Nuclear Magnetic Shieldings and Molecular Structure

NATO ASI Series

Advanced Science Institutes Series

A Series presenting the results of activities sponsored by the NATO Science Committee, which aims at the dissemination of advanced scientific and technological knowledge, with a view to strengthening links between scientific communities.

The Series is published by an international board of publishers in conjunction with the NATO Scientific Affairs Division

A Life Sciences	Plenum Publishing Corporation
B Physics	London and New York
C Mathematical	Kluwer Academic Publishers
and Physical Sciences	Dordrecht, Boston and London
D Behavioural and Social Sciences	
E Applied Sciences	
F Computer and Systems Sciences	Springer-Verlag
G Ecological Sciences	Berlin, Heidelberg, New York, London,
H Cell Biology	Paris and Tokyo
I Global Environmental Change	

NATO-PCO-DATA BASE

The electronic index to the NATO ASI Series provides full bibliographical references (with keywords and/or abstracts) to more than 30000 contributions from international scientists published in all sections of the NATO ASI Series.
Access to the NATO-PCO-DATA BASE is possible in two ways:

– via online FILE 128 (NATO-PCO-DATA BASE) hosted by ESRIN,
Via Galileo Galilei, I-00044 Frascati, Italy.

– via CD-ROM "NATO-PCO-DATA BASE" with user-friendly retrieval software in English, French and German (© WTV GmbH and DATAWARE Technologies Inc. 1989).

The CD-ROM can be ordered through any member of the Board of Publishers or through NATO-PCO, Overijse, Belgium.

Series C: Mathematical and Physical Sciences - Vol. 386

Nuclear Magnetic Shieldings and Molecular Structure

edited by

J. A. Tossell

Department of Chemistry and Biochemistry,
University of Maryland,
College Park, Maryland, U.S.A.

Springer Science+Business Media, B.V.

Proceedings of the NATO Advanced Research Workshop on
The Calculation of NMR Shielding Constants and Their Use in the Determination of
the Geometric and Electronic Structures of Molecules and Solids
College Park, Maryland, U.S.A.
July 20–24, 1992

ISBN 978-0-7923-2119-4   ISBN 978-94-011-1652-7 (eBook)
DOI 10.1007/978-94-011-1652-7

Printed on acid-free paper

TABLE OF CONTENTS

INTRODUCTION

In recent years a number of first principles or *ab initio* quantum mechanical methods have been developed which allow the accurate calculation of NMR shielding constant tensors for large molecules and for molecular models of solids. In addition to more efficient implementations of the conventional common-origin coupled Hartree-Fock perturbation theory and finite perturbation theory methods which were applied to small molecules in the 1960's, new methods have been developed based upon localized molecular orbitals and methods employing so-called gauge-including atomic orbitals have been improved. It has also become feasible to go beyond the Hartree-Fock level, including the effects of electron correlation using many body perturbation theory, polarization propagator techniques or multi-configuration SCF approachs. Greater understanding of the basis set dependence of the calculated shieldings has also improved their accuracy and reliability and has allowed their extension both to larger molecules and to many different types of magnetic nuclei, including transition metals and heavy main-group elements. As a result of these improvements in methodology, coupled with dramatic increases in the speed and storage capacity of routinely available computers, it has now become possible for many groups, using many different methods, to accurately calculate NMR shielding constant tensors for fairly complicated molecules.

The status of the various methods was reviewed at this workshop, and it was agreed that the results obtained with the various localized orbital and gauge including orbital methods were quite similar. The gauge-including atomic orbital method seems to have slightly superior accuracy if very small basis sets are used, but all the localized and gauge-including methods seem to converge to the same result using larger bases. Common-origin methods with core electron corrections can also give reasonably good absolute shieldings, but are perhaps best used for small molecules (for which very large basis sets can be employed and core corrections are small) or, when applied to large molecules, for analysis of shielding trends (rather than absolute shieldings) and in the elucidation of the electronic mechanisms of the shielding and the relationship between the shielding and other electronic properties. Studies of correlation effects using the various methods show many cases in which such effects seem fairly small, but also some (e.g. compounds with both multiple bonds and lone pairs and compounds poorly described by a single determinant wavefunction) in which they seem quite large. There is not yet sufficient experience with such correlation effects to make possible a simple, reliable estimation of their magnitudes in all cases.

Substantial advances have also recently occurred in experimental NMR spectroscopy, with improved methhods for determining the orientations of NMR shielding tensors in liquids and solids, with the use of many-dimensional NMR techniques to uniquesly assign resonances to structural sites and with studies of changes in the shielding tensor with changes of temperaute, phase or crystal structure. Some new NMR techniques uniquely relevant to the determination of the relationship between the shielding tensor and the geometric structure were reviewed in the workshop. Changes in shielding constants with changes in local geometry were also reviewed for minerals, glasses, semiconductors and for the compounds of many main group and transition elements. In many cases, predominantly experimental talks also incorporated first principles calculations to help in the assignment of shieldings to sites and in the determination of shielding tensor orientations. These calculations provided rigorous tests for the efficiency and accuracy of these methods when applied to large molecules.

In addition to the discussions of first principles calculations, there were several presentations stressing the utility of qualitative molecular orbital or semiempirical analyses of shielding trends. Some connections were made between these approachs and those based on first principles calculations, and the validity of some of the semiempirical approachs was discussed. The importance of validating the assumptions of the semiempirical approachs by appeal to first principles calculation or other electronic structural data was stressed, but it was concluded that such approachs still had great value for systematizing shielding constant data, interpolating high precision values of shielding constants from strucural data and identifying correlations for futher study by first principles methods.

Overall, the conclusion of the workshop was that present and emerging methods, along with expected computational advances, would make possible calculations at post-Hartree-Fock accuracy on quite large molecules in the near future. It was not so clear exactly how these capabilites could best be used to update and improve the present non-specialist's understanding of NMR shielding constants or to deepen our overall understanding of molecular electronic strcture. These remain important general topics for future discussion.

The Organizing Committe for the workshop consisted of:T. D. Bouman (Southern Illinois Univ. at Edwardsville, USA, deceased); P. Lazzeretti (Univ. of Modena, Italy); G. E. Maciel (Cororado State Univ., USA); J. A. Tossell (Univ. of Maryland, USA); and G. A. Webb (Univ. of Surrey, UK). Generous financial support was provided by the NATO Scientific Affairs Division. Additional support was provided by Biosym Technologies,

Chemagnetics, Cray Research, Inc., Hytec Electronics, Ltd. and by the Dept. of Chemistry and Biochemistry of the University of Maryland at College Park.

To give more of the flavor of the workshop the papers are presented in the order given, with the names of the lecturers marked with an asterix. Informal discussions were also held on the topics: (1) "How does the compound in the computer relate to that in the NMR spectrometer?", (2) "Do other electronic properties correlate with NMR shieldings?", and (3) "How do correlation and relativity affect NMR shieldings?". A review of the methods used for shielding constant calculations and an evaluation of their accuracy is given by G. A. Webb in the first chapter, entitled "An Overview of Nuclear Shielding Calculations" and a brief summary of the meeting lectures and discussions, along with some suggestions for future research, is given by C. J. Jameson in the final chapter, entitled "Overview and Directions for the Future". Also included are one-page abstracts of most of the posters presented at the workshop.

J. A. Tossell
College Park, MD, USA
October, 1992

LIST OF PARTICIPANTS

Michael Barfield
Dept. of Chemistry
University of Arizona
Tucson, AZ 85721

Ad Bax
Laboratory of Chemical Physics
NIDDK, National Institutes of health
Bethesda, MD 20892

Michael Buhl
Center for Computational Quantum
Chemistry
University of Georgia
Athens, GA 30602

David Case
Dept. of Molecular Biology
The Scripps Research Institute
10666 Torrey Pines Rd.
La Jolla, CA 92037

Donald Chesnut
Dept. of Chemistry
Duke University
Durham, NC 27706

Angel deDios
Dept. of Chemistry
University of Illinois at Chicago
P.O. Box 4348
Chicago, IL 60680

Joseph Delhalle
Dept. of Chemistry
Facultes Univeresitaires Notre-Dame de
la Paix
Rue de Bruxelles, 61
B-5000 Namur
Belgium

Ross Dickson
Dept. of Chemistry
Queens University
Kingston, Ontario K7L 3N6
Canada

James Duchamp
Dept. of Chemical Engineering
Cornell University
120 Olin Hall
Ithaca, NY

Ray Dupree
Dept. of Physics
University of Warwick
Coventry CV4 7AL
UK

Clifford E. Dykstra
Dept. of Chemistry
Indiana university-Purdue University at
Indianapolis
1125 E. 38th St.
Indianapolis, IN 46205

Hellmut Eckert
Dept. of Chemistry
University of California
Santa Barbara, CA 93106

Klaus Eichele
Dept. of Chemistry
Dalhousie University
Halifax, Nova Scotia B3H 4J3
Canada

Paul Ellis
Dept. of Chemistry
University of south Carolina
Columbia, SC 29036

Julio Facelli
Dept. of Chemistry
University of Utah
1320 HEB
Salt Lake City, UT 84112

Thomas Farrar
Dept. of Chemistry
University of Wisconsin
1101 University Avenue
Madison, WI 53706

Ulrich Fleischer
Lehrstuhl fur Theoretische Chemie
Ruhr-Universitat Bochum
4630 Bochum
Germany

Eugene Fleishcmann
Cray Research, Inc.
4041 Powder Mill Rd.
Calverton, MD

Jan Geertsen
Dept. of Chemistry
Odense University
Campusvej 55
Odense N 5230
Denmark

Corine Gerardin
Dept. of Geological and Geophysical
Sciences
Princeton University
Guyot Hall
Princeton, NJ 08566

David Grant
Dept. of Chemistry
University of Utah
1320 HEB
Salt Lake City, UT 84112

Arnd-Rudiger Grimmer
Projektgruppe Festkorper-NMR
KAI e.v. Berlin
Rudower Chaussee 5
D-O-1199 Berlin
Germany

Aage Hansen
Dept. of Chemistry
H.C. Orsted Institute
University of Copenhagen
Copenhagen DK-2100 CPH 0
Denmark

Marc Henry
Dept.Chimie de la Matiere Condensee
Universite P. et M. Curie
4 Place Jussieu
75252 Paris Cedex 05
France

Judith Herzfeld
Dept. of Chemistry
Brandeis University
415 South St.
Waltham, MA 02254-9110

James Hinton
Dept. of Chemistry
University of Arkansas
Fayetteville, AK 72701

Milan Hoodscek
MGSL
DCRT/NIH
Bethesda, MD 20892

A. Keith Jameson
Dept. of Chemistry
Loyola University Chicago
6525 N. Sheridan St.
Chicago, IL 60626

Cynthia Jameson
Dept. of Chemistry
University of Illinois at Chicago
P.O. Box 4348
Chicago, IL 60680

Todd Keith
Dept. of Chemistry
McMaster University
1280 Main St. West
Hamilton, Ontario L8S 4M1
Canada

R. James Kirkpatrick
Dept. of Geology
University of Illinois
1301 W. Green St.
Urbana, IL 61801

Jacek Klinowski
Dept. of Chemistry
University of Cambridge
Lensfield Rd.
Cambridge CB2 1EW
UK

Werner Kutzelnigg
Dept. of Chemistry
Ruhr-Universit Bochum
4630 Bochum-Querenburg
Germany

Yiu-Fai Lam
Dept. of Chemistry
University of Mayrland
College Park, MD 20742

Paolo Lazzeretti
Dept. of Chemistry
University of Modena
Via Campi 183
41100 Modena
Italy

Hongbiao Le
Dept. of Chemistry
University of Illinois
505 S. Mathews Ave.
Urbana, IL 61801

Michael Lumsden
Dept. of Chemistry
Dalhousie University
Halifax, Nova Scotia B3H 4J3
Canada

Gary Maciel
Dept. of Chemistry
Colorado State Univ.
Ft. Collins, CO 80523

Vladimir Malkin
Dept. of Chemistry
Universite de Montreal
C.P. 6128, Succ. A
Montreal, Quebec H3C 3J7
Canada

Joan Mason
Dept. of Chemistry
The Open University
Milton Keynes MK7 6AA
UK

Hiroshi Nakatsuji
Dept. of Synthetic Chemistry
Faculty of Engineering
Kyoto University
Kyoto 606
Japan

Jens Oddershede
Dept. of Chemistry
Odense University
Campusvej 55
Odense DK-5230
Denmark

Eric Oldfield
Dept. of Chemistry
Univ of Illinois
505 S. Matthews
Urbana, IL 61801

Victor Parziale
Technical Sales Dept.
Chemagnetics
2555 Midpoint Dr.
Fort Collins, CO 80525

John Pearson
Dept. of Chemistry
University of Illinois
505 S. Matthews
Urbana, IL 61801

William Power
Dept. of Chemistry
Yale University
225 Prospect St.
New Havem , CT 06511

Peter Pulay
Dept. of Chemistry and Biochemistry
University of Arkansas
Fayetteville, AK 72701

Marie-Francoise Quinton
ESPCI, Lab. de Physique Quantique
10 rue Vauquelin
75231 Paris Cedex 05
France

William Raynes
Dept. of Chemistry
Univ. of Sheffield
Sheffield S3 7HF
UK

Gotthard Saghi-Szabo
Dept. of Chemisty and Biochemistry
Univ. of Mayland
College Park, MD 20742

Paul v.R. Schleyer
Institut fur Organische Chemie
Friedrich-Alexander-Universitat
Erlangen-Nurnberg
Henkestr. 42
D-8520 Erlangen
Germany

Ulrich Sternberg
Physikalisch-Astronomische Fakultat
Friedrich-Schiller-Universitat Jena
Max-Wien-Platz 1
O-6900 Jena
Germany

Walter Stevens
Center for Advanced Research in
Biotechnology
Maryland Biotechnology Institute
9600 Gudelsky Dr.
Rockville, MD 20850

Francis Taulelle.
Dept.Chimie de la Matiere Condensee
Universite P. et M. Curie
4 Place Jussieu
75252 Paris Cedex 05
France

Jack Tossell
Dept. of Chemistry and Biochemistry
Univ. of Maryland
College Park, MD 20742

John Trudeau
Dept. of Chemistry
University of Wisconsin
1101 University Ave.
Madison, WI 53706

Drazen Vikic-Topic
LCP
NIDDK/NIH
Bethesda, MD 20892

Roderick Wasylsihen
Dept. of Chemistry
Dalhousie University
Halifax, Nova Scotia B3H 4J3
Canada

Graham Webb
Dept. of Chemistry
University of Surrey
Guildford, Surrey GU2 S4H
UK

Dennis Winkler
Dept. of Chemistry and Biochemistry
Univ. of Maryland
College Park, MD 20742

Renate Wolff
Analytisches Zentrum
Rudower Chaussee 5
D-1199 Berlin
Germany

Krysztof Wolinski
Dept. of Chemistry
University of Arkansas
Fayetteville, AK 72701

Ping Yip
Science Dept.
Biosym Technologies
9685 Scranton Rd.
San Diego, CA 92121

Riccardo Zanasi
Dept. of Chemistry
University of Modena
Via Campi 183
41100 Modena
Italy

Kurt Zilm
Dept. of Chemistry
Yale University
225 Prospect St.
New Haven, CT 06511

AN OVERVIEW OF NUCLEAR SHIELDING CALCULATIONS

G. A. Webb
Department of Chemistry
University of Surrey
Guildford
Surrey, England

ABSTRACT. An account is given of molecular orbital calculations of nuclear shielding. Both ab initio and semi–empirical calculations are considered. Correlation effects relativistic contributions and solvent effects are included.

> Whoever, in the pursuit of science, seeks after immediate practical utility, may generally rest assured that he will seek in vain.
> H.L.F. von Helmholtz

This quotation epitomises the story of nuclear shieldings calculations. The initial paper on the theory of nuclear shielding in generally agreed to be that by Ramsey in 1950.[1] Thus it has been known for over forty years how, in principle, nuclear shieldings may be calculated. However, it is only in recent years that it has been possible to perform such calculations with a meaningful accuracy. Some of the earlier results, although possibly indicative of shielding trends in series of closely related molecules, are often unsuitable for predictive purposes. Approximations are still part of current calculations, these are considered in the present account.

It seems reasonable to begin this, somewhat eclectic report with a consideration of the basic concepts of nuclear shielding after making reference to recent reviews of this area.[2-5]

Basic Principles

NMR active nuclei have magnetic moments which may interact with an applied r.f. magnetic field to produce the NMR experiment in an external magnetic field B. The energy, E, of a magnetic moment, μ, in the external field is:–

$$E = \mu \cdot B + \mu \cdot \sigma \cdot B$$
$$= -\mu \cdot (1-\sigma) \cdot B \qquad (1)$$

the first scalar product corresponds to the direct interaction between the magnetic moment and the applied field. The second scalar interaction includes the nuclear shielding tensor σ and describes the interaction of μ with the effective magnetic field $-\sigma \cdot B$ which is produced by the moderation of B by

1

J. A. Tossell (ed.), Nuclear Magnetic Shieldings and Molecular Structure, 1–25.
© 1993 Kluwer Academic Publishers.

the molecular electrons in the vicinity of the nucleus. Since σ may be of either sign, the electrons may produce either an increase or a decrease in the magnetic field experienced by the nuclei. A positive value of σ results in an increase in nuclear shielding sometimes referred to as a diamagnetic effect. If σ is negative nuclear deshielding occurs; this corresponds to a paramagnetic effect.

The chemical shift, δ, with respect to a given standard is given by

$$\delta_x = \sigma_{std.} - \sigma_x \tag{2}$$

thus an increase in chemical shift corresponds to a decrease in shielding and vice-versa.

In general, σ, is an unsymmetric second-rank tensor, i.e. $\sigma_{ij} \neq \sigma_{ji}$. The number of independent components required to fully specify σ depends upon the nuclear site symmetry[6]. It increases from one for cubic symmetry to a maximum of nine for the C_1 and C_i point groups. In contrast with experiment all of the tensor components are usually provided by theoretical calculations. For comparison with experimental results the tensor is usually decomposed into its symmetric and antisymmetric components, σ^s and σ^a respectively

$$\sigma^s_{ij} = \tfrac{1}{2} \left(\sigma_{ij} + \sigma_{ji} \right) \tag{3}$$

$$\sigma^a_{ij} = \tfrac{1}{2} \left(\sigma_{ij} - \sigma_{ji} \right) \tag{4}$$

σ^a_{ij} enters the Zeeman Hamiltonian only in terms which are second order in the magnetic field and thus it is not normally observed in standard NMR experiments. Consequently σ^a tends to be neglected and the principal axes of the shielding tensor are determined by diagonalization of σ^s. The diagonal elements are then used to give the isotropic shielding, σ_{av}, the shielding anisotropy, $\Delta\sigma$, and the asymmetry, η.,

Where $\sigma_{av} = \tfrac{1}{3} \left(\sigma_{11} + \sigma_{22} + \sigma_{33} \right)$ \hfill (5)

and $\Delta\sigma = \sigma_{33} - \tfrac{1}{2} \left(\sigma_{11} + \sigma_{22} \right)$ \hfill (6)

and $\eta = \dfrac{\left(\sigma_{22} - \sigma_{11} \right)}{\left(\sigma_{33} - \sigma_{av} \right)}$ \hfill (7)

Where the usual convention is that $\sigma_{33} \geq \sigma_{22} \geq \sigma_{11}$.

Normally from NMR studies on liquids only σ_{av} is compared with experimental data. However from NMR measurements on aligned molecules, for example in solids or liquid crystals, it is possible to obtain information about the individual tensor components. Thus a comparison may be made with the calculated values of $\Delta\sigma$ and η. In general the error in the experimental results for $\Delta\sigma$ and η tend to be larger than those for σ. For example an error of up to 50% in the measured value of η has been estimated[7]. Consequently caution should be exercised in comparing the theoretical and experimental data.

In the interpretation of low temperature solid state NMR spectra, ab initio calculations of the nuclear shielding tensor components have been very helpful[8,9]. Comparison with the calculated data being the basis for the assignment of the principal axes of many ^{13}C shielding tensors. Taking the ^{13}C tensor of cyclopropane as an example the calculated components of δ. with respect to TMS, are $\delta_{11} = 27$, $\delta_{22} = 9$ and $\delta_{33} = -40$ ppm, in comparison with the experimental values of 22, 2 and -36 ppm respectively[10]. Thus theoretical calculations can be successfully used to predict the orientation of the principal axes of the shielding tensor.

Some calculations of σ^a have also appeared.[7,11] Large values of σ^a are predicted for the nitrogen atom shielding in diazirine and for the ^{13}C shielding in cyclopropene[11]. Both of these molecules have a double bond which is highly perturbed by the third ring atom. It would be interesting to have some suitable experimental data for comparison purposes.

Quantum Mechanical Approach

Nuclear shielding is produced by the influence on the total electronic energy, E, of a molecule due to the nuclear magnetic moment μ, and the applied uniform magnetic field B. Thus the shielding tensor components can be defined by

$$\sigma_{\alpha\beta} = \left[\frac{\delta^2 E}{\delta\mu_\alpha \delta B_\beta}\right]\mu, B=0 \tag{8}$$

Where $\alpha,\beta = $ X,Y,Z the Cartesian coordinate components. Since the electron induced nuclear shielding is usually a fairly small contributor to the total molecular Hamiltonian, σ is evaluated by means of perturbation theory[2]. The resulting picture is that the wave function of the molecule is perturbed by B the effect of which is to induce preferred currents which then interact with μ. Consequently to be able to calculate σ we need to know the effect of the perturbation on the wavefunctions caused by each of the components of B.

Ramsey [1,12,13] used the perturbation approach to produce an expression for $\sigma_{\alpha\beta}$ comprising four terms

$$\sigma_{\alpha\beta} = \sigma_{\alpha\beta}^d + \sigma_{\alpha\beta}^{dg} + \sigma_{\alpha\beta}^p + \sigma_{\alpha\beta}^{pg} \tag{9}$$

Where d and p denote the diamagnetic and paramagnetic components respectively. The diamagnetic terms depend only on the wavefunction of the electronic ground state. Whereas the paramagnetic terms depend upon wavefunctions of the molecular excited states including those of the continuum. In the above expression g signifies gauge and refers to the need to define an origin for the vector potential of B. This is called the gauge origin, G. Terms with the g superscript depend upon this choice of origin.

This is shown by the presence of R in equations (11) and (13) where R is the separation of G from the origin of the coordinates, C, which is the point at which the shielding is normally evaluated. The electron k is taken to be at a distance of r_k from C.

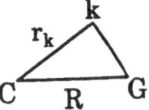

Expressions for the four terms in equation (9) are given as

$$\sigma_{\alpha\beta}^{d} = \frac{\mu_o e^2}{8\pi m} <o \mid \sum_k \frac{r_k^2 \delta_{\alpha\beta} - r_{k\alpha} r_{k\beta}}{r_k^3} \mid o> \qquad (10)$$

$$\sigma_{\alpha\beta}^{dg} = \frac{\mu_o e^2}{8\pi m} <o \mid \sum_k \frac{R_\alpha r_{k\beta} - R_\gamma r_{k\gamma} \delta_{\alpha\beta}}{r_k^3} \mid o> \qquad (11)$$

$$\sigma_{\alpha\beta}^{p} = \frac{\mu_o e^2}{8\pi m^2} \sum_n{}' (E_o - E_n)^{-1} \left[<o \mid \sum_k \frac{\ell_{k\alpha}}{r_k^3} \mid n><n \mid \sum_k \ell_{k\beta} \mid o> \right.$$
$$\left. + <o \mid \sum_k \ell_{k\beta} \mid n><n \mid \sum_k \frac{\ell_{k\alpha}}{r_k^3} \mid o> \right] \qquad (12)$$

$$\sigma_{\alpha\beta}^{pg} = \frac{\mu_o e^2}{8\pi m^2} \epsilon_{\beta\gamma\delta} R_\gamma \sum_n{}' (E_n - E_o)^{-1} \left[<o \mid \sum_k \frac{\ell_{k\alpha}}{r_k^3} \mid n><n \mid \sum_k P_{k\delta} \mid o> \right.$$
$$\left. + <o \mid \sum_k P_{k\delta} \mid n><n \mid \sum_k \frac{\ell_{k\alpha}}{r_k^3} \mid o> \right] \qquad (13)$$

Where μ_o is the permeability of free space, P_k is the linear momentum operator of the kth electron, $\ell_k = r_k \times P_k$ is the orbital angular momentum, $\sum'$ implies the summation over all states except the ground state (n≡0), but including the continuum of excited states, $\delta_{\alpha\beta}$ is the substitution tensor, $\delta_{\alpha\beta}$ = 1 if $\alpha=\beta$ and $\delta_{\alpha\beta}=0$ if $\alpha\neq\beta$. $\epsilon_{\beta\gamma\delta}$ is the alternating tensor, $\epsilon_{\beta\gamma\delta}=1$ if $\beta\gamma\delta$ is an even permutation of X,Y,Z $\epsilon_{\beta\gamma\delta} = -1$ if $\beta\gamma\delta$ is an odd permutation and $\epsilon_{\beta\gamma\delta}=0$ if any two of the labels β, γ or δ are identical.

If the gauge origin is chosen to be at the nucleus of interest then R=O and
$$\sigma_{\alpha\beta}^{dg} = \sigma_{\alpha\beta}^{pg} = O \qquad (14)$$
for approximate wavefunctions such as those usually employed in finite basis set coupled Hartree–Fock (CHF) Calculations[14]. If a complete basis set were

employed then the CHF results would be independent of the choice of gauge[15]. However, for approximate wavefunctions it is usually found that $\sigma_{\alpha\beta}^{dg} \neq \sigma_{\alpha\beta}^{pg}$ when the gauge origin is not taken at the nucleus of interest.

Since σ is a molecular property it must clearly be invariant to changes of gauge origin, i.e. it is gauge independent. Thus the question of gauge origin is of great importance in quantum mechanical calculations of nuclear shielding. There have been several approaches to dealing with this difficulty.

In ab initio calculations there appear to be two commonly adopted avenues. One is to employ a sufficiently complete basis set so that the consequences of the choice of gauge origin on the calculated value of σ are minimal. The second method is to introduce gauge factors into either the atomic orbitals of the basis set or into the molecular orbitals of a Coupled Hartree–Fock (CHF) calculation in such a manner that the results are independent of gauge origin even though the calculation is an approximate one. The former of these two avenues has been explored in calculations on small molecules. Whereas the second approach is finding more widespread acceptance for larger molecules.

Inclusion of gauge factors in the atomic orbitals used in a calculation is most often accomplished by means of gauge included atomic orbitals (GIAOs)[16]. Thus the complex orbitals $\aleph$ are used instead of the usual, real, atomic orbitals, φ, located at R, Hence,

$$\aleph_n = \varphi_n \exp\left[-\frac{ie}{\hbar} A(R_n) \cdot r\right] \tag{15}$$

Where the complex phase factor contains the vector potential $A(R_n)$ from the magnetic field evaluated at the point R_n.

This approach has been widely adopted by Ditchfield and others[17-19]. More recently the GIAO method has been used in conjunction with analytical derivative theory[20]. In this approach the magnetic field perturbation is treated in an analogous way to the perturbation produced by changing the nuclear coordinates.

Instead of using gauge factors to premultiply the normal atomic orbitals in a shielding calculation it is possible to use complex exponential gauge factors in the molecular orbitals[21-24].

This method uses individual gauge origins for different localised molecular orbitals (IGLO). The molecular orbitals, ψ_k, used in the shielding calculations are related to the molecular orbitals φ_k, by $\varphi_k = \exp(i\lambda_k)\psi_k$ (16) Where λ_k is a local multiplicative operator proportional to the applied field B, and φ_k are expanded in powers of B.

In addition to overcoming the gauge origin problem this approach provides a way of defining contributions to the nuclear shielding arising from atomic and bonding electrons. Thus the specific shielding contributions produced by the lone pairs on the atom in question and by orbitals centered on neighbouring atoms can be determined. Another, similar, method has been proposed in which different gauge origins are used for different pairs of orbitals, these may be either atomic or molecular orbitals[25].

Rather than use the CHF procedure for calculating second order magnetic properties an alternative approach is to use the equations of motion (EOM) method[26]. A well established approximation to the EOM procedure is the random phase approximation (RPA). Three methods of calculating the nuclear shielding within the RPA, have been proposed[27-29]. In one of these localised molecular orbitals with local origins (LORG) are used[27], while a common origin is employed in the other two methods[28,29]. It is perhaps comforting to realise that the CHF and RPA methods of calculating nuclear shielding are equivalent[27].

In LORG–RPA calculations of nuclear shielding as in the IGLO ones the occupied molecular orbitals are localised and each is associated with an origin vector relative to the magnetic nucleus. A consequence of the localization of the molecular orbitals, and the use of local origins, is that the total shielding can be decomposed into individual local bond contributions, involving electrons in bonds directly attached to the nucleus in question, and bond–bond contributions involving all the other bonds.

Complex orbitals, including gauge factors, are not introduced into the LORG RPA shielding calculations. However, due to an expansion of angular momentum terms relative to a local origin for each orbital, and the use of RPA properties, the resulting shielding expressions contain no gauge dependence.

The GIAO, IGLO and LORG –RPA calculations all employing local origins can be performed with different levels of basis set quality. The results are dependent on the choice of basis set used. However, all of these methods are reasonably successful in calculating nuclear shielding tensors. Their success appears to arise from the omission of large terms of opposite sign which would have been exactly cancelling in the limit of a complete basis set[5]. Consequently for a given basis set, the local origins provide a damping of basis set errors in long range shielding contributions. Thus the agreement found with the experimental data is better than that produced by the results of conventional CHF or finite perturbation theory calculations using a common gauge origin[5].

Some Results of Local Origin Shielding Calculations

In table 1 the results of some ^{13}C shielding calculations obtained from various local origin procedures, are compared with experimental data. In general there is a respectable agreement between the theoretical and experimental results. The calculations considered tend to employ medium sized basis sets; for example sets of triple zeta quality with a set of d polarisation functions for the heavy atoms. If a double zeta basis set is employed, together with polarization functions, then the ^{13}C shielding results are less accurate[22,32]. They tend to be more shielded by about 20 ppm. The results presented show that the IGLO, LORG and GIAO methods all produce reliable results of a comparable nature.

Table 2 shows the results of some local origin calculations of nuclear shielding for ^{15}N, ^{17}O and ^{19}F nuclei in various molecules.

TABLE 1

Comparison of some local origin calculated and experimental ^{13}C shieldings and their anisotropies (in ppm)

	IGLO CHF	LORG RPA[27]	GIAO CHF[30]	GIAO FPT[31]	EXP[5] σ_0
CH_4	196.7[23] 215[32]	196		193	195.1
HCN $\Delta\sigma$	72.9[24] 306[24]	77.3 301		74.8 304.8	82.1 316.3 ± 1.2
HC≡CH $\Delta\sigma$	116.4[24] 243.3[24]	122.3 235		118.3 241	117.2 240 ± 5
CO	−6[32] −23.9[24]		−19.3	−21.3	1.0
$\Delta\sigma$	420[32] 442.2[24]		435.5	439.0	405.5 ± 1.4 406 ± 30
CH_3CH_3 $\Delta\sigma$	183.5[23] 13.5[23]	184.7 8		181.2 11.3	180.9
CH_3OH $\Delta\sigma$	157.2[23]	145.7 60.5		134.6 77.3	136.6 63
$H_2C=O$ $\Delta\sigma$	−3.8[23] 183.8[23]	4 183		2.6 196.5	−8

TABLE 2

Comparison of some local origin calculated and experimental ^{15}N, [70]O and ^{19}F shieldings and their anisotropies (in ppm)

	IGLO[22,24] CHF	GIAO[30] CHF	GIAO[31] FPT	EXP[a]
N_2	−108.	−113.4	−109.4	−61.6
$\Delta\sigma$	670.8	677.8	662.3	676 ± 20
NH_3	265.4		265.2	264.5
$\Delta\sigma$	−37.7		−45.2	−40
CO	−84.1	−106.3	−88.0	−42.3
$\Delta\sigma$	742.0	775.2	748.1	676.1
H_2O	327.4		332.0	344.0
$\Delta\sigma$	55.8		39.0	
F_2	−172.5	−219.9	−181.5	−219
$\Delta\sigma$	991.6	1062	1005.3	1050 ± 50
CH_3F			484.5	471
$\Delta\sigma$			−71.5	−60.8 ± 15
HF	413.5	411.6	412.4	419.7
$\Delta\sigma$	102.3	106.7	105.8	108 93.7

[a]   From spin–rotation constants, except for N_2 and H_2O

Some Results of Shielding Calculations obtained without Incorporating a Local Origin

It has been known for almost thirty years that very accurate values of nuclear shielding can be obtained from CHF theory provided a very large atomic orbital basis set is employed[14,33]. There are many examples of successful calculations of this kind in the literature and the results of some ^{13}C and ^{1}H shielding calculations are given in Table 3, for some hydrocarbons[34,35]. For the ^{13}C shielding calculations the best gauge origin is found to be at the ^{13}C nucleus and for the ^{1}H calculations it is reported to be close to the ^{13}C nucleus. The large basis sets used include three d sets on the carbon and two p sets on the hydrogen atoms in agreement with those used in similar calculations[36,37]. Shielding calculations on second row hydrides, without using a local origin procedure require the addition of further d functions for the heavy atom, such that four or five d sets are required to reduce gauge dependence of the shielding results[35].

TABLE 3

Comparison of some non local origin Calculated and Experimental ^{1}H and [13]C Shieldings,[34] (in ppm)

[1]H Shieldings

	σ_{xx}	σ_{yy}	σ_{zz}	σ_{av}	σ_{av} (exp)
CH_4	27.90	27.90	38.36	31.39	30.61
C_2H_6	34.78	26.13	30.99	30.63	29.26
C_2H_4	24.66	23.13	28.47	25.42	25.43
C_2H_2	24.23	24.23	41.09	29.85	29.86

[13]C Shieldings

	σ_{xx}	σ_{yy}	σ_{zz}	σ_{av}	σ_{o} (exp)
CH_4	195.8	195.8	195.8	195.8	195.1
C_2H_6	187.7	182.7	193.1	186.2	180.9
C_2H_4	177.9	−81.1	84.3	60.4	64.5
C_2H_2	39.0	39.0	279.4	119.1	117.2

The unique molecular axis is taken to be the Z axes, e.g. in C_2H_4 Z is along the C–C bond and X is perpendicular to the molecular phase.

By taking a reasonable choice of gauge origin satisfactory shielding results are obtained with significantly smaller basis sets.

Electron Correlation Effects

Sum–over–States (SOS) perturbation theory and Finite Perturbation Theory (FPT) are alternatives to the CHF procedure for calculating nuclear shielding[38,39]. In an attempt to include some electron correlation effects in the SOS calculations Configuration Interaction (CI) may be used[40]. SOS–CI calculations on first and second row hydrides, with the centre of mass as the gauge origin, reveal that the results obtained are comparable to those produced by the CHF method using the same quality basis set but requiring less computer time[41]. It has also been demonstrated that the SOS–CI procedure is equivalent to the FPT method and an exact reproduction of the shielding results for HF, H_2O, NH_3 and CH_4 as test molecules, has been found for the two approximate methods[42,43]. Thus the SOS–CI procedure appears to be a convenient alternative to the FPT and CHF methods, at least with basis sets of 4–31G quality.

CHF calculations with extended basis sets of gaugeless gaussian orbitals have been used to calculate ^{31}P [44,45] and ^{29}Si [46–49] shieldings in a number of small molecules and ions. In general good agreement with the experimental results is found. IGLO calculations have also been reported for a number of the same ^{31}P and ^{29}Si containing species[50]. It is found that smaller basis sets are satisfactory for the ^{29}Si calculations than those required for ^{31}P. In the latter case three or four sets of d functions are found to be necessary. This indicates that it is more difficult to produce satisfactory shielding calculations for nuclei with non–bonding electrons than for those with fully saturated valence shells. A similar conclusion is reached from IGLO calculations of nitrogen shieldings[51]. In particular, nitrogen atoms involved in N–N double bonds have calculated shieldings which are too small even when near HF quality basis sets are used. It seems very likely that the lone pairs are not adequately described at the HF limit and that to obtain a more satisfactory shielding calculation it would be necessary to introduce correlation effects.

Similar disagreements between the experimental and calculated nuclear shieldings are found for O_3[24], and for the O and N shieldings of SO_2 and NSF[5]. It is very likely that the problems arising in the CHF method are associated with the existence of low lying excited electronic slates which are coupled to the ground state by the magnetic perturbation. These excited states play a dominant role in the calculation of the paramagnetic part of the shielding tensor. An accurate account of the shielding contribution arising from these low–lying states is not available within the single determinant perturbation scheme.

Calculations Including Correlation Effects

The results of some calculations on the shielding of N_2, using various methods of including correlation effects, are shown in Table 4.

TABLE 4

Nuclear Shielding of N_2 (in ppm) Comparison of Calculated and Experimental Values

CHF Value		Correlated value		Method	Ref.
σ	$\Delta\sigma$	σ	$\Delta\sigma$		
−110.0	672.9	−41.1	571.1	MCRPA	53
−106.5	668.1	−72.2	618.9	SOPPA	54
−72.2		−62.5		MBPT	55

Experimental value $\sigma = -61.6$, $\Delta\sigma = 603 \pm 28$ 56
MCRPA = Multiconfiguration Random Phase approximation
SOPPA = Second order Polarisation Propagator approximation.
MBPT   = Many Body Perturbation Theory

The MBPT calculation uses SCF results which appear to be far from convergence with respect to the basis set extension. Thus the apparent agreement between the MBPT result and experiment is probably fortuitous. The MCRPA calculations use a basis set of 124 CGTO's and the SOPPA method employs a basis of 76 CGTO's. Thus these results appear to be more reliable.

It is interesting to note that problems in shielding calculations arising from basis sets which are too small and thus give rise to gauge dependent results on the one hand and those due to electron correlation are distinguishable , and may be separated[57]. It appears that correlation effects are essentially independent of the effects of gauge choice.

Some EOM common origin calculations of ^{13}C and ^{29}Si shieldings provide results which are not in good agreement with experiment[58,59]. In principle these calculations should include some account of correlation effects on the shielding. Unfortunately, it seems that only modest basis sets are employed and the results are not sufficiently accurate to give a reliable estimate of any correlation contributions to the shielding that may be present. In the case of ^{13}C and ^{29}Si such effects are normally expected to be small.

Table 5 shows a comparison of various ^{31}P shielding tensor data obtained both from calculations and experiment. The SOLO method includes both correlation effects, as in the SOPPA procedure, and local origins by means of the LORG approach.

As can be seen the inclusion of second order correlation effects results in a better agreement with experiment than that produced by CHF calculations. In most cases the inclusion of correlation effects results in a significant change in the calculated value of σ_{av}, an increase for PN and a decrease for PF_3. In addition it is noteworthy that the sign of the correlation contribution to the shielding is preserved when localised origins are introduced into the calculations.

Geometry effects on Nuclear Shielding

Experimental NMR measurements are usually performed at around 300 K. At such temperatures the molecules in gaseous and liquid samples are normally quite mobile and collisions permit a molecule to have the possibility of populating excited vibrational and rotational states on the NMR timescale.

TABLE 5

Comparison of some ^{31}P Shielding tensor Components (in ppm) obtained by calculation and experiment[5,57].

Molecule	CHF	IGLO[3]	LORG	SOPPA	SOLO	Expt.[97]
PN σ_{11}	966	1011	.966	966	966	970
$\sigma_{\perp}$	−452	−617	−452	−383	−368	−406
σ_{av}	20	− 74	20	66	76	53
PH_3 σ_{11}	582	558	582	576	576	557
$\sigma_{\perp}$	606	584	606	602	602	613
σ_{av}	598	576	598	594	594	594
PF_3 σ_{11}	533	460	487	454	410	343
$\sigma_{\perp}$	217	148	182	145	110	162
σ_{av}	322	352	284	248	210	223
P_4 σ_{11}	642	593	603	604	572	610
$\sigma_{\perp}$	1135	1026	1023	1095	999	1015
σ_{av}	971	881	883	931	856	880

In contrast to this, shielding calculations are usually performed at a rigid nuclear framework, most often the equilibrium configuration is assumed. Thus in comparing experimental and calculated shieldings some account of rotational–vibrational effects may be necessary[60]. For ^{19}F shieldings corrections due to these effects range from −9.5 to −18.0 ppm for some haloethanes at 300 K and −6.8 to −16.4 ppm for some halomethanes[61,62]. A correction of −13.1 ppm for the ^{17}O shielding of H_2O, provides a further example of the importance of these effects[63]. Clearly rotation–vibration corrections are becoming more significant in comparing theoretical and experimental data as the accuracy and reliability of calculated nuclear shieldings increases. Consequently the observation of gas phase shieldings as a function of temperature can be interpreted in terms of the geometrical dependence of the shieldings.

A further aspect for consideration is that if a strong geometrical dependence of the shieldings exists then the choice of geometry for calculations becomes critical. If substantial differences in the calculated shielding values arise between the HF and experimental geometries then the latter are normally preferred[50]. However, experimental geometries are often unknown.

As shown in Table 6 for some molecules containing ^{15}N, [19]F [29]Si and ^{31}P, the shieldings of these nuclei depend significantly upon geometry. A negative value for $\delta\sigma/\delta r_{AX}$ shows that the shielding is predicted to decrease as the bond length increases. It is found that the dominant contribution to the shielding change, as the bond length increases, is due to the paramagnetic term. An increase in bond length results in a decrease in the relevant electronic excitation energies thus to an increase in the contribution by the paramagnetic component of the shielding tensor[50]. Unfortunately, direct comparison between the calculated data given in Table 6 and experiment is not possible since the experimental results involve assumptions which may not be valid[50]. However, the agreement between the Table 6 results and some relevant experimental data[60] is not close.

Other authors[31] report the results of calculated first derivatives of the shielding for a variety of small molecules in which the observed NMR nuclei are ^{13}C, ^{15}N, [17]O and [19]F. The first derivatives are found to be negative and in reasonable agreement with the available experimental data[60]. As bond distances normally increase with increasing temperature the negative first derivatives imply a decrease in shielding as the temperature rises. It is also observed that the shielding derivatives become more negative as the nuclear shielding decreases, again in agreement with experiment[60].

TABLE 6

Some first derivatives of IGLO calculated nuclear shieldings with respect to bond length (r_{AX}) and bond angle (α_{XAX}) for ^{15}N, [19]Si and [31]P.

	$\dfrac{\delta\sigma}{\delta r_{AX}}$	$\dfrac{\delta\sigma}{\delta\alpha_{XAX}}$
NH_3	− 433	− 6
NF_3	− 921	−107
HF	− 486	
LiF	54	
SiH_4	− 66	
SiF_4	80	
P_2	−2055	
PH_3	− 464	−332
PF_3	− 243	− 54

Ab Initio Calculations of Heavy Metal Shieldings

Calculations including a common origin approach within FPT have been used to investigate the metal shieldings of compounds containing Cu, Zn, Ag, Cd and Mo. Although modest basis sets are used the results obtained reproduce a number of the experimental trends[64-68].

As shown in Fig. 1. for the complexes $[MoO_{4-n}S_n]^{2-}$, n=0–4, and $MoSe_4^{2-}$ it is the d→d* excitations which are the dominant factors in determining the ⁹⁵Mo nuclear shielding[68]. The magnetically allowed excitation energies reported are calculated by the symmetry adapted cluster–configuration interaction (SAC–CI) theory[68].

LORG calculations of ⁹⁵Mo nuclear shieldings have also been reported[69]. Again the large ⁹⁵Mo deshielding which occurs in the oxothiomolybdate anions as sulphur replaces oxygen is well reproduced[69].

Analysis of the LORG calculated ⁹⁵Mo shieldings show the importance of the contributions arising from all of the bonds and lone pairs of electrons in the molecules concerned. The calculations were performed using four basis sets and it is found that the sensitivity of the calculated shieldings to the quality of the basis sets is no worse than that found for many first– and second–row nuclei. The reasonable agreement between the observed and calculated ⁹⁵Mo shielding trends is obtained without the inclusion of either electron correlation or relativistic effects.

Other heavy metal ab initio shielding calculations include those on titanium[70,71] tin[72] and cadmium[73]. The major shielding variations observed experimentally are reasonably well accounted for by the calculated data. The CHF calculations on some titanium oxides and halides are performed using polarised double–zeta basis sets[71]. Basis sets of comparable quality are used in the calculations of tin[72] and cadmium[73] nuclear shieldings calculations of ⁶⁷Zn shieldings[94,95] use Huzinaga MIDI basis sets for all atoms augmented by two additional valence p functions and one diffuse d function on Zn as well as d functions on the first and second row ligand atoms. The resulting agreement with the available experimental data is reasonable.

Semi empirical Calculations

All of the calculations of nuclear shielding so far mentioned in this account relate to those based on ab initio molecular orbital procedures. It is only in about the past ten years that such calculations have become increasingly popular due in large part, to significant increases in the widespread availability of computer power.

In contrast semi empirical molecular orbital calculations have provided the basis for some understanding of the various electronic contributions to nuclear shielding for about thirty years.

The more illuminating of these calculations are based upon Pople's GIAO model[74-77]. As an example, when it is appropriate to consider only s and p electrons on an atom A of interest then the expressions for the rotationally averaged values of the local diamagnetic, σ_A^d, and paramagnetic, σ_A^P, terms are given by

$$\sigma_A^d = \frac{\mu_0 e^2}{12\pi m} \sum_\mu^A P_{\mu\mu} <\mu \mid r_{av}^{-1} \mid \mu> \tag{17}$$

$$\sigma_A^p = -\frac{\mu_0 e^2 h^2}{6\pi m} <r^{-3}>_{np} \sum_j^{occ} \sum_k^{unocc} (E_k - E_j)^{-1}$$

$$+ \left[(C_{y,Aj}C_{z\,Ak} - C_{z,Aj}C_{y,Ak}) \sum_B^k (C_{y,Bj}C_{z,Bk} - C_{z,Bj}C_{y,Bk}) \right.$$

$$+ (C_{z,Aj}C_{x,Ak} - C_{x,Aj}C_{z,Ak}) \sum_B (C_{x,Bj}C_{y,Bk} - C_{y,Bj}C_{x,Bk})$$

$$+ \left. (C_{x,Aj}C_{y,Ak} - C_{y,Aj}C_{x,Ak}) \sum_B (C_{x,Bj}C_{y'Bk} - C_{y,Bj,}C_{x,Bk}) \right] \tag{18}$$

Where $P_{\mu\mu}$ is the charge density in the atomic orbital μ which is at an average distance of $r_{A\mu}$ from nucleus A, $C_{x',Aj}$ is the LCAO coefficient of the P_x orbital on atom A in the molecular orbital j etc. The summation over B includes A. Thus it is apparent that σ_A^P will be zero unless both A and B possess valence P electrons. From the ordering of the LCAO coefficients in Eqn. (18) it is apparent that $\pi \rightarrow \pi^*$ transitions do not contribute to $\sigma_A P$.

E_k and E_j refer to the energies of the molecular orbitals and k and j which are unoccupied and occupied respectively . The excitation energies $E_k - E_j$ are those for excited singlet states which are mixed with the ground state by the external magnetic field used in the NMR experiment. From this brief account it may be apparent that semi empirical nuclear shielding calculations can provide an entre to an interpretation of chemical shifts in terms of molecular electronic structure. More extensive accounts of these calculations are given elsewhere[5, 78-82].

Within the framework of semiempirical calculations attempts have been made to identify the importance of relativistic effects on the shielding of heavy nuclei in molecules and to consider possible medium effects on nuclear shielding [5,81-82].

Relativistic Effects for Heavy Nuclei

It seems that two such effects should be considered in discussions of the NMR parameters of heavy nuclei. The first arises from changes in the radial parts of the wavefunctions of heavier nuclei. Due to a relativistic electronic mass increase the radii of the innermost orbits tend to contract. This contraction may lead to a greater screening of the nuclear charge for the outer electrons such that the radii of their orbitals may expand. As shown in equations (17) and (18) changes in the separation between a given nucleus and the surrounding electrons are expected to significantly change both the diamagnetic and paramagnetic shielding contributions.

The second relativistic effect to be considered arises from the breakdown of Russell–Saunders coupling in favour of j–j coupling of electronic angular momenta. This can result in the introduction of some spin triplet character in the ground spin singlet state of a diamagnetic molecule. Thus a local magnetic field is produced at a given nucleus which gives rise to a change in nuclear shielding.

Nuclear shielding expressions have been reported for a relativistic analogue of Ramsey's theory which is based on the Dirac model[83-88]. So far only semi empirical molecular orbital calculations of the contributions from relativistic effects to the heavy atom chemical shift have appeared[89]. For the proton shieldings of the hydrogen halides these are found to be significant and arise from spin–orbit induced changes in the wavefunction[89]. Similar REX shielding calculations on PbH_3 reveal that a significant paramagnetic shielding contribution to lead arises from its 6S electrons. In the non–relativistic limit these electrons do not contribute to the paramagnetic shielding expression[98] as shown in Fig. 2. A comparison of some CHF and REX calculated shieldings, with the experimental results for the series CX and CX_2 where X = O, S, Se show some predicted relativistic contributions to the paramagnetic shielding terms for the heavier nuclei[99].

Medium Effects on Nuclear Shielding

Both specific and nonspecific interactions may occur between solute and solvent molecules and influence the nuclear shielding. Specific interactions such as protonation, hydrogen bonding and complex formation are well known to produce nuclear shielding changes and have been widely treated by both ab initio and semi empirical molecular orbital calculations[5].

Nonspecific solvent effects may be considered with respect to two types of model. The solvent may be taken as a continuum characterised by its macroscopic dielectric constant, ϵ. With this model electrostatic interactions between the solute and solvent molecules are accounted for in terms of the reaction field theory[90].

The second type of model permits the shielding changes to be interpreted in terms of interactions between pairs of molecules[91]. Unfortunately, precise mathematical expressions appear to be unavailable for many nonbonded interactions. Thus continuum models tend to be more popular for investigating solvent effects on nuclear shielding. The most widely used of these appears to be the solvaton model[92]. Within the INDO/S–SOS framework, the solvaton model predicts solvent induced shifts which are proportional to $(\epsilon-1)/2\epsilon$. In the case of nitromethane the nitrogen shielding is predicted to decrease by about 10 ppm as ϵ increases from about 2 to about 46. This is in very good agreement with the observed changes in the nitrogen shielding for nitromethane in a series of non–hydrogen bonding solvents[93]. The largest solvent induced nitrogen shielding variation found to date is that of 1,2–diazine where a value of 48 ppm is reported[96]. It appears that non–specific solute–solvent interactions are almost as important as those arising from hydrogen–bonding in determining this range.

Such large solvent induced shielding changes suggests caution in comparing solution and solid state nitrogen chemical shifts.

Nuclear Shielding of Macromolecules

Molecular orbital calculations on monomeric species are not readily transferable to polymers on account of electronic structure differences. In polymers electrons may not be constrained to finite regions of space as they are in small molecules. To account for this, and other features of the electronic structures of polymers, tight–binding (TB) molecular orbital calculations using semi empirical parameters have been reported for the nuclear shieldings of a number of polymers[100]. In general INDO/S parameters appear to be more suitable than the CNDO/2 sets. Reasonable agreement between observed ^{13}C shielding tensor components and those calculated using the TB–INDO/S procedure are given for the cis– and trans–polyacetylenes as shown in Fig. 3. It seems unlikely that ab initio calculations of the nuclear shielding of polymers will be available for sometime.

References

1. N.F. Ramsey, Phys. Rev., 1950, 77, 567.

2. D.B. Chesnut, Annual Reports on NMR Spectroscopy, 1989, 21, 51.

3. W. Kutzelnigg and M. Schindler, NMR Basic Principles and Progress, 1991, 23, 1.

4. A.E. Hansen and T.D. Bouman, Annual Reports on NMR Spectroscopy, 1992, 26, in the press.

5. C.J. Jameson in Specialist Periodical Reports on NMR Vol. 9, 1980 to Vol. 21, 1992.

6. A.D. Buckingham and S.M. Malm, Mol. Phys., 1971, 22, 1127.

7. J.C. Facelli, A.M. Orendt, D.M. Grant and J. Michl, Chem. Phys. Lett., 1984, 112, 147.

8. A.M. Orendt, J.C. Facelli, A.J. Beeler, K. Reuter, W.J. Horton, P. Cutts, D.M. Grant and J. Michl, J. Amer. Chem. Soc., 1988, 110, 3386.

9. J.C. Facelli, D.M. Grant and J. Michl, Acc. Chem. Res., 1987, 20, 152.

10. A.M. Orendt, J.C. Facelli, D.M. Grant, J. Michl, F.H. Walker, W.P. Dailey, S.T. Waddell, K.B. Wiberg, M. Schindler and W. Kutzelnigg, Theor. Chim. Acta, 1985, 68, 421.

11. A.E. Hansen and T.D. Bouman, J. Chem. Phys., 1989, 91, 3552.

12. N.F. Ramsey, Phys. Rev., 1950, 78, 699.

13. N.F. Ramsey, Phys. Rev., 1957, 86, 243.

14. N.N. Lipscomb in Adv. Magn. Reson., 1966, 2, 137.

15. S.T. Epstein, J. Chem. Phys., 1965, 42, 2897.

16. F. London, J. Phys. Radium, 1937, 8, 397.

17. R. Ditchfield, J. Chem. Phys., 1972, 56, 5688.

18. R. Ditchfield, Chem. Phys. Lett., 1972, 15, 203.

19. R. Ditchfield, Mol. Phys., 1974, 27, 789.

20. K. Wolinski, J.F. Hinton and P, Pulay, J. Amer. Chem. Soc., 1990, 112, 8251.

21. W. Kutzelnigg, Isr. J. Chem., 1980, 19, 193.

22. M. Schindler and W. Kutzelnigg, J. Chem. Phys., 1982, 76, 1919.

23. M. Schindler and W. Kutzelnigg, J. Amer. Chem. Soc., 1983, 105, 1360.

24. M. Schindler and W. Kutzelnigg, Mol. Phys., 1983, 48, 781.

25. B. Levy and J. Ridard, Mol. Phys., 1981, 44, 1099.

26. D.J. Rowe, Rev. Mod. Phys., 1968, 40, 153.

27. A.E. Hansen and T.D. Bouman, J. Chem. Phys., 1985, 82, 5035.

28. V. Galasso and G. Fronzani, J. Chem. Phys., 1986, 34, 3215.

29. P. Lazzeretti and R. Zanasi, Phys. Rev. A., 1985, 32, 2607.

30. H. Fukui, K. Miura and H. Shinbori, J. Chem. Phys., 1985, 83, 907.

31. D.B. Chesnut and C.K. Foley, J. Chem. Phys., 1986, 84, 852.

32. A.J. Beeler, A.M. Orendt, D.M. Grant, P.W. Cutts, J. Michl, K.W. Zilm, J.W. Downing, J.C. Facelli, M.S. Shindler and W. Kutzelnigg, J. Amer. Chem. Soc., 1984, 106, 7672.

33. R.M. Stevens, K.M. Pitzer and W.N. Lipscomb, J. Chem. Phys., 1963, 38, 550.

34. R. Höller and H. Lischka, Mol. Phys., 1980, 41, 1017.

35. R. Höller and H. Lischka, Mol. Phys., 1980, 41, 1041.

36. P. Lazzeretti and R. Zanasi, J. Chem. Phys., 1978, 68, 832.

37. P. Lazzeretti and R. Zanasi, J. Chem. Phys., 1978, 68, 1523.

38. R. Ditchfield, D.P. Miller and J.A. Pople, J. Chem. Phys., 1970, 53, 613.

39. R. Ditchfield, D.P. Miller and J.A. Pople, J. Chem. Phys., 1971, 54, 4186.

40. H. Nakatsuji, J. Chem. Phys. 1974, 61, 3728.

41. V. Galasso, Theor. Chim. Acta., 1983, 63, 35.

42. H. Fukui, H. Yoshida and K. Miura, J. Chem. Phys. 1981, 74, 6988.

43. H. Fukui, H. Yoshida and K. Miura, J. Chem. Phys. 1982, 77, 5259.

44. P. Lazzeretti and J.A. Tossell, J. Phys. Chem, 1987, 91, 800.

45. J.A. Tossell and P, Lazzeretti, J. Chem. Phys., 1987, 86, 4066.

46. J.A. Tossell and P. Lazzeretti, Chem. Phys. Lett., 1986, 128, 420.

47. J.A. Tossell and P. Lazzeretti, J. Chem. Phys., 1986, 84, 369.

48. J.A. Tossell and P. Lazzeretti, Chem. Phys. Lett, 1986, 132, 464.

49. J.A. Tossell, J. Non Cryst. Solids, 1990, 120, 13.

50. A. Fleischer, M. Schindler and W. Kutzelnigg, J. Chem. Phys., 1987, 86, 6337.

51. M. Schindler, J. Amer. Chem. Soc., 1987, 109, 5950.

52. M. Schindler, J. Chem. Phys., 1988, 88, 7638.

53. M. M. Jaszunski, A. Rizzo and D.L. Yeager, Chem. Phys., 1989, 136, 385.

54. J. Oddershede and J. Geertsen, J. Chem. Phys., 1990, 92, 6036.

55. E.C. Vauthier, M. Comean, S. Odiot and S. Fliszar, Can. J. Chem., 1988, 66, 1781.

56. C.J. Jameson, A.K. Jameson, D. Oppusunggu, S. Wille, P.M. Burrell and J. Mason, J. Chem. Phys., 1981. 81, 74.

57. T.D. Bouman and A.E. Hansen, Chem. Phys. Lett., 1990, 175, 292.

58. G. Fronzoni and V. Galasso, Chem. Phys. 1986, 103, 29.

59. V. Galasso and G. Fronzoni, J. Chem. Phys., 1986, 84, 3215.

60. C.J. Jameson and H.J. Osten, Annual Reports on NMR Spectroscopy, 1986, 17, 1.

61. C.J. Jameson and H.J. Osten, J. Chem. Phys., 1985, 83, 5425.

62. C.J. Jameson and H.J. Osten, Mol. Phys., 1989, 55, 383.

63. P.W. Fowler and W.T. Raynes, Mol. Phys., 1981, 43, 65.

64. H. Nakatsuji, K. Kanda, K. Endo and T. Yanezawa, J. Amer. Chem. Soc., 1984, 106, 4653.

65. K. Kanda, H. Nakatsuji and T. Yonezawa, J. Amer. Chem. Soc., 1984, 106, 5888.

66. H. Nakatsuji and M. Sugimoto, Inorg. Chem., 1990, 29, 1221.

67. H. Nakatsuhi, T.Inoue and T. Nakao, Chem. Phys. Lett., 1990, 167, 111.

68. H. Najatsuji, M. Sugimoto and S. Saito, Inorg. Chem., 1990, 29, 3095.

69. J.E. Combariza, M. Barfield and J.H. Enemark, J. Phys. Chem., 1991, 95, 5463.

70. H. Nakatsuji and T. Nakao, Chem Phys. Lett., 1990, 167, 571.

71. J.A. Tossell, J. Magn. Reson., 1991, 94, 301.

72. H. Nakatsuji, T.Inoue and T. Nakao, Chem. Phys. Lett., 1990, 167, 111.

73. H. Nakatsuji, T. Nakao and K. Kanda, Chem. Phys., 1987, 118, 25.

74. J.A. Pople, Discuss. Faraday Soc., 1962, 34, 7.

75. J.A. Pople, J. Chem. Phys., 1962, 37, 53.

76. J.A. Pople, J. Chem. Phys., 1962, 37, 60.

77. J.A. Pople, Mol. Phys., 1965, 7, 301.

78. K.A.K. Ebraheem and G.A. Webb, Prog. NMR Spectroscopy, 1978, 11, 149.

79. G.A. Webb, in "The Multinuclear Approach to NMR Spectroscopy", Ed. J.B. Lambert and F.G. Riddell, Reidel Co. Dordrecht, 1982, p.29.

80. G.A. Webb, in "NMR of Newly Accessible Nuclei", Vol. 1, Ed. P. Laszlo, Academic Press, London, 1983, p.79.

81. I Ando and G.A. Webb, "Thoery of NMR Parameters", Academic Press, London, 1983, p.47.

82. G.A. Webb, in "NMR of Living System", Ed. T. Axenrod and G. Ceccarelli, Reidel Co, Dordrecht, 1986,p.19.

83. P. Pyykkö, Chem. Phys., 1983, 74, 1.

84. N.C. Pyper, Chem. Phys. Lett., 1983, 96, 204.

85. N.C. Pyper, Chem. Phys. Lett., 1983, 96, 211.

86. Z.C. Zhang and G.A. Webb, J. Mol. Struct., 1983, 104, 439.

87. N.C. Pyper, Mol. Phys., 1988, 64, 933.

88. Z.C. Zhang and N.C. Pyper, Mol. Phys., 1988, 64, 963.

89. P. Pyykkö, A. Görling and N. Rösch. Mol. Phys., 1987, 61, 195.

90. C. Böttcher, "Theory of Electric Polarization", Elsevier, Amsterdam, Vol. 1 (1973).

91. W.T. Raynes, A.D. Buckingham and H.J. Bertstein, J. Chem., Phys., 1961, 36, 3481.

92. I. Ando and G.A. Webb, Org. Magn. Reson., 1981, 15, 111.

93. M. Witanowski, L. Stefaniak, B. Na Lamphun and G.A. Webb, Org. Magn. Reson., 1981, 16, 57.

94. J.A. Tossell, J. Phys. Chem., 1991, 95, 366.

95. J.A. Tossell, Chem. Phys. Lett., 1990, 169, 145.

96. M. Witanowski, W. Sicinska, S. Biernat and G.A. Webb, J. Magn. Reson., 1991, 91, 789.

97. C.J. Jameson, A.C. de Dios and A.K. Jameson, Chem. Phys. Lett., 1990, <u>167</u>, 575.

98. M. Edlund, T. Lejon, P. Pyykkö, T.K. Venkatacholos and E. Buncel, J. Amer. Chem. Soc., 1987, <u>109</u>, 5982.

99. J. Jokisaari, P, Lazzeretti and P. Pyykkö, Chem. Phys., 1988, <u>123</u>, 339.

100. I. Ando, T. Yamanobe, H. Kurosu and G.A. Webb. Annual Reports on NMR Spectroscopy, 1990, <u>22</u>, 205.

GAW/CEA
17.2.92

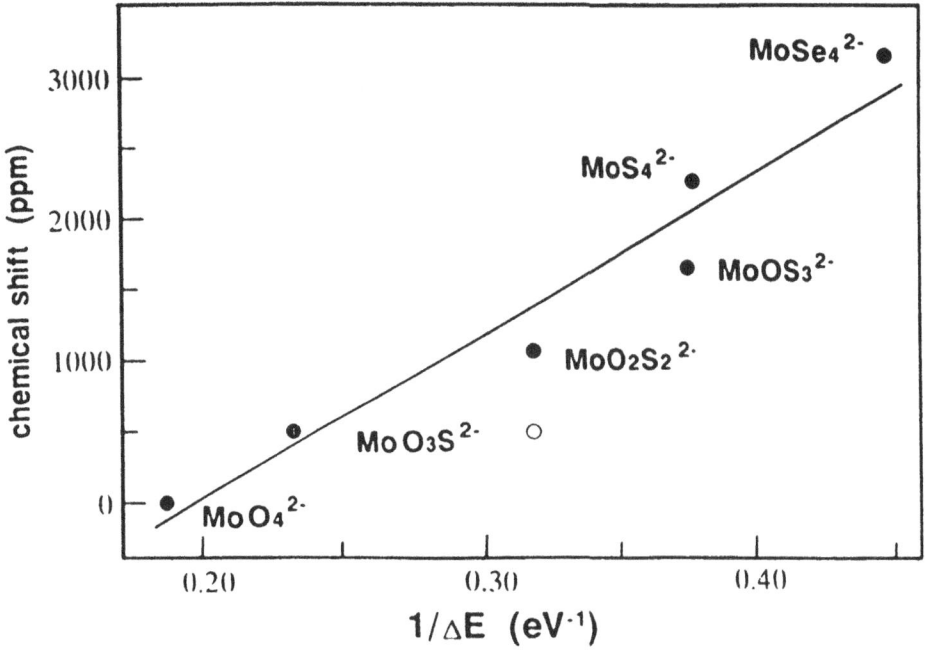

Fig. 1.   Relationship between the inverse of the lowest magnetic dipole
allowed transition energies, ΔE and Mo chemical shifts.
Reproduced from
M. Nakatsuji, M. Sugimoto and S. Saito, Inorg. Chem., 1990, 29,
3096.

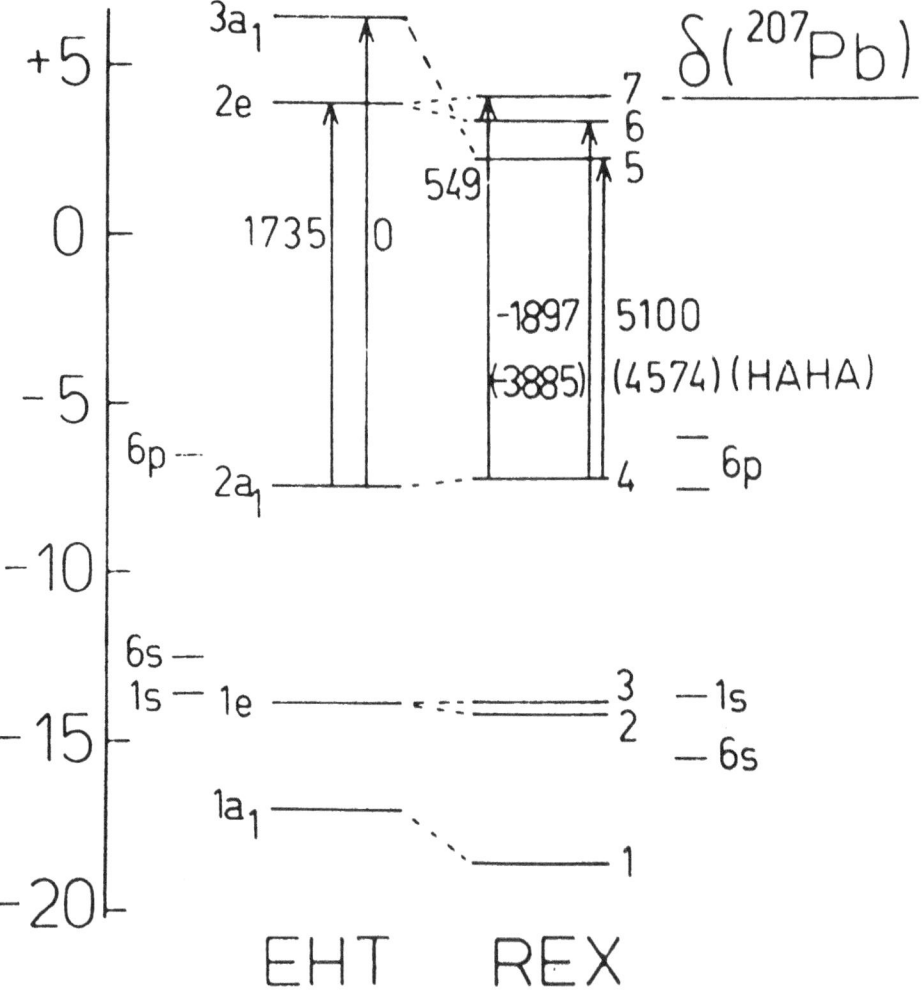

Fig. 2. Paramagnetic contributions (in ppm) to the [207]Pb chemical shift in PbH₃⁻ according to relativistic (REX) and nonrelativistic (EHT) calculations. The contributions from the Pb 6s orbitals are given in parenthesis.
Reproduced from
U. Edlund, T. Lejon, P. Pyykko, T.K. Venkatacholan and E. Buncel, J. Amer. Chem. Soc., 1987, 109, 5982.

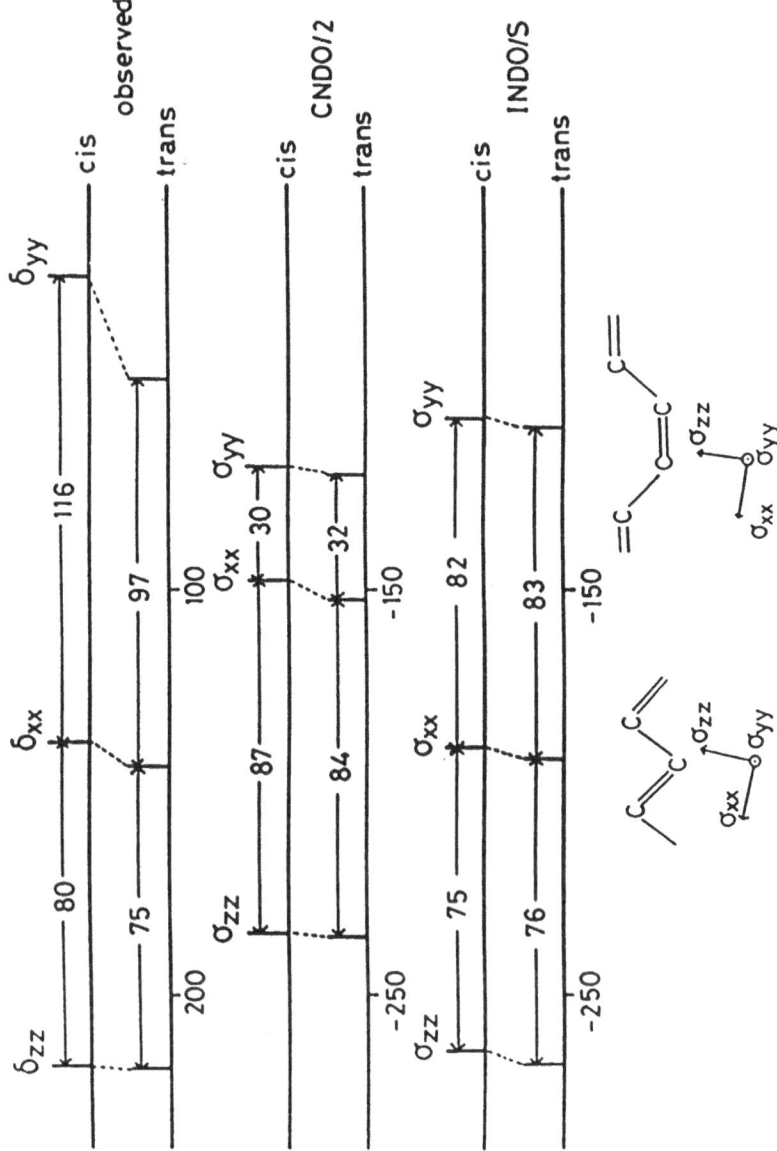

Fig. 3.   The observed and calculated components of ^{13}C NMR shielding
tensors of cis and trans–polyacetylenes.
Reproduced from
I. Ando, T. Yamanobe, H. Kurosu and G.A. Webb. Annual Reports
on NMR Spectroscopy 1990, 22, 220.

Experimental and Theoretical Studies of the Chemical Shift Tensors of Phosphites and Fluorophosphates

Thomas C. Farrar and Jon D. Trudeau
Department of Chemistry
University of Wisconsin–Madison
Madison, Wisconsin 53706

Abstract

Experimental NMR lineshape and relaxation time measurements of phosphite and monofluorophosphate salts have been carried out for both solid state and solution state samples. The chemical shift tensors for the phosphorus, proton and fluorine nuclei were measured as a function of cation, pH, solvent, temperature, concentration and extent of hydration. The P–F, P–O, and P–H bond distances all decrease as the pH is decreased (i.e. as the oxygen atoms are protonated) or as the crystalline sample becomes more hydrated. The isotropic J–coupling, the isotropic chemical shifts and the chemical shift anisotropies for the proton, phosphorus and fluorine nuclei are all dependent on the bond distance between the spin pairs (P–F or P–H). The chemical shift tensors have also been calculated from first principles for these same molecular systems using GAUSSIAN and the LORG method. It was discovered that if a sufficiently large basis set is used in the calculations, there is excellent agreement between the results obtained from the *ab initio* calculations and that from experimental observations for a sample of Na_2PHO_3. For the $HPHO_3^-$ anion and phosphorus acid, H_2PHO_3, however, the agreement is poor.

J. A. Tossell (ed.), Nuclear Magnetic Shieldings and Molecular Structure, 27–48.

1 INTRODUCTION

Over the past several decades enormous advances have been made in both the development of *ab initio* calculations and NMR theory. These developments have been accompanied by equally enormous advances in sophisticated, multinuclear NMR spectrometers and in the power, speed and cost/performance of computers. With these advances, a greater interest in theoretical calculations and experimental measurements of NMR parameters such as chemical shift tensors, dipolar coupling tensors (both direct and indirect) and quadrupole coupling tensors has arisen. This is due to the fact that a knowledge of these parameters can provide a wealth of information about the geometry and electronic structure of molecules.

A major part of our research over the past several years has been focused on obtaining a better understanding of molecular interactions, especially solvent/solute interactions, and how the molecular environment affects molecular and electronic structure and molecular dynamics. Since the chemical shift tensor is very sensitive to changes in the molecular and electronic structure it may be used as a probe to monitor molecular changes. To use this method of investigation to its fullest extent, it is of fundamental importance to have a complete understanding of how changes in structure are related to changes in the chemical shift tensor. In particular, it is important to provide answers to the following questions:

1. What methods are available for the measurement of the chemical shift tensor[1] in the solid state (for single crystal and polycrystalline samples), in the liquid state and in the gas phase? How accurate and/or precise are these measurements?

2. Does the chemical shift tensor change appreciably as a function of physical state and/or environment in a given physical state?

3. To which molecular parameters are the components of the chemical shift tensor most sensitive?

4. How much do intermolecular interactions affect molecular structure and/or chemical shift tensors?

---

[1]Although we direct our attention here to the chemical shift tensor, the same questions are relevant for indirect dipolar coupling and electric field gradient (quadrupole coupling) tensors.

5. How accurate are *ab initio* calculations of chemical shift tensors?

We have chosen two model systems for our investigations, phosphites and fluorophosphates. These two systems have some physical and chemical properties which are very similar, but the chemical shift tensors and the dipolar coupling constants for the two systems are quite different. Both systems are small enough that *ab initio* calculations may be carried out in reasonable time periods. The molecular structures for both systems may be changed in a systematic way such that we may gain some insight into how molecular structure and the values of chemical shift tensor elements are related. The NMR experiments and the *ab initio* calculations reported here are for the anions $PHO_3^=$, $HPHO_3^-$, $PFO_3^=$, and $HPFO_3^-$ and for the molecules H_2PHO_3 and H_2PFO_3. Studies have been carried out in the solid state and the liquid state. Unfortunately, we have not yet been able to carry out gas phase experiments for these molecules.

As will be seen below, the NMR experiments indicate that one of the important molecular parameters in both the phosphites and the fluorophosphates is the PH or the PF bond distance. Since this parameter is so important, we have also carried out x–ray and neutron diffraction studies as well as infrared studies of these molecular systems in order to get complementary information about how the PH and PF bond distances change as a function of physical state, pH, and so on.

2   Experimental

2.1   NMR

For the pH, temperature, solvent and concentration dependent studies of the isotropic J–coupling and chemical shifts, Na_2PHO_3 and Na_2PFO_3 salts were purchased from Pfaltz and Bauer and Johnson Matthey Electronics, respectively. The ethylene glycol was purchased from Aldrich Chemical Company. All water used in solution preparation was deionized via ion exchange columns. The pH values of the different samples were adjusted by adding appropriate amounts of HCl or NaOH. All chemicals were used as obtained from the manufacturer without further purification. A Fisher Accumet Model 925 pH meter and buffered reference solutions were used for the pH measurements.

For the NMR relaxation time studies of sodium phosphite, $Na_2PHO_3 \cdot 5H_2O$ was obtained from Reidel Chemical Co. The waters of hydration were removed by heating to 90°C under vacuum at 10^{-3} Torr. The dried Na_2PHO_3 was then dissolved in D_2O and redried in order to exchange any residual waters of hydration. This Na_2PHO_3 was dissolved in a 40/60 mixture (by volume) of d_6–ethylene glycol/D_2O to make a 0.88 M solution. NaOD was added to adjust the pH to 12.4. The amount of residual HDO was checked by proton NMR measurements. The HDO peak intensity was less than that of either of the proton resonance lines in the $PHO_3^=$ spectrum. For the solid state NMR studies, D_2O solutions of phosphite salts were evaporated to dryness.

Similarly, to perform NMR relaxation time studies of sodium monofluorophosphate, anhydrous Na_2PFO_3 (Alfa Catalog Chemicals) was dissolved in D_2O (99.996% D, Aldrich Chemical Co.) to make a 1.0 M solution. The solution pH was maintained at 8.0 by the addition of 0.1 M NaOD which was prepared by the addition of sodium metal to D_2O. At pH 8 the monofluorophosphate is 99% in the form of the doubly charged anion, $PFO_3^=$, and is quite stable. The Na_2PFO_3 solution was added to deuterated ethylene glycol (99.8% deuterium, MSD Isotopes) to make a final working solution of 0.6 M Na_2PFO_3 in a 60/40 mixture of d6–ethylene glycol/D_2O. For the solid state NMR studies of $Na_2PHO_3 \cdot 5D_2O$, D_2O solutions of sodium fluorophosphate were evaporated almost to dryness in a desiccator.

All NMR solution samples for the relaxation time studies, were pipetted into 8mm NMR tubes which were cleaned before use and treated with sodium EDTA solutions to remove any trace of paramagnetic ion contaminants. The tubes were filled to a height of 40mm. They were degassed by at least eight freeze–pump–thaw cycles on a high–vacuum line to remove any dissolved gases and sealed under high vacuum. Measurements made over a period of 36 months show no change in the relaxation times for a given sample. The 8mm tube is placed and held concentric in a 10mm tube filled with d_6–acetone. This allows field frequency locking at temperatures where the sample d_6–ethylene glycol/D_2O resonances become too broad. The sample has a freezing point of 235 K.

In the liquid state relaxation experiments performed here temperature stability is of paramount importance. For this reason we have designed and built a temperature controller which is capable of holding the temperature constant to better than $\pm$ 0.035°C for periods up to 36 hours or longer. This device has been described elsewhere [1]. Typical phosphorus 90° pulse widths were 25 μs and fluorine 90° pulse widths were 20 μs. All spin inversions were accomplished with composite pulses designed to compensate for RF inhomogeneity ($90_x - 180_y - 90_x$) [2]. Relaxation delays of ten times the relaxation time for the most slowly relaxing of the four spectral lines were used. A spectral width of approximately 1000 Hz was used (the value of J_{PF} varies with pH from 860 Hz at pH 13 to 980 Hz at pH 2). The FIDs were zero filled prior to Fourier transformation to provide at least 15 data points above the half height for each peak. Peak heights were used as a measure of the magnetization.

2.2 *Ab Initio* Calculations

Self-Consistent Field (SCF) wave functions were calculated using GAUSSIAN 86[3] on a VAX 8650. For molecular structure optimizations, we used the Berny optimization method[4] and the default convergence limits for forces (3E-4 Hartree/Bohr) and displacements (1.2E-3 Bohr). An HF/6-31G* basis set {(31)=[2s] for hydrogen and (631/31/1)=[3s,2p,1d] for heavy atoms} was used with default scale factors and exponents. Geometry optimizations were performed assuming proper symmetry and varying all of the bond lengths and angles except for the dihedral angle (when present). Molecular properties were then in general calculated at the optimized geometry.

The chemical shift tensor was calculated using a program written by Hansen and Bouman[5] called RPAC which involves a local origin variant of the coupled Hartree-Fock method in which the random phase approximation (RPA) is applied[6, 7]. These calculations were run on a Cray–Y–MP/8128 where the two–electron integral file needed as input for this program was obtained using GAUSSIAN 90[8]. The 6–311++G(2d,1p) basis was used for these calculations which gives similar results to those of IGLO[9] and Hartree–Fock theories using large basis sets. These types of chemical shift calculations, although not perfect, have provided excellent results for carbon and hydrogen atoms and satisfactory results for many other atoms [10]. As seen below, such calculations also provide excellent agreement with the experimental results for the phosphite ion.

3 Theory

3.1 NMR Relaxation

The theory of spin–lattice relaxation in coupled spin systems has been treated in detail by a number of authors [11–18]. In this treatment, the populations of the energy levels are represented by the diagonal elements of the reduced density matrix, σ_{ii}. Spin–lattice relaxation is then described by the time rate of change of the diagonal matrix elements:

$$\frac{d\sigma_{ii}^*}{dt} = \sum_j R_{iijj}[\sigma_{jj}^* - \sigma_{jj}^*(\infty)], \tag{1}$$

where the asterisk denotes that the tensors are in the interaction representation. In the above equation, σ_{ii}^* and σ_{jj}^* refer to the instantaneous populations of spin levels i and j respectively, $\sigma_{jj}^*(\infty)$ is the equilibrium population of spin level j, and R is the relaxation matrix. The elements of the relaxation matrix, R_{iijj}, are calculated by the theory developed by Redfield [19]. Two terms, A and B, (for ^{31}P and ^1H, respectively) were included in the analysis to account for the residual intermolecular dipolar interactions with solvent deuterium and sodium cations, interactions with trace impurities of paramagnetic ions, and/or the relaxation due to spin–rotation[18]. The random field approximation was used to describe these interactions. The random terms take the form $\gamma_i^2 H_i^2 \tau_c$, where H_i is the randomly fluctuating field produced by spins external to $PHO_3^=$. These random terms depend upon temperature because of the implicit dependence on τ_c.

Experimentally, one does not observe the actual populations of energy levels but population differences which are equal to the line intensities. In order to use line intensities directly in the analysis, the relaxation matrix was transformed to the line intensity basis[20]. The equation describing the spin–lattice relaxation of two–spin systems such as $PHO_3^=$ in line intensity mode is

$$\frac{d\mathbf{I}}{dt} = \mathcal{R}'\mathbf{I}, \tag{2}$$

where $\mathbf{I}$ is a column vector containing, in the present case, the four line intensities, I_1, I_2, I_3, and I_4, and $\mathcal{R}'$ is the relaxation matrix in the line intensity basis given by the Redfield equations. Elements of $\mathcal{R}'$ (for a system with C_{3v} symmetry) contain the following six parameters: dipolar parameter $(p = \gamma_H \gamma_P \hbar / 2r_{PH}^3)$, phosphorus chemical shift anisotropy (CSA), $(\Delta\sigma_P = \sigma_\parallel^P - \sigma_\perp^P)$, proton CSA, $(\Delta\sigma_H = \sigma_\parallel^H - \sigma_\perp^H)$, correlation time (τ_c), A, and B which are the random field terms for the phosphorus and proton nuclei, respectively (see above). The matrix elements of $\mathcal{R}'$ are given in Table 1. The solution to equation 2 is given by

$$I_i(t) = c_1 e^{\lambda_1 t} v_{i1} + c_2 e^{\lambda_2 t} v_{i2} + c_3 e^{\lambda_3 t} v_{i3} + c_4 e^{\lambda_4 t} v_{i4} + I_i(\infty), \qquad (3)$$

where the λ_i and v_{ij} are the eigenvalues and eigenvectors of $\mathcal{R}'$. The c_i's are constants determined by the boundary conditions which are reflected by the initial intensities (or boundary parameters), $I_i(0)$, used in the experiment. The parameter $I_i(\infty)$ represents the equilibrium intensity of peak i. The intensities as measured with a 90° pulse are assumed to be the population differences of the energy levels involved in the nuclear spin transition. There are four such equations corresponding to the four lines in the spectrum in this case.

In the analysis program the relaxation matrix is calculated based on estimated values of $\Delta\sigma_P$, $\Delta\sigma_F$ and/or $\Delta\sigma_H$, p, τ_c, A, and B. This is then numerically diagonalized to give eigenvalues λ_1 through λ_4 and eigenvectors v_1 through v_4. Note that since the number of nuclei is constant, one of the eigenvalues is zero and the relaxation is described by a triple exponential. Using the initial conditions (which depend upon the pulse experiment) and the equilibrium line intensities, the constants c_1 to c_4 are obtained and the line intensities are then calculated. An iterative, non–linear least–squares method is used to find the set of best fit parameters to the experimental line intensities as a function of the relaxation delay. This procedure is a very common one and, in fact, the mathematics of this fitting process is identical to the one used in many laboratories to analyze data from 2D–FT–NMR NOESY experiments to obtain interproton distances in very complex molecules for which the number of coupled differential equations may be much greater than the four equations which we deal with for our system of two coupled spins of 1/2. It is most important to remember, however, that if the values of many parameters are to be estimated simultaneously, the data from a combination of experimental boundary conditions must be fit simultaneously. For example, in the analysis of our experiments which require the estimation of six experimental parameters, almost any reasonable guess of the six parameters will give a reasonable fit if only one set of boundary conditions is used in the fitting process. This is due to the fact that for a single set of boundary conditions, the six NMR parameters are highly correlated and, consequently, many combinations of the parameters will give a good fit. To obtain a meaningful, accurate set of parameters it is essential to fit the data obtained from two sets of experiments simultaneously using two very different boundary conditions (recall that for each set of experiments there are four initial conditions, the intensities of the four lines in the spectra).

The components of the chemical shift tensor, $\sigma_\perp$ and $\sigma_{\parallel}$, may be obtained using the isotropic chemical shift and the chemical shift anisotropy (CSA). The coupled relaxation time measurements give information about the CSA, $(\Delta\sigma = \sigma_{\parallel} - \sigma_\perp)$, and the high resolution experiment gives the isotropic chemical shift, $[\sigma_p(iso) = (1/3)(\sigma_{xx} + \sigma_{yy} + \sigma_{zz}) = (1/3)(2\sigma_\perp + \sigma_{\parallel}]$. Using these identities, one obtains

$$\begin{aligned} \sigma_\perp &= 3\sigma_p(iso) - \Delta\sigma \\ \sigma_{\parallel} &= (2/3)\Delta\sigma + \sigma_p(iso). \end{aligned} \tag{4}$$

The molecular motions in solution are described by τ_c, the correlation time. This parameter is defined to be the time required for typical molecule to move appreciably and is of course dependent upon temperature. For many simple molecular or ionic systems, it has been found that the temperature dependence of the correlation time may be accurately described by an Arrhenius equation

$$\tau_c = \tau_o\, exp(E_a/RT), \tag{5}$$

where E_a is the activation energy for isotropic reorientation, T is the absolute temperature, and τ_o is a pre–exponential factor which is often interpreted as an inverse collision frequency.

Using the ethylene glycol/water solutions describe above, equation 5 is an accurate representation of the temperature dependence of τ_c (or $\tau_\perp$) for temperatures down to 240 K, somewhat above the T_1 minimum (where $\omega_o\tau_c \simeq 1$). Note that $\eta_{\parallel}$, in the present case, does not contribute to the relaxation. At temperatures below the T_1 minimum the solution becomes glassy and the behavior is no longer Arrhenius. The data over the entire tempērature range is thus no longer accurately described by equation 5. Instead, we have found that the equation

$$\tau_c = \tau_0 \, exp[E_a/R(T - T_g)] \tag{6}$$

provides an accurate representation of the decoupled experiments over the entire range of temperatures for the relaxation data presented in this paper where the parameter T_g is sometimes referred to as the glass transition temperature. As the sample approaches T_g, its viscosity increases much more rapidly than predicted by the Arrhenius equation. This equation has frequently been used to describe molecular motions in glasses.

3.2 Line Shapes

NMR line shape theory was worked out in detail many years ago and there are now a variety of computer programs which carry out least squares fitting procedures of the line shape of solid polycrystalline powder samples[21]. The frequency of any system whose Hamiltonian contains a single second–rank tensor (such as chemical–shift or quadrupolar coupling) can be calculated from the relation

$$\omega(\theta,\phi) = (\sin^2\theta\,\cos^2\phi)T_{11} + (\sin^2\theta\,\sin^2\phi)T_{22} + (\cos^2\theta)T_{33} \tag{7}$$

where T_{11}, T_{22}, and T_{33} are the principal values of the tensor of interest, in frequency units, and θ and ϕ are the spherical coordinates of the magnetic field vector in the axis system of the tensor. In the event that dipolar coupling is also present, a doublet is observed. The separation between these doublet lines is given by

$$\Delta\omega(\theta,\phi) = D(1 - 3[\sin\beta\,\sin\theta\,\cos(\phi - \alpha) + \cos\beta\,\cos\theta]^2), \tag{8}$$

where D is the dipolar constant $(\gamma_I\gamma_S\hbar/r_{IS}^3)$ and β and α are the polar and azimuthal angles, respectively, which describe the orientation of the I–S internuclear vector in the principal axis system of the chemical shift tensor. The complete equation for the frequency spectrum is then given by

$$\omega_{\pm}(\theta,\phi) = \omega(\theta,\phi) \pm \Delta\omega(\theta,\phi)/2. \tag{9}$$

We have seen that unless there are very sharp and distinctive features in the NMR line shapes for the powder samples, it is difficult to obtain accurate and meaningful values for the various NMR parameters. For this reason the data from the phosphites samples which were prepared with D_2O give much more accurate results than those prepared with H_2O, since the deuterated water of hydration contributes very little to the over all powder pattern through intermolecular interactions and the resulting spectra have rather sharp features.

4 Results

The isotropic phosphorus and proton chemical shifts and the chemical shift anisotropies have been calculated as a function of the basis set for the $PHO_3^=$ anion. The 6–31G* optimized geometry [r(PH)=147 pm, r(PO)=151.8 pm, $\angle OPH$=103.3°] was used except for one calculation in which the experimentally measured geometry [r(PH) = 142 pm, r(PO) = 152.8 pm, an OPO angle of 107°] of $MgPHO_3 \cdot 5H_2O$ from neutron diffraction data was used. These are compared to results obtained using NMR relaxation studies in solution as well as solid state NMR on $Na_2PHO_3 \cdot 5D_2O$. The results are summarized in Table 2. Note that the calculated shielding components are given on the absolute scale relative to the bare nucleus. The isotropic chemical shift of the phosphorus has been reported on the absolute scale assuming the absolute shielding of 85% aqueous H_3PO_4 is 328.35 ppm while the isotropic proton chemical shift is reported on the delta scale relative to TMS. As can be seen from the table, only for the largest basis set used, at the 6–311++G(2d,1p) level, does one begin to obtain good agreement with the experimentally observed results. The smaller basis sets give poor results without seeming to converge to the experimental values. As was stated above, the last computational entry in the table used the experiment geometry. The results are very similar to the results obtained when using the optimized geometry.

Using the large basis set, the components of the phosphorus and proton chemical shift tensors of the $HPHO_3^-$ anion and phosphorus acid were also calculated; the results are shown in Table 3. Again the calculated values are reported on the absolute shielding scale. Experimental data obtained by solid state NMR experiments performed on $Na_2PHO_3 \cdot 5D_2O$ are also shown given on the delta scale with respect to TMS and 85% H_3PO_4 for phosphorus; results for proton are with respect to TMS on the delta scale. As can be seen, although there is good agreement between the experimental and calculated results for the doubly charged anion, the agreement for the singly charged anion and for the acid is rather poor. This seems to indicate that, at least for the systems studied here, there are significant differences between the structure of anions or the acid in the solution and in the solid state and their structure in the gaseous state, assuming that the *ab initio* calculations faithfully represent the chemical shift tensor in the gas phase.

The pH dependence of the isotropic chemical shifts and J–coupling for phosphite and fluorophosphate are shown in Table 4 (and in Figure 1). The scalar coupling constants for both the phosphites and the fluorophosphates exhibit typical titration curves. This is also the case for the isotropic proton chemical shift in the phosphites and for isotropic chemical shifts of both the phosphorus and the fluorine nuclei in the fluorophosphates. The isotropic phosphorus chemical shift for the $HPHO_3^-$ ion, however, shows a clear minimum in the titration curve. It appears that the pH dependence of the isotropic phosphorus chemical shift in $PHO_3^=$ is a particularly demanding test of the theory. Even with the largest basis set, this behavior is not predicted correctly.

5   Discussion

From these experiments we have discovered that accurate values of chemical shift tensors can be obtained from solid powder samples of simple spin systems. The most accurate results are obtained if a static, decoupled spectrum is obtained for a two spin system. This provides accurate values for all components of the chemical shift tensor. Using this value for the chemical shift tensor, line shape analysis programs can provide accurate information about the dipolar coupling and hence the internuclear bond distance. It is also important to minimize intermolecular dipolar interactions. In our experience static line shape experiments give more accurate information than magic angle spinning experiments or second moment analysis.

Relaxation time studies of two spin systems in the liquid state provide accurate information about the chemical shift tensors of the two spins. This is a particularly good way to obtain information about the proton chemical shift tensor where in the solid state there are too many resonances in the spectrum to analyze. It is essential that data is obtained from enough independent experiments with different boundary conditions to ensure that the correlation between the various NMR parameters is minimal. For this reason, experiments with systems of low molecular symmetry may be quite difficult.

An important part of the development of any theory is a test of the accuracy of the theoretical predictions. As can be seen from Table 2, at the 6–311G++(2d,1p) level the agreement for the $Na_2PHO_3 \cdot 5D_2O$ sample in the solid and solution states is quite good. The agreement is even better comparing experimental data from the solid state sample of anhydrous Na_2PHO_3 and K_2PHO_3. From Table 5, however, we see that the phosphorus CSA for the phosphite anion changes from −95.1 ppm for the phosphorus acid to −147 ppm for the anhydrous K_2PHO_3 sample. This change of about 50 ppm is somewhat smaller than 80 ppm change predicted by the *ab initio* calculations. We should also point out that Table 5 shows a large dependence of the chemical shift tensor on the cation bound to the molecule which is probably changing the cell structure in the single crystal. This dependence is not taken into account in the theoretical calculations.

There is an interesting correlation between the various chemical shift parameters and the directly bonded P–H bond distance. If for the phosphorus nucleus in the phosphite salts $(\sigma_{11} + \sigma_{22})/2$ is plotted as a function of bond distance, one observes a linear dependence with a slope of about +2.5 ppm/pm. A similar linear relationship is seen for the dependence of the phosphorus CSAs and σ_{33} values as a function of bond distance. The slope is −4.3 ppm/pm for the CSA and −2.1 ppm/pm for σ_{33}.

6   Conclusions

For the systems studied here there is a wide variation in the components of the chemical shift tensor as a function of pH, cation, extent of hydration, and physical state. Preliminary results indicate that there is a linear relation between the components of the chemical shift tensor and the bond distance in an isolated two spin system. *Ab initio* calculations provide considerable insight into the relation between molecular structure and the components of the chemical shift tensor, but the accuracy of such calculations is still limited, even for very large basis sets, . At least part of the difference between the calculated values and the experimentally observed values is due to an incomplete and/or inaccurate knowledge of the values of the structural parameters in the liquid or solid state.

7 Acknowledgements

We would like to thank the National Science Foundation, grant number CHE–9102674, for the support of this research. Dr. Ilene Locker–Carpenter from Cray Research kindly performed the *ab initio* calculations at the 6–311++G(2d,1p) level.

References

[1] Farrar, T.C.; Sidky, E.Y.; Decatur, J.D., J. Magn. Reson., **1990**, 86, 605.

[2] Levitt, M.H., J. Magn. Reson., **1982**, 48, 234.

[3] Gaussian 86, Frisch, M. J.; Binkley, J. S.; Schlegel, H. B.; Ragharachari, K.; Melius, C. F.; Martin, R. L.; Stewart, J. J. P.; Bobrowicz, F. W.; Rohlfing, C. M.; Kahn, L. R.; Defrees, D. J.; Seeger, R.; Whiteside, R. A.; Fox, D. J.; Fleuder, E. M.; Pople, J. A., Carnegie-Mellon Quantum Chemistry Publishing Unit, Pittsburgh PA, **1984**.

[4] Schlegel, H. B. *J. Comp. Chem.* **1980**, *3*, 214.

[5] Hansen, A. E.; Bouman, T. D. *J. Chem. Phys.* **1985**, *82*, 5035.

[6] Hansen, A. E.; Bouman, T. D. *Adv. Chem. Phys.* **1980**, *44*, 545.

[7] Bouman, T. D.; Hansen, A. E.; Voigt, B.; Rettrup, S. *Int. J. Quant. Chem.* **1983**, *23*, 595.

[8] Gaussian 90, Revision F, M. J. Frisch, M. Head-Gordon, G. W. Trucks, J. B. Foresman, H. B. Schlegel, K. Raghavachari, M. Robb, J. S. Binkley, C. Gonzalez, D. J. Defrees, D. J. Fox, R. A. Whiteside, R. Seeger, C. F. Melius, J. Baker, R. L. Martin, L. R. Kahn, J. J. P. Stewart, S. Topiol, and J. A. Pople, Gaussian, Inc., Pittsburgh PA, **1990**.

[9] Schindler, M. and Kutzelnigg, W., *JACS*, Vol. 105, **1983**, pp 1360.

[10] Chesnut, D. B. *Annual Reports on NMR Spectroscopy* **1989**, *21*, 51.

[11] Farrar, T.C. and Locker, I.C.; *J. Chem. Phys.*, **1987**, *87*, 3281.

[12] Redfield, A.G.; *Advances in Magnetic Resonance*, Waugh, J.S. Ed; Academic: New York, **1965**; Vol. 1.

[13] Shimizu, H.; *J. Chem. Phys.* **1964**, *40(11)*, 3357.

[14] Mackor, E.L. and MacLean, C.; *Prog. in NMR Spect.*, **1967**, *3*, 129.

[15] Werbelow, L.G. and Grant, D.M.; *Adv. Mag. Res.* **1977**, *9*, 189.

[16] Blicharski, J.S.; *Z. für Naturforschung*, **1972**, *A27*, 1355.

[17] Koenigsberger, E.; Sterk, H.; *J. Chem. Phys.*, **1985**, *83*, 2723.

[18] Vold, R. L.; Vold, R. R.; *Prog. NMR Spect.*, **1978**, *12*, 79.

[19] Redfield, A.G.; *Advances in Magnetic Resonance*, Waugh, J.S. Ed; Academic: New York, **1965**; Vol. 1.

[20] Locker, I.; Ph.D. Thesis, University of Wisconsin–Madison, **1987**.

[21] Oas, T.G.; Drobny, G.P.; Dahlquist, F.W.; J. Magn. Reson., **1985**, *78*, 408.

Table 1:

Elements of the Relaxation Matrix in the line intensity basis for a system with two coupled spins of 1/2 and C_{3v} symmetry (i.e. the assymetry parameter is zero). Note that here the principal axis frames of the dipolar and chemical shift interactions are coincident.

$$\mathcal{R}'_{1111} = -\frac{2}{3}(3p - \Delta\sigma_s)^2\, k_s - 6p^2 k_+ - p^2 k_- - A$$

$$\mathcal{R}'_{1122} = -6p^2 k_+ - p^2 k_-$$

$$\mathcal{R}'_{1133} = -\frac{1}{3}(3p - \Delta\sigma_I)^2\, k_I - 6p^2 k_+ + p^2 k_- - \frac{B}{2}$$

$$\mathcal{R}'_{1144} = \frac{1}{3}(3p + \Delta\sigma_I)^2\, k_I - 6p^2 k_+ + p^2 k_- + \frac{B}{2}$$

$$\mathcal{R}'_{2222} = -\frac{2}{3}(3p + \Delta\sigma_s)^2\, k_s - 6p^2 k_+ - p^2 k_- - A$$

$$\mathcal{R}'_{2233} = \frac{1}{3}(3p - \Delta\sigma_I)^2\, k_I - 6p^2 k_+ + p^2 k_- + \frac{B}{2}$$

$$\mathcal{R}'_{2244} = -\frac{1}{3}(3p + \Delta\sigma_I)^2\, k_I - 6p^2 k_+ + p^2 k_- - \frac{B}{2}$$

$$\mathcal{R}'_{3311} = -\frac{1}{3}(3p - \Delta\sigma_s)^2\, k_s - 6p^2 k_+ + p^2 k_- - \frac{A}{2}$$

$$\mathcal{R}'_{3322} = \frac{1}{3}(3p + \Delta\sigma_s)^2\, k_s - 6p^2 k_+ + p^2 k_- + \frac{A}{2}$$

$$\mathcal{R}'_{3333} = -\frac{2}{3}(3p - \Delta\sigma_I)^2\, k_I - 6p^2 k_+ - p^2 k_- - B$$

$$\mathcal{R}'_{4411} = \frac{1}{3}(3p - \Delta\sigma_s)^2\, k_s - 6p^2 k_+ + p^2 k_- + \frac{A}{2}$$

$$\mathcal{R}'_{4422} = -\frac{1}{3}(3p + \Delta\sigma_s)^2\, k_s - 6p^2 k_+ + p^2 k_- - \frac{A}{2}$$

$$\mathcal{R}'_{4444} = -\frac{2}{3}(3p + \Delta\sigma_I)^2\, k_I - 6p^2 k_+ - p^2 k_- - B$$

$$\mathcal{R}'_{2211} = \mathcal{R}'_{3344} = \mathcal{R}'_{4433} = \mathcal{R}'_{1122}$$

where $\Delta\sigma_i = \gamma_i B_0(\sigma_\parallel - \sigma_\perp)_i,$ $p = \gamma_I \gamma_S \hbar/2r_{IS}^3$ and

$$k_0 = \frac{\tau_c}{5}$$

$$k_i = \frac{\tau_c}{5(1 + \omega_i^2 \tau_c^2)} \qquad i = \text{I or S}$$

$$k_{2i} = \frac{\tau_c}{5(1 + 4\omega_i^2 \tau_c^2)} \qquad i = \text{I or S}$$

$$k_+ = \frac{\tau_c}{5[1 + (\omega_I + \omega_S)^2 \tau_c^2]}$$

$$k_- = \frac{\tau_c}{5[1 + (\omega_I - \omega_S)^2 \tau_c^2]}$$

Table 2:

Chemical shift tensor components of $PHO_3^=$ as a function of basis set. The structural parameters were obtained from an energy minimization using 6-31G* basis [r(PH)=147 pm, r(PO)=151.8 pm, OPH=103.3°]. The isotropic chemical shifts are relative to the bare nucleus where an upfield shift is positive. The experimental values reported are an average from results obtained using NMR relaxation studies in solution as well as solid state NMR on $Na_2PHO_3 \cdot 5H_2O$. The isotropic chemical shift values are given relative to the bare nucleus.

Basis	σ_o^P	$\Delta\sigma^P$	σ_o^H	$\Delta\sigma^H$	# contracted gaussians
4-31G	498.87	−69.7	31.17	7.98	42
6-31G	532.25	−72.93	31.98	6.62	42
6-31++G	513.09	−67.23	30.23	7.73	59
6-311G	201.90	−123.34	31.36	8.30	63
4-31G*	490.64	−109.61	40.56	27.29	66
4-31G**	630.15	−102.98	28.07	7.76	69
6-31G**	642.89	−107.19	27.97	7.20	69
DZD	587.44	−108.24	29.52	6.09	70
DZP	587.63	−110.76	27.80	7.58	73
6-311++G	200.38	−105.71	29.93	8.26	80
6-31++G**	633.78	−108.42	26.82	7.38	86
6-311G**	400.41	−112.53	42.01	21.19	86
6-311++G(2d,1p)	360.8	−154.3	26.08	8.24	183
6-311++G(2d,1p)	362.6	−147.5	26.14	8.65	183[†]
experiment (1)	360[††]	−130 ± 1	5.94[†††]	5.4 ± 0.3	

[†]This result was obtained for r(PH) = 142 pm, r(PO) = 152.8 pm and an OPO angle of 107°, the experimental values obtained from neutron diffraction study of $MgPHO_3 \cdot 5H_2O$.

[††]Reported on the absolute scale assuming the isotropic phosphorus chemical shift of 85% H_3PO_4 is 328.35 ppm.

[†††]Reported on the delta scale relative to TMS.

Table 3:

Comparison of chemical shift tensors between *ab initio* calculations and experimental measurements. The calculations were performed using 6–311++G(2d,1p) at the 6–31G* optimized geometries on a Cray-Y-MP/8128 computer by Ilene Locker–Carpenter. The experimental data was obtained by solid state NMR experiments and/or solution state relaxation time experiments of $Na_2PHO_3 \cdot 5D_2O$. The calculated values are reported on the absolute shielding scale. The experimental values for the phosphorus have been converted to the absolute shielding scale. The proton values were all done on solution samples and are all reported on the delta scale with respect to TMS.

Molecule		^1H Theory	Exp	^{31}P Theory	Exp
$PHO_3^=$	$\sigma_0 =$	26.08	5.95	360.79	323
	$\sigma_\parallel =$	31.57		257.94	237
	$\sigma_\perp =$	23.33		412.22	366
	$\Delta\sigma =$	8.24	5.40	−154.28	− 130
	$\eta =$			0.12	
$HPHO_3^-$	$\sigma_0 =$	26.12	6.10	382.25	323
	$\sigma_{11} =$	23.33		261.37	401
	$\sigma_{22} =$	23.36		350.16	323
	$\sigma_{33} =$	31.69		535.12	244
	$\Delta\sigma =$	8.34		229.44	118
	$\eta =$	0.006		0.581	0.98
H_2PHO_3	$\sigma_0 =$	25.94	6.20	389.86	318
	$\sigma_{11} =$	23.06		279.59	379
	$\sigma_{22} =$	23.13		351.34	322
	$\sigma_{33} =$	31.64		538.65	255
	$\Delta\sigma =$	8.54		223.19	95.5
	$\eta =$	0.012		0.482	0.9

Table 4:

NMR parameters as a function of pH as measured by NMR relaxation experiments of solution samples. The J-coupling values are in units of Hz and the chemical shift values are in ppm. The internuclear distances have been corrected for vibrational averaging and are in excellent agreement with x-ray and neutron diffraction values.

NMR Parameter	$PHO_3^=$	$HPHO_3^-$	H_2PHO_3
$J_{iso}(PH)$	565	630	> 680
$\sigma_{iso}(H)$	5.95	6.10	6.20
$\sigma_{iso}(P)$	3.8	3.3	5.4
CSA(H)	5.5 ± 0.4		
CSA(P)	−128 ± 5		
$\Delta\sigma_{iso}(H)/\Delta T$	−0.004		
$\Delta\sigma_{iso}(P)/\Delta T$	+0.006		
r_{PH}	146.0 ± 0.5		

NMR Parameter	$PFO_3^=$	$HPFO_3^-$	H_2PFO_3
$J_{iso}(PF)$	−870	−909	< −920
$\sigma_{iso}(F)$	4.20	3.50	
$\sigma_{iso}(P)$	1.5	−4.5	
CSA(F)	+125.7 ± 1.0		
CSA(P)	−127.9 ± 3.0		
$\Delta\sigma_{iso}(F)/\Delta T$	0.016		
$\Delta\sigma_{iso}(P)/\Delta T$	0.010		
r_{PF}	166.0 ± 0.5		

Table 5:

Phosphorus chemical shift tensor components of $PHO_3^=$ as a function of cation and pH using solid state NMR on non–crystalline samples. The components are given assuming that the resonance of 85% H_3PO_4 is 0.0 ppm on the σ scale where upfield is positive. The chemical shift anisotropy (CSA) is defined to be $\sigma_{33} - (\sigma_{11} + \sigma_{22})/2$ where $|\sigma_{33} - \sigma_{iso}| \geq |\sigma_{22} - \sigma_{iso}| \geq |\sigma_{11} - \sigma_{iso}|$. The asymmetry parameter, η, is given by $(\sigma_{22} - \sigma_{11})/(\sigma_{33} - \sigma_{iso})$. The numbers in parenthesis are the estimated error in the least significant digit.

Molecular System	CSA(P)	η	σ_{33}	σ_{22}	σ_{11}
H_2PHO_3	−95.1	−0.90	−73.1(3)	−6.5(2)	50.6(2)
$MgPHO_3 \cdot 6H_2O$	−111.9	0.0	−82.6(2)	29.3(5)	29.3(5)
$MgPHO_3 \cdot 6D_2O$	−112.3	0.0	−83.1(3)	29.3(2)	29.3(2)
$MgPHO_3$	−106.3	−0.42	−74.1(2)	17.1(3)	47.2(4)
$Mg(DPHO_3)_2$	−101.6	−0.93	−72.2(4)	−0.1(3)	60.8(4)
$CaPHO_3 \cdot H_2O$	−107.6	−0.27	−77.3(6)	21(2)	40(2)
$Ca(HPHO_3)_2$	−115.4	−0.89	−81.5(3)	−0.2(2)	68.2(2)
$Ca(DPHO_3)_2$	−115.6	−0.91	−79.2(2)	1.3(2)	71.4(2)
$Na_2PHO_3 \cdot 5H_2O$	−127.5	−0.17	−90.0(4)	30.4(8)	44.7(8)
$Na_2PHO_3 \cdot 5D_2O$	−129.5	−0.12	−91.7(4)	33(1)	43(1)
Na_2PHO_3	−139.8				
$NaDPHO_3$	−118.0	−0.98	−84.2(5)	−4.8(4)	72.5(5)
$Na(H_{1.5}PHO_3)_2$	−85.1	0.0	−70.1(7)	25.9(2)	25.9(2)
K_2PHO_3	−147.1	0.0	−96.4(4)	50.7(5)	50.7(5)
$KHPHO_3$	−141.8	0.83	−83.6(4)	−5.4(3)	97.3(4)
$KDPHO_3$	−138.4	0.84	−82.7(6)	−5.2(4)	94.4(5)

48

Figure 1.   The isotropic scalar coupling, J_{PH}, as a function of pH for Na_2PHO_3 in H_2O solution at 298 K.

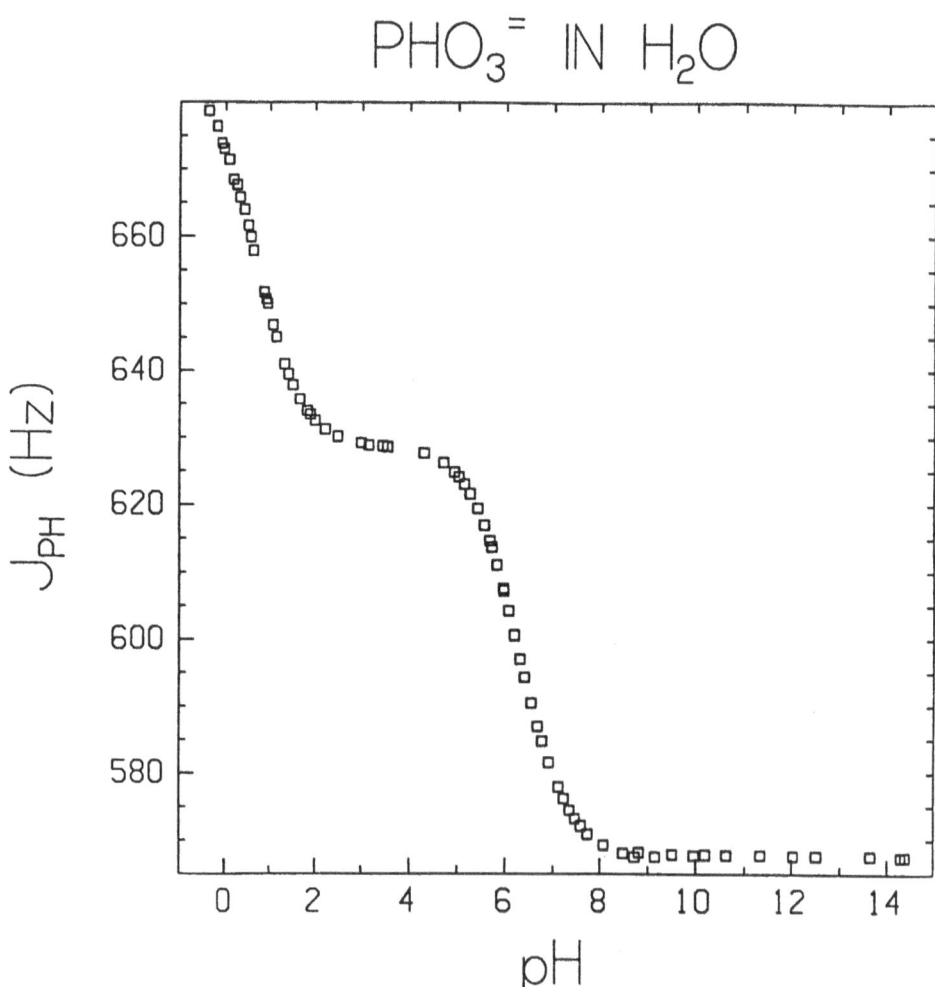

SOLID STATE NMR CHEMICAL SHIFTS OF CHALCOGENIDES AND PNICTIDES

HELLMUT ECKERT, KELLY MORAN, DEANNA FRANKE,
AND CHRISTOPHER HUDALLA

Department of Chemistry
University of California
Santa Barbara, CA 93106

ABSTRACT. Chemical shift data available in the literature on binary II-VI and III-V semiconductors and semiconductor alloys and new data obtained on II-IV-V_2 chalcopyrite semiconductors and their respective alloys are reviewed and discussed in terms of current molecular orbital descriptions for solids. The chemical shift trends observed for the ^{33}S, ^{77}Se, and ^{125}Te resonances in binary II-VI compounds are satisfactorily explained in terms of the Bond Orbital Model (BOM) developed by Harrison, whereas the ^{113}Cd chemical shift trends observed for the cadmium chalcogenides disagree with BOM predictions. For the III-V compounds, the BOM fails to account for the experimentally observed chemical shift trends. Here, the two-electron bond orbital model including the effect of inter-electronic interactions provides a much better description. In terms of this model, the major chemical shift differences seen along homologous series are due to differences in anion-cation overlap rather than bond ionicity. Applications of this model for explaining chemical shift trends in II-IV-V_2 chalcopyrite semiconductors and II-VI, III-V, and II-IV-V_2 semiconductor alloys are discussed.

1. Introduction

Solid chalcogenides and pnictides comprise an important class of materials with unique and technologically promising optical and electronic properties. There has been substantial recent interest in the full characterization of new phase diagrams in such systems [1], the development of new synthetic strategies [2], the identification of unusual bonding environments [3], and the development of a deeper

49

J. A. Tossell (ed.), Nuclear Magnetic Shieldings and Molecular Structure, 49–73.
© 1993 *Kluwer Academic Publishers.*

understanding of their physical properties on the basis of structural and chemical bonding characteristics [4]. Modern solid state nuclear magnetic resonance techniques are greatly contributing to this development [5], particularly by providing detailed information about local bonding arrangements in disordered and amorphous systems. This is so because the exact resonance frequency of the nuclei in a magnetic field is influenced by their electronic environments, which in turn reflect coordination number and symmetry, the hybridization state and the electron density of the atom to which the nucleus studied belongs. This phenomenon, known in the literature as the "chemical shift", is largely responsible for the pivotal significance of solid (and liquid) state NMR in structural chemistry. Unfortunately, the precision with which chemical shifts can be measured exceeds greatly the accuracy with which they are calculable from first principles. Thus, structural applications of chemical shift data in glasses generally utilize one or a combination of the following approaches: (1) experimental chemical shift comparisons with crystallographically characterized model compounds, (2) semi-empirical predictions based on correlations of chemical shifts with local structural and bonding parameters, and (3) ab initio MO calculations of chemical shifts for simple model structures.

The systems of interest in this study comprise the sulfides, selenides, tellurides, phosphides, arsenides, and antimonides of main group and post transition-metal elements. Due to their complexity these systems have thus far eluded the ab initio approach, although there is some hope that the chemical shift properties of simpler systems will become calculable with reasonable accuracy in the near future [6]. Nevertheless, many attempts have been made to understand the chemical shift properties of simple cubic II-VI and III-V semiconductors on the basis of tight-binding MO calculations in close feedback with experimental measurements of dielectric constants and spectroscopic energy gaps [4,7]. The present paper will review the current state of this field and discuss new data obtained on II-IV-V$_2$ chalcopyrite semiconductors in the light of this theory. We will finally identify some common threads in NMR studies of semiconductor alloys, which provide unique information about the bonding properties in these materials.

2. Solid-State MO Theoretical Predictions for Chemical Shifts in Cubic Average-Valence-IV Compounds.

It is generally accepted that the large variations of the chemical shifts observed for a given atom bonded to atoms within homologous series are due to changes in the (dominant) paramagnetic term of Ramsey's formula [8] for chemical shifts. This term describes the paramagnetic shift arising from the angular momentum of excited electronic states which are mixed into the ground state wavefunction. Following Jameson and Gutowsky [9], only shielding effects arising from electrons centered on the atom in question need to be considered, and σ_{para} can be expressed in terms of a LCAO-MO framework as arising from valence electron imbalances P_u and D_u in the p and d orbitals, respectively:

$$\sigma_{para} = -(\mu_o e^2 h^2/6\pi m^2)\, \Delta^{-1}[<r^{-3}>_p P_u + <r^{-3}>_d D_u]. \qquad \{1\}$$

In this equation, m and e denoted the electron mass and charge, μ_o is the magnetic permeability constant and Δ is an average excitation energy involving summation over all excited states. For simple cubic tetrahedrally coordinated II-VI and III-V compounds the d-electron contribution is generally neglected. Harrison and coworkers [10] have developed a simple solid-state molecular orbital approach that allows the imbalances P_u to be evaluated: Four sp^3 wave functions of the anions and cations $|h^a>$ and $|h^c>$, respectively, are combined to form four bonding orbitals, b:

$$b = u_a|h^a> + u_c|h^c>$$

Minimization of the bond energy

$$E_b = <b|\mathcal{H}|b>/<b|b>$$

by variation of the coefficients u_a and u_c yields the result [10]:

$$u_c^2 = \frac{1}{2} \left[\frac{1 - S(1 - \alpha_p^2)^{1/2}}{1 - S^2} - \frac{\alpha_p}{(1 - S^2)^{1/2}} \right] \qquad \{2a\}$$

$$u_a^2 = \frac{1}{2} \left[\frac{1 - S(1 - \alpha_p^2)^{1/2}}{1 - S^2} + \frac{\alpha_p}{(1 - S^2)^{1/2}} \right] \qquad \{2b\}$$

where the plus and the minus signs refer to the anion and the cation, respectively. The total bonding energy is separated into a homo- and a heteropolar contribution V_2 and V_3, respectively. Harrison's MO approach does not yield values of the average excitation energy Δ, however, this parameter is expected to correlate with the total binding energy defined as

$$E_b = 2(V_2^2 + V_3^2)^{0.5} \qquad \{3\}$$

Furthermore, a bond polarity α_p is defined as the heteropolar contribution of the total binding energy, i.e.

$$\alpha_p = V_3/(V_2^2 + V_3^2)^{1/2} \qquad \{4\}$$

While the overlap $S = \langle h^a | h^c \rangle$ between the anion and cation wave functions is frequently neglected for strongly ionic compounds, it can become quite large for more covalent systems. Huang et al. have presented further refinement of Harrison's one-electron BOM considering inter-electronic interactions ('two-electon BOM') [11]. Their analysis reveals that the bond polarities between the nine III-V materials are more similar than estimated by Harrison's method, and vary with atomic number in an unsystematic manner. The overlap integrals range between 0.6 and 0.66 and show systematic trends with atomic number: for a given cation, S increases with increasing size of the counter-anion, whereas for a given anion, S decreases with increasing size of the counter-cation.

In the LCAO-MO framework, eq. 1 can be written as

$$\sigma_{para} = -(\mu_0 e^2 h^2/4\pi m^2) \, \Delta^{-1} \{ \langle r^{-3} \rangle_p [u^2(1 - u^2)] \}, \qquad \{5\}$$

where only p-electron imbalances have been considered.

The $<r^{-3}>_p$ average is calculated from Slater orbitals according to:

$$<r^{-3}> = (Z_{eff})^3/n_{eff}^3 l(l+1/2)(l+1). \tag{6}$$

Z_{eff} and n_{eff} are the effective nuclear charge and principal quantum numbers, respectively and $l = 1$ for p-electrons [12]. The effective nuclear charge, in turn, depends on the polarity α_p and can be calculated according to Slater's rules [13]. The results are:

Group II elements: $Z_{eff} = 3.2 - 1.4\ u_i^2$
Group III elements: $Z_{eff} = 4.2 - 2.1\ u_i^2$
Group IV elements: $Z_{eff} = 5.2 - 2.8\ u_i^2$
Group V elements: $Z_{eff} = 5.5 - 2.1\ u_i^2$
Group VI elements: $Z_{eff} = 5.8 - 1.4\ u_i^2$ {7}

Here the u_i^2 values are taken from eq. {2a} for Group II, III, and IV elements, which are positively polarized in the compounds under study and from eq. {2b} for the Group V and VI elements, which are negatively polarized. Thus, for cationic constituents a decrease in α_p decreases Z_{eff} and thus reduces $<r^{-3}>$, whereas for anionic constituents the opposite will be true.

Equation {5} states that chemical shift trends within homologous or solid solution series could be due to variations in either Δ^{-1}, $<r^{-3}>$, or P_u (D_u). For the negatively polarized (anionic) bonding partner, a decrease in bond polarity α_p causes an increase in σ_p, both due to the effect on P_u (D_u) and $<r^{-3}>$. This will be true for any given value of S. It is further important to realize that for homologous series of II-VI, III-V, and II-IV-V$_2$ semiconductors, a decrease in α_p is generally accompanied by a decrease in the binding energy (and thus presumably an increase in Δ^{-1}). Therefore, for the negatively polarized bonding partner, all variables in eq. {5} change in the same direction, resulting in increasingly paramagnetic chemical shifts with decreasing bond polarity. As discussed further below, these considerations are supported by the majority of the actual experimental results.

For the positively polarized (cationic) bonding partner the effect of decreasing bond polarity on P_u (or D_u) and Δ^{-1} on the one hand and on $<r^{-3}>$ on the other hand go in opposite directions. Based on interpretations in terms of the ionicity scales by Phillips and Harrison, previous workers have reported observations of increased σ_p with increasing polarity within homologous series. It has thus been suggested that the effect of polarity on $<r^{-3}>_p$ is predominant [14,15]. Since both parameters depend on u_i^2, this question can be easily resolved, comparing the terms in eqs. {2} and {7} for homologous series of compounds.

3. Survey of Chemical Shifts in Binary and Ternary Average-Valence IV Compounds.

Below we summarize all of the chemical shift data available in the literature for cubic tetrahedral semiconductors, including some new data for several tetragonal II-IV-V_2 chalcopyrites, which are also characterized by tetrahedral bonding. Unless noted otherwise, the chemical shifts are referenced to the binary compound that resonates most upfield and downfield shifts are listed as positive. Besides the simple binary compounds, several solid-solution systems have been studied in detail. For the latter it is often possible to observe resolved spectra characterizing individual local environments. For instance, [125]Te NMR spectra of $Cd_{1-x}Zn_xTe$ and $Hg_{1-x}Cd_xTe$ alloys show five partially resolved resonances attributable to the five possible $TeZn_nCd_{4-n}$ ($TeCd_nHg_{4-n}$) nearest neighbor configurations ($0 \leq n \leq 4$) [16-19]. Similar site resolution has recently been demonstrated in $In_{1-x}Ga_xP$ alloys [20], while others show only unstructured single-line resonances [21]. To facilitate the discussion of these chemical shift in terms of Bond orbital considerations, Table 1 lists Phillips ionicities, the Harrison parameters α_p and E_g, values of α_p and S given by Huang for III-V systems, and calculated values of $u^2(1-u^2)$ and Z_{eff}^3 obtained from this information.

3.1. NMR CHEMICAL SHIFTS IN BINARY II-VI, III-V, AND II-IV-V_2 COMPOUNDS

3.1.1. *Binary II-VI Systems.* Table 2 lists the [33]S, [77]Se, [125]Te, and [113]Cd chemical shifts obtained by various groups on the post-transition metal chalcogenides. These values are contrasted in Figure 2 with the parameter $u^2(1-u^2)$ calculated for the Zn and Cd chalcogenides. As predicted by eq. {5}, the more ionic cadmium chalcogenides are shifted upfield with respect to their zinc analogs. This effect is amplified further by the effect of polarity on Z_{eff} and

Table 1: Phillips ionicities f_i, Harrison polarities α_p (values from [11a] in parentheses), anion-cation overlap values S [11a], total energy gap E_g (values from [11a] in parentheses), and calculated parameters relevant for σ_p in eq. {5}.

Compound	f_i	α_p	S	$E_g(H)^a$	anion $u^2(1-u^2)$	cation $u^2(1-u^2)$	$Z_{eff}^3(A)$	$Z_{eff}^3(C)$
ZnS	0.623	0.73		6.36	0.1168	0.1168	96.64	27.30
CdS	0.685	0.77		6.16	0.1018	0.1018	94.88	28.07
ZnSe	0.676	0.72		6.24	0.1204	0.1204	97.08	27.11
CdSe	0.699	0.77		6.10	0.1018	0.1018	94.88	28.07
HgSe		0.77		6.32^b	0.1018	0.1018	94.88	28.07
ZnTe	0.546	0.72		5.56	0.1204	0.1204	97.08	27.11
CdTe	0.675	0.76		5.44	0.1056	0.1056	95.32	27.87
HgTe	0.65	0.76		5.70^b	0.1056	0.1056	95.32	27.87
AlP	0.307	0.47(0.53)	0.633	5.00(5.52)	0.1977	0.0422	62.55	69.29
GaP	0.374	0.52(0.51)	0.629	5.10(5.35)	0.2069	0.0490	64.66	68.49
InP	0.421	0.58(0.52)	0.602	4.84(5.04)	0.2073	0.0523	64.78	68.11
AlAs	0.274	0.44(0.55)	0.645	4.84(5.44)	0.1851	0.0339	60.00	70.25
GaAs	0.310	0.50(0.54)	0.637	4.94(5.30)	0.1921	0.0384	61.39	69.73
InAs	0.357	0.53(0.56)	0.617	4.60(4.95)	0.1873	0.0373	60.42	69.85
AlSb	0.426	0.54(0.54)	0.659	4.66(4.76)	0.1862	0.0334	60.22	70.31
GaSb	0.261	0.44(0.52)	0.654	4.32(4.61)	0.1974	0.0402	62.50	69.52
InSb	0.321	0.51(0.56)	0.644	4.08(4.36)	0.1800	0.0314	59.07	70.54

^{a}in eV, ^{b}less reliable, see text

$<r^{-3}>$. It should be noted that the ^{77}Se and ^{125}Te resonances of the corresponding mercury compounds are again paramagnetically shifted with respect to those of the cadmium compounds. Unfortunately, however, there are no reliable α_p-values, because the dielectric constants have not been measured for cubic HgY compounds. However, the Coulson polarity scale [8] does produce the correct chemical shift ordering for the II-VI chalcogenides.

Table 2. Chalcogen {22,23] and ^{113}Cd [15] NMR shifts (ppm) in II-VI semiconductors.

	^{33}S	^{77}Se	^{125}Te
ZnY	56	126	200
CdY	0	0	0
HgY		378	216

	S	Se	Te
^{113}Cd	409	267	0[a]

[a]similar trends observed in ref. [15]

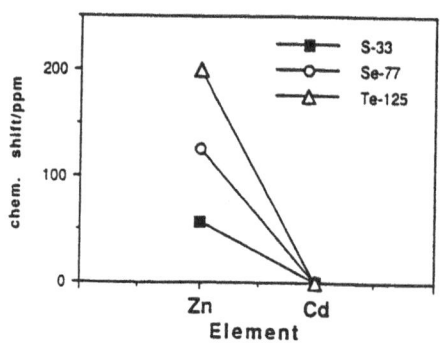
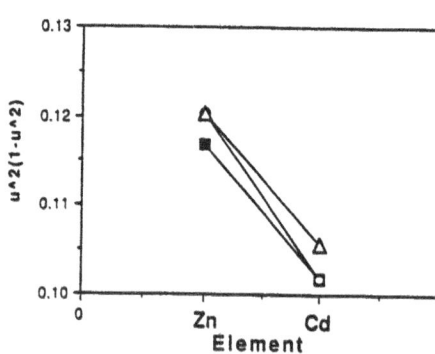

Figure 1. Chalcogen chemical shifts in Zn and Cd chalcogenides. Left: experimental data; right: parameter $u^2(1-u^2)$ calculated from Harrison's α_p-values, assuming S=0.

Experimental [113]Cd NMR studies in CdS, CdSe, and CdTe reveal a decrease of σ_p with increasing atomic number of the chalcogen. This trend cannot be due to changes in $u^2(1-u^2)$ as the α_p values for these three compounds are essentially identical [10]. According to the Phillips scale CdSe is slightly more ionic than CdS, in agreement with the [113]Cd chemical shift trend observed [4]. However, the significant upfield shift observed for CdTe is unexpected.

3.1.2. *Binary III-V Systems.* The chemical shifts of III-V semiconductors have been the subject of repeated investigations. Where available, the most recent values, determined by magic-angle spinning techniques are deemed to be most accurate and are listed in Table 3.

Table 3. Cation and Anion NMR shifts (ppm) in III-V Semiconductors [26-29].

	[31]P	[75]As	[121]Sb
AlY	-210[a]	0[a]	0[c]
GaY	-143[a,b]	228[c]	320[c]
InY	-147[a,b]	208[c]	120[c]

	P	As	Sb
[27]Al[d]	72	60[e]	0
[69]Ga[d]	353	262	0
[115]In[d]	776	598	0

[a]referenced to 85% H_3PO_4, this work, [b]ref.[26], [c]ref.[14,27], [d]ref. [28]. Additional data in [30] and [31]. [e]$Al_xGa_{1-x}As$ samples show a 7 ppm downfield [27]Al shift from x=1.0 to x=0.25 [29].

Figure 2 compares the experimental chemical shifts with the MO predictions for $u^2(1-u^2)$ for the pnictogen atoms. Note that Harrison's method with the assumption S=0 fails completely to reproduce the chemical shift trend within the homologous series, except for the antimonides. In contrast, by using the parameters pertinent to the two-electron bond orbital method [11], the agreement is greatly improved. In particular, this more sophisticated theory predicts the upfield shifts of the aluminum

58

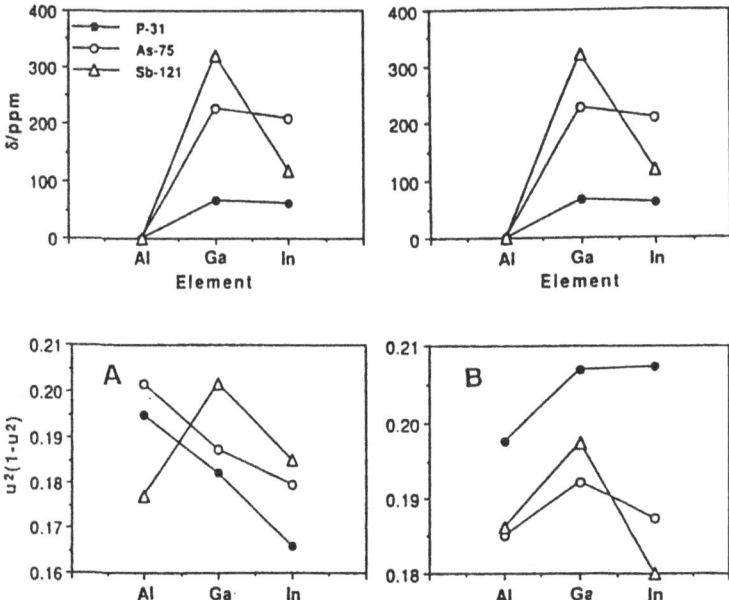

Figure 2: Experimental and predicted chemical shift trends for the anionic constituents in III-V semiconductors. A: one-electron BOM [10]. B: two-electron BOM [11].

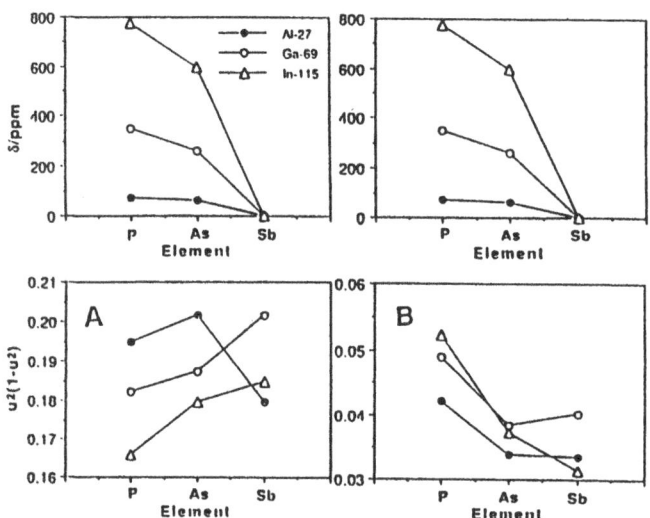

Figure 3. Experimental and predicted chemical shift trends for the cationic constituents in III-V semiconductors. A: one-electron BOM; B: two-electron BOM.

compounds relative to their respective homologues. Differences in Z_{eff} (relevant for the term $<r^{-3}>$ in eq. {5}) due to differences in polarity produce only minor corrections.

Figure 3 summarizes the cation chemical shifts observed in III-V semiconductors, contrasting them with $u^2(1-u^2)$ values calculated from both models. Again, the two-electron bond orbital method produces vastly superior results, particularly with regard to the ordering of the phosphides versus the arsenides. With one exception, the model predicts the correct order of the chemical shifts within homologous series, it does not, however, reproduce the strong upfield shifts observed for the antimonides. Again, the differences in Z_{eff} as calculated from eq. {4} have only a very small influence.

3.1.3. *Nanosized II-VI and III-V particles.* Semiconductor particles with sizes on the order of 10-100 nm show quantum confinement effects, resulting in physical properties that are intermediate between bulk semiconductors and truly isolated molecules. One consequence of the small particle size is an incomplete band structure and hence an increase in the average band gap. [111,113]Cd and ^{77}Se NMR studies of nanosized CdS [32] and CdSe [33,34], as well as ^{31}P and [115]In MAS-NMR studies of GaP [35] and InP [36] samples consistently show substantial upfield chemical shift effects. Since the bonding character and symmetry remains essentially unchanged in these materials, these upfield shifts must be attributed to increases in the average excitation energy parameter Δ as a result of the particle size effect.

3.1.4. *Ternary II-IV-V$_2$ Chalcopyrites.* Table 4 summarizes chemical shift measurements on II-IV-V$_2$ chalcopyrites. These materials are closely analogous to the binary tetrahedral II-VI and III-V zincblende-type systems, except that the presence of two different metal atoms causes a small tetrahedral lattice distortion and small local distortions at the P and metal atom sites. Each group V atom is bonded to two group II and two group IV atoms, while each of the metal atoms is surrounded by four group V atoms. Figure 4 illustrates the experimental chemical shift trend. As pointed out in ref. [26], the trend σ_p(Si compounds) < σ_p (Sn compounds) < σ_p (Ge compounds) correlates with the ordering of the electronegativities on different scales. The ordering σ_p(Cd compounds) > σ_p(Zn compounds) is opposite to the situation in the chalcogenides, is, however, in agreement with Pauling's electronegativity scale.

Table 4. ^{31}P and ^{113}Cd shifts (ppm) in II-IV-V$_2$ semiconductors.

	^{31}P	^{113}Cd
ZnSiP$_2$	-147[b]	
ZnGeP$_2$	-60[a,b]	
ZnSnP$_2$	-92[b]	
CdSiP$_2$	-131[a]	201[a]
CdGeP$_2$	-32[a]	237[a]
CdSnP$_2$	-74[c]	

[a]this work, ^{31}P shifts referenced to 85% H$_3$PO$_4$; ^{113}Cd shifts relative to Cd(CH$_3$)$_2$.
[b]ref.[26], [c]ref.[37]

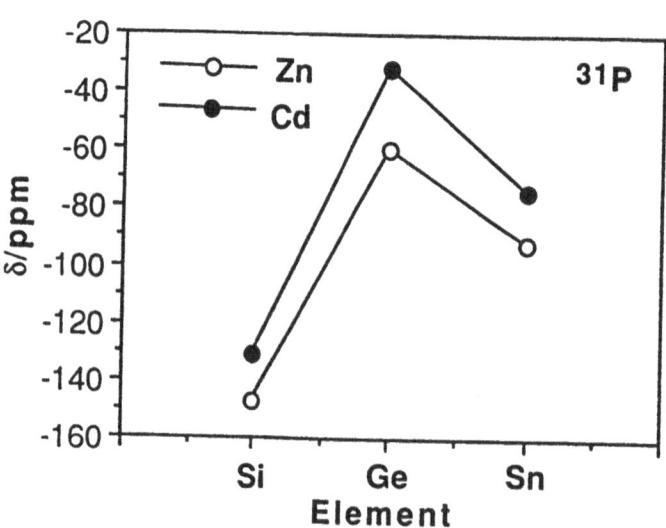

Figure 4. Experimental chemical shift trends in chalcopyrite-type II-IV-V$_2$ semiconductors.

The extension of the Bond Orbital Model from binary zincblende-type compounds to chalcopyrite-type compounds has to take into consideration that two different types of bonds (II-V and IV-V) are present in the latter. We assume that their respective contributions to $u^2(1-u^2)$ are simply additive. Each compound is now characterized by parameter pairs E_g, E_g', α_p, α_p, and S, S', for the II-V and the IV-V bonds, respectively. Values for the first two pairs are easily determined, using the procedure delineated by Lines and Waszczak [38]. With this set of parameters (which is not self-consistent), the calculation fails to reproduce most of the experimentally observed chemical shift trends. Specifically, the BOM is unable to explain the strong upfield shift observed for the silicon compounds and the relative ordering observed for the Cd and Zn compounds. We point out that the α_p values for the IV-V bonds are probably unreliable, since they are derived from the hybrid energies ε_h given only as estimates for Si, Ge, Sn in Table II of ref. [10]. Furthermore, Harrison's method is likely to exaggerate the α_p difference between the Zn-P and the Cd-P bonds. As is the case for the III-V compounds, it is more likely that the decisive difference here lies with the anion-cation overlap integrals S, even if there is a slight difference in polarities. Based on the dependence of S on atomic number alluded to earlier, we expect S(P-Cd) < S(P-Zn). According to eq. {2} the factor $u^2(1-u^2)$ increases with decreasing overlap at constant α_p thus reproducing the experimental observation that σ_p (Cd compounds) > σ_p(Zn compounds). It would certainly be possible to fit the experimentally observed NMR chemical shift data to a reasonable parameter set of α_p' and S'; however, in view of the large number of adjustable parameters the physical insights of such a fit would be limited.

The [113]Cd and [29]Si chemical shifts in CdMP$_2$ and CdMAs$_2$ (M = Ge, Si) show without exception upfield [113]Cd chemical shift displacements when phosphorus is replaced by arsenic. Again, this phenomenon can be understood in terms of the greater Cd-As overlap integral as compared to the Cd-P bond.

3.2. CHEMICAL SHIFTS IN SEMICONDUCTOR ALLOYS

3.2.1. *II-VI Semiconductor Alloys*. At present, data are available only for two systems, Zn$_x$Cd$_{1-x}$Te, and Cd$_x$Hg$_{1-x}$Te (see Tables 5 and 6).

Table 5. ^{125}Te,^{113}Cd NMR shifts (ppm) in $Cd_xZn_{1-x}Te$ alloys [16][a]:

x	Cd	TeCd$_4$	TeCd$_3$Zn	TeCd$_2$Zn$_2$	TeCdZn$_3$	TeZn$_4$
0.1	93			80	145	220
0.5	50	-90	10	95	160	250
0.7	28	-70	25	120	215	260
0.88	13	-10	75	170		
0.93	8	0	108			

[a]shifts are quoted for the five individual ^{125}Te sites observed

Table 6. ^{125}Te NMR shifts (ppm) in $Cd_xHg_{1-x}Te$ alloys [18][a]:

x	TeCd$_4$	TeCd$_3$Hg	TeCd$_2$Hg$_2$	TeCdHg$_3$	TeHg$_4$
0.05					250
0.10					275
0.25			500	650	300
0.50	100	300	500	600	
0.75	50	250	400		
0.90	0	200			

[a]estimated from a figure published in [18]; additional data are published in [39]

The feature common to these systems is the presence of multiple mixed nearest neighbor environments for the anions. These individual $TeZn_nCd_{4-n}$ and $TeHg_nCd_{4-n}$ sites are partially resolved by ^{125}Te MAS-NMR. For both systems, the chemical shift ordering of these five sites can be predicted from the shifts of the pure endmembers: σ_p increases with n, reflecting the differences between Zn,Cd, and Hg with respect to ionicity and overlap as expected on the basis of the trends in α_p and S discussed earlier in connection with Table 1. In the $Zn_xCd_{1-x}Te$ system σ_p for each individual site decreases with increasing substitution level x (this trend is less clear in the mercury containing system, where peak assignments are less straightforward). On the other hand, for the ^{113}Cd resonance σ_p *increases* with x. As discussed below, this observation cannot be understood in terms of simple polarity or overlap effects in individual bonds.

Eqs. {2} and {5} predict that for α_p values around 0.75, the term $u^2(1-u^2)$ increases with increasing S, if S > 0.27. (The exact value of S is not known). Due to the shrinkage of the unit cell with increasing x, the Te-Cd overlap is expected to increase, resulting in a reduced σ_p contribution to the ^{125}Te chemical shift as observed experimentally. Likewise, the Te-Zn overlap integral in a lattice dilated by cadmium atoms is smaller than in pure ZnTe, resulting in an enlarged σ_p contribution compared to the bulk compound.

To explain the opposite compositional dependence of the ^{113}Cd chemical shift, it is necessary to assume a net charge transfer from cadmium to zinc, resulting in more covalency at the cadmium site [16] in the mixed alloy as compared to pure CdTe. This charge transfer can be viewed as an "averaging" of the Zn-Te and Cd-Te bond covalencies due to the simultaneous presence of both cations. A similar effect has been observed for the ^{113}Cd resonance in $Hg_{1-x}Cd_xTe$ alloys [39], although in this case the downfield trend is equally well explained by the reduced Cd-Te overlap in the lattice dilated by Hg. (The tremendous chemical shift effects observed for the ^{199}Hg resonance in that system [39] cannot be explained by any of these effects, however).

3.2.2. III-V Semiconductor Alloys.

The system $Ga_xIn_{1-x}P$ studied recently by Tycko and coworkers is a particularly intriguing example for the changes in chemical bonding properties in semiconductor alloys and for the ability of ^{31}P spectroscopy to detect such changes [20]. Although the chemical shifts of pure GaP and InP differ by only 4 ppm, the five individual $PIn_{4-n}Ga_n$ sites in $In_{0.5}Ga_{0.5}P$ span a chemical shift range of more than 100 ppm, resulting in excellent site resolution by MAS NMR (see Table 7). The In_4P sites are found 50 ppm upfield and the Ga_4P sites ca. 50 ppm downfield from their respective locations in the endmembers. The situation is in perfect analogy to the $Zn_xCd_{1-x}Te$ system. Compression of the InP lattice by Ga substitution increases the In-P and Ga-P overlap integrals, resulting in a monotonic upfield displacement of the ^{31}P resonances observed for all individual sites.

Finally, $Al_xGa_{1-x}As$ alloys show a small upfield displacement of the ^{27}Al resonance with increasing Al content [29]. The weakness of the effect in this case can be understood with the help of Table 1, showing α_p and S to be very similar for AlAs and GaAs. The slight increase in σ_p with Ga content probably reflects a slight decrease in Al-As overlap due to the expansion of the lattice.

Table 7. ^{31}P NMR shifts (ppm) in $Ga_xIn_{1-x}P$ alloys [20][a]:

x	PIn_4	PIn_3Ga	PIn_2Ga_2	$PInGa_3$	PGa_4
0	-147				
0.14	-162	-135	-105	-88	
0.5	-195	-170	-140	-112	-88
1					-143

[a]shifts observed for resolved sites, estimated from a graph published in ref. [20]

3.2.3. *II-IV-V$_2$ alloys*.   Table 8 summarizes chemical shift data obtained by our group for II-IV-V$_2$ alloys.

The [113]Cd and [29]Si chemical shifts in alloys of composition $CdBAs_{2-x}P_x$ (B = Ge, Si) increase with increasing phosphorus substitution. While in principle we expect five distinct $CdAs_{4-x}P_x$ nearest neighbor configurations to be present, MAS-NMR offers no clear site resolution in this case. Again, the chemical shift trend reflects the expected decrease in cation-anion overlap when Cd-As (and Si-As) bonds are partially replaced by Cd-P (Si-P) bonds. Note that in this system the direction of this effect is opposite to that expected from the lattice contraction with increasing x.

The ^{31}P resonances in the systems $ZnGeAs_{2-x}P_x$, $CdGeAs_{2-x}P_x$, and $CdSiAs_{2-x}P_x$ show a strong upfield shift trend with increasing phosphorus substitution. In this case the data are explained well by the increased Cd-P overlap concomitant with the compression of the lattice.

Finally, we discuss results for the system $Cd_{1-x}Zn_xGeP_2$. Although in principle three ^{31}P sites are expected to be present, most of the spectra show only a single broad line, the chemical shift of which corresponds to the compositionally weighted average of the pure endmember values. Again, the decrease of the [31]P σ_p value with increasing x reflects the increase in overlap from the P-Cd to the P-Zn bond and the overall increase in overlap due to the lattice contraction. We cannot currently explain why in samples with high Zn content the mixed PGe_2ZnCd site (detected selectively by

Table 8. ^{31}P, ^{29}Si and ^{113}Cd NMR shifts (ppm) in II-IV-V$_2$ semiconductor alloys.

x	ZnGeAs$_{2-x}$^{31}P$_x$	CdGeAs$_{2-x}$^{31}P$_x$	CdSiAs$_{2-x}$^{31}P$_x$
0.25	1	34	
0.50	-10	23	-99
0.75	-20	14	
1.00	-29	3	-114
1.25	-35	-8	
1.50	-47	-15	-123
1.75	-52	-24	
2.00	-60	-32	-131.4

x	^{113}CdGeAs$_{2-x}$P$_x$	^{113}CdSiAs$_{2-x}$P$_x$	Cd^{29}SiAs$_{2-x}$P$_x$
0.0	151	89	
0.25	158		
0.50	161	113	-4
0.75	172		
1.00	186	147	6
1.25	190		
1.50	209	179	13
1.75	220		
2.00	237	202	15

x	Cd$_{1-x}$Zn$_x$Ge^{31}P$_2$	^{31}PGe$_2$ZnCd sites[a,b]	^{113}Cd$_{1-x}$Zn$_x$GeP$_2$
0.00	-32		237
0.125	-32		238
0.25	-33		242
0.375	-37		272
0.50	-42		278
0.625	-47		300
0.75	-54	-65	319
0.875	-58	-70	330
1.00	-60		

[a] all ^{31}P shift relative to 85% H$_3$PO$_4$, all ^{113}Cd shifts relative to Cd(CH$_3$)$_2$.

[b] partial site resolution observed and peak assignment made by ^{31}P-^{113}Cd double quantum filtering

heteronuclear double quantum filtering) resonates upfield of the symmetric PGe_2Zn_2 majority sites. The [113]Cd resonances move consistently downfield with increasing Zn substitution. In complete analogy with the situation in the system $Cd_{1-x}Zn_xTe$ this result cannot be explained by the lattice compression effect and requires the invocation of charge transfer effects.

3.3. COMMON THREADS OF CHEMICAL SHIFT TRENDS IN SEMICONDUCTOR ALLOYS.

A survey of the chemical shift data reviewed earlier shows three principal reasons for systematic chemical shift changes in semiconductor alloys:

-- nearest-neighbor substitution, affecting anion-cation overlap and/or bond polarity in a generally predictable fashion. This effect is expected to influence cation shifts in common cation systems (anion substitution) and anion shifts in common-anion systems (cation substitution).

-- electron density re-distribution ("charge transfer") expected to result in averaging of bond polarities. This effect is expected to influence cation shifts in common-anion systems and anion shifts in common-cation systems.

-- lattice compression (expansion), affecting anion-cation overlap, resulting in upfield (downfield) shift effects. This effect influences both cation and anion shifts in both common-cation and common-anion systems. Figures 5-8 present the currently available database in terms of the chemical shift effect caused by substitution of a smaller ion into a given lattice, suggesting rather uniform behavior:

    1. In common anion systems, nearest neighbor substitution of a heavier cation by its lighter homologue increases the overlap and shifts the anion resonance upfield, whereas the cation resonance is shifted downfield as the result of charge transfer effects (Figures 5 and 6).

    2. In common cation systems, the substitution of a heavier anion by its lighter homologue decreases the overlap and shifts the cation resonance downfield, whereas the anion resonance is shifted upfield, due to charge transfer effects (Figures 7 and 8).

    3. In all substitution systems studied so far, the effect on the cation and the anion chemical shifts is opposite in direction,

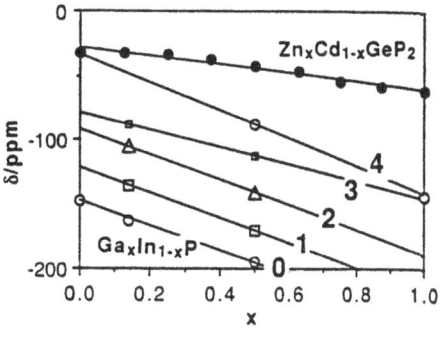

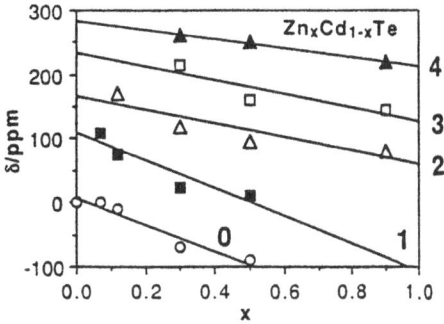

Figure 5: Compilation of anionic chemical shift data in common-anion systems, as a function of cation substitution level x. The numerals denote the number n in the individually resolved $TeZn_nCd_{4-n}$ and PGa_nIn_{4-n} sites.

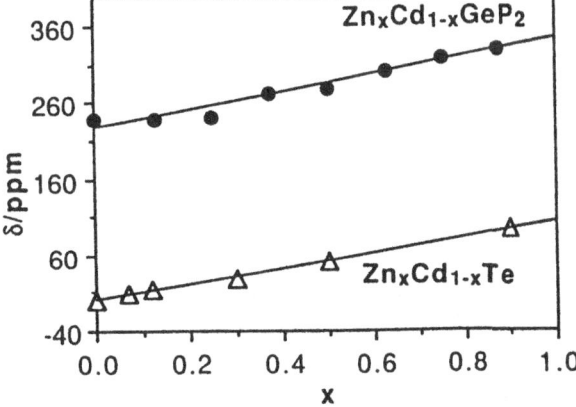

Figure 6: Compilation of cation chemical shift data in common-anion systems, as a function of cation substitution level x.

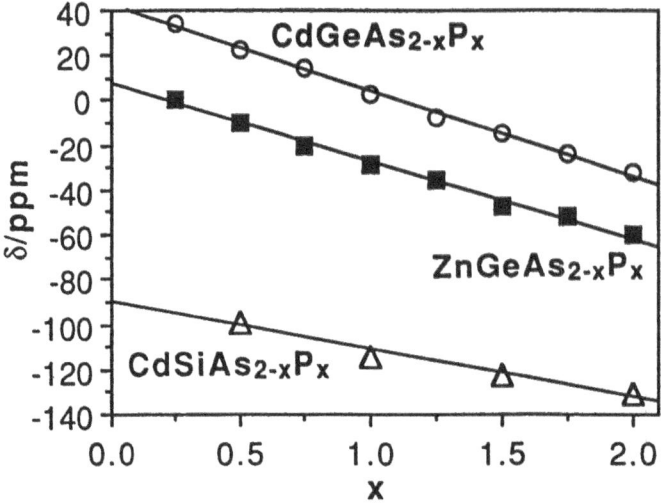

Figure 7. Compilation of anion chemical shifts in common-cation systems, as a function of anion substitution level x.

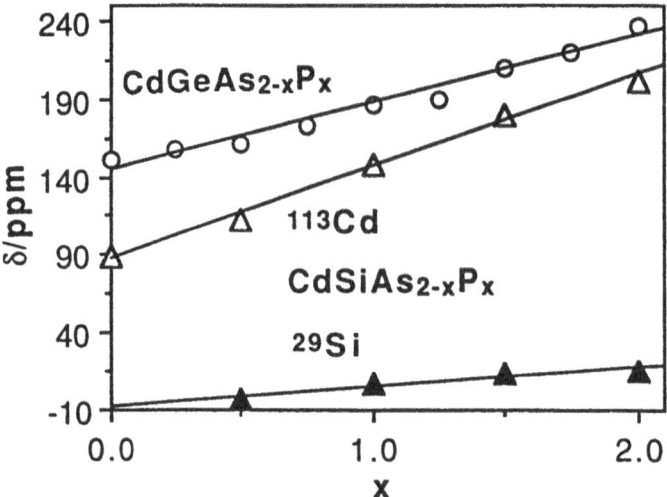

Figure 8. Compilation of cation chemical shift data in common-anion systems, as a function of substitution level x.

indicating that the lattice compression (expansion) effect is smaller than the other two factors. The only case in which the lattice compression effect appears to be decisive is the $Al_{1-x}Ga_xAs$ system, presumably since according to reference [11] there is very little difference between the Al-As and the Ga-As overlap integrals.

4. Conclusions

The strict correlation between anion and cation shift trends in the semiconductor alloys studied so far warrants further detailed investigations. It strongly suggests that the nearest neighbor substitution around the common ion induces the re-distribution of charge density around the ion that is subject of the substitution. Further insights will arise as an increasing number of semiconductor substitution systems will become available. Of particular interest with regard to the postulated charge transfer process are systems in which all three elemental species involved are amenable to solid state NMR investigations. Systems to be studied in the future thus will include $CdTe_{1-x}Se_x$, $Hg_{1-x}Cd_xSe$, and $Al_{1-x}Ga_xP$. In addition, studies under experimental variation of the overlap S by externally applied pressure or temperature variation will add new insights into the nature of chemical shifts in these materials.

Finally, the foregoing discussion has disregarded the effects of the average excitation energy parameter Δ in eq. {5}. As illustrated by the results on the size-quantized semiconductor clusters this may not be justified, and the issue will need to be addressed eventually. Unfortunately, Δ is experimentally not easily accessible. For the present compounds, particularly the chalcopyrites, simple correlations of σ_p with energy observables representative of the entire solid (such as Harrison's E_g values or the published bandgaps [41] cannot be established. In future studies it might be possible to extract Δ from XPS studies in conjunction with band structure calculations.

5. Acknowledgments

This research was supported by National Science Foundation grant no. DMR-8913738.

6. References

[1] Kennedy, J. H. (1989) 'Ionically conductive glasses based on silicon disulfide', Mater. Chem. Phys. 23, 29-50

[2] Bonneau, P. R., Jarvis, R. F., and Kaner (1991) R. B. 'Rapid solid state synthesis of materials from molybdenum disulfide to refractories', Nature (London) 349, 510-512.

[3] (a) Liao, J. H., Kanatzidis, M. G. (1990) 'Hydrothermal Synthesis of Metal Polychalcogenides. Structural Characterization of $[Mo_{12}Se_{56}]^{12-}$. A Cluster of Clusters.', J. Am. Chem. Soc. 112, 7400-7402.

(b) Kanatzidis, M. G., Huang, S. P. (1989) 'Synthesis of $[(Ph_4P)AgSe_4]_n$. A Novel One-Dimensional Inorganic Polymer', J. Am. Chem. Soc. 111, 760-761.

[4] (a) Van Vechten, J. A. (1969)' Quantum dielectric theory of electronegativity in covalent systems. II. Ionization potentials and interband transition energies', Phys. Rev. 187, 1007-1020.

(b) Van Vechten, J. A. (1969) 'Quantum dielectric theory of electronegativity in covalent systems. I. Electronic dielectric constants', Phys. Rev. 182, 891-905.

(c) Phillips, J. C. (1973) 'Bonds and Bands in Semiconductors', Academic Press, N.Y.

(d) Phillips, J. C. (1968) 'Dielectric definition of electronegativity', Phys. Rev. Lett. 20, 550-553.

(e) Phillips, J. C. (1969) 'Dielectric theory of cohesive energies of tetrahedrally coordinated crystals', Phys. Rev. Lett. 22, 645-647.

(f) Phillips, J. C. (1969) 'Resonating bond theory of tetrahedrally coordinated crystals', Phys. Rev. Lett. 23, 482-484.

[5] H. Eckert (1989) 'Structural Characterization of Non-Oxide Chalcogenide Glasses Using Solid State NMR', Angew. Chem. 101, 1723-1732.

[6] J. A. Tossell, private communication

[7] (a) Coulson, C. A., Redei, L. B., and Stocker, D. (1962), 'The electronic properties of tetrahedral intermetallic compounds', Proc. R. Soc. London 270, 357-372.

(b) Lanoo, M., Decarpigny, J. N. (1973) 'Simple tight binding calculation of the tranverse effective charges in III-V, II-VI, and IV-IV compound semiconductors', Phys. Rev. B 8, 5704-5710.

[8] Ramsey, N. F. (1950) 'Magnetic shielding of nuclei in molecules', Phys. Rev. 78, 699-704.

[9] Jameson, C. J., and Gutowski, H. S. (1964),'Calculation of Chemical Shifts. I. General formulation and the Z-dependence', J. Chem. Phys. 40, 1714-1724.

[10] (a) Harrison, W. A., and Ciraci, S. (1974), 'Bond-orbital model, II, Phys. Rev. B 10, 1516-1527.

(b) Chadi, D. J., White, R. M., and Harrison, W. A. (1975) 'Theory of magnetic susceptibility of tetrahedral semiconductors', Phys. Rev. Lett. 35, 1372-1375.

(c) Harrison, W. A. (1973), 'Bond-orbital model and the properties of tetrahedrally coordinated solids', Phys. Rev. B 8, 4487-4498.

[11] (a) Huang, C.; Moriarty, J. A., and Sher (1976), A. 'Two electron bond-orbital model, II' Phys. Rev. B 14, 2539-2558.

(b) Huang, C., Moriarty, J. A., Sher, A., and Breckenridge, R. A. (1975) 'Two electron bond-orbital model I', Phys. Rev. B 12, 5395-5406.

[12] Pauling, L, and Wilson, E. B. (1935), 'Introduction to Quantum Mechanics', Dover Publications Inc., p. 145.

[13] Slater, J. C. (1930) 'Atomic Shielding Constants', Phys. Rev. 36, 57-65.

[14] (a) Lütgemeier, H. (1964) 'Die chemische Verschiebung der Kernresonanzlinien in $A^{III}B^V$ Verbindungen', Z. Naturforsch. 19a, 1297-1300.

(b) Hübner, K. (1971) 'Berechnung der chemischen Verschiebung der Kernresonanz in A^{III}-B^V Halbleitern, Phys. Status Solidi (b) 45, 619-632.

[15] Look, D. C. (1972) 'Nuclear magnetic resonance chemical shifts in CdS, CdSe, and CdTe', Phys. Status Solidi (b) 50, K97-K100.

[16] Beshah, K., Zamir, D., Becla, P., Wolff, P. A., Griffin, R. G. (1987), 'Te and Cd nuclear magnetic resonance study of local structure and bonding in $Cd_{1-x}Zn_xTe$', Phys. Rev. B 36, 6420-6425.

[17] Vieth, H. M., Vega, S., Yellin, Y., and Zamir, D. (1991) 'Temperature dependence of the NMR line shifts and T_1 relaxation

times of [125]Te in the semiconductor alloys $Hg_{1-x}Cd_xTe$', J. Phys. Chem. 95, 1420-1424.

[18] Zax, D. B., Vega, S., Yellin, N, and Zamir, D. (1987), 'Study of structural ordering in $Hg_{1-x}Cd_xTe$ by [125]Te NMR', Chem. Phys. Lett. 138, 105-109.

[19] Zamir, D., Beshah, K., Becla, P., Wolff, P. A., Griffin, R. G., Zax, D., Vega, S., Yellin, N (1988) ' Nuclear magnetic resonance studies of II-VI semiconductor alloys, J. Vac. Sci. Technol. A6, 2612-2613.

[20] Tycko, R., Dabbagh, G., Kurtz, S. R., and Goral, J. P. (1992) 'Quantitative study of atomic ordering in $Ga_{0.5}In_{0.5}P$ thin films by ^{31}P nuclear magnetic resonance', Phys. Rev. B 45, 13452-13456.

[21] Franke, D., Banks, K., Maxwell, R., and Eckert, H. (1992) 'Atomic distribution in crystalline II-IV-V_2 semiconductor allyos. [31]P and [113]Cd magic-angle spinning, spin echo and [31]P-[113]Cd spin echo double resonance NMR studies of the systems $ZnGeAs_{2-x}P_x$ and $CdGeAs_{2-x}P_x$', J. Phys. Chem. 96, 1906-1915.

[22] Eckert, H., and Yesinowski, J. P. (1986) 'Sulfur-33 NMR at natural abundance in solids', J. Am. Chem. Soc. 108, 2140-2146.

[23] Koch, W., Lutz, O, and Nolle, A. (1978) '[77]Se and [125]Te nuclear magnetic resonance investigations in II-VI and IV-VI compounds', Z. Physik A 289, 17-20.

[24] Nolle, A. (1978), 'Isotropic and anisotropic nuclear magnetic shielding of [113]Cd in cadmium halides, cadmium chalcogenides and in cadmium carbonate, Z. Naturforsch. 33a, 666-671.

[25] Willig, A., and Sapoval, B.(1977) 'NMR orbital shifts of [125]Te in solids. Dependence on ionicity and local structure', J. Physique - Lettres 38, L57-60.

[26] Vanderah, T. A., and Nissan, R. A. (1988) ^{31}P MAS NMR of II-IV-V_2 chalcopyrite type series', J. Phys. Chem. Solids 49, 1335-1338.

[27] Bogdanov, V. L., and Lemanov, V.V. (1968) 'Quadrupole relaxation and chemical shift in A^{III}-B^V crystals', Sov. Phys. Solid State 10, 223-224.

[28] Han, Oc Hee, Timken, H. K. C., and Oldfield, E. (1988) 'Solid state magic angle sample spinning nuclear magnetic resonance spectroscopic study of group III-V (13-15) semiconductors' J. Chem. Phys. 89, 6046-6052.

[29] Akimoto, J., Mori, Y., and Kojima, C. (1987), 'Effective electron-density varianion and atomic configuration of Al in $Al_xGa_{1-x}As$', Phys. Rev. B 35, 3799-3803.

[30] (a) Sears, R. E. J. (1981) '[11]B nuclear magnetic shielding in BN and BP', Phys. Rev. B 24, 4072-4074.
(b) Humphries, L. J., and Sears, R. E. J. (1975) '[31]P chemical shifts in BP, GaP, and InP', J. Phys. Chem. Solids 36, 1149.
[31] Sears, R. E. J. (1980) '[27]Al nuclear magnetic shielding in aluminum group V semiconductors', Phys. Rev. B 22, 1135-1140.
[32] Herron, N., Wang, Y., and Eckert, H. (1990), 'The synthesis and characterization of surface capped, size quantized CdS clusters. Chemical control of cluster size', J. Am. Chem. Soc. 112, 1322-1326.
[33] Thayer, A. M., Steigerwald, M. L., Duncan, T. M., and Douglass, D. C. (1988), 'NMR study of semiconductor molecular clusters', Phys. Rev. Lett. 60, 2673-2676.
[34] Steigerwald, M. L., Alivisatos, A. P., Gibson, J. M., Harris, T. D., Kortan, R., Muller, A. J., Thayer, A. M., Duncan, T. M., Douglass, D. C., and Brus, L. E. (1988) 'Surface derivatization and isolation of semiconductor cluster molecules', J. Am. Chem. Soc. 110, 3046.
[35] Mac Dougall, J. E., Eckert, H., Stucky, G.D., Herron, N., Wang, Y., Möller, K., Bein, T., and Cox, D. (1989) 'Synthesis and characterization of group III-V semiconductor clusters: gallium phosphide (GaP) in zeolite-Y', J. Am. Chem. Soc. 111, 8006-8007.
[36] Moran, K. L., Mac Dougall, J. E., Stucky, G. D., Eckert, H., Fagan, P. J., Herron, N. and Wang, Y.,'NMR of semiconductor clusters', to be published.
[37] Ryan, M. A., Peterson, M. W., Williamson, D. L., Frey, J. S., Maciel, G. E., and Parkinson, B. A. (1987) 'Metal site disorder in zinc tin phosphide', J. Mater. Res. 2, 528-537.
[38] Lines, M. E., Waszczak, J. V. (1977) 'A bond-orbital interpretation of the linear dielectric and magnetic properties of the ternary chalcopyrites", J. Appl. Phys. 48, 1395-1403.
[39] Willig, A., Sapoval, B., Leibler, K., and Verie, C. (1976), 'Chemical shift of NMR in HgTe, CdTe and their alloys', J. Phys. C. 9, 1981-1989
[40] Becker, K. D., (1978), 'Temperature dependence of NMR chemical shifts in cuprous halides', J. Chem. Phys. 68, 3785-3793.
[41] Shaukat, A. (1990) 'Composition dependent band gap variation of mixed chalcopyrites, J. Phys. Chem. Solids 51, 1413-1418.

Intra- and Intermolecular Electrial Effects on Nuclear Magnetic Resonance, Nuclear Quadrupole Resonance and Infra-Red pectroscopic Parameters from Ab Initio Calculation and Experiment: From CO to Proteins

Joseph D. Augspurger and Clifford E. Dykstra
Department of Chemistry
Indiana University-Purdue University at Indianapolis
1125 East 38th Street
Indianapolis, IN 46205, USA

Eric Oldfield and John G. Pearson
Department of Chemistry
University of Illinois at Urbana-Champaign
505 South Mathews Avenue
Urbana, IL 61801, USA

ABSTRACT. We examine the role of electrostatic influences on chemical shifts, quadrupole couplings, and vibrational frequencies in proteins as well as small prototype molecules. The goal is to test the notion that the primary change in electronic structure is that of electrical polarization. It is shown that the effects of these interactions on molecular properties are accurately determined by properties which can be calculated by analytical differentiation at the SCF level with Derivative Hartree-Fock (DHF) theory. Several applications of this model are compared with experiment.

Introduction

In this paper, we discuss progress towards obtaining a better understanding of the origins of chemical shifts and infra-red vibrational frequencies in proteins. The origins of the chemical shift nonequivalences due to folding a protein (or nucleic acid) into its native conformation are poorly understood, a somewhat surprising situation given the central importance of folding-induced shielding changes in permitting application of multi-dimensional NMR studies of protein structure [1]. Early workers noted in ^{1}H NMR that a number of residues in e.g. hen egg white lysozyme (EC 3.2.1.17) could be resolved [2], and in 1973, Allerhand et al. [3] observed very large chemical shift non-equivalences, due to folding, in the ^{13}C NMR spectra of lysozyme. For C^{γ} of the six Trp residues, folding produced a ~ 6 ppm range of chemical shifts, which was almost completely removed upon protein denaturation [3].

Other workers have also observed very large chemical shift ranges for other nuclei in proteins, about 12 ppm for [19]F [4-6] and up to ~25 ppm for [15]N [7]. Similarly, we have observed about a 10 ppm chemical shift range for [17]O in $C^{17}O$ labelled heme proteins[8], and very recently, we have observed an unprecedented 15 ppm chemical shift range for

J. A. Tossell (ed.), Nuclear Magnetic Shieldings and Molecular Structure, 75–94.

fluorotryptophan-labelled hen egg white lysozyme (17 ppm in the presence of the inhibitor $(NAG)_3$), due almost entirely due to folding (9).

Given the increasing body of chemical shift data becoming available on proteins, there has been a renewed interest in interpreting the chemical shift changes caused by folding, with most emphasis being placed on the ^1H nucleus (10-21). One general conclusion of these studies is that CH^α proton resonances are shifted downfield in β-sheet regions, while upfield shifts (from the random coil values) are seen for α-helices. All of these investigations have employed empirical correlations between shielding and some structural parameter, and many interesting observations have been made. Of particular utility is the "digital" (or tri-state) approach of Wishart et al. (21) to predicting secondary structure, in which observed shifts are assigned, based on their relationship to the random coil shift, a "-1", "0", or "1" index, corresponding to helix, (random) coil or sheet structure. Applying some simple rules, excellent correlations with known secondary structure were observed. However, the *origins* of these shielding effects have not yet been fully explained. For ^1H NMR, although very large data bases are available, there are many contributions to the observed shielding, including:

- ring current effects
- magnetic susceptibility anisotropies (principally of peptide CO groups)
- electric field effects (helix dipoles, fixed charges, "hydrogen bonding", etc.)

Several groups have used multi-parameter optimization methods in which initial estimates of the electric (E) and magnetic (B) field shifts are iterated to give a best fit with experiment, but there are uncertainities associated with this approach. 10 or 11 parameters need to be optimized, using even semi-classical approaches, which do not include **electronic** structural changes due to helix/sheet conformations, and in addition, there are many possible **models** for the E-field effects - uniform fields, $(field)^2$ terms, field gradients, etc. (22,23). Similar empirical correlations between secondary structure and chemical shielding have also recently been reported for $^{13}C^\alpha$ (24), but a detailed quantum chemical analysis of these shieldings has not yet been reported. Moreover, the empirical approaches have been much less successful for e.g. NH protons, and for ^{15}N or ^{19}F, essentially no predictions have yet been made, even though the chemical shift ranges are ≈15-25 ppm. Thus, more rigorous approaches to the shielding of the heavier elements, as well as hydrogen, need to be developed.

The focus of our current work is to develop models of long range effects upon chemical shielding, due to **electrostatic** fields, using a combination of quantum chemical, molecular dynamics and electrostatics calculations. To date, only simple electrostatic (Coulomb) models have been employed for H^α and HN, and we believe that more realistic modelling of the effects of electric fields should lead to better agreement with experiment - especially for ^{13}C, ^{15}N and ^{19}F. As we describe below, the effects of fields, field gradients and $(field)^2$ terms may all need to be investigated, combined with energy minimization/molecular dynamics methods, using a variety of force fields, and electrostatic modelling methods. We suggest below that long range, weak electrical interactions are large for the heavier elements, and these weak interactions may influence infra-red vibrational frequencies (as well as pure quadrupole frequencies).

Calculation of NMR Parameters by Ab Initio Methods

The first aspect of our efforts to develop models for shielding is the direct ab initio calculation of shielding and related properties for small prototype molecules. Numerous molecular properties are derivatives of the energy of the molecular eigenstate. For example, the dipole polarizability is the second derivative with respect to the strength of an applied electric field. Magnetic susceptibilities, harmonic vibrational frequencies, and chemical shielding tensors are still other examples of properties that are formally defined as second energy derivatives. Hyperpolarizabilities, hypersusceptibilities and so on, are still higher order derivatives. The analytical evaluation of these energy derivatives follows from differentiating the Schrödinger equation and solving the resulting equations, which are generally inhomogeneous differential equations. If energy and wavefunction in the presence of a perturbation are expressed as Taylor series, then the coefficients of the series are the derivatives. Explicit, recursive expressions for these derivatives are given as well by Rayleigh-Schrödinger perturbation theory (RSPT) (25), and it is possible, in principle, to organize the solutions of the RSPT inhomogeneous equations in an open-ended manner. Alternatively, derivative quantities may be obtained by finite differences, but usually an analytical procedure is preferred because of the possible numerical errors in finite difference procedures and/or the substantial number of differences required for higher order property values.

A direct differentiation approach can be applied to the Hartree-Fock equation to evaluate derivatives (and therefore properties) of the SCF energy (26-29). One approach to analytic differentiation, derivative Hartree-Fock (DHF) theory, has been developed (30,31) to be completely open-ended with respect to both the number of parameters and the order of differentiation. It is general with respect to real or imaginary perturbations, and whether the basis is perturbation dependent (e.g. geometrical derivatives) or not. It has the further advantage of a fully implemented 2N/2N+1 energy derivative evaluation, where only Nth order derivatives of the wavefunction are required to calculate 2N and 2N+1 order energy derivatives. Since the wavefunction derivatives are the computationally intensive step, this enhances computational efficiency. DHF requires as input the SCF wavefunction coefficient matrix, the one- and two- electron operator integrals, integrals over derivatives of the hamiltonian, and in the case of basis dependent perturbations derivatives of the one- and two- electron integrals. We have developed associated algorithms for readily obtaining all such integrals (32,33).

The chemical shielding of a particular nucleus is formally defined as the second derivative of the energy with respect to the strength of a uniform external magnetic field and the magnetic dipole moment of the nucleus. A particular element of the chemical shielding tensor is

$$\sigma_{\alpha\beta} = \frac{\partial^2 E}{\partial \mu_\alpha \partial B_\beta} \tag{1}$$

These elements can be calculated at the SCF level with DHF. But in addition, there are higher order properties which are relevant to our ultimate goal. To consider the effects of an external electric field on the chemical shift, $\sigma_{\alpha\beta}$, one can first consider the derivative of $\sigma_{\alpha\beta}$ with respect to the electric field strength, a third derivative property designated A:

$$A_{\alpha\beta,\gamma} = \frac{\partial \sigma_{\alpha\beta}}{\partial V_\gamma} = \frac{\partial^3 E}{\partial \mu_\alpha \partial B_\beta \partial V_\gamma} \tag{2}$$

We may term the quantity $A_{\alpha\beta,\gamma}$ a dipole shielding polarizability, as it is the first derivative of the chemical shielding with respect to an external electric field, analogous to the dipole polarizability being the first derivative of the dipole moment with respect to an electric field. Extending the terminology, the dipole shielding hyperpolarizability, B, is the second derivative of the shielding with respect to the electric field, which is a fourth derivative property. The effects of non-uniform fields can be examined as well by differentiating with respect to a field gradient to calculate the quadrupole shielding polarizability, C, and so on.

In calculating these electro-magnetic properties, basis set completeness is perhaps the foremost concern. One would like to use the smallest basis which achieves the desired accuracy in determining molecular structures or properties. Often, practical considerations (computation time, memory, disk space, etc.) place constraints on the basis size. One means of evaluating completeness is to systematically enlarge bases and test the property for convergence with respect to further basis enlargement. When incorporating the effects of an external magnetic field on a molecule, the additional issue of gauge dependence arises (27). With a complete basis, any magnetic property, such as magnetic susceptibility or chemical shielding, is invariant to the choice of gauge. With a truncated basis, this invariance may not result, and so gauge invariance sometimes is taken as another criterion for judging basis completeness.

In practice, the choice of gauge is often that which appears to lead to the most correct property values for a truncated basis. The center of charge, the center of mass, or a nuclear center have been common choices. Chan and Das (34) showed that choosing the electronic centroid of charge minimized the paramagnetic susceptibility. Ditchfield (35) developed an approach to Hartree-Fock perturbation theory using gauge invariant atomic orbitals (GIAO). These incorporate the magnetic vector potential into the basis functions. However, it is the properties which are invariant, and so this type of basis is better described as gauge-dependent (36). Epstein (37) subsequently demonstrated that the resultant GIAO wavefunction is not invariant to gauge transformation, even though the energy is. He termed this "enforced gauge invariance" for it results not from basis completeness, but from the basis always being returned to the original gauge by the built-in dependence on the magnetic vector potential. Sadlej (38) proposed a criterion for choosing the gauge origin which minimizes the error with respect to gauge origin. This criterion is the quantity $[<0|A(R)B(R)|0> - \sum_K <0|A(R)|K><K|B(R)|0>]$, where $|0>$ is the ground state SCF wavefunction, $|K>$ the other Slater determinants which arise from the SCF orbitals, and where A and B are operators which depend on the gauge origin. A similar proposal was made by Yaris (39). Kutzelnigg and Schindler have introduced an approach where an individual gauge for each localized orbital (IGLO) is used in incorporating dependence on the vector potential into the basis so that molecular properties can be built up from atomic and bond contributions (40,41). Hansen and Bouman (42) have applied a conceptually similar local orbital/local origin (LORG) approach. Lazzeretti et. al. (43-45) have likewise introduced a multiple-origin gauge method, this one based on computing susceptibilities from nuclear electric shieldings which they term "distributed origin gauge with origin at the nuclei (DOGON)" (45). Geertsen (46,47) has shown that a polarization propagator based method leads to gauge invariant susceptibilities for limited basis sets. Hinton and Pulay have recently found means for improving the efficiency of GIAO calculations (48, 49). In many ways, these are attempts to achieve proper results from conventional-sized bases versus extended bases. We may also examine the issue of basis set completeness vis-a-vis gauge sensitivity through very extended bases, and in particular to look at a wider class of properties, magneto-electrical response properties. Gauge sensitivity for these properties has been considered rarely, perhaps the earliest case being the calculations of Day and Buckingham on hydrogen fluoride (50).

There have been quite a number of important studies of basis set requirements for accurate calculation of purely magnetic second order properties, e.g. susceptibilities and chemical shieldings. Lazzeretti, et al. (51,52) examined several small molecules, using basis sets that ranged from minimal up to better than triple zeta, doubly polarized. Kutzelnigg and Schindler (41) carried out similar tests using the IGLO method with similar bases, and also went beyond to include triply polarized sets, with a single 4f function on first row atoms. More recently, Chesnut has conducted tests of small basis sets using Ditchfield's GIAO approach (35) for calculating the chemical shieldings of small molecules of first row atoms (53), larger basis calculations for molecules of second row atoms (54), and use of locally dense basis sets (55), where the nucleus for which the shielding is to be calculated has a larger basis set than other atoms in the molecule. We may expect that the basis set requirements to be somewhat similar for magneto-electric properties, and our results show to what extent this holds.

We calculated the chemical shielding and the shielding polarizability of carbon monoxide with a number of basis sets at several gauge origins on the molecular axis to evaluate the gauge dependence as a function of basis set size and also to examine the property convergence (56). The valence sets used were Dunning's double zeta (DZ) (9s5p/4s2p) and triple zeta (TZ) (10s6p/6s3p) contractions (57) of Huzinaga's primitive sets (58). An initial finding was that added flexibility in the valence p set had a noticeable effect, and so two other valence sets were constructed by uncontracting the two most

TABLE 1. Exponent Values of the Uncontracted, Diffuse and Polarization Gaussian Functions that Were Used to Augment Double-Zeta and Triple Zeta Sets[a].

Basis Designation[b]	s	p	d	f
DZP/DZ'P			0.75/0.80	
TZP/TZ'P			0.75/0.80	
TZ'2P			0.9	
			0.18	
TZ$^+$3P[c] and	0.05/0.06	0.03 /0.05	0.9	
TZ'$^+$3P		0.005/0.007	0.13	
			0.02	
TZ'$^+$3Pf	0.05/0.06	0.03 /0.05	0.9	0.4
		0.005/0.007	0.13	
			0.02	

[a]Where the exponents of the augmenting functions differ for carbon and oxygen, it is the carbon function exponent that is given first. The oxygen exponent value follows the "/". If there is one value, then the same exponent was used for both carbon and oxygen centers.

[b]The + superscript designates the addition of two uncontracted diffuse p functions and one uncontracted diffuse s function.

[c]This is the Electrical Properties (ELP) basis described in Ref. 59.

diffuse primitive functions of Dunning's contracted gaussian. These are designated DZ' (9s5p/4s4p) and TZ' (10s6p/6s5p). To these valence sets were added one, two and three sets of polarization functions (in some cases augmented with diffuse valence functions as well), up to the largest basis tested which also included one 4f function. Table 1 describes the diffuse valence and polarization functions which augmented the valence basis sets. In these calculations the C atom was placed at the origin, the center of mass at x=1.218Å, and the O atom at x=2.132Å.

The results for the chemical shielding and shielding polarizability of oxygen are given in Fig. 1:

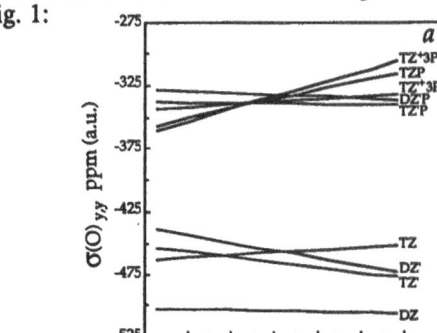 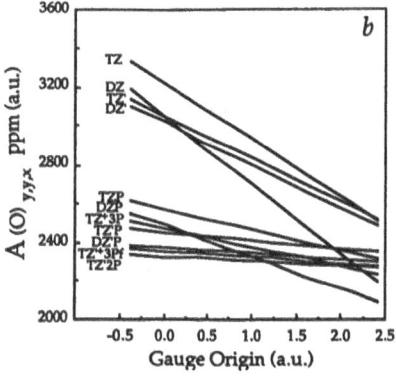

Figure 1. *a)* $\sigma_{y,y}$ and *b)* $A_{yy,x}$ (ppm/a.u.) for ^{17}O as a function of gauge origin along the molecular axis (56). On the scale of the graph for $\sigma_{y,y}$, the DZP basis result is essentially coincident with TZ^+3P, as are TZ'2P and TZ'^+3P with TZ'^+3Pf. The nearly coincident curves are not displayed. For $A_{yy,x}$, where the TZ'^+3P and TZ'^+3Pf bases are essentially coincident, only one curve is shown. The carbon atom lies at the origin, the center of mass at 1.218, and the oxygen atom at 2.132.

Since for linear molecules the axial component of the shielding tensor (σ_{xx}) is independent of the gauge origin position on the molecular axis, we concentrated on the off-axis component (σ_{yy}). For the shielding, the unpolarized bases are shown to give results which fall into a group quite separate from the larger bases. It is significant that even the unpolarized bases give shieldings which are quite insensitive to the gauge origin, but which are 100 to 200 ppm lower than the larger basis results. The increased flexibility of TZ versus DZ has a greater effect for the shieldings than for the susceptibility. On the other hand, there is little difference between DZ' and TZ'. Additional polarization functions augmenting the DZ and TZ bases leads to successively less gauge sensitivity (DZP to TZP to TZ^+3P). For the more flexible DZ' and TZ' bases, adding a single set of polarization functions leads to significant improvement while adding further polarization functions leads to much smaller improvements. There is little difference between the TZ'2P, TZ'^+3P, and TZ'^+3Pf bases. For the ^{17}O shieldings, there is a variance of about 15 ppm over the range of 2.5 a.u. tested, which is a variation of 10 and 5 percent respectively. The results for the shielding polarizabilities show much the same trends.

We also carried out similar tests of the magnetic susceptibility, and the results are given in Fig. 2:

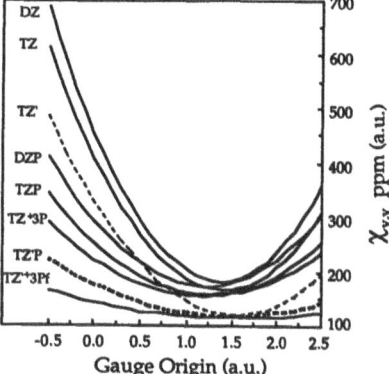

Figure 2. $\chi_{y,y}$ as a function of gauge origin along the molecular axis. On this scale, the DZ' basis results are essentially coincident with the TZ' results, as are those of DZ'P with TZ'P, and of TZ'2P and TZ'⁺3P with TZ'⁺3Pf; the nearly coincident curves are not displayed. Broken lines are used for clarity only. The carbon atom lies at the origin, the center of mass at 1.218, and the oxygen atom at 2.132.

The susceptibility is evaluated as a derivative similarly to the chemical shielding, only it is the pure second derivative of the energy with respect to the magnetic field.

$$\chi_{\alpha\beta} = \frac{\partial^2 E}{\partial B_\alpha \partial B_\beta} \tag{3}$$

The most obvious feature of the susceptibility curves is that they exhibit extrema (60), and that basis set enlargement leads fairly smoothly to lessening of gauge dependence. The additional valence flexibility of the TZ valence set versus the DZ valence set leads to a negligible improvement in gauge invariance and a small lowering in the vicinity of the curves' minima. The significant improvement in the valence basis is the de-contraction of the p-set: All DZ' bases are overall better than DZ and TZ' bases are better than TZ. The TZ'P basis shows a marked decrease in gauge dependence over TZ' as evidenced by a flatter curve in Fig. 2. Use of the second set of polarization functions shows a lesser effect, and the additional flexibility of TZ'⁺3P and TZ'⁺3Pf shows a continuing but smaller effect.

These results show that basis set tests at a single gauge origin may be misleading: They may suggest convergence when it has not been achieved. Likewise, gauge invariance is not a useful criterion for basis quality; it may be achieved when there is still a considerable error in the property. What seems the best means of assessing the reliability in the range of magneto-electrical properties we have considered is to examine basis set convergence for a range of gauge choices. From the body of results collected here, the TZ'2P basis appears optimal in limiting basis size yet achieving reliable values. Smaller bases such as DZ'P may be appropriate if a slightly larger error is acceptable. Table 2 is a collection of A values for a number of small molecules. These values and higher order properties have been used to calculate changes in chemical shieldings from external electrical potentials

Table 2. Shielding polarizabilities for several small molecules in ppm/a.u. (31)

Molecule	Nucleus	$\bar{A}_x$ [a]	$\bar{A}_{xx}$ [b]
H_2	H	-50.3	5.8
HCCH	H	-67.2	-109.5
HCN	H	-54.1	89.5
H_2CO	H	0.1	5.0
HF	H	-81.5	61.8
H_2O	H	-47.3	3.5
NH_3	H	-27.7	-18.1
CH_4	H	-45.1	-2.1
SiH_4	H	-35.8	-1.7
H_2CO	C	-697.4	746.0
CO	C	-374.5	532.7
HCN	C	-422.6	512.1
HCCH	C	-733.9	-351.8
CH_4	C	0.0	-54.1
NH_3	N	50.8	333.3
HCN	N	1910.1	603.6
H_2O	O	401.1	-190.8
CO	O	1526.7	1044.0
H_2CO	O	7019.0	1377.8
HF	F	636.5	-741.7
SiH_4	Si	0.0	456.5

[a] $\bar{A}_x = 1/3 [A_{xx,x} + A_{yy,x} + A_{zz,x}]$.

[b] $\bar{A}_{xx} = 1/3 [A_{xx,xx} + A_{yy,xx} + A_{zz,xx}]$

NMR and IR Results on Proteins

Our ideas about electrical interactions in proteins have developed from recent work on CO-labelled heme proteins, where we observed ^{13}C and ^{17}O chemical shifts for over a dozen proteins, and were also able to determine ^{17}O nuclear quadrupole coupling constants for the bound CO. When we compared each of these three NMR parameters with the CO infra-red vibrational frequency, v_{CO}, we found excellent correlations between $\delta_i(^{13}C)$, $\delta_i(^{17}O)$, e^2qQ/h (^{17}O) and v_{CO}, and we explained our results in terms of a weak electrical interaction model (61) in which changes in the vibrational frequency of CO, the ^{13}C and ^{17}O chemical shifts, as well as the ^{17}O nuclear quadrupole coupling constants, were all interpreted as changes due to polarization of CO by large electric fields from the protein. Our model is based on the demonstration by Dykstra that the primary electronic structure change upon weak interaction is electrical polarization (62-64). A typical experimental result showing the relation between $v(C-O)$ and $\delta_i(^{17}O)$ for CO bound to a number of different heme proteins is shown in Figure 3, and Figure 4 shows

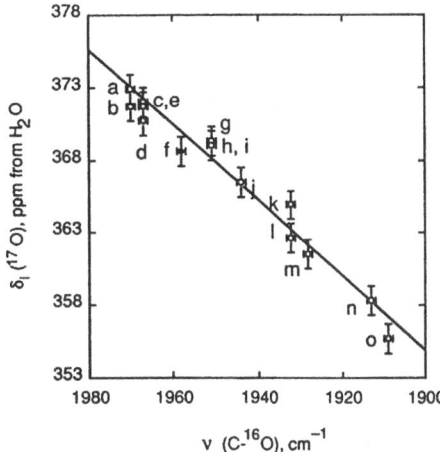

Figure 3. Graph showing relation between (infrared) CO vibrational stretch frequency [ν(C-O), cm^{-1}] and ^{17}O NMR isotropic chemical shift, $\delta_i(^{17}O)$, for heme proteins. Data points for proximal His as well as proximal Cys (chloroperoxidase) containing proteins fall on the same curve. Letters correspond to proteins, as follows: a, *Glycera dibranchiata* hemoglobin; b, synthetic *Physeter catodon* myoglobin His E7 → Val; c, picket fence porphyrin; d, *P. catodon* myoglobin, pH "low"; e, synthetic *P. catodon* myoglobin, His E7 → Phe; f, chloroperoxidase, pH = 6.0; g, rabbit hemoglobin, β chain; h, human adult hemoglobin, α chain; i, human adult hemoglobin, β chain; j, *P. catodon* myoglobin, pH = 7.0; k, horseradish peroxidase isoenzyme C, pH = 10.5; l, horseradish peroxidase isoenzyme A, pH = 9.5; m, rabbit hemoglobin, α chain; n, horseradish peroxidase isoenzyme C, pD = 7.0; o, horseradish isoenzyme A, pD = 4.5.

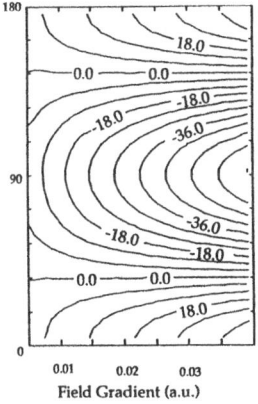
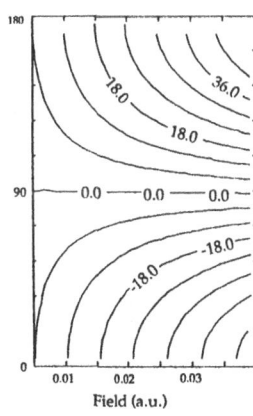

Figure 4. Contours of (a) the isotropic chemical shift at the oxygen nucleus of CO as a function of the strength of an applied field (horizontal axis) and the angle of orientation of the field with respect to the molecular axis (vertical axis), and (b) the isotropic chemical shift at the oxygen nucleus of CO as a function of the strength of an applied field gradient (horizontal axis) and the angle of orientation of the field gradient with respect to the molecular axis (vertical axis) for the ground state of CO.

the perturbation of the [17]O chemical shift of free CO due to a uniform field or applied field gradient, obtained by using DHF to calculate the electrical response properties (moments, polarizabilities and hyperpolarizabilities) as well as the chemical shift polarizabilities as a function of the CO separation distance. With these properties, the CO potential energy curve and chemical shift were calculated for a number of different axial electric field/ field gradient strengths, and from these the ground vibrational state chemical shift and fundamental transition frequency calculated. Figure 5a shows the

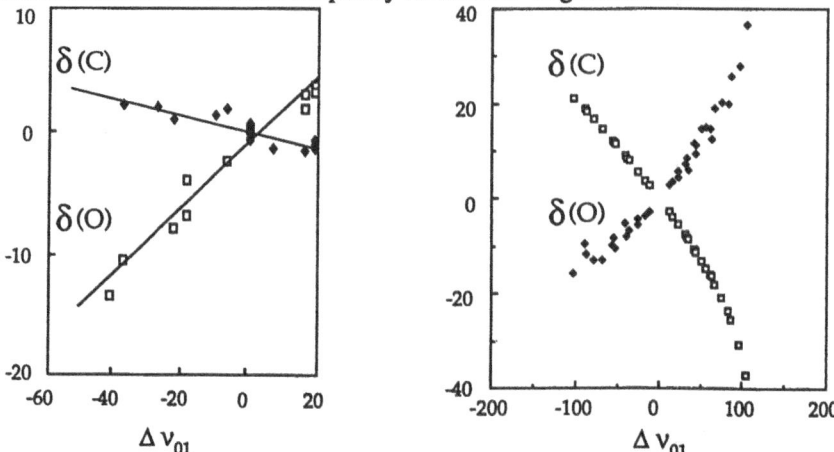

Figure 5. a) Plots of the isotropic chemical shifts for [17]O and for ^{13}C versus the change in the fundamental vibrational frequency measured experimentally for a range of carbon monoxyheme proteins; b) Plots of the isotropic chemical shifts for [17]O and for ^{13}C versus the change in the fundamental vibrational frequency calculated for various positions of an axial dipole consisting of two point charges.

experimental [17]O and ^{13}C chemical shifts of CO bound to heme proteins as a function of CO vibrational frequency (using 1950 cm[-1] as a nonperturbed zero-frequency reference), while Figure 5b shows the results of DHF calculations of the field and field gradient perturbation of a free CO molecule, again taking the field-free ν_{CO} as a zero reference point.

Although not perfect, the sign and magnitude of the observed NMR and IR frequency shifts are quite well reproduced by the calculation, which tends to support the idea that the uniform component of the fields generated by the protein are responsible for the observed chemical shift changes, and the large (~60 cm[-1]) range of CO vibrational frequencies. The range of fields required to cause the experimental shifts is ~0.008 au (~4x10[7] Vcm[-1]), a reasonable value for the interior of a protein, where dielectric constants are generally low.

To further test the idea that electrical interactions in proteins make a significant contribution to the observed shieldings, we have computed the $\bar{A}_x$ values for ^{1}H, [13]CO, [13]C(aromatic), C[17]O and ^{19}F in a number of model compounds (see Table II), and compared these $\bar{A}_x$ values with the *observed chemical shielding ranges in proteins*, as shown in the following Figure:

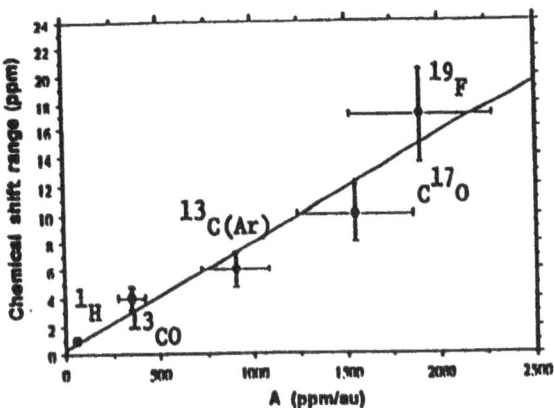

Figure 6. Graph showing relationship between the isotropic dipole shielding polarizabilities computed by using derivative Hartree-Fock theory, and the observed (maximum) chemical shift ranges, for ^1H, ^{13}C, ^{17}O and ^{19}F nuclei in proteins.

All of these results can be fit to a linear equation, assuming the dominance of the dipole shielding polarizability, as:

$$\Delta\delta(ppm) = 0.00765\ \bar{A}_x + 0.32 \tag{4}$$

where the slope, ΔV_x, represents the range of effective fields experienced by the nuclei in question. Although Equation (4) is only strictly applicable for highly symmetric systems, the slope we find, $\Delta V_x = 0.00765$ au, appears to be a very reasonable value, based upon previous electrostatic calculations of electric fields in proteins (65). For example, Dao-Ping et al. (65) found an electrostatic field in the active site cleft in lysozymes of up to 13.7×10^6 Vcm^{-1} (for an interior dielectric of 4); for $\varepsilon=2.5$ this is 0.0043 au (~2.2×10^7 Vcm^{-1}). Thus, the effective range of fields in proteins of ~3.9×10^7 Vcm^{-1} determined by this NMR approach is close to the ~4×10^7 Vcm^{-1} value we deduced previously from our IR and NMR results (61) and the work of others. However, these fields cannot always be expected to be very uniform, and further work is underway on calculating the non-uniform field contributions to shielding in several systems.

Given these results, it is clear that the next step is to try to interpret individual, assigned chemical shifts in proteins. This is expected to be a lengthy process, since questions about solution-crystal structural differences, dynamical averaging, non-uniform field contributions, and of course, protein electrostatics, all come into play. Nevertheless, preliminary results using a variety of electrostatics models show promise for analyzing the ^{19}F NMR spectra of the *Escherichia coli* galactose binding protein, as shown in Figure 7. Fluorine should be a particularly sensitive probe of protein electrostatics, since its dipole shielding polarizability, $\bar{A}_x = 1885$ ppm/au (in fluorobenzene) is the largest we have encountered for any of the lighter elements.

Using a simple Coulomb point-charge model, and an energy minimized (500 steps using Biosym Technologies Discover program) GBP structure, we obtain the theoretical spectrum shown in Figure 7B (7A is the experimental spectrum, in the form of a stick diagram).

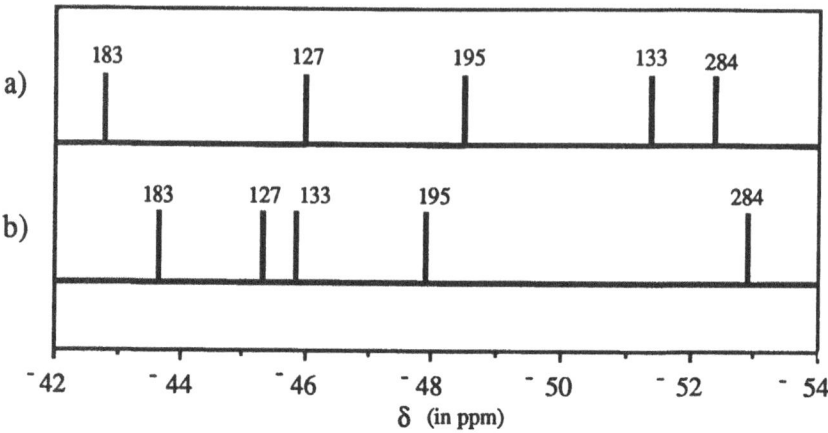

Figure 7. Uniform field point charge calculation of ^{19}F NMR spectrum for 5-F-Trp GBP. Calculation assumes AMBER charges on protein atoms, Ca(2+), STO-3G charges for glucose atoms, and a dielectric constant of 2.0. A 'random coil shift' of -47.6 ppm was determined by comparing the calculated and experimental spectra. Experimental spectrum based on Luck and Falke (1991; Ref. 4).

There is good general agreement between theory and experiment, for example, the overall range of shielding of ~9 ppm is in accord with the experimental range of 11 ppm, and four of the five resonances are in their correct sequence. However, the spectrum does vary considerably with minimization conditions (e.g. F charge, Ca^{2+} charge, number of steps) and it seems likely that molecular dynamics averaging over long trajectories will be required for more definitive calculations, as will the effects of field gradient (and possibly higher order) terms.

We now return to the topic of infra-red vibrational frequencies in proteins, and their electrical perturbation.

For about twenty-five years, it has been known that the infra-red vibrational frequencies of CO ligands *in a given protein* may vary, as a function of temperature, pressure, pH, and so forth. Given the ideas put forth above, we thought that these frequency shifts might be due to weak electrical interactions. Of particular importance here is the observation by Caughey (66) that there are always "four, and only four" v_{CO} peaks, even when investigating over twenty protein systems.

It is clear from Figure 8 that four different electric fields can be generated in proteins containing distal histidine residues by the simple processes of H$^{\delta 1}$ ↔ H$^{\epsilon 2}$ tautomerism, combined with a 180° C$^{\beta}$-C$^{\gamma}$ ring flip. The CO is close enough to be influenced by the histidines' dipole and quadrupole moments, so we carried out *ab initio* electronic structure calculations on the model system CO-methylimidazole, and found that at the four orientations shown in Figure 8, the field due to the quadrupole happens to be

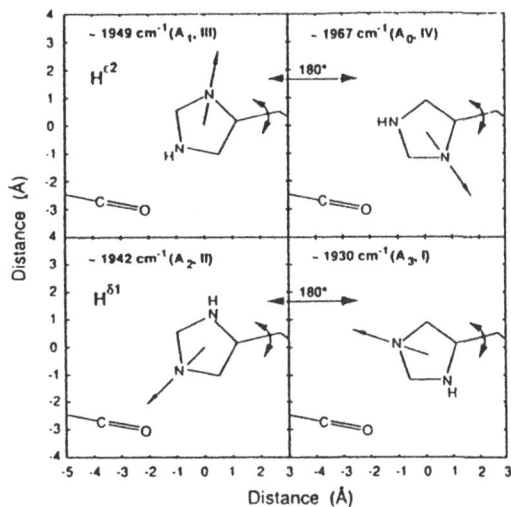

Figure 8. Schematic diagram illustrating the $H^{\delta 1}$ and $H^{\varepsilon 2}$ tautomers of the C^{β}-C^{γ} ring-flip isomers of the distal histidine in HbCO or MbCO, together with the FeCO fragment of the heme.

essentially proportional to the field due to the dipole (67). Thus, it is reasonable for a cosine dependence of just a dipole interaction to represent the deviation of the fundamental vibrational frequency, $\Delta\bar{\nu}$, on θ, the angle between E_x and the CO bond axis. For simplicity, we write the fundamental vibrational frequency, $\bar{\nu}$, in the carbonmonoxyhemeproteins as:

$$\bar{\nu} = \bar{\nu}_0 + a \cos\left[\frac{n360}{5} + b\right] \qquad (5)$$

where $\bar{\nu}_0$ and a are variables that describe the field-free fundamental vibrational frequency and the histidines' local charge field, which based on dipole moment and other considerations is expected to be $\sim 4 \times 10^7$ Vcm^{-1}. n takes the values -1,0, 1,2 by symmetry, and b reflects in a simple way the orientation of the CO bond axis with respect to the imidazole, as shown in Figure 8.

We obtain excellent agreement with the experimental observations on carbonmonoxyhemoglobin (HbCO) using the following parameters: $\bar{\nu}_0 = 1954$ cm^{-1}, a = -23 cm^{-1}, b = -11° and these and other results are shown for several proteins graphically in Figure 9.

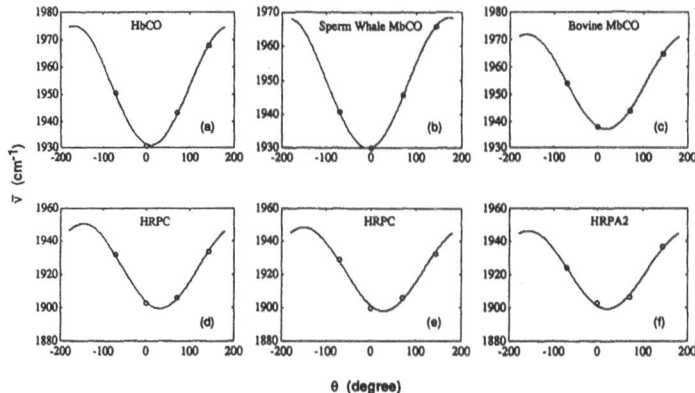

Figure 9. Comparison between experimental data (circles) and Eq. 5 (solid curve) for (a) HbCO; (b) sperm whale MbCO; (c) bovine heart MbCO; (d) horseradish peroxidase (mainly C isoenzyme); (e) horseradish peroxidase, C isoenzyme; and (f) horseradish peroxidase, A_2 isoenzyme.

Thus, the electric field generated by the distal histidine, which is immediately adjacent to the CO ligand, has four projections (cos -82°, cos -10°, cos 62°, cos 134°) due to the presence of a large dipole moment vector which rotates by 360/5 = 72° in the $H^{\delta 1}$ and $H^{\epsilon 2}$ tautomers of the two, 180° ring-flip isomers. The resultant fields are then $\approx 5.5 \times 10^6$ Vcm^{-1}, 3.9×10^7 Vcm^{-1}, 1.9×10^7 Vcm^{-1} and -2.8×10^7 Vcm^{-1}. Fields of this size significantly affect vibrational stretching frequencies. However, since we have used three adjustable parameters to fit four vibrational frequencies, the question of the adequacy of this fitting procedure needs to be addressed. Can we fit anything?

There are 4!/2=12 possible combinations of the four dipole moment vector orientations for the distal histidine residue. We have investigated several different proteins to determine whether all of these combinations can be fit by judicious choice of v_0, a and b. For e.g. mouse HbCO, the 12 solutions yield the following r.m.s. deviations between theory and experiment: 0.1, 0.8, 3.2, 4.1, 4.1, 6.1, 6.9, 8.8, 8.9, 9.9, 10.7 and 11.8 cm^{-1}. Clearly, there are at most two solutions, having b = -11, -61°, and only the first one is close to that anticipated from x-ray studies of CO hemoglobins. Similar behavior is found for other proteins, so it appears that varying v_0, a and b is only capable of producing a small number of acceptable solutions.

Electrostatic Effects on Quadrupole Coupling Constants

As mentioned in the previous section, in our recent work on CO-labelled heme proteins (61) we also found from the experimental results a correlation between the quadrupole coupling constants of the ^{17}O nucleus in CO and v_{CO}. The experimental results, given in Fig. 10A, show an essentially linear correlation between the e^2qQ/h and v_{CO}. We calculated the electric field gradient experienced by [17]O from well correlated, ab initio wave functions using a large basis, and the effect of an external electric field of various strengths was explicitly included. The results are given on Fig. 10B. As for the chemical shifts, the relative changes due to application of a uniform electric field to CO are quite similar to those seen between CO experiencing the different environments of the various heme proteins. The correlation is well reproduced by these calculations; the slopes of the lines in Figs. 10A and 10B are 0.011 and 0.017 MHz/cm^{-1}, respectively.

We can extend that analysis and apply DHF to calculate the effects of external fields on the quadrupole coupling just as we have for the chemical shielding. In a series of papers (68-70), Sternheimer presented an analysis of the effect of an external point charge on the field gradient due to the electron distribution at a nucleus of an atom. The original analysis gives, in effect, the extent to which a nuclear quadrupole polarizes an isolated atom's electron distribution (68). Subsequently, it was shown that to second order it was equivalent to view the perturbation as either that of the nuclear quadrupole inducing a quadrupole moment in the atom's electrons, or an external charge inducing a field gradient at the nucleus. Simply, the field gradient at a nucleus is taken to be a "shielded" external field gradient:

$$V_{zz} = V_{zz}^{ext} (1 - \gamma) \qquad (6)$$

where γ is the shielding factor. In [83]Kr, for example, $\gamma = -77.5 \pm 15$ (71), according to a least squares fit of experimental values of V_{zz} vs. V_{zz}^{ext} for a number of Kr–X dimer

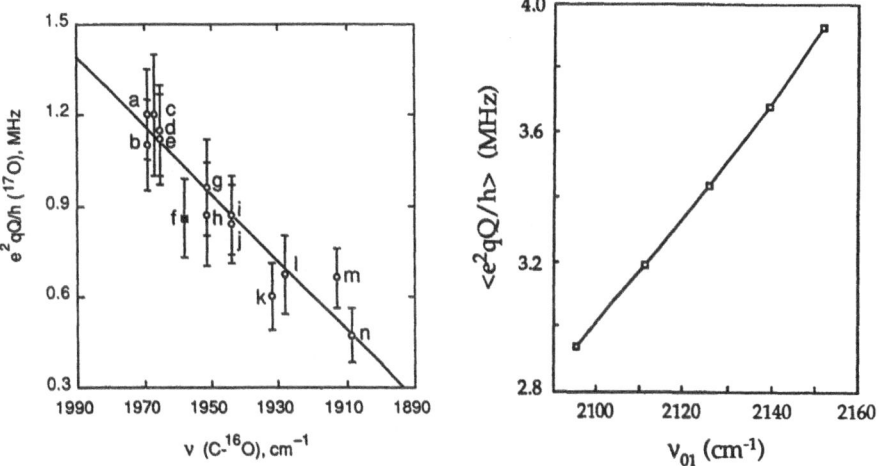

Figure 10. A.e^2q Q/h versus ν_{CO} for CO in heme proteins. B. $\langle e^2qQ/h \rangle$ for the ground vibrational state of CO (in MHz) calculated for different axial electric fields, versus the vibrational frequency shifts that are calculated to arise due to the fields. The upper rightmost point is the field-free data point. Note that ν_{CO} decreases from left to right in A while it increases in B.

complexes, where V_{zz} was found from the experimental χ value, and V_{zz}^{ext} was found from the low order electrical moments of the partner molecule.

The analysis was extended to ionic diatomic molecules (70) to include the added effect of the induced dipole moment. Buckingham (72) formally wrote out the expression to explicitly include the effects of uniform fields on ions, in effect by adding a term to Eqn. (6)

$$V_{zz} = V_{zz}^{ext}(1-\gamma)+\varepsilon V_z^2 \tag{7}$$

where ε is the second order field gradient response due to the uniform field. Engström, et. al. (73) applied these ideas to covalent molecules, using a perturbation approach. Legon and Millen (74) have used this approach phenomenologically to experimentally determine for the first time the first order response of the field gradient due to an external electric field for HCl.

Table 3. Field Gradient and Field Gradient Polarizabilities of Several Small Molecules (a.u.)

Molecule		q^a	q_z	q_{zz}	q_{xx}	$q_{z,z}$	$q_{x,x}$	$q_{zz,zz}$	$q_{xx,xx}$
CH$_3$F	C	0.5947	4.649	-2.228	-2.049	-2.030	-13.65	-11.38	-21.19
	F	3.0950	-13.043	5.998	-0.797	70.216	-2.76	99.44	25.02
HF	F	2.9052	-8.569	10.448	-5.068	52.943	-37.17	73.03	-43.04
	H	0.5803	-0.745	0.856	0.214	2.716	0.55	1.93	0.05
HCCH	C	-0.3570	5.854	4.695	-1.955	-11.268	-27.23	33.95	-49.65
	H	0.3650	0.872	1.566	0.239	1.814	-0.14	7.41	-1.56
HCN	N	-1.1470	7.126	4.229	-1.966	10.249	-15.69	63.99	-29.44
	C	-0.4618	-5.364	4.555	-0.939	1.764	-11.28	46.46	-19.46
	H	0.3443	-0.767	1.467	0.271	1.930	0.58	8.04	0.24
CH$_4$	C	0.0	0.0	1.410	-0.705	19.482	-9.74	34.34	-17.17
	H	0.0	0.259	-0.077	0.039	-2.203	1.10	-4.81	2.40
H$_2$O	O	-1.8617	-2.155	6.794	-5.197	64.868	-52.69	94.83	-93.86
NH$_3$	N	-0.9665	0.161	9.728	-1.624	131.236	13.38	234.08	-5.86
CO	O	-0.6469	6.068	5.793	-3.289	-2.162	-7.44	-90.87	-17.44
	C	-1.1226	-2.603	1.541	1.209	22.923	11.66	62.11	14.28
HCl	Cl	3.6050	-17.399	23.877	-16.670	67.352	-135.24	229.33	-222.70
	H	0.3076	-0.748	0.919	0.259	2.252	0.32	3.16	-1.15

[a]Following convention, we use q to denote V_{zz}^{Nuc}, so that, e.g., $q_{x,yy} = \dfrac{\partial^2 V_{zz}^{Nuc}}{\partial V_x \partial V_{yy}}$.

Except for H$_2$O, the z-axis is the molecular symmetry axis. For H$_2$O, the z-axis is perpendicular to the molecular plane.

[b]This property is q_x. q_z is identically zero by symmetry.

With the analytical capability of DHF, we approach this problem head-on and calculate the derivatives of the field gradient at a nucleus with respect to the elements of the external electrical potential.

$$V_{zz}^{Nuc} = V_{zz,0}^{Nuc} + V_{zz,x}^{Nuc} V_x + V_{zz,y}^{Nuc} V_y + V_{zz,z}^{Nuc} V_z + \frac{1}{2} V_{zz,xx}^{Nuc} V_{xx}^2 + ... \qquad (8)$$

(V_{zz}^{Nuc} is the electric field gradient at the particular nucleus.) Both Bacskay, et al. (75,76) and Fowler (77) have presented a derivative approach to calculating certain of the terms in Eqn. (6). We can obtain all such values, and in Table 3 we present results for several small molecules. The results can be used to anticipate the effect on quadrupole coupling constants of the electrical influence of nearby molecules.

Summary

Overall, our results suggest the following:
1) For small isolated molecules, SCF level calculations (e.g. DHF) can give accurate representations of the chemical shielding tensors.
2) For larger molecules, long-range contributions to shielding in many instances contain calculable contributions from weak electrical interactions, which may be computed by using the multipole shielding polarizabilities such as we have obtained from DHF theory.
3) In CO heme proteins, there are strong correlations between $\delta_i(^{13}C)$, $\delta_i(^{17}O)$, $e^2qQ/h(^{17}O)$ and v_{C-O}, and these experimental trends are reproduced in DHF theory.
4) The four major conformational substates evidenced in the IR spectra of heme proteins are likely due to 180° ring flips about C^β-C^γ of the two imidazole tautomers.
5) For ^{19}F NMR of (labelled) proteins, weak electrical interactions describable by the dipole shielding polarizability and the uniform components of the field appear to make significant contributions to the observed spectral shifts.
6) For ^1H, ^{13}C, ^{17}O and ^{19}F NMR in proteins, there is a good correlation between known shielding ranges and $\bar{A}_x$, suggesting a range of uniform fields of ≈ 0.008 au, in basic agreement with fields calculated using a variety of electrostatic modelling approaches.
7) The range of fields in proteins as deduced from our NMR and IR results on a dozen heme proteins, from calculations on GBP, from the relation between $\bar{A}_x$ and the overall shielding ranges for several nuclei, and from electrostatics calculations by others, are all rather similar, and support the idea that both NMR and IR frequency shifts in proteins (and other macromolecules) will be influenced by such long range interactions.

Acknowledgements

This work was supported in part by the US National Institutes of Health (grants HL-19481 and GM-40426), the US National Science Foundation (grant CHE 9107317), and by the American Heart Association, with funds contributed by the AHA, Illinois Affiliate, Inc.

J. Pearson is a National Institutes of Health Postdoctoral Fellow (grant GM-14545).

92

References

1. Bax, A., Ikura, M., Kay, L., Torchia, D.A., and Tschudin, R. (1991) J. Magn. Reson. 86, 304; Clone, G. M. and Groenenborn, A. M. Progr. NMR Spectroscopy 23, 43-92 .
2. McDonald, C. C. and Phillips, W. D. (1969) J. Amer. Chem. Soc. 91, 1513.
3. Allerhand, A., Childers, R. F., and Oldfield, E. (1973) Biochemistry 12, 1335; Sykes, B. D. and Weiner, J. H. (1980) Magn. Res. Biol. 1, 171.
4. Gerig, J. T. (1989) Meth. Enzym. 177, 3.
5. Millett, F. and Raftery, M. A. (1972) Biochem. Biophys. Res. Commun. 47, 625.
6. Luck, L. A. and Falke, J. J. (1991) Biochemistry 30, 4248.
7. Glushka, J., Lee, M., Coffin, S., and Cowburn, D. (1989) J. Am. Chem. Soc. 111, 7716.
8. Park, K. D., Guo, K., Adebodun, F., Chiu, M. L., Sligar, S. G., and Oldfield, E. (1991) Biochemistry 30, 2333.
9. Oldfield, E., Montez, B., Patterson, J., Harrell, S., Lian, C., and Le, H., unpublished results.
10. Hoch, J. C., Ph.D. Thesis (1983) Harvard University, University Microfilms International, Ann Arbor, MI, #8322365.
11. Pardi, A., Wagner, G., and Wüthrich, K. (1983) Eur. J. Biochem. 137, 445.
12. Redfield, C. and Dobson, C. M. (1990) Biochemistry 29, 7201.
13. Oldfield, E. Abstracts, American Chemical Society Southeastern-Southwestern Meeting, New Orleans, LA, December 7, 1990.
14. Williamson, M. P. and Asakura, T. (1991) J. Magn. Res. 94, 557.
15. Williamson, M. P., Asakura, T., Nakamura, E., and Demura, M. J. Biomol. NMR, in press.
16. Asakura, T., Nakamura, E., Asakawa, H., and Demura, M. (1991) J. Magn. Res. 93, 355.
17. Wagner, G., Pardi, A., and Wüthrich, K. (1983) J. Am. Chem. Soc. 105, 5948.
18. Ösapay, K. and Case, D. A. (1991) J. Am. Chem. Soc. 113, 9436.
19. Dalgarno, D. C., Levine, B. A., and Williams, R. J. P. *Bioscience Reports*, **3**, 443 (1983); A. Pastore and V. Saudek, *J. Magn. Res.*, **90**, 165 (1990); M. P. Williamson, *Biopolymers*, **29**, 1423 (1990); I. D. Kuntz, P. A. Kosen and E. C. Craig, *J. Am. Chem. Soc.*, **113**, 1406 (1991).
20. Wishart, D. S., Sykes, B. D., and Richards, F. M. (1991) J. Mol. Biol. 222, 311.
21. Wishart, D. S., Sykes, B. D., and Richards, F. M. (1992) Biochemistry 31, 1647.
22. Buckingham, A. D. (1960) Can. J. Chem. 38, 300.
23. Batchelor, J. G. (1975) J. Am. Chem. Soc. 97, 3410.
24. Pera, S. and Bax, A. (1991) J. Am. Chem. Soc. 113, 5490.
25. Bartlett, R. J. and Brandas, E. J. (1972) J. Chem. Phys. 56, 5467; Bartlett, R. J. and Shavitt, I. (1977) Chem. Phys. Lett. 50, 190.
26. McWeeny, R. (1961) Phys. Rev. 126, 1028.
27. Stevens, R. M., Pitzer, R. M. and Lipscomb, W. N. (1963) J. Chem. Phys. 38, 550.
28. Gerratt, J. and Mills, I. M. (1968) J. Chem. Phys. 49, 1719.
29. Pulay, P. (1969) Molec. Phys. 17, 197.
30. Dykstra, C. E. and Jasien, P. G. (1984) Chem. Phys. Lett. 109, 388.
31. Augspurger, J. D. and Dykstra, C. E. (1991) J. Phys. Chem. 95, 9230.
32. Augspurger, J. D. and Dykstra, C. E. (1990) J. Comp. Chem. 11, 105.

33. Augspurger, J. D., Bernholdt, D. E., and Dykstra, C. E. (1990) J. Comp. Chem. 11, 972.
34. Chan, S. I. and Das, T. P. (1962) J. Chem. Phys. 37, 1527.
35. Ditchfield, R. (1972) J. Chem. Phys. 56, 5688.
36. Ditchfield, R. (1974) Mol. Phys. 27, 789.
37. Epstein, S. T. (1973) J. Chem. Phys. 58, 1592.
38. Sadlej, A. J. (1975) Chem. Phys. Lett. 36, 129.
39. Yaris, R. (1976) Chem. Phys. Lett. 38, 460.
40. Kutzelnigg, W. (1980) Isr. J. Chem. 19, 193.
41. Schindler, M. and Kutzelnigg, W. (1982) J. Chem. Phys. 76, 1919.
42. Hansen, A. E. and Bouman, T. D. (1985) J. Chem. Phys. 82, 5035; (1989) 91, 3552.
43. Stephens, P. J., Jalkanen, K. J., Lazzeretti, P., and Zanasi, R. (1989) Chem. Phys. Lett. 156, 509.
44. Stephens, P. J. and Jalkanen, K. J. (1989) J. Chem. Phys. 91, 1379.
45. Lazzeretti, P., Malagoli, M., and Zanasi, R. (1991) Chem. Phys. 150, 173.
46. Geertsen, J. (1989) J. Chem. Phys. 90, 4892.
47. Oddershede, J. and Geertsen, J. (1990) J. Chem. Phys. 92, 6036.
48. Wolinsky, K., Hinton, J. F., and Pulay, P. (1990) J. Am. Chem. Soc. 112, 8251.
49. Hinton, J. F., Guthrie, P., Pulay, P., and Wolinski, K. (1992) J. Am. Chem. Soc. 114, 1604.
50. Day, B. and Buckingham, A. D. (1976) Mol. Phys. 32, 343.
51. Lazzeretti, P., Zanasi, R., and Cadioli, B. (1977) J. Chem. Phys. 67, 382.
52. Lazzeretti, P. and Zanasi, R. (1977) Int. J. Quan. Chem. 12, 93; (1978) J. Chem. Phys. 68, 1523.
53. Chesnut, D. B. and Foley, C. K. (1985) Chem. Phys. Lett. 118, 316.
54. Chesnut, D. B. and Foley, C. K. (1986) J. Chem. Phys. 85, 2814.
55. Chesnut, D. B. and Moore, K. D. (1989) J. Comp. Chem. 10, 648.
56. Augspurger, J. D. and Dykstra, C. E. (1991) Chem. Phys. Lett. 183, 410.
57. Dunning, T. H. (1970) J. Chem. Phys. 53, 2823; (1971) 55, 716.
58. Huzinaga, S. (1965) J. Chem. Phys. 42, 1293.
59. Dykstra, C. E., Liu, S.-Y., and Malik, D. J. (1989) Adv. Chem. Phys. 75, 37.
60. Note that the convention of Eqn. 3 corresponds to a diamagnetic susceptibility being positive.
61. Augspurger, J. D., Dykstra, C. E., and Oldfield, E. (1991) J. Am. Chem. Soc. 113, 2447.
62. Dykstra, C. E. (1989) J. Am. Chem. Soc. 111, 6168.
63. Dykstra, C. E. (1987) J. Phys. Chem. 91, 6216.
64. Gutowsky, H. S., Germann, T. C., Augspurger, J. D., and Dykstra, C. E. (1992) J. Chem. Phys. 96, 5808.
65. Dao-Ping, S., Lao, D.-I., and Remington, S. J. (1989) Proc. Natl. Acad. Sci. USA 86, 5361; Hol, W. G. J. (1985) Prog. Biophys. Molec. Biol. 45, 149; Sharp, K. A. and Honig, B. (1990) Ann. Rev. Biophsy. Chem. 19, 301; Warshel, A. and Åqvist, J. (1991) Annu. Rev. Biophys. Biophys. Chem. 20, 267; Lee, F. S. and Warshel, A. submitted to J. Chem. Phys.
66. Potter, W. T., Hazzard, J. H., Choe, M. G., Tucker, M. P., and Caughey, W. S. (1990) Biochemistry 29, 6283.
67. Oldfield, E., Guo, K., Augspurger, J. D., and Dykstra, C. E. (1991) J. Am. Chem. Soc. 113, 7537.

68. Sternheimer, R. (1950) Phys. Rev. 80, 102; (1951) 84, 244; (1952) 86, 316; (1954) 95, 736; (1956) 102, 731; (1966) 146, 140.
69. Sternheimer, R. M. and Foley, H. M. (1953) Phys. Rev. 92, 1460.
70. Foley, H. M., Sternheimer, R. M., and Tycko, D. (1954) Phys. Rev. 93, 734.
71. Campbell, E. J., Buxton, L. W., Keenan, M. R., and Flygare, W. H. (1981) Phys. Rev. A24, 812.
72. Buckingham, A. D. (1962) Trans. Farad. Soc. 58, 1277.
73. Engström, S., Wennerström, H., Jönsson, B., and Karlström, G. (1977) Mol. Phys. 34, 813.
74. Legon, A. C. and Millen, J. D. (1988) Chem. Phys. Lett. 144, 136; (1988) Proc. Roy. Soc. A 417, 21.
75. Bacskay, G. and Gready, J. E. (1988) J. Chem. Phys. 88, 2526.
76. Bacskay, G., Kerdraon, D. I., and Hush, N. S. (1990) Chem. Phys. 144, 53.
77. Fowler, P. W. (1989) Chem. Phys. Lett. 156, 494; (1990) Chem. Phys. 143, 447.

THE NUCLEAR SHIELDING SURFACE:
THE SHIELDING AS A FUNCTION OF MOLECULAR GEOMETRY
AND INTERMOLECULAR SEPARATION

CYNTHIA J. JAMESON
University of Illinois at Chicago
Department of Chemistry M/C-111
Chicago, Illinois 60680

ANGEL C. de DIOS
University of Illinois at Urbana
Department of Chemistry
Urbana, Illinois 61801

(dedicated to the memory of Thomas D. Bouman 1940-1992)

ABSTRACT. In making the connection between theoretical shielding values and experiments, it becomes necessary to consider medium effects and rovibrational averaging. The rovibrational averaging which takes the shielding for a molecule at its rigid equilibrium geometry into a thermal average shielding also provides the temperature dependence in the zero-pressure limit and the changes upon isotopic substitution which are experimentally accessible. The averaging of ^{15}N and ^{31}P shielding in the NH_3 and PH_3 molecules uses *ab initio* shielding surfaces and intramolecular potential surfaces or their derivatives. General trends are observed in comparing the shielding surfaces in these two molecules with H_2O and CH_4. A proper account of medium effects requires the knowledge of the intermolecular shielding surfaces, which are explored here for model systems ^{39}Ar in Ar...Ar, Ar...Ne, Ar...Na$^+$, Ar...NaH, and for ^{21}Ne in Ne...Ne and Ne...He, employing *ab initio* calculations (LORG and SOLO). The shapes of the shielding function $\sigma(R)$ for two atoms are shown to be similar for intra-and intermolecular shielding.

1. The Connection Between Experimental and Theoretical Shielding Values

Theoretical calculations of nuclear magnetic shielding provide the entire shielding tensor on an absolute basis, i.e., with respect to a bare nucleus. On the other hand, a measurement in the laboratory provides a chemical shift δ between the sample and the reference substance, i.e., a difference in shielding such as

$$\delta_A = \sigma(^{13}C, \text{TMS in CDCl}_3 \text{ soln. with A, } x_A, x_{TMS}, x_{CDCl_3}, 300K)$$
$$- \sigma(^{13}C, \text{A in CDCl}_3 \text{ soln., } x_A, x_{TMS}, x_{CDCl_3}, 300K) .$$

There are changes in shielding in going from the isolated molecule in its rigid equilibrium geometry to the single molecule at its ground vibrational state at 0 K, to the molecule in the gas at the zero-pressure-limit at 300 K, to the molecule in a gas at a given density, to the molecule in a neat liquid or a solution or a single crystal or powder.[1] As indicated in Fig. 1, these changes are not necessarily small.[2] They are not usually the same for

95

J. A. Tossell (ed.), Nuclear Magnetic Shieldings and Molecular Structure, 95–116.
© 1993 *Kluwer Academic Publishers.*

96

the sample and the reference substance. Therefore, comparisons between theoretically calculated values of the absolute shielding tensor and the published chemical shift data in condensed phases can not be properly made. It is essential that experimental gas phase data be made available, especially for those nuclei in the first and second row of the periodic table, since the *ab initio* calculations are becoming sufficiently accurate for many of the molecules containing these nuclei. The absolute shielding scale is set by a measured spin rotation tensor for the nucleus in a (primary standard) molecule.[2-10] These are ^{13}C in CO molecule, ^{19}F in HF, ^{31}P in PH_3, ^{17}O in CO, ^{15}N in NH_3, ^{29}Si in SiH_4, etc. The measured chemical shift between molecule A and the primary standard molecule in the same sample in the gas phase provides the absolute isotropic shielding value for the nucleus in molecule A.

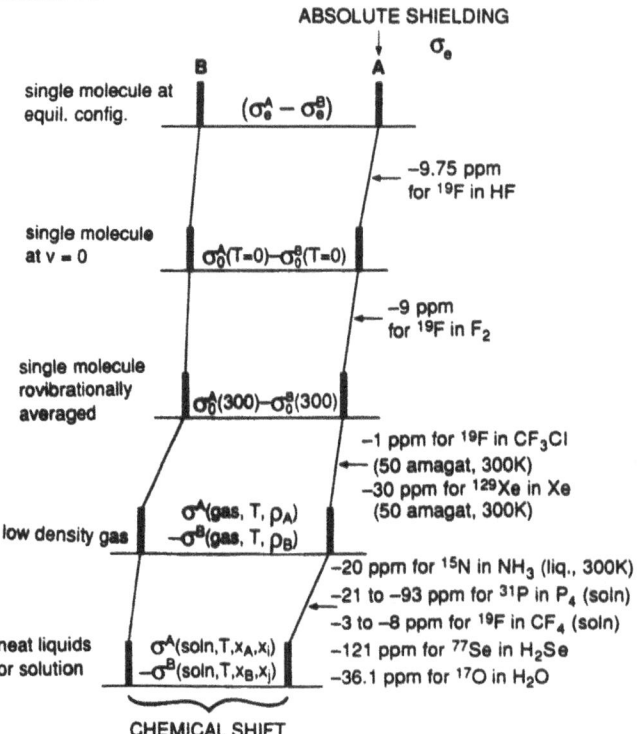

Fig. 1 Relation between the chemical shift and the absolute shielding at the equilibrium geometry

Ab initio calculations published in the last few years and those calculations being reported in this conference are reaching the level of accuracy such that even comparison with gas phase data may require some corrections for intermolecular effects and rovibrational averaging. We start with a gas phase experiment in which the nuclear shielding in a molecule in the pure gas is given by [11,12]

$$\sigma(T,\rho) = \sigma_0(T) + \sigma_1(T)\rho + \sigma_2(T)\rho^2 + ... \tag{1}$$

Experimentally, one can measure the temperature dependence of the shielding in the molecule, $[\sigma_0(T) - \sigma_0(300\ K)]$, in the zero-pressure-limit.[13] The theoretical considerations of this are given in Section 2. Experimentally, one can also measure the density dependence of the shielding for the molecule in the gas phase; the terms $\sigma_1(T)\rho + \sigma_2(T)\rho^2 + ...$ can be obtained.[14] For example, the second virial coefficient of nuclear shielding for a pair of molecules can be determined by

$$\sigma_1(T) = -\left\langle \lim_{\rho \to 0} \left[\frac{v(T,\rho) - v_0(T)}{v_0(300\ K)} \right] \frac{1}{\rho} \right\rangle \tag{2}$$

from measurements in gas mixtures. Here, $v_0(T)$ is the resonance frequency in the limit of zero density. In a mixture of gases, both the $\sigma_1(T)$ for molecule A in the A-A pair and in the A-B pair can be determined.[15,16] The theoretical considerations of this are given in Section 3.

In making the corrections that bridge experiment with theory, we need companion mathematical surfaces. For *intermolecular* or medium effects, e.g., virial coefficients of shielding in the low density gas, or gas-to-liquid shifts, or solvent shifts, we need (a) the shielding surface as a function of the intermolecular coordinates, (b) the intermolecular potential energy surface in terms of the same intermolecular coordinates, and (c) the means of carrying out the averaging by some statistical mechanical means. For the dilute gas in the limit of density approaching zero, we need only the first term in the pair distribution function, $\exp[-V(R)/kT]$. For example, for ^{129}Xe in xenon gas where $V(R)$ is the Xe-Xe potential, the function $\sigma(R)$ is the shielding surface describing the ^{129}Xe shielding as a function of Xe-Xe distance, and $\sigma(\infty)$ is the value for an isolated atom, we have

$$\sigma_1(T) = \int_0^\infty 4\pi R^2\, dR[\sigma(R) - \sigma(\infty)]\exp[-V(R)/kT] . \tag{3}$$

For condensed phases or adsorbed species we need to carry out a grand canonical ensemble average, for example, and perhaps even include the many-body corrections.

For *intramolecular* rovibrational averaging effects, [17] e.g., the (v,J,K) dependence of shielding, or the temperature dependence of the thermal average shielding, and the mass dependence of shielding (isotope shifts), [18] we need (a) the shielding surface which is a function of molecular geometry, or as a function of nuclear displacement coordinates, (b) the intramolecular potential energy surface, or at least its higher derivatives with respect to displacements away from the equilibrium molecular geometry, and (c) the means of carrying out the averaging over the (v,J,K) states, e.g., by using the full anharmonic vibrational wave functions, [19] or by some other approach such as perturbation theory or by using a contact transformation which allows the averages to be expressed in terms of derivatives of the shielding surface and the derivatives of the potential energy surface.[17,20] Thermal averages are thereafter obtained by a statistical average over the thermally populated (v,J,K) states.[19,20] For example, the rovibrational average is

$$\langle \sigma \rangle_{vJK} = \int \psi_{vJK}^*(S_1, S_2, ...)\, \sigma(S_1, S_2, ...)\psi_{vJK}(S_1, S_2, ...)dS_1\, dS_2 ... \tag{4}$$

for the shielding surface $\sigma(S_1, S_2, ...)$ in terms of nuclear displacement symmetry coordinates $S_1, S_2, ...$. The thermal average is given by:

$$\sigma_0(T) = \frac{\sum\limits_{vJK} (2J+1)\, g_{NS}\, \langle\sigma\rangle_{vJK}\, \exp[-E_{vJK}/kT]}{\sum\limits_{vJK} ((2J+1)\, g_{NS}\, \exp[-E_{vJK}/kT]} \qquad (5)$$

In other words, the interpretation of experimental data at a fundamental level requires a knowledge of the shielding as a function of intermolecular distances and molecular geometry. We show some examples of each in this paper and in accompanying posters.

2. Ab Initio Calculations of Shielding Surfaces

Previous calculations of shielding in molecules at selected geometries displaced from equilibrium have been used to determine first and second derivatives of nuclear shielding in a wide variety of molecules.[21-26] However, only very few systematic studies of the shielding surface have so far been carried out. The first report of a full shielding surface was that of ^1H in the H_2^+ molecule by Hegstrom.[27] Ab initio calculations of ^{13}C and ^{17}O shielding in the CH_4 and in the H_2O molecules in the vicinity of their equilibrium geometries yield values of the surface along a variable symmetry displacement coordinate while maintaining the other symmetry displacements at zero.[21,28] When all such traces of the shielding surface are obtained, full rovibrational averaging can be carried out

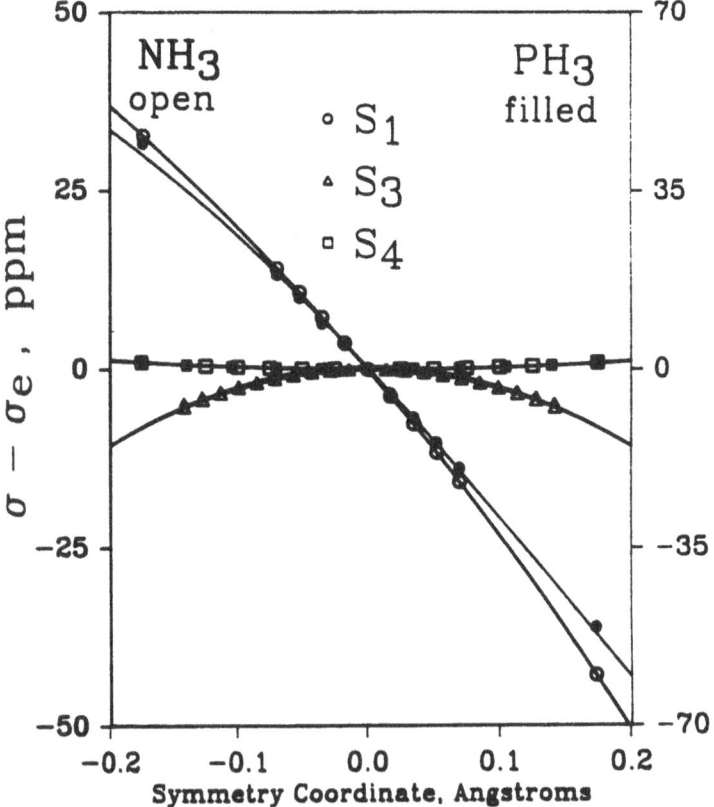

Fig. 2 Traces on the ^{15}N and ^{31}P shielding surfaces for NH_3 and PH_3.

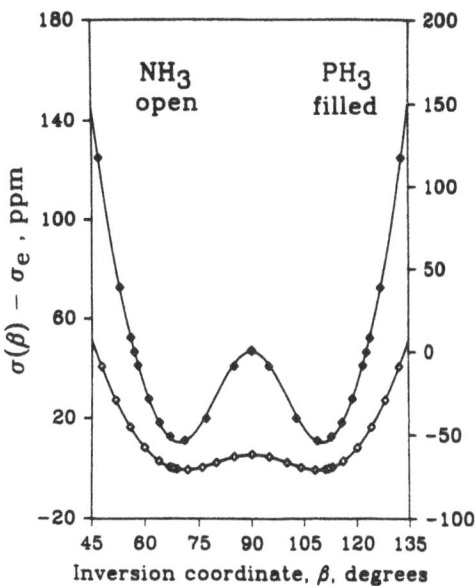

Fig. 3 Variation of the ^{15}N and ^{31}P shielding with inversion in NH_3 and PH_3.

to yield thermal average values, $\sigma_0(T)$, which can be compared with experiment.[19,29,30] Furthermore, the same rovibrational averaging in each of several isotopomers yields the shielding differences due to isotopic substitution (isotope shifts) which can also be compared with experiment.[19,29,30]

2.1 SHIELDING SURFACES FOR NH_3 and PH_3

We have recently reported the shielding surfaces of ^{15}N in NH_3 and ^{31}P in PH_3.[19,30] These are shown in Fig. 2 and 3. The symmetry coordinates are $S_1 = \left(1/\sqrt{3}\right)(\Delta r_1 + \Delta r_2 + \Delta r_3)$, $S_{3a} = \left(1/\sqrt{6}\right)(2\Delta r_1 - \Delta r_2 - \Delta r_3)$, $S_{4a} = \left(1/\sqrt{6}\right)r_e(2\Delta \alpha_1 - \Delta \alpha_2 - \Delta \alpha_3)$. The coordinates forming a degenerate pair with these (S_{3b} and S_{4b} respectively) provide exactly the same traces as S_{3a} and S_{4a}. The remaining coordinate S_2 is the symmetric bend. Since this corresponds to the inversion coordinate over which full averaging takes place at room temperature in NH_3, a more extensive portion of the surface has been calculated so as to include angles from 45° to 135° for the NH or PH bond relative to the C_3 symmetry axis of the molecule.

In this work the shielding calculations were performed with RPAC Version 8.5 program [31] by Hansen and Bouman [32] which employs the localized orbital local origin (LORG) method while GAUSSIAN88 provided the necessary SCF information.[33] The computations were carried out in an IBM 3090/300J/vector facility. The basis set employed is relatively large: three sets of d-type polarization functions on the apex atom and two sets of p-type polarization functions on the hydrogen were added to the 6-311+G basis of Pople.

Fig. 2 shows that when the vertical scales (^{31}P and ^{15}N shieldings) are related by the ratio of $\langle a_0^3/r^3 \rangle_{np}$ for P and N atoms, the surfaces are very close to being superposable, except for the inversion coordinate. The traces on the shielding surface with respect to the inversion coordinate in Fig. 3 have similar shapes to the potential energy surface. However, the barrier in the PH_3 molecule is much much higher than that for NH_3 so that in the averaging over the inversion coordinate all the shielding values from 45° to 135° need to be considered in NH_3 whereas the S_2 coordinate in PH_3 can be treated in the normal way as with all other coordinates, assuming small displacements. The NH_3 inversion wavefunctions were obtained from numerical solution of the Schrödinger equation for inversion, [19] using the empirical potential surface that was fit to high resolution spectra of all isotopomers of NH_3 by Spirko et al.[34] The results of the thermal averaging are in very good agreement with the experimental data for the $[\sigma_0(T) - \sigma_0(300\ K)]$ measured recently for NH_3 and PH_3, [19,30] and the isotope shifts $[\sigma(PD_3) - \sigma(PH_3)]$ and $[\sigma(ND_3) - \sigma(NH_3)]$ are in reasonable agreement with experiment. Details of this work can be found in ref [19] and [30].

2.2 GENERAL TRENDS

We can compare our results for NH_3 and PH_3 in Fig. 2 with the other shielding surfaces for CH_4 and H_2O, that have been previously reported by W. T. Raynes and

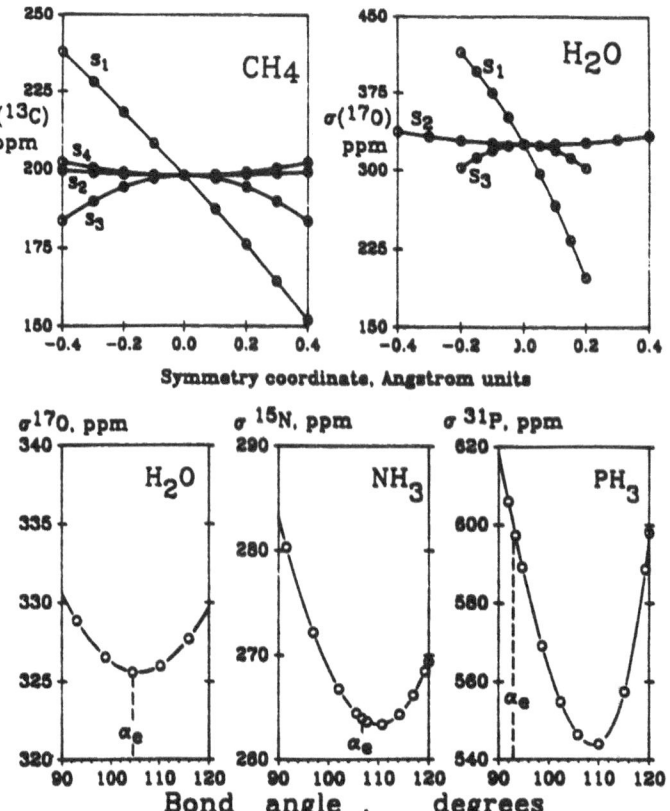

Fig. 4 Traces on the shielding surfaces of CH_4 and H_2O. Comparison of bond angle dependence

coworkers.[21,29] The latter are shown in Fig. 4. For the hydrides, at least, the shielding surfaces look strikingly similar. The symmetric stretch (S_1) leads to deshielding of ^{15}N, ^{31}P, ^{13}C, and ^{17}O in these molecules. The trace along the asymmetric stretch is concave downward (deshielding), whereas the trace along the asymmetric bend is concave upward (toward increased shielding). The asymmetric stretch corresponds to larger changes in shielding than the asymmetric bend but since the vibrational frequencies are greater for the stretch than the bend, the effects on the thermal average nearly cancel. The shielding trace along the symmetric bond angle change (or inversion) in OH_2, NH_3, and PH_3 are all concave upward and the shielding value is minimum at very nearly the tetrahedral angle in all cases. However, the equilibrium bond angle is close to tetrahedral in OH_2 and NH_3 but is 93° in PH_3. Therefore, in the former, the symmetric bond angle changes lead to uniformly higher shielding whereas in PH_3, the linear term in this coordinate leads to *deshielding* while the quadratic term leads to increased shielding with increasing temperature.

Another point of similarity is that in all these systems the largest contribution to the isotope shift, $[\sigma^X(XD_n) - \sigma^X(XH_n)]$, comes from the symmetric stretch. The thermal averages in Fig. 5 show very similar trends. In these hydrides, the thermal average shielding, $\sigma_0(T)$ decreases with increasing temperature, although much less pronounced in CH_4 and NH_3. The bond angle contributions in all 4 molecules make the calculated

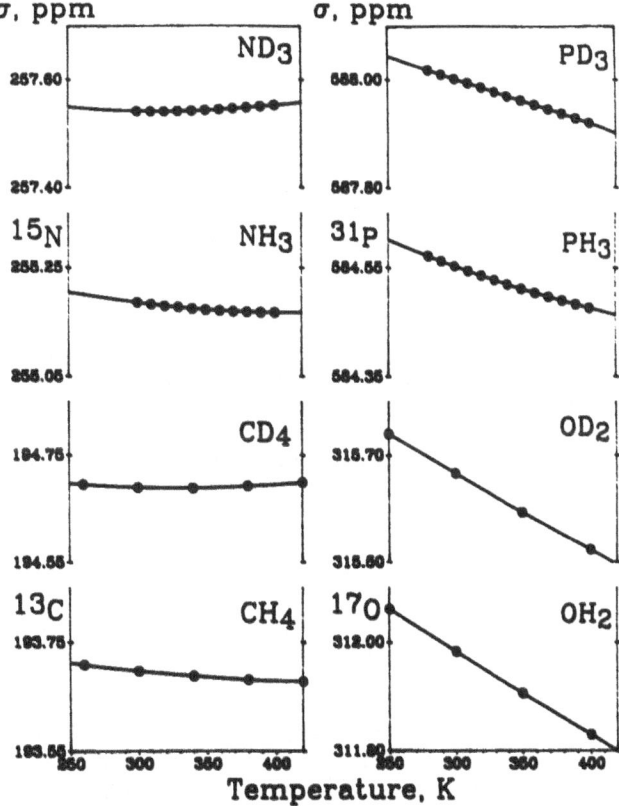

Fig. 5 Thermal average shielding $\sigma_0(T)$

$\sigma_0(T)$ functions concave upward[†] and this becomes more pronounced for the deuterated versions.

Although the theoretical shielding values decrease with bond extension in most molecules, and most of the experimental evidence (isotope shifts and temperature dependence of shielding) point toward deshielding upon bond extension, there are, however, theoretical cases of increasing shielding upon bond extension. The first example was reported for LiH by Stevens and Lipscomb[36] and Ditchfield.[25] Furthermore, D. B. Chesnut[22] has discovered a fascinating variation of $(\partial\sigma^X/\partial R)_{eq}$ in XH_n across the periodic table, positive for hydrides LiH and BeH_2, becoming increasingly negative toward FH. Similarly it is positive for NaH, MgH_2, and AlH_3, becoming increasingly negative toward ClH. A smooth algebraic change in $(\partial\sigma^X/\partial R)_{eq}$ occurs as X moves across the periodic table, tracing out an arch in each row with X = Be and Mg at the most positive value. At the end of this paper we will offer a reasonable explanation for this fascinating trend.

3. Intermolecular Shielding

Excepting molecular beam studies, the application of NMR to the elucidation of molecular structure and mechanisms of molecular reactions nearly always involves observations of the molecules in some medium, whether in gas, liquid, solid, or adsorbed phases. An understanding of the intermolecular effects on nuclear magnetic shielding is crucial to taking proper account of solvent effects, or conversely, to using the NMR chemical shift as a probe of intermolecular interactions. There is a large body of gas phase data [37] which provides information on the intermolecular contributions to the chemical shift; the observed signs, magnitudes, and the dependence on temperature need to be accounted for. Apart from the bulk susceptibility shifts which are present in all samples, there are true intermolecular shifts that can be attributed to polar and hydrogen-bonding effects and magnetic anisotropy of neighbor molecules (as in aromatic solvents). These have well-understood mechanisms and will not be discussed further here. The enigmatic part of the intermolecular shielding is that which gives rise to what had been termed van der Waals shifts.[38] This is the subject of this Section.

The study of the intermolecular shielding function, the change in the nuclear shielding as a function of the distance between interacting molecules and their orientation, is of fundamental importance in its own right. The intermolecular shielding function in general has important dependence on orientations of the interacting molecules. However, we start with the simplest case of two interacting rare gas atoms and focus attention on the dependence of the shielding on the intermolecular separation R. Therefore, within this article, we will refer to the intermolecular shielding function as $\sigma(R)$.

3.1 ^{29}Ar AS A MODEL SYSTEM

In principle, the behavior of the intermolecular shielding function of two rare gas atoms at the two limits is well known. The separated atoms is the reference limiting situation; the

---

[†]This is not generally the case. In the diatomic molecules (such as ^1H in H_2^+ or H_2, ^{19}F in HF or F_2) and for ^{19}F in a large number of fluorides, [35] the $\sigma_0(T)$ functions are concave downward and strongly decreasing (deshielding) with increasing temperature. The shielding of end atoms are largely dependent on the bond stretch and much less sensitive to bond angle changes.

diamagnetic shielding in a free atom is well known. When the interacting atoms correlate with a united atom S state, as Ar + Ar does with the Kr ground state, the shielding of the united atom is also well known, a purely diamagnetic shielding.[39] Thus, [σ(R) - σ(∞)] the intermolecular shielding relative to the separated system was known for Ar-Ar to be [σ(free Kr atom) - σ(free Ar atom)] at R = 0, and was known to be zero at R = ∞. In the intermediate region, the general shape of the function was unknown.

For calculations of intermolecular shielding, it is even more important to damp out the errors which result from incomplete cancellation of large positive and negative long-range contributions. In calculations of magnetic properties, this is known as a gauge origin problem. Furthermore, since the long-range interaction between two rare gas atoms is a result of electron correlation, we need to investigate any additional contributions from electron correlation beyond the coupled Hartree-Fock or random phase approximation (RPA) level of theory. For these reasons, we choose a local origin method that effectively damps out the errors in calculating the long-range contributions to the diamagnetic and paramagnetic terms and which can be extended to include second-order correlation contributions. We used the localized orbital local origin (LORG) method developed by Hansen and Bouman [32] and the second order LORG (SOLO) approach, in which the energy matrices and transition moments are evaluated such as to retain all terms through second order in electron correlation, i.e., using $\Omega^{(2)}$ and $t^{(2)}$.[40] The second-order effects on the diamagnetic terms have not been included in this calculation. These have been shown to be small, even for molecules such as CO and N_2 for which a correct description requires more than one determinant. For these molecules, the diamagnetic shielding results from configuration interaction (CI) calculations differ by less than 2 ppm from the self-consistent field (SCF) results for all nuclei.[41,42]

The SOLO calculations were carried out with the help of Professor T. D. Bouman using GAUSSIAN 90 [33] to generate the SCF molecular orbitals, and the RPAC 9.0 molecular properties programs system [43] implemented on the Cray 2 of the National Center for Supercomputing Applications at the University of Illinois, Urbana.

The use of counterpoise or ghost orbitals [44] would reduce the bias that results if the pair basis represents the wave function of Ar_2 in the simultaneous fields of the external magnet and the nuclear moment more accurately than the single atom basis represents the wave function of an isolated Ar atom in the presence of the two fields. Since the chemical shielding is a localized property (the shielding field produced by an electron is proportional to the inverse cube of the distance of the electron from the observed nucleus), basis set superposition errors are expected to be negligible even in this case where the intermolecular effects on shielding are being sought. Nevertheless, to verify this expectation, we carried out calculations of the single Ar atom at the 6-311G ($3d$) level with and without an equal number of ghost orbitals at various internuclear distances. We have established that the ghost orbital contribution is entirely negligible in this system.

The LORG and SOLO results for Ar_2 [45] are shown in Fig. 6. A numerical comparison at selected distances is shown in Table 1 for the LORG vs. SOLO results. This is the first *ab initio* intermolecular σ(R) function that has been calculated for a range of values of internuclear separations wide enough to show the shape of the function, and which is consistent with the correct limit at R = 0. The behavior at extremely short distances ~1Å or shorter can become very complicated because of avoided curve crossings with other states of the same symmetry. In any case, these regions are not sampled at the modest temperatures and pressures at which the gas phase experiments have been carried out.

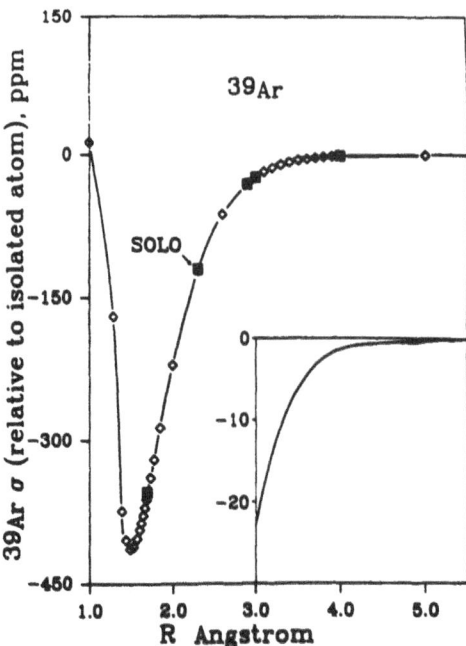

Fig. 6 ^{39}Ar shielding in the Ar...Ar system

TABLE 1. ^{39}Ar shielding in Ar$_2$ at two levels of calculation, with (SOLO) and without (LORG) second order correlation. [45]

$R_{Ar\text{-}Ar}(\text{Å})$	$[\sigma(R) - \sigma(\infty)](ppm)$	
	LORG	SOLO
1.70	-356.90	-353.49
2.30	-121.39	-119.64
2.90	-30.32	-30.68
3.00	-23.53	-23.91
4.00	-1.30	-1.40
5.00	-0.04	-0.00

TABLE 2. ^{39}Ar shielding in Ar...Na$^+$ [45]

$R_{Ar\text{-}Na^+}(\text{Å})$	$[\sigma(R) - \sigma(\infty)](ppm)$	
	LORG	SOLO
1.50	-42.60	-62.85
1.56	-39.78	-60.78
3.20	-4.77	-5.55

3.2 THE R-DEPENDENCE OF INTERMOLECULAR SHIELDING

SOLO includes both dispersion and overlap/exchange contributions and their interaction. However, at large distances, the intermolecular shielding is the small difference between two large terms and thus is difficult to obtain accurately. Therefore, the exact R-dependence at long range, the coefficients of the R^{-6} and R^{-8} terms in the *ab initio* shielding function $[\sigma(R) - \sigma(\infty)]$ cannot be determined accurately. For intermediate distances, a plot of log $[(\sigma(\infty) - \sigma(R))/ppm]$ vs. log $(R/\text{Å})$ in the range of distances (0.5 - 1.0) times the r_{min} of the Ar_2 potential function gives a slope equal to -6.67, i.e., $[\sigma(R) - \sigma(\infty)]$ behaves as $R^{-6.67}$. In this range of separations $[\sigma(R) - \sigma(\infty)]$ for ^{39}Ar in Ar ... Ne behaves as $R^{-6.79}$, ^{21}Ne in Ne... Ne behaves as $R^{-6.74}$, and ^{21}Ne in Ne... He goes as $R^{-7.41}$.[46]

A simple physical explanation for the R dependence of $[\sigma(R) - \sigma(\infty)]$ from *ab initio* calculations in rare gas pairs over the range of R values from $0.7r_0$ to $3-4r_0$ is as follows: The dispersion interaction causes a buildup of charge in the region between the two nuclei. The R^{-6} dispersion energy arises from the classical electrostatic attraction of each nucleus to its own dipolar electron cloud.[47] At the same time the R^{-6} shielding term due to dispersion arises from the paramagnetic shielding of a dipolar electron cloud. Similarly, the mutual distortion of the two atoms due to overlap leads to dipolar electron clouds, and the overlap dipole varies exponentially with R at very short range. The R^{-6} to R^{-8} dependence of shielding due to overlap arises from the paramagnetic shielding of a dipolar electron cloud.

There is a somewhat different behavior of the shielding with separation when electrical charges are involved. For example, let us compare ^{39}Ar shielding in Ar...Ne with that in Ar...Na$^+$ in Fig. 7. We find that approach of $\sigma(R)$ to the separated atoms value is

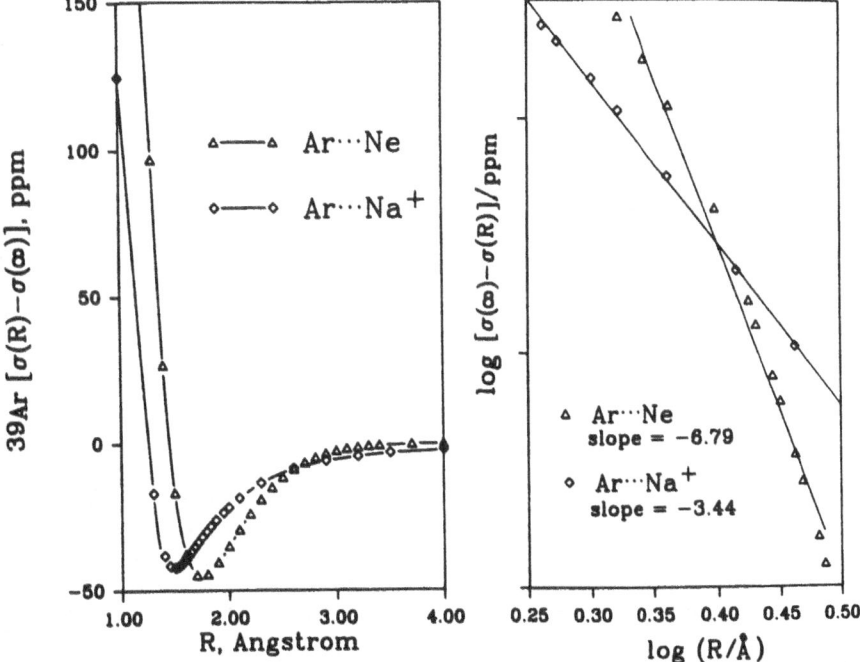

Fig. 7 ^{39}Ar shielding surfaces for the isoelectronic systems Ar...Ne and Ar...Na$^+$

approximately $R^{-3.44}$ for ^{39}Ar in Ar...Na$^+$ for $R \geq 1.8$Å whereas it is the typical (for rare gas pairs) approximately $R^{-6.79}$ for ^{39}Ar in Ar...Ne for $R \geq 2.0$Å. An understanding of this different R dependence at large R can be gained from the expected functional forms of $\sigma(R)$ in the long-range limit.

The long-range limiting behavior in these systems can be modeled by considering the effects of the neighboring atom or ion on the ^{39}Ar shielding in terms of the electric fields F_γ created by the neighbor at the position of the ^{39}Ar nucleus. In the presence of an applied electric field the shielding can be written as: [48-50]

$$\sigma_{\alpha\alpha} = \sigma_{\alpha\alpha}^{(0)} + \left(\frac{\partial\sigma_{\alpha\alpha}}{\partial F_\gamma}\right)_{F_\gamma=0} F_\gamma + \left(\frac{\partial^2\sigma_{\alpha\alpha}}{\partial F_\gamma \partial F_\beta}\right)_{F_\beta=F_\gamma=0} F_\gamma F_\beta + .. \qquad (6)$$

We may further include the effects of electric field gradients $F_{\gamma\beta}$. For an Ar atom, the shielding first derivatives with respect to the electric field (the shielding polarizability, a term coined by Dykstra [51,52]) vanish by symmetry,

$$\left(\frac{\partial\sigma_{\alpha\alpha}}{\partial F_\gamma}\right)_{F_\gamma=0} = 0. \qquad (7)$$

If we are observing only the isotropic shielding, then the 1-2 pair at a fixed separation R is considered to be tumbling freely in the magnetic field B_0 and the contributions of electric field gradients to $[\sigma_{iso} - \sigma_{iso}^{(0)}]$ for an Ar atom vanish also.

Therefore, in the mean field model of Raynes, Buckingham, and Bernstein, [12] where the neighbor atom effects on the shielding are considered in terms of the mean square electric fields $\overline{F^2}$ it produces at the nucleus in question,

$$\left[\sigma_{iso} - \sigma_{iso}^{(0)}\right]_{\substack{long-\\range}} \approx \frac{1}{3}\left\{\frac{1}{2}\sigma_{zz,z,z}^{(2)} + \sigma_{zz,x,x}^{(2)}\right\}_{Ar} \overline{F^2} \qquad (8)$$

where $\qquad \sigma_{zz,x,x}^{(2)} \equiv \left(\frac{\partial^2\sigma_{zz}}{\partial F_x^2}\right)_{F_x=0} \qquad (9)$

and $\qquad \overline{F^2} = q_2^2\, R^{-4} + \mu_2^2\, R^{-6}(3\cos^2\theta+1) + \frac{9}{8}\, \Theta_2^2 R^{-8}\, (1-2\cos^2\theta+5\cos^4\theta)$

$$+ \left[-\frac{C_6}{\alpha_{Ar}(0)}\, R^{-6} - \frac{R^{-8}}{\alpha_{Ar}(0)}\left(C_8 - 5C_6\frac{C_2}{\alpha_2(0)}\right) - ... \right] \qquad (10)$$

Here, for Ar with a linear molecule, for example, q_2, μ_2, and Θ_2 are the electric charge, dipole moment, and quadrupole moment of the neighbor (2), and $\alpha_2(0)$ and C_2 are its static electric dipole polarizability and quadrupole polarizability.[53] The angle between the axis of molecule 2 and the intermolecular axis is θ. The dispersion energy coefficients for the 1-2 pair are C_6 and C_8. The dispersion term in the mean square field (in square brackets) is that given by Buckingham and Clarke.[54]

Indeed, we find the appropriate interpretation for the R^{-4} tail shown by the *ab initio* ^{39}Ar shielding in the Ar...Na$^+$ system. On the other hand in Fig. 8, as expected, a near $\sim R^{-6}$ tail is found for Ar...NaH (collinear). Note too how the additional electrons in NaH makes a difference in the short/intermediate range behavior of the shielding of Ar.

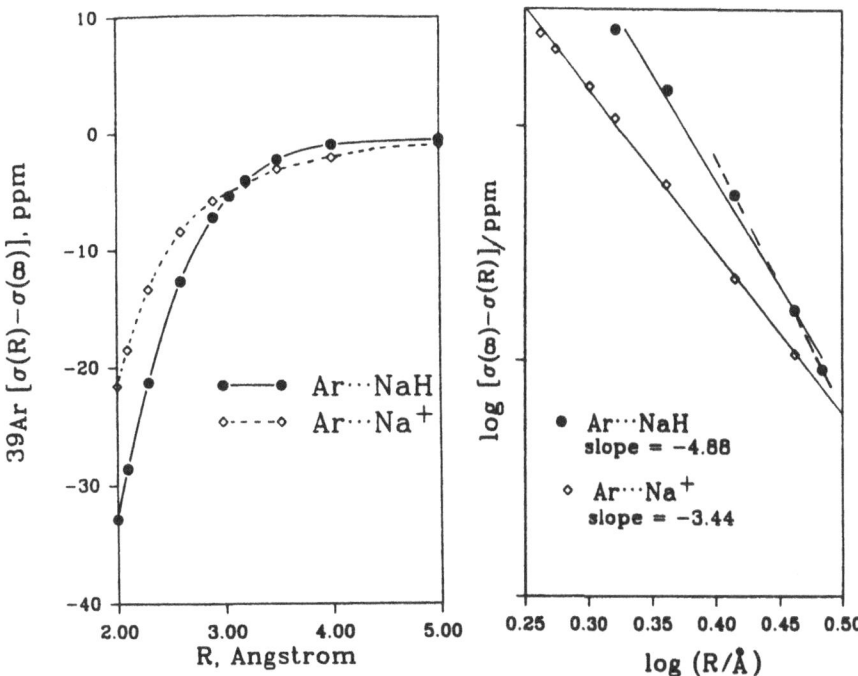

Fig. 8 ^{39}Ar shielding surface for Ar...Na$^+$ compared with Ar...NaH (collinear)

Only *ab initio* calculations can provide the behavior of σ(R) over a wide range of separations. However, by using simple physical models, we conclude that we have a reasonably good understanding of the R-dependence of the intermolecular shielding function.

3.3 SCALING OF THE INTERMOLECULAR SHIELDING FUNCTION

A possible scaling of intermolecular shielding functions for the rare gas pairs is suggested by the long-range limit form of the function.[45] In the mean field model, the mean square effective field, created at 1 due to the dispersion interaction between 1 and 2, can be related to the dispersion energy of the pair of interacting atoms as in Eq. (10):[54]

$$\overline{F^2} = - \frac{C_6}{\alpha_1(0)}\, R^{-6} - \frac{R^{-8}}{\alpha_1(0)}\left(C_8 - 5C_6\,\frac{C_2}{\alpha_2(0)}\right) - \ldots \tag{11}$$

In the simpler London approximation, the mean square field at 1 due to 2 is [55,56]

$$\overline{F^2} \approx \frac{3}{2}\,\alpha_2(0)\,\frac{U_1 U_2}{U_1 + U_2}\,R^{-6} \tag{12}$$

where U_1 and U_2 are the ionization potentials and $\alpha(0)$ is the static polarizability. Although the mean field model may not be strictly valid at distances shorter than r_{min} of the potential function, the factors appearing in the above expression suggest that an approximate scaling of the shielding functions may be possible. Furthermore, the nature of the shielding hyperpolarizability,

$$\left(\frac{\partial^2 \sigma}{\partial F^2}\right)_{F=0} = \frac{1}{3}\left\{\frac{1}{2}\,\sigma_{zz,z,z}^{(2)} + \sigma_{zz,x,x}^{(2)}\right\} \tag{13}$$

suggests scaling factors $\langle a_0^3/r^3\rangle \cdot \alpha_1(0)$. It is well known that the magnitudes of NMR chemical shifts of nuclei across the Periodic Table scale as $\langle a_0^3/r^3\rangle_{np}$ for the free atom in question, [57] a quantity which can be derived directly from the spin-orbit splittings observed in atomic spectra. In other words, the sensitivity of the nuclear shielding to changes in the electronic environment of the nucleus, e.g., in going from one molecule to another, has been found to be reflected by $\langle a_0^3/r^3\rangle_{np}$. The distortion of the electronic distribution of an atom in the presence of a static uniform electric field depends on the static electric dipole polarizability of the atom. Thus, the shielding hyperpolarizability is expected to scale as $\langle a_0^3/r^3\rangle_{np} \cdot \alpha(0)$. We have therefore suggested that the scaling factors relating intermolecular shielding functions of rare gas pairs are [45]

$$\langle a_0^3/r^3\rangle_1 \,\alpha_1(0) \cdot \frac{3}{2}\,\alpha_2(0)\,\frac{U_1 U_2}{U_1 + U_2}\,.$$

For example, at long-range separations,

$$[\sigma(^{129}\text{Xe in Xe ... Kr}) - \sigma(\infty)] \approx \frac{\langle a_0^3/r^3\rangle_{Xe}\,\alpha_{Xe}(0)\,\alpha_{Kr}(0)}{\langle a_0^3/r^3\rangle_{Ar}\,\alpha_{Ar}(0)\,\alpha_{Ar}(0)}\,\frac{U_{Xe}\,U_{Kr}}{U_{Xe} + U_{Kr}}\,\frac{2}{U_{Ar}}$$

$$\times\,[\sigma(^{39}\text{Ar in Ar...Ar}) - \sigma(\infty)]_{\substack{ab \\ initio}} \tag{14}$$

There are no adjustable parameters here.

We have some evidence that these factors provide a successful scaling of the intermolecular shielding in rare gas pairs, in the range of values $R \geq 0.7\ r_0$. One is the nearly conformal R dependence of the shielding for all rare gas pairs for which we have carried out *ab initio* calculations: ^{39}Ar in Ar...Ar, Ar...Ne, ^{21}Ne in Ne...Ne, and Ne...He exhibit $\sim R^{-6.67}$, $\sim R^{-6.79}$, $\sim R^{-6.74}$, and $\sim R^{-7.41}$ behavior, respectively in this range of separations.[46] This indicates that the shielding functions are conformal and scaling is possible. Secondly, the *ab initio* ^{21}Ne shielding in Ne...Ne is closely reproduced by the *ab initio* ^{39}Ar shielding in Ar...Ar when the latter is multiplied by the scaling factors given above. This is shown in Fig. 9. Thirdly, the range of gas-to-solution shifts of ^{129}Xe, ^{83}Kr, and ^{21}Ne atoms in the same set of solvents spanning a range of dielectric constants do form a straight line passing through the origin when plotted against $\langle a_0^3/r^3\rangle \cdot \alpha(0) \cdot U$.[45] Finally, the scaling factors have been used to convert the *ab initio* $[\sigma(R) - \sigma(\infty)]$ for ^{39}Ar in Ar...Ar into the corresponding values for ^{129}Xe in Xe...Ar, Xe...Kr, and Xe...Xe. These can be used to calculate the second virial coefficients of shielding

$$\sigma_1(T) = \int_0^\infty 4\pi R^2\,dR[\sigma(R) - \sigma(\infty)]\,\exp\,[-V(R)/kT].$$

These values are compared with the gas phase experimental data in Fig. 10. The agreement is satisfactory. The experimental signs, orders of magnitude, and temperature dependences are well-reproduced. The appropriate scaling of the internuclear distance is unknown but we chose r_0 for convenience, as suggested by the conformal interaction potential energy functions for rare gas pairs which are scaled by r_0 and ε.[58]

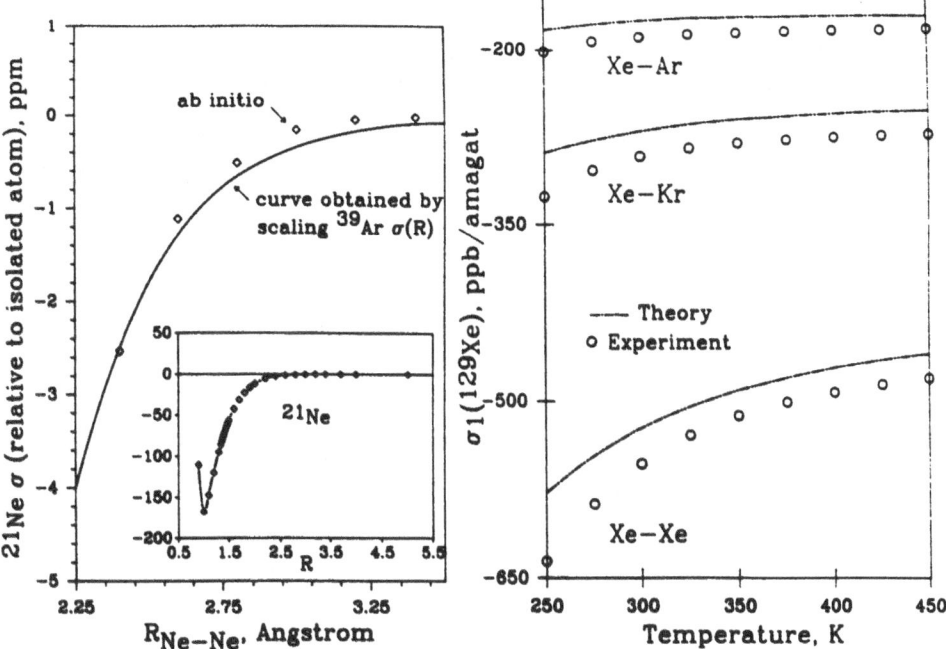

Fig. 9 ^{21}Ne shielding in Ne...Ne compares well with scaled Ar...Ar

Fig. 10 Second virial coefficients of ^{129}Xe shielding from ^{129}Xe $\sigma(R)_{Xe-X}$ obtained by scaling $\sigma(R)_{Ar-Ar}$.

4. The General Shape of the Shielding Function

All the *ab initio* intermolecular shielding surfaces that we have found (Ar...Ar, Ar...Ne, Ne...Ne, Ne...He, Ar...Na$^+$, and Ar...NaH) have the general shape like that of a potential energy function for rare gas pairs.[46] We propose that this shape is the general qualitative shape of an intermolecular shielding function. Some facts concerning $\sigma(R)$ require the existence of a negative minimum in the function $[\sigma(R) - \sigma(\infty)]$ for Ar$_2$ and predominantly negative values in the well region of the V(R). (a) The united atom is Kr which has a known large diamagnetic shielding which is 2.6 times that of σ for an isolated Ar atom.[39] (b) The observed second virial coefficients of shielding in nearly all systems studied in the gas phase is *negative* with few exceptions.[37] (c) The distortion asssociated with the long-range dispersion forces provides a negative $[\sigma(R) - \sigma(\infty)]$.

In every case the R dependence in the range of separations inside of r_{min} is close to R^{-6} to R^{-8} (except in Ar...Na$^+$ where there is an obvious long-range ~R^{-4} dependence that is expected for the system). In the presence of an electric dipole (the Ar...NaH system) the R dependence approaches R^{-6} once again, not unexpected from Buckingham's mean square field approach.[12] It is interesting that for the intermediate range of distances (between 0.8 r_0 and r_{min}) the R dependence is still about the same as might be expected from the dispersion contributions predicted by the mean square field model of Buckingham. SOLO calculations which include second order electron correlation contributions to shielding provide only a small correction in this range of distances.[45]

The shapes of the intermolecular shielding functions are similar to the *intra*molecular shielding function for ^1H in H_2^+ molecule. Interestingly (we don't know the fundamental basis for it yet), the very accurate proton shielding in the H_2^+ molecule has an $R^{-6.0}$ behavior for R values greater than that which corresponds to the mininum in the shielding function, for distances 8 a.u up to 20 a.u. Ergo, we propose that all two-atom shielding functions involving ^1S states have the *same qualitative* shape, that is, if we do not include electron spin or the natural paramagnetism of separated atom states which have non-zero angular momentum. (For example, if HF were to separate to the F^- and H^+ ions then we would have ^1S separated atom states.)

What do shielding functions in diatomic molecules (other than ^1H in H_2^+) look like? *Ab initio* calculations in the vicinity of the equilibrium geometry are well-known for H_2, HF, HCl, NaH, LiH, CO, N_2, but a wide range of R values have not been investigated. If indeed a universal qualitative shape is that which we have found for intermolecular shielding, we should be able to locate a shielding minimum. For most molecules the shielding derivative with respect to bond extension is *negative*. Thus the shielding minimum, if it exists, will be found at R values much greater than the equilibrium bond length. Unfortunately, at very large separations the orbital paramagnetism of the non- ^1S separated atom states comes into play. This paramagnetism is of course not included in the frequently quoted diamagnetic shielding for the free atom.[39] When the separated atom states are not ^1S, the shielding just keeps on being larger negative with increasing bond length.

We can however, search for a shielding minimum at R values shorter than the equilibrium bond length, if the derivatives of the shielding at the equilibrum geometry is positive. Possible candidates are NaH and LiH. We did not reach the shielding minimum in LiH even at R = 0.5Å. However, we were able to calculate ^{23}Na shielding in the NaH molecule over a wide range of R values, [46] and what we found is shown in Fig. 11. This too is a shielding function of the same general shape that we have seen before.

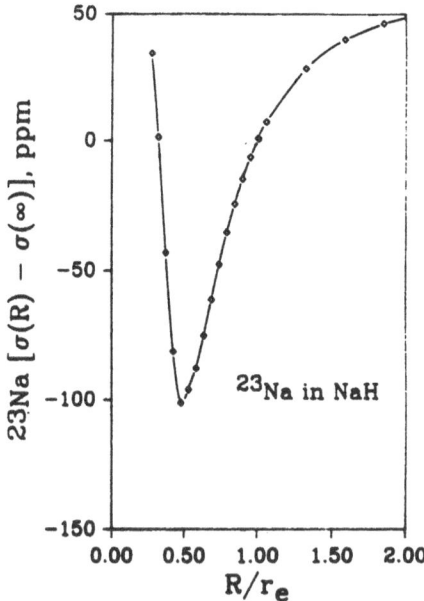

Fig. 11 The ^{23}Na shielding surface in the NaH molecule

The shielding that is observed experimentally depends on the relevant region over which the averaging takes place. In Fig. 12 we attempt to show the relationships between the portion of the $\sigma(R)$ function that is of importance, and the potential energy function $V(R)$ which governs the averaging over the appropriate range of R values. The unit of energy in each subfigure is Kelvins, to indicate which portion of the potential function becomes relevant in the sampling of $\sigma(R)$ at any given temperature. The progression of the "active" range of R values across the $\sigma(R)$ function in going from ^1H in the H_2^+ molecule, to ^{23}Na in the NaH molecule, to ^{39}Ar in $ArNa^+$ and in Ar_2 provides different samplings of generally the same shape of $\sigma(R)$ function.

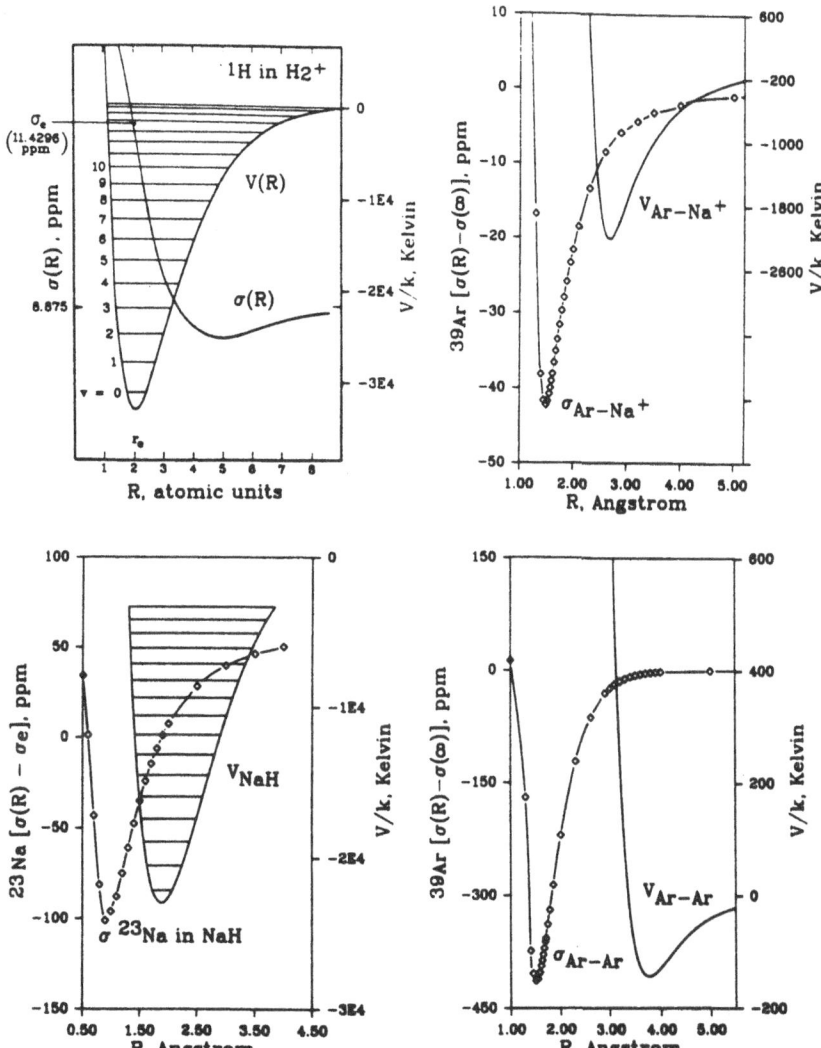

Fig. 12 Comparison of inter- and intra-molecular shielding surfaces

Although we cannot calculate the entire shielding function for nuclei in most diatomic molecules (for reasons stated earlier), we believe that they *all* nevertheless have shapes similar to those shown in Fig. 12 (excluding contributions from orbital paramagnetism in non-S systems). We propose that even for those diatomic molecules for which we are unable to calculate $\sigma(R)$ over a wide enough range of separations (without encountering the problems cited above), the shape of the shielding function is qualitatively similar to that we found for the intermolecular case of Ar in Ar_2. Evidence for this assertion is provided by the shape of the shielding function in the vicinity of the equilibrium geometry: The second derivatives in both the extremes (that is, large negative first derivative and positive derivatives corresponding to the extreme ends of a row in the Periodic Table) are negative. ^{39}Ar in Ar_2 does have negative second derivatives at the small R side (where it has a large negative first derivative) and the large R side of the shielding minimum (where it has a positive first derivative). In the immediate vicinity of r_e the *ab initio* (LORG) shielding functions for diatomic molecules are remarkably similar for analogous diatomic molecules. If we scale them by $\langle a_0^3/r^3 \rangle \cdot r_e$, the comparison can be made directly. We find that $[\sigma(R) - \sigma(\infty)]/\langle a_0^3/r^3 \rangle r_e$ for ^{35}Cl in ClF and ^{19}F in F_2 are nearly exactly superposable in the range of R values corresponding to the classical turning points of the ground vibrational states of these molecules.[46] The same remarkable superposability is found for ^{35}Cl in HCl and ^{19}F in FH, and even ^{23}Na in NaH and ^7Li in LiH are nearly superposable.

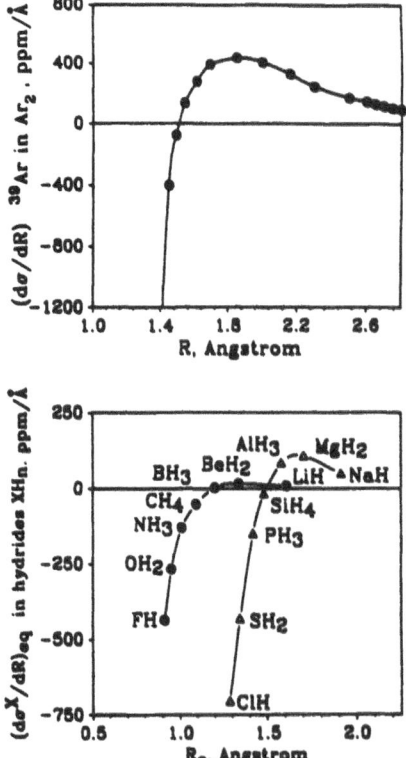

Fig. 13 Derivatives of the X nuclear shielding in XH_n vary with R_e in the same way as the shielding derivative function in Ar...Ar

Finally, we offer as evidence the variation of the sign of the derivative of the shielding at r_e, $(d\sigma/dR)_{eq}$ among the hydrides of the first two rows of the periodic table. The change in sign was first noted by Chesnut.[22] For both first- and second-row hydrides, the shielding derivative starts out to be positive and small at the alkali end of the periodic table. A slight increase in the shielding derivative is observed upon going to the hydrides of the alkaline earth metals. Then a decrease in the derivative is observed upon going to boron and aluminum hydrides. This drop in the shielding derivative continues on, such that at the halogen end of the Periodic Table, it is generally negative and large. Our explanation for these trends suggests itself when the equilibrium bond lengths in these hydrides are compared.[46] In Fig. 13 we plot the shielding derivatives $(d\sigma^X/dR)_{eq}$ vs. the equilibrium X-H bond length. The similarity between *inter*molecular and *intra*molecular shielding functions is clearly depicted by the similarity between the derivative function $d\sigma/dR$ vs. R for ^{39}Ar in Ar$_2$ and the $(d\sigma^X/dR)_{eq}$ vs. r_e for the XH$_n$ hydrides. We propose that the shielding derivatives provide glimpses into the $\sigma(R)$ function at the specific values of R which are the equilibrium bond lengths, and these glimpses trace out the same shape as the one we have already found for the intermolecular $\sigma(R)$ and for ^{23}Na in NaH molecule.

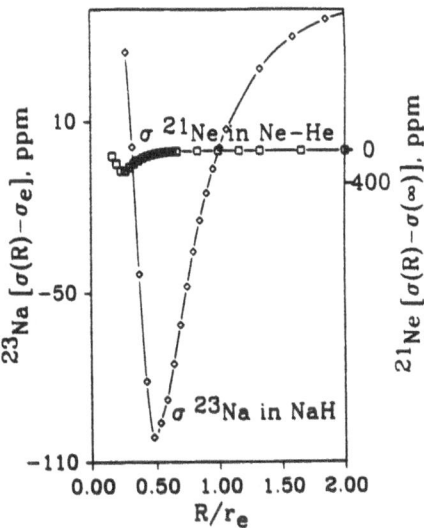

Fig. 14 Shielding surfaces in the isoelectronic systems NaH and Ne...He. The vertical scales are in the same ratio as $\langle a_0^3/r^3 \rangle_{np}$ for ^{23}Na and ^{21}Ne atoms

In closing, we provide the proper perspective on the relative magnitudes of the changes in shielding with changes in interatomic distance in two-atom systems, the two can be compared if the $\langle a_0^3/r^3 \rangle$ scaling of shielding sensitivity is taken into account. This is shown in Fig. 14 for the isoelectronic systems Ne...He and NaH. When ^{21}Ne is drawn on the ^{23}Na scale by using the respective $\langle a_0^3/r^3 \rangle$ for the Ne and Na atoms for the vertical scales, we find that the intermolecular shielding changes are indeed much smaller than the intramolecular shielding changes.

ACKNOWLEDGMENT. The research described in this talk has been supported by the National Science Foundation (Grants CHE89-01426 and CHE92-10790). Professor

Thomas D. Bouman provided assistance in the SOLO calculations described here. His untimely death is a loss to all those interested in the elucidation of the nuclear magnetic shielding tensor.

References

1. Jameson, C. J., (1980) 'Effects of intermolecular interactions and intramolecular dynamics on nuclear resonance', Bull. Magn. Reson. 3, 3-28.
2. Jameson, C. J., Jameson, A. K., and Burrell, P. M. (1980) '^{19}F nuclear shielding scale from gas phase studies', J. Chem. Phys. 73, 6013-6020.
3. Hindermann, D. K. and Cornwell, C. D. (1968) 'Fluorine and proton NMR study of gaseous hydrogen fluoride', J. Chem. Phys. 48, 2017-2024.
4. Jameson, C. J., Jameson, A. K., Oppusunggu, D., Wille, S., Burrell, P. M., and Mason, J. (1981) '^{15}N nuclear magnetic shielding scale from gas phase studies', J. Chem. Phys. 74, 81-88.
5. Jameson, A. K., and Jameson, C. J. (1987) 'Gas phase ^{13}C chemical shifts in the zero-pressure limit: refinements to the absolute shielding scale for ^{13}C', Chem. Phys. Lett. 134, 461-466.
6. Wasylishen, R. E., Mooibroek, S., and Macdonald, J. B. (1984) 'A more reliable ^{17}O absolute chemical shielding scale', J. Chem. Phys. 81, 1057-1059.
7. Davies, P. B., Neumann, R. M., Wofsy, S. C., and Klemperer, W. (1971) 'Radiofrequency spectrum of phosphine (PH_3)', J. Chem. Phys. 55, 3564-3568.
8. Jameson, C. J., de Dios, A., and Jameson, A. K. (1990) 'Absolute shielding scale for ^{31}P from gas-phase nmr studies', Chem. Phys. Lett. 167, 575-582.
9. Wasylishen, R. E., Connor, C., and Friedrich, J. O. (1984) 'An approximate absolute ^{33}S nuclear magnetic shielding scale', Can. J. Chem. 62, 981-985.
10. Jameson, C. J. and Jameson, A. K. (1988), 'Absolute shielding scale for ^{29}Si', Chem. Phys. Lett. 149, 300-305.
11. Buckingham, A. D. and Pople, J. A. (1956) 'Electromagnetic properties of compressed gases', Faraday Soc. Discuss. 22, 17- 21.
12. Raynes, W. T., Buckingham, A. D., and Bernstein, H. J. (1962) 'Medium effects in proton magnetic resonance. I. Gases', J. Chem. Phys. 36, 3481-3488.
13. See for example, Jameson, C. J., Jameson, A. K., and Oppusunggu, D. (1986) 'Temperature dependence of ^{77}Se, ^{125}Te, and ^{19}F shielding and M-induced isotope shifts in MF_6 molecules', J. Chem. Phys. 85, 5480-5483.
14. Jameson, A. K., Jameson, C. J., and Gutowsky, H. S. (1970) 'Density dependence of ^{129}Xe chemical shifts in mixtures of xenon and other gases', J. Chem. Phys. 53, 2310-2321.
15. Jameson, C. J., Jameson, A. K., and Cohen, S. M. (1975) 'Temperature and density dependence of ^{129}Xe chemical shift in rare gas mixtures', J. Chem. Phys. 62, 4224-4226.
16. Jameson, C. J., Jameson, A. K., and Cohen, S. M. (1976) 'Second virial coefficient of ^{129}Xe chemical shielding in mixtures of Xe with spherical top molecules, CH_4, CF_4, and SiF_4', J. Chem. Phys. 65, 3401-3406.
17. Jameson, C. J. (1991) 'Rovibrational averaging of molecular electronic properties', in Z. B. Maksic (ed.), Theoretical Models of Chemical Bonding, Part 3. Molecular Spectroscopy, Electronic Structure, and Intramolecular Interactions, Springer-Verlag, Berlin, pp. 457-519.
18. Jameson, C. J. and Osten, H.-J. (1986) 'Theoretical aspects of the isotope effect on nuclear shielding', in G. A. Webb (ed.), Annual Reports on NMR Spectroscopy, Academic Press, London, Vol. 17, pp. 1-78.

19. Jameson, C. J., de Dios, A. C., and Jameson, A. K. (1991), 'Nuclear magnetic shielding of nitrogen in ammonia', J. Chem. Phys. 95, 1069-1079.

20. Riley, G., Raynes, W. T., and Fowler, P. W. (1979) 'The variation of molecular properties with vibration-rotation', Mol. Phys. 38, 877-892.

21. Fowler, P. W., Riley, G., and Raynes, W. T. (1981), 'Dipole moment, magnetizability, and nuclear shielding surfaces for the water molecule', Mol. Phys. 42, 1463-1481.

22. Chesnut, D. B. (1986) 'NMR chemical shift bond length derivatives of the first- and second-row hydrides', Chem. Phys. 110, 415-420.

23. Chesnut, D. B. and Wright, D. W. (1991) 'Chemical shift bond derivatives for molecules containing first-row atoms', J. Computational Chem. 12, 546-559.

24. Chestnut, D. B. and Foley, C. K. (1986) 'Chemical shifts and bond modification effects for some small first-row-atom molecules', J. Chem. Phys. 84, 852-861.

25. Ditchfield, R. (1981) 'Theoretical studies of the temperature dependence of magnetic shielding tensors: H_2, HF, and LiH', Chem. Phys. 63, 185-202.

26. Fleischer, U., Schindler, M., and Kutzelnigg (1987) 'Magnetic properties in terms of localized quantitities. VI. Small hydrides, fluorides, and homonuclear molecules of phosphorus and silicon', J. Chem. Phys. 86, 6337-6347.

27. Hegstrom, R. A. (1979) 'g factors and related magnetic properties of molecules. Formulation of theory and calculations for H_2^+, HD^+, and D_2^+', Phys. Rev. A 19, 17-30.

28. Lazzeretti, P., Zanasi R., Sadlej, A. J., and Raynes, W. T. (1987) 'Magnetizability and carbon-13 shielding surfaces for the methane molecule', Mol. Phys. 62, 605-616.

29. Raynes, W. T., Fowler, P. W., Lazzeretti, P., Zanasi, R., and Grayson, M. (1988) 'The effects of rotation and vibration on the carbon-13 shielding, magnetizabilities and geometrical parameters of some methane isotopomers', Mol. Phys. 64, 143-162.

30. Jameson, C. J., de Dios, A. C., and Jameson, A. K. (1991) 'The ^{31}P shielding in phosphine', J. Chem. Phys. 95, 9042-9053.

31. RPAC version 8.5, Thomas D. Bouman, Southern Illinois University at Edwardsville, and Aage E. Hansen, H. C. Oersted Institute, Copenhagen, Denmark.

32. Hansen, A. E. and Bouman, T. D. (1985) 'Localized orbital/local origin method for calculation and analysis of NMR shieldings. Applications to ^{13}C shielding tensors', J. Chem. Phys. 82, 5035-5047.

33. GAUSSIAN88, M. J. Frisch, M. Head-Gordon, H. B. Schlegel, K. Raghavachari, J. S. Binkley, C. Gonzalez, D. J. Fox, R. A. Whiteside, R. Seeger, C. F. Melius, J. Baker, R. Martin, L. R. Kahn, J. J. P. Stewart, E. M. Fluder, S. Topial, and J. A. Pople, Gaussian, Inc., Pittsburgh, PA. 1988.

34. Spirko, V., Stone, J. M. R., and Papousek, D. (1976) 'Vibration-inversion-rotation spectra of ammonia: centrifugal distortion, coriolis interactions, and force field in $^{14}NH_3$, $^{15}NH_3$, $^{14}ND_3$, and $^{14}NT_3$', J. Mol. Spectrosc. 60, 159-178.

35. Jameson, C. J. and Osten, H. J. (1985) 'Systematic trends in the variation of ^{19}F nuclear magnetic shielding with bond extension in halomethanes', Mol. Phys. 55, 383-395.

36. Stevens, R. M., Pitzer, R. M., and Lipscomb, W. N. (1963) 'Perturbed Hartree-Fock calculations I. Magnetic susceptibility and shielding in the LiH molecule', J. Chem. Phys. 38, 550-560.

37. Jameson, C. J. (1991) 'Gas phase NMR spectroscopy', Chem. Rev. 91, 1375-1395.

38. Rummens, F. H. A. (1975) 'Van der Waals forces and shielding effects', in P. Diehl, E. Fluck, and R. Kosfeld (eds.) NMR Basic Principles and Progress, Springer, Berlin, Vol. 10, pp. 1-118.

39. Malli, G. and Froese C. (1967) 'Nuclear magnetic shielding constants calculated from numerical Hartree-Fock wave functions', Intl. J. Quantum Chem. 1S, 95-98.

40. Bouman, T. D. and Hansen, A. E. (1990) 'NMR shielding calculations beyond coupled Hartree-Fock: Second order correlation effects in localized-orbital/local-origin calculations of molecules containing phosphorus', Chem. Phys. Lett. 175, 292-299.

41. Amos, R. D. (1980) 'SCF and CI calculations of the one-electron properties, polarizabilities and polarizability derivatives of the nitrogen molecule', Mol. Phys. 39, 1-14.

42. Amos, R. D. (1979) 'SCF and CI calculations of the one-electron properties of carbon monoxide as a function of internuclear distance', Chem. Phys. Lett. 68, 536-539.

43. RPAC Version 9.0 T. D. Bouman, Southern Illinois University, Edwardsville, and A. E. Hansen, H. C. Oersted Institute, Copenhagen, Denmark.

44. Boys, S. F. and Bernardi, F. (1970) 'The calculation of small molecular interactions by the differences of separate total energies', Mol. Phys. 19, 553-566.

45. Jameson, C. J. and de Dios, A. C. (1992) 'Ab initio calculations of the intermolecular chemical shift in nuclear magnetic resonance in the gas phase and for adsorbed species', J. Chem. Phys. 97, 417-434.

46. Jameson, C. J. and de Dios, A. C. (1992) 'The NMR shielding as a function of internuclear separation', to be published.

47. Buckingham, A. D. (1978) 'Basic theory of intermolecular forces: applications to small molecules', in B. Pullman (ed.), Intermolecular Interactions: From Diatomics to Biopolymers, John Wiley, New York pp. 1-67.

48. Buckingham, A. D. (1960) 'Chemical shifts in the nuclear magnetic resonance spectra of molecules containing polar groups', Can. J. Chem. 38, 300-307.

49. Marshall, T. W. and Pople, J. A. (1958) 'Nuclear magnetic shielding of a hydrogen atom in an electric field' Mol. Phys. 1, 199-202.

50. Raynes, W. T. and Ratcliffe, R. (1979) 'Nuclear site symmetry and nuclear magnetic shielding in a uniform electric field', Mol. Phys. 37, 571-578.

51. Augspurger, J. D., Dykstra, C. E., and Oldfield, E. (1990) 'Correlation of carbon-13 and oxygen-17 chemical shifts and the vibrational frequency of electrically perturbed carbon monoxide: A possible model for distal ligand effects in carbonmonoxyheme proteins', J. Am. Chem. Soc. 113, 2447-2451.

52. Augspurger, J. D. and Dykstra, C. E. (1991) 'Electromagnetic properties of molecules from a uniform procedure for differentiation of molecular wave functions to high order', J. Phys. Chem. 95, 9230-9238.

53. Buckingham, A. D. (1967) 'Permanent and induced molecular moments and long-range intermolecular forces', in J. O. Hirschfelder (ed.), Adv. Chem. Phys. Vol. 12, pp. 107-142.

54. Buckingham, A. D. and Clarke, K. L. (1978) 'Long-range effects of molecular interactions on the polarizability of atoms', Chem. Phys. Lett. 57, 321-325.

55. London, F. (1930) 'Über einige Eigenschaften und Anwendungen der Molekularkräfte', Z. physik. Chem. (Leipzig) B, 11, 222-251.

56. London, F. (1937) Trans. Faraday Soc. 33, 8-

57. Jameson, C. J. and Mason, J. (1987) 'The chemical shift', in J. Mason (ed.) Multinuclear NMR, Plenum, New York, pp. 51-88.

58. Maitland, G. C., Rigby, M., Smith, E. B., and Wakeham, W. A. (1981) Intermolecular Forces their Origin and Determination, Clarendon Press, Oxford.

AB-INITIO CALCULATION and ANALYSIS of NUCLEAR MAGNETIC SHIELDING TENSORS: the LORG and SOLO APPROACHES

AAGE E. HANSEN
Chemical Laboratory IV, Department of Chemistry
H.C. Ørsted Institute, Universitetsparken 5
DK-2100 Copenhagen Ø, Denmark
and
THOMAS D. BOUMAN[†]
Department of Chemistry, Southern Illinois
University, Edwardsville, Illinois 62026, USA

ABSTRACT.
The localized orbital/local origin approach labelled LORG to first order and SOLO to second order in electron correlation is reviewed in the general context of ab-initio calculations of nuclear shielding tensors. The performance is illustrated and discussed for a range of molecules containing ^{13}C, ^{15}N, and ^{19}F nuclei. We discuss tensorial aspects of nuclear shielding emphasizing the shielding antisymmetry for nuclei at sites of low symmetry, and show how a decomposition of the paramagnetic contributions to the shielding tensor elements into symmetry species can assist the identification of electronic structures reponsible for antisymmetry. For first row nuclei the trend in the effects of second order electron correlation is towards increased isotropic shieldings and decreased anisotropies, generally improving agreements with experiments significantly.

---

† It is with deep sorrow that AaEH must report that Professor Thomas D. Bouman died on April 12, 1992.

J. A. Tossell (ed.), Nuclear Magnetic Shieldings and Molecular Structure, 117–140.
© 1993 Kluwer Academic Publishers.

1.   Introduction

The energy of a nucleus of dipole moment μ_N in an isolated molecule subjected to a magnetic field $\mathbf{B_0}$ is given as

$$W = -\mu_N \cdot (1 - \sigma) \cdot \mathbf{B_0} \tag{1}$$

in the absence of coupling to other magnetic nuclei. σ is the shielding tensor, and retaining terms bilinear in μ_N and $\mathbf{B_0}$ in a second order perturbation expansion, Ramsey [1] obtained the following expression for the i,j^{th} Cartesian component of σ

$$\sigma_{ij} = \sigma_{ij}^{d} + \sigma_{ij}^{p} = (1/2c^2) \langle 0 | [\mathbf{u}_j \times (\mathbf{r-R})] \cdot [\mathbf{u}_i \times \mathbf{r}] r^{-3} | 0 \rangle$$

$$- (1/c^2) \sum_{q \neq 0} [\mathbf{u}_i \cdot \langle 0 | \boldsymbol{\ell}/r^3 | q \rangle] (E_q - E_0)^{-1} [\mathbf{u}_j \cdot \langle q | (\mathbf{r-R}) \times \mathbf{p} | 0 \rangle] \tag{2}$$

(atomic units). σ_{ij}^{d} and σ_{ij}^{p} are dia- and paramagnetic contributions, $\mathbf{u}_i$ and $\mathbf{u}_j$ are Cartesian unit vectors, and the angular momentum $\boldsymbol{\ell}$ and position vectors $\mathbf{r}$ are referred to the magnetic nucleus. $\mathbf{R}$ locates an arbitrary origin for the magnetic gauge [1], and for complete sets of eigenstates $|q\rangle$ and energies E_q of the field-free molecule, the results obtained from eq. (2) are invariant to the choice of $\mathbf{R}$ [1].

In practice a central issue has been the formulation of approaches that remove or minimize errors arising from an unbalance of the $\mathbf{R}$-dependent terms in approximate calculations. Following a summary of ab-initio approaches to the magnetic origin problem, Section 2 outlines our localized orbital/local origin approach labelled LORG [2] to first order in electron correlation and SOLO [3] to second order. In Section 3 we discuss the ^{13}C and ^{19}F shielding in methylfluoride and, to illustrate the performance for aromatic systems, the ^{13}C and ^{15}N shielding in benzene and the various azines.

The shielding tensor enters the energy, eq. (1), in a bilinear form, and is hence not necessarily symmetric in the

Cartesian indices [4-6]. However, the observed resonance frequency is governed almost entirely by the symmetric part of σ. The antisymmetric part affects the resonance frequency only to second order, but is in principle obtainable from relaxation mechanisms [4,6-10]. In Section 4 we outline some general aspects of tensorial shielding, and illustrate how ab-initio computations can assist in analyzing the relationship between molecular structure and characteristics of the shielding tensor, emphasizing in particular the antisymmetric component. Section 5 contains concluding remarks.

2.   Computational Schemes

2.1   Common Origin Approaches

Coupled Hartree-Fock (CHF) theory for nuclear shielding can be expressed in various forms [2,11-14], of which early versions [11,12] assumed canonical (delocalized) molecular orbitals. The following sum-over-states type CHF expression is equally valid in canonical and localized orbital bases [2]

$$\sigma_{ij}(\text{CHF}) = \sigma_{ij}{}^{d}(\text{CHF}) + \sigma_{ij}{}^{p}(\text{CHF}) =$$

$$c^{-2} \sum_{\alpha} \langle \alpha | \{ (\mathbf{r}-\mathbf{R}) \cdot \mathbf{r} \, \delta_{ij} - [\mathbf{u}_i \cdot (\mathbf{r}-\mathbf{R})][\mathbf{u}_j \cdot \mathbf{r}] \} / r^3 | \alpha \rangle$$

$$- 2c^{-2} \sum_{\alpha m} \sum_{\beta n} [\mathbf{u}_i \cdot \langle \alpha | \ell / r^3 | m \rangle] \, \Omega_{\alpha m, \beta n} \, [\mathbf{u}_j \cdot \langle n | (\mathbf{r}-\mathbf{R}) \times \mathbf{p} | \beta \rangle] \qquad (3)$$

Here orbitals α and β are doubly occupied in the field-free (spin-singlet) Hartree-Fock ground state wave function $|\Delta_o\rangle$, m and n are virtual orbitals, and Ω is an inverse energy matrix

$$\Omega = (\mathbf{A} - \mathbf{B})^{-1} \left\{ \begin{array}{l} A_{\alpha m, \beta n} = \langle \alpha \rightarrow m | H - E_o | \beta \rightarrow n \rangle \\[1em] B_{\alpha m, \beta n} = \langle \alpha \rightarrow m, \ \beta \rightarrow n | H | \Delta_o \rangle \end{array} \right. \qquad (4)$$

E_o is the energy of $|\Delta_o\rangle$. The CHF expression formulated by

Kutzelnigg [13,14] is also applicable to localized as well as canonical orbitals. In terms of the order expansion formulated by Oddershede et al [15-17], eqs. (3,4) are consistent through first order in the fluctuation potential relative to $|\Delta_o\rangle$, and the CHF level can hence be considered consistent through first order in electron correlation .

Eqs (1,3) contain a single origin located at **R** and are referred to as Common Origin expressions [18]. In complete orbital bases (i.e. in the Hartree-Fock limit) the **R**-dependent terms in eq. (3) cancel identically, see e.g. ref. [2], and a direct route to gauge insensitive shielding results is hence to use basis sets large and flexible enough to approach the Hartree-Fock limit. Early CHF calculations of this kind were published by Höller and Lischka [19] for small molecules, while Lazzeretti et al [20] recently approached the Hartree-Fock limit in CHF calculations for benzene by expanding the orbitals in sets of up to almost 400 contracted atomic basis functions. Large basis set Common Origin calculations including terms to second order in electron correlation have been published by Oddershede et al [16,17] in the Second Order Polarization Propagator Approximation (SOPPA) [15] (see Section 2.2.2). Alternatively, Geertsen [21] has proposed a formulation of the diamagnetic contribution that makes the gauge dependent contributions from the dia- and paramagnetic terms cancel identically for the isotropic part of the shielding tensor, regardless of basis set quality and computational approach. Applications of this formulation are included in papers by Geertsen and Oddershede et al presented at this Workshop.

2.2 Local Origin Approaches

The majority of shielding calculations for larger molecules over the past ten years have been based on local (or distributed [18]) origin approaches. Here the common origin **R** is replaced by local reference points distributed throughout the molecule. This idea can be traced to Kirkwood's treatment of optical rotatory power [22,23], where magnetic transition moments are expanded relative to local reference points, and to London's introduction of a complex local gauge factor in a theory of diamagnetic susceptibilities [24].

The earliest implementation of this at the CHF level is by Hameka [25] and Ditchfield [26]. Here molecular orbitals are expanded in a basis of atomic orbitals where each atomic orbital carries a complex gauge factor referred to the position of the nucleus carrying that atomic orbital. This is the socalled GIAO approach, the acrynom originally referring to Gauge Invariant Atomic Orbitals [25]; as a more accurate alternative that retains the acronym we suggested the term Gauge Including Atomic Orbitals [2]. The method has been applied quite extensively, and has proven successful even when small basis sets are employed (see ref. [27] and references therein). This advantage was offset to some extent by the need to compute and transform two-electron integrals in the complex basis, and for many years applications were restricted to relatively small molecules. However, an efficient new GIAO implementation at the CHF level by Wolinsky, Hinton and Pulay [28] has expanded the applicability significantly. Gauss [29] has implemented a GIAO approach at the level of second order Møller-Plesset perturbation theory (MP2), and Helgaker and Jørgensen [30] have developed an electronic Hamiltonian in which origin independence is ensured through a basis of GIAO's; the latter formulation allows origin independent calculation of magnetic properties at any approximate ab-initio level.

A very different application of local gauge factors was developed by Kutzelnigg and Schindler [13,14,31] in their Individual Gauge for Localized Orbitals (IGLO) CHF approach, where complex gauge factors now refer to centroids of localized occupied orbitals. Here complex gauge factors enter only in the formal developments leading to the working equations, and do not introduce the double book-keeping required in the GIAO approaches. The IGLO method has been applied to large classes of compounds [32], and an extension to higher order in electron correlation has been presented at this Workshop.

2.2.1. LORG. Our original development of the LORG (Localized Orbital/Local Origin) method [2] utilizes a Kirkwood [22,23] type expansion of the angular momentum terms in eq. (3) within the Random Phase Approximation [15,33,34]. For static properties the latter is identical to CHF; a later derivation [35] is closer to conventional CHF theory. The resulting working equation for an element of the shielding tensor is

$$\sigma_{ij}(\text{LORG}) = \sigma_{ij}{}^{d}(\text{LORG},1) + \sigma_{ij}{}^{d}(\text{LORG},2) + \sigma_{ij}{}^{p}(\text{LORG})$$

$$= c^{-2} \sum_{\alpha} \langle \alpha | \{ (\mathbf{r}-\mathbf{R}_{\alpha}) \cdot \mathbf{r}\, \delta_{ij} - [\mathbf{u}_i \cdot (\mathbf{r}-\mathbf{R}_{\alpha})] [\mathbf{u}_j \cdot \mathbf{r}] \} / r^3 | \alpha \rangle$$

$$+ ic^{-2} \sum_{\alpha} \sum_{\beta} [\mathbf{u}_i \cdot \langle \alpha | \boldsymbol{\ell}/r^3 | \beta \rangle] [\mathbf{u}_j \cdot | (\mathbf{R}_{\alpha}-\mathbf{R}_{\beta}) \times \langle \beta | \mathbf{r} | \alpha \rangle |]$$

$$- 2c^{-2} \sum_{\alpha m} \sum_{\beta n} [\mathbf{u}_i \cdot \langle \alpha | \boldsymbol{\ell}/r^3 | m \rangle]\, \Omega_{\alpha m, \beta n}\, [\mathbf{u}_j \cdot \langle n | \boldsymbol{\ell}^{(\alpha)} | \beta \rangle] \tag{5}$$

where $\mathbf{R}_{\alpha}$ is an origin associated with localized orbital α, and $\boldsymbol{\ell}^{(\alpha)} \equiv (\mathbf{r}-\mathbf{R}_{\alpha}) \times \mathbf{p}$ is the angular momentum relative to this local origin. The diamagnetic label d now simply refers to the fact that these terms are computed from occupied orbitals, while the paramagnetic term labelled p involves a sum-over-states. The primary result is that all explicit references to an origin at the arbitrary position $\mathbf{R}$ are replaced by internal distance vectors.

Localization of the occupied orbitals is based on the Foster-Boys procedure [36,37]. However, the ability of localized molecular orbital approaches to ensure correct representation of molecular symmetry was questioned recently [28] since, as is well-known, straightforward application of the Foster-Boys procedure produces banana multiple bonds, Kekule structures for simple aromatics, and quite complicated bonds for heteroaromatics or strongly polar units. In our implementation of the LORG method in Program RPAC [38-40], we utilize an invariance of the Random Phase Approximation, and hence of CHF, with respect to unitary transformations within the occupied orbitals [15,37]. This feature allows options where banana bonds can be rotated to yield conventional σ and π bonding orbitals, and where groups of orbitals can be treated separately, for example localizing core and σ frame separately, while leaving π-orbitals delocalized in planar aromatics. The latter procedure manifestly retains the full symmetry of the electronic structure and hence ensures against artificial symmetry breaking of the shielding tensor. Localization of the virtual orbitals is immaterial for the evaluation of eq. (5), but can be introduced [38] if analysis of the paramagnetic terms into localized orbital excitations is desired.

For the choice of origins we use the centroids

$$\mathbf{R}_\alpha = <\alpha|\mathbf{r}|\alpha> \tag{6}$$

for lmo's not bonded to magnetic nucleus, while for lmo's bonded to the magnetic nucleus we retain two options [2,41]:

LORG $\qquad \mathbf{R}_\alpha = 0$ (i.e. position of nucleus) $\hspace{2cm}$ (7a)

FULL LORG $\qquad \mathbf{R}_\alpha = <\alpha|\mathbf{r}|\alpha>$ $\hspace{3.3cm}$ (7b)

The compromise represented by eqs. (6,7a) seems to provide a good balance of short and long range shielding contributions for ecomony-size basis sets [2]; for extended basis sets the difference between LORG and FULL LORG is small (see below).

For larger basis sets explicit evaluation of $\sigma_{ij}^P(\text{LORG})$, eq. (5), is time-consuming and impractical, and following refs. [42,43] we introduce instead the linear equations [44]

$$\sum_{\alpha,m} (\mathbf{A}-\mathbf{B})_{\beta n,\alpha m} \langle\alpha|\rho|m\rangle = i\langle\beta|\mathbf{p}|n\rangle \tag{8a}$$

$$\sum_{\alpha,m} (\mathbf{A}-\mathbf{B})_{\beta n,\alpha m} \langle\alpha|\eta|m\rangle = i\langle\beta|\ell|n\rangle \tag{8b}$$

and solve iteratively for the implied matrix elements $\langle\alpha|\rho|m\rangle$

and $\langle \alpha | \eta | m \rangle$. The paramagnetic Lorg term then becomes [44]

$$\sigma_{ij}{}^{P}(\text{LORG}) =$$

$$(-2ic^{-2}) \sum_{\alpha} \sum_{m} [\mathbf{u}_i \cdot \langle \alpha | \ell / r^3 | m \rangle][\mathbf{u}_j \cdot \{\langle m | \eta | \alpha \rangle - \mathbf{R}_\alpha \times \langle m | \rho | \alpha \rangle\}]$$

$$\equiv \sum_{\alpha} \sigma_{ij}{}^{P}(\text{LORG}, \alpha) \tag{9}$$

which is now a sum of bond contributions. In the Hartree-Fock limit, $\rho = \mathbf{r}$ [2,33,45], while the $\langle \alpha | \eta | m \rangle$'s provide a partial representation of an operator with the formal property $[\eta, H] = i\ell$ [45]. No coordinate representation of η seems known.

2.2.2. SOLO.   In order to extend the LORG approach to second order in electron correlation, we rewrite the paramagnetic term in eq. (1) as follows

$$\sigma_{ij}{}^{P} = (1/c^2) \langle\langle \{\mathbf{u}_i \cdot \ell / r^3\}; \{\mathbf{u}_j \cdot (\mathbf{r} - \mathbf{R}) \times \mathbf{p}\} \rangle\rangle_{E=0} \tag{10}$$

where $\langle\langle A; B \rangle\rangle_{E=0}$ is the static limit of the polarization propagator for operators A and B [15,34]. In the socalled particle-hole form this becomes

$$\sigma_{ij}{}^{P} = (1/c^2) \, \mathbf{t}[\mathbf{u}_i \cdot \ell / r^3] \, \mathbf{P}(E=0)^{-1} \, \mathbf{t}[\mathbf{u}_j \cdot (\mathbf{r} - \mathbf{R}) \times \mathbf{p}] \tag{11}$$

in terms of transition moments $\mathbf{t}$ and the principal propagator $\mathbf{P}(E)$ [15,17]. The abstract notation of eqs. (10,11) can be brought into contact with the notation used above by writing the paramagnetic part of eq. (5) in the form [46]

$$\sigma_{ij}{}^{P}(\text{LORG})$$

$$= -2c^{-2} \sum_{\alpha m} \sum_{\beta n} [\langle \alpha | \mathbf{u}_i \cdot \ell / r^3 | m \rangle] \, \Omega_{\alpha m, \beta n} \, [\mathbf{u}_j \cdot \langle n | \ell^{(\alpha)} | \beta \rangle]$$

$$= -2c^{-2} \sum_{\alpha m} \sum_{\beta n} t_{\alpha m}{}^{(1)} [\mathbf{u}_i \cdot \ell / r^3] \, \Omega_{\alpha m, \beta n}{}^{(1)} \, t_{n\beta}{}^{(1)} [\mathbf{u}_j \cdot \ell^{(\alpha)}] \tag{12}$$

and recalling that Ω is in fact an inverse energy matrix, eq. (4). The superscripts indicate that CHF/LORG is consistent to first order in electron correlation, as discussed above.

Following this line of approach we <u>define</u> [3] the working equations of a Second Order LORG (SOLO) method as follows

$$\sigma_{ij}(\text{SOLO}) = \sigma_{ij}{}^d(\text{LORG},1) + \sigma_{ij}{}^d(\text{LORG},2) + \sigma_{ij}{}^p(\text{SOLO})$$

$$= c^{-2} \sum_{\alpha} \langle \alpha | \{(\mathbf{r}-\mathbf{R}_\alpha) \cdot \mathbf{r} \, \delta_{ij} - [\mathbf{u}_i \cdot (\mathbf{r}-\mathbf{R}_\alpha)][\mathbf{u}_j \cdot \mathbf{r}]\}/r^3 | \alpha \rangle$$

$$+ ic^{-2} \sum_{\alpha} \sum_{\beta} [\mathbf{u}_i \cdot \langle \alpha | \boldsymbol{\ell}/r^3 | \beta \rangle] [\mathbf{u}_j \cdot \{ (\mathbf{R}_\alpha-\mathbf{R}_\beta) \times \langle \beta | \mathbf{r} | \alpha \rangle \}]$$

$$- 2c^{-2} \sum_{\alpha m} \sum_{\beta n} t_{\alpha m}{}^{(2)} [\mathbf{u}_i \cdot \boldsymbol{\ell}/r^3] \, \Omega_{\alpha m, \beta n}{}^{(2)} \, t_{n\beta}{}^{(2)} [\mathbf{u}_j \cdot \boldsymbol{\ell}^{(\alpha)}] \quad (13)$$

This differs from eq. (5) only in the last term, in which all contributions now include second order corrections to ensure a consistent second order evaluation of the entire paramagnetic term. The required order expansions for the contributions to energy matrices and to transitions moments are available in refs. [15,47]. Note that in Common Origin $\mathbf{R}_\alpha = \mathbf{R}_\beta = \mathbf{R}$ for all α and β, and $\sigma_{ij}{}^d(\text{LORG},2)$ vanishes identically. The remaining terms in eq. (13) then reproduce the working equations of the Second Order Polarization Propagator (SOPPA) approach employed by Oddershede et al [16,17], and we follow their approach in assuming that the diamagnetic terms can be evaluated directly from the Hartree-Fock ground state. Eq. (13), and eqs. (8,9) modified similarly through second order, are implemented in Program RPAC, version 9.0 [40], except that we neglect a W_4 term that appears in the full SOPPA approach [15]. In our first application of the SOLO approach we considered a number of ^{31}P containing compounds [3], and found second order effects for the isotropic shielding ranging up to about 75 ppm, generally improving agreements with experiments significantly.

2.3 Long Range Shielding and IGLO/LORG Comparison

As stated above, the computationally important feature of local origin approaches is that basis set errors associated with shielding contributions from distant groups are quenched very effectively [2,13]. To indicate how this is achieved the following expression provides a relation between the shielding tensor in Common Origin CHF and in LORG and IGLO [35]

$$\sigma_{ij}(\text{LORG/IGLO}) = \sigma_{ij}(\text{CHF}, \mathbf{R}=0)$$

$$- c^{-2} \sum_{\alpha} \langle \alpha | \{ \mathbf{R}_{\alpha} \cdot \mathbf{r} \, \delta_{ij} - [\mathbf{u}_i \cdot \mathbf{R}_{\alpha}][\mathbf{u}_j \cdot \mathbf{r}] \} / r^3 | \alpha \rangle$$

$$+ ic^{-2} \sum_{\alpha} \sum_{\beta} [\mathbf{u}_i \cdot \langle \alpha | \ell / r^3 | \beta \rangle] [\mathbf{u}_j \cdot (\mathbf{R}_{\alpha} - \mathbf{R}_{\beta}) \times \langle \beta | \mathbf{r} | \alpha \rangle]$$

$$- 2ic^{-2} \begin{cases} \sum_{\alpha m} [\mathbf{u}_i \cdot \langle \alpha | \ell / r^3 | m \rangle] [\mathbf{u}_j \cdot \{ \mathbf{R}_{\alpha} \times \langle m | \rho | \alpha \rangle \}] & \text{(LORG)} \\[2ex] \sum_{\alpha m} [\mathbf{u}_i \cdot \langle \alpha | \ell / r^3 | m \rangle] [\mathbf{u}_j \cdot \{ \mathbf{R}_{\alpha} \times \langle m | \mathbf{r} | \alpha \rangle \}] & \text{(IGLO)} \end{cases}$$

$$(14)$$

(see eq. (9)). In the Hartee-Fock limit, orbital completeness and the fact that $\rho = \mathbf{r}$ in complete bases as noted above, make the correction terms on the right hand side of eq. (14) cancel identically in LORG as well as IGLO [35], proving that CHF, IGLO, LORG, and in fact GIAO [26], converge to the same result in this limit. Eq. (14) is not used in the implementations of these methods [3,32,44], but in this representation the difference between (FULL) LORG and IGLO [48] lies in the appearance of electric dipole transition moments as the matrix elements $\langle m | \rho | \alpha \rangle$ and $\langle m | \mathbf{r} | \alpha \rangle$, respectively. The convergence characteristics of these matrix elements are hence important for the performance of the two methods for a given basis set.

In modest basis sets the paucity of virtual orbitals, and the fact that $\langle \alpha | \ell / r^3 | \beta \rangle$ is large only when α and β are both close to the magnetic nucleus, imply that the leading contribution to the long range damping comes from the first correction term on the right hand side of eq. (14). The effect of this term is precisely to damp out long range diamagnetic terms which are not properly counterbalanced by paramagnetic contributions in CHF calculations with such basis sets [1,2]. The compromise origin choice in LORG, eqs. (6,7a), tends to focus on long range contributions, since all the correction terms in eq. (14) that refer to orbitals bound to the magnetic nucleus vanish identically. Numerical examples illustrating these features are given in Table I of ref. [2]. Closely parallel considerations apply in second order (SOLO).

3. Methyl-Fluoride, Benzene and Azines

As illustrated in Table 3 below for the benzene molecule, the performance of the local origin approaches is often quite similar, although the GIAO approach can appear formally less demanding than LORG and IGLO in terms of basis set size. In practice this is counterbalanced by the need to operate in a dual space in GIAO due to the complex nature of the atomic basis. The shielding in methylfluoride, Table 1, provides an interesting case study. Here the ^{19}F shielding is anomalous since its experimental anisotropy, $\Delta\sigma = \sigma_\parallel - \sigma_\perp$, is large and <u>negative</u>, while it is <u>positive</u> for ^{13}C as commonly expected. However, the double-zeta LORG and IGLO results do not reproduce the correct sign for the ^{19}F anisotropy, in contrast to the GIAO results at this level. For the ^{13}C nucleus the shielding is not as sensitive to basis set, although bases 4-31G and A do not perform particularly well in LORG and IGLO. For triple-zeta polarized basis sets and beyond, the various approaches yield very similar results. We note that second order effects, as displayed by the differences between CHF/LORG and SOPPA/SOLO results, are small, but the trend is here that the isotropic shielding decreases, while the magnitude of the anisotropy increases at second order.

The LORG decomposition into bond contributions shown in Table 2 traces the basis set problem for ^{19}F specifically to the lone-pair orbitals, whose paramagnetic contributions to the perpendicular shielding change by 80 ppm between the two basis sets included in the table. All other bond contributions are quite stable, and it is tempting to relate the difference between LORG/IGLO and GIAO to the sparcity of the virtual orbital space around the F atom at the double-zeta level. Lone-pair orbitals have no virtual counterparts at this level for the real basis sets used in LORG and IGLO, while the dual basis set introduced by the gauge factor on the individual atomic orbitals in the GIAO formulation may be viewed as adding higher angular momentum (i.e.polarization) components [28].

Table 1. ^{13}C and ^{19}F shielding (in ppm) in CH_3F

Method	N[a]	^{13}C		^{19}F	
		$\bar{\sigma}$	$\Delta\sigma$[b]	$\bar{\sigma}$	$\Delta\sigma$[b]
CHF[c]					
[3s2p/2s]	24	168	10	425	-7
[3s3p1d/2s]	40	135	65	461	-71
[3s3p1d/2s1p]	49	133	78	464	-73
- - SOPPA	-	130	75	452	-82
LORG					
4-31G[c]	24	165	77	403	30
[3s2p/2s][d]	24	129	69	406	19
[3s3p1d/2s][d]	40	129	74	455	-63
[3s3p1d/2s1p][d]	49	127	78	461	-68
- - SOLO[d]	-	123	86	448	-75
IGLO					
A[c]	26	164	72	465	26
C[c]	56	147		488	-47
III[e]	100	122	92	474	-66
GIAO[f]					
4-31G	24	142	85	489	-75
6-311G(d,p)	56	128	88	483	-68
6-311G+(2d,2p)	81	126	91	483	-67
Exp. gas[g]		117		471	-67
solid,liq.cr.[h]		117	90	471	-90±4

a: Contracted basis functions. b: $\Delta\sigma = \sigma - \sigma$.
c: ref. [35]   d: Present results.   e: ref. [49].
f: ref. [28]. g: refs. [50,51].   h: ref. [52].

Table 2. **LORG Bond Contributions to ^{19}F shielding (in ppm) in Methyl Fluoride**

Bond		σ a^b	b^c	σ a	b	$\Delta\sigma^a$ a	b
F_{1s}		312	312	312	312	0	0
F-C		24	24	91	98	-67	-74
$C-H_3$		1	-2	32	25	-31	-27
	dia	144	144	126	126	18	18
$F-lp_3$	par	-63	-63	**-162**	**-82**	**99**	**19**
	sum	82	81	-35	43	117	36
All		418	414	400	476	**18**	**-63**

a: $\Delta\sigma=\sigma - \sigma$ b: [3s2p/2s](N=24). c: [3s3p1d/2s](N=40)

Turning to aromatic systems, results for benzene are given in Table 3. The LORG and SOLO results labelled (σ,π) are obtained with separately localized core and sigma orbitals and delocalized π orbitals, as discussed in Section 2, while the (Kek) calculations used local σ and π orbitals corresponding to a Kekule structure; even at double-zeta level, the sensitivity to localization scheme is small. The trend is clearly that the local origin approaches converge faster to the Hartree-Fock limit than the Common Origin calculations, and that LORG, IGLO and GIAO performances are now quite similar for comparable basis sets. In second order, local origin results (SOLO and FULL SOLO) are in closer agreement with experiments than the Common Origin results (SOPPA) for this triple-zeta polarized basis set, while the difference between SOLO and FULL SOLO is marginal. The magnitude of the second order effect is about the same regardless of origin approaches. Note that the estimated CHF limit for $\bar{\sigma}$ is 5 ppm lower than the gas phase experiment, while the second order results suggest that this level of electron correlation will suffice to close the gab by increasing the shielding by just about that amount.

Table 3. ^{13}C shielding (in ppm) in benzene

Method	N[a]	$\bar{\sigma}$	δ[b]	σ_{11}	σ_{22}	σ_{33}	Ω[c]
CHF[3s2p/2s][d]	66	144	49	0	139	298	298
LORG(Kek)	-	58	135	-67	52	189	256
LORG(σ,π)	-	62	131	-60	57	187	247
IGLO(DZ)[e]	72	88	131				211[f]
CHF[3s3p1d/2s][g]	114	82	114	-37	72	210	247
SOPPA	-	88	108	-23	79	209	232
LORG(σ,π)	-	64	132	-50	53	188	238
SOLO(σ,π)	-	69	127	-36	56	187	222
FULL SOLO	-	69	127	-34	55	186	220
GIAO[h]							
6-31G	66	74	133	-40	60	202	242
6-31G(d,p)	120	73	129	-38	61	195	233
6-311G(d,p)	144	59	138	-62	49	188	250
6-311G+(2d,2p)	216	58	139	-63	49	189	252
CHF[i]							
Basis(I)	198	61		-63	51	194	257
Basis(III)	300	56		-68	47	189	257
Basis(V)	396	55		-69	46	187	256
HF-limit,est.		53					
Exp. gas/liq.[j]		57/58	138				
solid (T= 14K)[k]		65		-32	44	184	216
(T=223K)[l]		53			-7	173	180[f]

a: Contracted basis functions. b: $\delta = \sigma(CH_4) - \sigma(C^*)$.
c: $\Omega = \sigma_{33} - \sigma_{11}$.   d: ref. [2].   e: ref. [31].
f: $\Delta\sigma = \sigma_{33} - (\sigma_{11} + \sigma_{22})/2$. g: ref. [53]. h: ref. [28].
i: ref. [20]. j: ref. [50]. k: ref. [54]. l: ref. [55].

The largest shielding (σ_{33}) is perpendicular to the molecular plane, σ_{22} is tangential to the ring, and σ_{11} lies along the C-H bond, in all approaches and basis sets. However, the table shows that basis set, origin approach and correlation have quite different effects on individual tensor components. For example, the resulting correlation effects are largest for the paramagnetic component (σ_{11}), as can be anticipated from eq. (13), and the the span, $\Omega = \sigma_{33} - \sigma_{11}$, hence decreases with second order correlation for this system.

In the six-membered heteroaromatic azines [53] the ^{13}C shieldings follow the pattern found for benzene; the relatively small second order effects tend to increase the isotropic shielding and decrease the span of the principal shieldings. Table 4 summarizes the results for ^{15}N shieldings in these molecules, and we have again used separate localization of core

Table 4. ^{15}N **shieldings (in ppm) in azines**

Molecule	Method	$\bar{\sigma}$	$\Omega=\sigma_{33}-\sigma_{11}$
Pyridine	SOLO[a]	-72 (-73)[c]	664
	LORG[a]	-94	691
	IGLO[b]	-104	772
Pyrazine	SOLO	-102 (-90)	728
	LORG	-136	774
	IGLO	-121	756
Pyrimidine	SOLO	-45 (-51)	566
	LORG	-58	574
	IGLO	-71	602
Pyridazine	SOLO	-197 (-156)	858
	LORG	-235	911
	IGLO	-240	909
s-Triazine	SOLO	-28 (-39)	487
	LORG	-33	481
	IGLO	-41	489
s-Tetrazine	SOLO	-159 (-141)	746
	LORG	-213	841
	IGLO	-221	857
1,2,4-Triazine			
N-1	SOLO	-207 (-178)	887
	LORG	-255	971
N-2	SOLO	-151 (-134)	723
	LORG	-171	732
N-4	SOLO	-42 (-54)	618
	LORG	-76	669

a: [3s3p1d/2s] ref. [53]. b: [5s4p1d/3s1p] ref. [56].
c: Exp., assuming $\bar{\sigma}(NH_3,liq.) = 244.5$ ppm, ref. [57].

and σ orbitals, while the π orbitals are delocalized. σ_{33} is perpendicular to the ring, but in contrast to ^{13}C the most deshielded direction (σ_{11}) is now tangential to the ring [53]. Except for pyridine LORG and IGLO agree quite well, and in accord with the range of ^{15}N shieldings across these molecules, the second order effects vary quite drastically. For the most deshielded nuclei the isotropic shielding and the span are changed by about 50 and 100 ppm, respectively, and again the second order effects increase the isotropic shieldings and decrease the span, generally improving agreements with experiments significantly.

4.   Analysis of the Shielding Tensor

For a spin-½ system of gyromagnetic ratio γ, the resonance frequency derived from eq. (1) is [9,58]

$$\omega(\mathbf{B}) = B_0 \, |\gamma| \, [\mathbf{u_B}^\dagger \cdot (\mathbf{1} - \sigma^\dagger)(\mathbf{1} - \sigma) \cdot \mathbf{u_B}]^{1/2} \qquad (15)$$

or to first order in the components of the shielding tensor

$$\omega(\mathbf{B}) = B_0 \, |\gamma| \, (1 - \mathbf{u_B}^\dagger \cdot \sigma^s \cdot \mathbf{u_B}) \qquad (16)$$

Here $\mathbf{u_B}$ is a unit vector along $\mathbf{B_0}$, $\sigma^s \equiv$ ½$(\sigma + \sigma^\dagger)$ is the symmetric part of the tensor, and eqs. (15,16) express the NMR frequency as function of field direction relative to the molecule. The elements of the shielding tensor rarely exceed 10^{-3}, and by the first order expression in eq. (16) the resonance is then governed almost entirely by the quantity [9]

$$T_B = \mathbf{u_B}^\dagger \cdot \sigma \cdot \mathbf{u_B} = \mathbf{u_B}^\dagger \cdot \sigma^s \cdot \mathbf{u_B} \qquad (17)$$

also called the secular shielding [58]. Eqs. (16,17) hence provide the rationale for reporting principal shieldings σ_{11}, σ_{22} and σ_{33}, obtained by diagonalization of σ^s. In [9,39] we show that contour plots of T_B are a valuable alternative to principal axis ellipsoids [58] for illustrating the shielding variation relative to the molecular structure.

Observation of an antisymmetric component, $\sigma^a \equiv$ ½$(\sigma - \sigma^\dagger)$, requires determination of second order contributions in eq. (15), or careful analysis of relaxation contributions [7,8,10]. Only one experimental estimate of the magnitude of σ^a, namely for the olefinic ^{13}C nucleus in tetrachloro-cyclopropene, seems available in the literature [10]. From the theoretical side, symmetry selection rules for the elements of σ^a and σ^s were derived by Buckingham and Malm [5], but since only relatively few computed results are available, see e.g. [2,8,9,44], little is known about systematic correlations between magnitude of the shielding antisymmetry and molecular and electronic structural features. At the same time theoretical calculations are in fact in a rare position to be able to provide insight into an intriguing aspect of molecular physics for which empirical data apparently are quite hard to come by.

In this context LORG and SOLO offer an advantage in the exploitation of symmetry, since the paramagnetic terms can be built up one symmetry block at a time [2,9,40] utilizing that the (A-B) matrix, eq. (4), and hence eqs. (8), are block diagonal in canonical bases. Eqs. (8) can then be solved separately within irreducible representations, before being transformed to a localized basis [2,37]. The canonical (A-B) matrix is block diagonal with respect to the full molecular point group while a site group, if different from the full group, must be a subgroup. An analysis can hence be carried out in terms of overall molecular symmetry as well as local nuclear site symmetry.

This is illustrated in Table 5 for one of the olefinic ^{13}C nuclei in cyclopropene; the molecule is in the xz-plane with the apex along the positive z-axis and the ^{13}C nucleus under consideration (C_2) along the positive x-axis [9]. We show the overall LORG results and decomposition into contributions both according to the full C_{2v} molecular point group and according to the C_s site group for this nucleus. Also shown are typical excitations transforming as the various species, and the SOLO results are included for comparison. Diagonalization of the symmetric part of the tensors yields the LORG(SOLO) results δ = $\overline{\sigma}(CH_4)-\overline{\sigma}(C_2)$ = 119(111) ppm and Ω = $\sigma_{33}-\sigma_{11}$ = 303(257) ppm, while IGLO [8] yields δ = 119 ppm and Ω = 285 ppm. Solid state experiments yield δ = 108 ppm and Ω = 234 ppm [59]. SOLO again increases the shielding (lowers the shift) and decreases the span, in both cases towards experiment.

Table 5. **Cartesian Elements of the Shielding Tensor for Carbon Atom 2 in Cyclopropene (in ppm)**

		xx	yy	zz	xz	zx	
LORG[a,b]							
dia		297	296	294	8	3	
para	A′ A_1	0	24	0	0	0	$\sigma \rightarrow \sigma^*$
	B_1	0	-116	0	0	0	$\pi \rightarrow \pi^*$
	A″ A_2	0	0	-386	171	0	$\sigma \rightarrow \pi^*$
	B_2	-199	0	17	-9	-34	$\pi \rightarrow \sigma^*$
Total		99	204	-74	171	-30	
SOLO[a,c]		93	201	-39	138	-34	

a: [3s3p1d/2s1p]. b: ref. [9]. c: Present results.

In accord with the selection rules [5] for the C_s site of C_2, the $\frac{1}{2}(\sigma_{zx} - \sigma_{xz})$ element of σ^a is non-vanishing, and LORG and SOLO yield -101 ppm and -85 ppm for this element. IGLO [8] yields 100 ppm for the magnitude, while in tetrachloro-cyclo-propene the magnitude is experimentally estimated to be larger than 50 ppm [10]. Hence theory and experiment agree that the shielding antisymmetry of this olefinic carbon atom is quite large, approaching 1/3 of the span of the principal shieldings.

The diamagnetic contribution is almost isotropic, and the paramagnetic contributions again reflect the general symmetry rules [5]; the A_1 and B_1 (A′) species contribute only to the secular shielding perpendicular to the molecular plane, while the A_2 and B_2 (A″) species contribute to in-plane secular shielding as well as antisymmetry. However the decomposition clearly identifies the leading antisymmetry contribution as A_2 terms. Since A_2 species are electric-dipole forbidden, these terms become origin independent, and the dissymmetry, ($\sigma_{xz}(A_2)=171$ ppm, $\sigma_{zx}(A_2)=0$), then follows from the b_1 and a_2 transformation of the angular momentum operators ℓ_x and ℓ_z (eq. (2)). $\sigma \rightarrow \pi^*$ and $\pi \rightarrow \sigma^*$ excitations are typical A_2 species.

In ref. [9] we also study the diazirine molecule, which derives from cyclopropene by replacing each of the olefinic C-H groups by a nitrogen atom. Here the LORG result for the antisymmetric shielding element $\frac{1}{2}(\sigma_{zx} - \sigma_{xz})$ for the N_2 nucleus is 262 ppm, compared to a span Ω = 895 ppm. In fact, the in-plane Cartesian anisotropy, $\sigma_{xx} - \sigma_{zz}$, is only 120 ppm, making the shielding tensor so dissymmetric that diagonalization of σ itself produces two complex eigenvalues [9]. The predicted resonance frequencies, eqs. (15,16), of course remain real. The symmetry analysis of the LORG contributions for N_2 is similar to the results shown in Table 5; the A_2 contributions are about the same, while B_2 terms now contribute even more than A_2 to the antisymmetry. Studies of second order correlation effects and basis set stability of LORG and SOLO results for diazirine are presently underway.

5.   Concluding Remarks

At the CHF level the local origin approaches generally converge faster towards the complete basis set limit than common origin calculations, as illustrated in Table 3. Apart from special cases, such as the ^{19}F nucleus in methyl-fluoride discussed in Section 3, there does not seem to be a clear-cut difference in the performance of the local origin approaches, and a choice between them may well depend on local facilities and on the options available in the packages. The present post-SCF implementation of the LORG and SOLO approaches [40] requires storage of transformed 2-electron integrals for the construction of the **A** and **B** matrices and the second order terms. However once the 2-electron files are available experimentation with localization and origin choices, and with analyses into bond and symmetry terms, are inexpensive restart options. As shown in Section 4, symmetry analysis can provide valuable insight into structural correlations.

Shielding calculations including electron correlation to second order are presently being pursued, both in common origin [16,17] and in local origin [3,17,29,53] methodology, and such results are presented also in this and other communications at this Conference. In common origin approaches [16,17] (see also contributions by Geertsen and Oddershede at this Conference), basis sets approaching saturation seem to be important for reliable second order results. The corresponding basis set requirements for second order local origin approaches have not yet been studied in detail. However in local origin methods, the basis set requirements for proper representation of close and long range effects of correlation may in fact be different, and in this context the use of the attenuated (locally dense) basis sets proposed by Chesnut et al [60] offer an interesting possibility for the simulation of saturation.

The second order results obtained so far for first row nuclei [17,29,53] suggest a trend towards increased isotropic shieldings and decreased anisotropies and spans, as illustrated in Tables 4-6, although exceptions such as methylflouride (Table 1) have been found. The effects are generally small for ^{13}C [17,53], but can be quite large for ^{15}N [17,53] and ^{17}O [29]. The shielding of the ^{31}P nucleus [3] shows both up-field and down-field second order effects ranging up to almost 100 ppm. The second order methods generally improve the agreement with experiments, increasing significantly the range of nuclei and molecules for which ab-initio approaches can offer reliable shielding results.

Acknowledgment

This work was supported by grants to TDB from the U.S. National Science Foundation (CHE-9007809) and from the Donors of the Petroleum Research Fund of the American Chemical Society (23469-B6). Also acknowledged is a travel grant (CRG 910126) from the Scientific Affairs Division of NATO.

References

[1]   N.F. Ramsey, Phys. Rev. 86 (1952) 243.

[2]   Aa.E. Hansen and T.D. Bouman, J. Chem. Phys. 82 (1985) 5035.

[3]   T. D. Bouman and Aa. E. Hansen, Chem. Phys. Letters 175 (1990) 292.

[4]   R.F. Schneider, J. Chem. Phys. 48 (1968) 4905.

[5]   A.D. Buckingham and S.M. Malm, Mol. Phys. 22 (1971) 1127.

[6]   R.G. Griffin, J.D. Ellett Jr., M. Mehring, J.G. Bullitt, and J.S. Waugh, J. Chem. Phys. 57 (1972) 2147.

[7]   H.W. Spiess, in **NMR Basic Principles and Progress** 15 (Springer Verlag, Berlin, 1978).

[8]   J.C. Facelli, A.M. Orendt, D.M. Grant and J. Michl, Chem. Phys. Letters 112 (1984) 147.

[9]   Aa.E. Hansen and T.D. Bouman, J. Chem. Phys. 91 (1989) 3552.

[10]   F.A.L. Anet, D.J. O'Leary, C.G. Wade and R.J. Johnson, Chem. Phys. Letters 171 (1990) 401.

[11]   W.N. Lipscomb, in **Advances in Magnetic Resonance** 2, edited by J.S. Waugh (1966) 137.

[12]   H. Nakatsuji, J. Chem. Phys. 61 (1974) 3728.

[13]   W. Kutzelnigg, Isr. J. Chem. 19 (1980) 193.

[14]   W. Kutzelnigg, Theochem 61 (1989) 11.

[15]   J. Oddershede, P. Jørgensen, and D. L. Yeager, Computer Phys. Rept. 2 (1984) 33.

[16] J. Oddershede and J. Geertsen, J. Chem. Phys. 92 (1990) 6036.

[17] I. Paidarova, J. Komasa and J. Oddershede, Mol. Phys. 72 (1991) 559.

[18] P.J. Stephens, J. Phys. Chem. 91 (1987) 1712.

[19] R. Höller and H. Lischka, Mol. Phys. 41 (1980) 1017, and 41 (1980) 1041.

[20] P. Lazzeretti, M. Malagoli and R. Zanasi, Theochem 234 (1991) 127.

[21] J. Geertsen, Chem. Phys. Letters 179 (1991) 479.

[22] J.G. Kirkwood, J. Chem. Phys. 5 (1937) 479, and 7 (1939) 139.

[23] Aa.E. Hansen and T.D. Bouman, J. Am. Chem. Soc. 107 (1985) 4828.

[24] F. London, J. Phys. Rad. 8 (1937) 397.

[25] H.F. Hameka, **Advanced Quantum Chemistry** (Addison-Wesley, Reading Mass., 1965) and references therein.

[26] R. Ditchfield, in **MTP International Review of Science, Physical Chemistry Series I. Molecular Structure and Properties** vol II (Butterworths, London, 1972) and references therein.

[27] Chesnut, D.B. in **Annual Reports on NMR Spectroscopy** 21 (Academic Press, London, 1989).

[28] K. Wolinski, J.F. Hinton and P. Pulay, J. Am. Chem. Soc. 112 (1990) 8251.

[29] J. Gauss, Chem. Phys. Letters 191 (1992) 614.

[30] T. Helgaker and P. Jørgensen, J. Chem. Phys. 95 (1991) 2595.

[31] M. Schindler and W. Kutzelnigg, J. Chem. Phys. 76 (1982) 1919.

[32] W. Kutzelnigg, U. Fleischer and M. Schindler, in **NMR Basic Principles and Progress** 23 (Springer Verlag, Berlin, 1990).

[33] Aa.E. Hansen and T.D. Bouman, Mol. Phys. 37 (1979) 1713.

[34]   P. Jørgensen and J. Simons, **Second Quantization-Based Methods in Quantum Chemistry** (Academic Press, New York, 1981).

[35]   J.C. Facelli, D.M. Grant, T.D. Bouman, and Aa.E. Hansen, J. Comput. Chem. $\underline{11}$ (1990) 32.

[36]   J.M. Foster and S.F. Boys, Rev. Mod. Phys. $\underline{32}$ (1960) 300.

[37]   T.D. Bouman, B. Voigt and Aa.E. Hansen, J. Am. Chem. Soc. $\underline{101}$ (1979) 550.

[38]   T.D. Bouman, Aa.E. Hansen, B. Voigt and S. Rettrup, Int. J. Quant. Chem. $\underline{23}$ (1985) 595.

[39]   T.D. Bouman and Aa.E. Hansen, Int. J. Quant. Chem. $\underline{Symp.}$ $\underline{23}$ (1989) 381.

[40]   T. D. Bouman and Aa. E. Hansen, RPAC Molecular Properties Package, Version 9.0, 1991.

[41]   The criterion for the application of eq. (7a) is based on the distance between the magnetic nucleus and the centroid for a given orbital. Experimentation can be advisable for non-standard localization choices.

[42]   G.L. Bendazzoli, S. Evangelisti, P. Palmieri and S. Rettrup, J. Chem. Phys. $\underline{85}$ (1986) 2015.

[43]   W.A. Parkinson and M.C. Zerner, Chem. Phys. Letters $\underline{139}$ (1987) 563.

[44]   T.D. Bouman and Aa.E. Hansen, Chem. Phys. Letters $\underline{149}$ (1988) 510.

[45]   Aa.E. Hansen, P.J. Stephens and T.D. Bouman, J. Phys. Chem. $\underline{95}$ (1991) 4255.

[46]   Apart from the sign difference caused by the definition of the propagator [15,34], eq. (14) differs from eq. (13) in a factor of two arising from the summation over double occupied orbitals in the latter expression.

[47]   J. Geertsen and J. Oddershede, Chem. Phys. $\underline{90}$ (1984) 301.

[48]   As presently implemented [31,32], IGLO identifies all local reference points with orbital centroids; direct comparison hence is between FULL LORG and IGLO.

[49]   U. Fleischer and M. Schindler, Chem. Phys. $\underline{120}$ (1988) 103.

[50] A.K. Jameson and C.J. Jameson, Chem. Phys. Lett. <u>134</u> (1987) 461.

[51] C.J. Jameson and H.J. Osten, Mol. Phys. <u>56</u> (1985) 1083.

[52] D.B. Chesnut and C.G. Phung, J. Chem. Phys. <u>91</u> (1989) 6238.

[53] T.D. Bouman and Aa.E. Hansen, Chem. Phys. Lett. <u>197</u> (1992) 59.

[54] M. Linder, A. Höhener and R.R. Ernst, J. Magn. Res. <u>35</u> (1979) 379.

[55] A. Pines, M.G. Gibby and J.S. Waugh, Chem. Phys. Lett. <u>15</u> (1972) 373.

[56] M. Schindler, Magn. Res. Chem. <u>26</u> (1988) 394.

[57] M. Witanowski, L. Stefaniak and G.A. Webb in **Annual Reports on NMR Spectroscopy** <u>21</u> (Academic Press, London, 1989).

[58] M. Mehring, **Principles of High Resolution NMR in Solids** (Springer Verlag, Berlin, 1983).

[59] K.W. Zilm, R.T. Conlin, D.M. Grant and J. Michl, J. Am. Chem. Soc. <u>102</u> (1980) 6672.

[60] D.B. Chesnut and K.D. Moore, J. Comp. Chem. <u>10</u> (1989) 648.

The IGLO method. Recent developments

*W. Kutzelnigg, Ch. van Wüllen, U. Fleischer, R. Franke, T. v. Mourik**
Lehrstuhl für Theoretische Chemie
Ruhr-Universität Bochum
Universitätsstr. 150

D-4630 Bochum, Germany

Abstract

After some historical comments on IGLO and related methods, a new derivation of the IGLO method based on a non-local gauge transformation is presented, which is the basis of a multiconfigurational generalization (MC-IGLO) of IGLO. Some results obtained both with the (old) IGLO (SCF-IGLO) method and with MC-IGLO are presented, e.g. on the dependence of chemical shifts on geometry, on idealized annulenes and the limits of the Spiesecke-Schneider relation, on the heavy-atom effect, the shielding along the potential curve of He_2, and generally on a comparision of MC-IGLO with SCF-IGLO.

1. Introduction

The limited space does not allow a detailed introduction to the general problems associated with the ab-initio calculation of NMR chemical shifts and of magnetic susceptibilities, including a comparison of the methods that are at present available. We can only give some historical remarks on the IGLO method with occasional comments on other methods (sec. 2).

One of the big achievements in recent years has been the implementation of methods that take care of correlation effects on magnetic properties. Again we cannot review the literature and we have to concentrate on the multiconfiguration generalization of IGLO, the MC-IGLO method. It is not possible to present MC-IGLO in all details, but we want, at least, to outline its basic idea. For this it is necessary to say a few words on non-local gauge transformations and on a rederivation of IGLO in the spirit that has been developed for MC-IGLO. This is done in sec. 3.

We then show some new results, both obtained with the 'old' SCF-IGLO method (sec. 4) and the new MC-IGLO (sec. 5). Indirectly we gain some insight concerning relativistic effects on chemical shifts.

2. Historical remarks

The IGLO formalism was presented in 1980 [1], the first applications were published in 1982 [2]. At that time the methods available for the calculation of chemical shifts (and of magnetic susceptibilities) were coupled Hartree-Fock (CHF) [3] with a common gauge origin and GIAO [4,5]. Both methods were limited to rather small molecules and were quite computer-time consuming. Compared to these existing methods IGLO was a real progress

J. A. Tossell (ed.), Nuclear Magnetic Shieldings and Molecular Structure, 141–161.

and allowed for the first time ab-inito calculations of magnetic properties of molecules of the size of, say, naphtalene. A summary on applications has recently been published [6].

In the very early days of IGLO a fruitful collaboration was started with Josef Michl and coworkers on experimental and theoretical studies of the full shielding tensors of a series of interesting molecules including propellane [7]. The IGLO method turned out to be an important help for the assignment of the observed tensor components of the shifts. In the course of this collaboration J. Downing from Josef Michl's laboratory spent some time at Bochum to prepare a special version of the IGLO program for the computer at Salt Lake City. This first IGLO program given away from Bochum was used at Salt Lake City and at some other places for several years, although meanwhile a much improved (and much accelerated) workstation version of IGLO was made generally available and distributed all over the world. Unfortunately a paper on a comparison of the performance of IGLO and LORG [8] was based on the old Salt-Lake version of the IGLO program and a just newly written version of the LORG method, and gave so a wrong impression on the real virtues and drawbacks of the two methods. The LORG method presented in 1985 [9] looked at first glance very different from IGLO, while on closer inspection [10] it can be regarded as a simplified version of IGLO. The manipulations related to the introduction of different gauge origins for different MOs look similar, but LORG does not satisfy a stationarity condition for the magnetic susceptibility. Nevertheless IGLO and LORG results for chemical shifts are usually very close. The main advantage of LORG for users has been that it can be obtained as an 'extra' to the GAUSSIAN package.

The IGLO program that is generally available from the authors on request (so far free for non-profit-making organizations) is based on and includes − as the original version − a Gaussian-lobe integral program [11]. Meanwhile most of the IGLO calculations of the Bochum group use cartesian Gaussians, but we have not yet distributed this program, since we have no copyright for the integral routines. There are plans to include IGLO into MOTECC, and we also consider to make a version available that has a direct interface to existing integral and SCF-programs like GAUSSIAN. Gaussian lobes are actually still used in the CYBER 205 version, for which we have an effectively vectorized integral program [12]. There is also a 'direct-IGLO' program (DIGLO) [13] interfaced to the semi-direct SCF program of Ahlrichs et al [14], in which two-electron integrals are not stored but recalculated. This program was especially designed for rather large molecules. Since it has been developed in cooperation with BAYER AG, there are some restrictions concerning its distribution.

Various cooperations started in recent years. Among these that with Paul Schleyer and his group has been particularly effective, as is demonstrated by many joint papers [15] especially on the elucidation of the structures of the carbonium ions and boranes.

In very recent years programs based on the GIAO method have been so effectively improved [16] that GIAO calculations became competitive with IGLO calculations. Under comparable conditions IGLO is still faster than GIAO by a factor 2-3, mainly because in IGLO no extra two-electron integrals in addition to those needed in a SCF calculation are required. GIAO has some advantages for highly symmetric molecules [17] because for these it is easier than in IGLO to take full advantage of the molecular symmetry.

All methods mentioned so far are based on a single Slater determinant wave function. It is somewhat surprising how good methods of this type are. The reason is probably that the perturbing operators for magnetic properties are all one-electron operators. Nevertheless there are cases where the one-determinant approximation definitely fails. For these cases an MC-SCF version of IGLO has been prepared [18].

Other methods to take care of correlation effects on chemical shifts have been proposed [19], but cannot be reviewed here.

3. Theory

3.1 THE BASIC IDEAS OF STATIONARY PERTURBATION THEORY [10,20]

In the conventional derivation of perturbation theory it is assumed that the unperturbed problem is solved exactly. Since this is almost never the case, it is better to base perturbation theory on the variation principle. We hence want to make the expectation value

$$< \Psi|U^{-1}HU|\Psi > \tag{3.1.1}$$

stationary, where Ψ is the wave function determined by the stationary condition, U is an infinitesimal unitary transformation being an element of the *variational group* $\mathcal{G}$. Condition for stationarity of (3.1.1) is that

$$< \Psi|[H,X]|\Psi >= 0; \; \forall X \in \mathcal{L} \tag{3.1.2}$$

where $\mathcal{L}$ is the Lie algebra associated with the variational group $\mathcal{G}$.

We consider the case (realized for magnetic perturbations) that

$$H = H_0 + \lambda H_{10} + \lambda^2 H_{20} + \mu H_{01} + \lambda\mu H_{11} \tag{3.1.3}$$

We describe the dependence of the wave function Ψ on λ and μ in the form

$$\Psi = e^{Y(\lambda,\mu)}\Psi_0; \; Y(\lambda,\mu) = -Y^\dagger(\lambda,\mu) \in \mathcal{L} \tag{3.1.4}$$

Expanding in powers of λ and μ

$$Y(\lambda,\mu) = \sum_{k,l=0}^{\infty} \lambda^k \mu^l Y_{kl}; \; (k = l = 0 \text{ excluded}) \tag{3.1.5a}$$

$$E(\lambda,\mu) = \sum_{k,l=0}^{\infty} \lambda^k \mu^l E_{kl}; \; E_{00} \equiv E_0 \tag{3.1.5b}$$

one gets [10,20]

$$E_0 =< \Psi_0|H_0|\Psi_0 > \tag{3.1.6a}$$

$$E_{10} =< \Psi_0|H_{10}|\Psi_0 > \tag{3.1.6b}$$

$$E_{20} =< \Psi_0|H_{20} + \frac{1}{2}[H_{10}, Y_{10}]|\Psi_0 > \tag{3.1.6c}$$

$$E_{01} =< \Psi_0|H_{01}|\Psi_0 > \tag{3.1.6d}$$

$$E_{11} =< \Psi_0|H_{11} + [H_{01}, Y_{10}]|\Psi_0 > \tag{3.1.6e}$$

with Ψ_0 determined by the condition

$$< \Psi_0|[H_0, X]|\Psi_0 >= 0; \; \forall X \in \mathcal{L} \tag{3.1.7}$$

and Y_{10} satisfying

$$< \Psi_0|[[H_0, X], Y_{10}] + [H_{10}, X]|\Psi_0 >=< \Psi_0|[[H_0, Y_{10}], X] + [H_{10}, X]|\Psi_0 >= 0 \tag{3.1.8}$$

Eqn. (3.1.8) can also be interpreted as the condition that the generalized Hylleraas functional

$$F(Y_{10}) = < \Psi_0 | \frac{1}{2}[[H_0, Y_{10}], Y_{10}] + [H_{10}, Y_{10}] + H_{20} | \Psi_0 > \qquad (3.1.9)$$

is stationary with respect to variations of Y_{10}. To solve (3.1.8) one expands Y_{10} in a basis of $\mathcal{L}$

$$Y_{10} = \sum_k b_k X_k; \ X_k \in \mathcal{L} \qquad (3.1.10)$$

and gets a linear system of equations for the coefficients b_k

$$\sum_l H_{kl} b_l + V_k^* = 0 \qquad (3.1.11)$$

with

$$H_{kl} = < \Psi_0 | [[X_k^\dagger, H_0], X_l] | \Psi_0 > = < \Psi_0 | [X_k^\dagger, [H_0, X_l]] | \Psi_0 > \qquad (3.1.12a)$$

$$V_k = < \Psi_0 | [H_{10}, X_k] | \Psi_0 > \qquad (3.1.12b)$$

The matrix with elements H_{kl} is called the (electronic) Hessean.

3.2 Coupled Hartree Fock (CHF) as stationary perturbation theory

We now specify that Ψ_0 is a single Slater determinant, henceforth symbolized as Φ. We write the Hamiltonian in 2nd quantization form as [21]

$$H = h_q^p a_p^q + \frac{1}{2} g_{rs}^{pq} a_{pq}^{rs} \qquad (3.2.1)$$

In (3.2.1) h_q^p and g_{rs}^{pq} are the matrix elements of the one-electron and two-electron part of the Hamiltonian respectively in terms of a given orthonormal spin-orbital basis $\{\chi_p\}$,

$$h_q^p = < \chi_q | h | \chi_p > \qquad (3.2.2a)$$

$$g_{rs}^{pq} = < \chi_r(1) \chi_s(2) | r_{12}^{-1} | \chi_p(1) \chi_q(2) > \qquad (3.2.2b)$$

the a_p^q and a_{pq}^{rs} are the one and two-particle excitation operators respectively that are expressible in terms of the annihilation operators a_p and creation operators $a^q = a_q^\dagger$.

$$a_q^p = a^p a_q; \ a_{rs}^{pq} = a^p a^q a_s a_r \qquad (3.2.3)$$

In (3.2.1) the Einstein summation convention over repeated indices is implied.

The Lie algebra $\mathcal{L}$ of the variational group is taken as that spanned by the elements a_p^q that generate rotations in the spin-orbital space. The stationarity condition (3.1.7) – known as Brillouin condition – becomes (for all matrix elements now expressed in terms of the Hartree-Fock spin orbitals ψ_p)

$$f_q^p(n_p - n_q) = 0 \qquad (3.2.4)$$

$$f_q^p = h_q^p + (g_{qr}^{pr} - g_{rq}^{pr}) n_r \qquad (3.2.5)$$

where f_q^p is the matrix element of the Fock operator $\hat{f}$, which consists of the one-electron operator $\hat{h}$, and the Coulomb and exchange operators. n_r is the occupation number of the orbital ψ_r in Φ, which can be equal to 1 or 0. The expressions $g_{qr}^{pr}n_r$ and $g_{rq}^{pr}n_r$ are recognized as matrix elements of the Coulomb and exchange operator respectively.

We first choose the spin-orbitals such that (3.2.4) is satisfied i.e. we solve the Hartree-Fock equations. Then we choose the expansion (3.1.10) appropriate for coupled Hartree-Fock

$$Y_{10} = b_q^p a_p^q; \quad (b_q^p)^* = -b_p^q \tag{3.2.6}$$

implying again the Einstein summation convention. The V_k in (3.1.11) gets two indices and the H_{kl} four indices, so that instead of (3.1.12) we have

$$
\begin{aligned}
H_{qs}^{pr} =&< \Psi_0|[a_q^p, [H_0, a_s^r]]|\Psi_0 >= (\delta_q^r f_s^p - \delta_s^p f_q^r)(n_s - n_r) \\
&+ (g_{qs}^{pr} - g_{sq}^{pr})(n_q - n_p)(n_s - n_r)
\end{aligned}
\tag{3.2.7a}
$$

$$(V_q^p)^* = - < \Psi_0|[H_{10}, a_p^q]|\Psi_0 >= (H_{10})_q^p(n_q - n_p) \tag{3.2.7b}$$

The perturbation that we want to consider comes from the interaction with an external magnetic field of the field strength $\vec{B}$ (first label in 3.1.3) and with the magnetic field of a nucleus with magnetic moment $\vec{\mu}$ at position $\vec{\varrho}$ (second label in 3.1.3). The explicit expressions for the H_{kl} are (ignoring spin, which does not contribute for singlet states)

$$H_{10} = \frac{e}{mc}\vec{A}\cdot\vec{p} = \frac{e}{2mc}\{\vec{B}\times(\vec{r} - \vec{R})\}\cdot\vec{p} \tag{3.2.8a}$$

$$H_{20} = \frac{e^2}{2mc^2}\vec{A}^2 = \frac{e^2}{8mc^2}\{\vec{B}\times(\vec{r} - \vec{R})\}^2 \tag{3.2.8b}$$

$$H_{01} = \frac{e}{mc}\vec{A}_\mu\cdot\vec{p} = \frac{e}{mc}\frac{\vec{\mu}\times(\vec{r} - \vec{\varrho})}{|\vec{r} - \vec{\varrho}|^3}\cdot\vec{p} \tag{3.2.8c}$$

$$H_{11} = \frac{e^2}{mc^2}\vec{A}\cdot\vec{A}_\mu = \frac{e^2}{2mc^2}\frac{\{\vec{B}\times(\vec{r} - \vec{R})\}\cdot\{\vec{\mu}\times(\vec{r} - \vec{\varrho})\}}{|\vec{r} - \vec{\varrho}|^3} \tag{3.2.8d}$$

In (3.2.8) $\vec{A}$ is the vector potential due to the external field with field strength $\vec{B}$, $\vec{A}_\mu$ the vector potential of the magnetic field of the nucleus. $\vec{R}$ is an arbitrary gauge origin.

Since H_{10} and H_{01} are purely imaginary (note that $\vec{p} = \frac{\hbar}{i}\nabla$) and since we only consider non-degenerate states with a real wave function, E_{10} and E_{01} vanish. Except for trivial factors E_{20} as to be evaluated from (3.1.6.c) is the magnetic susceptibility and E_{11} given by (3.1.6e) the chemical shift of NMR.

The so-called uncoupled-Hartree-Fock method (UCHF) deserves only historical interest, since it is now hardly used. It is obtained from CHF, if in (3.2.7a) one ignores the terms that explicitly contain matrix elements g_{qs}^{pr} of the electron interaction operator, i.e. if one replaces (3.2.7a) by

$$F_{qs}^{pr} =< \Psi_0|[a_q^p, [H_{HF}, a_s^r]]|\Psi_0 >= (\delta_q^r f_s^p - \delta_s^p f_q^r)(n_s - n_r) \tag{3.2.9a}$$

$$H_{HF} = f_q^p a_p^q - \frac{1}{2}(g_{pq}^{pq} - g_{qp}^{pq})n_p n_q \tag{3.2.9b}$$

where H_{HF} is the Hartree-Fock Hamiltonian which has Φ as eigenfunction with eigenvalues E_{HF}.

If, moreover, one choses the spin orbitals so that they diagonalize the Fock operator, i.e.

$$f_q^p = \varepsilon_p \delta_q^p \tag{3.2.10}$$

which is no loss of generality, (3.2.9) simplifies to

$$F_{rs}^{pq} = \delta_q^r \delta_s^p (\varepsilon_s - \varepsilon_r)(n_s - n_r) \tag{3.2.11}$$

The Hylleraas functional (3.1.9) becomes then a sum of contributions of the occupied spin-orbitals ψ_i

$$F(Y_{10}) = \sum_i < \psi_i | \frac{1}{2} [[\hat{f}, Y_{10}^{(i)}], Y_{10}^{(i)}] + [H_{10}, Y_{10}^{(i)}] + H_{20} | \psi_i > \tag{3.2.12a}$$

$$Y_{10}^{(i)} = \sum_a \{ b_a^i a_i^a + b_i^a a_a^i \} \tag{3.2.12b}$$

and the system of equations for the determination of the $Y_{10}^{(i)}$ decouple completely (whence the name uncoupled Hartree-Fock).

3.3 GAUGE TRANSFORMATIONS

According to (3.2.8) the Hamiltonian does not directly contain the magnetic field strength $\vec{B}$, but rather its vector potential $\vec{A}$. There are many possible different choices of $\vec{A}$ compatible with the same $\vec{B}$. We only consider the freedom to choose the gauge origin $\vec{R}$ for the external field arbitrarily. For the field created by the magnetic moment of the nucleus its position $\vec{\varrho}$ is a 'natural gauge origin' that we keep there. A change of $\vec{R}$ modifies H_{10}, H_{20} and H_{11}. To understand the corresponding change in the wave function we realize that we can define a (unitary) so-called gauge transformation

$$e^{\lambda \Lambda_k(\vec{r})}; \quad \Lambda_k(\vec{r}) = \frac{ei}{2c\hbar} \{ (\vec{R}_k - \vec{R}) \times \vec{B} \} \cdot \vec{r} \tag{3.3.1}$$

We apply the gauge transformation (3.3.1) to the Hamiltonian and use the Hausdorff formula

$$H^{(k)} = e^{-\lambda \Lambda_k} H e^{\lambda \Lambda_k} = H + \lambda [H, \Lambda_k] + \frac{1}{2} \lambda^2 [[H, \Lambda_k], \Lambda_k] + \dots \tag{3.3.2}$$

Inserting the form (3.1.3) of the Hamiltonian and collecting powers of λ and μ leads to

$$H^{(k)} = H_0 + \lambda H_{10}^{(k)} + \lambda^2 H_{20}^{(k)} + \mu H_{01} + \lambda \mu H_{11}^{(k)} \tag{3.3.3}$$

$$H_{10}^{(k)} = H_{10} + [H_0, \Lambda_k] \tag{3.3.4a}$$

$$H_{20}^{(k)} = H_{20} + [H_{10}, \Lambda_k] + \frac{1}{2} [[H_0, \Lambda_k], \Lambda_k] \tag{3.3.4b}$$

$$H_{11}^{(k)} = H_{11} + [H_{01}, \Lambda_k] \tag{3.3.4c}$$

For the further evaluation we need the explicit expressions (3.2.8) for the H_{pq} and the following auxiliary results (in which it is used that Λ_k is local and commutes with all local operators, such that only the commutator with the kinetic energy T, and with H_{01} and H_{10} are non-vanishing)

$$[H_0, \Lambda_k] = [T, \Lambda_k] = \frac{e}{2mc}\{\vec{B} \times (\vec{R} - \vec{R}_k)\} \cdot \vec{p} \tag{3.3.5a}$$

$$\frac{1}{2}[[H_0, \Lambda_k], \Lambda_k] = \frac{e^2}{8mc^2}\{\vec{B} \times (\vec{R} - \vec{R}_k)\}^2 \tag{3.3.5b}$$

$$[H_{10}, \Lambda_k] = \frac{e^2}{4mc^2}\{\vec{B} \times (\vec{r} - \vec{R})\} \cdot \{\vec{B} \times (\vec{R} - \vec{R}_k)\} \tag{3.3.5c}$$

$$[H_{01}, \Lambda_k] = \frac{e^2}{2mc^2}\frac{\{\vec{B} \times (\vec{R} - \vec{R}_k)\} \cdot \{\vec{\mu} \times (\vec{r} - \vec{\varrho})\}}{|\vec{r} - \vec{\varrho}|^3} \tag{3.3.5d}$$

By virtue of (3.3.5) one sees that the $H_{pq}^{(k)}$ defined by (3.3.3) are of the same form as the H_{pq} given by (3.2.8), just with $\vec{R}$ replaced by $\vec{R}_k$.

The message of eqn. (3.3.3) is that a shift of the gauge origin is equivalent to a unitary transformation of the Hamiltonian. This means that the eigenvalues are not changed and the wave function is just affected by a unitary transformation – all this, of course, provided that one solves the Schrödinger equation exactly.

Let us now consider the effect of a gauge transformation in stationary perturbation theory. We want to make the Hylleraas functional (3.1.9) stationary, but we want to express this functional in terms of the gauge transformed Hamiltonian $\tilde{H}$ given by (3.3.3). We hence express H_{10} etc. in terms of $H_{10}^{(k)}$ etc. via (3.3.4) and get after some rearrangement

$$F(Y_{10}) = <\Psi_0|\frac{1}{2}[[H_0, Y_{10} - \Lambda_k] + [H_{10}^{(k)}, Y_{10} - \Lambda_k]$$
$$+ H_{20}^{(k)} - \frac{1}{2}[H_0, [X_k, Y_{10}]]|\Psi_0 > \tag{3.3.6}$$

If we ignore the double commutator in (3.3.6), this $F(Y_{10})$ is formally equivalent to the original $F(Y_{10})$ with the H_{pq} replaced by $H_{pq}^{(k)}$ and Y_{10} by $\tilde{Y}_{10}^{(k)}$ with

$$\tilde{Y}_{10}^{(k)} = Y_{10} - \Lambda_k \tag{3.3.7}$$

Now we can formulate the condition under which $F(Y_{10})$ is gauge invariant with respect to the gauge transformation generated by Λ_k which means that the functional form of F remains the same if we replace the H_{pq} by $H_{pk}^{(k)}$ and Y_{10} by $\tilde{Y}_{10}^{(k)}$. This is the case, if the variational group is so chosen that

$$Y_{10} \in \mathcal{L}; \; \tilde{Y}_{10}^{(k)} \in \mathcal{L} \tag{3.3.8a}$$

Then Λ_k (regarded as an operator) is an element of $\mathcal{L}$, or explicitly $\Lambda_k\psi_p$ is – for all p – an element of the one-electron basis $\{\psi_p\}$, from which we have constructed $\mathcal{L}$. In this case

$$[\Lambda_k, Y_{10}] \in \mathcal{L} \tag{3.3.8b}$$

and the last term in (3.3.6) vanishes.

If the condition (3.3.8) for gauge invariance is satisfied, the second order energy E_{20} – as given by (3.1.6c) in terms of the Y_{10} which makes (3.1.9) stationary – is independent of the gauge. Even then the two contributions $E_{20,d}$ and $E_{20,p}$ to E_{20}, referred to as *diamagnetic* and *paramagnetic* contributions

$$E_{20,d} = <\Psi_0|H_{20}|\Psi_0> \tag{3.3.9a}$$

$$E_{20,p} = \frac{1}{2} <\Psi_0|[H_{10}, Y_{10}]|\Psi_0> \tag{3.3.9a}$$

are *not* gauge independent, as is easily seen if we make the same decomposition for H_{k0} replaced by $H_{k0}^{(k)}$ and Y_{10} by $\tilde{Y}_{10}^{(k)}$

$$
\begin{aligned}
E_{20,d}^{(k)} &= <\Psi_0|H_{20}^{(k)}|\Psi_0> = <\Psi_0|H_{20}|\Psi_0> \\
&+ <\Psi_0|[H_{10}, \Lambda_k] + \frac{1}{2}[[H_0, \Lambda_k], \Lambda_k]|\Psi_0>
\end{aligned}
\tag{3.3.10a}
$$

$$
\begin{aligned}
E_{20,p}^{(k)} &= \frac{1}{2} <\Psi_0|[H_{10}^{(k)}, \tilde{Y}_{10}^{(k)}]|\Psi_0> \\
&= \frac{1}{2} <\Psi_0|[H_{10}, \tilde{Y}_{10}^{(k)}]|\Psi_0> + \frac{1}{2} <\Psi_0|[[H_0, \Lambda_k], \tilde{Y}_{10}^{(k)}|\Psi_0>
\end{aligned}
\tag{3.3.10b}
$$

Although it would be nice to have a gauge invariant formalism at hand, in practice this is prohibitive (because too large basis sets would be required). It is rather recommended not to insist on gauge invariance, but to choose a particular gauge and accordingly a Lie algebra $\mathcal{L}$ such that $Y_{10} \in \mathcal{L}$ for this gauge - or at least that Y_{10} is well approximated within $\mathcal{L}$.

For atoms there is a natural gauge origin, namely the position of the nucleus. For this choice Y_{10} vanishes and one simply gets

$$E_{20} = E_{20,d} = <\Psi_0|H_{20}|\Psi_0> \tag{3.3.11}$$

However for a different gauge origin, $E_{20,p}^{(k)}$ does not vanish and cancels with the extra contribution to $E_{20,d}^{(k)}$ provided that the conditions for gauge invariance hold. Then $Y_{10} = -\Lambda_k$. Now, one sees easily that if this is not the case, $E_{20,d}^{(k)}$ is much more easily evaluated accurately than $E_{20,p}^{(k)}$, since it is a simple expectation value, while for the evaluation of $E_{20,p}^{(k)}$ one need first to construct $\tilde{Y}_{10}$, which is correctly possible only if the basis is so large that gauge invariance is guaranteed.

It is, of course, not very clever to construct a complicated operator $\tilde{Y}_{10}$ to take care of the inappropriate gauge origin, what can much more easily done by means of a gauge function Λ_k.

Similar observations are made for the 'diamagnetic' and 'paramagnetic' contributions to the chemical shift

$$E_{11,d} = <\Psi_0|H_{11}|\Psi_0> \tag{3.3.12a}$$

$$E_{11,p} = <\Psi_0|[H_{01}, Y_{10}]|\Psi_0> \tag{3.3.12b}$$

In atomic calculations the gauge-invariance problem can — as we have seen — be completely avoided by using the appropriate gauge origin. Then Y_{10} vanishes and the choice of the Lie algebra $\mathcal{L}$ is irrelevant for the construction of E_{20} and E_{11} (it is, of course, relevant for the unperturbed Hartree-Fock problem).

3.4 THE IGLO METHOD GENERATED BY A NON-LOCAL GAUGE TRANSFORMATION

Let us, for a moment, consider UCHF rather than CHF. Since in UCHF the Hylleraas functional to be made stationary is a sum of contributions to the occupied spin-MOs ψ, as manifest in (3.2.12), we can take each term in the sum (3.2.12a) separately and evaluate the $Y_{10}^{(k)}$ independently from each other. This also means that we can perform independent gauge transformations for these contributions, trying to find a optimum gauge for each occupied MO. This is not so straightforward as it seems at first glance, namely for the following reasons.

1. Commutators of the Fock operator $\hat{f}$ with Λ_k are not simply reduced to commutators with the kinetic energy operator $\hat{T}$ as needed in (3.3.5), but there is also a contribution of the commutator with the exchange operator, since this is non-local.

2. In order to apply an optimal gauge transformation one should use localized rather than canonical orbitals, however (3.1.9) decouples only in terms of canonical orbitals.

3. If one takes the different contributions to $F(Y_{10})$ independently, these 'don't know' which other ψ_l are occupied and Y_{10} also contains contributions $b_q^p a_p^q$ with both ψ_q and ψ_p occupied.

If there is no perfect decoupling into MO-contributions, one cannot proceed as naively as done in the beginning of this section, one must rather find a global transformation that has the same effect as individual gauge transformations for the MO contributions. We do this now and switch again to CHF rather than UCHF. This leads to a new convincing derivation of the IGLO method, different from that given originally [1,2]. The desired generalized gauge transformation is of the form

$$H \rightarrow \tilde{H} = e^{-\lambda Z} H e^{\lambda Z}; \quad Z = Z_q^p a_p^q; \quad Z = -Z^\dagger \tag{3.4.1}$$

$$Z_q^p = (\Lambda_p)_q^p \text{ if } n_p = 1, n_q = 0 \tag{3.4.2a}$$

$$Z_q^p = (\Lambda_q)_q^p \text{ if } n_p = 0, n_q = 1 \tag{3.4.2b}$$

$$Z_p^q = \frac{1}{2}(\Lambda_p + \Lambda_q)_q^p \text{ if } n_p = n_q = 1 \tag{3.4.2c}$$

$$Z_q^p = 0 \text{ if } n_p = n_q = 0 \tag{3.4.2d}$$

$$(\Lambda_p)_q^p = <\psi_q|\Lambda_p|\psi_p> \tag{3.4.2e}$$

with Λ_p defined by (3.3.1). The transformed Hamiltonian is then (cf. (3.3.3,4))

$$\tilde{H} = H_0 + \lambda \tilde{H}_{10} + \lambda^2 \tilde{H}_{20} + \mu H_{01} + \lambda\mu \tilde{H}_{11} \tag{3.4.3}$$

$$\tilde{H}_{10} = H_{10} + [H_0, Z] \tag{3.4.4a}$$

$$\tilde{H}_{20} = H_{20} + [H_{10}, Z] + \frac{1}{2}[[H_0, Z], Z] \tag{3.4.4b}$$

$$\tilde{H}_{11} = H_{11} + [H_{01}, Z] \tag{3.4.4c}$$

The explicit evaluation of the operators (3.4.4) is relatively tedious. Let us first note that Y_{10} will now be replaced by $\tilde{Y}_{10}$ which is solution of

$$< \Psi_0|[[H_0, X], \tilde{Y}_{10}] + [\tilde{H}_{10}, X]|\Psi_0 >= 0; \quad \forall X \in \mathcal{L} \tag{3.4.5}$$

and that E_{20} and E_{11} are given as

$$E_{20} =< \Psi_0 | \frac{1}{2}[\tilde{H}_{10}, \tilde{Y}_{10}] + \tilde{H}_{20} | \Psi_0 > \qquad (3.4.6a)$$

$$E_{11} =< \Psi_0 | [H_{01}, \tilde{Y}_{10}] + \tilde{H}_{11} | \Psi_0 > \qquad (3.4.6b)$$

The evaluation of (3.4.6b) is relatively easy since

$$
\begin{aligned}
< \Psi_0 | [H_{01}, Z] | \Psi_0 > &= (H_{01})_p^i Z_i^p - (H_{01})_i^p Z_p^i = (H_{01})_a^i (\Lambda_i)_i^a - (H_{01})_i^a (\Lambda_i)_a^i \\
&= (H_{01})_p^i (\Lambda_i)_i^p - (H_{01})_k^i (\Lambda_i)_i^k - (H_{01})_i^p (\Lambda_i)_p^i + (H_{01})_i^k (\Lambda_i)_k^i \\
&= (H_{01} \Lambda_i)_i^i - (H_{01})_k^i (\Lambda_i)_i^k - (\Lambda_i H_{01})_i^i + (H_{01})_k^i (\Lambda_k)_i^k \\
&= [H_{01}, \Lambda_i]_i^i + (H_{01})_k^i (\Lambda_k - \Lambda_i)_i^k \qquad (3.4.7)
\end{aligned}
$$

where i, k count occupied spin orbitals, a, b virtual ones, while p, q go over a complete set of spin-orbitals. For (3.4.6b) we hence get using (3.3.4c)

$$
\begin{aligned}
< \Psi_0 | \tilde{H}_{11} | \Psi_0 > &=< \Psi_0 | H_{11} + [H_{01}, Z] | \Psi_0 > \\
&= \sum_i < \psi_i | H_{11}^{(i)} | \psi_i > + \sum_{i,k} (H_{01})_k^i (\Lambda_k - \Lambda_i)_i^k \qquad (3.4.8)
\end{aligned}
$$

We can, in fact, take care of $\tilde{H}_{11}$ in chosing $H_{11}^{(i)}$ for the respective ψ_i. There is an additional double sum, that is related to the fact that both $Y_i^j = 0$ and $\tilde{Y}_i^j = 0$, but that $Z_i^j \neq 0$.

While obviously the first sum in (3.4.8) represents local diamagnetic contributions and the first term in (3.4.6b) paramagnetic ones, it is a matter of taste whether the double sum in (3.4.8) is regarded as paramagnetic or diamagnetic or something else. We have usually (for historical reasons) referred to it as $p0$-term, and often added it to the paramagnetic term. In the light of the present derivation it looks more like part of the diamagnetic term.

Somewhat more complicated is the evaluation of (3.4.5). We choose

$$X = a_s^r \qquad (3.4.9a)$$

and get in view of (3.4.4a)

$$
\begin{aligned}
< \Psi_0 | [\tilde{H}_{10}, a_s^r] | \Psi_0 > &= (H_{10})_s^r (n_s - n_r) + Z_p^q H_{qs}^{pr} \\
&= (n_s - n_r) \{ (H_{10})_s^r + Z_p^r f_s^p - Z_s^q f_q^r + Z_p^q (g_{qs}^{pr} - g_{sq}^{ps})(n_q - n_p) \} \qquad (3.4.9b)
\end{aligned}
$$

We note that this is non-vanishing only if $n_s = 1, n_r = 0$ or $n_s = 0, n_r = 1$. We consider the first case and use the labels i, j, k, l for occupied and a, b, c, d for unoccupied orbitals. Then

$$
\begin{aligned}
< \Psi_0 | [\tilde{H}_{10}, a_i^a] | \Psi_0 > &= (H_{10})_i^a + Z_j^a f_i^j - Z_i^b f_b^a \\
&+ Z_b^i (g_{ji}^{ba} - g_{ij}^{ba}) - Z_j^b (g_{bi}^{ja} - g_{ib}^{ja}) \qquad (3.4.10)
\end{aligned}
$$

where we have used that f is block diagonal, i.e. that matrix elements f_a^i or f_i^a vanish. Next we insert (3.4.2) and use that

$$f_i^j (\Lambda_i)_j^a - \Lambda_i^b f_b^a = [f, \Lambda_i]_i^a = [T, \Lambda_i]_i^a - [K, \Lambda_i]_i^a \qquad (3.4.11)$$

$$K = g_{kq}^{pk} a_p^q \tag{3.4.12}$$

Since Λ_i is a local operator it commutes with the nuclear attraction and the Coulomb operators contained in f, only the commutators with the kinetic energy operator T and the exchange operator K don't vanish. By virtue of (3.3.4a) we achieve that in the evaluation of (3.4.10) the term $(H_{10}^{(i)})_i^a$ appears. However there are correction terms. The first of these is due to the fact that in the 2nd term in (3.4.10) Λ_j appears although Λ_i is needed to use (3.4.11). This term is

$$(\Lambda_j - \Lambda_i)_j^i f_i^j \tag{3.4.13}$$

It has been called 'resonance correction term', because it vanishes if the Fock matrix is diagonal, i.e. if $f_j^i = f_i^i \delta_{ij}$. It also vanishes, if the two spinorbitals ψ_i and ψ_j, have the same gauge origin. There are further correction terms due to the commutator of the exchange operator in (3.4.11) and the terms involving matrix elements g_{rs}^{pq} of the electron interaction in (3.4.10). These are referred to as exchange correction terms. They shall not be discussed here in detail. Even more correction terms arise in the evaluation of (3.4.6a).

3.5 COUPLED MULTICONFIGURATION SCF

In multiconfiguration SCF one wants to make the energy expection value (3.1.1) stationary for Ψ a linear combination of Slater determinants ϕ_μ with respect to both the expansion coefficients of the CI and the spin-orbitals from which these determinants are constructed.

Generally one considers three sets of orbitals, the occupied ones (which are occupied in all ϕ_μ), the active ones (which are occupied in some ϕ_μ) and virtual ones (occupied in none of the ϕ_μ). The CAS-SCF case is realized if all possibilities to construct Slater determinants from the chosen number of active orbitals are taken care of.

The power of the stationary perturbation theory as outlined in sec. 3.1 is that the basic equations of coupled MC-SCF are formulated rather easily. The main difference to ordinary coupled Hartree-Fock as discussed in sec. 3.2 is that we consider stationarity of the expectation value (3.1) with respect to two variational groups, one of orbital rotations as in CHF theory (this is generated by the operators a_q^p), and another group corresponding to full-CI within the active space − we consider the CAS-SCF case − generated by the shift operators

$$|\phi_\mu >< \phi_\nu| \tag{3.5.1}$$

for ϕ_μ and ϕ_ν Slater determinant contained in Ψ. Again, after having solved the unperturbed problem by means of CAS-SCF, we have to solve a linear system of equations of type (3.1.11), with a Hessean like (3.1.12a) and an inhomogenity (3.1.12b). The only difference is now that basis operators are taken from two independent Lie algebras [20]. The linear system (3.1.11) gets then a block form

$$\underline{\underline{H}}^{(11)} \vec{b}^{(1)} + \underline{\underline{H}}^{(12)} \vec{b}^{(2)} + \vec{V}^{(1)} = 0 \tag{3.5.2a}$$

$$\underline{\underline{H}}^{(21)} \vec{b}^{(1)} + \underline{\underline{H}}^{(22)} \vec{b}^{(2)} + \vec{V}^{(2)} = 0 \tag{3.5.2b}$$

The first block $\underline{\underline{H}}^{(11)}$ of the Hessean describes coupling between orbital rotations, the last block $\underline{\underline{H}}^{(22)}$ CI rotations, the other two blocks describe coupling between the two types of rotation.

The explicit expressions are rather lengthy and will not be given here.

3.6 THE MC-IGLO APPROACH

The step from coupled MC-SCF to MC-IGLO is analogous to that from coupled Hartree-Fock to IGLO. Having solved the 0th order problem, i.e. the MC-SCF equations, we transform both the occupied and the active orbitals independently to localized orbitals. We then apply a non-local gauge transformation as in sec. 3.4 to the Hamiltonian, which leads to a theory with individual gauge origins for the various localized orbitals and with coupling terms. The latter are much more complicated than in the ordinary IGLO case, but the theory is straightforward. Details will be given elsewhere [18].

4. Results from SCF IGLO calculations

4.1 DEPENDENCE OF THE CHEMICAL SHIFTS OF HYDROCARBONS ON THE CONFORMATION.

In the following discussion one must keep in mind that a different sign convention is used for absolute shifts (σ) and relative shifts (δ). $\sigma > 0$ means shielding with respect to the bare nucleus, while $\delta > 0$ means deshielding relative to the reference value.

It is well-known that chemical shifts change with variations of the geometry, especially with respect to changes of the molecular conformation. IGLO calculations furnish direct informations on these changes that can then be used to interpret experimental results. A simple example is the dependence of the ^{13}C shift of ethane on the angle ϑ of internal rotation. The results are shown on fig. 1. The relaxation of the other geometric parameters has some effect, but not a very strong one. The difference of $\sigma(C)$ between the staggered and eclipsed conformations is almost 5 *ppm*. Qualitatively the curves on fig. 1 are almost indistinguishable from the dependence of the energy on ϑ.

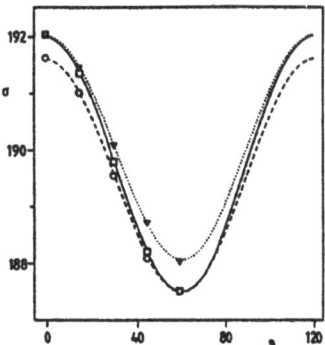

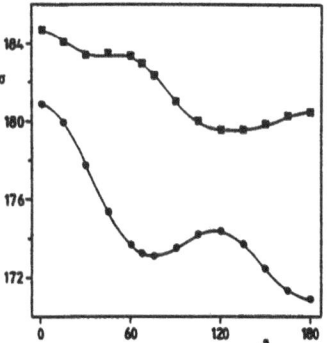

fig. 1
The absolute ^{13}C shift in ethane vs. the dihedral angle. Full line: all geometry parameters relaxed.

fig. 2
The absolute ^{13}C shift of C_1 (—•—) and C_2 (—■—) vs. the rotation angle around the C_2-C_3 bond of n-butane.

A somewhat more interesting example is that of the dependence of the ^{13}C shifts in n-butane on an internal rotation around the bond in the middle of the molecule. The results are on fig. 2. The curve for atom C_1 shows the γ-gauche effect. Here $\vartheta = 60°$ corresponds to the gauche conformation and $180°$ to the trans-conformation of the two external CH_3 groups. The difference between the gauche and trans values, calculated as ~ 3 *ppm* represents the γ-gauche effect [23]. Similar IGLO calculations with smaller basis sets and less optimization were published by Barfield et al. [24a]. More refined calculations by Barfield et al. can be found in this book [24b].

Unfortunately a simple qualitative interpretation of the γ-gauche effect can, so far, not be given. One can only refute previous attempts to explain this effect.

4.2 RING CURRENTS IN BENZENE. CHEMICAL SHIFTS IN IDEALIZED ANNULENES

For the explanation of magnetic properties of cyclic conjugated hydrocarbons the ring current picture of London [25] has played a central role, although it has occasionally been questioned [26,27]. We have constructed current density plots in order to get some insight into this problem [28].

On fig. 3 the current density in benzene in a plane parallel to the molecular plane for a magnetic field perpendicular to the plane is plotted, both the total density and the separate π and σ-contributions. One sees that the current density due to the π-system is exactly as one imagines a London type ring current, namely a diamagnetic (clockwise) one.

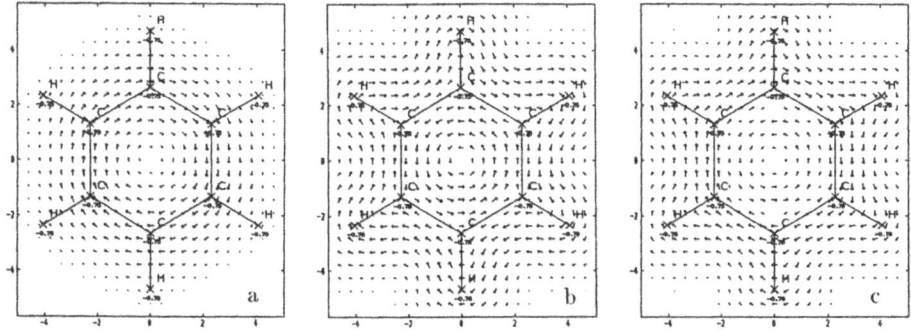

fig. 3 Current density induced in the benzene molecule in a plane parallel to the molecular plane at a distance $0.75a_o$. (a) contribution of π-MOs, (b) contribution of σ-MOs, (c) total.

The current density of the σ-system shows a behaviour that is also encountered in saturated ring systems. One namely finds a paramagnetic ring current inside the ring and a diamagnetic one outside of it. This can easily be rationalized as the sum of diamagnetic ring currents around all individual atoms. If one now adds the σ and π contributions the nice picture of the π-system alone disappears. Nevertheless on a circle through the C atoms there is still a diamagnetic current, only further inside it becomes paramagnetic.

Some information on the effect of ring currents on the proton shifts can be gained from a comparison of benzene with related molecules, especially with a C_{3v} distorted benzene that would correspond to a (nonconjugated) cyclohexatriene [29].

By quantum chemical methods one can study molecules that are not accessibly to experiment or existing molecules in non-equilibrium configurations. For a comparative study of Hueckel annulenes it is e.g. possible to compute these in the idealized geometries of

Table 1
Relative chemical shifts in idealized Hückel annulenes[a]

	$\delta(C)$			$\delta(H)$		
	DZ	II	exp.[b]	DZ	II	exp.[b]
$C_3H_3^+$	177.7	165.7	177	13.0	10.2	10.80
$C_4H_4^{2+}$	262.5	256.1		16.3	13.1	
$C_4H_4^{2-}$	69.3	60.1		0.8	1.8	
$C_5H_5^-$	102.8	100.4	102	5.8	4.0	5.44
C_6H_6	128.4	129.3	129	9.0	6.8	7.27
$C_7H_7^+$	149.9	154.2	155	11.1	8.7	9.18
$C_8H_8^{2+}$	161.2			12.8		
$C_8H_8^{2-}$	93.7		85	4.3		5.68
$C_9H_9^-$	109.6		109	7.1		6.80
$C_{10}H_{10}$[c]	123.3			9.3		
$C_{14}H_{14}$[c]	122.2			11.0		

a) with respect to TMS; the absolute shielding
 σ_C(TMS) is 218.1 and 192.7 *ppm* for DZ and Basis II.
 Basis sets DZ (double zeta), and II as in ref. 6.
 Structures optimized in D_{nh} Geometry on DZd SCF level.
b) H. O. Kalinowski, S. Berger, S. Braun, ^{13}C-NMR
 Spectroscopy, G. Thieme, Stuttgart, 1984
 H. Günther, NMR Spectroscopy, G. Thieme, Stuttgart, 1983
 D. G. Farnum, Adv. Phys. Org. Chem. (1975) 11, 123
c) CH and CC bond lengths as in benzene.

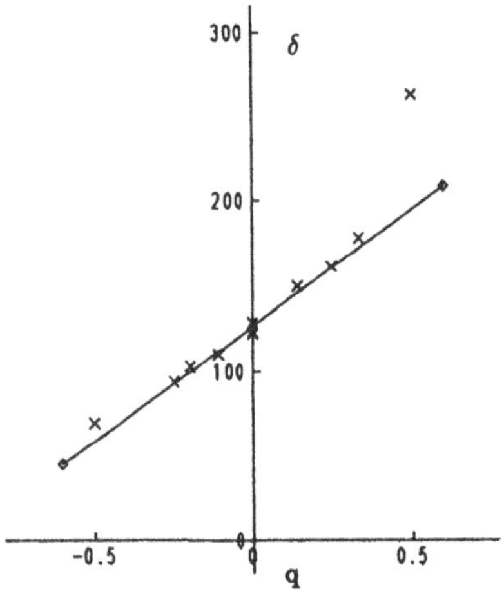

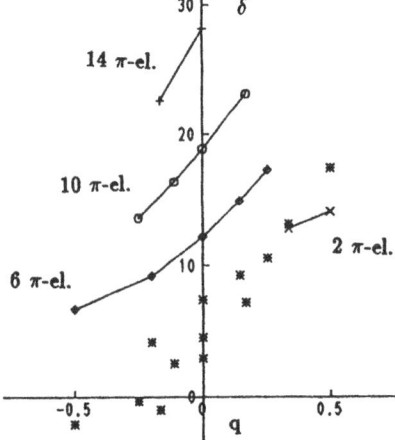

fig. 4
$\overline{\delta(^{13}C)}$ vs. the formal charge of
a CH unit in Hückel annulenes

fig. 5
$\overline{\delta(^1H)}$ vs. the formal charge of
a CH unit in Hückel annulenes
isolated points: $\sigma_\perp$
connected points: $\sigma_\parallel$

regular polygons. The results are in table 1 and compare well with experiment for the known molecules or ions.

On fig. 4 the ^{13}C chemical shifts are plotted against the formal charge per CH unit. The relation found by Spiesecke and Schneider [30] is essentially confirmed.

It must be stressed, however, that this is the only case of all that we have studied where there is a clear correlation between formal charge and chemical shift. For the ^{1}H shifts of the same systems there is no good correlation with the formal charges. However if one looks at $\sigma_\parallel$ (i.e. out of plane) and $\sigma_\perp$ (i.e. average of the two inplane components) separately (fig. 5) one finds that $\sigma_\perp$ correlates rather well with the formal charge, while for $\sigma_\parallel$ there is obviously not only a dependence of the charge, but also of the number of π-electrons. Taking 2-, 6-, 10 and 14-electron systems separately one finds a good correlation with the charge for each set.

4.3 HEAVY-ATOM EFFECTS

In table 2 the absolute shieldings of the hydrogen halides calculated for various basis sets are compared with the experimental values. While the agreement is satisfactory for HF and HCl, there is a discrepancy of 5 ppm for HBr and 13 ppm for HI. MC-IGLO calculations indicate that these discrepancies are not due to correlation effects. It is plausible to conclude indirectly that the strong shielding effect of a heavy halogen for the neighboring H atom is a relativistic effect.

Table 2
Absolute hydrogen shieldings in hydrogen halides (in ppm)

molecule	IGLO				experiment[a]
basis	DZ	DZd	II	'large'	
HF	35.4	32.6	28.2	28.1	28.8
HCl	35.3	33.0	31.3	30.8	31.4
HBr	34.5	31.7	30.4	30.4	35.3
HI	34.2	32.2	31.1	31.4	44.2

a) K. H. Hellwege, A. M. Hellwege, Landoldt-Börnstein,
   Zahlenwerte und Funktionen, Neue Serie II/6, 14, Springer,
   Berlin, 1987

There is also direct evidence on the relativistic origin of the heavy atom effect from semiempirical calculations by Pyykkö et al. [31].

However, table 2 indicates that not the entire heavy-atom effect, i.e. the observed increased shielding on going for F, via Cl, Br to I substitution, can have a relativistic origin. The increase from F to Cl is, at least, fully recovered on the non-relativistic IGLO level.

A similar observation can be made on table 3 where the relative ^{13}C shifts of halogen substituted hydrocarbons are given. The heavy-atom effect of Cl vs. F is well accounted for by IGLO calculations, it is hence non-relativistic. The heavy-atom effect of Br has, at least for CH_3Br and CH_2Br_2 in part a non-relativistic origin, but there must be a strong relativistic contribution, not accounted for by IGLO calculations. For I finally relativistic effects appear to dominate, but even there is a significant non-relativistic contribution.

Table 3
Relative carbon chemical shifts in halomethanes[a]

molecule	IGLO			experiment[b]
	DZ	DZd	II	
CH_3F	55.5	58.8	66.2	71.6
CH_2F_2	89.4	93.4	991.1	111.2
CHF_3	103.4	106.4	106.3	118.8
CF_4	112.8	113.3	109.5	119.9
CH_3Cl	25.1	11.9	24.4	25.1
CH_2Cl_2	43.7	31.2	53.2	54.2
$CHCl_3$	69.9	49.4	80.6	77.7
CCl_4	102.4	70.8	110.1	96.70
CH_3Br	10.5	7.9	14.3	10.2
CH_2Br_2	28.9	25.4	37.5	21.6
$CHBr_2$	55.3	49.6	68.0	12.3
CBr_4	91.0	82.7	104.1	-28.5
CH_3I	-3.7	-2.6	-6.7	-24.0
CH_2I_2	1.6	0.8	2.2	-53.8
CHI_3	8.9	6.9	12.4	-139.7
CI_4	13.7	10.5	20.1	-292.2

a) with respect to TMS; the absolute shielding
σ_C(TMS) is 218.1, 244.9 and 192.7 *ppm* for
DZ, DZd and Basis II
b) Kalinowski et al., see table 1

5. Results from MC-IGLO calculations

5.1 BH AS A TEST CASE

One of the theoretically best-studied molecules, at least as far as its magnetic properties are concerned, is boron monohydride BH. These are not known from experiment, but BH is interesting as the simplest prototype of a Van-Vleck paramagnetic molecule. The van Vleck paramagnetism is due to a low lying excited configuration $1\sigma^2 2\sigma^2 3\sigma 1\pi$. The $3\sigma/1\pi$ near degeneracy implies that there is a strong mixing of the 'excited' configuration $1\sigma^2 2\sigma^2 1\pi^2$ to the ground configuration $1\sigma^2 2\sigma^2 3\sigma^2$. BH is hence a good candidate for the nonvalidity of the single-determinant approximation and hence a good test case for MC-IGLO. On table 4 the MC-IGLO results for the chemical shift of B for various active spaces in the MC-IGLO calculation are displayed [18].

One first observes that $\sigma_{\parallel}$ remains almost unchanged. It only consists of a diamagnetic contribution, a simple expectation value, which is not much affected by electron correlation. However, $\sigma_{\perp}$ (and hence $\sigma_{iso} = (2\sigma_{\perp} + \sigma_{\parallel})/3$) change drastically with the active space.

Table 4
Chemical shift $\sigma(B)$ in BH from MC-IGLO
calculations for different active spaces
Basis: $11s, 7p, 3d, 1f; 6s, 3p, 1d/7s, 6p, 3d, 1f;$
$4s3p1d$

occupied MOs [a]	$\sigma_\perp(B)$	$\sigma_\parallel(B)$
3σ [b]	-493.71	198.81
$4\sigma, 1\pi$ [c]	-308.62	199.70
$4\sigma, 1\pi, 1\delta$	-343.55	199.69
$6\sigma, 2\pi, 2\delta$	-325.12	199.67
$6\sigma, 3\pi$	-368.39	199.68
$6\sigma, 3\pi, 2\delta$	-390.55	199.70
$7\sigma, 3\pi$	-334.96	199.66
$7\sigma, 3\pi, 1\delta$	-335.14	199.68
$10\sigma, 5\pi$	-341.04	199.69
$10\sigma, 5\pi, 3\delta$	-361.42	199.72
$12\sigma, 6\pi, 3\delta$	-362.10	199.71

a) 1σ is kept doubly occupied
   in all configurations
b) one-configuration SCF
c) full-valence CAS

The smallest possible active space of three σ-AOs corresponds to one-determinant IGLO (henceforth referred to as SCF-IGLO). We get $\sigma_\perp = -494\ ppm$. Full-valence CAS, which leads to correct dissociations of the BH bond, consists of four σ and one π (pair) MO. On this level $\sigma_\perp$ is reduced to $-309\ ppm$. Further increase of the active space raises $|\sigma_\perp|$ again, though in a not very systematic way. The result for a space of 12σ, 6π, 3δ of 362 ppm appears to be close to convergence. An experimental value is not known. Compared to this limit the IGLO value is in error by $-132\ ppm$, the MC-IGLO for full valence CAS by $+53\ ppm$. The result is improved for full-valence CAS, but one overshoots somewhat. Unfortunately one can usually not go beyond full valence-CAS, and if one could there is no clear recipe how one should extend the active space. For still higher accuracy one should probably take care of non-dynamic correlation effects by means of MC-IGLO and to use a different recipe for dynamical correlation effects.

5.2 THE POTENTIAL CURVE OF HE₂

He$_2$ is a somewhat academic system, because its chemical shift can hardly be measured, but it is a model system for analogous types of interaction. Fig. 6 shows the chemical shifts of He for a range of distances, and this on SCF-IGLO level for the configurations $1\sigma_g^2 1\sigma_u^2$ and $1\sigma_g^2 2\sigma_g^2$ and from MC-IGLO.

Let us first consider the configuration $1\sigma_g^2 1\sigma_u^2$ which dominates at all physically reasonable distances. The chemical shift $\sigma(\text{He})$ from an IGLO calculation remains nearly constant from ∞ down to a distance of $\sim 2.5\ a_0$ (the van der Waals minimum is at $\sim 5.6a_0$) and decreases very rapidly at shorter internuclear distances, going through zero at $\sim 1.1a_0$ and approaching $-\infty$ ar $R = 0$. The united atom limit is obviously $1s^2 2p^2$, and the lowest singlet state of this configuration, i.e. ^1D is orbit degenerate and (genuinely) paramagnetic. On the way to this limit He$_2$ becomes strongly van-Vleck paramagnetic.

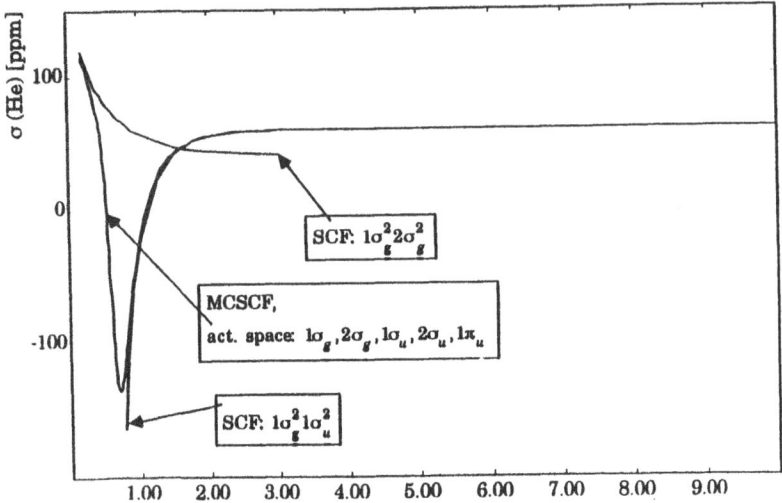

fig. 6 The shielding function in He_2.

This is, however, not the final truth. Somewhere near 0.55 a_0 there is an avoided crossing and the configuration $1\sigma_g^2 2\sigma_g^2$ has the lower energy. For $R \to 0$ it approaches the ground configuration $1s^2 2s^2$ of the united atom. This has the consequence that $\sigma(\text{He})$ from MC-SCF differs very little from the results for $1\sigma_g^2 1\sigma_u^2$ for distances larger than ~ 1.5 a_0. The decrease for small R is a little less steep for MC-IGLO than for $1\sigma_g^2 1\sigma_u^2$, MC-IGLO reaches a minimum near ~ 0.7 a_0 and rises then again to approach a positive finite value at the united atom limit. That $\sigma(\text{He})$ reaches a minimum and goes up again for very small R, will not have any influence at observable quantities except in high-energy collisions, because at the distance of the σ- minimum the repulsion energy is as large as 1.7 Hartree $\sim 46 eV \sim 1100 kcal/mol$.

5.3 TYPICAL CLOSED-SHELL MOLECULES

Since IGLO has been rather successful for many 'normal' molecules it would be disappointing if MC-IGLO calculations indicated significant correlation effects for these. This is, fortunately not the case. We illustrate this for two examples, namely CH_4 and PH_3.

For both molecules the changes from IGLO to MC-IGLO are of the same order as those on going from basis II to basis IV. For the largest basis IV (that is supposed to be close to saturation) the MC-IGLO value agree slightly better with experiment. For $\sigma(\text{H})$ there is hardly any correlation effect. Only the anisotropy of $\sigma(P)$ is affected and improved by MC-IGLO, but it is even more basis-set dependent.

Note that relative shifts, i.e. differences of σ between different molecules are usually much less basis-set dependent in SCF calculations than absolute shifts. Something similar is expected for correlation effects in typical closed-shell molecules.

Table 5

Absolute chemical shifts (in ppm) from SCF-IGLO (roman)
and MC-IGLO (italic) calculations for various basis sets

Molecule	basis	II	III	IV	experiment/Ref.
CH$_4$	σ(C)	198.47	193.85	193.82	
		201.13	*198.85*	*198.39*	198.7[a]
	σ(H)	31.07	31.26	31.22	
		31.01	*31.18*	*31.13*	30.61[b]
PH$_3$	σ(P)	575.54	559.45	583.39	
		581.68	*575.50*	*598.16*	594.40[c]
	$\Delta\sigma$(P)	25.79	24.68	38.94	
		33.48	*44.18*	*48.43*	56.0±2[d]
	σ(H)	29.66	29.67	29.43	
		29.88	*29.87*	*29.65*	28.28[c]
F$_2$	σ	-163.84	-176.64	-165.32	
		-201.75	*-215.82*	*-204.32*	-192.8[e]
	$\sigma_\perp - \sigma_\parallel$	-978.57	-997.81	-980.37	
		-1035.62	*-1056.75*	*-1039.04*	-1050.0± 50[f]
CO	σ(C)	-21.59	-23.56	-23.40	
		+17.14	*+13.62*	*+13.39*	+3.0±0.9[a]
	$\sigma_\perp - \sigma_\parallel$	-436.84	-441.62	-441.76	
		-384.23	*-388.59*	*-389.32*	-415.0[g]
	σ(O)	-86.87	-85.85	-83.86	
		-37.43	*-38.05*	*-36.66*	-36.7[h]
	$\sigma_\perp - \sigma_\parallel$	-765.95	-742.19	-742.45	
		-678.59	*-667.49*	*-670.48*	-653.0[g]
O$_3$	σ(O$_c$)	-2928.57	-2794.31	-2730.10	
		-662.43	*-663.07*	*-657.66*	-724.0[i]
	σ(O$_t$)	-3054.14	-2905.76	-2816.66	
		-1190.26	*1176.88*	*-1151.85*	-1290.0[i]

a) A. K. Jameson and C. J. Jameson, Chem. Phys. Lett. (1987) 134, 461
b) W. T. Raynes, Nucl. Magn. Reson. (1977) 7, 1
c) P. B. Davies et al., J. Chem. Phys. (1971) 55, 3564
d) C. J. Jameson et al., Chem. Phys. Lett. (1990) 167, 575
e) C. J. Jameson et al., J. Chem. Phys. (1980) 73, 6013
f) D. E. O'Reilley et al., Chem. Phys. Lett. (1970) 8, 470
g) W. H. Flygare, Chem. Rev. (1974) 74, 653
h) R. E. Wasylishen et al., J. Chem. Phys. (1984) 81, 1057
   C. J. Jameson, Chem. Rev. (1991) 91, 1375
i) I. J. Solomon et al., J. Am. Chem. Soc. (1968) 90, 5408

5.4 MOLECULES NOT WELL DESCRIBABLE BY A SINGLE SLATER DETERMINANT

An experimentally well-investigated simple molecule for which the agreement between SCF-IGLO theory and experiment has not been fully satisfactory, is F_2. The bond in F_2 is rather weak and at the equilibrium distance there is a significant admixture of the configuration with the antibonding rather than the bonding MO doubly occupied. MC-IGLO should hence improve its description. This is the case, as is seen from table 5 both for the isotropic part and the anisotropy. It is somewhat unexpected and opposite to what one usually finds, that correlation enhances the deshielding.

CO is a prototype of molecules with multiple bonds for which SCF-IGLO results were often not satisfactory. Table 5 indicates that MC-IGLO leads to significant improvement, in particular for $\sigma(O)$. For $\sigma(C)$ the agreement with experiment is not yet perfect. Like one often finds (see the discussion of BH) it appears that MC-IGLO overshoots the correlation effects somewhat.

Let us finally have a look at a molecule where the correlation effect is really spectacular, namely ozone O_3. It must be stressed that O_3 is a real exception as far as the failure of SCF-IGLO is concerned [32]. Even the isovalence-electronic molecules SO_2 or NSF are much better behaved. The results in table 5 are self explanatory.

6. Acknowledgement

The computations reported here were performed on the Cyber 205 of the Computer Center of the Ruhr-University, on the CRAY YMP of the KFA Jülich, the Siemens-Fujitsu S 400 of the regional Computer Center Aachen and on IBM 6000 risc work stations.

References

*Present adress: Vakgroep Theoretische Chemie, Rijksuniversiteit , Utrecht, Netherlands

1 W. Kutzelnigg, Isr. J. Chem. (1980) 19, 193
2 M. Schindler, W. Kutzelnigg, J. Chem. Phys. (1982) 76, 1919
3 R.M. Stevens, R.M. Pitzer, W. Lipscomb, J. Chem. Phys. (1963) 38, 550
4 H. Hameka, Mol. Phys. (1962) 1, 203
5 R. Ditchfield, J. Chem. Phys. (1972) 56, 5688
6 W. Kutzelnigg, U. Fleischer, M. Schindler, in NMR, Basic Principles and Progress 1990, Vol. 23, Springer Berlin
7a A.M. Orendt, J.C. Facelli, D.M. Grant, J. Michl, F.H. Walker, W.P. Dailey, S.T. Waddell, K.B. Wiberg, M. Schindler, W. Kutzelnigg, Theoret. Chim. Acta (1985) 68, 421
 b A.J. Beeler, A.M.Orendt, D.M. Grant, P.W. Cutts, J. Michl, K.W. Zilm, J.W. Downing, J.C. Facelli, M. Schindler, W. Kutzelnigg, J. Am. Chem. Soc. (1984) 106, 7672
8 J.C. Facelli, D.M. Grant, T.D. Bouman, A.E. Hansen, J. Comp. Chem. (1990) 11, 32
9 T.D. Bouman, P.E. Hansen, J. Chem. Phys. (1985) 82, 5035
10 W. Kutzelnigg, J. Mol. Struct. Theochem. (1989) 202, 11
11 R. Ahlrichs, Theoret. Chim. Acta (1974) 33, 1957
12 U. Meier, Diplomarbeit, Bochum 1984
   W. Kutzelnigg, M. Schindler, W. Klopper, S. Koch, U. Meier and H.Wallmeier, in: Supercomputer Simulations in Chemistry, M. Dupuis ed., Lecture Notes in Chemistry (1986) Vol. 44, p. 55, Springer, Berlin 1986
13 U. Meier, Ch. van Wüllen, M. Schindler, J. Comput. Chem. (1992) 13, 551

14 R. Ahlrichs, M.Bär, M. Häser, H.Horn, C.Kölmel, Chem. Phys. Letters (1989) 162, 165
15 P. v. R. Schleyer, W. Koch, B. Liu, U. Fleischer, J. Chem. Soc. Chem. Commun. (1989) 1098
   M. Bremer, K. Schötz, P. v. R. Schleyer, U. Fleischer, M. Schindler, W. Kutzelnigg, W. Koch, P. Pulay, Angew. Chem. (1989) 101, 1063; Angew. Chem. Int. Ed. Engl. (1989) 28, 1042
   M. Bremer, P. v. R. Schleyer, K. Schötz, M. Kausch, M. Schindler, Angew. Chem. (1987) 99, 795; Angew. Chem. Int. Ed. Engl. (1987) 26, 702
   M. Bremer, P. v. R. Schleyer, U. Fleischer, J. Am. Chem. Soc. (1989) 111, 1147
   P. v. R. Schleyer, M. Bühl, U. Fleischer, W. Koch, Inorg. Chem. (1990) 29, 153
   M. Bühl, N.J.R. v. E. Hommes, P. v. R. Schleyer, U. Fleischer, W. Kutzelnigg, J. Am. Chem. Soc. (1991) 113, 2459
16 K. Wolinski, J.F. Hinton and P. Pulay, J. Am. Chem. Soc. (1990) 112, 8251
   See also the contribution by the same authors in this book
17 M. Häser and R. Ahlrichs, Theoret. Chim. Acta in press
18 Ch. van Wüllen, W. Kutzelnigg, to be published
19 M. Jaszunski and A. Sadlej, Theoret. Chim. Acta (1975) 40, 157
   G.T. Darborn, N.C. Handy, Mol. Phys. (1983) 49, 1277
   J. Oddershedde, P. Jørgensen and D.L. Yaeger, Comp. Phys. Rep. (1984) 2, 33
   J. Geertsen and J. Oddershedde, Chem. Phys. (1984) 90, 301
   T.D. Bouman and A.E. Hansen, Chem. Phys. Lett. (1990) 175, 292
   J. Gauss, Chem. Phys. Lett. (1992) 191, 614
20 W. Kutzelnigg, Theoret. Chim. Acta, in press
21 W. Kutzelnigg, J. Chem. Phys. (1981) 77, 3081
   W. Kutzelnigg, in: 'Aspects of Many-Body Effects in Molecules and Extended Systems', D. Mukherjee, ed., Lecture Notes in Chemistry (1989), Vol. 50, p. 35
22 R. Franke, Diplomarbeit, Bochum 1990
23 D.M. Grant and E.G. Paul, J. Am. Chem. Soc. (1964) 86, 2984
   D.M. Grant and B.V. Cheney, J. Am. Chem. Soc. (1967) 89, 5315
24a M. Barfield and S.H. Yamamura, J. Am. Chem. Soc. (1990) 112, 4747
  b M. Barfield, this book
25 F. London, J. Phys. Radium (1937) 8, 397
   L. Pauling, J. Chem. Phys. (1936) 4, 673
26 J.I. Musher, J. Chem. Phys. (1965) 43, 4081; (1967) 46, 1219
27 P. Lazzeretti, E. Rossi and R. Zanasi, J. Chem. Phys. (1982) 77, 3129
28 T. v. Mourik, U. Fleischer, W. Kutzelnigg, to be published
29 P. Lazzeretti, U. Fleischer and W. Kutzelnigg, to be published
30 H. Spiesecke and W.G. Schneider, Tetrahedron Letters (1961) 41, 468
31 P. Pyykkö, P. Gerling, A. Rösch, Mol. Phys. (1987) 61, 195
32 M. Schindler and W. Kutzelnigg, Mol. Phys. (1983) 48, 781

ELECTRONIC CURRENT DENSITY INDUCED BY MAGNETIC FIELDS AND MAGNETIC MOMENTS IN MOLECULES

P. LAZZERETTI, M. MALAGOLI, and R. ZANASI
Dipartimento di Chimica dell'Università degli Studi di Modena
Via Campi 183, 41100 Modena, Italy

ABSTRACT. Mathematical tools for solving the real autonomous system of linear differential equations for the trajectories of electrons in a molecule in the presence of magnetic field are analyzed in Sec. 2. Methods based on the Runge-Kutta integration have been developed for quantum mechanical current density evaluated via approximate coupled Hartree-Fock perturbation theory. Shubnikov theory of magnetic groups has been examined in Sec. 3 to predict the essential features of the current density vector field. The results of large basis set calculations on benzene and borazine are reported in Sec. 4.

1. Introduction

In the presence of an external, time-independent and spatially uniform, magnetic field $\mathbf{B}$, and/or an intrinsic magnetic dipole $\boldsymbol{\mu}$, a charge current density $\mathbf{J} = -e\mathbf{j}$ is induced within the electron cloud of a molecule smeared out with charge density $\rho = -e\gamma$ (denoting by $-e$ the unit of electron charge). This quantity can be written $\mathbf{J} = \rho\mathbf{v}$; $\gamma(\mathbf{r})$ and $\mathbf{j}(\mathbf{r})$ are probability density and probability current density functions, depending on the position vector $\mathbf{r}$, and $\mathbf{v}(\mathbf{r})$ is the local velocity.

In the sequel we will limit ourselves to considering closed shell molecules, i.e., diamagnetic systems symmetric under time reversal, where $\mathbf{J}$ is vanishing in the absence of external perturbation.

The inducing magnetic field is assumed vanishingly small, so that perturbation theory can be applied. In addition, we will only study linear problems in which the current density depends on the first power of $\mathbf{B}$. A detailed knowledge of the basic patterns of electron circulation can provide essential information for understanding molecular magnetic properties, say NMR spectroscopy data. In Sec. 2 the theory of real autonomous systems of linear differential equations [1,2] is reviewed to analyze the behavior of streamlines in the current density vector field. In Sec. 3 we will see how far one can go, simply relying upon symmetry arguments to determine the fundamental features of electron flow *a priori*, that is, without actual calculation of current density maps. In Sec. 4 the results of approximate quantum mechanical calculations are presented, showing pictures of electron circulation in benzene and borazine.

J. A. Tossell (ed.), Nuclear Magnetic Shieldings and Molecular Structure, 163–190.
© 1993 *Kluwer Academic Publishers.*

2. Trajectories in the Current Density Field

Within classical mechanics the velocity vector field satisfies the equations

$$\mathbf{v} = \frac{d\mathbf{r}}{dt}, \quad \frac{dv_x}{x} = \frac{dv_y}{y} = \frac{dv_z}{z},$$

where t is used to denote time. Analogous equations can be written for the stationary $\mathbf{j}$ trajectories. The streamlines of the probability current density vector field $\mathbf{j}$ are found as solutions of the real autonomous system of differential equations

$$\mathbf{r}' \equiv \frac{d\mathbf{r}}{d\tau} = \mathbf{j}(\mathbf{r}), \tag{2.1}$$

where the derivatives $dx/d\tau$, $dy/d\tau$ and $dz/d\tau$ have been arranged in a column $\mathbf{r}'$, and $\tau = \tau(\mathbf{r})$ is any convenient coordinate along the trajectory. In the maps a (time) arrow will be used to indicate the direction of the current. Eq. (2.1) is equivalent to

$$\frac{dx}{j_x} = \frac{dy}{j_y} = \frac{dz}{j_z}. \tag{2.1'}$$

In general the three-dimensional vector field can be studied by analyzing the current regime over properly chosen planes, which is usually sufficient to understand the three-dimensional particle flow. When a two-dimensional problem is studied, say the trajectories in the z_0 plane, the differential equation (2.1') becomes

$$\frac{dx}{j_x(x, y, z_0)} = \frac{dy}{j_y(x, y, z_0)}. \tag{2.1''}$$

In the domains where $\mathbf{j}$ does not vanish, this problem can be easily solved, in many practical cases, by standard numerical procedures, as one and only one trajectory passes through any non singular point of the vector field [1]: the equation can be recast in the form

$$\frac{dy}{dx} = F(x, y) \equiv \frac{j_y}{j_x}, \tag{2.2}$$

which is well-suited for numerical applications, say Euler and Runge-Kutta integration.

The most interesting characteristics of the current density vector field are connected with its singularities, which are also called fixed, or equilibrium, or stagnation points. These are the loci in which the modulus of $\mathbf{j}$ goes to zero. However, in the vicinity of stagnation points, eq. (2.2) has no meaning, as it just the indetermined form $0/0$. The problem can be tackled, according to the theory of linear differential equations [1,2], by means of a perturbative approach. Let us introduce the Taylor expansion of the charge current density about the singularity $\mathbf{r}_0$. In dyadic notation,

$$\mathbf{J}(\mathbf{r}) = (\mathbf{r} - \mathbf{r}_0) \cdot \left[\nabla \mathbf{J}\right]_{\mathbf{r}_0} + \frac{1}{2}(\mathbf{r} - \mathbf{r}_0) \cdot \left[\nabla \nabla \mathbf{J}\right]_{\mathbf{r}_0} \cdot (\mathbf{r} - \mathbf{r}_0) + \dots. \tag{2.3}$$

Within the linear approximation, this series can be truncated to first order in $(\mathbf{r} - \mathbf{r}_0)$, that is

$$J_\alpha(\mathbf{r}) \approx \mathcal{J}_{\beta\alpha}(r_\beta - r_{0\beta}). \tag{2.4}$$

The analysis of the Jacobian matrix $\mathcal{J} = [\nabla \mathbf{J}]_{\mathbf{r}_0}$ and its eigenvalues leads to the identification of the type of singularity and its phase portrait, that is, the peculiar pattern of streamlines in its environment. The continuity equation for stationary states, $\nabla \cdot \mathbf{J} = 0$, is a constraint on the eigenvalues of the Jacobian. This means that, in most cases, we can limit ourselves to studying the current flow in two-dimensional maps: the behavior in three dimensions can be figured out bearing in mind the continuity equation as a physical condition on the eigenvalue relative to the out-of-plane component.

Using eq. (2.4), the differential equation for the trajectories about the singularity on the origin in the (x, y) plane becomes, to first order,

$$\frac{dx}{\mathcal{J}_{11}x + \mathcal{J}_{21}y} = \frac{dy}{\mathcal{J}_{12}x + \mathcal{J}_{22}y}, \tag{2.5}$$

and, using the transposed Jacobian matrix, this is equivalent to the system of differential equations in matrix form

$$\mathbf{r}' = \tilde{\mathcal{J}}\mathbf{r}, \qquad \det \mathcal{J} \neq 0. \tag{2.6}$$

It is expedient to carry out a change of coordinate system and the associated similarity transformation to get a Jacobian $\mathcal{D}_{\alpha\beta}$, having a simpler *canonical* form. In planar problems the transposed Jacobian $\tilde{\mathcal{D}}$ can be reduced to one of the following real matrices [2]

$$\tilde{\mathcal{D}}_1 = \begin{pmatrix} \lambda & 0 \\ 0 & \lambda \end{pmatrix}, \quad \lambda \neq 0; \qquad \tilde{\mathcal{D}}_2 = \begin{pmatrix} \lambda & 0 \\ 0 & \mu \end{pmatrix}, \quad \mu < \lambda < 0 \text{ or } 0 < \mu < \lambda;$$

$$\tilde{\mathcal{D}}_3 = \begin{pmatrix} \lambda & 0 \\ \gamma & \lambda \end{pmatrix}, \quad \lambda \neq 0, \gamma > 0; \qquad \tilde{\mathcal{D}}_4 = \begin{pmatrix} \lambda & 0 \\ 0 & \mu \end{pmatrix}, \quad \lambda < 0 < \mu;$$

$$\tilde{\mathcal{D}}_5 = \begin{pmatrix} \alpha & \beta \\ -\beta & \alpha \end{pmatrix}, \quad \alpha \neq 0, \beta \neq 0, \qquad \tilde{\mathcal{D}}_6 = \begin{pmatrix} 0 & \beta \\ -\beta & 0 \end{pmatrix}, \quad \beta \neq 0.$$

The solutions of the two-dimensional system (2.6) are denoted by $\phi = (\phi_1, \phi_2)$ and it is convenient to introduce the polar functions ρ and ω, associated with the solutions ϕ, defined by

$$\rho(\tau) \equiv (\phi_1^2(\tau) + \phi_2^2(\tau))^{\frac{1}{2}}, \qquad \omega(\tau) \equiv \tan^{-1}\frac{\phi_2(\tau)}{\phi_1(\tau)}.$$

These functions are to be distinguished from the polar coordinates r and θ in the (x, y) plane, defined by

$$r = (x^2 + y^2)^{\frac{1}{2}}, \qquad \theta = \tan^{-1}\frac{y}{x},$$

just as the solution coordinate functions (ϕ_1, ϕ_2) are to be distinguished from the Cartesian coordinates (x, y) in the plane.

Corresponding to canonical form (I)$\equiv \tilde{\mathcal{D}}_1$, one has the autonomous system

$$\frac{dx}{d\tau} = \lambda x, \quad \frac{dy}{d\tau} = \lambda y,$$

then, if (c_1, c_2) is any initial point except $(0, 0)$, a solution through this point is given by

$$\phi_1(\tau) = c_1 \exp(\lambda \tau), \qquad \phi_2(\tau) = c_2 \exp(\lambda \tau).$$

Eliminating the parameter τ the trajectories on the (x, y) plane are

$$\phi_2 = \frac{c_2}{c_1} \phi_1,$$

i.e., the orbits through (c_1, c_2) are open half lines passing through this point and with an end at the origin. Then $\rho(\tau) \to 0$ as $\tau \to +\infty$, if $\lambda < 0$, and $\rho(\tau) \to 0$ as $\tau \to -\infty$, if $\lambda > 0$. The singularity corresponding to $\tilde{\mathcal{D}}_1$ is referred to as a *proper* (or *degenerate*) *node*: the trajectories, see Fig. 1, are locally straight lines originating or ending at the singularity. The arrows indicate the direction of increasing τ. Accordingly the node is also called a *sink* or an *attractor* for $\lambda < 0$, a *source* or a *repellor* for $\lambda > 0$. The origin is asymptotically stable, (*vide infra* for a definition of stability) in the case $\lambda < 0$ and unstable when $\lambda > 0$.

The autonomous system corresponding to canonical form (II) is

$$\frac{dx}{d\tau} = \lambda x, \quad \frac{dy}{d\tau} = \mu y,$$

then the solution is given by

$$\phi_1(\tau) = c_1 \exp(\lambda \tau), \qquad \phi_2(\tau) = c_2 \exp(\mu \tau).$$

This curve passes through $(c_1, c_2) \neq (0, 0)$ at $\tau = 0$. Let us consider the case $\mu < \lambda < 0$, see Fig. 2a, then as $\tau \to +\infty, (\phi_1(\tau), \phi_2(\tau)) \to (0, 0)$, and, if $c_1 \neq 0$,

$$\frac{\phi_2(\tau)}{\phi_1(\tau)} = \frac{c_2}{c_1} \exp[(\mu - \lambda)\tau] \to 0, \qquad \tau \to +\infty.$$

If $c_1 = 0$ and $c_2 \neq 0$, then $\phi_1(\tau) = 0$ and $\phi_2(\tau) = c_2 \exp(\mu \tau)$, which is either the open positive or negative y axis, depending on $c_2 > 0$ or $c_2 < 0$. Every orbit, except one, has the same limiting direction at the singularity, which is asymptotically stable for $\mu < \lambda < 0$. The origin is unstable for $0 < \mu < \lambda$, see Fig. 2b. The singularity is called an *improper node*.

Form (III) $\equiv \tilde{\mathcal{D}}_3$, see Fig. 3, is also that of an improper node. The equations to solve in this case are

$$\frac{dx}{d\tau} = \lambda x, \quad \frac{dy}{d\tau} = \gamma x + \lambda y.$$

If (c_1, c_2) is the initial point for $\tau = 0$, then the solution through this point is

$$\phi_1(\tau) = c_1 \exp(\lambda\tau), \qquad \phi_2(\tau) = (c_2 + c_1\gamma\tau)\exp(\lambda\tau).$$

For $\lambda<0$, see Fig. 3a, $\phi_1(\tau)$ and $\phi_2(\tau) \to 0$ for $\tau \to +\infty$. If $c_1 \neq 0$,

$$\frac{\phi_2(\tau)}{\phi_1(\tau)} = \frac{c_2}{c_1} + \gamma\tau \to \pm\infty, \quad \tau \to \pm\infty.$$

If $c_1>0$ (<0), then $\phi_2(\tau)>0$ (<0) for τ positive and large enough. If $c_1 = 0$, then $\phi_1(\tau) = 0$ and $\phi_2(\tau) = c_2 \exp(\lambda\tau)$: the orbit becomes a half y axis. Also, if $c_1 \neq 0$,

$$\frac{\phi_2'(\tau)}{\phi_1'(\tau)} = \frac{\gamma}{\lambda} + \frac{\phi_2(\tau)}{\phi_1(\tau)} \to \pm\infty, \quad \tau \to \pm\infty.$$

Thus every trajectory has the same limiting direction at the singularity.

A *saddle point*, see Fig. 4, is observed for form (IV) $\equiv \tilde{D}_4$. The autonomous system in this case is

$$\frac{dx}{d\tau} = \lambda x, \qquad \frac{dy}{d\tau} = \mu y,$$

then, if (c_1, c_2) is any initial point except $(0,0)$, a solution through this point is given by

$$\phi_1(\tau) = c_1 \exp(\lambda\tau), \qquad \phi_2(\tau) = c_2 \exp(\mu\tau),$$

where $\lambda<0$ and $\mu>0$. If $|\lambda| = |\mu|$ the orbits would be rectangular hyperbolae, more generally the streamlines resemble these hyperbolae. If $(c_1, c_2) \neq (0,0)$, then $\phi_1(\tau) \to 0$, and $\phi_2(\tau) \to \pm\infty$, according as $c_2>0$ or $c_2<0$. The asymptotes are a couple of straight lines ending and originating at the singularity (which is never crossed by any orbit), that is, the (time) arrow is reversed on passing through the singularity.

Corresponding to (V) $\equiv \tilde{D}_5$ the equations to solve are

$$\frac{dx}{d\tau} = \alpha x + \beta y, \qquad \frac{dy}{d\tau} = -\beta x + \alpha y,$$

and the solution through the initial point (c_1, c_2) is given by

$$\phi_1(\tau) = \exp(\alpha\tau)(c_1 \cos(\beta\tau) + c_2 \sin(\beta\tau)),$$

$$\phi_2(\tau) = \exp(\alpha\tau)(-c_1 \sin(\beta\tau) + c_2 \cos(\beta\tau)).$$

If $\rho_0^2 = c_1^2 + c_2^2$, this solution may be written

$$\phi_1(\tau) = \rho_0 \exp(\alpha\tau) \cos(\beta\tau - \delta), \qquad \phi_2(\tau) = -\rho_0 \exp(\alpha\tau) \sin(\beta\tau - \delta),$$

where $\cos\delta = c_1/\rho_0$ and $\sin\delta = c_2/\rho_0$. The polar functions for this solution are

$$\rho(\tau) = \rho_0 \exp(\alpha\tau), \qquad \omega(\tau) = -\beta\tau + \delta,$$

so that

$$\rho = C \exp(-\frac{\alpha}{\beta}\omega), \qquad C = \rho_0 \exp(\frac{\alpha}{\beta}\delta),$$

i.e., a spiral. The singularity is referred to as *focus* or *spiral point*. The streamlines, see Fig. 5, are locally spirals which approach, or leave, the region of the singularity depending on the conditions $\alpha < 0$, $\beta < 0$ or $\alpha > 0$, $\beta > 0$ respectively.

Case (VI) is a special case of (V) where $\alpha = 0$. Thus the solution through (c_1, c_2) at $\tau = 0$ becomes

$$\phi_1(\tau) = c_1 \cos(\beta\tau) + c_2 \sin(\beta\tau), \qquad \phi_2(\tau) = -c_1 \sin(\beta\tau) + c_2 \cos(\beta\tau),$$

so that $\rho(\tau) = \rho_0$, which is the equation of a circumference of radius ρ_0 centered on the origin. The eigenvalues of $\tilde{\mathcal{D}}_6$ are pure imaginary $\pm ib$. The singularity is called a *center*. The streamlines are locally circumferences and the phase portrait is that of a *vortex* [2]. An arbitrarily small variation of the elements of $\tilde{\mathcal{D}}_6$ can cause the roots $\pm ib$ to have a real part: the center evolves into a focus [2].

From the six cases considered above a definition of *stability* of a singularity can be given: stable singularities are those corresponding to canonical Jacobian matrices whose eigenvalues are characterized by negative or zero real part.

Another interesting phase portrait is the *limit cycle* [2]. This is a special attractor (a close orbit): all internal and external streamlines in the environment of the limit cycle are spirals. The limit cycle can be detected by a first-return Poincarè map, according to the recipe (i) cross the ϕ orbits along a transversal l, (ii) construct a map $\phi(P)$ for every point P at the intersection between the orbits ϕ and l, (iii) determine the limit cycle C as $\phi_C(P) = P$.

The analysis of eq. (2.3) can be extended to non linear problems, but in most practical cases the first-order approach (2.4) is sufficient. It is useful to recall that some of the phase portraits previously examined may be not physical: they can however be found in approximate calculations which do not fulfill the continuity equation.

The fixed points of $\mathbf{J}$ can also be classified according to the $(rank, signature)$ index of the singularity, which has been adopted by Collard and Hall, [3] and by Bader et al. [4] in their studies of the trajectories of $\nabla\rho(\mathbf{r})$ in molecules. This approach is particularly useful to gather information on the spatial patterns in the $\mathbf{J}$ field by studying the Jacobian in three dimensions.

The rank r is defined as the number of non zero eigenvalues of the Jacobian, the signature s is the excess of positive over negative eigenvalues, if they are real or pure imaginary. If the eigenvalues are complex we may define the signature as the difference between the number of eigenvalues having positive real part and the number of eigenvalues having negative real part.

Owing to the continuity equation for stationary flow, the Jacobian is traceless and its eigenvalues are not independent (i) in real physical cases and (ii) in those ideal cases where ρ and $\mathbf{J}$ have been calculated from exact eigenfunctions to a model Hamiltonian. This restricts the number of possible (r, s) to a few physically meaningful cases [5]:

$(3, \pm 1)$ points corresponding to isolated singularities. Two eigenvalues satisfy $\xi_3 = -\Re(\xi_1 + \xi_2)$. If ξ_1 and ξ_2 are real (they may also be $\xi_1 = \xi_2$), a saddle point is observed, if they are complex conjugate a focus is found. Usually, owing to symmetry, $(3, +1)$ and $(3, -1)$ points occur as coupled singularities.

$(2, 0)$ points; eigenvalues $\xi_3 = 0, \xi_1 = -\xi_2$, where ξ_1 and ξ_2 are obtained by diagonalizing either (i) the canonical form $\tilde{\mathcal{D}}_6$, that is, $\xi_{1,2} = \pm ib$, or (ii) the form $\tilde{\mathcal{D}}_4$, with $\xi_1 = \lambda = -\mu$. These are respectively vortex and saddle points. The eigenvectors $\mathbf{t}_1$ and $\mathbf{t}_2$, corresponding to ξ_1 and ξ_2, are real in the case of a saddle (they give the direction of the asymptotes through the singularity) and imaginary in the case of a center. Saddle- and vortex-stagnation lines are continuous manifolds of $(2,0)$ points. As we will see, normally these stagnation lines are symmetry determined and lie entirely on symmetry planes of a molecule. The eigenvector $\mathbf{t}_3$ is locally tangent to the stagnation line. $(2,0)$ points can be open lines (this is the case of axial vortices) or form close loops [6-17]. A particularly interesting pattern is the toroidal vortex flowing around a closed vortex line of $(2,0)$ points.

$(0,0)$ points; they belong to $(2,0)$ manifolds and are found in correspondence of three zero eigenvalues of the Jacobian. These loci are called *branching* or *transition points*. The reason for this denomination will be immediately clear. Consider, for instance, a molecule of D_{nh} symmetry in a magnetic field $\mathbf{B}$ along the highest symmetry axis C_n. In general, in the outer reaches of the molecular domain, the induced electronic current density is diamagnetic, that is, resembling that occurring in atoms. In the vicinity of the North and South poles, at large distance from the molecular center of charge, the axis of the diamagnetic vortex, that is the primary $(2,0)$ stagnation line, is parallel to C_n and $\mathbf{B}$. Closer to the center of charge, that is, in the regions where the electron density $\rho(\mathbf{r})$ is higher, different regimes may take place: transition from vortex to saddle flow can be found, a stagnation line may branch into several stagnation lines. In fact, the primary vortex line may branch into saddle and vortical lines. From the mathematical point of view, the regime variation is due to an exchange between the canonical forms $\tilde{\mathcal{D}}_4$ and $\tilde{\mathcal{D}}_6$. As it is a transition between pure imaginary and pure real eigenvalues, the branching points must necessarily occur in correspondence of three zero eigenvalues. Also $(0,0)$ points may be due to molecular symmetry, *vide infra*.

Gomes [18] has introduced the idea of *stagnation graph* and proved an index conservation theorem. The index of a saddle (vortex) line is $-1(+1)$. When a stagnation line of index i_0 branches out into m new lines, the sum of the relative indices is

$$\sum_{k=1}^{m} i_k = i_0. \qquad (2.7)$$

For instance, a vortex line may bifurcate giving rise to two new vortex lines and one saddle line.

170

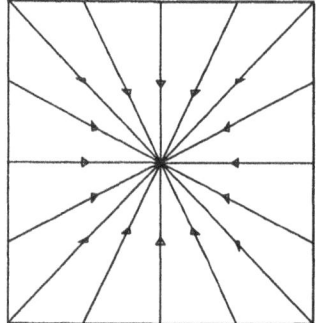

Fig 1a: Proper node, $\lambda < 0$

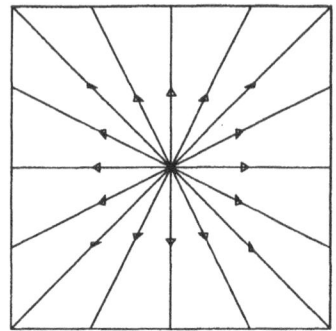

Fig 1b: Proper node, $\lambda > 0$

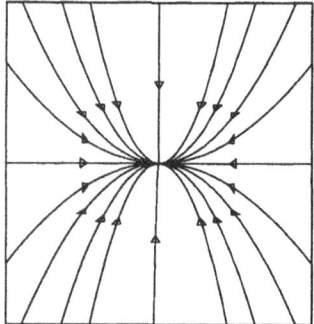

Fig 2a: Improper node, $\mu < \lambda < 0$

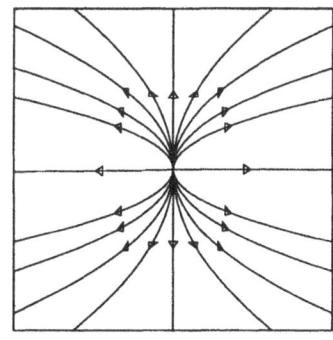

Fig 2b: Improper node, $0 < \mu < \lambda$

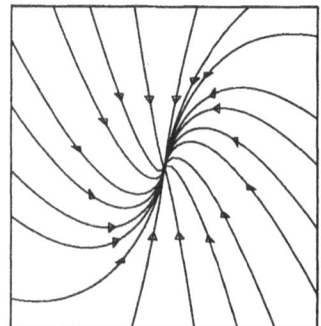

Fig 3a: Improper node, $\lambda < 0$

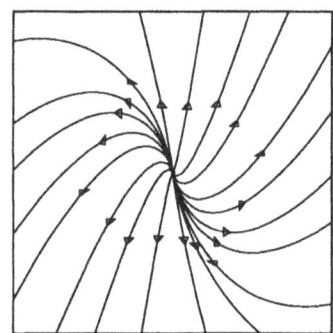

Fig 3b: Improper node, $\lambda > 0$

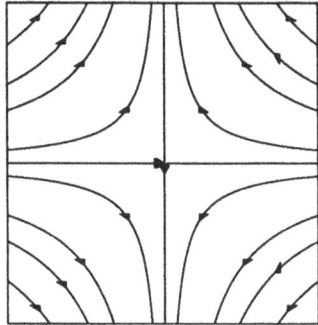

Fig 4: Saddle point

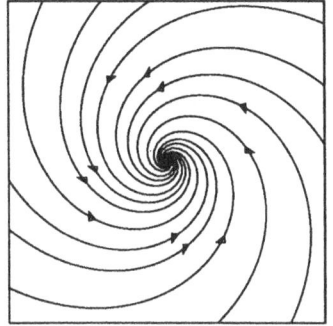

Fig 5a: Spiral point, $\alpha < 0, \beta < 0$

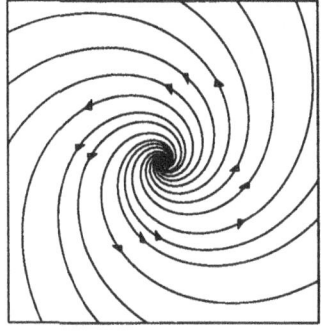

Fig 5b: Spiral point, $\alpha > 0, \beta < 0$

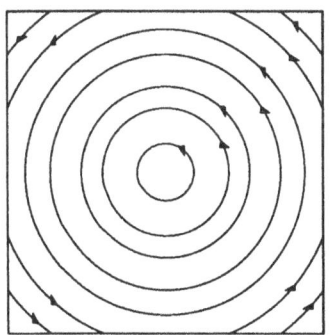

Fig 6a: Center, $\beta < 0$

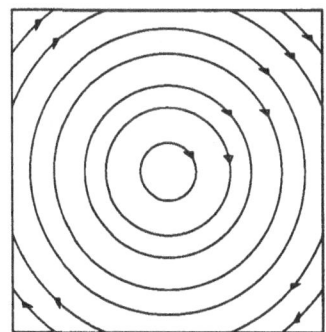

Fig 6b: Center, $\beta > 0$

3. Magnetic Groups and Magnetically Induced Current Density

The time averaged electron density $\rho(\mathbf{r})$ in a molecule possesses symmetry properties which are related to the geometrical shape of the molecule itself, considered as a rigid system in its equilibrium state. Accordingly, we may consider the 32 conventional point groups [19] as mathematical tools describing the possible point symmetry of ρ. This statement holds in the absence of (stationary) electron current density $\mathbf{J}(\mathbf{r})$.

For the vast majority of species $\mathbf{J} = \mathbf{0}$; this applies to nonmagnetic, as well as diamagnetic substances, whereas ferromagnetic, antiferromagnetic and pyromagnetic species must have $\mathbf{J} \neq \mathbf{0}$.

In the presence of magnetic field or nuclear magnetic dipoles, electronic current density is induced in the molecule in question. In this case the 32 classical point groups, which are all admissible symmetry groups when $\mathbf{J} = \mathbf{0}$, are no longer suitable to describe the actual symmetry properties of the system. Any symmetry transformation can now be made up of an operation acting on the geometrical shape of the molecule and another transformation of $\mathbf{J}$ which does not affect the space coordinates: there is the need for magnetic groups, equivalent to Shubnikov color groups [20,21].

An operation peculiar to vector fields is the reversal in direction. In the case of current density this is equivalent to time reversal R [19], since $\mathbf{J}$, as well as ρ, is a subobservable [17] and can be considered as a completely classical quantity. Then, in the presence of current density, the symmetry transformations will have the form $P = RQ$, where Q is either a rotation or a reflection. The operator R does not act on space coordinates $\mathbf{r}$, as it commutes with rotations and reflections. Accordingly, it does not affect $\rho(\mathbf{r})$, but reverses the sign of $\mathbf{J}(\mathbf{r})$ at each $\mathbf{r}$, that is,

$$R\rho(\mathbf{r}) = \rho(\mathbf{r}), \qquad R\mathbf{J}(\mathbf{r}) = -\mathbf{J}(\mathbf{r}). \tag{3.1}$$

A quantum mechanical operator T is antilinear [22,23] if

$$T\phi = \sum_k c_k^* T\psi_k, \qquad \phi = \sum_k c_k \psi_k, \tag{3.2}$$

T is antiunitary [22,23] if it is antilinear and

$$TT^\dagger = T^\dagger T = E, \tag{3.3}$$

so that

$$|<\psi|\chi>| = |<T\psi|T\chi>|. \tag{3.4}$$

The operator for time reversal is an antiunitary operator [22,23]. In the present context, however, R can be merely regarded as an operator which changes the sign of t and such that $R^2 = E$. It is not itself a symmetry operator in any molecular system characterized by a non vanishing current density: if a group contained R, that would imply $\mathbf{J} = -\mathbf{J} = 0$ everywhere.

In fact not every P can be a symmetry transformation, also excluded are those $P = RQ$ for which Q is of odd cycle g, as it would imply

$$P^g = R^g Q^g = RE = R. \qquad (3.5)$$

If P is a legitimate symmetry operation, then P^g is also allowed, which contradicts (3.5): it follows that the possibility of RC_3 axes is ruled out. Instead, RS_3 is an allowed operation which appears, for instance, in $C_{3h}(C_3)$ group, *vide infra*.

The set of symmetry transformations $\{P\}$ forms a group in the usual mathematical acceptation [19]. It is called a *magnetic group*. This is the same as one of Shubnikov *color* groups [20,21] in which the color change is replaced by the time-reversal operator.

Both Q and $P = RQ$ cannot occur in magnetic group M, since M would then contain Q^{-1} and $PQ^{-1} = R$, which, as we have seen, is not allowed in a system characterized by $\mathbf{J} \neq 0$. On the basis of this property, for any magnetic group of order q, we can label the elements as $P_k = RQ_k, k = 1, 2, \ldots m, Q_i, i = m + 1, m + 2, \ldots q$. Assume that $P_k = RQ_k$ and Q_i form a group. Replacing R with E, that is, $P_k = Q_k$, one finds that $\{Q_k\}, \{Q_i\}$ form a group belonging to one of the classical 33. If $P_k = RQ_k$ and Q_i form a group, $\{Q_i\}$ is a subgroup. Indeed,

$$Q_{i'} Q_{i''} = Q_{i'''} \in \{Q_i\} \neq \{P_k\}$$

and $\{Q_i\}$ is one of the usual 32 conventional groups.

In fact, in any finite magnetic group M, half of the symmetry elements are unitary and half antiunitary. It follows from the fact that the unitary operations form a stable subgroup H of index 2, that is, of order $m = q/2$, as the product of two antiunitary operators is unitary.

These properties lead to the Tavger-Zaitsev theorem [24]: a necessary and sufficient condition for the elements $\{Q_i\}$ and $\{P_k\} = \{RQ_k\}$ to form a group is that H is of index 2 in G.

In fact any unitary group G can be expressed in terms of left cosets with respect to subgroup H of index 2:

$$G = H + Q_k H, \qquad Q_k \in G - H, \qquad Q_i \in H, \qquad Q_i Q_k \in G - H.$$

The set

$$M = H + P_k H, \qquad P_k = RQ_k$$

is a group, since both H and $P_k H$ are stable:

$$H \cdot H = H$$

$$P_k H \cdot H = P_k H \in M - H$$

$$P_k H \cdot P_k H = RQ_k H \cdot RQ_k H = H.$$

Viceversa, if $\{Q_i\}$ and $\{P_k\}$ form a group M, then multiplying the m elements $\{Q_i\}$ by P_k gives the different elements of type RQ_k. Multiplying $P_{k'} \cdot P_{k''}$ we obtain the different elements of type $\{Q_i\}$. Thus H has index 2 in M and G as well.

Therefore, for each unitary group G, possessing a subgroup H of index 2 and hence a decomposition in terms of left cosets with respect to H, we can construct a magnetic group M according to the TZ algorithm: for any symmetry group G

(i) select a subgroup $H \equiv \{Q_i\}$ of index 2 in G

(ii) form the set $G - H$ of operators $\{Q_k\}$

(iii) construct $R\{G - H\}$, consisting of $\{P_k\} = \{RQ_k\}$

(iv) obtain the magnetic group M as $H + R\{G - H\}$, consisting of $\{P_k, Q_i\}$.

If G runs over the 32 symmetry classes, 58 different groups M are obtained, for a total of 90 magnetic groups. Note that M and G are isomorphic. Accordingly, it is often appropriate to use the representations of the unitary group G isomorphic with M in solving some problems of classical physics. For instance, some singularities of the current density vector field in molecules are symmetry determined and their existence can be easily guessed a priori by analyzing the unitary group G.

Within the Schönflies notation, a magnetic group can be given a symbol which makes explicit the H subgroup of index 2 used in the TZ algorithm, compare, for instance, the $D_{nh}(C_{nh})$ magnetic groups.

It is also expedient to introduce the International notation, according to which n is the symbol for an n-fold rotation axis C_n, m is a $\sigma_{v,d,h}$ symmetry plane, a bar over a symbol means product with space inversion. Thus, for instance, $C_s \equiv m$, $S_6 \equiv \bar{3}$, $C_{3h} \equiv \bar{6}$. A / stands for a symmetry plane perpendicular to the rotation axis, then $C_{2h} \equiv 2/m$, but $D_{2h} \equiv mmm$. In fact symmetry planes not perpendicular to a rotation axis are recorded without any additional marking. Other examples: $C_{3v} \equiv 3m$, $C_{4v} \equiv 4mm$, $D_{4h} \equiv 4/mmm$, $D_2 \equiv 222$. A bar under a symbol means product with time reversal, e.g., $\underline{n} \equiv RC_n$, $\underline{m} \equiv R\sigma$.

For the sake of clarity we exemplify the procedure for building up $D_{nh}(C_{nh})$ magnetic groups [5]. Assuming $\mathbf{B}$ along the C_n axis, parallel to the z direction, we select $G \equiv D_{nh}$, and choose C_{nh} as the corresponding H of index 2. Then we construct the set $R\{D_{nh} - C_{nh}\}$ and obtain the magnetic groups $D_{nh}(C_{nh})$ as $C_{nh} + R\{D_{nh} - C_{nh}\}$. Thus, for n=2,

$$D_{2h} \equiv mmm \equiv \{E\, C_2(z)\, C_2(x)\, C_2(y)\, i\, \sigma_h(xy)\, \sigma(xz)\, \sigma(yz)\},$$
$$C_{2h} \equiv 2/m \equiv \{E\, C_2(z)\, i\, \sigma_h(xy)\},$$
$$D_{2h}(C_{2h}) \equiv \underline{mmm} \equiv \{E\, C_2(z)\, RC_2(x)\, RC_2(y)\, i\, \sigma_h(xy)\, R\sigma(xz)\, R\sigma(yz)\};$$

for n=3,

$$D_{3h} \equiv \bar{6}m2 \equiv \{E\, 2C_3\, 3C_2\, \sigma_h\, 2S_3\, 3\sigma_v\},$$
$$C_{3h} \equiv \bar{6} \equiv \{E\, 2C_3\, \sigma_h\, 2S_3\},$$
$$D_{3h}(C_{3h}) \equiv \bar{6}\underline{m}2\{E\, 2C_3\, 3RC_2\, \sigma_h\, 2S_3\, 3R\sigma_v\};$$

for n=6,

$$D_{6h} \equiv 6/mmm \equiv \{E\, 2C_6\, 2C_3\, C_2\, 3C_2'\, 3C_2''\, i\, 2S_3\, 2S_6\, \sigma_h\, 3\sigma_d\, 3\sigma_v\},$$
$$C_{6h} \equiv 6/m \equiv \{E\, 2C_6\, 2C_3\, C_2\, i\, 2S_3\, 2S_6\, \sigma_h\},$$
$$D_{6h}(C_{6h}) \equiv 6/m\underline{mm} \equiv \{E\, 2C_6\, 2C_3\, C_2\, 3RC_2'\, 3RC_2''\, i\, 2S_3\, 2S_6\, \sigma_h\, 3R\sigma_d\, 3R\sigma_v\}.$$

Some interesting features of the **J**-field are a direct consequence of the magnetic group symmetry:

(i) A σ plane cannot be crossed by the streamlines

(ii) $R\sigma$ planes can be crossed only by trajectories perpendicular to them, which is the usual case of vortical regime. If a streamline approaches the $R\sigma$ plane forming an angle different from 90°, it is scattered away, compare for the phase portrait of a saddle, Fig. 4. Therefore open or closed, vortex or saddle stagnation lines, i.e., (2,0) manifolds, may lie on, but not pass through, an $R\sigma$ plane. Vortical stagnation lines may pass through perpendicularly, but not lie on, σ planes; only saddle lines may exist on σ planes. One might consider the possibility of a (0,0) point, lying on an $R\sigma$ plane, from which originates a couple of vortices rotating in different directions on either side of that plane. The $R\sigma$ plane is *not* crossed by the vortical stagnation lines perpendicular to it: there are two distinct (2,0) manifolds separated by the (0,0) point. Such a situation might occur in $C_{nh}(C_n)$ groups. As a consequence, stagnation lines are, quite frequently, fully contained in $R\sigma$ and σ planes.

(iii) As the in-plane components of the **J** vector vanish all over $R\sigma$ planes by symmetry, the continuity equation for stationary flow $\nabla \cdot \mathbf{J} = 0$ is necessarily fulfilled for the perpendicular component, even if **J** has been evaluated via approximate quantum mechanical methods.

(iv) The symmetry axes C_n, parallel to the inducing magnetic field **B**, lying on $R\sigma_v$ planes are necessarily stagnation lines.

(v) According to (i), all over a $\sigma(xy) \equiv z = 0$ plane, the perpendicular component $J_z(x, y, 0)$ vanishes for all the points $P(x, y, 0)$ on that plane, and consequently

$$\left[\frac{\partial J_z}{\partial x}\right]_{x,y,0} = \left[\frac{\partial J_z}{\partial y}\right]_{x,y,0} = 0.$$

Similar considerations can be made for the derivatives of the current density over $R\sigma$ planes.

From these simple rules one can obtain further interesting results. Thus, for instance, consider the case of water molecule in the presence of magnetic field **B** parallel to the $C_2(z)$ direction (in the sequel we will always assume a coordinate system with origin on the molecular center of mass). The symmetry of the molecule in the field is $C_{2v}(C_2) \equiv 2mm$, isomorphic with C_{2v}, with symmetry elements

$$E, C_2, R\sigma(xz), R\sigma(yz),$$

see Table 1. The molecule lies on the $R\sigma(yz)$ plane, therefore

$$J_x(x, 0, z) = J_y(0, y, z) = J_z(x, 0, z) = J_z(0, y, z) = 0,$$

and $C_2(z)$ is a stagnation axis. In addition

$$\frac{\partial}{\partial x} J_x(x, 0, z) = \frac{\partial}{\partial x} J_z(x, 0, z) = \frac{\partial}{\partial z} J_x(x, 0, z) = \frac{\partial}{\partial z} J_z(x, 0, z) = 0,$$

$$\frac{\partial}{\partial y} J_y(0, y, z) = \frac{\partial}{\partial y} J_z(0, y, z) = \frac{\partial}{\partial z} J_y(0, y, z) = \frac{\partial}{\partial z} J_z(0, y, z) = 0.$$

Hence, for all the points lying on the $C_2(z)$ axis, the Jacobian has the form

$$[\nabla J]_{0,0,z} = \begin{pmatrix} 0 & \nabla_x J_y & 0 \\ \nabla_y J_x & 0 & 0 \\ 0 & 0 & 0 \end{pmatrix}.$$

It can be seen [5] that this is a superposition of canonical forms (IV) and (VI): either a vortex or a saddle are expected to occur about the center of mass and any other point on the two-fold axis.

A similar situation is found for ethylene molecule in the presence of magnetic field perpendicular to the $\sigma_h \equiv \sigma(xy)$ molecular plane and parallel to the $C_2(z)$ axis. The symmetry of the system is $D_{2h}(C_{2h}) \equiv \underline{mmm}$, isomorphic with $D_{2h} \equiv mmm$, see Table 2. For this molecule one has the same conditions already seen for the components of the Jacobian matrix within the $C_{2v}(C_2)$ symmetry. In addition $J_z(x, y, 0) = 0$ and

$$\frac{\partial}{\partial x} J_z(x, y, 0) = \frac{\partial}{\partial y} J_z(x, y, 0) = 0,$$

all over the σ_h plane. However, also in this case the Jacobian for the points of the C_2 axis is a superposition of canonical forms (IV) and (VI), as in the previous case of $C_{2v}(C_2)$ symmetry.

One can easily show that the current regime is vortical about a C_3 axis parallel to the external magnetic field. Consider, for instance, the magnetic symmetry $C_{3v}(C_3) \equiv \underline{3m}$ isomorphic with $C_{3v} \equiv 3m$ (a typical example is ammonia molecule in magnetic field). Let us choose a non orthogonal basis set of unit vectors $\epsilon = \{\epsilon_1 \epsilon_2 \epsilon_3\}$, so that $\epsilon_3 \parallel \mathbf{B}$, and ϵ_2 lies at $120°$ from ϵ_1, on the same plane perpendicular to the direction of the magnetic field. The magnetic group $C_{3v}(C_3)$ possesses the symmetry elements

$$E, C_3, C_3^2, R\sigma_{13}, R\sigma_{23}, R\sigma_{34}.$$

All over the symmetry plane $R\sigma_{13}$, spanned by ϵ_1 and ϵ_3, the in-plane components of the current density vanish for any points $P(x_1, 0, x_3)$, i.e.,

$$J_1(x_1, 0, x_3) = J_3(x_1, 0, x_3) = 0.$$

Similarly, all over $R\sigma_{23}$, spanned by ϵ_2 and ϵ_3, for any point $P(0, x_2, x_3)$,

$$J_2(0, x_2, x_3) = J_3(0, x_2, x_3) = 0.$$

Accordingly, the C_3 axis is an open stagnation line. In addition

$$\frac{\partial}{\partial x_1} J_1(x_1, 0, x_3) = \frac{\partial}{\partial x_3} J_1(x_1, 0, x_3) = \frac{\partial}{\partial x_1} J_3(x_1, 0, x_3) = \frac{\partial}{\partial x_3} J_3(x_1, 0, x_3) = 0,$$

and

$$\frac{\partial}{\partial x_2} J_2(0, x_2, x_3) = \frac{\partial}{\partial x_3} J_2(0, x_2, x_3) = \frac{\partial}{\partial x_2} J_3(0, x_2, x_3) = \frac{\partial}{\partial x_3} J_3(0, x_2, x_3) = 0.$$

The symmetry plane $R\sigma_{34}$, spanned by ϵ_3 and ϵ_4, where $\epsilon_4 = -\epsilon_1 - \epsilon_2$, is a geometrical locus such that $x_1 = x_2$ for all its points $P(x_1, x_2, x_3)$. Thus the rate of change of an arbitrary function $f(x_1, x_2, x_3)$ with respect to x_1 and x_2 is the same all over this plane, i.e., $\partial f/\partial x_1 = \partial f/\partial x_2$. On the other hand, all over $R\sigma_{34}$, $J_1(x_1, x_2, x_3) = -J_2(x_1, x_2, x_3)$, and therefore $\nabla_1 J_2 = -\nabla_2 J_1$. Consequently, for any point lying on the C_3 axis, the Jacobian is characterized by only one nonvanishing quantity, $(\nabla \times \mathbf{J})_3$, which transforms according to the totally symmetric representation of $C_{3v}(C_3)$ (the same way as the product RR_3 between time inversion R and the rotation about the C_3 axis). Therefore

$$[\nabla J]_{0,0,x_3} = \begin{pmatrix} 0 & \nabla_1 J_2 & 0 \\ \nabla_2 J_1 & 0 & 0 \\ 0 & 0 & 0 \end{pmatrix}, \quad \nabla_2 J_1 = -\nabla_1 J_2.$$

This is related to canonical form VI and a vortex is accordingly expected to take place about the C_3 axis. A similar situation occurs for $D_{3h}(C_{3h}) \equiv \bar{6}m2$ systems.

A simple method for analyzing the symmetry of the Jacobian can be developed by splitting it up in terms of irreducible tensors. An irreducible tensor of rank l and degree k is the basis for an irreducible representation of degree k of the 3-dimensional rotation group. Only $d = 2k + 1$ of the 3^l components of a Cartesian tensor of rank l and degree k are independent. The Jacobian is therefore split according to

$$\nabla_\alpha J_\beta = S_{\alpha\beta} + A_{\alpha\beta} + t_{\alpha\beta},$$

$$t_{\alpha\beta} = \frac{1}{3}\nabla_\gamma J_\gamma \delta_{\alpha\beta}, \qquad l = 2, k = 0, d = 1,$$

$$S_{\alpha\beta} = \frac{1}{2}(\nabla_\alpha J_\beta + \nabla_\beta J_\alpha) - t_{\alpha\beta}, \qquad l = 2, k = 2, d = 5,$$

$$A_{\alpha\beta} = \frac{1}{2}(\nabla_\alpha J_\beta - \nabla_\beta J_\alpha) \equiv A_\gamma = \frac{1}{2}\epsilon_{\alpha\beta\gamma}\nabla_\alpha J_\beta, \qquad l = 2, k = 1, d = 3.$$

The transformation properties of these quantities are related to those of spherical harmonics Y_{00}, Y_{2m} and Y_{1m} respectively. The isotropic contribution $t_{\alpha\beta}$ is identically vanishing due to the continuity equation (this is no longer true if $\mathbf{J}$ is a theoretical current density evaluated via approximate quantum mechanical methods; however, the magnitude of $t_{\alpha\beta}$ gives information on the accuracy of a calculation and on discrepancies between theoretical and physical current density fields).

The number of independent components of the symmetric traceless $S_{\alpha\beta}$ second-rank tensor and of the pseudovector A_γ is equal to the dimension of the subspaces (stable under the operations of the magnetic group M) that they span. These components are easily identified, as they behave as basis functions belonging to the totally symmetric representation of M.

A few simple rules can be given for determining the representation spanned by $S_{\alpha\beta}$, A_γ and $t_{\alpha\beta}$ in the magnetic group M:
(i) x_α coordinates and R_α rotations transform as in the isomorphic group G,
(ii) J_α vector components transform as Rx_α,
(iii) $S_{\alpha\beta}$ symmetric components transform as the sum of products $x_\alpha J_\beta + x_\beta J_\alpha$,

(iv) $(\nabla \times \mathbf{V})_\alpha$, for any polar vector V_α left unchanged by R, transform as the components R_α of the axial rotation vector in the isomorpous group G,

(v) $A_\alpha \equiv (1/2)(\nabla \times \mathbf{J})_\alpha$ antisymmetric components transform as the operator product RR_α.

As an example consider any point lying on the two-fold symmetry axis in a molecule of $C_{2v}(C_2)$ magnetic symmetry, see Table 1 for group characters. One easily finds the sytem of characters for the reducible representation carried by the symmetric components $S_{\alpha\beta}$ of the Jacobian matrix,

$$\chi(\mathbf{P}) = \{6 \quad 2 - 2 - 2\}$$

and by the antisymmetric components A_γ,

$$\chi(\mathbf{P}) = \{3 \ - 1 \quad 1 \quad 1\}.$$

The frequency of the totally symmetric representation is 1 in both cases: one symmetric and one antisymmetric components fully characterize the Jacobian and the C_2 axis is therefore a vortex or saddle line, confirming the analysis of p. 13.

More powerful methods are provided by the approach developed by Bradley and Davis [25] to analyze tensor symmetry. As we will see, there is a substantial advantage related to this procedure: one uses G instead of the isomorphic M.

In the sequel the classification introduced by Birss [26] will be adopted: tensors symmetric and antisymmetric under R are i and c respectively. Four classes are considered according to Bradley and Davies [25], namely

(i) Polar i-tensors $\equiv$ s-tensors, transforming according to Γ_s,

(ii) Polar c-tensors $\equiv$ m-tensors, transforming according to Γ_m,

(iii) Axial i-tensors $\equiv$ p-tensors, transforming according to Γ_p,

(iv) Axial c-tensors $\equiv$ $(m \times p)$-tensors, transforming according to $\Gamma_{m \times p} \equiv \Gamma_m \times \Gamma_p$,

where Γ_s is the totally symmetric representation of G isomorphic to M, Γ_m is the one-dimensional representation of G in which all the operations of the subgroup common to G and M, i.e., H in the Tavger-Zaitsev partition, are represented by $+1$, Γ_p is the pseudoscalar representation of G in which all proper (improper) rotations are represented by $+(-1)$. In certain cases some of these representations coincide.

A t-tensor $\mathbf{d}^{(t)}$ of rank j in M, with $t = s, m, p, m \times p$, is defined to be one whose components transform

$$d^{(t)\prime}_{ab...j} = \chi_t(P) P_{al} P_{bm} \ldots P_{ju} d^{(t)}_{lm...u}, \tag{3.6}$$

under the active transformation $\mathcal{P} \in M$, isomorphic with $P \in G$ having matrix representation $\mathbf{P}$ over the basis set of (Cartesian) unit vectors $\epsilon = \{\epsilon_1 \epsilon_2 \epsilon_3\}$, i.e.,

$$P\epsilon = \epsilon \mathbf{P}.$$

The representation carried by ϵ, i.e., the collection $\Gamma(\mathbf{P})$ of matrices, is usually reducible. The relative system of characters is $\chi(\mathbf{P})$. The relevant irreducible representations of G are denoted $\Gamma_t(\mathbf{P}), (t = s, m, p, m \times p)$ and the corresponding characters are $\chi_t(\mathbf{P})$.

The number ν_t of unique components for a tensor $d_{ab...j}^{(t)}$ invariant under M is equal to the order of the space spanned by components which are left unchanged by all $\mathcal{P} \in M$ and which transform as basis functions for the totally symmetric Γ_s. In other words, ν_t is the frequency of Γ_s in the direct product $\Gamma_t \times \Gamma^j, \Gamma^j = \Gamma \times \Gamma \times \ldots \Gamma, j$ times:

$$\nu_t = \frac{1}{|G|} \sum_{P \in G} \chi_t(P) [\chi(P)]^j, \tag{3.7}$$

where $|G|^{-1} = |M|^{-1}$ is the order of the group. This is also the number of times that Γ_t appears in the reduction of Γ^j, as χ_t is real.

The explicit form of a $d_{ab...j}^{(t)}$ tensor invariant under the magnetic group M can be found by projecting out the components transforming as basis functions for the totally symmetric representation with the help of the operator

$$T_s = \frac{1}{|M|} \sum_{\mathcal{P} \in M} \chi_s(\mathcal{P}) \mathcal{P}, \qquad \chi_s(P) \equiv \chi_s(\mathcal{P}). \tag{3.8}$$

Alternatively, we can operate with the projector

$$T_t = \frac{1}{|M|} \sum_{\mathcal{P} \in M} \chi_t(P) \mathcal{P}, \tag{3.9}$$

on a corresponding polar i-tensor of rank j, since the former imposes on the tensor components $d_{ab...j}^{(t)}$ exactly the same relationships as are imposed by the latter on $d_{ab...j}^{(s)}$. One could also say that a t-tensor invariant under M suffers the same restrictions as a polar i-tensor transforming under Γ_t.

Since a typical polar i-tensor of rank j is obtained by the *ordered* product of j coordinates $x_a x_b \ldots x_j$, one can apply the simple projector

$$T_t = \frac{1}{|M|} \sum \chi_t(P) P, \tag{3.10}$$

The number of unique nonvanishing $\nabla_\alpha J_\beta$ components is found from the frequency formula

$$\nu_m = \frac{1}{|G|} \sum_{P \in G} \chi_m(P)[\chi(P)]^2, \qquad (3.11)$$

see the general expression (3.7).

Consider, for instance, $C_{3v}(C_3) \equiv 3\underline{m}$ magnetic symmetry, see Table 3. The character system for the reducible representation carried by the Cartesian coordinates is $\chi(\mathbf{P}) = \{3\,0\,1\}$. From (3.11) one finds $\nu_m = 1$, i.e., only one parameter fully determines the Jacobian for the points on the $C_3 \equiv z$ axis.

By applying

$$T_m = \frac{1}{6}\left(E + C_3 + \overline{C}_3 - \sigma_1 - \sigma_2 - \sigma_3\right)$$

on the ordered product xy we find $T_m xy = \frac{1}{2}(xy - yx)$, therefore $\left(\nabla \times \mathbf{J}\right)_z$ is totally symmetric under $3\underline{m}$ and an axial vortex is expected to rotate about the three-fold axis. This result is consistent with the previous discussion on ammonia molecule.

We will also examine other examples which are relevant to further discussion in Sec. 4. Thus, in the case of borazine and benzene in the presence of magnetic field perpendicular to the molecular plane, e.g., species belonging respectively to $D_{3h}(C_{3h})$ and $D_{6h}(C_{6h})$ symmetries, see Tables 4 and 5, the system of characters are respectively

$$\chi(\mathbf{P}) \equiv \{3\ \ 0 - 1\ \ 1 - 2\ \ 1\}, \quad \chi(\mathbf{P}) \equiv \{3\ \ 2\ \ 0 - 1 - 1 - 1 - 3 - 2\ \ 0\ \ 1\ \ 1\ \ 1\}.$$

In both cases $\nu_m = 1$, and a vortex circulating about the higher symmetry axis in the proximity of the center of mass is expected. The regions surrounding an H-C bond and an H-C=C-H fragment in benzene have local $C_{2v}(C_s)$ symmetry, see Table 1. The symmetry analysis of the Jacobian gives two nonvanishing components, one symmetric and one antisymmetric: either a vortex or a saddle can appear about the local RC_2' and RC_2'' axes.

These examples also show that it is sometimes convenient to exploit lower subgroup symmetry instead of the full symmetry to predict the existence of singularities in the current density field.

Another interesting feature occurs in the case of C_{4v} symmetry in the presence of magnetic field $\mathbf{B}$. This group contains three subgroups of index 2, i.e., C_4 and two C_{2v} groups having different setting with respect to a set of fixed axes, although crystallographically indistinguishable. Using the TZ algorithm, one can build up either $C_{4v}(C_{2v})$ or $C_{4v}(C_4)$ magnetic groups, see Table 6. The first case corresponds to $\mathbf{B} \parallel C_4(z)$. The second one might occur in a C_{4v} molecule, say AB_4, where intrinsic magnetic dipoles $\boldsymbol{\mu}_{B_i} \parallel C_4(z)$, with $i=1, 2, 3$, and 4, are placed on B_i nuclei, in such a way that the couple of nuclear moments lying on $R\sigma_{va}$ has opposite direction with respect to those lying on $R\sigma_{vb}$. Operating on $x_\alpha x_\beta$ with the corresponding T_m projectors (3.10), one finds

$$T_s^{C_{4v}(C_{2v})} S_{xy} = T_s^{C_{4v}(C_{2v})} S_{zz} = 0, \quad T_s^{C_{4v}(C_{2v})} S_{xx} = \frac{1}{2}(S_{xx} - S_{yy}),$$

$$T_s^{C_{4v}(C_4)} S_{xx} = T_s^{C_{4v}(C_4)} S_{yy} = T_s^{C_{4v}(C_4)} S_{zz} = 0, \quad T_s^{C_{4v}(C_4)} \nabla_x J_y = \frac{1}{2}(\nabla_x J_y - \nabla_y J_x).$$

Accordingly, the C_4 axis is a saddle stagnation line for $C_{4v}(C_{2v})$ and a vortex stagnation line for $C_{4v}(C_4)$.

From the series of examples discussed previously one can deduce the simple rules:

(i) *if the operator R_z for the rotation about the z axis transforms according to the irreducible representation Γ_m of a unitary group G, then a vortex will take place about the $C_n(z)$ axis of the magnetic group M isomorphic with G.*

(ii) *if the $x^2 - y^2$ or xy functions transform according to Γ_m, then saddle circulation will occur about the $C_n(z)$ axis.*

M groups as $D_n(C_n)$, $C_{nv}(C_n)$, $D_{nh}(C_{nh})$, $D_{nd}(S_{2n})$, for $n \geq 3$, provide examples of the first case, groups as $D_4(D_2)$, $C_{4v}(C_{2v})$, $C_{4h}(C_{2h})$, $D_{4h}(D_{2h})$ of the latter. Groups as $D_{2h}(C_{2h})$ are characterized by the presence of *both* R_z and xy in the same irreducible representation and can accordingly show either saddle or vortex circulation about the $C_2(z)$ axis.

Groups as C_{nh}, where R_z belongs to the totally symmetric representation, are magnetic groups describing the symmetry of electron circulation in molecules characterized by vortical circulation about the symmetry axis $C_n \parallel \mathbf{B}$.

Table 1. $C_{2v}(C_2) \equiv 2\underline{mm}$ and $C_{2v}(C_s) \equiv \underline{2}mm$

$C_{2v}(C_2)$		E	$C_2(z)$	$R\sigma_v$	$R\sigma'_v$		
		E	$RC_2(z)$	$R\sigma_v$	σ'_v	$C_{2v}(C_s)$	
$\Gamma_1 \equiv \Gamma_s \equiv \Gamma_{m\times p}$	$z, A_z, x^2, y^2, z^2, S_{xy}$	1	1	1	1	$\Gamma_1 \equiv \Gamma_s$	J_y, A_x, S_{yz}
$\Gamma_2 \equiv \Gamma_m \equiv \Gamma_p$	R_z, J_z, xy	1	1	-1	-1	$\Gamma_2 \equiv \Gamma_p$	J_x
Γ_3	x, R_y, J_y	1	-1	1	-1	$\Gamma_3 \equiv \Gamma_{m\times p}$	
Γ_4	y, R_x, J_x, yz	1	-1	-1	1	$\Gamma_4 \equiv \Gamma_m$	J_z

Table 2. $D_{2h}(C_{2h}) \equiv \underline{mmm}$

	E	$C_2(z)$	$RC_2(x)$	$RC_2(y)$	i	σ_{xy}	$R\sigma_{xz}$	$R\sigma_{yz}$	
$\Gamma_1 \equiv \Gamma_s$	1	1	1	1	1	1	1	1	$A_z, x^2, y^2, z^2, S_{xy}$
$\Gamma_2 \equiv \Gamma_p$	1	1	1	1	-1	-1	-1	-1	J_z
$\Gamma_3 \equiv \Gamma_m$	1	1	-1	-1	1	1	-1	-1	R_z, xy
$\Gamma_4 \equiv \Gamma_{m\times p}$	1	1	-1	-1	-1	-1	1	1	z
Γ_5	1	-1	-1	1	1	-1	1	-1	R_y, xz
Γ_6	1	-1	-1	1	-1	1	-1	1	y, J_x
Γ_7	1	-1	1	-1	1	-1	-1	1	R_x, yz
Γ_8	1	-1	1	-1	-1	1	1	-1	x, J_y

Table 3. $C_{3v}(C_3) \equiv 3\underline{m}$

	E	$2C_3(z)$	$3R\sigma_v$	
$\Gamma_1 \equiv \Gamma_s \equiv \Gamma_{m\times p}$	1	1	1	$z, \left(\nabla \times \mathbf{J}\right)_z$
$\Gamma_2 \equiv \Gamma_m \equiv \Gamma_p$	1	1	-1	R_z, J_z
Γ_3	2	-1	0	$(x,y), (R_x, R_y), (J_x, J_y)$

Table 4. $D_{3h}(C_{3h}) \equiv \bar{6}m2$

	E	$2C_3$	$3RC_2$	σ_h	$2S_3$	$3R\sigma_v$	
$\Gamma_1 \equiv \Gamma_s$	1	1	1	1	1	1	$(\nabla \times \mathbf{J})_z$
$\Gamma_2 \equiv \Gamma_p$	1	1	1	-1	-1	-1	J_z
$\Gamma_3 \equiv \Gamma_m$	1	1	-1	1	1	-1	R_z
$\Gamma_4 \equiv \Gamma_{m\times p}$	1	1	-1	-1	-1	1	z
Γ_5	2	-1	0	2	-1	0	$(x,y),\ (J_x,J_y)$
Γ_6	2	-1	0	-2	1	0	(R_x,R_y)

Table 5. $D_{6h}(C_{6h}) \equiv 6/mmm$

	E	$2C_6$	$2C_3$	C_2	$3RC_2'$	$3RC_2''$	i	$2S_3$	$2S_6$	σ_h	$3R\sigma_d$	$3R\sigma_v$	
$\Gamma_1 \equiv \Gamma_s$	1	1	1	1	1	1	1	1	1	1	1	1	$(\nabla \times \mathbf{J})_z$
$\Gamma_2 \equiv \Gamma_m$	1	1	1	1	-1	-1	1	1	1	1	-1	-1	R_z
Γ_3	1	-1	1	-1	1	-1	1	-1	1	-1	1	-1	
Γ_4	1	-1	1	-1	-1	1	1	-1	1	-1	-1	1	
Γ_5	2	1	-1	-2	0	0	2	1	-1	-2	0	0	(R_x,R_y)
Γ_6	2	-1	-1	2	0	0	2	-1	-1	2	0	0	
$\Gamma_7 \equiv \Gamma_p$	1	1	1	1	1	1	-1	-1	-1	-1	-1	-1	J_z
$\Gamma_8 \equiv \Gamma_{m\times p}$	1	1	1	1	-1	-1	-1	-1	-1	-1	1	1	z
Γ_9	1	-1	1	-1	1	-1	-1	1	-1	1	-1	1	
Γ_{10}	1	-1	1	-1	-1	1	-1	1	-1	1	1	-1	
Γ_{11}	2	1	-1	-2	0	0	-2	-1	1	2	0	0	$(x,y),\ (J_x,J_y)$
Γ_{12}	2	-1	-1	2	0	0	-2	1	1	-2	0	0	

Table 6. $C_{4v}(C_{2v}) \equiv \underline{4}m\underline{m}$ and $C_{2v}(C_s) \equiv 4\underline{mm}$

	$C_{4v}(C_{2v})$	E	$2RC_4(z)$	$C_2(z)$	$2\sigma_v$	$2R\sigma_d$		
		E	$2C_4(z)$	$C_2(z)$	$2R\sigma_v$	$R\sigma_d$	$C_{4v}(C_4)$	
$\Gamma_1 \equiv \Gamma_s$	$z,\ x^2+y^2,\ z^2,\ S_{xx}-S_{yy}$	1	1	1	1	1	$\Gamma_1 \equiv \Gamma_s \equiv \Gamma_{m\times p}$	$(\nabla \times \mathbf{J})_z$
$\Gamma_2 \equiv \Gamma_p$	R_z	1	1	1	-1	-1	$\Gamma_2 \equiv \Gamma_m \equiv \Gamma_p$	J_z
$\Gamma_3 \equiv \Gamma_m$	$J_z,\ x^2-y^2$	1	-1	1	1	-1	Γ_3	
$\Gamma_4 \equiv \Gamma_{m\times p}$	xy	1	-1	1	-1	1	Γ_4	
Γ_5	$(x,y),\ (R_x,R_y),\ (J_x,J_y),\ (xz,\ yz)$	2	0	-2	0	0	Γ_5	(J_x,J_y)

4. Planar cyclic molecules: benzene and borazine

Coupled Hartree-Fock perturbation theory [5] has been used to calculate the current density field in benzene and borazine. A large basis set of 396 contracted Gaussian functions, which furnishes near Hartree-Fock theoretical magnetic properties [27] has been adopted for benzene. A *polarized* basis set from Sadlej's tables [28], containing 210 contracted Gaussians, have been used for borazine. The differential equations (2.2) for the trajectories were solved using the Runge-Kutta quadrature. This method was found more accurate than Euler's, previously employed in similar maps [5,29]. The requirements of $D_{6h}(C_{6h})$ and $D_{3h}(C_{3h})$ magnetic symmetries examined in the previous section are fulfilled.

The behavior of streamlines on a plane parallel to the molecular plane of benzene, lying 0.52 bohr above, is shown in Fig. 7a. The electron circulation taking place in the outer regions of the molecular domain is diamagnetic and atom-like. These delocalized circulations are sustained by both σ- and π-electrons, as separate plots, see also Ref. 29, clearly show. As it is evident from the corresponding map Fig. 7b, the intensity of the J field rapidly decreases increasing the distance from the molecular center of charge. Very high intensities only occur in the environment of carbon nuclei and, to a lesser extent, of protons.

Getting closer to the nuclear skeleton, the primary diamagnetic whirlpool branches out in seven vortices: six diamagnetic vortices, separated by saddle stagnation lines, circulate above C=C bonds, and a central vortex rotates about the six-fold axis. It is paramagnetic, as expected on the basis of continuity requirements for the J field.

The transition point lies at 2.461 bohr from the center of charge, much higher than the region of maximum π-electron density, approximately 0.75 bohr over the molecular plane. It has been detected by careful analyis of the Jacobian for the points on the C_6 axis, i.e., three zero eigenvalues are found for that point.

Toroidal vortices occur in the vicinity of carbon atoms, see also Ref. 29. They are fully encased in separatrices which are topological spheres [18].

Quite different features are observed for borazine molecule, see Figs. 8a and 8b. In the tail regions the circulation is delocalized all over the molecular system. The transition point lies at 8.682 bohr, much higher than in benzene (this estimate should however be confirmed via larger basis set calculations). Four vortices originate there: three are diamagnetic and circulate over the nitrogen nuclei, the fourth one, rotating about the three-fold axis, is paramagnetic as in benzene.

The presence of localized axial vortices about nitrogen nuclei and the absence of bond vortices like those occurring in benzene could be related to polarity of the B-N bonds, due to the fact that nitrogen is more electronegative than boron. This seems to be confirmed by the trend of intensities, see Fig. 8b, showing much higher values over nitrogen nuclei. Toroidal vortices are found in the region of a B-H bond: they are observed in the map as juxtaposed paramagnetic (on B) and diamagnetic (on H) lines of current, where the torus is cut by the σ_h molecular plane. $(3, \pm 1)$ singularities are placed respectively at the North and South poles of each toroidal vortex. They look like isolated saddle points on the plane of the plot, but they would appear as nodes on planes through the singular point perpendicular to the molecular plane. Three $(2,0)$ stagnation lines occur in the regions between the axial vortices over nitrogen nuclei and the central paramagnetic whirlpool, in accordance with the Gomes theorem (2.7).

Fig 7a: Trajectories of the current density vector field at 0.52 bohr above the molecular plane of benzene.

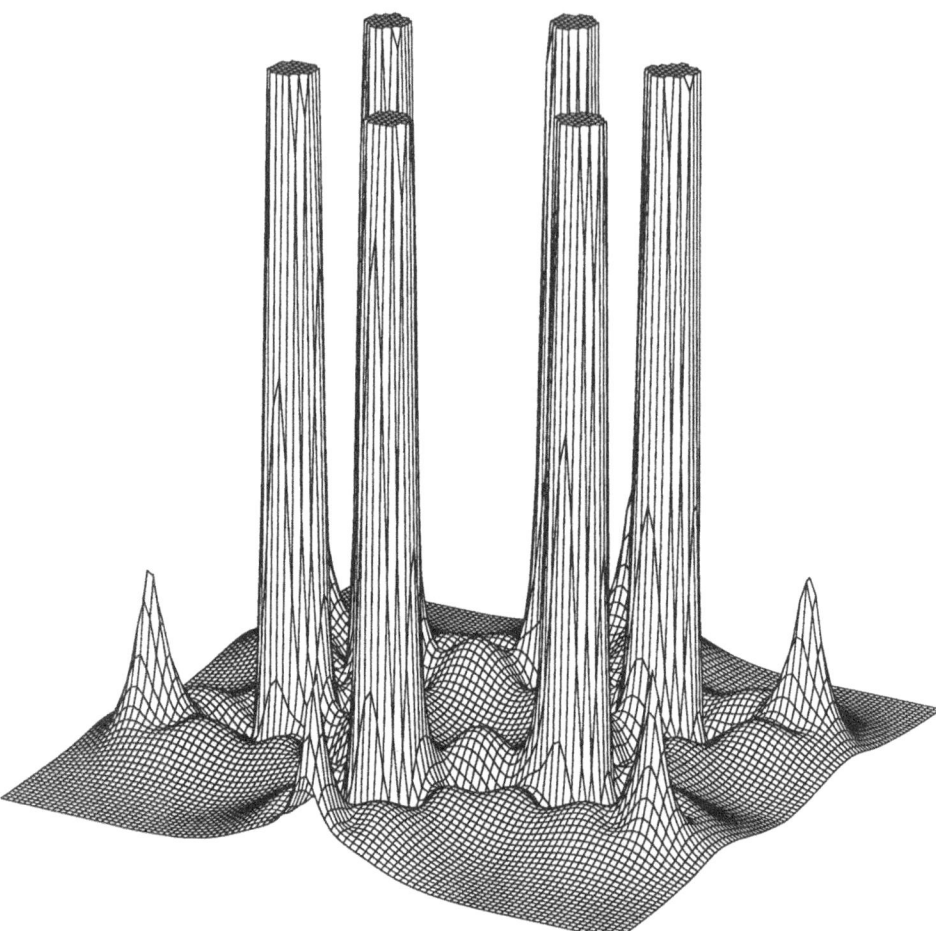

Fig. 7b: 3-D representation of the modulus of the current density on the molecular plane of benzene. Peaks of intensity higher than 7.3×10^{-3} a.u. are cut out.

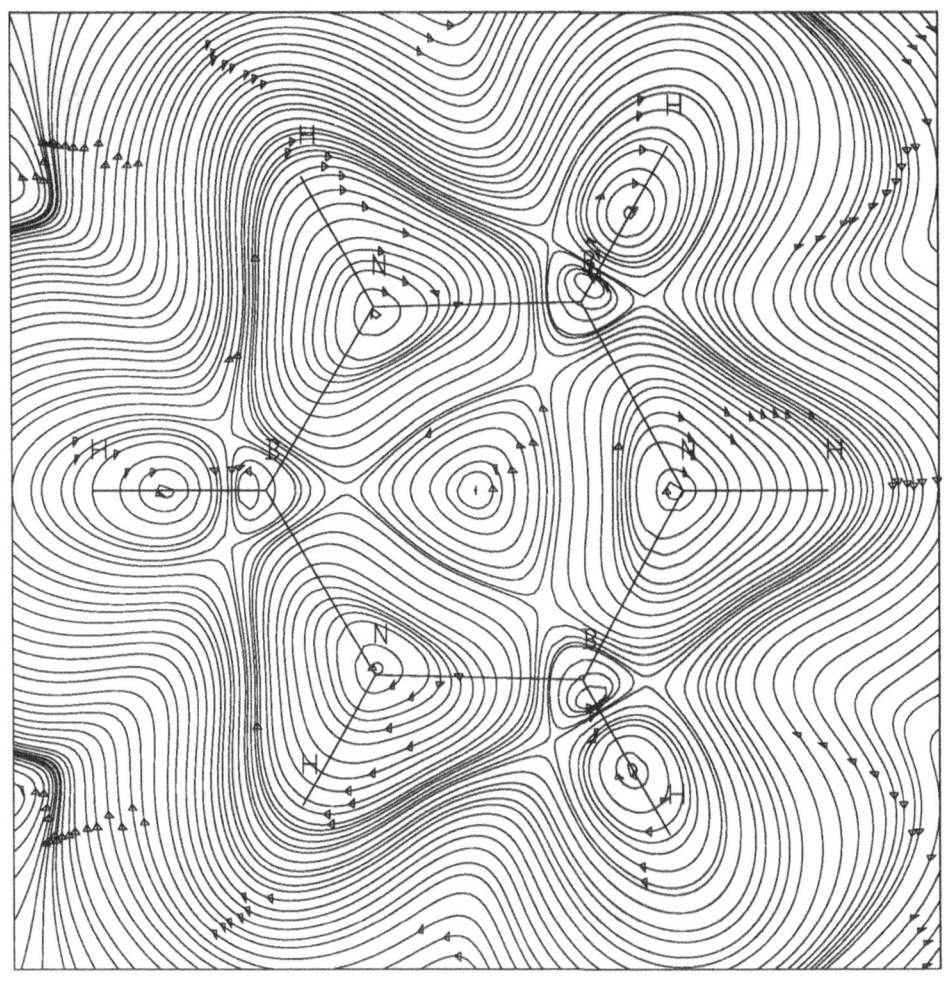

Fig 8a: Trajectories of the current density vector field at 0.7 bohr above the molecular plane of borazine.

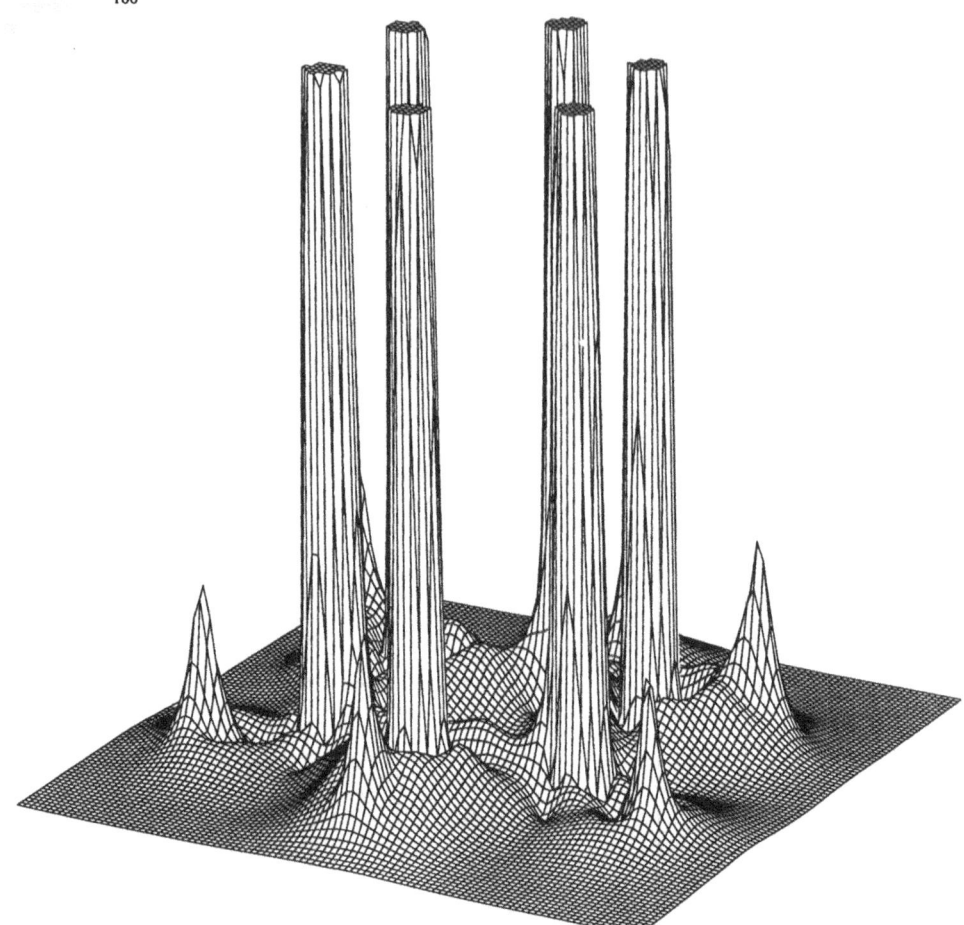

Fig. 8b: 3-D representation of the modulus of the current density on the molecular plane of borazine. Peaks of intensity higher than 1.0×10^{-2} a.u. are cut out.

References

[1] Smirnov, V. (1970) *Cours de Mathematiques Superieurs*, Tome II, Mir, Moscow

[2] Coddington, E. A. and Levinson, N. (1955) *Theory of Ordinary Differential Equations*, McGraw-Hill, New York

[3] Collard, K. and Hall, G. G. (1977) *Int. J. Quantum. Chem.* **12**, 623

[4] Bader, R. F. W and Nguyen-Dang, T. T. (1981) *Adv. Quantum. Chem.* **14**, 63, and references therein

[5] Lazzeretti, P. Rossi, E., and Zanasi, R. (1984) *Int. J. Quantum Chem.* **25**, 929; (1984) *ibid.* **25**, 1123

[6] Riess, J. (1970) *Ann. Phys.* **57**, 301

[7] Riess, J. (1971) *Ann. Phys.* **67**, 347

[8] Riess, J. (1970) *Phys. Rev. D* **2**, 647

[9] Riess, J. (1976) *Phys. Rev. B* **13**, 3862

[10] Riess, J. and Primas, H. (1968) *Chem. Phys. Letters* **1**, 545

[11] Hirschfelder, J. O., Cristoph, A. C., and Palke, W. E. (1974) *J. Chem. Phys.* **61**, 5435

[12] Hirschfelder, J. O., Goebel, C. J., and Bruch, L. W. (1974) *J. Chem. Phys.* **61**, 5456

[13] Hirschfelder, J. O. and Tang, K. H. (1976) *J. Chem. Phys.* **64**, 760

[14] Hirschfelder, J. O. and Tang, K. H. (1976) *J. Chem. Phys.* **65**, 470

[15] Heller, D. F. and Hirschfelder, J. O. (1977) *J. Chem. Phys.* **66**, 1929

[16] Hirschfelder, J. O. (1977) *J. Chem. Phys.* **67**, 5477

[17] Hirschfelder, J. O. (1978) *J. Chem. Phys.* **68**, 5151

[18] Gomes, J. A. N. F. (1983) *Phys. Rev.* **A28**, 559; (1983) *J. Chem. Phys.* **78**, 4585

[19] Hamermesh, M. (1972) *Group Theory and its Applications to Physical Problems*, Addison-Wesley, London

[20] Shubnikov, A. N. (1951) *Symmetry and Antisymmetry of Finite Figures*, U.S.S.R. Academy of Sciences, Moscow

[21] Shubnikov, A. N. and Belov, N. V. (1964) *Coloured Symmetry*, Pergamon Press Ltd., London

[22] Wigner, E. P. (1971) *Group Theory and its Application to the Quantum Mechanics of Atomic Spectra*, Academic, New York

[23] Messiah, A. (1973) *Quantum Mechanics*, North-Holland, Amsterdam,

[24] Tavger, B. A. and Zaitsev, V. M. (1956) *Sov. Phys. JETP* **3**, 430

[25] Bradley, C. J. and Davies, B. L. (1968) *Rev. Mod. Phys.* **40**, 359

[26] Birss, R. R. (1963) *Rep. Progr. Phys.* **26**, 307; (1964) *Symmetry and Magnetism*, North Holland Publ. Co., Amsterdam

[27] Lazzeretti, P., Malagoli. M., and Zanasi, R. (1990) *J. Molec. Struct. THEOCHEM* **234**, 127

[28] Sadlej, A. J. (1988) *Collect. Czech. Chem. Commun.* **53**, 1995

[29] Lazzeretti, P., Rossi, E., and Zanasi, R. (1982) *J. Chem. Phys.* **77**, 3129

SHIELDING TENSOR DATA AND STRUCTURE: THE BOND-RELATED CHEMICAL SHIFT CONCEPT

A.-R.GRIMMER
Projektgruppe Festkörper-NMR , WIP-KAI e.V.
Rudower Chaussee 5, Haus 4.1
O-1199 Berlin-Adlershof, Germany

ABSTRACT: Using symmetry relations it is possible to measure "bond-related chemical shifts" along an individual bond direction as a new property to characterize chemical bonds.

1.Introduction

During the last decade high-resolution solid-state NMR became an essentially experimental method in the field of solid state chemistry. In most applications of solid-state NMR isotropic chemical shift (chs) data have been used as analytical fingerprint to characterize solid materials or to follow pathways of synthesis from precursors to final products. Inasmuch as the information is extracted from chs data the situation is comparable to the role of the classical liquid-state NMR in chemistry in general. But exceeding liquid-state NMR the solid-state NMR is capable of "portraying" the chemical bonding around the resonance nucleus more in detail. The basic phenomenon for this sort of investigation is the chs anisotropy measured as magnetic shielding tensor (mst).

It is the aim of the present paper to demonstrate the use of mst data for detailed insights in the structure of solid silicates. The fundamental techniques for structure analysis of solids are diffraction methods, but situated at the high-energy end of the electromagnetic spectrum diffraction methods have different physical background. Fig.1 illustrates the situation: both methods look at the investigated solid from very different point of views in a literal sense.

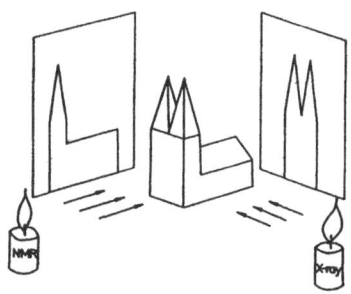

Fig.1 Schematical representation of the different features of the same object given by different experiments

J. A. Tossell (ed.), Nuclear Magnetic Shieldings and Molecular Structure, 191–201.
© 1993 *Kluwer Academic Publishers.*

2. Methodical background

2.1 CHEMICAL SHIFT TENSOR OF POLYCRYSTALLINE SAMPLES: LOSS OF INFORMATION

To a good approximation the anisotropy of chs can be described by a symmetric mst σ with six independent values σ_{rs} with $r,s = 1,2,3$ and $\sigma_{rs} = \sigma_{sr}$ or alternatively three principal values σ_{11}, σ_{22} and σ_{33} plus three Eulerian angles ϕ, φ and ψ. At least in principle all these six data are measurable by investigating single crystals. Due to the well-known general problems of NMR like low natural abundance and/or long T1 times such measurements of small-sized crystals are extremely time consuming. A second problem is to get suitable single crystals of sufficient quality and usually the NMR spectroscopist restricts themselves to the more general case of polycrystalline samples.

The study of polycrystalline materials, like powder samples, has an essential consequence: due to the statistic distribution of the orientations of the crystallites with respect to the magnetic field B_0 the information on the Eulerian angles is lost, and we measure only the mentioned three principal values of the mst σ.

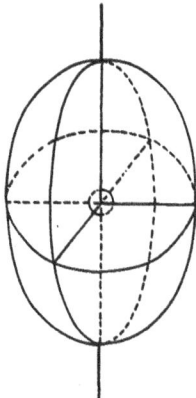

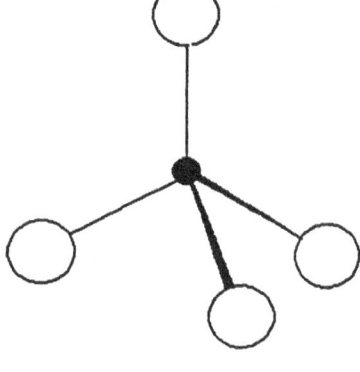

Fig.2a Ellipsoid representation of the mst given in inverse shieldings

Fig.2b Tetrahedral representation of the SiO_4 unit used as axis system

Fig.2a gives a pictorial representation of the mst, the three principal values correspond to the three axes of the ellipsoid. Even in this restricted situation, the three principal values contain enough useful qualitative information to be worth extracting.

Tab.1 shows the more or less qualitative correlation between the mst data and the connectivity of SiO4 tetrahedra given by the index n (Qn notation: Q0 - isolated unit, Q1 - one Si-O-Si bridging bond per unit etc.)

Tab.1: Connectivity of SiO_4 units and mst data given as isotropic shielding σ, axiality $\Delta\sigma$ and asymmetry parameter η

	Q0	Q1	Q2	Q3	Q4
σ/ppm	≈65	<	increasing	<	≈130
$\Delta\sigma$/ppm	≈0	<0	>0	>0	≈0
η	medium	large	very large	large	small

An example for the semi-quantitative application of mst data gives the investigation of Belinite $4Ca_2SiO_4 \cdot MgCl_2$ [1]. 29Si MAS NMR spectra shows an axially symmetric 29Si mst (σ = 75.6 ppm and $\Delta\sigma$ = - 40.6 ppm) typically for an unit (Q"0.5") intermediate between Q0 and Q1. This leads to a structure with fourfold coordinated magnesium surrounded by four SiO_4 units with weak Mg···O-Si bonds.

2.2 "REVIVAL" OF INFORMATION: LOCAL SYMMETRY AND TENSOR ORIENTATION

To make some steps in the direction of a more quantitative use of the tensor data we have to reconstruct or to "revive" the information about the orientation of the tensor (Fig.2a) with respect to a given axis system. For the sake of simplicity in the silicates we prefer the system of the four Si-O bonds of the given SiO_4 tetrahedron (Fig.2b).

Fig. 3a shows 29Si NMR spectra (static and MAS) of TetraCalciumSilicate Hydrate $Ca_6(Si_2O_7)(OH)_6$ (TCSH), a Q1 silicate.

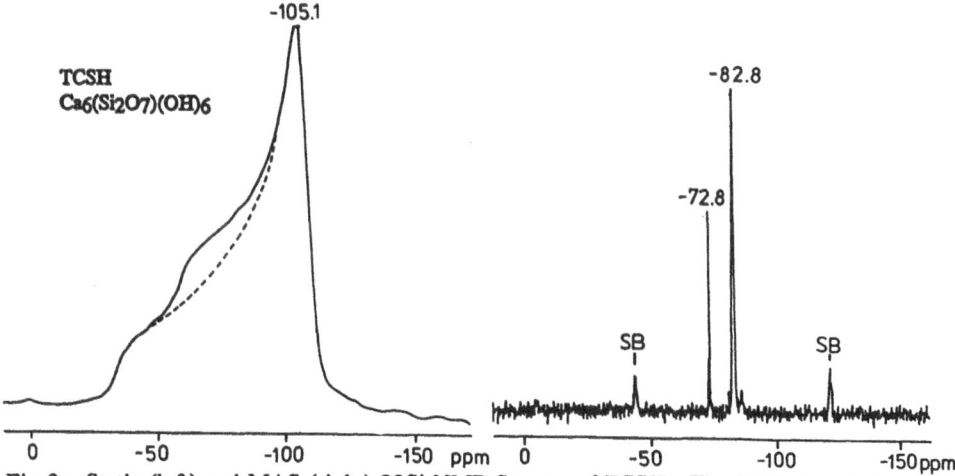

Fig.3a Static (left) and MAS (right) 29Si NMR Spectra of TCSH . The MAS signal at -72.8 ppm is due to an impurity of Q0 silicate visible in the staic spectrum as central deviation of the observed lineshape from the typically axial lineshape.

Obviously the tensor is axially symmetric with an negative axiality , schematically represented in Fig 3b (left). The x-ray results are shown in Fig. 3b (right), they are typically for Q1 units. Within the limits of error the local symmetry around the Si position is C_{3v} and agrees with the NMR results.

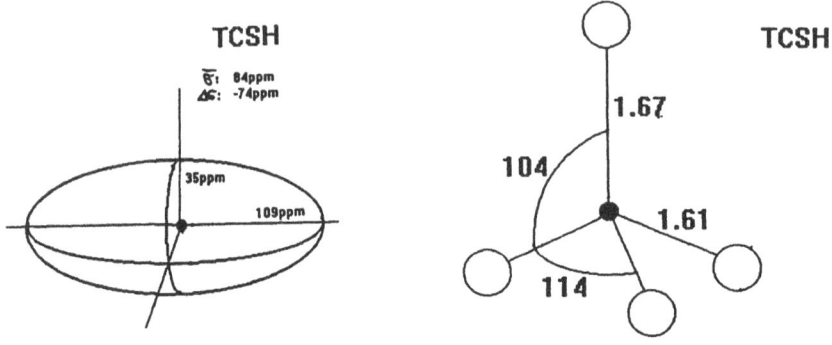

Fig.3b mst (left) and SiO_4 tetrahedron of TCSH (right, distances/A and angles/°)

The inverse situation is encountered in Fig.4a showing the static 29Si NMR spectrum of the silicate TetraMethylAmmoniumSilicate $(CH_3)_4N_8Si_8O_{20}.69H_2O$ (TMAS), a pure Q3 silicate and the precursor of the well-known secondary standard for 29Si MAS NMR $(CH_3)_3Si_8 Si_8O_{20}$ (Q8M8).

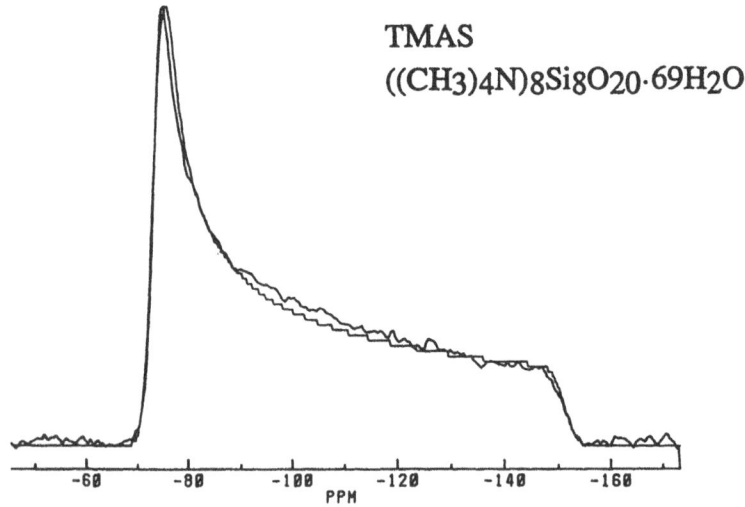

Fig.4a 29Si NMR Spectrum of TMAS (waving line: experiment, terraced line: simulation)

Again the chemical shift tensor is axially symmetric, but in this case with a positive axiality (Tab.2), schematically represented in Fig 4b (left), the corresponding x-ray data are shown in Fig. 4b (right).

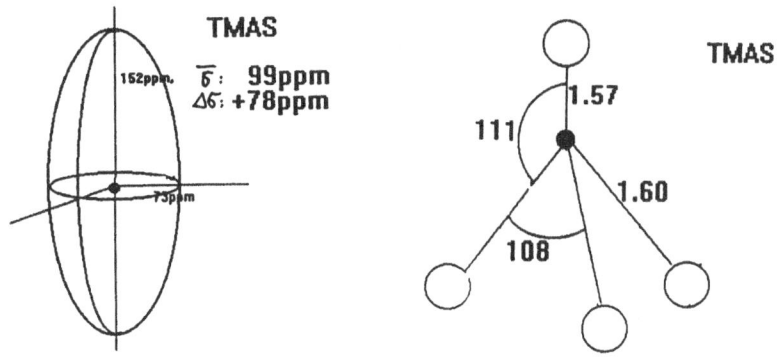

Fig.4b mst (left) and SiO_4 tetrahedron of TMAS (right, distances/Å and angles/°)

The local symmetry around the Si-atom is C_{3v} and agrees again with the NMR result. Due to the axial symmetry of the chemical shift tensor as well as of the SiO_4 tetrahedron the direction of the bond must be coincident with the symmetry axis of the tetrahedron.Now we know the orientation of the chemical shift tensor with respect to the binding system. Fig. 5 shows as an example the situation for the Q3 units in TMAS.

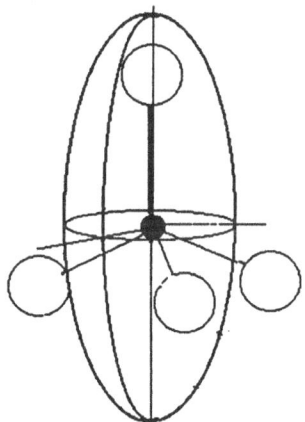

Fig. 5 Orientation of the mst with respect to the Si-O bond system

Applying the mentioned procedure to cases of axial symmetry it is possible to measure the magnetic shielding σ(Si-O) along a defined individual Si-O bonding. In the following we call this special parameter "*bond-related chemical shift*".

It should be mentioned in the degenerated case of spheric (isotropic) chemical shift tensors corresponding to regular SiO_4 tetrahedra that all three (equivalent) shielding constants obviously represent the bond related chemical shift. Many Q0 and Q4 units fulfill this condition to a good approximation (Tab. 1).

In the cases discussed we used x-ray data as a source for the additional needed information. It is also possible to extract additional information from thr NMR data provided that there is an independent dipolar interaction such as found in PO_3F units.

3. Results

3.1. BOND RELATED CHEMICAL SHIFTS *VERSUS* CORRESPONDING BOND LENGTHS

To test the validity of the bond-related shift concept possible correlations have been studied between the new property $\sigma(Si\text{-}O)$ and classical properties attributed to chemical bonds like bond lengths or force constants. Tab. 2 compares $\sigma(Si\text{-}O)$ with the corresponding bond lengths $d(Si\text{-}O)$ of silicates with fourfold coordinated silicon.

Tab.2 mst data and Si-O bond lengths of silicates [2]

	Qn	$d_{\parallel}(Si\text{-}O)/A$	$\sigma_{\parallel}(Si\text{-}O)/ppm$	$d_{\perp}(Si\text{-}O)/A$	$\sigma_{\perp}(Si\text{-}O)/ppm$
Ca_3SiO_5	Q0	1.635	73 [a]	1.635	73 [a]
$Ca_2Si_2O_7$ [b]	Q1	1.667	30.3	1.605	97.8
$BaSi_2O_5$ [c]	Q3	1.581	158.9	1.631	63.6
TMAS [d]	Q3	1.569	152	1.608	73
Q8M8 [d]	Q3	1.588	131.3	1.600	98.5
SiO2 t-quartz	Q4	1.609	107.4	1.609	107.4

[a] average of nine signals, [b] rankinite, [c] sanbornite, [d] see text

It is evident that higher shielding corresponds to shorter Si-O bond length and *vice versa*. Fig.6 demonstrates the linear correlation between shielding and distance data, and the corresponding regression equations are

$$\sigma_{\parallel}(Si\text{-}O)/ppm = 2070 - 1.222 * 10^3\ d(Si\text{-}O)\ /\text{Å} \qquad (1a)$$
$$d(Si\text{-}O)/\text{Å} = 1.694 - 8.18 * 10^{-4}\ \sigma_{\parallel}/ppm \qquad (1b)$$

with R = 0.999 and N = 6.

In contrast to the parallel tensor component $\sigma_{\parallel}$ the second component $\sigma_{\perp}$ does not coincide with any Si-O bond direction, because the *three* orthogonal axis of the principal axis system can not fit simultaneously the *four* Si-O bond directions of the SiO_4 tetrahedron. For SiO_4 tetrahedra typically O-Si-O angles are between 101° and 117°, therefore the deviation of the three Si-O bond directions equals to (19°±8°) (Fig.7).

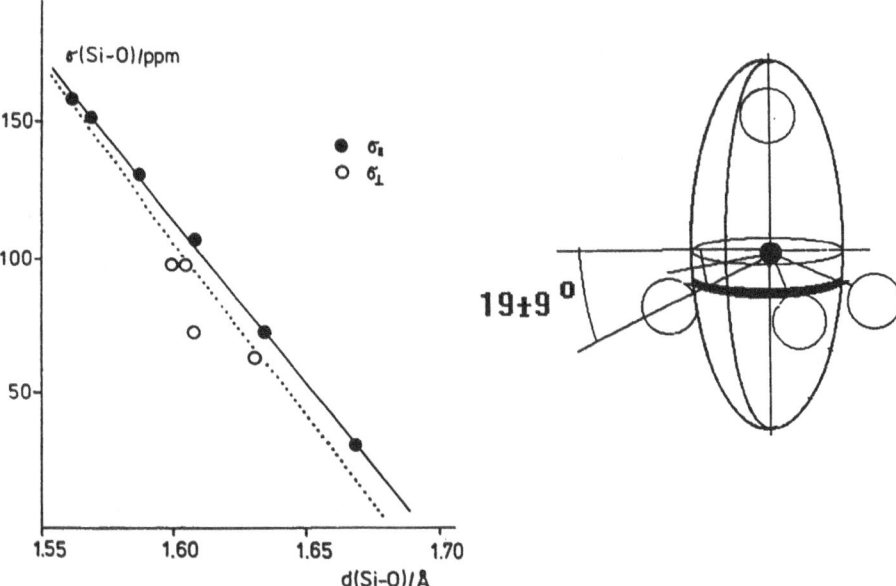

Fig.6 σ(Si-O) *versus* d(Si-O), eqs.(1)   Fig.7 Deviation of the three Si-O bond
straight line , eqs.(2) dotted line   from the σ plane

Neglecting this angular deviation the $\sigma_\perp$ values were added to the data set (Tab.2). As a consequence the correlation bond related chemical shift *versus* bond length changes to

$$\sigma_{total}(Si\text{-}O)/ppm = 2250 - 1.334 * 10^3 \, d(Si\text{-}O)/\text{Å} \qquad (2a)$$
$$d(Si\text{-}O)/\text{Å} = 1.687 - 7.50 * 10^{-4} \, \sigma_{total}(Si\text{-}O)/ppm \qquad (2b)$$

with $R = 0.963$ and $N = 10$.

The small numerically differences between eqs (1) and eqs (2) suggest that the observed
short bond length ↔ high shielding
is valid for **all** Si-O bond distances in SiO_4 units.

3.2. ISOTROPIC CHEMICAL SHIFT DATA *VERSUS* MEAN ("ISOTROPIC") BOND LENGTHS

The presented bond related chemical shift concept has been developed by the comparison of individual bond lengths with individual shielding constants given as principal values of the shielding tensor.

Due to the popularity of the MAS technique the majority of magnetic shielding data consists not of principal values, but of mean values of the three principal values in form of isotropic chemical shift . This gives rise to the question how to use these data. Obviously these data should be compared with the "isotropic" averaged Si-O bond length d of a given SiO_4 unit

$$d = \sum_{i}^{i=4} d(Si\text{-}O)_i.$$

It is well known that increasing the connectivity index n corresponds to a shorter mean bond length d. Two typical examples are shown in Tab.3 where the given isotropic shift data confirm the trend

short bond $\leftrightarrow$ high shielding.

Tab. 3   Comparison of isotropic shielding and distance data

	n	d/Å	σ/ppm
Mg_2SiO_4	0	1.635	63
Q-SiO_2 t-quartz	4	1.609	107.4

Encouraged by Tab.3 and other examples we correlated the known σ with d as follows. The stepwise procedure is due to the fact that the two populations of silicates with known structures on one hand and silicates with measured NMR data on the other hand are overlapping but not completely conincident.

Step 1: d *versus* n gives
$$d/Å = 1.638 - 7.5 * 10^4\, n$$
with R = 0.83 and N = 128
as an updated version of the Bailey-Smith correlation (1963) [3]
$$d/Å = 1.632 - 4.8 * 10^4\, n.$$

Step 2: *versus* n gives
$$\sigma/ppm = 69.8 + 8.9\, n$$
with R = 0.90 and N = 56.

The combination of step 1 and step 2 gives the following correlation :

$$\sigma /ppm = 2014 - 1.187 * 10^3\, d/Å \qquad (3a)$$
$$d/Å = 1.697 - 8.42 * 10^{-4}\, /ppm \qquad (3b)$$

Later investigation of Higgins and Woessner [4] confirmed this correlation, they found as a special case for Q4 units

$$\sigma/ppm = 2312 - 1.371 * 10^3\, d/Å \qquad (4a)$$
$$d/Å = 1,685 - 7.29 * 10^{-4}\, \sigma/ppm \qquad (4b)$$

The satisfactory agreement of eqs.(3) derived from the isotropic mean data set with eqs.(1) derived independently from the individual data set is a good argument for the general validity of the bond related chemical shift concept for silicates.

Using this concept and eqs.(1) calculated values for the bond distances Si1-O and Si2-O in Na [(CH$_3$)$_4$N]$_7$Si$_8$O$_{20}$.54H$_2$O have been obtained very recently by Wiebcke and Koller [5] "which are in excellent agreement with those determined crystallographically " (Tab.4).

Tab.4   mst data and Si-O bond lengths in Na [(CH$_3$)$_4$N]$_7$Si$_8$O$_{20}$.54H$_2$O (Wiebcke and Koller [5])

	σ (Si$_i$-O)/ppm	d (Si-O)/Å NMR	d (Si-O)/Å XRD
Si1	132(3)	1.586(3)	1.589(2)
Si2	130(3)	1.588(3)	1.588(2)

3.3 SILICATES WITH SIXFOLD COORDINATED SILICON

The structural units of silicates are usually SiO$_4$ tetrahedra with fourfold coordinated silicon, but there are also silicates with five- and sixfold coordinated silicon. The comparison of isotropic chemical shifts and structural data gives for sixfold coordinated silicates [6]

$$\sigma \text{ /ppm} = 4682 - 2536 \text{ d/Å} \qquad (5a)$$
$$\text{d/Å} = 1.846 - 3.9 * 10^{-4} \sigma \text{ /ppm,} \qquad (5b)$$

there are only a few data for the corresponding mst´s.

Taking into consideration the differences in the character of the Si-O bond, indicated by the large difference in the fictive Si-O bond lengths d$_0$ normalized for σ = 0 equals 1.685Å (eqs.3) for Si[4] and 1.846Å (eqs.5) for Si[6] resp. it is not expected *a priori* that even for sixfold coordinated silicon shorter bond length correspond to higher shielding. This empirical observation can be taken as an additional argument in favour of the discussed concept.

4. Discussion

4.1 SENSITIVITY OF NMR DATA TO STRUCTURAL FEATURES

The relationship between chemical shift and bond length give rise to consequences in the field of applied spectroscopy as well as of theoretical considerations [7]. They can be used for prediction of individual and mean bond length and *vice versa* of individual and mean chemical shifts, for refinement of diffraction data and for assignment of NMR signals of ambiguous origin.

Here the extreme sensitivity of chemical shift tensor data to structural features should be emphasized . The slope of eq.(1) equals

$$d \text{ d(Si-O)}/d \text{ (Si-O)} = -8.18 *10^{-4} \text{ Å/ppm} \approx -0.001 \text{ Å/ppm}$$

indicating that changes of the Si-O bond length in the order of magnitude of 0.001 Å corresponds to chemical shift differences of 1 ppm. Using a good spectrometer, well-crystallized silicates give 29Si MAS NMR signals with line-widths in the order of ≤0.1 ppm (≈10 Hz), suggesting the structural resolution with respect to bond lengths is 0.0001 Å (or 10 fm). This is considerably better than for diffraction methods. As a consequence, x-ray data measured before the appearance of four-circle diffractometers (≈1975) are not significantly accurate enough to be used by NMR spectroscopists.

4.2. EQUAL WEIGHTING OF STRUCTURAL PARAMETERS

The basic structural parameter used in the bond related chemical shift concept is the (individual) Si-O bond length. On the other hand the geometry of any SiO_4 unit is defined not only by one bond length but by at least 10 structural data, usually four Si-O-bond lengths plus six O-Si-O angles or alternatively six O-O distances.

Therefore it may not be justified to restrict the complete structural feature of the SiO_4 unit to bond lengths - why should we not also use bond angles to charcterize the geometry of the tetrahedron? The answer is as follows: indeed, all ten structural parameters may have equal importance, but they are not independent. If we change **one** bond length all the other nine parameters, bond angles and bond lengths, will change also.

Fig.9 gives a schematical representation of the effect. Starting with a regular tetrahedra (left) lengthening of one Si-O(B) bond corresponds to shortening of the three Si-O(T) bonds plus narrowing (<109°) of the three bond angles O(T)-Si-O(B) plus widening (>109°) of the three bond angles O(T)-Si-O(T) (Fig.9 right). These changes are changes of the first (silicon-oxygen) coordination sphere.

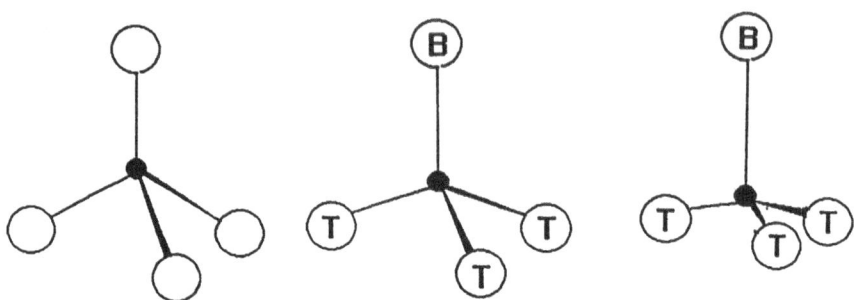

Fig. 9 Changes of shape of a SiO_4 tetrahedron as consequence of the lengthening of one silicon xygen (bridging) bond (B: bridging oxygen, T: terminal oxygen)

Furthermore: the second coordination sphere around silicon characterized by usually higher coordinated (≥4) cations and/or lower coordinated (≤4) atoms like Al or Si itself is significantly reflected by the first sphere. Therefore the discussion of the first silicon-oxygen coordination sphere includes *implicite* the discussion of second sphere effects.

This discussion has two consequences:

1. The intimate interaction of all ten parameters justifies focusing on one of them.

2. There is a wealth of correlations other than the present between chemical shift and bond length. Comparable correlations exist between σ and the mean Si-O-Si bond angles [8,9,10], it should be possible to correlate to a certain degree all structural parameters of silicates with shielding data.

The equal weighting of structural parameters offers also an explanation for the observation of Wolff and Radeglia [11] that in silicate units with C_{3v} symmetry the tensor components perpendicular to the symmetry axis play an important role.

Finally it should be emphasized that the bond related chemical shift concept is applicable to families of more or less closely related compounds. The present limits of a possible generalization are demonstrated by the fact that the "isotropic" Si-O bond lenghts in fourfold coordinated silicates are shorter than in sixfold coordinated ones, but the isotropic shielding for $Si^{[6]}$ is higher than for $Si^{[4]}$. As above mentioned only few [12] is known about mst data for five- or sixfold coordinated silicates, systematic investigations of the mst´s of $Si-O_6$ octa-hedra are needed to overcome these difficulties.

The bond related chemical shift concept does not claim to be more than an empirical tool given to NMR spectroscopists to help bridge over the gap between structural and spectroscopic data. Hiding manyfold structural information on the solid chemical shift anisotropy speaks loudly and clearly to the experimentalists - but till now our understanding of this new and rich language is limited. What we need is a dictionary NMR↔structure and perhaps the presented concept can be helpful to add new pages to this book.

5. Acknowledgements

This work was funded by the Fonds der Chemischen Industrie. I also acknowledge critical reading by the referee and helpful discussions during the NATO ARW 1992 College Park,

6.References

[1]   Grimmer, A.-R. and von Lampe, F., Chem. Comm. 1986 (1986) 219

[2]   Grimmer, A.-R., Fechner, E.and Peter,R., Chem.Phys.Letters 77(1981) 331
        Grimmer, A.-R. and Radeglia, R., Chem.Phys.Letters 106 (1984) 262

[3]   Smith, J.V. and Bailey, S.W., Acta Cryst.,16 (1963) 801

[4]   Higgins, D.B. and Woessner, E.E., EOS 63 (1982) 1139

[5]   Wiebcke, M. and Koller, H., IVth ECSC, Dresden, 7.-9.9.1992, abstracts p. 239

[6]   Grimmer, A.-R., von Lampe, F., and Mägi, M., Chem.Phys.Letters 132 (1986) 549

[7]   Tossell, J.A. and Lazzeretti, J.Chem.Phys. 84 (1986) 369

[8]   Smith, J.V. and Blackwell, C.S., Nature 303 (1983) 819

[9]   Thomas, J.M., Klinowski, J. Ramdas, S., Hunter, B.K. and Tennakoon, D.T.B.,
        Chem.Phys.Letters 102 (1983) 158

[10]   Engelhardt, G. and Radeglia, R., Chem.Phys.Letters 108 (1984) 271

[11]   Wolff. R. and Radeglia. R., (this volume)

[12]   Grimmer, A.-R., Wieker, W., von Lampe, F., Fechner, E., Peter, R. and
        Molgedey,G., Z.Chem. 20 (1980) 453

INVESTIGATION OF STRUCTURAL PHASE TRANSITIONS IN OXIDE MATERIALS USING HIGH-RESOLUTION NMR SPECTROSCOPY

Brian L. Phillips[1]
R. James Kirkpatrick
Yuehui Xiao
Department of Geology
University of Illinois
Urbana, IL 61801, U.S.A.

John G. Thompson
Research School of Chemistry
Australian National University
Canberra, A.C.T. 2601 Australia

ABSTRACT. The structural changes associated with reversible, non-quenchable structural phase transitions (SPT's) in crystals and the dynamical behavior of the crystals during these transitions can be effectively investigated using NMR methods. This paper presents a review of our recent work on SPT's of oxide materials which illustrate the information obtainable from, especially, the isotropic chemical shift obtained using magic-angle spinning (MAS) methods at elevated temperatures. The materials investigated are Sr_2SiO_4, the cristobalite polymorphs of SiO_2 and $AlPO_4$, and the Ca-feldspar anorthite ($CaAl_2Si_2O_8$). Depending on the material, these methods provide the temperature dependence of the isotropic chemical shift (which can be compared to theoretical and empirical relationships between chemical shifts and structure), transition temperatures, the fraction of each phase coexisting at a given temperature in the transition interval and therefore a quantitative picture of any hysteresis, a measure of the order parameter for analysis using Landau theory, and a quantitative measure of the soliton density for incommensurate phases. In addition, chemical shift measurements are readily combined with more traditional measurements of quadrupole coupling constants and relaxation rates for both $I=1/2$ and quadrupolar nuclides.

1. Present address: Earth Sciences Department, L-219, Lawrence Livermore National Laboratory, P.O. Box 808, Livermore, CA 94550 USA

J. A. Tossell (ed.), Nuclear Magnetic Shieldings and Molecular Structure, 203–220.

1. INTRODUCTION

Many crystalline materials undergo reversible structural phase transitions (SPT's) with varying temperature. Often the stable, high-temperature phases cannot be quenched to room temperature. Thus, the structure and dynamical behavior of the high temperature phases and the nature of the SPT must be studied in situ at elevated temperatures. Over the past approximately 20 years, NMR has been used effectively to study SPT's and has contributed greatly to our understanding of them (see Blinc, 1981; Rigamonti, 1984; Blinc et al., 1986, for reviews). Most of this work has involved measurement of quadrupole coupling constants, spin-lattice (T_1) relaxation rates, and quadrupole shifts for special orientations of single crystals.

We present here a review of our recent high-resolution NMR work on SPT's in oxide systems which emphasizes measurement of the isotropic chemical shift. The magic-angle spinning (MAS) methods used provide significantly enhanced spectral resolution compared to single-crystal techniques for nuclides with spin I = 1/2 (e.g., ^{29}Si, ^{31}P) and sometimes also for such quadrupolar nuclides as ^{23}Na and ^{27}Al. Thus, the isotropic chemical shift can be used as an effective probe of the structural changes at SPT's. Such experiments have the advantage that they require only powdered samples and thus are not limited to materials which can be obtained as large single crystals. They are also readily combined with more traditional T_1 and QCC measurements, which we also illustrate.

Depending on the structures of the phases involved and the nature of the SPT, the following types of information can be obtained from high-resolution NMR measurements. 1) The temperature dependence of the isotropic chemical shifts and their change across the transition. These values can then be interpreted in terms of empirical and theoretical models of the relationship between the chemical shift and structure to yield structural information. 2) Transition temperatures and for SPT's with significant first-order character the fraction of each phase coexisting in the transition interval. 3) Any thermal hysteresis associated with the transition. 4) The spatial variation of the modulation wave and the soliton density in incommensurate phases. Such phases often occur stablely between high-temperature, disordered phases and low-temperature, ordered phases. 5) The temperature dependence of the order parameter for analysis of the transition in terms of Landau theory. 6) Information concerning the dynamical behavior of, especially, the high-temperature phases through line-shape analysis.

We illustrate these points with our recent results for the SPT's of Sr_2SiO_4, the cristobalite polymorphs of SiO_2 and $AlPO_4$, and the Ca-feldspar anorthite ($CaAl_2Si_2O_8$). Similar studies have investigated highly siliceous zeolites (Fyfe et al., 1985, Hay et al., 1985, Strobl, et al., 1987), Na site exchange in nepheline (Stebbins et al., 1989), and SiO_2-cristobalite (Spearing et al., in press).

2. Sr_2SiO_4

Sr-orthosilicate undergoes a series of SPT's from an ordered, low temperature, ferroelastic phase (β, space group $P2_1/n$)) through an incommensurately modulated phase (α'_L) at about 90°C to a high temperature, paraelastic phase (α'_H, space group Pmnb) at about 500°C (Withers et al., 1987). The crystal structure of Sr_2SiO_4 has the topology of the β-K_2SO_4 structure. It is isomorphous with some other alkaline earth orthosilicates (e.g., Ca-orthosilicate) and the well-studied class of improper ferroelectrics which includes Rb_2ZnBr_4, Rb_2ZnCl_4 and K_2SeO_4, which undergo a similar set of transitions (Hyde et al., 1986; Blinc, 1986).

The incommensurate α'_L phase has been observed in electron diffraction experiments by the appearance of two sets of satellite Bragg reflections with wave vectors $q_1=0.303b^*$ and $q_2=0.396b^*$, where q_2 is a second harmonic of q_1 (Withers et al., 1987). These wave vectors do not change with temperature through the α'_L-phase.

We have examined the β to incommensurate transition using ^{29}Si MAS NMR and a 95 mg sample enriched to 95% ^{29}Si (Phillips et al., 1991). The results illustrate the utility of high-resolution MAS NMR for detecting such a transition, characterizing its hysteresis, and determining the soliton density in incommensurate phases. The chemical shifts for this material are reported to 0.01 ppm but are precise to only about ±0.05 ppm due to changes in the homogeneity of the H_0 field with temperature and magnet drift with time.

The resonance of the β-phase at room temperature is a single, approximately Lorentzian peak with a chemical shift of -69.42 ppm and a full-width at half-height (FWHH) of 0.24 ppm (Figure 1). The chemical shift becomes less negative with increasing temperature to -69.20 ppm at 80°C. This single peak is consistent with the single tetrahedral Si-position in the β-Sr_2SiO_4 structure (Figure 2; Catti and Gazzoni, 1983; Catti et al., 1983). The chemical shift is similar to that of other orthosilicates.

With increasing temperature, the appearance of the incommensurate α'_L-phase is first detected at 83° by the appearance of a broad asymmetrical doublet (Figure 1). The most intense peak in the doublet is at the same position as the peak for the β-phase just below the onset of the transition, and the shift of the less intense peak is less negative (less shielded). With increasing temperature the intensity of the less shielded peak increases but its position remains constant at about -68.1 ppm. The position of the larger peak changes from -69.26 ppm at 83°c to -69.08 ppm at

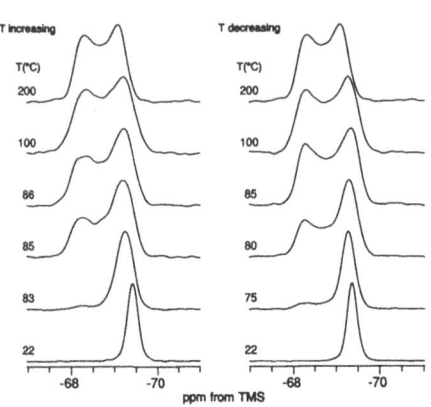

1. ^{29}Si MAS NMR spectra of Sr$_2$SiO$_4$ taken through the β to incommensurate transition. After Phillips et al., 1991.

200°C. With decreasing temperature from 200°C, the changes in the spectra are reversed, except that the doublet persists to below 70°C, corresponding to a thermal hysteresis of at least 10°. The relative intensity of the small peak still present at 70° does not increase on reheating to 85°. The spectrum of the sample taken at 25° after the heating and cooling sequence is identical to the original room temperature spectrum.

The broad doublet observed for the α'_L-phase is characteristic of incommensurately modulated phases (see Blinc, 1986, for a review and introduction) and is due to a continuum of Si-environments and thus a quasi-continuous distribution of ^{29}Si chemical shifts. The observed peak maxima are maxima in the spectral density of this distribution.

On an atomic level the ^{29}Si chemical shifts can be related to the rotation angle of the Si-tetrahedron (Figure 2). In the incommensurate phase, this rotation angle varies spatially with a periodicity that is not an integral number of unit cells and is one manifestation of the incommensurate structural modulation occurring parallel to b*. The more shielded maximum corresponds to regions in the structure in which tetrahedra are near the maximum (positive or negative) rotation angle, corresponding approximately to the two orientational variants of the β-phase (Figure 2). Because these two orientational variants are related only by a mirror symmetry operation, they have the same chemical shift. The less shielded peak corresponds to regions having a near-zero rotation angle and is expected to be at the same position as the peak for the orthorhombic α'_H-phase, although this has not been confirmed.

The non-symmetrical shape of the observed NMR spectra and the presence of satellite reflections with higher-order harmonics in the electron diffraction patterns (Withers et al., 1987) both indicate that the phase of the incommensurate modulation is non-linear, i.e., that the modulation is not a simple sine-wave.

Using the methods described by Blinc (1986), we have expanded the ^{29}Si chemical shift in terms of the local displacement and modeled the spectra with a multi-soliton solution to the sine-Gordon equation (Bruce et al., 1978). This equation minimizes the free energy density of the

A.

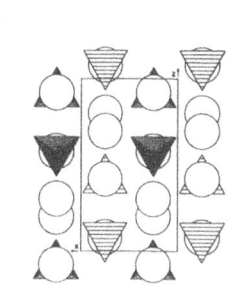

Figure 2. Projection of the crystal structure of (A) α'_H Sr_2SiO_4 and (B) two orientational variants of $\beta-Sr_2SiO_4$ onto (010). Large circles are Sr, and the tetrahedra are SiO_4 groups.

B.

incommensurate phase with respect to the phase of the modulation. Simulation of the spectra yields the soliton density (soliton width divided by inter-soliton spacing), the amplitude of the modulation and its initial phase. In this context the term soliton is used to describe a region of the crystal in which the phase of the structural modulation (e.g., the tetrahedral rotation angle) changes rapidly. The crystal structure, then, can be described by solitons alternating with nearly commensurate regions having approximately the structure of the β-phase (Figure 3). In Sr_2SiO_4 this modulation repeats along the b*-axis with a periodicity that is not equal to an integral number of unit cell lengths in that direction. The soliton density, then, is simply the fraction of the crystal occupied by regions of rapidly changing structure (phase of the modulation) and varies from zero for a commensurate phase to one for a plane-wave modulation.

For Sr_2SiO_4 the soliton density as determined from our NMR spectra increases rapidly through the β to incommensurate transition and shows an approximately 10° hysteresis (Figure

4). This behavior is generally similar to that observed for Rb$_2$ZnBr$_4$ and Rb$_2$ZnCl$_4$ (Blinc et al., 1986, and references therein). The presence of a modulated structure at temperatures below T_c in our down-temperature experiments might be due to metastablely preserved solitons which are randomly pinned, probably by impurities or defects (see Blinc et al., 1986).

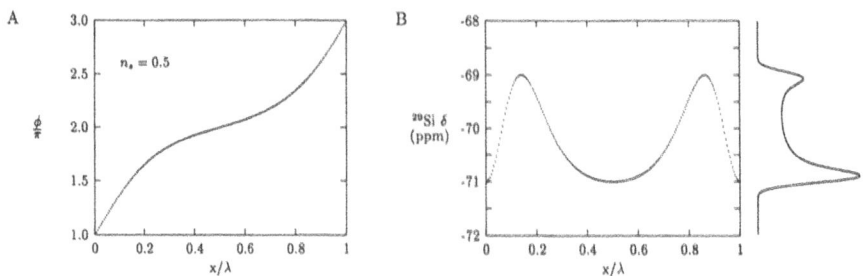

Figure 3. A) Spatial variation of the phase over one period of the incommensurate modulation wave in Sr$_2$SiO$_4$ for a soliton density, n$_s$, of 0.5. B) Grid of ^{29}Si NMR chemical shifts computed for the same soliton density. The spectral density function at the right is computed by convolution with a Gaussian linebroadening function and is essentially identical with the observed spectrum.

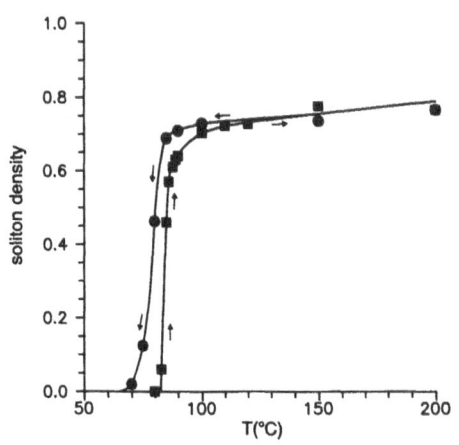

Figure 4. Temperature variation of the soliton density in incommensurate Sr$_2$SiO$_4$ obtained by least-squares fit to the ^{29}Si MAS NMR spectra, which shows an approximately 10° hysteresis between up-temperature and down-temperature runs. After Phillips et al., 1991.

The difference in [29]Si chemical shifts between the two singularities for the incommensurate phase is related to the amplitude of the modulation wave, which goes to zero at the incommensurate to α'_H (paraelastic) transition and is a measure of the order parameter for this transition. Although our data are quite far from the temperature of this transition (ca. 500°C), fitting the temperature dependence of the observed splitting yields a critical exponent of 0.31, which is very similar to those observed for Rb_2ZnBr_4 and Rb_2ZnCl_4 (Blinc et al., 1984, 1986b). The fitted critical temperature, T_i, is 496°C, essentially identical to that observed by electron diffraction (Withers et al., 1987).

3. SiO_2 AND $AlPO_4$ CRISTOBALITE

Phases with the cristobalite structure are the stable high-temperature polymorphs of both SiO_2 and $AlPO_4$. Both persist metastablely to lower temperatures and undergo reversible, first-order SPT's near 200°C. The structures of the α (low-temperature) and β (high-temperature) phases and the properties of the α to β transitions are similar for both materials (SiO_2-c and $AlPO_4$-c), differing primarily by the lower symmetry of the $AlPO_4$-c caused by ordering of Al and P onto the tetrahedral sites in the structure (Wright and Leadbetter, 1975; Leadbetter and Wright, 1976; Leadbetter and Smith, 1976).

The mechanism of the α to β transitions and the structures and dynamical behavior of the high-temperature β-phases are not fully understood (see Hatch and Ghose, 1991). For both materials the average symmetry of the β-phase as determined by X-ray diffraction is cubic. The primary unresolved questions are whether the actual local symmetry of the β-phase is cubic and whether the apparent disorder of its structure is due to dynamical averaging caused by atomic motion (and if so what kind) or to small static domains of varying orientations over which diffraction averages to produce an apparently ordered structure. The β-phase cannot have a long-range, static, cubic structure, because this structure requires linear T-O-T bonds and unrealistically short T-O distances. The disorder in the β-phase is reflected in the structure refinements as a distribution of oxygen atoms over a set of partially filled, symmetry-related positions (the h-sites located on an annulus about the Si-Si vector in Figure 5 and an analogous set of 3 sites in $AlPO_4$-c; Peacor, 1973, Wright and Leadbetter, 1976; Ng and Calvo, 1976).

We have obtained ^{29}Si, ^{27}Al and ^{31}P MAS NMR spectra, ^{27}Al static NMR spectra and ^{27}Al and ^{31}P T_1 relaxation rates through the SPT's of both SiO_2-c and $AlPO_4$-c (Phillips et al., submitted). The results indicate that the local (unit-cell scale) structure of the β-phase is not cubic and that its average cubic symmetry probably results from a dynamical and/or spatial averaging of orientational variants. They are also consistent with first-order character of the transitions, and demonstrate that dynamical effects can cause the mean Si-O-Si bond angles obtained from structure refinements to differ significantly from the time-averaged Si-O-Si angles estimated from ^{29}Si chemical shifts.

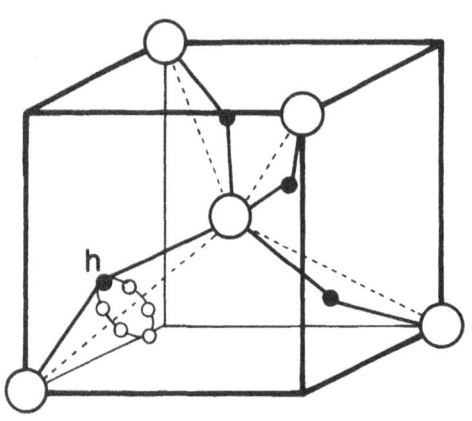

Figure 5. Structure of SiO_2-c (after Wright and Leadbetter, 1975 and Welberry et al., 1989). In the low-temperature α-phase, the oxygens are ordered onto one of the h-sites, leading to Si-O-Si bond angles of about 150°. In the ideal, so-called C9 structure of the β-phase the oxygens would occupy the sites in the center of the annulus of h-sites along the Si-Si vector, but the Si-O distances would be unreasonable short, and the Si-O-Si bond angle would be 180°. Available evidence suggests that in the β-phase the oxygens are dynamically disordered over the h-sites, that the Si-O-Si bond angle remains near 150°, and that the local symmetry remains orthorhombic on time scales up to ca. 25 μs.

Upon heating of a well ordered sample of SiO_2-c from room temperature, the ^{29}Si chemical shift of the low-temperature phase changes linearly from -108.5 ppm at 23°C to -111.0 at 246°C, the β-phase first appears at 249°C, the α- and β-phases coexist over a temperature interval of 7°, and the ^{29}Si chemical shift of the high-temperature phase varies only from -113.9 at 249°C to -114.1 ppm to 325°C (Figures 6 and 7). Similar results have been obtained by Spearing et al. (in press). Upon cooling from 325°C in our experiments, the low-temperature phase reappears at 254°C, but the high-temperature phase persists to 224°C, 25° below its first appearance during heating. In TEM studies, the α to β transition for each crystallite occurs rapidly (Hua et al., 1988), and thus the coexistence of the α- and β-phases indicated by the NMR spectra is likely due to a distribution of transition temperatures.

A more disordered, ^{29}Si-enriched sample of SiO_2-c shows generally similar behavior, except that its chemical shifts for the α-phase are about 0.8 ppm more negative (shielded) and upon heating the β-phase first appears at 172°C and the two phases coexist over 32°, and upon cooling the α-phase first appears at 186°C, and the β-phase persists to 145°C. Apparently the extent of stacking disorder has a significant effect on T_c and the magnitude of the hysteresis.

The ^{27}Al and ^{31}P MAS NMR spectra of $AlPO_4$-c show that its

SPT behaves in an analogous manner and that for our sample there is an approximately 20° hysteresis in the transition upon cooling (Figure 8). The isotropic ^{27}Al and ^{31}P chemical

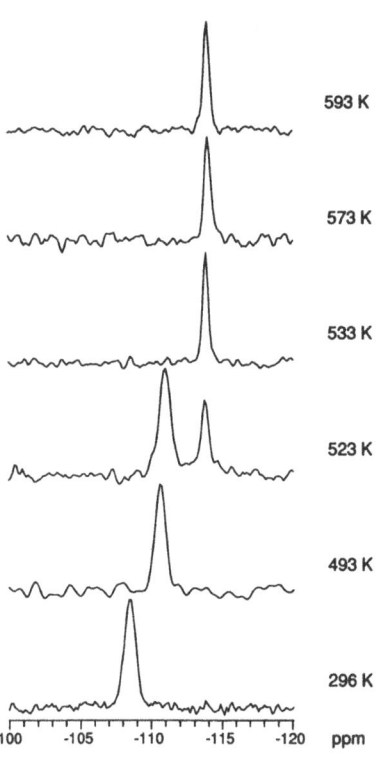

Figure 6. ^{29}Si MAS NMR spectra of SiO$_2$-c at the indicated temperatures. The peak between about −108.5 and −111 ppm is for the α-phase, and the peak near −113.5 ppm is for the β-phase.

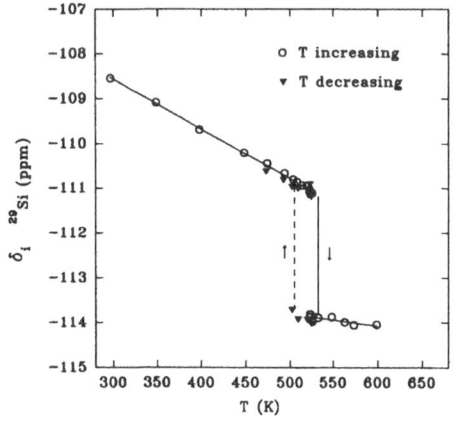

Figure 7. Temperature variation of the ^{29}Si chemical shift of SiO$_2$-c in up-temperature and down-temperature heating cycles.

shifts of the low-temperature phase likewise become more shielded with increasing temperature but remain nearly constant for the high-temperature phase (Figure 9). For both there is a stepwise increase in shielding from the low temperature phase to the high temperature phase of about 3 ppm, essentially the same as for ^{29}Si in SiO_2-c.

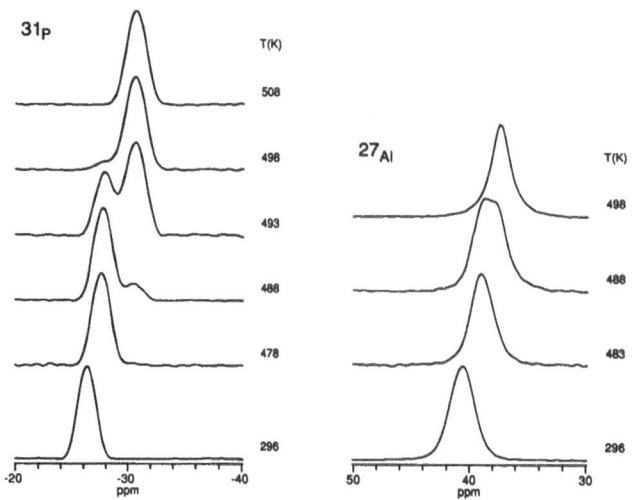

Figure 8. ^{27}Al and ^{31}P MAS NMR spectra of $AlPO_4$ through the α - β transition. After Phillips, et al., submitted.

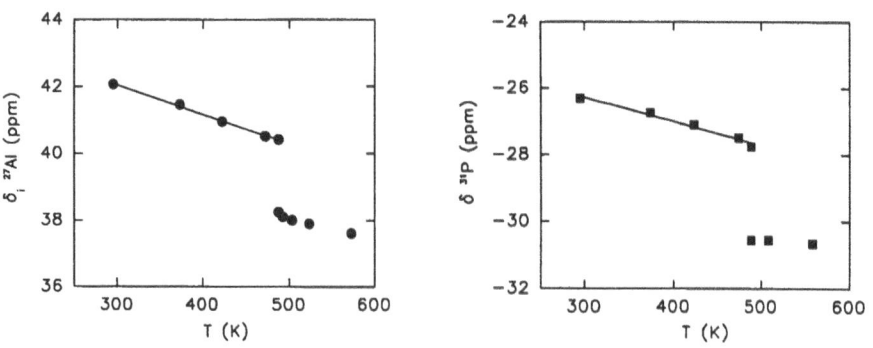

Figure 9. Temperature variation of 27Al and ^{31}P isotropic chemical shifts for $AlPO_4$-c. After Phillips et al., submitted.

The static ^{27}Al NMR peak of $AlPO_4$-c narrows considerably from the low temperature to the high temperature phase, consistent with a large decrease of the quadrupole coupling constant (QCC) from the α- to β-phases obtained from the ^{27}Al satellite transitions (Figure 10). The QCC of the β-phase is, however, not zero as required for a truly cubic phase, but is 60(2) kHz with an asymmetry parameter of 0.55(5).

The α to β transitions are also marked by sharp increases in the ^{27}Al and ^{31}P T_1 relaxation rates for $AlPO_4$ (Figure 11) and also for ^{29}Si and ^{17}O in SiO_2-c (Spearing et al., in press).

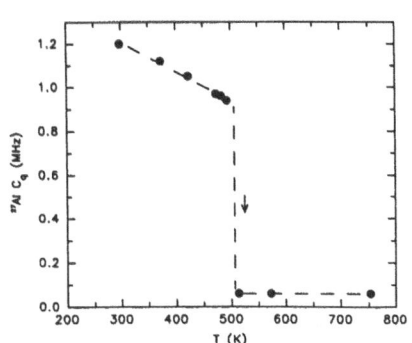

Figure 10. Temperature variation of ^{27}Al QCC for $AlPO_4$-c. After Phillips et al., submitted.

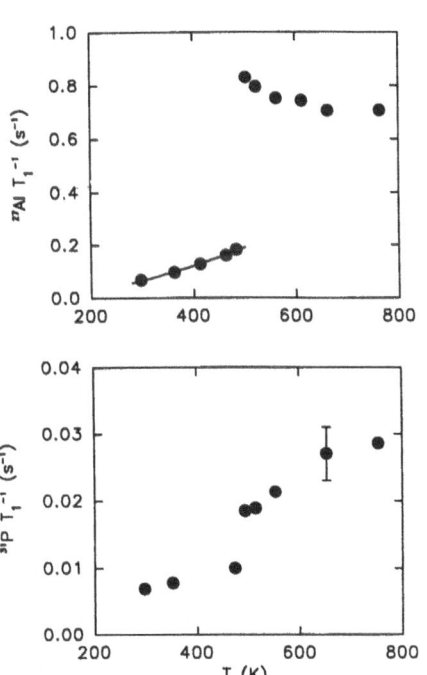

Figure 11. Temperature variation of ^{27}Al and ^{31}P T_1 relaxation rates for $AlPO_4$-c. After Phillips et al., submitted.

The linear changes of the ^{27}Al, ^{29}Si and ^{31}P chemical shifts in the α phases and the discontinuous changes across the α-β transitions can be readily interpreted in terms of changes in the mean T-O-T bond angle per tetrahedron, $\langle\theta\rangle$, (T=Si for SiO_2 and T=Al,P for $AlPO_4$). In the range of $\langle\theta\rangle$ values of interest here (140 - 160°), the chemical shifts of all three nuclides correlate linearly with $\langle\theta\rangle$ with a proportionality constant of about 0.6ppm/° (Figure 12; Müller, et al., 1984, Pettifer et al., 1988, Phillips et al., submitted).

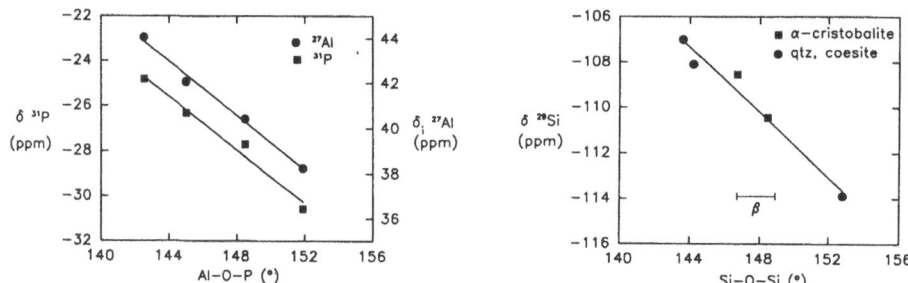

Figure 12. Correlations of the ^{27}Al and ^{31}P isotropic chemical shifts with mean Al-O-P bond angle per tetrahedron for the quartz, α-$AlPO_4$-c (RT and 180°C) and β-$AlPO_4$-c polymorphs of $AlPO_4$ and of the ^{29}Si chemical shifts with mean Si-O-Si bond angle per tetrahedron for the quartz, α-SiO_2-c (RT and 200°C) and coesite polymorphs of SiO_2, along with chemical shifts for the β-phase of SiO_2-c. After Phillips et al., submitted.

For $AlPO_4$-c, the ^{27}Al and ^{31}P chemical shifts correlate well with the $\langle\theta\rangle$ values obtained for the α-phase by XRD at room temperature (145.5°; Mooney, 1956) and at 180°C (148.5°; Ng and Calvo, 1977) and for the β-phase (151.9°; Ng and Calvo, 1977), but give a value 6° larger than obtained for the β-phase by Wright and Leadbetter (1976). The three points for $AlPO_4$ fall on a linear correlation between chemical shift and mean Al-O-P bond angle (Figure 12) that also includes the results for the quartz-structure polymorph fo $AlPO_4$ (Müller et al., 1984).

For SiO_2-c the ^{29}Si chemical shifts are consistent with the $\langle\theta\rangle$ for the α-phase from refinements at room temperature and 200°C (146.7° and 148.5°; Pluth et al., 1984). A linear correlation that includes these points and similar data for quartz and coesite (Figure 12) predicts a $\langle\theta\rangle$ of about 152.5±1° for the β-phase. This value is significantly greater than the values of 145.9° to 148.9° obtained from refinements of diffraction data (Peacor, 1968; Wright and Leadbetter, 1976).

The discrepancy between the mean Si-O-Si angles obtained

from diffraction data and predicted from the ^{29}Si chemical shifts for the β-phase are most likely due to difficulties in determining true T-O-T bond angles from diffraction data for phases in which the atoms have large thermal vibrations. For cristobalite, the oxygens in particular have large thermal ellipsoids, and correlated motion of the oxygens and their bonded Si-atoms may cause the time-averaged Si-O-Si angle to deviate from the angle between the mean atom positions.

In addition, the apparent disorder of the oxygen atoms in structure refinements of the β-cristobalite phases described above, the presence of diffuse x-ray and electron scattering for the β-phases (Hua et al., 1988), and the loss of twinning in the β-phase (Withers et al., 1987) all suggest that dynamical processes beyond normal lattice vibrations are integral to the structures of the β-cristobalite phases. These processes could be fluctuations of ordered domains between twin-related orientations (Hatch and Ghose, 1989) or normal-mode vibrations perhaps involving coupled rotations of rigid TO$_4$ tetrahedra (rigid-unit modes, RUM's, Dove et al., in press; Hua et al., 1989). For either model, the observed non-zero ^{27}Al QCC has important implications.

In the average cubic structure of β AlPO$_4$-c (space group F$\bar{4}$3m) Al occupies a position having point symmetry $\bar{4}$3m (Wright and Leadbetter, 1976; Ng and Calvo, 1977). Thus, by symmetry its QCC must = 0. The observed QCC of about 60 kHz indicates that the average local symmetry remains non-cubic for times as long as the inverse of the NMR linewidth (ca. 25 μs). Therefore, any realistic model for the dynamics of the β-cristobalite structure should not average the Al-site over this time scale.

The reduction of the ^{27}Al QCC from ca. 900 kHz in the α-phase to ca. 60 kHz in the β-phase does, however, provide additional evidence for dynamical averaging in the β-phase. Relative to the so called Wyckoff C9 structure of the β-phase (Al-O-P angle = 180°), both the QCC and the deviation of the Al-O-P angle from 180° are measures of the order parameter, both being zero in the C9 structure. Across the α - β transition the QCC decreases 94%, whereas the difference of the Al-O-P bond angle from 180° decreases only 12%. For static structures one expects a correlation between QCC and the chemical shift, because both correlate with the mean Al-O-P angle. In fact, for the α-phase such a correlation exists (Phillips et al., submitted) and predicts a QCC of ca. 600 kHz for the β-phase, compared to the observed value of 60 kHz. This difference, thus, suggests that the EFG tensor experiences orientational averaging whereas the average Al-O-P bond angle remains significantly less than 180° (Figure 5).

4. P$\bar{1}$ - I$\bar{1}$ TRANSITION OF ANORTHITE

Anorthite (CaAl$_2$Si$_2$O$_8$) is one end-member of the plagioclase feldspar series, the most common mineral in the earth's crust, and undergoes a reversible triclinic to triclinic SPT from space group P$\bar{1}$ (low temperature) to I$\bar{1}$ (high temperature) at about 240°C (Brown et al., 1963). Single crystal ^{27}Al NMR measurements were essential for confirming that the high-temperature phase is body-centered rather than a superposition of small P$\bar{1}$ domains (Staehli and Brinkmann, 1974). Understanding this transition is essential to understanding the structure and thermodynamic behavior of the plagioclase feldspars, and its importance to mineralogy and geology is underscored by the large amount of recent work (see, e.g., Van Tendeloo et al., 1989, Angel et al., 1989, and references in those papers).

Spontaneous strain measurements from X-ray diffraction data at 25 to 700°C indicate that the P$\bar{1}$-I$\bar{1}$ transition is tricritical for the well ordered Val Pasmeda sample (Redfern and Salje, 1987). A tricritical transition lies on the boundary between a first-order transition and a second-order transition.

The structure of anorthite consists of a framework of corner-shared Al and Si tetrahedra with feldspar topology and Ca atoms in large sites between the tetrahedra (Ribbe, 1983). The P$\bar{1}$ phase has 16 crystallographically distinct tetrahedral sites, 8 occupied by Si and 8 by Al in the ordered structure. The transition to the I$\bar{1}$ structure adds a body-centering sub-cell translation, 1/2[111], which halves the number of Si and Al sites. In perfectly ordered anorthite the Si and Al tetrahedra alternate regularly, and all the Si-sites are Q^4(4Al).

We have investigated the P$\bar{1}$-I$\bar{1}$ transition for the Val Pasmeda sample using ^{29}Si MAS NMR from 25° to 500°C (Phillips, 1990). These results complement previous results for this SPT in two ways. 1) Direct observation of the Si nuclei provides a view of the transition from the Al,Si framework, whereas in X-ray diffraction experiments a large fraction of the scattering comes from the Ca atoms. 2) NMR is sensitive to only local structure and can thus distinguish between a true P$\bar{1}$ structure or an average I$\bar{1}$ structure composed of small, static P$\bar{1}$ domains. In addition, the frequency separation of pairs of peaks due to sites related by the body-centering sub-cell translation 1/2[111] yields the order parameter for this transition, allowing investigation of the nature of the SPT.

The observed room temperature ^{29}Si MAS NMR spectrum of the Val Pasmeda sample (Figure 13) consists of 6 peaks, which we have previously assigned to the 8 Si-sites of P$\bar{1}$ anorthite

(Phillips et al., 1992). With increasing temperature in the PĪ phase these peaks converge, and at 241°C the spectrum contains only 4 peaks which we assign to the 4 Si-sites of the IĪ structure based on the mean Si-O-Al angles obtained in a recent neutron diffraction study (Ghose et al., in press).

Simulation of the spectra to yield the chemical shifts of the 4 or 8 Si-sites and demonstrates that peaks for sites related by the body-centering do converge with increasing temperature

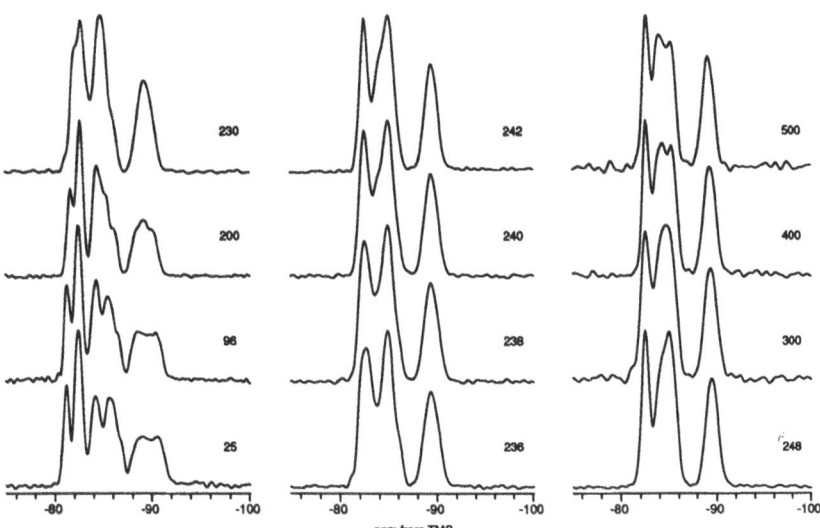

13. ²⁹Si MAS NMR spectra of Val Pasmeda anorthite through the PĪ - IĪ transition. Temperature in °C. After Phillips, 1990.

for the PĪ phase and that only four peaks are present for the IĪ phase (Figure 14).

These results clearly demonstrate that the average Al,Si framework of IĪ anorthite does not consist of small domains with PĪ symmetry. If this were true, the IĪ phase would still yield 8 ²⁹Si NMR peaks. The ²⁹Si spectra alone, however, cannot distinguish whether this IĪ structure has static IĪ symmetry or whether it is a dynamical average of PĪ domains.

The splitting of the pairs of peaks for the PĪ anorthite which converge at the PĪ-IĪ transition can be used to follow the temperature evolution of the order parameter within the context of Landau theory. To second order, the frequency

218

separation, $\Delta\nu$ is given by,

$$\Delta\nu = A_1Q + A_2Q^2,$$

where A_1 and A_2 are constants and Q is the order parameter. In Landau theory the temperature dependence of the order parameter, Q, is given by

$$Q = \left[\frac{T_c - T}{T_c}\right]^\beta$$

where T is temperature in K, T_c the critical temperature, and β is the critical exponent which is 0.5 if the transition is second-order and 0.25 if the transition is tricritical. The splitting of the best resolved pair of peaks due to sites related by body-centering (T1mzo and T1mzi), which are at -88.6 and -90.5 ppm at room temperature, was obtained by fitting the spectra such that only the peak positions vary with temperature (Figure 15). The temperature dependence of this splitting gives $T_c = 241\pm1°C$ and $\beta = 0.29\pm0.05$. The critical exponent for a tricritical transition is 0.25, and our results are thus consistent with this transition being tricritical or nearly so. Salje (1987) concluded on the basis of diffraction data that this SPT is tricritical.

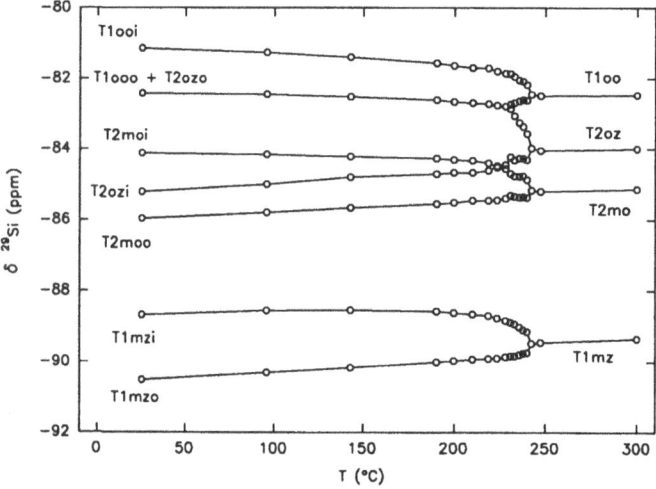

Figure 14. Temperature variation of the fitted ^{29}Si chemical shifts for Val Pasmeda anorthite showing convergence of peaks for sites related by body-centering at the P1̄ - I1̄ transition.

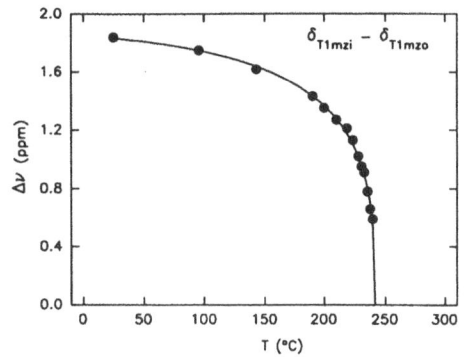

Figure 15. Temperature variation of the separation of the ^{29}Si chemical shifts of the T1mzo and T1mzi sites in anorthite. This splitting is related to the order parameter for the P$\bar{1}$ to I$\bar{1}$ transition. The fitted curve yields a T_c of 241±1°C and a critical exponent of 0.29 ±0.05, consistent with a tricritical or nearly tricritical SPT. After Phillips, 1990.

5. ACKNOWLEDGEMENTS

This paper was supported by NSF Grant 90-04260, R.J.K. P.I.

6. REFERENCES

Angel, R. J., Redfern, S. A. T., and Ross, N., L., 1989, Phys. Chem. Mineral., 16, 539.
Blinc, R., 1981, Phys. Rep., 79, 331.
Blinc, R., Milia, F., Topic, B., and Zumer, S., 1984, Phys. Rev. B, 29, 4173.
Blinc, R., Prelovsek, P. Rutar, V., Seliger, J., and Zumer, S.,1986a, in Incommensurate Phases in Dielectrics, 1.
Fundamentals, Blinc, R. and Levanyuk, A. P., eds., North-Holland Publishing, Amsterdam.
Blinc, R., Rutar, V., Topic, B., Milia, F., and Rasing, T., 1986b, Phys. Rev. B, 33, 1721.
Brown, W. L., Hoffmann, W. and Laves, F., 1963, Naturwiss., 50, 221.
Bruce, A. D., Cowley, R. A., and Murray, A. F., 1978, J. Phys. C, 11, 3591.
Catti, M., and Gazzoni, G., 1983, Acta Cryst. B, 39, 679.
Catti, M., Gazzoni, G., and Ivaldi, G., 1983, Acta Cryst. C, 39, 29.
Dove, M. Y., Giddy, A. P., and Heine, V., in press, Trans. Am. Crystallogr. Assn., 27.
Ghose, S., McMullan, R.K., and Weber, H.-P., in press, Zeit. Kristallogr.
Ghose S., Van Tendeloo, G., and Amelinckx, S., 1988, Science, 242, 1539.
Hatch, D. M., and Ghose, S., 1989, Phys. Chem. Mineral., 16,

614.

Hatch, D. M., and Ghose, S., 1991, Phys. Chem. Mineral., 17, 554.

Hua, G. T., Welberry, T. R., Withers, R. L., and Thompson, J. G., 1988, J. Appl. Cryst., 21, 458.

Hyde, B. G., Sellar, J. R., and Stenberg, L., 1986, Acta Cryst. B, 42, 423.

Leadbetter, A. J., and Wright, A. F., 1976, Philos. Mag., 33, 105.

Leadbetter, A. J., and Smith, T. W., 1976, Philos. Mag., 33, 113.

Müller, D., Jahn, E., Ladwig, G., and Haubenreisser, U., 1984, Chem. Phys. Lett., 109, 332.

Mooney, R. C. L., 1956, Acta Cryst., 9, 728.

Ng, H. N., and Calvo, C., 1977, Can. J. Phys., 55, 677.

Peacor, D. R., 1973, Z. Kristallogr., 138, 274.

Pettifer, R. F., Dupree, R., Farnan, I., and Sternberg, U., 1988, J. Non-cryst. Solids, 106, 408.

Phillips, B. L., 1991, Unpublished PhD Thesis, Department of Geology, University of Illinois at Urbana, Champaign.

Phillips, B. L., Kirkpatrick, R. J., and Carpenter, M. A., 1992, Amer. Mineralogist, 77, 484.

Phillips, B. L., Kirkpatrick, R. J., and Thompson, J. G., 1991, Phys. Rev. B, 43, 13,280 - 13,289.

Phillips, B. L., Xiao, Y., and Kirkpatrick, R. J., submitted, Phys. Chem. Minerals.

Pluth, J. J., Smith, J. V., and Faber, J. Jr., 1985, J. Appl. Phys., 57, 1045.

Redfern S. A. T., and Salje, E., 1987, Phys. chem. Minerals, 14, 189.

Ribbe, P. H., ed, 1983, Revs. in Mineralogy, Vol. 2, Feldspar Mineralogy, 2nd ed., Min. Soc. Am., Washington D. C.

Rigamonti, A., 1984, Adv. Phys., 33, 115.

Salje, E., 1987, Phys. Chem. Minerals, 14, 181.

Smith, J. V., and Blackwell, C. S., 1983, Nature 303, 223.

Spearing, D., Farnan, I., and Stebbins, J., in press, Phys. Chem. Minerals.

Staehli, J. L., and Brinkmann, D., 1974, Z. Kristallogr., 140, 360.

Van Tendeloo, G., Ghose, S., and Amelinckx, 1989, Phys. chem. Mineral., 16, 311.

Welberry, T., R., Hua, G. L., and Withers, R. L., J. Appl. Cryst., 22, 87

Withers, R. L., Hyde, B. G., and Thompson, J. G., 1987, J. Phys. C, 20, 1653.

Wright, A. F., and Leadbetter, A. J., 1975, Philos. Mag., 31, 1392.

A Model Study of Chemical Shielding in a Partially Hydrated Dipeptide

D. B. Chesnut and C. G. Phung
P. M. Gross Chemical Laboratory
Duke University
Durham, NC 27706, USA

Abstract: Chemical shift calculations of the amide group nuclei in several configurations of mono- and di-hydrated glycylglycine have been carried out using the gauge including atomic orbital (GIAO) method with 6-311G(d,p)/3-21G locally dense basis sets. Significant effects are noted for the amide carbonyl oxygen and hydrogen but only small effects for carbon and nitrogen. Shift parameters for the doubly hydrated species are very nearly the sum of those for the monohydrates. While the orientation of the shift tensors for oxygen, carbon, and nitrogen are little affected by hydration, the change of the amide proton tensor is significant and appears to follow the orientation of the local hydrogen bonding water. A larger model system appears necessary to properly reflect interactions in the solid state.

Introduction

Advances in computer facilities during the past decade have made possible first principle quantum mechanical calculations on larger and larger molecular systems. This has been true in general, and is true specifically in the determination of the nuclear magnetic resonance (NMR) chemical shift, a sensitive local probe of electronic structure. Nearly all shielding work has been concerned with isolated small molecules mimicking essentially a gas phase environment. This has been so because not only is it difficult to treat large chemical systems but also because one needs confirmation of the theoretical technique by comparing "isolated molecule" calculations to experimental gas phase data, data which is only available for relatively small molecules. There have been a few attempts to look at systems that are quite sizeable; among these are the calculations by Schindler [1] on some purine and pyrimidine bases by means of the IGLO approach [2] (systems containing from 8 to 11 heavy atoms), our own work on the neutral and zwitterionic forms of glycylglycine [3] (a 9 heavy atom problem), and most recently the work of Hinton et al. [4] on the $(H_2O)_{17}$ cluster employing an

J. A. Tossell (ed.), Nuclear Magnetic Shieldings and Molecular Structure, 221–241.

efficient implementation [5] of the gauge including atomic orbital (GIAO) method [6] and the locally dense basis set approach [7] developed in this laboratory.

Of greater practical and theoretical interest are non-gas-phase systems, in particular those where the molecule of interest has been solvated, or, in general, systems where intermolecular interactions affect the molecular property being studied. In the present work we extend our work on glycylglycine (GG) by including one or two waters forming hydrogen bonds with the amide group of the dipeptide. We focus our interest on the very important OCNH amide group of this simplest of all dipeptides and inquire into the changes in the chemical shifts of these nuclei upon hydration either at the carbonyl oxygen (WO), or at the amide proton (WN), or at both locations (2W). Following a discussion of the changes noted for a small selection of hydration configurations, we compare our calculated results to some recently observed for several species in the solid state [8-11].

Model Structures and Theoretical Methods

The various molecular configurations studied are discussed below where we use a special shorthand notation. P and R refer to the planar and rotated forms of the dipeptide, WO, WN, and 2W refer to hydration at the carbonyl oxygen, the NH proton, and at both sites, respectively, and s and o stand for the standard and angle-optimized forms of the hydrates.

The GG molecular geometry has been taken from our prior work [3] where it was optimized at the 6-311G(d,p) level [12]; the water monomer was also optimized at this level of basis. The monomer optimized geometries for both GG and H_2O were kept intact in the hydrates. As can be seen in the left hand portion of Figure 1, the carboxyl proton has been placed in an s-trans conformation and the amino lone pair in the plane of the amide group "cis" to the CO bond in order to mimic a trans-extended peptide structure. Linear hydrogen bonds were assumed with solid state average H-O bond lengths of the NH...O and OH...H bonds of 1.8650 A [13] and 1.8181 A [14], respectively. Both the GG planar form (P) and its rotated form (R) were studied, where the rotated form is obtained by a 90° rotation about the NC_α bond to place the plane of the carboxyl group 90° to the amide plane; it has been suggested [15] that dipeptides tend to exist either in the planar or carboxyl-rotated 90° form, which is the basis of our investigation of these two conformations. In the crystal structure of glycylglycine hydrochloride [16] these two planes are 77.9° apart.

For the planar and rotated forms of GG we have looked at two types of arrangements of the waters of hydration. In the first, indicated as our standard (s) model, the hydrogen bonds are placed so as to coincide with the "classical" lone pair orientations of the oxygen of the carbonyl group (sp^2 hybridization) and of the water hydrating the NH proton (sp^3 hybridization). In addition to these

GG(P)2W(s)　　　　　　　GG(R)2W(s)

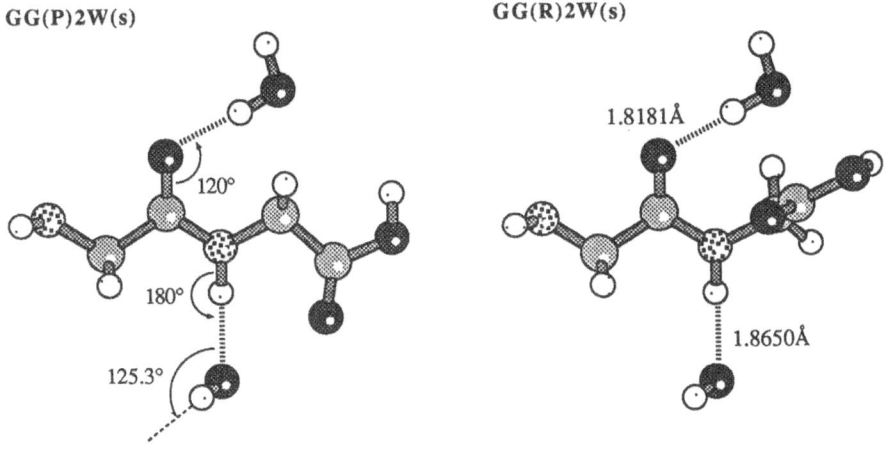

Figure 1. Drawings of the planar and rotated model hydrated glycylglycine systems studied. The standard (s) configurations are shown in which the hydrogen bonding lone pairs are classically oriented; the hydrogen bond lengths were always kept linear at the average experimental distances shown.

standard (s) configurations, we have also looked at those angle optimized structures (o) in which the water molecules of the monohydrated species were allowed to have a modified angular orientation while maintaining the hydrogen bond distances and the linearity of the hydrogen bonds.

The energy changes upon (mono)hydration are given in Table 1. The use of average hydrogen bond distances taken from the solid state results in several cases of a low or even positive (destabilizing) hydration energy. Clearly, many facets of the solid are involved in the determination of its structure, whereas in our case of a slightly hydrated, gas-phase-like glycylglycine the optimal structures are undoubtably quite different. We have looked at the optimal angle structures (o) as well as the standard structures (s) in an effort to see what difference, if any, partial energy optimization makes.

The determination of the hydration energies was carried out using a balanced 6-311G(d,p) basis for all atoms of the system; the polarization functions employed for heavy atoms consisted of the 6 d Cartesian gaussians, and basis set superposition errors were corrected for by the counterpoise method [17]. For chemical shift calculations, on the other hand, we employed our locally dense basis set approach [7] in which the important amide group nuclei (O,C,N, and H) were treated at the 6-311G(d,p) level, while all other atoms in the dipeptide and the hydrating waters were given the (attenuated)

Table 1

Energies of hydration (ΔE, kcal/mole) and orientation angle changes ($\Delta \varphi$, degrees) upon angular optimization for the monohydrate glycylglycines. Positive angles are counterclockwise rotations viewing the systems as pictured in Figure 1.

WO and WN Structures	ΔE	$\Delta \varphi$
GG(P)WO(s)	-1.04	
GG(P)WO(o)	-6.39	15.4˙
GG(R)WO(s)	5.32	
GG(R)WO(o)	-5.69	19.8˙
GG(P)WN(s)	7.03	
GG(P)WN(o)	4.03	-14.5˙
GG(R)WN(s)	-2.19	
GG(R)WN(o)	-2.47	12.8˙

3-21G basis. We have shown previously [3,7] the accuracy and usefullness of the locally dense basis set approach for determining chemical shifts, an approach which is successful largely due to the essentially local nature of the shift phenomenon. Quite recently Hinton et al. [4] have used this approach to study the $(H_2O)_{17}$ cluster with good results. The locally dense basis set approach allows one to look at key nuclei in molecules whose size might preclude a balanced approach, and also requires much less cpu time for a given system. The use of this approach is justified by its success, as demonstrated in our previous work on glycylglycine [3] which showed that shifts for the amide group nuclei were basically unchanged upon moving from a balanced to a locally dense approach.

Ditchfield's GIAO approach [6] was used for the chemical shift calculations, all of which were carried out on a CRAY Y-MP 8/464 supercomputer located at the North Carolina Supercomputing Center.

Energetics

Table 1 lists the various hydration energies. Note that these energies refer to hydration of a particular monohydrate and not the dihydrate. The optimizations which allowed the hydrogen bond angle to vary while maintaining a constant hydrogen bond distance were also done for the monohydrates, although both waters of hydration are shown in Figure 1.

For the planar (P) model the water hydrating the carbonyl oxygen leads to a small stabilization of the system, while that water hydrating the NH proton is destabilizing by 7.03 kcal/mol. The instability of this latter hydrate is due to the fact that the carboxyl double bond oxygen atom is very close (2.56 Å) to the hydrating water oxygen, leading to significant repulsion. If each water, in turn, is allowed to angle-optimize (o), the NH water moves away from that carboxyl oxygen by -14.5° (a clockwise rotation viewing the system as pictured in Figure 1) and, while still leading to a repulsion of 4.03 kcal/mol, the repulsive energy is reduced. For the CO water, a change of 15.4° (a counterclockwise rotation) is made, allowing a significant increase in stability to -6.39 kcal/mol for that hydrate.

In the case of the rotated molecule (R) hydrates, the rotation of the carboxyl group away from the NH water removes the repulsive behavior and, in the case of the standard (s) molecule, leads to a stability of -2.19 kcal/mol. If this water is allowed to angle-optimize, the stability is increased only slightly to -2.47 kcal/mol.

Rotating the carboxyl group brings one of its alpha-carbon hydrogens very close to the CO water (1.53 Å to the bonding water hydrogen, 1.75 Å to the water oxygen) in its standard orientation, resulting in a net repulsion of 5.32 kcal/mol. If this water is allowed to relax, it moves some 19.8° away and results in a stabilized, -5.69 kcal/mol, hydrate.

Accordingly, our standard models are ones clearly not of minimum energy. The angle-optimized structures in both the planar and rotated forms are partially optimized in-so-far as the angular orientation is concerned. However, the energetics are not our principle concern here since we are really trying to mimic the situation in the solid state. Nonetheless, it must be kept in mind when looking at the various shift changes upon hydration that in some instances we are working with a basically unstable configuration, while in others the presence of the hydrating water leads to an increased stability.

Chemical Shift Tensor Changes Upon Hydration

The chemical shifts for the unhydrated planar (P) and rotated (R) forms as well as changes upon hydration at either the carbonyl oxygen, the NH hydrogen, or at both sites, are given in Table 2 for each of the amide group nuclei. The principal values of the pertinent shift tensors are given in the table under the labels XX, YY, and ZZ. XX refers to that principal value whose principal direction lies close to an adjacent bond; for oxygen and carbon the appropriate bond is the CO bond, for nitrogen the NC bond (of the carbonyl group), and for hydrogen it is the HN bond. YY represents that principal direction perpendicular to XX and is either in or virtually in the plane of the amide group. The XX and YY directions lie in the plane of the amide group for the planar (P) form of the molecule, and the direction ZZ is perpendicular to that plane. Since no molecular plane of symmetry is present for the rotated form of the molecule, no constraining directions are imposed upon the shift tensor; however, even in this case the XX and YY directions lie essentially in the amide plane and the ZZ direction nearly perpendicular to it. The variable A (bond) represents the angle

Table 2

Calculated properties of the shift tensors of the amide group nuclei in the various glycylglycine systems. The isotropic shift (σ), anisotropy ($\Delta\sigma$), range (R), and tensor components (XX, YY, ZZ) are given in ppm. The angle A that the XX component makes with the principal bond (indicated in parentheses) direction and the tilt angle theta are given in degrees.

A. Oxygen

	A(CO)	theta	XX	YY	ZZ	σ	$\Delta\sigma$	R
a. Unhydrated								
GG(P)	14.0°		-399.7	-209.4	390.1	-73.0	694.6	789.8
GG(R)	17.7°	0.8°	-384.2	-205.4	404.9	-61.6	699.6	789.1
b. Changes upon hydration at the carbonyl oxygen								
GG(P)WO(s)	0.0°		95.4	53.4	-10.7	46.0	-85.1	-106.1
(R)	0.3°	-0.4°	97.1	54.6	-15.6	45.4	-91.4	-112.7
GG(P)WO(o)	-0.3°		79.7	46.7	-8.3	39.4	-71.5	-88.0
(R)	0.1°	-0.2°	73.8	45.5	-9.5	36.6	-69.2	-83.5
c. Changes upon hydration at the NH hydrogen								
GG(P)WN(s)	-1.1°		15.1	7.9	0.2	7.8	-11.2	-14.9
(R)	-1.1°	0.1°	16.9	7.4	-2.0	7.5	-14.0	-18.9
GG(P)WN(o)	-1.2°		14.7	7.9	0.1	7.5	-11.2	-14.6
(R)	-1.1°	0.1°	17.7	7.6	-2.4	7.6	-15.2	-20.3
d. Changes upon dual hydration								
GG(P)2W(s)	-1.4°		110.4	61.7	-11.5	53.5	-97.5	-121.9
(R)	-1.1°	-0.3°	114.2	62.2	-18.1	52.8	-106.1	-132.3
GG(P)2W(o)	-1.8°		94.6	55.1	-8.7	47.0	-83.5	-103.3
(R)	-1.2°	-0.1°	91.9	53.4	-11.9	44.4	-84.7	-104.0

(Table 2 continued)

B. Carbon

	A(CO)	theta	XX	YY	ZZ	σ	Δσ	R
a.	**Unhydrated**							
GG(P)	-4.3°		39.2	-70.3	107.3	25.4	122.9	177.6
GG(R)	-5.5°	1.7°	41.2	-72.3	108.0	25.6	123.5	180.3
b.	**Changes upon hydration at the CO oxygen**							
GG(P)WO(s)	0.2°		-18.1	7.5	1.2	-3.1	6.5	-6.3
(R)	0.2°	-0.3°	-20.0	8.4	-0.6	-4.0	5.2	-9.0
GG(P)WO(o)	-0.1°		-15.1	7.0	1.2	-2.3	5.2	-5.8
(R)	-0.3°	-0.3°	-14.9	7.4	0.0	-2.5	3.8	-7.4
c.	**Changes upon hydration at the NH proton**							
GG(P)WN(o)	3.7°		-2.0	2.7	-0.1	0.2	-0.6	-2.8
(R)	3.4°	0.0°	-2.7	3.1	-0.3	0.0	-0.5	-3.4
GG(P)WN(o)	4.0°		-1.4	3.0	-0.1	0.5	-0.9	-3.1
(R)	3.3°	0.0°	-3.3	3.0	-0.3	-0.2	-0.2	-3.3
d.	**Changes upon dual hydration**							
GG(P)2W(s)	5.3°		-19.7	10.2	0.9	-2.8	5.6	-9.3
(R)	4.1°	-0.3°	-22.2	11.4	-0.9	-3.9	4.6	-12.3
GG(P)2W(o)	3.6°		-16.2	10.0	1.1	-1.7	4.1	-8.9
(R)	3.9°	-0.3°	-18.1	10.5	-0.3	-2.6	3.5	-10.8

(Table 2 continued)

C. Nitrogen

	A(CN)	theta	XX	YY	ZZ	σ	$\Delta\sigma$	R
a. Unhydrated								
GG(P)	5.8°		288.7	100.2	176.8	188.6	150.2	188.5
(R)	10.6°	1.1°	260.5	97.7	191.3	183.2	116.1	162.8
b. Changes upon hydration at the CO oxygen								
GG(P)WO(s)	0.3°		-9.5	-13.6	1.5	-7.2	-3.5	4.1
(R)	-0.6°	-0.4°	-5.9	-13.7	-0.6	-6.8	1.2	7.8
GG(P)WO(o)	0.1°		-6.2	-11.2	1.1	-5.5	-1.1	5.0
(R)	-0.3°	-0.1°	-4.0	-10.4	0.0	-4.8	1.0	6.4
c. Changes upon hydration at the NH proton								
GG(P)WN(s)	-1.0°		-5.2	-9.2	6.5	-2.7	-3.8	4.0
(R)	-1.6°	0.0°	-3.4	-9.5	7.1	-2.0	-2.3	6.1
GG(P)WN(o)	-0.7°		-7.1	-8.3	6.8	-2.9	-6.4	1.2
(R)	-2.0°	0.3°	-1.7	-11.1	4.8	-2.7	1.4	9.4
d. Dual hydration								
GG(P)2W(s)	-0.7°		-15.7	-23.5	7.9	-10.5	-7.9	7.8
(R)	-2.1°	0.0°	-10.2	-24.0	6.6	-9.2	-1.6	13.8
GG(P)2W(o)	-0.6°		-14.1	-20.1	7.8	-8.8	-7.9	6.0
(R)	-2.3°	0.2°	-6.5	-22.0	5.0	-7.9	1.9	15.5

(Table 2 continued)

D. Hydrogen

	A(HN)	theta	XX	YY	ZZ	σ	Δσ	R
a. Unhydrated								
GG(P)	19.6°		33.90	31.54	19.75	28.40	8.25	14.15
(R)	-13.9°	5.7°	34.72	30.50	21.62	28.94	8.66	13.10
b. Changes upon hydration at the CO oxygen								
GG(P)WO(s)	-1.2°		-0.06	-0.25	-0.16	-0.16	0.15	0.10
(R)	1.5°	-0.5°	0.10	-0.17	-0.08	-0.04	0.23	0.18
GG(P)WO(o)	-1.1°		-0.09	-0.32	-0.24	-0.22	0.19	0.15
(R)	1.2°	-0.2°	-0.04	-0.26	-0.20	-0.16	0.20	0.16
c. Changes upon hydration at the NH proton								
GG(P)WN(s)	-19.8°		4.39	-6.87	-7.20	-3.23	11.43	11.59
(R)	8.0°	-1.9°	4.73	-7.46	-7.32	-3.34	12.12	12.05
GG(P)WN(o)	-29.8°		4.02	-6.69	-7.35	-3.34	11.04	11.37
(R)	17.2°	-1.6°	4.45	-6.99	-7.33	-3.28	11.61	11.78
d. Changes upon dual hydration								
GG(P)2W(s)	-19.7°		4.31	-7.08	-7.38	-3.39	11.54	11.69
(R)	8.4°	-2.1°	4.82	-7.62	-7.35	-3.38	12.31	12.17
GG(P)2W(o)	-29.4°		3.88	-6.97	-7.60	-3.57	11.17	11.48
(R)	17.3°	-1.7°	4.43	-7.26	-7.55	-3.45	11.83	11.98

(measured counterclockwise) between the XX direction and the pertinent bond. The (Euler) angle theta is the amount of tilt of ZZ away from the perpendicular to the amide plane. While theta is zero by symmetry in the planar (P) form, a glance at the table shows that the theta values are very small for the rotated form, indicating that the amide plane effectively constrains the chemical shift tensors.

Table 2 also contains the isotropic shift, σ, the shift anisotropy, $\Delta\sigma$, and a quantity indicated as R, the "range" of the spectrum, representing the largest difference among the principle elements of the shift tensor (that is, $R = \sigma_{ii}^{max} - \sigma_{jj}^{min}$), all in ppm.

The molecular configuration notation in the tables and figures is that which has been introduced previously. Note that the data in Section a of each part of Table 1 represent <u>absolute</u> chemical shifts and angles, while the data in Sections b, c, and d represent <u>changes</u> upon hydration at the various sites.

Major Effects of Hydration

Inspection of Table 2 shows that while small changes in shielding occur for all configurations of hydration, the large effects occur for the oxygen atom when the carbonyl oxygen is hydrated, and for the proton of the NH group when that hydrogen is hydrated, as one might have expected. The changes for the carbon and nitrogen of the amide group are fairly small and may well lie within the noise level of the theoretical calculation. What perhaps might not have been anticipated is the large change of the anisotropy (and the range) of the spectrum. In the case of oxygen the decrease in both $\Delta\sigma$ and R is due to the large increase (positive change) of the XX-component, the component nearly parallel to the CO bond. Although carbon shows relatively small changes, it has a relatively large decrease of nearly 20 ppm in the same direction. The changes for nitrogen are generally small, although, interesting enough, the nitrogen is affected more by the water of hydration at the carbonyl oxygen than by the water at the NH proton.

The changes at the amide proton are large, σ showing the typical hydrogen bonding downfield shift of 3-4 ppm, and the anisotropy showing a huge increase of the order of 12 ppm. The range of the powder spectrum for the amide proton in the standard, planar form is calculated to increase from 14.15 ppm to 25.74 ppm upon hydration at the amide proton. For the amide hydrogen, the largest changes (negative) take place not along the XX component (close to the NH bond) but rather in the YY and ZZ directions. While changes in the angles A and theta for oxygen, carbon, and nitrogen are generally small, such is not the case for the amide proton; the hydrogen angle changes are discussed below.

Additivity of Shift Changes

Careful inspection of the data in Table 2 shows that the angle and shift changes seen upon double hydration are virtually the sum of those seen for the monohydrates. That is, the effects of hydration appear to be additive. One finds that the deviation from additivity for the angles is equal to or less than 1.4°, while the changes in shift values (the individual shift tensors elements, σ, $\Delta\sigma$, and R) are all less than or equal to 1.2 ppm for the "heavy" atoms (oxygen, carbon, and nitrogen), and less than or equal to 0.06 ppm for hydrogen. There is no a priori reason to expect this additivity other than in those cases where the perturbation is small. However, the changes in the proton shifts are hardly small in terms of the observed range of proton chemical shifts and shift anisotropies. This additivity suggests that the size of the shift changes observed experimentally may reflect the total degree of hydration of the amide group as a whole.

Planar vs Rotated Forms

The data in Table 2 is grouped by planar (P) and rotated (R) forms of the standard (s) and optimized (o) hydrates. The table shows that the differences between the unhydrated and hydrated planar and rotated forms is small for oxygen, carbon, and nitrogen, but is noticable in the case of hydrogen. For this species the angle A and angle differences are quite different (vide infra) while the tensor elements in the unhydrated form differ by one or two ppm, a significant variation for hydrogen.

The differences between the standard (s) and angle-optimized forms (o) show up for oxygen, carbon and nitrogen in a slightly reduced shift change as one goes from the s to the o forms for the water of hydration at the carbonyl group; for hydration at the NH group, the relative changes are approximately the same. For hydrogen the changes are essentially the same for carbonyl group hydration and only very slightly reduced for hydration at the NH proton.

The relatively large orientation changes for the proton tensor are worthy of additional comment. Here it will prove useful to refer to the angle A itself rather than to angle changes in order to compare hydration of the planar and rotated forms. We discuss these effects for the important case of monohydration at the NH proton; the pertinent data from Table 2 is shown schematically in Figure 2. For the unhydrated planar form, A is 19.6°; upon hydration using our standard parameters this angle changes to -0.2°, XX thus aligning itself essentially along the NH...O bond. If the water of hydration at the NH group is allowed to angle-optimize, ΔE changes by -3.0 kcal/mole and the angle A moves further negative (clockwise) to -10.2°. For the carboxyl-group-rotated form (R), the unhydrated species has an angle A of -13.9°. Upon hydration using our standard parameters this angle is reduced to -5.9°, thereby being more closely aligned with the NH...O bond, as is the case for the standard

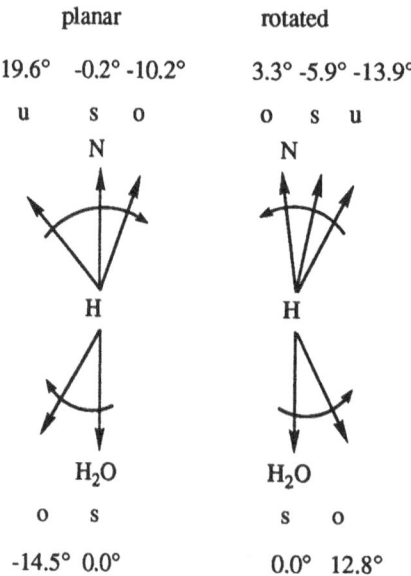

planar   rotated

19.6°   -0.2° -10.2°   3.3° -5.9° -13.9°

u   s   o   o   s   u

N   N

H   H

H_2O H_2O

o   s   s   o

-14.5° 0.0°   0.0°   12.8°

Figure 2. A schematic representation of the orientation of the proton XX (near bond) tensor and NH...O axis directions for the unhydrated (u), standard (s), and angle optimized (o) planar and rotated NH monohydrates.

hydrated planar form. If the water of hydration at the NH proton is allowed to angle optimize, only a small reduction in ΔE ensues (-0.28 kcal/mole), but the angle A changes further counterclockwise from -5.9˙ to +3.3˙. We note that the changes of the angle A going from the s to the o forms are opposite for the planar and rotated forms and that the change in the orientation of the water molecule itself is opposite to the A angle change in each individual case. The interesting observation is that the near-bond tensor component (XX) tends to "follow" the orientation of the NH...O hydrogen bond axis (as defined by the H...O direction). That is, the XX direction tends to reorient itself going from s to o configurations so as to try and remain parallel to the NH...O direction. In the two examples here, the two directions remain colinear within 10 degrees. It is a problem worthy of further study, for being able to determine the local hydration configuration from the orientation of the chemical shift tensor would be a valuable probe.

Comparison Between Calculation and Observation

We present in this section some data on experimental peptide shift tensors that have been reported and compare them with some of the results in this study.   Oas et al. have presented data for the carbonyl carbon [8] and amide nitrogen [9] chemical shift tensors for several dipeptides as determined from dipole-coupled chemical shift powder patterns.   They synthesized a homologous series of four end-protected dipeptides, N-acetyl-gylcyl-X-amide where X was the residue alanine, glycine, or tyrosine, and also present data on glycylglycine hydrochloride; Hartzel et al. [10] provide additional carbon and nitrogen data for alanylalanine.   These shifts were measured relative to internal standards (benzene for carbon and liquid ammonia for nitrogen) which we have converted to absolute shifts using the absolute scales of Jameson and co-workers [18,19].   In this experimental work one of the principal axes was assumed to be perpendicular to the amide plane, although the authors imply a directional error of 2-3°.   Disposition of the "in plane" tensor directions was determined relative to the CN bond.   We are able to relate this data to the CO bond for carbon (used in this study) by knowing the OCN angle of 124.5° assumed by these workers (compared to our model angle of 121.8°).   In addition to the above work, Teng and Cross [11] present nitrogen data for the doubly labled molecules (glycyl)$_2$(alanyl)$_3$-gramicidin A and (alanyl)$_3$(leucyl)$_4$-gramicidin A, each with a peptide bond labeled with ^{13}C and ^{15}N.   Their shifts were again measured relative to an internal standard (a saturated solution of NH$_4$NO$_3$) and, as with the other experimental data cited here, have been converted to an absolute scale using the absolute nitrogen scale of Jameson and co-workers. [18]

In Table 3A we consider the experimental data for the carbon tensors and compare them to that calculated here for the unhydrated, rotated form of glycylglycine (GG(R)) as well as the dihydrated, angle-optimized form (GG[R]2W[o]); similar data for nitrogen are shown in Table 3B.   Averages and standard deviations are given for the experimental data, not so much to presume that this data should be averaged, but simply to indicate the mean value found among a variety of peptides along with an estimate of the spread of these values.   These averages may be compared to our calculated results for both the unhydrated and the doubly hydrated model glycylglycine (section b of Table 3).   In addition, differences (section c of Table 3) are given between our dihydrated model results and the mean experimental results tabulated in the tables.

An inspection of the carbon data in Table 3A shows that in all cases the changes present going from our model unhydrated system to our model dihydrated system are in the proper direction to mimic the experimental results tabulated there.   A combination of experimental error in the angle determination plus the necessary crudeness of our model suggest that angle differences of 5° or 10° may not necessarily be significant.   Recall that the CO bond in our model calculation has a different orientation than that assumed experimentaly in that the OCN angles are different in the two cases, differing by approximately

Table 3

Calculated and (solid state) observed tensor properties of the amide group carbon and nitrogen atoms. See the caption of Table 2 for definitions.

A. Carbon Data

System	A(CO)	XX	YY	ZZ	σ	Δσ	R
a. Observed							
GlyGlyHCl	12.1°	7.0	-59.6	95.1	14.2	121.4	154.7
AcGlyGlyNH$_2$	0.0°	0.0	-58.8	93.0	11.4	122.4	151.8
AcGlyAlaNH$_2$	2.1°	-0.7	-57.9	94.2	11.8	123.5	152.1
AlaAla	3.6°	13.4	-59.8	89.2	14.3	112.4	149.0
AcGlyTyrNH$_2$	6.2°	18.7	-58.3	88.9	16.4	108.7	147.2
x(s)	4.80°(4.70°)	7.7(8.4)	-58.9(0.8)	92.1(2.9)	13.6(2.0)	117.7(6.7)	151.0(2.9)
b. Calculated							
GG(R)2W(o)	-1.6°	23.1	-61.8	107.7	23.0	127.0	169.5
GG(R)	-5.5°	41.2	-72.3	108.0	25.6	123.5	180.3
c. Differences	-6.4°	15.4	-2.9	15.6	9.4	9.3	18.5

(Table 3 continued)

B. Nitrogen Data

System	A(CN)	XX	YY	ZZ	σ	Δσ	R
a. Observed							
GlyGlyHCl	9°	183.7	31.0	181.2	132.0	77.6	152.7
AcGlyGlyNH$_2$	10°	200.3	30.4	176.8	135.8	96.7	169.9
AcGlyAlaNH$_2$	10°	196.4	11.6	155.9	121.3	112.6	184.8
AlaAla	16°	175.7	25.5	162.9	121.4	81.5	150.2
AcGlyTyrNH$_2$	8°	188.9	31.7	163.9	128.2	91.1	157.2
(Gly)$_2$(Ala)$_3$gramA	14°	186.8	17.8	160.8	121.8	97.5	169.0
(Ala)$_3$(Leu)$_4$gramA	15°	188.8	22.8	159.8	123.8	93.5	166.0
x(s)	12°(3°)	188.7(8.1)	24.4(7.6)	165.9(9.4)	126.3(5.8)	92.9(11.5)	164.3(12.0)
b. Calculated							
GG(R)2W(o)	8.4°	254.0	75.7	196.3	175.3	118.0	178.3
GG(R)	10.6°	260.5	97.7	191.3	183.2	116.1	162.8
c. Differences	-3.6°	65.3	51.3	30.4	49.0	25.1	14.0

3°. An examination of the differences (part c in Table 3A) shows that those cited for the XX and ZZ principal elements of the shift tensor of approximately 15-16 ppm are significant. The differences in σ and Δσ of 9 ppm are again probably significant as is the difference between the calculated and observed range of the spectrum (R). That is, to the extent that our model gas phase calculations mimic the true situation in the solid, we would expect to have somewhat better agreement between calculated and observed tensor data than that seen in Table 3A. On the other hand, it is good that the relative ordering of the tensor elements is correct, and that the angular orientation of the shift tensor agrees with that seen experimentally. We dissus these differences further below.

In Table 3B is shown comparable data for nitrogen. Here the angles calculated and observed (relative to the CN bond) are in very good agreement, while the individual principal values are high by an average of some 49 ppm. While nitrogen shifts are more difficult to calulate than those for carbon, we believe our calulations should be good to the order of 10-15 ppm and certainly the observed differences here of approximately three times this value need explanation. The relative ordering of the tensor elements and angles are again correct, and in all but the case of the ZZ principal value, the shifts move closer to experiment upon hydration; in the ZZ case the shift change is small and probably within the calculational noise level.

While the differences for the carbon values shown in Table 2A are borderline to being significant, the differences for the nitrogen shift results are definitely much greater than we would like to accept theoretically. If we accept the premise that, given an adequate model for what one observes experimentally, the calculated chemical shifts should more nearly agree, then we must conclude that our model is in some way _inadequate_. This is not surprising since we are trying to use a relatively simple dihydrate of an isolated dipeptide to mimic what is present in the solid (or solution) where there are many more nearest neighbors and the likelihood of many more interactions to affect the amide group nuclei.

We can, in fact, partially quantify these differences by looking at some gas phase calculations on relatively simple molecules and compare our calculated results to those observed both in the gas phase and in the neat liquid. Jameson et al. [18] have provided an absolute shift for nitrogen in HCN of -20.4 ppm in the gas phase, and have estimated a gas-to-liquid shift of approximately +12.0 ppm leading to a shift value for HCN liquid of -8.3 ppm. Our own calculations on HCN using a 6-311G(d,p) basis with a structure optimized in the same basis lead to a value of -31.1 ppm, in reasonable agreement with the Jameson group gas-phase value. However, while differing by some 10 ppm with the gas phase species, our calculations relative to the neat liquid differ by some 22 ppm; the intermolecular interactions in liquid HCN are obviously reflected in its chemical shift. Srinivasan and Lichter [20] report a nitrogen chemical shift for formamide of 112.4 ppm relative to liquid ammonia, which, when converted to an absolute value using the data of Jameson et al. [18] leads to an absolute chemical shift of 132.0 ppm. Using the same 6-311G(d,p) basis and a structure optimized at the same level, we calculate a nitrogen chemical shift of 176.5 ppm, a result which is some 44.5 ppm higher than neat formamide. The difference here of 44.5 ppm is very close to that found in our model dipeptide system when comparing it to dipeptides observed in the solid state (49.3 ppm). Obviously, the formamide group in such systems is quite susceptible to intermolecular effects such as would be found in the liquid or solid state but which in our simple model are apparently only poorly mimicked by the presence of only two, in-plane waters of hydration. If one were to "correct" our calculated nitrogen tensors by +44.5 for each element one would acheive an agreement between calculated and observed nitrogen tensor elements of the order of 15 ppm as an rms error, quite acceptable at the current level of theoretical capability. This would, however, be an artificial approach; it clearly would be better to derive a model which more nearly reflects the interactions in the liquid and solid states.

Data for carbon chemical shifts for formamide and HCN are also available from the compilations of Breitmaier and Voelter [21]. They report values for carbon in these two compounds relative to TMS which, again, when corrected to the Jamesons' absolute scale for carbon [19] lead to values in the neat liquids of formamide and HCN of 16.5 and 73.2 ppm, respectively. Our calculated isotropic shifts for these two species (again, at the 6-311G(d,p) level) are 35.4 and 81.6 ppm, respectively, leading to calculated-versus-observed differences of 18.9 ppm in the case of formamide and 8.4 ppm in the case of HCN. The gas phase value for HCN is 82.1 ppm, in excellent agreement with the calculated value; the difference of the calculated and neat observed values is not unreasonable in terms of our current theoretical noise level. The difference between calculated and neat observed values for formamide of nearly 19 ppm is, however, beyond that which we would presently accept theoretically. Here again, then, we see that the amide group is apparently especially sensitive to the nature of its surroundings, showing a large change in going from gas to the liquid or solid states.

Data for amide proton tensors and shifts is sparse. Reimer and Vaughan [22] present tensor elements for the amide proton in acetanilide relative to TMS which can be converted to absolute values using the reported shift of TMS of Emsley, Feeney, and Sutcliffe [23] relative to CH_4 [24] and Raynes' reported value for CH_4 [25]. Since this data was obtained from powder spectra no tensor orientation data is available. R. Gerald has kindly provided us with the unpublished crystal data of Haeberlen et al. [26] for the hydrated amide proton in N-acetyl-d,l-valine (denoted here as N-AcVal), a system in which the amide and carboxyl planes have a dihedral angle of 40 degrees (or 50 degrees, depending on definition of the dihedral angle). This data is given in Table 4 along with our calculated standard (s) planar and rotated (90°) forms of GG both unhydrated and doubly hydrated, as well as the comparable data for that form of GG in which the rotation of the COOH group relative to the amide group is 45 degrees. For the one piece of experimental orientation data (N-

Table 4

Observed and calculated shift tensor parameters for the amide proton. See the caption to Table 2 for definitions.

Hydrogen Data

System	A(NH)	XX	YY	ZZ	σ	Δσ	R
a. Observed							
N-AcVal	-8.8°	31.09	20.33	15.89	22.44	12.98	15.20
acetanilide	...	35.3	17.7	17.6	23.5	17.7	17.7
b. Calculated							
GG(P)	19.6	33.90	31.54	19.75	28.40	8.25	14.15
GG(P)2W(s)	-0.1	38.21	24.46	12.37	25.01	19.79	25.84
GG(45)	-13.3	33.76	31.18	21.45	28.80	7.44	12.31
GG(45)2W(s)	-4.3	38.45	23.75	13.98	25.39	19.58	24.47
GG(R)	-13.9	34.72	30.50	21.62	28.94	8.66	13.10
GG(R)2W(s)	-5.5	39.53	22.88	14.27	25.56	20.97	25.27

238

AcVal) our hydrated species compare favorably, differing at most by 8.7 degrees. Our calculated XX and YY (in plane) tensor values are larger than those observed, while our ZZ (out of plane) values are somewhat too low. Accordingly, σ, $\Delta\sigma$, and R are all calculated to be larger than observed for these two experimental cases. The disagreement here is likely due to two problems: our model is undoubtably oversimplified compared to experimental reality, and it remains generally difficult to accurately calculate proton chemical shifts (calculations generally being too large [27], as observed here). The calculations for our hydrated species are much closer to experiment than those for the unhydrated species, so the basic essentials of hydrogen bonding are being properly reflected but not yet in any quantitative manner. With the exception of the XX value for the 45-degree-rotated form, there is, as expected, a smooth trend in values and angles moving from planar to the completely rotated (90°) forms.

Summary

We reiterate here the key points made previously and point to future calculations that need to be done.

The largest effects of hydration are seen at the carbonyl oxygen and the amide proton, as would be expected. The oxygen resonance is calculated to move upfield by some 42 ppm on average, while the amide proton moves downfield by some 3.3 ppm; significant changes are seen in both cases for the anisotropy, oxygen showing a decrease of some 79 ppm on average while the proton exhibits an extremely big change (for hydrogen) of some 11.6 ppm. The oxygen, carbon, and nitrogen shift tensor orientations are minimally affected by hydration, while there are large changes for the amide proton which show that the proton XX tensor direction tends to follow the orientation of the hydrogen bonded water. While the amide plane is a plane of symmetry for the planar form of the hydrates, it also effectively contains the shift tensors in the rotated form. The shift tensors and orientation angles exhibit an additivity in that the doubly hydrated structures have values that are very nearly the sum of the individual monohydrates.

Can such model calculations provide helpful insights into the effects of hydration? The answer would appear to be "yes". The additivity effect indicates that the total degree of hydration may be exhibited in shift changes, and the amide proton following of the local water orientation, if general, would be most useful in determining the local hydration geometry. Do our present calculations properly mimic the experimental solid state? The answer here is clearly "no" in that the discrepancy between model calculated and experimental values is too large in terms of our current theoretical (and experimental) uncertainties. The disparity between gas phase calculated shifts for carbon and nitrogen in HCN and formamide and the values observed for the corresponding neat liquids shows clearly that interactions other than two in-plane hydrating waters are needed to properly reflect interactions in the solid. In particular, the pi system of the amide group is undoubtable sensitive to molecular interaction outside the amide group plane. We feel optimistic that with larger and faster computer facilities and use of techniques such as the locally dense basis set approach systems of the size necessary to mimic condensed molecular systems can be studied.

Acknowledgements

We are indebted to Mr. K.D. Moore of the North Carolina Supercomputing Center for performing the glycyglycine optimizations and for many helpful conversations, to Mr. Rex Gerald II for providing the proton tensor data for N-acyl-d,1-valine prior to publication, and to the North Carolina Supercomputing Center for providing the cpu time on the Center's Cray Y-MP 8/464 that allowed us to carry out the calculations.

References

1. M. Schindler, J. Am. Chem. Soc. 110, 6623-6630 (1988).
2. W. Kutzelnigg, Isr, J. Chem. 19, 193-200 (1980); M. Schindler and W. Kutgelnigg, J. Chem. Phys. 76, 1919-1933 (1982).
3. D. B. Chesnut and C. G. Phung, Chem. Phys. Lett. 183, 505-509 (1991).
4. J. F. Hinton, P. Guthrie, P. Pulay, and K.Wolinski, J. Am. Chem. Soc. 114, 1604-1605 (1992).
5. K. Wolinski, J. F. Hinton, and P. Pulay, J. Am. Chem. Soc. 112, 8251-8260 (1990).
6. R. Ditchfield, Mol. Phys. 27, 789-807 (1974).
7. D. B. Chesnut and K. D. Moore, J. Comp. Chem. 10, 648-659 (1989).
8. T. G. Oas, C. J. Hartzell, T. J. McMahon, G. P. Drobny, and F. W. Dahlquist, J. Am. Chem. Soc. 109, 5956-5962 (1987).
9. T. G. Oas, C. J. Hartzell, F. W. Dahlquest, and G. P. Drobny, J. Am. Chem. Soc. 109, 5962-5966 (1987).
10. C. J. Hartzell, M. Whitfield, T. G. Oas, and G. P. Drobny, J. Am. Chem. Soc. 109, 5966-5969 (1987).
11. Q. Teng and T. A. Cross, J. Magn. Reson. 85, 439-447 (1989).
12. Basis set notation used here is discussed in W. J. Hehre, L. Radom, P.v.R. Schleyer, and J. A. Pople, Ab Initio Molecular Orbital Theory, John Wiley and Sons, New York, 1986.
13. R. Taylor and O. Kennard, Acta Cryst. B39, 133-138 (1983).
14. C. Ceccarelli, G. A. Jeffrey, and R. Taylor, J. Mol. Structure 70, 255-271 (1981).
15. H. C. Freeman, G. Robinson, and J. C. Schoone, Acta Cryst. 17, 719-730 (1964); R. E. Marsh and J. P. Glusker, Acta Ctryst. 14, 1110-1116 (1961).
16. R. Parthasarathy, Acta Cryst. B25, 509-518 (1969).

17. S. F. Boys and F. Bernardi, Mol. Phys. 19, 553-566 (1970).

18. C. J. Jameson, A. K. Jameson, D. Oppusunggu, S. Wille, P. M. Burrell, and J. Mason, J. Chem. Phys. 74, 81-88 (1981).

19. A. K. Jameson and C. J. Jameson, Chem. Phys. Lett. 134, 461-466 (1987).

20. P. R. Srinivasan and R. L. Richter, J. Magn. Reson. 28, 227-234 (1977).

21. E. Breitmaier and W. Voelter, Carbon-13 NMR Spectroscopy, 3rd ed. (VCH, Weinheim, FRG, 1987).

22. J. A. Reimer and R. W. Vaughan, J. Magn. Reson. 41, 483-491 (1980).

23. J. W. Emsley, J. Feeney, and L. H. Sutcliffe, "High Resolution Nuclear Magnetic Resonance", Vol 2, Pergamon Press, New York, 1968

24. M. Schindler and W. Kutzelnigg, J. Am. Chem. Soc. 105, 1360-1370 (1983).

25. W. T. Raynes, "Nuclear Magnetic Resonance", 7, pg 25 (1978).

26. Private Communication, U. Haeberlen, T. Bernhard, S. Opella, J. Rendell, and R. Gerald II, 1992.

27. D. B. Chesnut and D. A. Egolf, unpublished data.

EFFICIENT IMPLEMENTATION OF THE GIAO METHOD FOR MAGNETIC PROPERTIES: THEORY AND APPLICATION

P. Pulay, J. F. Hinton and K. Wolinski
Department of Chemistry and Biochemistry
University of Arkansas
Fayetteville, Arkansas 72701 USA

ABSTRACT. An implementation of the gauge independent atomic orbital (GIAO) method for the ab initio self-consistent-field calculation of nuclear magnetic resonance chemical shifts is described. Significant improvement in the efficiency of the GIAO method has been achieved using techniques borrowed from analytical derivative methods. This improved GIAO method has been used to calculate the individual tensor components of the magnetic shielding of ^1H, ^{13}C, ^{15}N, ^{19}F, ^{29}Si and ^{33}S in a wide variety of molecules and molecular systems containing a relatively large number of heavy atoms. Very good agreement has been found between the theoretically determined magnetic shielding properties and those obtained experimentally.

1. Introduction

Nuclear Magnetic Resonance is the most widely used structural tool in modern chemistry and biochemistry. In spite of the large amount of empirical information accumulated over the years, correct signal assignment and the understanding of the relationship between molecular structure and isotropic chemical shifts may be difficult at times. Quantum chemical methods, especially *ab initio* calculations, are becoming accurate and affordable to be useful in solving these problems. NMR chemical shift tensors are less accessible experimentally and less well understood than the isotropic values, and theoretical values are particularly important in interpreting experimental results. The present paper summarizes our experience with the GIAO (Gauge Independent Atomic Orbital) method for calculating NMR chemical shifts and shift tensors. Although the overwhelming majority of chemically significant results have been obtained at the Hartree-Fock level, we will discuss the inclusion of electron correlation.

NMR chemical shifts, although the most important magnetic property, are not the only ones of importance. The simplest magnetic property, susceptibility, does have some chemical applications. More important are emerging techniques like vibrational circular dichroism and Raman optical activity. Although they appear quite different at first sight, they are in fact closely related to other magnetic properties such as NMR chemical shifts, and share some of the techniques and difficulties with them.

2. The GIAO Method

A common difficulty in calculating magnetic properties is the gauge problem [1]. In its simplest

J. A. Tossell (ed.), Nuclear Magnetic Shieldings and Molecular Structure, 243–262.
© 1993 *Kluwer Academic Publishers.*

form, this manifests itself in the (obviously unphysical) dependence of the calculated results on the position of the system in the Cartesian frame. One way to resolve this difficulty is to use basis functions which explicitly depend on the external homogeneous magnetic field **B**:

$$\chi_p(\mathbf{r},\mathbf{B})=\chi_p(\mathbf{r}) \exp[-i(\mathbf{B}\times\mathbf{R}_p)\bullet\mathbf{r}/2c] \tag{1}$$

Here the usual Coulomb gauge is being used, i.e. the vector potential is defined as $\mathbf{A}=\mathbf{B}\mathrm{x}\mathbf{R}/2$, $\chi_p(\mathbf{r})$ is a field-free basis function, usually a Gaussian, with its center at $\mathbf{R}_p$, and $\mathbf{r}$ is the electronic coordinate vector. These orbitals have been introduced by London [2], and have been thoroughly discussed by Ditchfield [3]. The usual term, gauge-independent atomic orbitals (GIAOs) is not very fortunate. The term gauge-dependent orbitals [4] is better but the GIAO abbreviation has been too deeply entrenched to change it. The expression gauge-including atomic orbitals, suggested by the Ahlrichs group, seems to be a good alternative.

The most important property of the GIAOs is not formal translational invariance. Rather, $\chi(\mathbf{r},\mathbf{B})$ is the correct first-order wave function for a central field problem in a magnetic field if $\chi(\mathbf{r})$ is the exact solution in the field-free case [5]. This statement is valid for arbitrary location of the center. (For higher angular momentum functions, the correct zeroth-order functions must be chosen, e.g. for a magnetic field in the z direction, the functions have to be eigenfunctions of $\mathbf{L}_z$, the z component of the angular momentum.) GIAOs thus incorporate the bulk of the effect of the magnetic field at the basis function level.

The GIAO method at the *ab initio* level was first adopted for NMR chemical shift calculations in the pioneering work of Ditchfield [6]. Later, this method was independently implemented by Giessner-Prettre *et al.* [7] and Fukui *et al.* [8]. This method has been used successfully for small and medium-sized molecules, e.g. [9-12]. Nevertheless, most of these programs required too much in computing resources to be routinely applicable to large molecules. Part of the reason for this was the limited capabilities of most computers in this period (1975-85). Another reason was that these early computer programs could not yet include many technical innovations which have been developed in meantime, mainly in the context of *ab initio* gradient and derivative theory. With the recent dramatic advances in computer power, particularly the power of workstations, calculations on large molecules, using good basis sets, became feasible. This, in turn, required a reinvestigation of the techniques used in MNR chemical shift calculations.

The gauge factor in Eq. (1) can be absorbed [13] in the coordinates of the orbital centers, $\mathbf{R}_p$ in Eq. (1). The latter then become complex and field-dependent. This picture provides a compelling analogy with *ab initio* gradient theory [14]. In the latter, the basis function centers depend on the nuclear positions, and usually travel with the atoms on which they are centered. In the GIAO method, the basis functions are twisted around the z axis in the *imaginary* xy plane (assuming a field in the z direction). In spite of the difference in the physics, the two phenomena are closely related at the mathematical level [14]. This is the basis for our efficient implementation of the GIAO method [15]. As emphasized in our paper [15], imaginary quantities present only a minor complication: no complex arithmetic is required in the fully analytical method, as all quantities involved are either real or pure imaginary.

The GIAO method is not the only one guaranteeing gauge invariance. Very large basis sets span the orbital space necessary to describe the first-order wave function, and thus lead to approximate gauge invariance without gauge factors. The formulation of this Coupled Hartree-Fock method [16] was an important milestone in the field. However, the requirements on the basis set become more and more severe as the distance of the atom from the origin increases. For this reason, we do not believe that this method can be readily extended for large systems, in spite of the accurate results obtained for smaller molecules (e.g. [17]).

Gauge factors do not have to be included at the AO level. Two recent successful methods, the Individual Gauge for Localized Orbitals (IGLO) method of Schindler and Kutzelnigg [18], and the Localized Orbital/Local Origin (LORG) method of Hansen and Bouman [19], include gauge factors at the level of localized molecular orbitals. These methods, particularly IGLO, started the era of modern NMR chemical shift calculations. In light of our comments on the GIAOs, both IGLO and LORG depend on the fact that localized molecular orbitals are somewhat similar to atomic orbitals. Both IGLO and LORG has been very successfully applied to a number of chemical problems (see e.g. [20]). Nevertheless, it appears to us that applying the gauge factors at the AO level, i.e. the GIAO method, has some advantages over the localized methods (IGLO and LORG), and, as we hope to demonstrate, they can be as efficient as they are. Our reasons for preferring the GIAO method are the following:

(a) Both localized methods involve the closure relation which is only approximately valid for finite basis sets.

(b) While GIAOs are exact first order wavefunctions of the one-center problem in the presence of magnetic field, this is only approximately true for the localized orbitals. This is supported by preliminary experience [15] which indicates that, using the same basis sets, GIAO calculations are somewhat more accurate in reproducing the NMR shifts, and particularly the tensor components, than IGLO calculations are.

(c) Some chemical systems have genuinely delocalized orbitals. The performance of the localized methods may suffer in these cases. More frequent are multiple solutions to localization, or symmetry violations which make the localized results slightly less well defined than the GIAO ones.

(d) Generalization to correlated wavefunctions, although possible in both techniques, is more straightforward in the GIAO method. A GIAO Møller-Plesset second order (MP2) implementation has been reported by Vauthier and Fliszar [21] but it has been criticized [22] as missing some terms. Gauss [22] has very recently derived and implemented the full equations. A second order approximation, SOLO, which is similar to but not identical with the Møller-Plesset second order expression, has been implemented for the LORG method [23].

(e) The GIAO method can be implemented without the four-index integral transformation.

3. SCF Theory with GIAO

In order to establish the notation, we summarize here the basic procedure for calculating NMR shieldings by the *ab initio* SCF technique. The magnetic shielding constant is the second derivative, with negative sign, of the electronic energy with respect to the external magnetic field B and the nuclear magnetic moment of a given nucleus, m_n. As the dimension of the magnetic moment is energy/field, this quantity is dimensionless. In general, it is a nonsymmetrical tensor but in most experiments only its symmetrical part is observable.

In the following, we will follow Ditchfield by denoting differentiation with respect to the external magnetic field and the nuclear magnetic moment, in this order, by superscripts. The tensor components, if important, will be denoted by a,b (a, b=x,y,z); if unimportant, we use the superscripts 0 (for unperturbed quantities) and 1 (for first-order quantities). In derivative Hartree-Fock theory, using atomic orbital basis everywhere, the shielding is

$$\sigma = \text{Tr}\{\mathbf{D}^{oo}\mathbf{h}_n{}^{ab}\} + \text{Tr}\{\mathbf{D}^{ao}\mathbf{h}_n{}^{ob}\} \tag{2}$$

Here $\mathbf{D}$ is the first-order reduced density matrix in AO basis, and $\mathbf{h}$ is the matrix of the one-electron part of the Hamiltonian. This formula is valid for both field-dependent and field-independent basis functions; the difference between the two is in the expression for the first-order density $\mathbf{D}^{ao}$ and $\mathbf{h}^{ab}$ elements. Note that Eq. (2) is an asymmetrical expression, and an alternative formalism could use the first-order density with respect to the magnetic moments instead of the external field. This alternative has not been seriously considered because, in general, the shielding for a number of nuclei are evaluated at once, necessitating the calculation of many first-order densities $\mathbf{D}^{ob}$, while the external field can have only three components. Nevertheless, the formulation in which the two perturbations are interchanged avoids the complications arising from the derivative two-electron integrals, since the basis functions in the GIAO formalism do not depend explicitly on the magnetic moments of the nuclei, only on external field.

The derivatives of the one-electron Hamiltonian are given in the GIAO formalism as

$$(\mathbf{h}^{ab})_{pq} = <p\,|\,h^{ab}q> + <p^a\,|\,h^{ob}\,|\,q> + <p\,|\,h^{ob}\,|\,q^a> \tag{3}$$

where the subscript $_n$, denoting the nucleus of interest, has been omitted. One-electron integrals are usually considered as a minor time consumer in *ab initio* programs but in the present case these one-electron integrals are computationally significant, as there are 9N matrices to be evaluated, one for each nucleus and a pair of spatial directions (ab). One way to improve efficiency in this stage is to calculate the integrals in batches for all nuclei which eliminates repeations in the logic. The operators h^{ab} and h^{ob} are similar to field gradient and spin-orbit coupling operators and their matrix elements over Gaussians can be evaluated using standard techniques; we use the Rys polynomial formulation of Dupuis *et al.* [24].

The first order density matrices in Eq. (2) are the solutions of the coupled-perturbed Hartree-Fock (CPHF) equations [16,3] with the external field as perturbation. Compared to the standard CPHF equations [16] with field-independent functions, there is perturbation in both the one-electron and the two-electron part of the Hamiltonian, due to the field dependence of the GIAOs. The problem is analogous to the calculation of force constants in SCF derivative theory, and has been formulated by Bratoz [25], Gerratt and Mills [26], and Pople *et al.* [27]. We prefer a density matrix based formalism [15]. This can be expressed as

$$\mathbf{D}^1 = (1/2)\mathbf{D}\mathbf{S}^1\mathbf{D} + \sum_{ia} (e_i-e_a)^{-1}\mathbf{C}_i{}^T\times(\mathbf{F}^1-e_i\mathbf{S}^1)\mathbf{C}_a\times(\mathbf{C}_i\mathbf{C}_a{}^T-\mathbf{C}_a\mathbf{C}_i{}^T) \tag{4}$$

In this equation, $\mathbf{C}_i$ and $\mathbf{C}_a$ denote the orbital coefficient vectors of the occupied and virtual orbitals, respectively, and e_i and e_a are the corresponding orbital energies. The superscript 1 denotes first order quantities with respect to the external field, $\mathbf{D}$ is the density matrix, $\mathbf{S}$ is the overlap matrix, and $\mathbf{F}$ is the Fock matrix. The first order Fock matrix depends on the first order density:

$$\mathbf{F}^1 = \mathbf{h}^1 + G(\mathbf{D}^1, g^o) + G(\mathbf{D}, g^1) \tag{5}$$

where

$$\mathbf{h}^1{}_{pq} = <p\,|\,h^1\,|\,q> + <p^1\,|\,h\,|\,q> + <p\,|\,h\,|\,q^1> \tag{6}$$

and

$$G(P,g)_{pq} = \sum_{rs} P_{rs}g_{pqrs} \tag{7}$$

The first order Hamiltonian h^1 is just $(1/2c)$ times the angular momentum operator, and the two-electron integral supermatrix is defined as

$$g_{pqrs} = (pq \mid rs) - (1/4)(pr \mid qs) - (1/4)(ps \mid qr) \tag{8}$$

Because of the implicit dependence of the right-hand side of Eq. (4) on D^1, it must be solved iteratively.

4. Efficiency

The major additional computational tasks in the SCF calculation of NMR shieldings are integral derivative evaluation and the solution of the CPHF equations. Our present program uses the Rys polynomial integral evaluation method [24]. However, we have recently implemented the Obara-Saika recursive technique, as modified by the Pople group [28] and Hamilton and Schaefer [29], and we have also introduced the calculation of the integrals in vectorizable blocks, removing thus the logic overhead. While we do not have full NMR chemical shift timings with the new program, preliminary results indicate that integral/derivative calculations are speeded up by about a factor of 3 relative to our old timings [15]. In addition to higher efficiency, the new code is also more flexible. It allows the use of general contractions [30], and higher angular momentum functions to an essentially unlimited value of angular momentum but at least to h type functions. In order to make the method more efficient for larger molecules, we also plan to include a simple form of the multiplicative integral approximation. The latter was implemented with considerable success by van Alsenoy [31], based on a suggestion of one of us (P. P.)

The new code also allows the use of semi-direct algorithms [32], allowing us thus to extend the calculations to very large systems. Although the use of a fully direct algorithm [33] leads to an algorithm which is essentially unlimited, it is often quite expensive compared to the traditional SCF method, particularly on workstations where the I/O speed is more in line with the CPU speed. Semidirect algorithms store the more expensive integrals and recalculate the cheap ones. According to our experience, one-fourth of the integrals is usually responsible for three-fourths of the computer time. Therefore, integral storage can be reduced by a factor of 4 with only a modest increase in computing costs. Most integrals can be stored economically in 4-byte, an even in 2-byte integers, with implicit labels, with the large integrals stored separately in 8-byte floating point words. Coupled with affordable multigigabyte disks, this allows very large calculations on modest workstations, with little integral recalculation.

Several early implementations of the GIAO method, and the first analytical force constant calculations, have stored the derivative two-electron integrals. As each two-electron integral has 3 derivatives, this method quickly leads to a bottleneck. Fortunately, this step can be easily eliminated, by contracting the integral derivatives with the density matrix elements immediately, and storing only the derivative Fock matrices [34].

An important feature of any program for large molecules is the neglect of small integrals which do not contribute significantly to the results. This is well understood for the two-electron integrals themselves, see e.g. [32]. It is more difficult to approximate the derivative integrals for a second order property like NMR shielding. The reason is that the contribution of the derivative integrals is not direct: they enter the formalism via the first order Fock matrix F^1 in Eq. (4), and their exact contribution becomes known only after the CPHF equations have been solved. Efficient neglect requires, of course, that the contributrion of an integral, or an approximation to it, be known *before* the integral is evaluated. This is possible by invoking the uncoupled Hartree-Fock approximation (UCHF) [35]. While this is considered unreliable by today's standards, we use it only to estimate the magnitude of the contributrion of an integral. In the UCHF approximation, the the first-order density matrix contribution in Eq. (5) is neglected. This allows a simple, non-iterative solution of Eq.(4) for the first order density matrix. Substituting this in the formula for the chemical shift, and extracting the coefficient of a particular two-electron integral, we arrive at an effective density matrix element for each integral which is similar in form to the Hartree-Fock two-particle density, i.e. it decomposes to one-particle densities. This quantity can be used to determine an approximate upper bound for the contribution of an integral. In practice, the upper bound for a whole shell of integrals is determined. The only disadvantage of this method is that it requires the magnetic dipole matrix elements, h^{o1}, before the two-electron integrals are evaluated. h^{o1} consists of three matrices for each nucleus for which the shielding is evaluated, and they need to be stored in this method, or else evaluated again after the first order density matrices from the CPHF step are available. If this device is not used, then an alternative program organization is preferable in which the CPHF equations are solved first, and the integrals h^{o1} are contracted with the first order density matrix directly after being evaluated, see Eq. (2). Although the device described above works quite efficiently for larger systems [15], the savings are more modest than those reported for energy and gradient evaluations. We do not yet quite understand the reason for this.

Another important computational expense is the coupled-perturbed Hartree-Fock procedure. The latter can be performed either in atomic orbital (AO) or in molecular orbital (MO) basis. Most formulations [25-27] use the MO form or a mixed AO-MO form [16]. Both of these require the partial transformations of the integrals to MO basis. The AO method, recommended by Pulay [34] and Osamura *et al.* [35]. avoids the cost of the integral transformation. The relative merits of the AO versus MO formalism deserves some discussion. The integral transformation is, without integral neglect and at constant basis set quality, an $O(N^5)$ procedure, while the rest of the calculation is $O(N^4)$. However, once the integrals are in MO form, the CPHF procedure is significantly less expensive: it scales like n^2V^2 where n is the number of occupied orbitals and V the virtuals. As the ratio n/N is usually less than 3, the CPHF iteration is at least an order of magnitude faster than the AO method. For very many perturbations, as in the case of the calculation of force constants, the MO method is probably preferable, particularly on vector supercomputers where the integral transformation is fast. However, for the magnetic perturbation, which has only three components, the AO method is preferable in our opinion. The transformation part requires extensive integral sorting steps, and these become very I/O intensive on workstation with smaller fast memories. Some of the larger calculations shown later would have been difficult to carry out on workstations in the MO formalism.

The CPHF step is one which significantly benefitted from research in analytical derivative theory. In particular, the seminal paper of Pople *et al.* [27] introduced an efficient combination of direct and iterative methods for the solution of the CPHF equations. Although the CPHF eqation (4) can be regarded as an algebraic equation in the components of the first-order density matrix or the related orbital coeffient derivatives, its solution by direct matrix inversion techniques, as in Ref. [26], is feasible only for very small systems, as it requires $O(N^6)$ operations

[36]. Iterating Eq. (4) is only $O(N^4)$ per iteration but convergence is frequently slow. The method of Pople *et al.* [26] starts with the simple iterative method but finds the solution in the iteratively generated space by finite matrix inversion, in the spirit of similar eigenvector methods [37,38]. Our modification of this method [15,39] is based on the insight of Wormer *et al.* [40] who has shown that the method of Ref. [26] is equivalent to the conjugate gradient method [41], and thus requires only the storage of three vectors and residues for each equation. This results in a simpler program organization and diminished storage but is less important in the present case than for force constant calculations where a large number of CPHF equations are to be solved.

5. Work in Progress

In this section, we discuss some developments which are being implemented or planned in our laboratory. Two projects will be considered: NMR chemical shift calculations using density functional methods, and the vibrational circular dichroism calculations.

Present experience indicates that the Hartree-Fock method, with flexible basis sets, performs well for the NMR chemical shifts of organic molecules, particularly for the chemical shift differences between atoms in similar chemical environment are considered. Nevertheless, significant correlation effects have been reported for the absolute shielding constants, particularly for molecules with multiple bonds [22,23] and for inorganic molecules. Only a few correlated calculations of NMR chemical shieldings have beeen reported, partly because the programs have been developed only recently, and partly because they require significantly more computation than the SCF results. Both the second order LORG (SOLO) approach of Bouman and Hansen [23], and the GIAO-MP2 method of Gauss [22] invokes the second order approximation; this is reasonable in view of the fact that correlation effects are not dramatic. The main drawback of these methods is that they have a steep $O(N^5)$ dependence on the molecular size, and are thus difficult to apply for large systems. We do not discuss MC-SCF calculations for very strongly correlated systems (e.g., ozone). Our emphasis is on large, ground-state molecules where dynamical correlation usually predominates.

Density functional theory (see, e.g. [42]) is able to account for most correlation effects with less computational effort than traditional correlation methods. This theory made remarkable advances in the last few years. Originally viewed skeptically by many in the chemical community because of the semiempirical nature of all practical implementations, numerical imprecision, and the overblown claims of some of its early protagonists, it became a remarkably useful tool in the past years. The main developments contributing to this were improved formulation [43], implementations which do not use the crude muffin-tin approximation [44,45], accurate exchange and correlation potentials from brute force calculations (e.g. [46]), the implementation of analytical derivative techniques [47,48], and improved non-local potentials [49,50]. In view of the good results obtained for molecular geometries, dissociation energies, and force constants, it is very likely that the predictions for NMR chemical shifts will be very good. Moreover, most NMR experiments are performed on ground state closed shell molecules where density functional theories perform best.

In principle the implementation of a density functional GIAO NMR chemical shielding program is straightforward. The modern formulation of density functional theory [43] is very close to Hartree-Fock theory, the only difference being the replacement of the exchange contribution by a semiempirical function of the electron density. Most existing programs also replace the Coulomb contribution by an equivalent but numerically less expensive expression. The main problems are numerical. Most existing density functional programs, at least in their default setting, do not appear numerically accurate enough for NMR chemical shielding calculations. The reason for this is probably the claim that density functional methods are significantly less expensive than equivalent Hartree-Fock methods, and have a lower asymptotic dependence on the molecular size. We are not quite convinced of these claims. Asymptotically, both methods should go as $O(N^k)$ with k somewhat below 3. Based on a limited comparison, we believe that the computational performance of the two methods should be comparable *if the same high numerical accuracy is imposed on both.*

The main feature of our current preliminary implementation of the density functional equations is its high numerical accuracy compared to other similar programs. We aim for an accuracy which is comparable to the SCF methods where two different programs return energies which are identical to at least 10^{-7} atomic units. This will probably lead to a modest loss of efficiency. However, the main reason preferring density functional theory to SCF theory may be not greater efficiency but higher accuracy.

A different but promising application of some of the aspects of the theory presented above is the calculation of vibrational circular dichroism (VCD). VCD is emerging as an important tool to ascertain the conformation of asymmetric centers in biological molecules. The theory of this third-order effect is fairly involved; the reviews [51-54] give an excellent introduction to the field.

In the present context the formulation of Stephens [55] is the most useful. In this formalism, the key quantity is the overlap of the first order wave functions with respect to the external magnetic field and the nuclear coordinates. This, and the usual dipole moment derivatives suffice to evaluate VCD rotatory strengths and signs. The calculation of the first order wave function with respect to the nuclear coordinates, and of the dipole moment derivatives is well known from analytical second derivative theory. The only problem is the calculation of the first order wave function under magnetic perturbation, the same problem we face in NMR chemical shift calculations. Existing methods [56,57] for VCR calculations do not use GIAOs. In view of the importance of the gauge factors for NMR shieldings, one expects that they are also important for VCD. The use of local (or distributed) gauge origins [58] has improved the non-GIAO calculations, and the results, using both the CPHF procedure [56] and a simpler, uncoupled-like approximation [57,59] are quite satisfactory for small molecules (e.g. [51-54]). However, it appears that accurate results for larger molecules will require GIAO calculations. Work along these lines is in progress.

6. APPLICATION OF THE ENHANCED GIAO METHOD

As part of the program to develop an efficient GIAO computer program for the calculation of NMR chemical shifts of nuclei in relatively large molecular systems, we have performed calculations with a number of nuclei (e.g., ^1H, ^{13}C, ^{15}N, ^{17}O, ^{19}F, ^{29}Si and ^{33}S). Before presenting the results for the chemical shift calculations, it is important to understand the criteria used for selecting the molecules and molecular systems to be studied. The ultimate test of the accuracy of the chemical shift calculation lies within the determination of the absolute value of the chemical shielding and the comparison of the calculated value with that obtained by experimental procedures. The absolute shielding value does not require the use of a reference compound containing the nucleus of interest. The chemical shielding calculation is performed typically on an isolated molecule at 0 K. Therefore, there is a limit to the expected agreement between the calculated and experimental absolute chemical shielding value. In the absence of experimentally determined value of the absolute chemical shielding for a particular nucleus of interest, the chemical shift anisotropy of the shielding tensor is next best chemical shielding parameter to use for a comparison of the calculated and experimental results. This chemical shielding parameter does not involve the experimental problems of referencing, bulk susceptibility corrections and magnetic field strength stabilization. The experimental value is customarily obtained from solid state NMR measurements, preferably with the solid at very low temperature to minimize contributions from molecular motion to the chemical shielding. Environmental effects (e.g., crystal packing forces and non-bonded interactions) may affect the chemical shielding. Again, if the calculation is performed on a single molecule at 0 K, the agreement between the experimental and calculated value may not be optimal. Finally, the isotropic chemical shift may be used to compare calculated and experimental chemical shielding values. This comparison depends upon having a suitable reference. In addition to the complicating factors of temperature and environment that may make the agreement between the calculated and experimental result poor, it has been observed that one may obtain fortuitously a reasonable value of the calculated isotropic chemical shift even with a calculation of inferior quality since the calculated isotropic value is the average of the components of the chemical shielding tensor.

Obviously, for the calculated results to be meaningful one must make every effort to have the same geometry for the molecule in the computer is that for which the experimental data was obtained. If an experimentally determined structure (e.g., a structure determined from X-ray crystallography) does not exist, one must resort to theoretical geometry optimization methods to obtain the geometry of the molecule used for the chemical shielding calculations.

In the initial testing phase of the enhanced GIAO chemical shift program chemical shielding parameters were determined, as a function of the quality of basis set, for ^1H, ^{13}C in methane, cyclopropane, cyclopropene, ethylene oxide and benzene; ^{13}C and ^{19}F in methyl fluoride; ^{13}C and ^{33}S in carbon disulfide, ^{13}C, ^{17}O and ^{33}S in dimethyl sulfide, dimethyl sulfoxide and dimethyl sulfone; and ^{17}O and ^{33}S in the sulfate and thiosulfate anions [15]. These molecules were selected for comparison with the results obtained using the localized (IGLO and LORG) methods. A detailed discussion of this comparison of results has been presented [15]. In general, the localized methods are more sensitive to the quality of the basis set employed or it may be stated that convergence of the calculated chemical shieldings is faster with the GIAO method. It was also found that the accuracy of the individual tensor components is often lower with the localized methods than with the GIAO method, even in cases where the isotropic average is correctly predicted by IGLO and LORG.

To further test the efficiency of the GIAO procedure, calculations have been performed on relatively large molecular molecules and molecular systems. The chemical shift anisotropy for the hydrogen atom in the water cluster $(H_2O)_{17}$ has been calculated. This system was chosen for study for a number of reasons: (1) The efficiency of the GIAO chemical shift program can be examined using large basis sets to describe a molecular system that contains a relatively large number of heavy atoms, 17; (2) Experimental NMR measurements have been made on ice to determine the chemical shift anisotropy of the hydrogen atom [60-63]. Significantly different values of the chemical shift anisotropy were obtained in these experimental measurements. A previous calculation, using a small basis set, for the water cluster $(H_2O)_5$ did not resolve the issue of different experimental values [10]; (3) The experimental values were obtained at very low temperatures, 173-195 K. Therefore, the affect of vibrational and librational motions on the chemical shift anisotropy are minimized in the comparison with the calculated

value for the water cluster $(H_2O)_{17}$ at 0 K; (4) Since the crystal structure of ice has been determined at 123 K [64], geometry optimization was not necessary to obtain the cluster structure necessary for the chemical shift calculation and; (5) The effect of a second sphere of solvent molecules on the chemical shift anisotropy can be determined from a comparison of the value for the central water molecule in the $(H_2O)_5$ and $(H_2O)_{17}$ clusters.

Table 1 contains the experimental and calculated values for the principal components of the chemical shift tensor and the chemical shift anisotropy for the water clusters at 0 K and ice. Several basis sets and basis set combinations were used to determine the size necessary for obtaining accurate chemical shift parameters. For the water cluster $(H_2O)_5$, chemical shift calculations were performed with two basis sets and a basis set combination (4-31G basis set on all atoms, 6-311G+1p+1d basis [65] on all atoms) and an *attenuated* basis set in which the 6-311G+1p+1d basis set was used to describe the central atoms (H-O-H) and the 4-31G basis set was used for all of the other atoms. The *attenuated* basis set concept has been described by Chesnut [66,67]. An analysis of the data in Table 1 shows that it is imperative to use a large basis set, the very least, to characterize the central hydrogen-bonded atoms.

For the water cluster $(H_2O)_{17}$ chemical shielding calculations were performed using two basis set combinations. With one calculation a 6-311G+1p+1d basis set was used for the central atoms involved in the hydrogen bond and a 4-31G basis set was used for all other atoms. For the second calculation the central water molecule and the four first sphere water molecules were described by a 6-311G+1p+1d basis set and the remaining twelve water molecules in the second hydration sphere were described by the 4-31G basis set. The values calculated for the chemical shift anisotropy (35.17 ppm and 34.83 ppm) are in good agreement with those experimental values of about 34 ppm. Therefore, these calculations seem to support the higher experimental values Although the two basis set combinations used in these calculations produced essentially the same value for the chemical shift anisotropy for the hydrogen atom in ice, the time required for the two calculations was significantly different A time of 14 hours was required for the calculation for the system in which the central water molecule and the four first sphere water molecules were described by the 6-311G+1p+1d basis set and 8 hours for the system in which only the central water molecule was described by the 6-311G+1p+1d basis.

The necessity for including a second sphere of water molecules in the calculation is evident when one compares the value of the chemical shift anisotropy for the central hydrogen in the $(H_2O)_5$ and $(H_2O)_{17}$ clusters. The addition of the second hydration sphere increases the chemical shift anisotropy by approximately 1 ppm. The chemical shift anisotropy increases by more than a factor of two upon completing the first hydration sphere. These results illustrate the potential importance of considering solvent effects in performing chemical shift calculation on molecules whose experimental values were obtained with the molecule in solution.

The data in Table 1 also show that the tensor components, σ_{11} and σ_{22}, are almost identical in the $(H_2O)_{17}$ cluster, being only different by 0.27 ppm. This result is in agreement with the prediction of approximate axial symmetry for the shielding associated with H-O - - H hydrogen bonds [68]. Such a small difference in principal component values would be difficult to observe experimentally and seems to explain why the powder pattern of ice appears to be axially symmetric. For the water cluster $(H_2O)_5$, these tensor components vary by 1.08 ppm. Again, the need to consider long-range solvent effects is evident.

The calculated values of the chemical shielding parameters for nitrogen atoms involved in multiple bonds, where correlation effects may be significant, frequently are not in good agreement with those obtained experimentally [69]. The GIAO method has been used to calculate the nitrogen chemical shielding parameters for the nitrogen atom in the (-C=N-) moiety found in benzylideneaniline and analogs of all-*trans*-retinylidenebutylimine. This moiety serves as a very demanding test for the GIAO procedure since it has a doubly bonded nitrogen atom with a lone pair of electrons. These molecules also contain a relatively large number of heavy atoms, which tests the efficiency of the procedure, and experimental values of the chemical shielding parameters for the imine atoms have been determined for benzylideneaniline (^{13}C and 15N), all-*trans*-retinylidenebutylimine (^{15}N) and the nitrogen-protonated form of all-*trans*-retinylidenebutylimine (^{15}N) by solid state NMR spectroscopy [70,71].

TABLE 1. Principal components of the chemical shift tensor (σ) and the chemical shift anisotropy [$\Delta\sigma = \sigma_{33} - (\sigma_{11} + \sigma_{22})/2$]

Water	σ_{11} (ppm)	σ_{22} (ppm)	σ_{33} (ppm)	$\Delta\sigma$ (ppm)
ice[a,b]	-17.5	-17.5	11.2	28.7±1
ice[c,b]	-12.5	-12.5	16	28.5±1
ice[d,b]	-19	-19	15	34±4
ice[e,b]	-6.5	-6.5	27.7	34.2±1
$(H_2O)_{17}$[f]	11.97	12.19	46.91	34.83
$(H_2O)_{17}$[g]	11.98	12.25	47.28	35.17
$(H_2O)_5$[h]	12.30	13.32	47.72	33.92
$(H_2O)_5$[g]	12.35	13.43	47.13	34.24
$(H_2O)_5$[i]	15.77	16.27	48.19	32.17
$(H_2O)_3$[i]	17.21	18.36	48.70	30.92
$(H_2O)_2$[i]	18.60	19.57	46.07	26.99
$(H_2O)_1$[i]	24.77	26.54	41.36	15.71

[a]Reference [63]. Temperature was 195 K.
[b]Liquid was used as a chemical shift reference.
[c]Reference [62]. Temperature was 195 K.
[d]Reference [61]. Temperature was 183 K.
[e]Reference [60]. Temperature was 173 K.
[f]In the present work a 6-311G+1p+1d basis set was used for all atoms in the central water molecule and those in the four water molecules in the first hydration sphere and a 4-31G basis set was used for the twelve water molecules in the second hydration sphere. The values listed are those of the chemical shielding and not the chemical shift with respect to a reference.
[g]In the present work a 6-311G+1p+1d basis set was used for all atoms.
[h]In the present work a 6-311G+1p+1d basis set was used for the central H-O - - H atoms and 4-31G on all others. The values listed are those for the chemical shielding not chemical shift with respect to a reference.
[i]Reference [10]. Calculations performed with a 4-31G basis set on all atoms. The values listed are those for the chemical shielding not chemical shift with respect to a reference.

The molecular geometries used for the calculations were obtained from X-ray crystallography for benzylideneaniline [72] and geometry optimization in the case of all-*trans*-retinylidenebutylimine analogs. The calculated geometries were obtained by full optimization using the 4-21G(*) basis set and empirical corrections in the form of *offset forces* [73].

Chemical shielding calculations were performed on two analogs of all-*trans*-retinylidenebutylimine, $CH_2(CH)_9CH=N-(n-butyl)$ and $CH_2(CH)_9CH=N(H^+)-(n-butyl)$. The two analogs differ from all-*trans*-retinylidenebutylimine by not having the cyclohexene ring with the attached methyl groups on the end of the molecule opposite to the nitrogen atom. Additional calculations were performed on three geometry-optimized *t*-butyl analogs, $CH_2(CH)_1CH+N-(t-butyl)$, $CH_2(CH)_3CH=N-(t-butyl)$ and $CH_2(CH)_9CH=N-(t-butyl)$ to determine the effects of the change from *n-butyl* to *t-butyl* and of chain length $(CH)_n$ on the nitrogen chemical shielding parameters.

TABLE 2. Principal components and isotropic value of the chemical shift tensor of benzylideneaniline and all-*trans*-retinylidenebutylimine

NUCLEUS	BASIS SET[a]	σ_{11} ppm[b]	σ_{22} ppm[b]	σ_{33} ppm[b]	σ_i
Benzylideneaniline $(C_6H_5CH=NC_6H_5)$					
^{13}C	1	208.2	106.9	52.8	122.6
	2	238.1	125.5	62.8	131.4
	3	258.7	132.7	72.4	154.6
	4	258.2	134.8	72.3	155.1
	Experiment	235	167	79	160
^{15}N	1	522.8	275.4	20.1	272.8
	2	595.9	314.5	30.6	313.7
	3	645.2	335.6	34.0	338.3
	4	649.2	339.1	35.5	341.3
	Experiment	610	321	65	332

[a] 1, 4-21G(2d,2p,diffuse) on the -CH=N- atoms and 4-21G on all other atoms;
   2, 4-31G(2d,2p,diffuse) on the -CH=N- atoms and 4-21G on all other atoms;
   3, 6-311G(2d,2p,diffuse) on the -CH=N- atoms and 4-21G on all other atoms;
   4, 6-2111G(2d,2p,diffuse) on the -CH=N- atoms and 4-21G on all other atoms.
[b] ^{15}N chemical shifts were referenced to NH3 (liq) at 298 K by determining the chemical shift difference between the compound and the calculated value for $(NH3)_g$ and correcting to the experimental value of $(NH3)_l$. Calculated ^{13}C chemical shifts are referenced to the calculated value of CH_4. TMS was the reference for the experimental data.

TABLE 2. (continued)

all-*trans*-retinylidenebutylimine analogs

CH$_2$(CH)$_9$CH=N-(*n-butyl*)

^{15}N	3	652.0	376.9	8.7	345.9
	Experiment[c]	644.8[d]	330.8[d]	32.7[d]	336.1
		650.9[e]	336.9[e]	38.8[e]	342.2

CH$_2$(CH)$_9$CH=N(H$^+$)-(*n-butyl*)

^{15}N	3	244.5	124.3	39.2	136.0
	Experiment[f]				
	HCl salt	317.5[d]	215.8[d]	48.5[d]	193.9
		323.6[e]	221.9[e]	54.6[e]	200.0
	HBr salt	307.8[d]	204.5[d]	47.9[d]	186.7
		313.9[e]	210.6[e]	54.0[e]	192.8
	HI salt	295.0[d]	185.0[d]	47.5[d]	175.8
		301.1[e]	191.1[e]	53.6[e]	181.9

[c] The sample was all-*trans*-retinylidenebutylimine
[d] Based on a chemical shift difference of 21.1 ppm between NH$_4$Cl (5.6M) and (NH$_3$)$_l$ at 298 K [76].
[e] Based on a chemical shift difference of 27.3 ppm between NH$_4$Cl (5.6M) and (NH$_3$)$_l$ at 298 K [76].
[f] The sample was the protonated form of all-*trans*-retinylidenebutylimine.

Table 2 contains the calculated and experimentally determined values of the principal components of the imine ^{13}C and/or ^{15}N chemical shielding tensor for the molecules described above. A comparison of the calculated values of the principal components of the chemical shielding tensor for the imine moiety of benzylideneaniline as a function of basis set shows that convergence is attained with the *attenuated* basis set, 6-311G(2p,2d,diffuse functions) used to describe the imine group atoms and the directly attached carbon atoms with their protons and a 4-31G used to describe all of the other atoms.

The calculated chemical shielding parameters for the ^{13}C nucleus of the imine group in benzylideneaniline are in good agreement with those determined experimentally. The isotropic chemical shift value ($\sigma_{11}+\sigma_{22}+\sigma_{33}$)/3 for the ^{13}C nucleus is 155 ppm and 160 ppm determined by calculation and experiment, respectively. However, experience in this laboratory with chemical shielding calculations suggests that a more valid comparison of the accuracy of the calculated values is obtained from the comparison of the individual tensor components or the chemical shielding anisotropy. The calculated value of the chemical shift anisotropy (σ_{11}-σ_{33}) is 187.3 ppm compared to an experimental value of 156 ppm. The difference chemical shift anisotropy values may reflect environmental and motional (vibrational and librational) effects in the solid. It is of interest to note that the calculated value (187.3 ppm) is in good agreement with the chemical shift anisotropy found experimentally for other carbon double-bonded species, for example, 196.5 ppm for the -C(CH$_3$)=N-OH moiety of (E)-acetophenone oxime [74] and approximately 200 ppm for olefinic carbons [75] and approximately 195 ppm for carbons of aldehydes and ketones [75].

A comparison of the data for the ^{15}N nucleus of the imine group of benzylideneaniline reveals good agreement between the calculated and experimentally determined chemical shift parameters The calculated isotropic chemical shift is 338 ppm and the experimentally determined value is 332 ppm. The calculated chemical shift anisotropy is 611 ppm compared to an experimentally determined value of 545 ppm. Again, the differences between the calculated and experimental values may be the result of vibrational and librational motion as well as environmental effects.

The data contained in Table 2 shows that the calculated value of the isotropic chemical shift for the nonprotonated form of all-*trans*-retinylidenebutylimine is 345.9 ppm compared to an experimentally determined value of 342.2 ppm. The chemical shift anisotropy is 643.3 ppm (calculated) and 612.1 ppm (experimental). This comparison of calculated and experimental value of chemical shielding parameters illustrates again the danger of using only the isotropic value as a the way to properly evaluate the accuracy of the calculation. The protonation of the imine nitrogen atom of all-trans-retinylidenebutylimine produces a dramatic decrease in the chemical shift anisotropy of the nitrogen atom and a large high-field shift in the isotropic chemical shift value compared to the nonprotonated form. The experimental values for the chemical shielding parameters for the HCl, HBr and HI salts of all-trans-rentiylidenebutylimine given in Table 2 show that the isotropic chemical shift and the chemical shift anisotropy of the nitrogen decrease in the order Cl > Br > I. This trend appears to reflect a decrease in the hydrogen-bond strength between the immonium proton and the anion [71]. The calculations were perform on the protonated nitrogen analog in the absence of an anion. Therefore, the calculated values would be expected to be lower that the experimental values but closer to the values of the HI salt. The calculated isotropic chemical shift value of 136 ppm for the protonated analog of all-*trans*-retinylidenebutylimine is also in quite good agreement with that of 148.3 ppm [71] for the Schiff base linkage observed in dark-adapted ε-[^{15}N]lysylbacteriorhodopsin. This is the single protein in purple membrane in which the anion is not closely associated with the protonated Schiff base.

The calculations showed that the chain length does not make a critical difference in the chemical shielding parameters. However, the replacement of the *n-butyl* group with the *t-butyl* has a significant effect on the chemical shielding parameters, increasing the value of the chemical shift anisotropy and producing a low-field shift in the isotropic chemical shift.

In summary, the enhanced GIAO procedure used for the calculation of chemical shielding parameters for the nitrogen atom having a lone pair of electrons and involved in a double bond appears to produce very satisfactory results. The agreement between the calculated and experimental results, however, must be viewed with with caution since rarely will the molecule in the computer be identical to the molecule or molecular system in the NMR spectrometer.

The chemical shielding parameters for the ^{29}Si nucleus in a variety of compounds have been calculated using the GIAO method. Molecules were chosen for study for which low temperature (-186 °C), solid state NMR spectroscopy had been used to obtain the chemical shielding parameters [77,78]. The molecules are: (1) Si(CH$_3$)$_4$; (2) (CH$_3$)$_3$SiOCH$_3$; (3) (CH$_3$)$_2$Si(OCH$_3$)$_2$; (4) (CH$_3$)Si(OCH$_3$)$_3$; (5) Si(OCH$_3$)$_4$; (6) [(CH$_3$)$_3$Si]$_2$O and (7) (CH$_3$)$_3$SiC$_6$H$_5$ [3]. The ^{29}Si chemical shielding parameters have also been determined for the compound, tetramesityldisilene; (8) [2,4,6-(CH$_3$)$_3$C$_6$H$_2$]$_2$Si=Si[2,4,6-(CH$_3$)$_3$C$_6$H$_2$]$_2$ [78]. The absolute shielding value has also been determined experimentally for the molecule, $Si(CH_3)_4$, in the liquid state [79].

For the molecule, $Si(CH_3)_4$, calculations were performed with molecular geometries; one obtained from gas phase electron diffraction [80] and the other from *ab initio* quantum mechanical geometry optimization. The geometry optimization was performed with a 4-31G basis set for C and H while a 3-21G(*) basis set was used to describe the Si atom. The chemical shielding calculations were performed with a 6-311G+p+d basis set for C and H atoms and a McLean-Chandler [6,5] 12s/6s (631111) 9p/5p (42111)+2d basis set [81] to describe the Si atom. The absolute chemical shielding value of 390.4 ppm and 393.9 ppm were obtained using the theoretically determined and gas phase electron diffraction determined geometries, respectively. The experimentally determined value of the absolute absolute chemical shielding for the Si atom in $Si(CH_3)_4$ is 368.5 $\pm$ 10 ppm [79]. The agreement between the theoretical and experimental values appears to be quite good. Previously, a calculated value of 447.9 ppm was obtained using a 6-31G basis set for geometry optimization and a 6-31G basis set for C and H and 6-31G + d functions basis set for Si in the chemical shielding calculation employing a GIAO method [82].

Table 3 contains a comparison of the experimentally [77] and theoretically determined chemical shielding parameters for the nine molecules previously listed. *Ab initio* quantum mechanical geometry optimization was used to obtain the molecular geometry of each molecule employing the basis set previously described. A comparison of the experimental and theoretical chemical shift values for the principal components of the chemical shielding tensor reveals good agreement between the two sets of values for each molecule. The experimental result for molecule 5, $Si(OCH_3)_4$, indicates cubic symmetry about the Si atom since the principal components of the chemical shielding tensor have the same value. However, a consideration of the molecular geometry of the molecule shows that this cannot be correct for the case of the isolated molecule, as theoretically studied. The difference between the experimental and theoretical values may be a manifestation of experimental error and/or environment effects.

Molecules containing a silicon-silicon double bond are rare [83-86]. One molecule, 8, does exist for which the principal components of the ^{29}Si chemical shift tensor have been determined experimentally [80]. This molecule has too many heavy atoms to permit a calculation to be performed. However, calculations were performed on an analog that was thought to be an adequate model of it, molecule 9 $[CH_2=CH]_2Si=Si[CH=CH_2]_2$. An X-ray structure analysis of molecule 8 revealed that the four C-Si bonds are approximately co-planar [87]. Geometry optimization was used to determine the structure of the model compound in the same planar conformation. The comparison of the data for molecules 8 and 9 reveals poor agreement between the chemical shift anisotropy but better agreement agreement for the isotropic chemical shift. There are several possible explanations for the discrepancy in the chemical shift anisotropy results: (1) The model molecule may not adequately represent molecule 8; (2) The basis set used may not satisfactorily describe the silicon-silicon double bond. The addition of electron correlation may be required for silicon-silicon double bonds [88,89]. These results illustrate the danger of relying exclusively upon the isotropic chemical shift as a measure of how accurately one is able to calculate accurately chemical shifts.

One area of NMR spectroscopy that might benefit significantly from the ability to calculate accurate chemical shielding parameters is ^{13}C chemical shift analysis of large, complex molecules. Calculations have now been performed on three relatively large molecules, corannulene ($C_{20}H_{10}$) [7]-circulene ($C_{28}H_{14}$), [8]-circulene ($C_{32}H_{16}$) and $C_{30}H_{10}$. The results obtained for these molecules suggest that very large molecule calculations should be possible in the near future. Corannulene ($C_{20}H_{10}$) is a polycyclic aromatic hydrocarbon that consist of five six-membered rings joined along their edges, resulting in a five-membered ring in the center of the molecule. The geometry of the molecule is that of a bowl. The carbon framework of corannulene can be thought of as one-third of a buckminsterfullerene molecule (C_{60}). There are three types of nonequivalent ^{13}C atoms in this molecule. Using a 6-31G(*) basis set to describe all of the atoms in the molecule, the enhanced GIAO procedure gave the following chemical shifts, relative to the carbon with the highest field chemical shift which serves as an internal reference, C(I)=0 ppm, C(II)=5.23 ppm and C(III)=9.62 ppm. The experimentally determined chemical shift values are C(I)=0 ppm, C(II)=4.97 ppm and C(III)=8.63 ppm [90]. The theoretical results appear to be quite satisfactory. [7]-Circulene ($C_{28}H_{14}$) is a polycyclic aromatic molecule with a circular arrangement of seven benzene rings. There are three nonequivalent ^{13}C atoms in this molecule. Again, using the carbon with the highest field chemical shift as an internal reference, the calculated chemical shift values are C(I)=0 ppm, C(II)=4.55 ppm and C(III)=10.62 ppm. The experimentally determined values are C(I)=0 ppm, C(II)=4.9 ppm and C(III)=8.5 ppm [91]. Agreement between the calculated and experimental values appear to be quite satisfactory. [8]-circulene is a polycyclic aromatic molecule with a circular arrangement of eight benzene rings. There are four nonequivalent carbon atoms whose relative chemical shifts are C(I)=0 ppm, C(II)=2.73 ppm, C(III)=11.52 ppm and C(IV)=12.70 ppm. Unfortunately, there are no experimental data available for making a comparison with the theoretical results. For the molecule, $C_{30}H_{10}$, there are four nonequivalent carbon atoms whose relative chemical shifts are C(I)=0 ppm, C(2)=-0.94 ppm, C(III)=4.64 ppm and C(IV)=18.85 ppm. Again, no experimental data is available for this molecule. *Ab initio* quantum mechanical geometry optimization was used to obtain the molecular geometry for corannulene, [7]- and [8]-circulene and $C_{30}H_{10}$ necessary for the chemical shift calculations.

Recently, ^{13}C chemical shift calculations performed in this laboratory were used to corroborate experimental evidence [92] that 2,3,5,6-tetramethylbicyclo [2.2.0]hexane] is not the product of photolyzing 7-oxa[2.2.1]hericene.

TABLE 3. Principal components, chemical shift anisotropy and isotropic chemical shift

Compound	σ_{11} ppm	σ_{22} ppm	σ_{33} ppm	$\Delta\sigma$ ppm	σ_i
1 Theory	390.4[a]	390.4[a]	390.4[a]	0	0[b]
2 Expt.[b]	-33 ± 3	-31 ± 3	8 ± 3	41 ± 3	-19 ± 3
2 Theory[b]	-30.6	-20.2	11.1	41.7	-13.2
3 Expt.[b]	-12 ± 3	-12 ± 3	35 ± 3	47 ± 3	4 ± 3
3 Theory[b]	-12.6	-11.2	27.5	40.1	1.2
4 Expt.[b]	23 ± 3	35 ± 3	68 ± 3	45 ± 3	42 ± 3
4 Theory[b]	10.8	29.8	56.1	45.3	32
5 Expt.[b]	80 ± 3	80 ± 3	80 ± 3	0	80 ± 3
5 Theory[b]	67.3	78.9	78.9	11.6	75
6 Expt.[b]	-16 ± 3	-8 ± 3	14 ± 3	30 ± 3	-3 ± 3
6 Theory[b]	-16.4	-8.2	7.5	23.9	-5.7
7 Expt.[b]	-2 ± 3	4 ± 3	29 ± 3	31 ± 3	10 ± 3
7 Theory[b]	-8.2	-5.4	18	26.2	1.5
8 Expt.[b]	180	27	-15	195	64
9 Theory[b]	131.6	86.6	11.3	120.3	76.5

[a] Absolute shielding value
[b] Chemical shift relative to compound 1 [$Si(CH_3)_4$]

1. $CH_3)_4Si$; 2. $(CH_3)_3Si)CH_3$; 3. $(CH_3)_2Si(OCH_3)_2$; 4. $(CH_3)Si(OCH_3)_3$; 5. $Si(OCH_3)_4$
6. $[(CH_3)_3Si]_2O$; 7. $(CH_3)_3SiC_6H_5$; 8. $[2,4,6\text{-}(CH_3)_3C_6H_2]_2Si=Si[2,4,6\text{-}(CH_3)_3C_6H_2]_2$
9. $[CH_2=CH]_2Si=Si[CH=CH_2]_2$

7. References

1. Hameka, H. F., Advanced Quantum Chemistry, Addison-Wesley, New York, 1963, p. 162.
2. London, F. (1937) J. Phys. Radium 8, 397.
3. Ditchfield, R. and Ellis, P. D. (1974), in G. C. Levy (ed.), Topics in Carbon-13 NMR Spectroscopy, Wiley, New York, pp. 1-51.

4. Pople, J. A. (1962) J. Chem. Phys. 37, 53.
5. Epstein, S. T. (1973) J. Chem. Phys. 58, 1592.
6. Ditchfield, R. (1974) Mol. Phys. 27, 789.
7. Ribas Prado, F., Giessner-Prettre, C., Daudey, J.-P., Pullman, A., Young, F., Hinton, J. F., and Harpool, D. (1980) J. Magn. Res. 37, 43.
8. Fukui, H., Miura, K., Yamazaki, H., and Nosaka, T. (1985) J. Chem. Phys. 82, 1410.
9. Ditchfield, R. (1977) Prog. Theor. Org. Chem. 2, 503 and references therein.
10. Hinton, J. F., Bennett, D. L., (1985) Chem. Phys. Lett. 116, 292.
11. McMichael Rohlfing, C., Allen, L. C. and Ditchfield, R. (1982) Chem. Phys. Lett. 86, 380.
12. Chesnut, D. and Foley, C. K. (1986) J. Chem. Phys. 84, 852.
13. Hall, G. G. (1973) Int. J. Quantum Chem. 7,15.
14. P. Pulay (1987) Adv. Chem. Phys. 69, 241.
15. Wolinski, K., Hinton, J. F., and Pulay, P. (1990) J. Am. Chem. Soc. 112, 8251.
16. Stevens, R. M., Pitzer, R. M., and Lipscomb, W. N. (1963) J. Chem. Phys. 38,550.
17. See e. g. Lazzaretti, P. and Tossell, J. A. (1987), J. Phys. Chem. 91, 800 and references therein.
18. Kutzelnigg, W. (1980) Isr. J. Chem. 19, 193; Schindler, M., Kutzelnigg, W. (1982) J. Chem. Phys. 76, 1919.
19. Hansen, A. E. and Bouman, T. D. (1985) J. Chem. Phys. 82, 5035; Hansen, A. E. and Bouman, T. D. (1989) J. Chem. Phys. 91, 3552.
20. Kutzelnigg, W., Fleischer, U., and Schindler, M. in NMR: - Basic Principles and Progress, Vol. 23, pp. 165, Springer, Berlin, 1991.
21. Vauthier, E., Comeau, M., Odiot, S. and Fliszar, S. (1988) Can. J. Chem. 66, 1781.
22. Gauss, J. (1992) Chem. Phys. Lett., in press.
23. Bouman, T. D. and Hansen, A. E. (1990) Chem. Phys. Lett. 175, 292.
24. Dupuis, M., Rys, J. and King, H. F. (1976) J. Chem. Phys. 65, 111.
25. Bratoz, S. (1958) Colloq. Int. C. N. R. S. 82, 2876.
26. Gerratt, J. and Mills, I. M. (1968) J. Chem. Phys. 49, 1719, 1730.
27. Pople, J. A., Raghavachari, K., Schlegel, H. B., and Binkley, J. S. (1979) Int. J. Quantum Chem. Symp. 13, 225.
28. Gill, P. M., Head-Gordon, M., and Pople, J. A. (1989) Int. J. Quantum Chem. Symp. 23, 269.
29. Hamilton, T. P. and Schaefer, H. F. (1991) Chem. Phys., 150, 163.
30. Raffenetti, R. C. (1973) J. Chem. Phys. 58, 4452.
31. van Alsenoy, C. (1988) J. Comput. Chem. 9, 620.
32. Häser, M. and Ahlrichs, R. (1989) J. Comput. Chem. 10, 104.
33. Almlöf, J. (1983) J. Comput. Chem. 3, 385.
34. Pulay, P. (1983) J. Chem. Phys. 78, 5043; Osamura, Y., Yamaguchi, Y., Saxe, P., Fox, D. Vincent, M. A., and Schaefer, H. F. III (1983) J. Mol. Struct. 103, 183.
35. Dalgarno, A. (1962) Adv. Phys. 11, 281.
36. Pulay, P. (1977) in H. F. Schaefer (ed.), Applications of Electronic Structure Theory, Plenum, New York, p. 153.
37. Brändas, E. J. and Goscinski, O. (1970) Phys. Rev. A 1, 552.
38. Davidson, E. R. (1975) J. Comput. Phys. 17, 87.
39. van Lenthe, J. H. and Pulay, P. (1990) J. Comp. Chem. 11, 1164.
40. Wormer, P. E. S., Visser, F. and Paldus, J. (1982) J. Comput. Phys. 48, 23.

41. Hestenes, M. R. and Stiefel, E. (1952) J. Res. Nat. Bur. Stand. 49, 498.
42. Parr, R. G. and Yang, W. (1989) Density Functional Theory of Atoms and Molecules, Oxford University Press, New York.
43. Kohn, W. and Sham, L. J. (1965) Phys Rev. A 140, 1133.
44. Baerends, E. J., Ellis, D. E. and Ros, P. (1973) Chem. Phys. 2, 41.
45. Dunlap, B. I., Connolly, J. W. D., and Sabin, J. R. (1979) J. Chem. Phys. 71, 3396.
46. Vosko, S. H., Wilk, L., and Nusair, M. (1980) Can. J. Phys. 58, 1200.
47. Versluis, L. and Ziegler, T. (1988) J. Chem. Phys. 88, 322.
48. Fournier, R. Andzelm, J., and Salahub, D. R. (1989) J. Chem. Phys. 90, 6371.
49. Perdew, J. P. (1986) Phys. Rev. B 33, 8822.
50. Becke, A. D. (1988) Phys. Rev. A 38, 3098.
51. Stephens, P. J. and Lowe, M. A. (1985) Ann. Rev. Phys. Chem. 36, 213.
52. Freedman, T. B. and Nafie, L. A. (1987) Topics in Stereochemistry 17, 113.
53. Polavarapu, P. L. (1989), in H. D. Bist, J. R. Durig and J. S. Sullivan (eds.), Vibrational Spectra and Structure, Esevier, New York, Vol. 17 B, p. 319.
54. Rauk, A. (1991), in P. G. Mezey (ed.) New Developments in Molecular Chirality, Kluwer, Amsterdam, p. 57.
55. Stephens, P. J. (1985) J. Phys. Chem. 89, 7489.
56. Amos, R. D., Handy, N. C., Jalkanen, K. J., and Stephens, P. J. (1987) Chem. Phys. Lett. 142, 153.
57. Nafie, L. A. and Freedman, T. B. (1983) J. Chem. Phys. 78, 7108.
58. Stephens, P. J. (1987) J. Phys. Chem. 91, 1712.
59. Dutler, R. and Rauk, A. (1989) J. Am. Chem. Soc. 111, 6957.
60. Ryan, L. M., Wilson, R. C., and Gerstein, B. C. (1977) Chem. Phys. Lett., 52, 341.
61. Pines, A., Ruben, D. J., Vegga, S., and Mehring, M. (1976) Phys. Rev. Lett., 36, 110.
62. Rhim, W. K. and Burum, D. P. (1979) J. Chem. Phys., 71, 3139.
63. Burum, D. P. and Rhim, W. K. (1979) J. Chem. Phys., 70, 3553.
64. Peterson, W. W. and Levy, H. A. (1957) Acta Crystallogr., 10, 70.
65. Krishnan, R., Brinkley, J. S., Seeger, R., and Pople, J. A. (1980) J. Chem. Phys., 72, 650.
66. Chesnut, D. B. and Foley, C. K. (1985) Chem. Phys. Lett., 118, 316.
67. Chesnut, D. B. and Moore, K. D. (1989) J. Comput. Chem., 10, 648.
68. Haeberlen, U. (1976) High Resolution NMR of Solids, Academic Press, New York, 159.
69. Chesnut, D. B. (1989) in G. Webb (ed.) Annual Reports on NMR Spectroscopy, Academic Press, vol. 21, 51.
70. Curtis, R. D., Penner, G. H., Power, W. P., and Wasylishen, R. E. (1990) J. Phys. Chem., 94, 4000.
71. Habison, G. S., Herzfeld, and Griffin, R. G. (1983) Biochemistry, 22, 1.
72. Burgi, H. B. and Dunitz, J. D. (1970) Helv. Chim. Acta, 52, 1747.
73. Pulay, P., Forgarasi, G., Zhou, X., and Taylor, P. W. (1990) Vibr. Spectrosc., 1, 159.
74. Wasylishen, R. E., Penner, G. H., Power, W. P., and Curtis, R. D. (1989) J. Am. Chem. Soc., 111, 6082.
75. Veeman, W. S. (1984) Prog. NMR Spectrosc., 216, 193.
76. Mason, J. (1987) in Multinuclear NMR, Plenum Press, London, 336.
77. Gibby, M. G., Pines, A., and Waugh, J. S. (1972) J. Am. Chem. Soc., 94, 6232.
78. Zilm, K. W., Grant, D. M., Michl, J., Fink, M. J., and West, R. (1983) J. Am. Chem. Soc., 1, 193.
79. Jameson, C. J. and Jameson, A. K. (1988) Chem. Phys. Lett., 149, 300.

80. Beagley, A D., Monaghan, J. J., and Hewitt, T. G. (1971) J. Mol. Struct., 8, 401.

81. McLean, A. D. and Chandler, G. S. (1980) J. Chem. Phys., 72, 5639.

82. Van Wazer, J. R., Ewig, C. S., and Ditchfield, R. (1989) 93, 2222.

83. West, R., Fink, M. J., and Michl, J. (1981) Science, 214, 1344.

84. Fink, M. J., Michalczyk, M. J., Haller, D. J., West, R., and Michl, J. (1983) J. Chem. Soc. Chem. Commun., 1010.

85. Masamune, S., Hanazawa, Y., Murakami, S., and Blount, T. (1982) J. Am. Chem. Soc., 104, 1150.

86. Boudjouk, P., Hans, B., and Anderson, K. R. (1982) J. Am. Chem. Soc., 104, 4992.

87. West, R., Fink, M. J., and Michl, J. (1981) 6th International Symposium on Organosilicon Chemistry, Budapest, Hungary.

88. Gordon, M. S., Thruong, T. N., and Bonderson, E. K. (1986) J. Am. Chem. Soc., 108, 1421.

89. Boatz, J. A., and Gordon, M. S. (1990) J. Phys. Chem., 94, 7331.

90. Scott, L. T., Hashemi, M. M., Meyer, D. T., and Warren, H. B. (1991) J. Am.Chem. Soc., 113, 7082.

91. Yamamoto, K., Harada, T., Okamoto, Y., Chikamatsu, H., Nakazaki, M., Kai, Y., Nakao, T., Tanaka, M., Harada, S., and Kasai, N. (1988) J. Am. Chem. Soc., 110, 3578.

92. Reynolds, J. H., Berson,J. A., Kumashiro, K. K., Duchamp, J.C., Zilm, K. W., Rubello, A, and Vogel, P. (1992) J. Am. Chem. Soc.,114, 763.

ACKNOWLEDGMENTS: We gratefully acknowledge the financial support from the IBM Corp and the National Science Foundation through grants DMB-9003671 (JFH) and CHE-8500487 (PP).

ELECTRONIC MECHANISMS OF METAL CHEMICAL SHIFTS FROM Ab Initio THEORY

Hiroshi Nakatsuji
Department of Synthetic Chemistry
Faculty of Engineering, Kyoto University, Kyoto 606, Japan
and
Institute for Fundamental Chemistry,
Nishi-Hiraki-cho, Kyoto 606, Japan

ABSTRACT.   A progress report on the study of the mechanisms of the metal chemical shifts carried out in this laboratory is given. The major mechanism is understood by the atomic electron configuration of the central metal: p- and d-mechanisms for $d^{10}s^{1-2}p^0$ metal complexes, d-excitation mechanism for d^n metal complexes, and p-excitation mechanism for s^2p^2 metal complexes. Though the paramagnetic term is the origin for most complexes, the chemical shifts of the Ga and In (s^2p^1) halides are primarily determined by the diamagnetic term, and therefore by the structural factors (geometry and nuclear charges) alone.

1. INTRODUCTION

This article is intended to be a progress report of the theoretical studies on the metal NMR chemical shifts of various metal complexes performed in my laboratory. Due to recent advances in multi-nuclear NMR technique, a lot of experimental observations of metal chemical shifts have been reported [1]. Since chemical shifts depend largely on the angular momenta of electrons around the observed nuclei, they reflect p and d electronic structures in the bondings of the metal complexes. The purpose of our series of studies is four-fold.
(1) to establish a reliable ab initio method for calculating metal chemical shifts.
(2) to clarify electronic origins and mechanisms of metal chemical shift by analyzing calculated results.
(3) to give a guiding concept to experimental chemists which is useful for understanding the trends in metal chemical shifts.
(4) to thus have a deeper understanding on the electronic structure of metal complexes.
We classify the metal complexes we have studied so far into four groups, according to the similarity in the mechanism of the chemical shifts. They are
(1) $d^{10}s^{1-2}p^0$ metals; Cu, Ag, Zn, Cd complexes [2]
(2) d^n metals; Ti, Nb, Mo, Mn complexes [3]
(3) s^2p^2 metals; Si, Ge, Sn compounds [4]
(4) s^2p^1 metals; Ga, In halides [5]

J. A. Tossell (ed.), Nuclear Magnetic Shieldings and Molecular Structure, 263–278.
© 1993 Kluwer Academic Publishers.

We briefly review our studies on the electronic mechanisms of the metal chemical shifts in these groups of compounds [6].

2. METHOD AND ANALYSIS

The chemical shift of the compound M is defined as a difference in the nuclear magnetic shielding constant σ relative to the reference compound as

$$\Delta\sigma_M \;=\; \sigma(\text{reference}) \;-\; \sigma(M). \tag{1}$$

The nuclear magnetic shielding constant σ is expressed as a sum of the diamagnetic term σ^{dia} and the paramagnetic term σ^{para},

$$\sigma \;=\; \sigma^{dia} \;+\; \sigma^{para}. \tag{2}$$

σ^{dia} and σ^{para} are the first and second order terms, respectively, in the perturbation theory [7];

$$(\sigma^{dia})_{xy} = -\frac{\mu_0 e^2}{8\pi m^2} < 0 \,|\, \sum_\nu \frac{r_\nu r_{A\nu}\delta_{xy} - r_{x\nu}r_{Ay\nu}}{r_{A\nu^3}} \,|\, 0 > \tag{3}$$

$$(\sigma^{para})_{xy} = -\frac{\mu_0 e^2}{8\pi m^2} \sum_n \frac{1}{E_n - E_0} [< 0 \,|\, \sum_\nu l_{x\nu} \,|\, n > < n \,|\, \sum_\nu \frac{l_{Ay\nu}}{r_{A\nu}^3} \,|\, 0 > \;+\; \text{c.c.} \;] \tag{4}$$

where $| \, 0 >$ and $| \, n >$ denote the ground and excited states, respectively, and $l_{Ay\nu}$ is the y component of the angular momentum operator of the νth electron around the nucleus A. The summation is taken over all the excited states.

The nuclear magnetic shielding constant is calculated by the Hartree-Fock/ finite perturbation method [8]. It is connected to and can be rewritten in the sum-over-state perturbation formula given by eq.(4) [9].

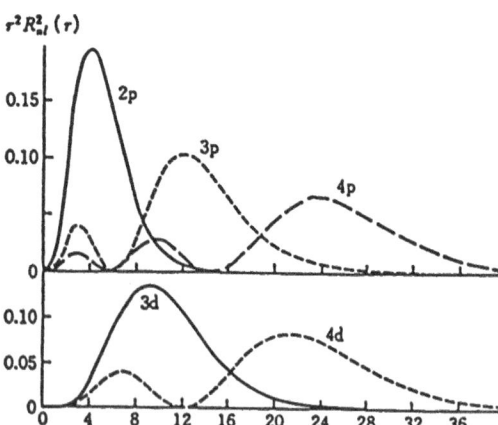

Figure 1. Radial distribution functions of the p- and d-orbitals of hydrogenic atom.

We analyze the magnetic shielding constant into AO contributions and MO contributions [2a]. They are defined by eqs.(10) and (11), respectively, of ref. 2a. The former is defined in the spirit of the Mulliken population analysis. It is important to note here that the shift in the paramagnetic term is dependent on the inner distributions of the valence electrons near the nucleus, since the NMR operator in eq.(4) involves the term $1/r_A^3$. Figure 1 shows the radial distribution functions of the np and nd orbitals of a hydrogenic atom. The 4p orbital, for example, has two small amplitudes in the 3p and 2p regions, which are important for the NMR operators. Namely, the valence electrons near the nucleus are observed through the NMR experiment.

In the following sections, we review our studies on the metal NMR chemical shifts. We do not discuss the geometries of the compounds, the basis sets, and some other computational details, which are explained in the original articles [2-5].

3. $d^{10}s^{1-2}p^0$ METAL COMPOUNDS; Cu, Ag, Zn, Cd COMPLEXES

The metal complexes belonging to this group are characterized by the electronic configuration of the central metal, $d^{10}s^{1-2}p^0$. The results of the ab initio finite perturbation calculation reproduce well the experimental chemical shifts: as an example, Figure 2 shows the correlation between theoretical and experimental values for the Cd chemical shift [2b]. Table 1 shows the analysis for the Cd shift

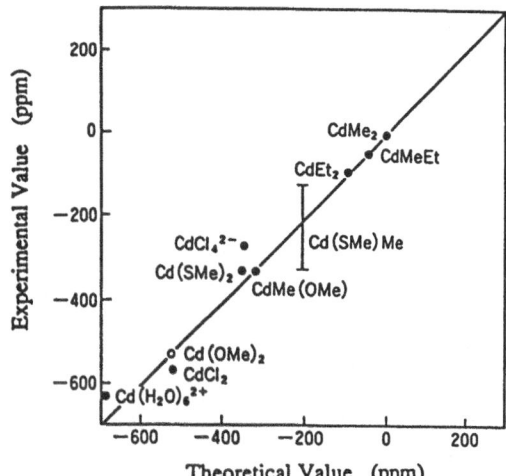

Figure 2. Comparison between experimental and theoretical values for the Cd chemical shifts of the cadmium complexes.

Table 1. Diamagnetic and paramagnetic contributions to the Cd magnetic shielding constant and their analysis into core and valence MO contributions (in ppm)

molecule	diamagnetic term σ^{dia}				paramagnetic term σ^{para}				magnetic shielding σ	
	core	valence	total	shift	core	valence	total	shift	total	shift
CdMe$_2$	4595	256	4851	0	-43	-1047	-1090	0	3761	0
CdMeEt	4602	272	4874	-23	-43	-1021	-1064	-26	3810	-48
CdEt$_2$	4607	289	4896	-45	-43	-995	-1038	-52	3857	-96
CdMe(OMe)	4602	276	4878	-27	-33	-755	-788	-302	4089	-328
Cd(OMe)$_2$	4609	296	4905	-54	-27	-590	-617	-473	4288	-527
CdMe(SMe)	4630	270	4900	-49	-36	-898	-934	-156	3965	-204
Cd(SMe)$_2$	4665	284	4949	-98	-31	-801	-832	-258	4117	-356
CdCl$_2$	4655	253	4908	-57	-25	-643	-668	-422	4240	-479
CdCl$_4^{2-}$	4731	314	5045	-194	-40	-875	-915	-175	4129	-368
Cd(H$_2$O)$_6^{2+}$	4626	378	5004	-153	-33	-442	-475	-615	4529	-768

into the diamagnetic and paramagnetic terms, which are further divided into core and valence MO contributions. The valence electron contribution to the paramagnetic term is the origin of the chemical shift. Table 2 further shows an analysis of the Cd paramagnetic term into Cd p and d AO contributions (s AO contribution is identically zero) and ligand contributions. We see that the p AO contribution is dominant, though the d AO contribution is not negligible. We conclude that for the Cd chemical shift, the valence p AO contribution is most important.

Table 2. Contributions to the paramagnetic term of the Cd nuclear magnetic shielding constant σ^{para} from the cadmium s, p and d AOs and the ligands (in ppm)

molecule	Cd					Ligand					
	s	p	shift	d	shift	Me	Et	OMe	SMe	Cl	H_2O
CdMe$_2$	0	-992	0	-68	0	-15	---	---	---	---	---
CdMcEt	0	-958	-34	-71	3	-15	-21	---	---	---	---
CdEt$_2$	0	-923	-69	-74	6	---	-21	---	---	---	---
CdMe(OMe)	0	-656	-336	-101	33	-14	---	-16	---	---	---
Cd(OMe)$_2$	0	-460	-532	-125	57	---	---	-16	---	---	---
CdMe(SMe)	0	-850	-142	-56	-12	-15	---	---	-14	---	---
Cd(SMe)$_2$	0	-766	-226	-38	-30	---	---	---	-14	---	---
CdCl$_2$	0	-291	-701	-124	56	---	---	---	---	-6	---
CdCl$_4{}^{2-}$	0	-604	-388	-52	-16	---	---	---	---	-7	---
Cd(H$_2$O)$_6{}^{2+}$	0	-723	-269	-166	98	---	---	---	---	---	-10

For Cu, Zn, Ag and Cd complexes, ab initio calculations are performed similarly and the mechanisms of the chemical shifts are investigated [2a]. Tossell reported ab initio calculations for Zn and Cd complexes [10].

Table 3 shows a summary of the mechanisms of the metal chemical shifts for the Cu, Ag, Zn and Cd complexes [2a]. For the complexes of the $d^{10}s^{1-2}p^0$ metals, the paramagnetic term, which is an origin of the chemical shift, is due to the electrons in the outer valence p orbitals and the holes in the valence d orbitals of the metal atom. The mechanisms of these electrons and holes being produced are shown in Figure 3. These mechanisms of the chemical shifts are referred to as p-mechanism or p-electron mechanism and d-mechanism or d-hole mechanism. For the Cd and Zn complexes, the p-mechanism is more important than the d-mechanism, so that the chemical shift would go lower field as the electron donating ability of the ligand increases. For the Cu complexes, on the other hand, the metal chemical shift is due primarily to the d-mechanism so that it increases with increasing electron-withdrawing ability of the ligand. For the Ag complexes, the p and d mechanisms are competitive, and therefore, both of the donating and withdrawing properties of the ligand are important. Note that the effect of the electron donating (or with drawing) ligand on the chemical shift will be opposite, depending on whether the d- or p-mechanism is more important.

Referring to Table 1, we see that the paramagnetic term for the Cd chemical shift becomes more positive in the order of the ligands, Me < Et < SMe < OMe, which is the order of the electron withdrawing ability. Thus, the Cd chemical shift moves to higher field in this order of the ligands, in accordance with Table 3.

Now, why do these differences in the mechanism of the metal chemical shift occur ? The answer may be given from the atomic energy levels of the Cu^+, Ag^+, Zn^{2+}, and Cd^{2+} ions shown in Figure 4. It shows the energy levels of the d^{10}, d^9s^1, and d^9p^1 configurations [11]; the d^9s^1 level is taken as a standard since this configuration is important for the bonding with the ligands. The s-d separation of Cu is smaller than the s-p separation. The reverse is true for Zn and Cd, but for

Table 3. Major mechanism of the metal chemical shift for Cu, Zn, Ag, Cd complexes and the nature of ligand which gives lower field shift

Metal complex	Major mechanism	Lower field shifting ligand
Cu	d-mechanism	electron-acceptor
Cd, Zn	p-mechanism	electron-donor
Ag	p- and d-mechanisms	electron-donor ‖. electron-acceptor

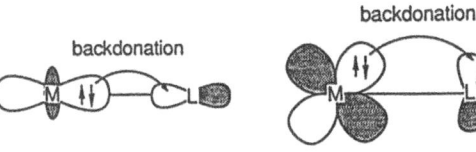

d-mechanism : holes in d shell

p-mechanism : electrons in p orbital

Figure 3. Illustration of the d-hole and p-electrom mechanisms of the chemical shifts of the 1B and 2B metal complexes. They are due to the metal-ligand interactions which produce holes in the valence d shell and eletrons in the valence p orbital, respectively.

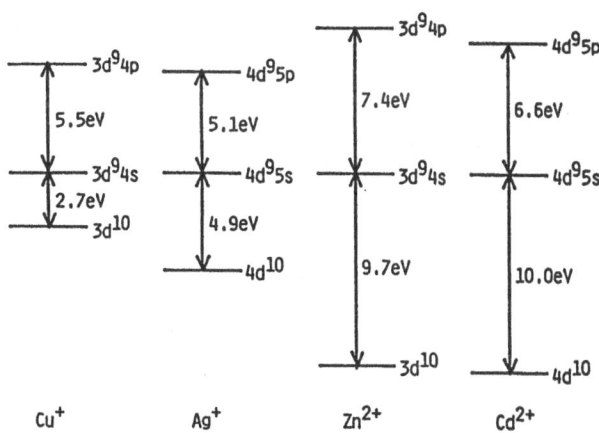

Figure 4. Atomic energy levels of the 1B (Cu, Ag) and 2B (Zn, Cd) metal ions. The energy levels of the d^9s^1 configurations are taken as a standard.

Ag the two separations are almost equal. Therefore, the d orbitals of Cu would more easily mix with the metal-ligand bonds than the p orbitals. On the contrary, the p orbitals of Zn and Cd would more easily mix with the metal-ligand bonds than the d orbitals. For Ag, the two tendencies should be almost equal. Thus, the major mechanisms of the metal chemical shifts of the Cu, Cd, Zn, and Ag complexes summarized in Table 3 are explained from the atomic energy levels of the central metal atoms: they are the intrinsic properties of the metals.

4. d^a METAL COMPOUNDS; Ti, Nb, Mo, Mn COMPLEXES

The transition metals, Ti, Nb, Mo, and Mn, are characterized by their open d subshells, d^2s^2, d^4s^1, d^5s^1, and d^5s^2, respectively: d-orbitals are active and split into both occupied and unoccupied MO's. The mechanism of the chemical shifts of these compounds is closely related to this open d-subshell nature of the central metal atoms and is commonly d-excitation or d-d* mechanism. We explain general features of the chemical shifts of these compounds, taking the Mo complexes, $MoO_{4-n}S_n^{2-}$ (n=0-4) and $MoSe_4^{2-}$, as an example [3b]. For Mo complexes, we also refer to the study of Combariza et al. [12].

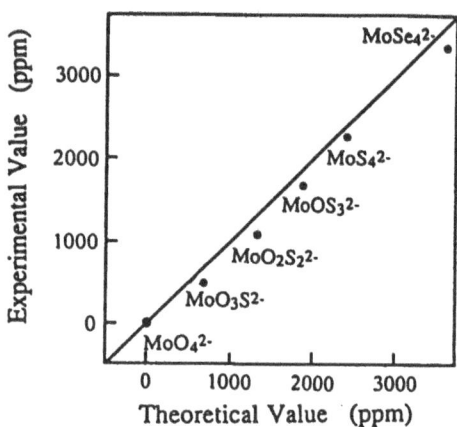

Figure 5. Comparison between theory and experiments for the Mo chemical shift.

Table 4. Diamagnetic and paramagnetic contributions to the Mo magnetic shielding constant and their analysis into core and valence MO contributions (in ppm)

molecule	diamagnetic term σ^{dia}				paramagnetic term σ^{para}				magnetic shielding σ		exptl
	core	valence	total	shift	core	valence	total	shift	total	shift	shift
MoO_4^{2-}	3968	195	4163	0	-172	-5429	-5601	0	-1438	0	0
MoO_3S^{2-}	3958	190	4148	15	-34	-6236	-6269	668	-2121	683	497
$MoO_2S_2^{2-}$	3948	185	4132	31	122	-6933	-6810	1209	-2801	1367	1066
$MoOS_3^{2-}$	3938	180	4117	46	306	-7728	-7422	1821	-3305	1867	1654
MoS_4^{2-}	3928	175	4102	60	505	-8443	-7938	2337	-3835	2397	2258
$MoSe_4^{2-}$	3928	169	4097	66	668	-9820	-9152	3550	-5055	3616	3339

The correlation between theory and experiment for the Mo chemical shift is shown in Figure 5. The ab initio Hartree-Fock/finite perturbation method reproduces reasonably well the Mo chemical shifts of the compounds studied. An analysis of the magnetic shielding constant into diamagnetic and paramagnetic contributions and further analysis into core and valence electron contributions are shown in Table 4. We see that the valence electron contribution to the paramagnetic term is the dominant origin of the chemical shift. In Table 5 we analyze the paramagnetic contribution into the molybdenum AO contribution and the ligand contribution and find that the change in the molybdenum d-AO contribution induced by the ligand substitution is the dominant origin of the chemical shift. Note that the Mo s-AO contribution is zero, since it does not have an angular momentum. Since Mo has an open d-subshell and since d-electrons have large angular momentum, the rotation of the d electrons around the Mo nucleus induced by the applied magnetic field gives an additional magnetic field at the nucleus. The ligand effect on this induced magnetic field is the origin of the chemical shift.

Table 5. AO contributions to the paramagnetic term of the Mo magnetic shielding constant (in ppm)

molecule	Mo				Ligand				σ^{para}
	s	p	d	total	O	S	Se	total	total
MoO_4^{2-}	0	-703	-4862	-5565	-8.9	---	---	-35	-5601
MoO_3S^{2-}	0	-679	-5566	-6264	-8.7	-1.1	---	-26	-6269
$MoO_2S_2^{2-}$	0	-661	-6127	-6793	-8.0	-0.8	---	-18	-6810
$MoOS_3^{2-}$	0	-671	-6746	-7413	-7.7	-0.5	---	-9	-7422
MoS_4^{2-}	0	-700	-7236	-7936	---	-0.4	---	-2	-7938
$MoSe_4^{2-}$	0	-656	-8489	-9145	---	---	-1.5	-6	-9151

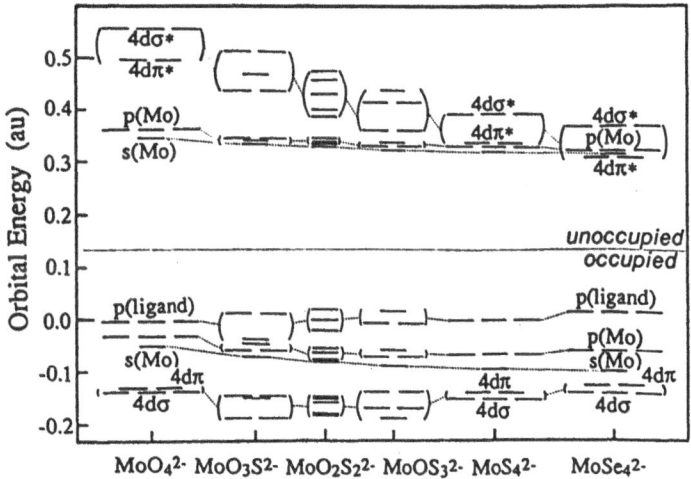

Figure 6. Molecular orbital energy diagram for the Mo complexes.

Now, how does the ordering of the chemical shifts among the $MoO_{4-n}S_n^{2-}$ (n=0 - 4) and $MoSe_4^{2-}$ arise ? The molecular orbital energy diagram shown in Figure 6 gives a solution to this question. Among the occupied and unoccupied valence MO's, the unoccupied $4d\pi^*$ and $4d\sigma^*$ MO's are much stabilized as the ligand is substituted from O to S and to Se, namely from hard to soft ligands. We therefore expect that the excitations from the $4d\sigma$ and $4d\pi$ MO's to the $4d\sigma^*$ and $4d\pi^*$ MO's are the most important terms in the sum-over-state perturbation formula given by eq.(4). The softer the ligand is, the more stabilized are the $4d\sigma^*$ and $4d\pi^*$ MO's and the smaller is the excitation energy from the $4d\sigma$ and $4d\pi$ MO's to the $4d\sigma^*$ and $4d\pi^*$ MO's. This change in the excitation energy, appearing as a denominator of eq.(4), leads to an increase in the chemical shift. Our analysis [3a,3b], has shown that the stabilization of the $4d\sigma^*$ and $4d\pi^*$ MO's is due to the stabilization and the mixing of the outer p orbitals of the ligands. As the ligand becomes softer, its outer p orbitals are stabilized and the Mo chemical shift increases.

If the above analysis is correct, we expect from eq.(4) that the Mo chemical shift is inversely proportional to the d-d* excitation energy, ΔE as [3c]

$$\Delta\sigma = A(1/\Delta E_{ref} - 1/\Delta E)$$

$$= \alpha + \beta/\Delta E. \qquad (5)$$

Here we have assumed that only one state mainly contributes to the magnetic shielding constant and that the factor A is roughly constant among the complexes.

We note that the excitation involved in eq.(5) is not necessarily an optically allowed transition but should be a magnetically allowed transition; the transition for which the numerator of eq.(4) is non-zero. For molecules with higher symmetry, like tetrahedral as MoX_4^{2-} (X=O, S, Se), the magnetically allowed excited states have T_1 symmetry, while the optically allowed excited states have T_2 symmetry. Thus, we can not expect an existence of the observed d - d* transition energies which are magnetically allowed. We therefore calculated the excitation energies of the Mo complexes using the symmetry adapted cluster-configuration interaction (SAC-CI) method [13]. The SAC-CI method has been shown to give excited states to a considerable accuracy within a reasonable amount of computational time [14].

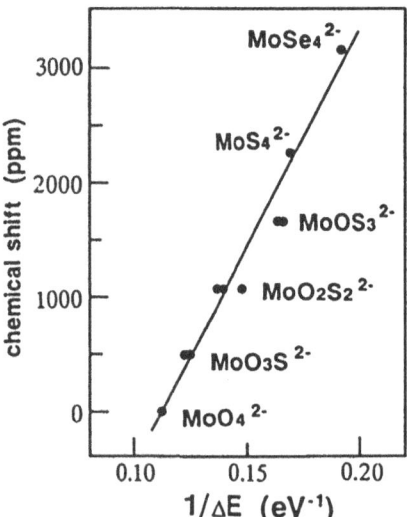

Figure 7. Relationship between the inverse of the lowest magnetically allowed d-d* excitation energy and the chemical shift. The chemical shifts are the experimental values and the excitation energies are the SAC-CI theoretical values.

Figure 7 shows a plot of the experimental Mo chemical shift against $1/\Delta E$ of the energy ΔE calculated for the $4d\sigma \rightarrow 4d\pi^*$ excitation, the lowest possible magnetically allowed d-d* transition. The two and three points for the C_{3v} and C_{2v} molecules, respectively, occur because T_1 splits into $A_2 + B_1 + B_2$ and $A_2 + E$, respectively. We see a very beautiful linear relationship, which justifies the validity of the mechanism of the Mo chemical shift discussed above. We call this mechanism of the chemical shift as d-excitation mechanism or d-d* mechanism.

From our systematic studies for the Ti, Nb, Mo, and Mn chemical shifts [3], it became clear that the d-excitation mechanism is common to these metal complexes. The origin of the d-excitation mechanism is attributed to the open d-subshell nature of these transition metals, so that we expect that the d-excitation mechanism is common to most of the transition metal complexes.

5. s^2p^2 METAL COMPOUNDS; Si, Ge, Sn COMPLEXES

The chemical shifts of the Si, Ge, and Sn complexes show interesting common behaviors on substitutions of ligands [1]. The compounds which are written as $MR_{4-x}R'_x$ with R and R' both representing organic ligands like H, Me, Ph, etc. show a linear dependence of the metal (M) chemical shift on the number of the ligands x. On the other hand, the complexes of the type $MR_{4-x}Y_x$ with Y being electronegative ligand like halogen, alkoxy, amino, etc., show U-shaped dependence on x. We here want to elucidate the electronic mechanism of the metal chemical shifts and the origins of the linear and U-shaped dependences, taking the Sn compounds, $SnMe_{4-x}H_x$ and $SnMe_{4-x}Cl_x$, as an example. The chemical shifts of some Si and Ge compounds are studied by Ditchfield et al. [15] and Fleischer et al. [16], respectively.

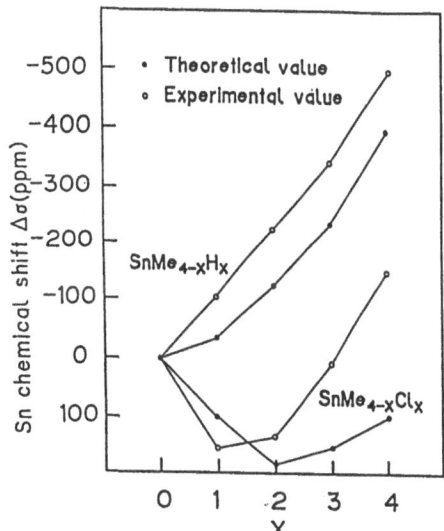

Figure 8. Comparison between theoretical and experimental values for the Sn chemical shifts of $SnMe_{4-x}H_x$ and $SnMe_{4-x}Cl_x$.

We show in Figure 8 the x dependence of the Sn chemical shift. The theoretical values roughly reproduce the linear and U-shaped dependences of the Sn chemical shifts in $SnMe_{4-x}H_x$ and $SnMe_{4-x}Cl_x$, respectively. The MO and AO analyses of the diamagnetic and paramagnetic terms shown in Table 6 reveal that the Sn chemical shift reflects mainly the change in the Sn valence 5p orbital induced by the ligand substitution. The Sn d AO contribution and the diamagnetic ligand contribution are of secondary importance and are compensating to each other.

Table 6. Diamagnetic and paramagnetic contributions to the Sn magnetic shielding constant and their analysis into Sn p- and d-AO contributions and ligand contributions (in ppm)

molecule	diamagnetic term σ^{dia}					paramagnetic term σ^{para}						
	Sn total	Ligand				Sn^a			Ligand			
		Me	Cl	H	total	p	d	total	Me	Cl	H	total
SnH_4	5073	---	---	6	24	-1432	-253	-1686	---	---	-2	-8
$SnMeH_3$	5072	38	---	6	56	-1581	-282	-1864	-16	---	-2	-23
$SnMe_2H_2$	5071	38	---	6	88	-1684	-308	-1992	-16	---	-2	-36
$SnMe_3H$	5070	38	---	6	120	-1768	-330	-2098	-16	---	-2	-50
$SnMe_4$	5069	38	---	---	152	-1795	-356	-2152	-16	---	---	-64
$SnMe_3Cl$	5069	38	69	---	183	-1927	-365	-2292	-16	-13	---	-61
$SnMe_2Cl_2$	5070	38	69	---	214	-2028	-378	-2407	-16	-13	---	-58
$SnMeCl_3$	5070	38	70	---	248	-2019	-392	-2411	-16	-13	---	-55
$SnCl_4$	5070	---	70	---	280	-1973	-421	-2394	---	-13	---	-52

a) s AO contribution is zero since it has no angular momentum.

The 5p AO contribution to Sn σ^{para} is determined by the two factors: the excitation energy in the denominator of eq.(4) and the integral terms in the numerator. The larger factor is the excitation energy from the Sn-L σ bonding MO to the antibonding MO. We call this mechanism as p-excitation or p-p* mechanism. Table 7 shows the experimentally observed excitation energies for SnH_4, $SnMe_4$, and $SnCl_4$ [17]. For $SnCl_4$, the lowest transition is the excitation of the Cl lone pair electron to the Sn-L antibonding MO, and the second peak observed at 7.80 eV is $\sigma \to \sigma^*$. We note that the Rydberg state mixes to some extent with the σ^* state. These excitations have T_2 symmetry, so that they are optically allowed but magnetically forbidden. However, we can expect a rough parallelism between them. We have confirmed that the experimental chemical shifts show a rough linear relationship with $1/\Delta E$ where ΔE is the $\sigma - \sigma^*$ excitation energy [4a]. We thus understand the ordering of the observed chemical shifts, SnH_4 <$SnCl_4$ <$SnMe_4$, though the ordering between $SnCl_4$ and $SnMe_4$ was not reproduced in our calculation.

Table 7. Observed excitation energy of the Sn compounds.

Compound	Nature	Excitation Energy (eV)
SnH_4	$3t_2 \to 3a_1$; $\sigma \to \sigma^*$	8.86
$SnMe_4$	$3t_2 \to 3a_1$; $\sigma \to \sigma^*$	6.63
$SnCl_4$	$3t_2 \to 3a_1$; $n(L) \to \sigma^*$	6.23
	$2t_2 \to 3a_1$; $\sigma \to \sigma^*$	7.80

The Sn AO contributions and the ligand contributions in the diamagnetic and paramagnetic terms are plotted in Figure 9 against x for the $SnMe_{4-x}Cl_x$ and $SnMe_{4-x}H_x$ series. It is clearly seen that the origin of the U-shaped dependence in the former compounds is the p-AO contribution to σ^{para}. The linear relationship for the $SnMe_{4-x}H_x$ series is understood from the large change in the excitation energy and the similarity between the H and Me ligands. On the other hand, the U-shaped relationship in the $SnMe_{4-x}Cl_x$ series is considered to be due to the change in both denominator and numerator of eq.(4). In the excitation energy term, two transitions showing opposite behaviors seem to exist: the mixing of these two transitions in the compounds with x = 1,2,3 is interesting. In the numerator, the anisotropy of the Sn p-AO density would be induced for less symmetric compounds by a large inductive effect of the Cl ligand and works to enlarge the paramagnetic term. For more details, we refer to Ref. 4a.

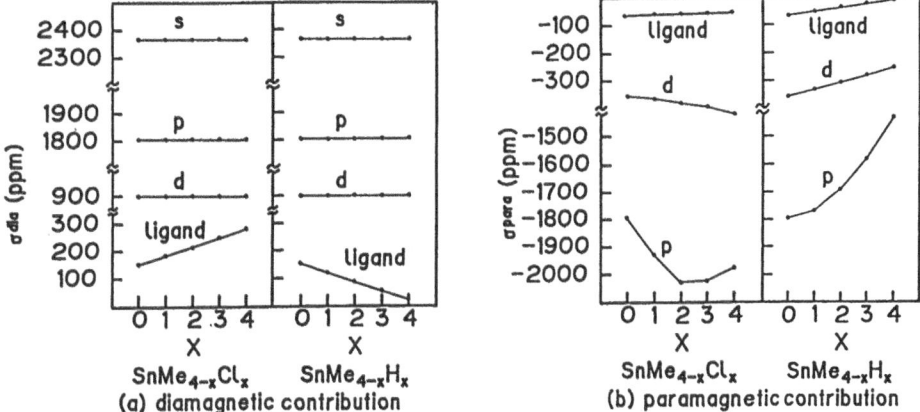

Figure 9. The change as a function of x in the (a) diamagnetic and (b) paramagnetic terms divided into the Sn s, p and d contributions and ligand contributions in $SnMe_{4-x}H_x$ and $SnMe_{4-x}Cl_x$.

6. s^2p^1 METAL COMPOUNDS; Ga, In HALIDES

All of the metal chemical shifts so far studied have been dominated by the paramagnetic term. The chemical shifts of gallium and indium halides are unique in that they are predominantly determined by the diamagnetic term [5]. Since the diamagnetic term depends only on the structural factors, like bond distance and ligand nuclear charge, as the Flygare-Goodisman equation [18] implies, the chemical shifts of these compounds can be calculated without the knowledge of the electronic structure. They show normal halogen dependence [1], in contrast to the reverse halogen dependence for the compounds dealt with in the previous sections. We briefly explain these facts in this section.

We compare in Figure 10 the experimental and theoretical values of the Ga chemical shifts for the compounds $GaCl_{4-x}Br_x^-$ ($x=0-4$). The agreement is excellent. Table 8 shows a breakdown of the magnetic shielding constant into the paramagnetic and diamagnetic terms. The diamagnetic term is three times larger than the paramagnetic one. Table 9 shows the atomic contributions to σ^{dia}. The individual atomic contributions are quite constant, so that σ^{dia} of the complex is written in a Pascal-rule like formula as

$$\sigma^{dia}(\text{complex}) = \sigma^{dia}(\text{M atom}) + \sum_L n_L \sigma^{dia}(L) \qquad (6)$$

where $\sigma^{dia}(\text{M atom})$, $\sigma^{dia}(L)$ and n_L are the free atom and ligand contributions and the number of the ligands, respectively. On the other hand, Flygare and Goodisman showed that σ^{dia} is written to a good approximation as

$$\sigma^{\text{dia}}(\text{complex}) = \sigma^{\text{dia}}(\text{M atom}) + \frac{e^2}{3mc^2} \sum_{\text{L}} Z_{\text{L}}/R_{\text{L}} \qquad (7)$$

where R_L and Z_L are metal-ligand distance and ligand nuclear charge, respectively.

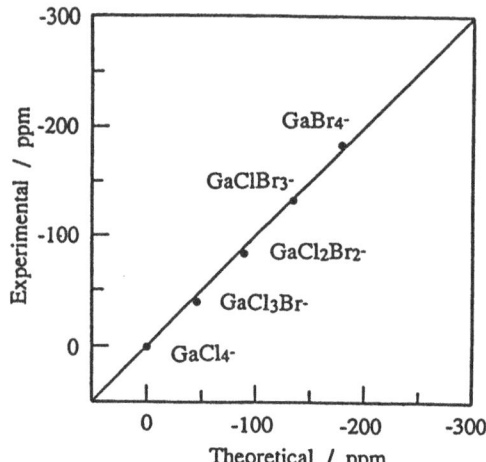

Figure 10. Correlation between experimental and theoretical values of the Ga chemical shifts in $GaCl_{4-n}Br_n^-$ ($n = 0 - 4$).

Table 8. Diamagnetic and paramagnetic contributions to the Ga magnetic shielding constant (in ppm)

molecule	diamagnetic term		paramagnetic term		magnetic shielding		exptl
	σ^{dia}	shift	σ^{para}	shift	σ	shift	shift
GaCl$_4^-$	2932.1	0.0	-888.7	0.0	2043.4	0.0	0.0
GaCl$_3$Br$^-$	3000.1	-68.0	-911.7	23.0	2088.4	-45.0	-39.8
GaCl$_2$Br$_2^-$	3068.4	-136.4	-936.0	47.4	2132.4	-89.0	-83.8
GaClBr$_3^-$	3136.1	-204.0	-958.5	69.9	2177.5	-134.1	-132.8
GaBr$_4^-$	3204.1	-272.0	-982.3	93.6	2221.8	-178.4	-183.8

Table 9. Metal and ligand contributions to the diamagnetic term σ^{dia}(in ppm)

molecule	Metal	Ligand		σ^{dia}
	Ga	Cl	Br	
GaCl$_4^-$	2630.6	75.4		2932.1
GaCl$_3$Br$^-$	2630.8	75.4	143.2	3000.1
GaCl$_2$Br$_2^-$	2631.1	75.5	143.2	3068.4
GaClBr$_3^-$	2631.3	75.3	143.1	3136.1
GaBr$_4^-$	2631.6		143.1	3204.1

We compare in Table 10 the values calculated by eq.(7) with those of the ab initio calculations. σ^{dia}(M atom) in eq.(7) is the free atom value and taken from Malli and Froese [19]. The agreement of the two methods is excellent, so that the diamagnetic term is determined solely by the structural factors, R_L and Z_L alone. For chemical shifts, σ^{dia}(M atom) cancels out. Thus, if we neglect the paramagnetic contribution σ^{para}, which is roughly one third of σ^{dia} with opposite sign, we can say that the Ga chemical shifts of the complexes $GaCl_{4-x}Br_x$ are determined by the structural factors, R_L and Z_L alone.

Table 10. Estimate of the diamagnetic term of the Ga magnetic shielding constant from Flygare-Goodisman equation compared with the ab initio results (in ppm)

molecule	Flygare-Goodisman eq.			ab initio result		
	Ga	ligands	σ^{dia}	Ga	ligands	σ^{dia}
$GaCl_4^-$	2638.6	296.0	2934.6	2630.6	301.6	2932.1
$GaCl_3Br^-$	2638.6	364.1	3002.7	2630.8	369.4	3000.1
$GaCl_2Br_2^-$	2638.6	432.2	3070.8	2631.1	437.4	3068.4
$GaClBr_3^-$	2638.6	500.3	3138.9	2631.3	504.6	3136.1
$GaBr_4^-$	2638.6	568.4	3207.0	2631.6	572.4	3204.1

It is interesting to examine this possibility for a wider class of compounds. Since σ^{dia} is very easily calculated when molecular geometry of a complex is known, we plot in Figure 11 the experimental values of the Ga chemical shifts of various halide complexes againt $\Delta\sigma^{dia}$, which is calculated from the second term of eq.(7). We see the points fall above and below the 45 degree line. These compounds are further classified into $GaCl_{4-x}Br_x^-$, $GaCl_{4-x}I_x^-$, $GaBr_{4-x}I_x^-$, and some mixed ones. We find an approximate linearity among each class of compounds, and the slope is steeper as the ligands become heavier. It is interesting to investigate the origin of this slope: is it explained without including the spin-orbit effect of the ligand ?

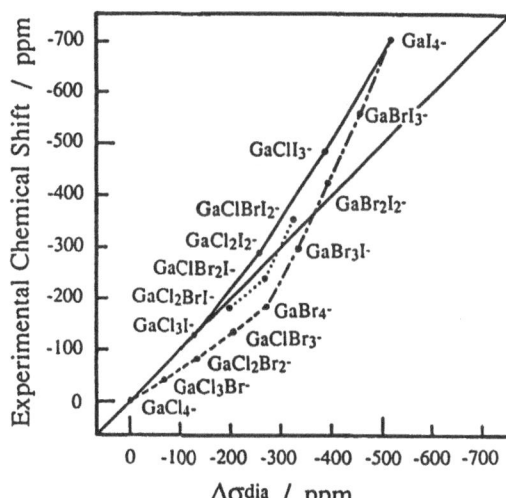

Figure 11. Correlation between experimental Ga chemical shift and diamagnetic shift value $\Delta\sigma^{dia}$ calculated by the Flygare-Goodisman equation.

Figure 12 is the plot of $\Delta\sigma^{dia}$ against the experimental chemical shifts for the indium complexes. The plots for light halides, $InCl_{4-x}Br_x^-$ lie nicely on the 45 degree line, but those for heavier halides deviate upwards. It is safe to conclude that the diamagnetic term is an important origin of the indium chemical shifts of these complexes.

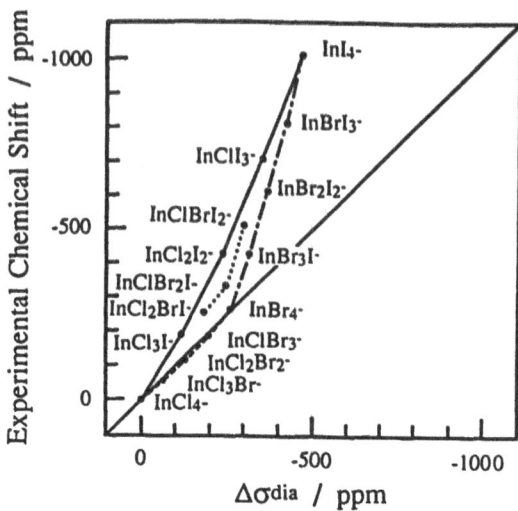

Figure 12. Correlation between experimental In chemical shift and diamagnetic shift value $\Delta\sigma^{dia}$ calculated by the Flygare-Goodisman equation.

One may expect that the aluminum chemical shift may also belong to this class of compounds. However, we have confirmed this is not the case. For aluminum chemical shifts, the paramagnetic term is important.

Now, why is σ^{para} small in this class of compounds ? The reason is that the single p electron of gallium or indium is tightly bound in the M-L bond, so that the excitation energy for this electron is large, leading to a small σ^{para} as eq.(4) implies.

7. CONCLUSION

In this progress report, we have emphasized that the mechanism of the metal chemical shifts of the complexes is closely related with the atomic electronic structure of the central metal atom. When the central metal atom has the valence electron configuration $d^k s^l p^m$ in its free atomic ground state, the open subshell nature of the d or p orbitals determines the major mechanism, since the angular momentum of the chemical shift operator is represented by that open subshell. The chemical shift is a measure of the ligand perturbation on this angular momentum. The s orbital and the full and empty orbitals do not give angular momentum.

In the $d^{10} s^{1-2} p^0$ complexes, the d subshell is full and the p subshell is empty. Therefore, the p electrons suplied by the ligand and the d holes produced by the electron withdrawing ligand give angular momenta and chemical shifts. These mechanisms of the chemical shifts are referred to as the p-electron mechanism (or p-mechanism) and the d-hole mechanism (or d-mechanism), respectively. The relative importance of the p- and d-mechanisms is primarily determined by the relative s-p and d-s spacings in the atomic spectrum of the metal atom: when the d-s spacing is smaller than the s-p spacing, the d-mechanism is more important and in the reverse situation, the p-mechanism is importnat. The Cu complex belongs to the former and the Zn and Cd complexes to the latter. In the Ag complexes, two spacings are close, so that the two mechanisms are competitive. Further, as being self-evident, the effect of the electron donating (or withdrawing) ligand on the metal chemical shift is opposite, depending on whether the p- or d-mechanism is important.

For the d^n complexes (n=2 ···· 8), the chemical shift is dominated by the d-excitation mechanism: the angular momentum is produced by the transition of electrons from the occupied d orbital to the unoccupied d orbital. Therefore, the induced angular momentum would be proportional to $1/\Delta E$ with ΔE being the d-d* excitation energy, as the perturbation theoretic formula eq.(4) implies. This is proved for the Mo complexes [3c]. Therefore, the chemical shift is measured by the magnitude of the ligand effect on the d-d* excitation energy. Generally speaking, the softer the ligand is, the ΔE is smaller and the chemical shift is larger: this mechanism is explained in our articles [3].

For the p^n complexes (n=2 ∼ 4), the p-excitation mechanism is important. The origin is similar to the d-excitation mechanism. An interesting behavior of the chemical shifts of this class of compounds is the existence of the linear and U-shaped dependences on the number of the ligands. The linear dependence is normal. The U-shaped dependence is caused by two factors; one is the existence of two competitive excitations, p-p* and n(L)-p* excitations (n(L) is the lone pair on the ligand) and the other is the anisotropy of the metal p-electron distribution in the non-symmetric compounds.

Though most chemical shifts originate from the paramagnetic term, those of the Ga and In halides are dominated by the diamagnetic term. The halogen dependence [1] is also clearly different. For the former case, most show the reverse halogen dependence, but the Ga and In complexes show the normal halogen dependence. Since the p electron of the s^{2}p^1 Ga and In complexes are tightly bound in the M-L bond, the paramagnetic term is small. Therefore, the Ga and In chemical shifts are determined mostly by the geometrical factors, the M-L bond length and ligand nuclear charge. This is quite unique among the metal complexes.

ACKNOWLEDGEMENT

The author thanks all the coworkers of this series of studies, Drs. K. Endo, K. Kanda, T. Yonezawa, T. Nakao and Messrs. M. Sugimoto, T. Inoue, S. Saito and M. Kanayama for active and enjoyable collaborations. He further thanks Mr. M. Sugimoto and Mrs. S. Furukawa for some helps in preparing this manuscript. He also thanks Professor J. H. Enemark for letting him know a numerous amount of experimental data on the Mo chemical shifts.

REFERENCES
[1] (a) R. K. Harris and B. E. Mann ed., "NMR and the Periodic Table", Academic Press, New York, 1978.
    (b) J. Mason, ed., "Multinuclear NMR", Plenum Press, New York, 1987.
    (c) E. A. Williams and J. D. Cargioli, Annu. Rept. NMR Spectr., 9, 221 (1977); E. A. Williams, ibid., 15, 235 (1983).
    (d) P. J. Smith and A. P. Tupčiauskas, Annu. Rept. NMR Spectr. 8, 291 (1978).
    (e) B. Wrackmeyer, Annu. Rept. NMR Spectr. 16, 73 (1985).
    (f) M. Minelli, J. H. Enemark, R. T. C. Brownlee, M. J. O'Connor, and A. G. Wedd, Coord. Chem. Rev. 68, 169 (1985).
[2] (a) H. Nakatsuji, K. Kanda, K. Endo, and T. Yonezawa, J. Am. Chem. Soc. 106, 4653 (1984).

(b) H. Nakatsuji, T. Nakao, and K. Kanda, Chem. Phys. 118, 25 (1987).

[3] (a) K. Kanda, H. Nakatsuji, and T. Yonezawa, J. Am. Chem. Soc. 106, 5888 (1984).

(b) H. Nakatsuji and M. Sugimoto, Inorg. Chem. 29, 1221 (1990).

(c) H. Nakatsuji, M. Sugimoto and S. Saito, Inorg. Chem. 29, 3095 (1990).

(d) H. Nakatsuji and T. Nakao, Chem. Phys. Letters, 167, 571 (1990).

(e) M. Sugimoto, M. Kanayama, and H. Nakatsuji, J. Phys. Chem. in press.

[4] (a) H. Nakatsuji, T. Inoue, and T. Nakao, Chem. Phys. Letters, 167, 111 (1990); J. Phys. Chem. in press.

(b) H. Nakatsuji and T. Nakao, to be published.

(c) T. Nakao, Desertation for Doctor of Engineering, Kyoto University, 1991.

[5] H. Nakatsuji, M. Sugimoto, and S. Kanayama, submitted for publication.

[6] (a) H. Nakatsuji in "Comparisons of Ab Initio Quantum Chemistry with Experiment: State of the Art", R. J. Bartlett, ed., Reidel, Dordrecht, The Netherlands, 1985, p. 409.

(b) H. Nakatsuji in "Modern Chemistry, Supplement 11, High Resolution NMR Spectroscopy", H. Saito and I. Morishima, eds., Tokyo Kagaku Dojin, Tokyo, Japan, 1987, p.237 (in Japanese).

(c) M. Sugimoto and H. Nakatsuji, Organometallic News, No.2, 63 (1992).

[7] N. F. Ramsey, Phys. Rev. 77, 567 (1950); 78, 699 (1950); 83, 540 (1951);86, 243 (1952); A. Saika and C. P. Slichter, J. Chem. Phys. 22, 26 (1954).

[8] H. D. Cohen and C. C. J. Roothaan, J. Chem. Phys., 43, s34 (1965); H. D. Cohen, J. Chem. Phys., 43, 3558 (1965); H. D. Cohen, J. Chem. Phys., 45, 10 (1966); J. A. Pople, J. W. McIver and N. S. Ostlund, Chem. Phys. Letters, 1, 46 (1967); J. A. Pople, J. W. McIver and N. S. Ostlund, J. Chem. Phys., 49, 2960 (1968). R. Ditchfield, D. P. Miller and J. A. Pople, 53, 613 (1970); H. Nakatsuji, K. Hirao and T. Yonezawa, Chem. Phys. Letters, 6, 541 (1970).

[9] H. Nakatsuji, J. Chem. Phys. 61, 3728 (1974).

[10] (a) J. A. Tossel, Chem. Phys. Lett., 169, 145 (1990); J. Phys. Chem. 95, 366 (1991).

(b) J. A. Tossel, VIIth International Congress on Quantum Chemistry, Proceeding, p.213, Menton, July 1991.

[11] C. E. Moore, "Atomic Energy Levels", National Bureau of Standard; Washington, 1971.

[12] J. E. Combariza, J. H. Enemark, M. Barfield, and J. C. Facelli, J. Am. Chem. Soc., 111, 7619 (1989); J. E. Combariza, M. Barfield, and J. H. Enemark, J. Phys. Chem. 95, 5463 (1991).

[13] H. Nakatsuji, Chem. Phys. Letters, 59, 362 (1978); 67, 329 (1979).

[14] (a) H. Nakatsuji, Intern. J. Quantum Chem. Symp. 17, 241 (1983).

(b) H. Nakatsuji, J. Chem. Phys. 80, 3708 (1984).

(c) O. Kitao and H. Nakatsuji, J. Chem. Phys. 87, 1169 (1987).

(d) H. Nakatsuji, Acta Chimica Hungarica, in press (1992).

[15] J. R. Van Wazer, C. S. Ewig and R. Ditchfield, J. Phys. Chem. 93, 2222 (1989).

[16] U. Fleischer, M. Schindler and W. Kutzelnigg, VII International Congress on Quantum Chemistry, Proceeding p.46, Menton, July 1991.

[17] G. C. Causley and B. R. Russell, J. Elec. Spectr. 11, 383 (1977); J. Fernandez, G. Lespes, and A. Dargelos, Chem. Phys. 103, 85 (1986); 111, 97 (1986).

[18] W. H. Flygare and J. Goodisman, J. Chem. Phys. 49, 3122 (1968).

[19] G. Malli and C. Froese, Int. J. Quantum Chem. 1, 95 (1967).

APPLICATIONS OF NMR SHIELDING CONSTANT CALCULATIONS IN MINERALOGY AND GEOCHEMISTRY

J. A. Tossell
Department of Chemistry and Biochemistry
University of Maryland
College Park, MD 20742
USA

Abstract

NMR has become a very powerful technique for the characterization of the local order present in amorphous solids, glasses and aqueous solutions. The NMR shielding constant depends upon the identity, number and distance(s) of nearest-neighbour atoms, upon the nature of the connection(s) between the nearest neighbour units and upon the identities, distances and angular relationships of the atoms in the second nearest-neighbour coordination sphere. *Ab initio* SCF calculations on carefully chosen molecular clusters have reproduced many of the trends in shielding constants observed in molecules and solids. For species for which no appropriate crystalline models exist, *ab initio* techniques have been used to calculate equilibrium structures and NMR shieldings have then been calculated at those structures, to assist in species identification. Changes in shielding constants have been correlated with changes in other electronic properties, so as to give a unified picture of the electronic structure of the species. Our calculations on O, Al, Si, P, Zn and Cd NMR shieldings, using both conventional common-origin coupled Hartree-Fock theory and the localized-orbital local-origin modification known as LORG, have reproduced many trends in both isotropic and anisotropic NMR shielding values and have, in some cases, substantiated qualitative interpretations of shielding trends. As examples of geochemical applications of such calculations, we consider the variation of Si NMR shieldings and O electric field gradients with <Si-O-Si in siloxanes and silicates, the change in Si shielding with coordination number, the anisotropy of the ^{29}Si NMR shielding tensor in olivine (Mg_2SiO_4), Al NMR shieldings and electric field gradients for various Al fluorides and oxyfluorides in F-bearing aluminosilicate glasses, and Zn and Cd NMR shieldings for chlorides and bisulfides in aqueous solution. Erroneously large predicted values of O shieldings from conventional CHF theory are corrected by the LORG approach or by a simple core electron correction, while highly polar compounds, such as the Al fluorides, are shown to present problems for LORG in its conventional implementation. Our results emphasize the need for simultaneous study of other properties which complement the NMR results, for improvements in methodology and computer capabilites which will allow the study of more realistic model systems or real bulk solids and for further study of the relationship between NMR shielding constants and other electronic properties.

J. A. Tossell (ed.), Nuclear Magnetic Shieldings and Molecular Structure, 279–296.
© 1993 *Kluwer Academic Publishers.*

Introduction

The development of cross-polarization and magic angle spinning techniques have made possible the characterization of chemically inequivalent Si sites in many minerals and glasses. Early work showed systematic changes in the Si NMR shielding, σ^{Si}, as a function of the degree of polymerization (Lippmaa et al., 1980), the <Si-O-T, where T is a tetrahedrally coordinated cation (Thomas et al, 1983) and the Si coordination number (Smith and Blackwell, 1983). Two semiempirical schemes were soon developed to rationalize some of these trends, one based on a localized bond orbital approach utilizing atomic electronegativies (Engelhardt and Radeglia, 1984: Radeglia and Engelhardt, 1985) and one based on a canonical molecular orbital approach and utilizing experimental excitation energies from x-ray spectra (Tossell, 1984). The CMO approach was a natural extension of that used to explain the photoemission and UV absorption spectra of silicates (Tossell, 1977). Both these approachs contained a number of unproven asumptions, required experimental input and were applicable to only limited classes of Si compounds. To establish a sounder basis for the interpretation and prediction of NMR shieldings in silicates, Tossell established a collaboration with Lazzeretti, who had recently calculated the Si NMR shielding in SiH_4 (Lazzeretti and Zanasi, 1980) using coupled Hartree-Fock perturbation theory in its conventional common-origin form, using a set of programs known as SYSMO. Many of the studies described herein were obtained using these programs.

Method

The method used in these studies in generally that of the conventional CHF theory, as described by Lipscomb (1966) and in the papers by Webb and others in this volume. In some of the studies we have also used the RPA LORG method of Hansen and Bouman (1985). In general, the Gaussian expansion basis sets used have been of polarized double or triple zeta type, with a single polarization function on each of the nuclei. For some of the smaller molecules, multiple polarization functions have been used on the magetic nucleus. Although we used experimental geometries in some early work, most of our later work uses energy optimized geometries obtained at the SCF level using polarized double zeta bases. In most cases in the CHF calculations the magnetic nucleus has been chosen as the gauge origin. Diamagnetic and paramagnetic shielding contributions, σ^d and σ^p, refer to this choice of origin. In our latest common-origin CHF studies we have also employed a correction for the effect of core electrons on distant atoms, described in detail for the SiF_n^{4-n} species below.

Results

SI OXIDES AND FLUORIDES

Early studies on Si oxides and fluorides (Tossell and Lazzeretti, 1986) supported the main assumption in Tossell (1984) of the dominant contribution of the Si-O or Si-F sigma-bonding MO's to the paramagnetic part of the shielding. These studies indicated that the Si NMR shielding of a highly charged anion, such as SiO_4^{-4} (even though clearly unstable against loss of electrons to the continuum) can be described with almost the same accuracy as that of isoelectronic SiF_4. The calculated decrease in shielding from SiF_4 to SiO_4^{-4} of 44 ppm was in good agreement with the experimental difference of about 40 ppm, and could be attributed to the shorter bond distance in SiF_4, which increased its diamagnetic shielding, and the smaller covalency, which decreased the magnitude of its paramagnetic shielding. The general success of the SiO_4^{-4} calculations can probably be attributed to the r^{-1} nature of the shielding operator, wieghting regions near the nucleus more heavily than distant regions, and to the small shielding contributions from the ligand non-bonding orbitals at the top of the valence region.

The calculated increase in Si shielding from SiF_4 to SiF_6^{-2} was about 110 ppm, considerably larger than the experimental value of about 75 ppm, partly an artifact of the imcomplete cancellation of diamagnetic and paramagnetic contributions from the core (and lone pair) electrons on the fluorines (a problem automatically eliminated in the localized-orbital methods such as IGLO, LORG and GIAO described in this volume), Such a cancellation can be artifically imposed (for the core electrons) by simply subtracting out the core electron diamagnetic contribution using the atom superposition model of Flygare and Goodisman (1968). This core correction is simply equal to the number of core electrons divided by the bond distance in Å and multiplied by 9.39 ppm. Values recently calculated (Tossell, 1992a) using this approach, which we call core-corrected CHF, are shown in Table 1.

Table 1. Calculated Si NMR shieldings for SiF_4, SiF_5^- and SiF_6^{-2}, compared with experiment

molecule	previous results[a]			present results[b]			
	R(Si-F)	σ(CHF)	σ(core-corrected CHF)	R(Si-F)	σ(CHF)	σ(core-corrected CHF)	$\Delta\sigma_{exp}$[b]
SiF_4	1.556	556	508	1.553	594	546	0
SiF_5^-	1.684,	619	561	1.628,	634	575	
	1.594			1.594			
SiF_6^{-2}	1.685	668	601	1.662	678	610	+74

a. Tossell and Lazzeretti (1984), assumed geometries, triple Si3d basis
b. Tossell (1992a), optimized geometries, 3-21G* basis

The core corrected CHF results are in reasonable agreement with experiment given that we have only a single value for the SiF_6^{-2} shielding, which might well depend slightly on the counter ion present within the solid. We have also calculated the shielding difference between Si three-, four- and five-coordinated to O, using very simple unprotonated SiO_n^{4-n} models, with calculated Si-O distances (Tossell, 1990a), and have obtained the expected trend, with the five coordinate species shielded with respect to the four coordinate by about 75 ppm, in reasonable agreement with experiment. These studies were, however, somewhat crude in terms of geometries assumed, basis sets used and lack of core corrections.

An interesting result of the Tossell and Lazzeretti (1986) study was the very small dependence of the Si NMR shielding on the bond distance for SiO_4^{-4}. A variation in Si-O distance of 0.10Å about the typical nesosilicate value of 1.634Å changed the shielding by less than 1 ppm (before core correction; the core corrected value would be 2 ppm), with longer distances giving (very slightly) larger shieldings. Early correlations had attributed the increase in Si NMR shielding with increasing polymerization and decreasing average bond length to a direct bond length effect, but our results supported the qualitative MO interpretation of Tossell (1984) and the empirical correlations developed by many other groups which attributed the increased shielding to the direct effect of polymerization, rather than to the small changes in average Si-O distance associated with it. Recently a correlation of Si NMR shielding with average bond length has been reported for a series of Ca silicates (Skibsted et al., 1990) which purports to support the Tossell and Lazzeretti (1986) calculations. According to this interpretation, however, the dependence of shielding on distance is much larger than that calculated, with a 0.02 Å decrease in Si-O distance leading to a deshielding of about 6 ppm Further study will be needed to determine whether this correlation is really a simple one between shielding and distance or whether it depends upon other effects arising in the second-coordination sphere which perturb both bond distances and shieldings..

AL FLUORIDES AND HYDROXYFLUORIDES

We have recently studied the Al NMR shieldings of various Al fluorides and hydroxyfluorides (Tossell, 1992a) to identify the Al species present in F-bearing aluminosilicate glasses (Kohn et al., 1991) Although the overall structure of silicate magmas is mainly determined by the "network-forming" tetradrally coordinated cations Al and Si, small amount of other network-formers such as B or P or small amounts of volatiles such as H_2O or F can substantially modify magma properties. NMR is perhaps the most powerful experimental tool for the study of such systems. Briefly, we find that an observed Al NMR peak with a shielding intermediate between those from well-characterized Al(O-)4 and AlF_6^{-3} species is best assigned on the basis of Al and F NMR shieldings and Al electric field gradient (EFG) to a species with three F and two O as nearest neighbour to Al. We have modeled this species as $AlF_3(OH)_2^{-2}$ and have calculated its minimum energy geometry, vibrational spectrum, EFG at Al and Al and F NMR shieldings . Calculated Al NMR shieldings are given in Table 2 and calculated EFG's in Table 3.

Table 2. Calculated Al NMR isotropic shieldings using 3-21G* optimized geometries, a 3-21G* basis and the program RPAC (6-31G* values in parentheses).

		CHF	Core-corrected CHF	LORG
AlF_4^-		638.4 (648)	592.6	655.5
AlF_5^{-2}	D_{3h}	661.3 (673)	606.8	603.4
	C_{4v}	660.7	606.3	597.9
AlF_6^{-3}		680.2	617.7	595.6
$Al(OH)_4^-, < = 143.3°$		628.4	585.3	
$Al(OH)_4^-, S_4$[a]		586.6	543.5	
AlO_4^{-5}, T_d		552.9	509.8	
$Al(OH)_5^{2-}, C_S$		632.6[c]	589.5	
AlO_5^{-7}, D_{3h}		600.4	549.5	
$AlF(OH)_3^-$		596.6	552.9	
$AlF_2(OH)_2^-$		612.4	568.0	
$AlF_3(OH)_2^{-2}$		645.4[b]	592.4	

a. 3-21G* equilibrium value, with <Al-O-H=106.5°
b. Calculated with the common origin coupled Hartree-Fock program SYSMO

Table 3. Calculated EFG's at Al in AlF_5^{2-}, $AlF_2(OH)_3^{2-}$, $AlF_3(OH)_2^{2-}$, $AlF(OH)_3^-$ and $AlF_2(OH)_2^-$ from 3-21G* bases at 3-21G* optimized geometries.

		$|q|$ (au)	e^2qQ/h (MHz)[a]
AlF_5^{2-}	C_{4v}	0.1366	4.2
	D_{3h}	0.0600	1.8
$AlF_2(OH)_3^{2-}$, D_{3h}		0.2815	8.7
$AlF_3(OH)_2^{2-}$, C_S		0.0375	1.2
$Al(OH)_5{-2}$, C_S		0.0856	2.6
$AlF(OH)_3^-$, C_{3v}		0.1484	4.6
$AlF_2(OH)_2^-$, C_{2v}		0.0377	1.2
AlF, $C_{\infty v}$		0.8578	26.4

a. Assuming $eQ = 1.5 \times 10^{-1}$ barns for ^{27}Al.

The narrowness of the intermediate shielding feature in the Al NMR spectrum in Kohn et al. (1991) argues for a small Al EFG. As seen in Table 3, the $AlF_3(OH)_2^{-2}$ species, with three short Al-F bonds in the equatorial plane and two longer Al-OH bonds in the apical position, indeed gives a very small EFG. Unfortunately, the calculated Al-F stretching frequences for this species do not differ significantly from those for AlF_6^{-3} More important, we cannot at present from the NMR properties characterize the second nearest neighbours about Al, i. e. determine what atoms are bonded to the O's, and thus cannot determine the role of this Al species within the overall polymeric network.

Note also that the LORG results in Table 2 show a trend opposite to experiment, with the shielding decreasing as the coordination number increses. In the conventional implementation of LORG one chooses a "pulling distance" such that for all the localized MO's whose centroids are within that pulling distance from the nucleus of interest, the shielding contributions are evaluated with that nucleus (rather than the LMO centroid) as gauge origin. This scheme works well for covalent systems, where bonding and nonbonding LMO's occupy distinct regions of space, but is problematic for the highly ionic AlF_n^{3-n} species, for which all the valence LMO's have centroids near the F's. It is very difficult to find a pulling distance that allows proper treatment of both bonding LMO's (which should be pulled to the nucleus) and the nonbonding LMO's (which should not be) and such a choice introduces an undesirable semiempirical component into the method.

SILOXANES AND POLYMERIZED SILICATES

We have also studied the effect of polymerization on Si and O NMR shieldings and upon the O EFG. In amorphous silicates and silicate glasses the main source of disorder is variation in the <Si-O-Si, which can assume values from about 120 to 180^0, although it is most commonly around 140-150°, Determination of the distribution of this (and related) angles is a very important topic in glass science. Many different correlations between Si NMR shielding and <Si-O-Si have been extracted from limited sets of experimental data. We first investigated this question by calculating properties for the model compounds disiloxane, $(SiH_3)_2O$, as a function of <Si-O-Si, with the Si-O distance held constant (Tossell and Lazzeretti, 1988). Disiloxane has been a favorite model for study of the Si-O-Si bridging bond for many years (see e. g. Gibbs, 1982 and Sauer, 1989). It is small enough that one can utilize fairly rigorous quantum mechanical methods for its description and its bond angle and bond-bending force constants appear to be similar enough to those in solid SiO_2 to give a good description of the bulk modulus and vibrational properties of SiO_2 and other silicates (Lasaga and Gibbs, 1987). Disiloxane is, of course, also a real molecule, stable and well-studied in both gas and solid phases.

We found that the Si NMR shielding calculated in disiloxane indeed decreased as the <Si-O-Si decreased, with only small variations in the rate of decrease depending upon our choice of Si-O distance, basis set and gauge origin. Comparison with experiment was difficult since the best experimental correlations were for tektosilicates, Q^4 species, in which all four oxygens of the silicate tetrahedron were shared, rather than for the one shared O in $(SiH_3)_2O$. Naively multiplying our calculated trend in σ^{Si} vs. <Si-O-Si by a factor of four gave changes in shielding about 25% larger than those from the empirical correlations. This was a reasonably significant accomplishment - a semiempirical MO approach was later shown to give exactly the opposite trend for shielding vs. angle (Malkin and Zhidomirov, 1990). Nonetheless, it was somewhat unsatisfying to calculate many millions of electron repulsion integrals to high precision and then be forced to multipy the final results by 4, assuming additivity of the briding bond effects. Comparing the $(SiH_3)_2O$ results with those for a monomeric model, e. g. SiH_3O^- or SiH_3OH, we found an increase in shielding with polymerization of perhaps 19 ppm. At that time the importance of the core electron correction had not been recognized by us. Correcting for the presence of the core electrons on the second Si in $(SiH_3)_2O$ could reduce this difference by as much as 30 ppm Clearly it was desireable to utilize better models to more firmly establish the effect of degree of polymerization and <Si-O-Si on the shielding.

There were also a number of other clear successes and failures of the Tossell and Lazzeretti (1988) calculations. First, the EFG at O was reproduced reasonably well and it was clearly calculated to decrease as the <Si-O-Si decreased. Lindsay and Tossell (1991) have supported and refined these results and recent experimental studies by Farnan et al. (1992) show the predicted trend of EFG with angle. On the other hand, for Si-O-Al linkages our calculated O EFG was somewhat too large and the Si was calculated to be shielded compared to the Si-O-Si case, exactly the opposite of the difference observed experimentally. This suggested that inclusion of counterions, e. g. the alkali metal ions, M^+, present in feldspars, would be necessary to properly describe the effect of the coupled

substitution of Al^{3+} and M^+ for Si^{4+}. We are presently still grappling with the methodolical and technical problems arising from theneed to include alkali metal ions. We also found in these studies that the calculated oxygen NMR shielding was too large and the shielding anisotropy was also larger than values estimated experimentally. These problems were the result of unbalanced diamagnetic contributions from the Si core electrons, which could be eliminated by our core correction approach or by using a local origin method. For example, the CHF value of 464 ppm for the O isotropic shielding in $(SiH_3)_2O$, <Si-O-Si=180°, was reduced to 351 by the core correction and to 336 in a LORG calculation (Tossell, 1991 unpublished results, with slightly different distances and basis set) in reasonable agreement with an experimental value around 324 ppm The calculated shielding anisotropy was reduced from about 200 ppm to about 100 ppm by core correction and the corresponding LORG value was 106 ppm.

In later work we utilized $(SiH_2O)_n$, n=2-4, models to describe small rings occuring in siloxanes and silicates (particularly silicate glasses) Our work and that of others established that for n≥4, the <Si-O-Si were much like those in silicates and stabilities per SiO unit were comparable. On the other hand, for n=3 the <Si-O-Si was smaller (133°) and the stability was lower, while n=2, $Si_2H_4O_2$, was an edge sharing (rather than a corner sharing) dimer, with a 91° Si-O-Si angle and was highly unstable. Calculated Si NMR shieldings (using conventional CHF theory, a polarized triple zeta basis and Si as the gauge origin) were 541, 526 and 472 ppm for the n=4, 3 and 2 cases, respectively. The calculated deshielding of the n=3 or "three-ring" case, was consistent with the assignment of a deshielded peak to such species in silica gels (Brinker et al., 1990) The calculated frequency for the symmetric oxygen breathing mode in $Si_3H_6O_3$ was also consistent with that observed for the so-called "D2" defect band in the Raman spectrum of silica gel, which had been shown to correlate in intensity with the assigned NMR peak At that time no values were available for shielding in "two-rings" - indeed such species have only recently been synthesized . The cross-polarization magic angle spinning solid state NMR of $Si_2R_4O_2$, where R=2,4,6-trimethylphenyl (McKillop et al., 1992), gives a Si shift (compared to tetramethyl silane) of about -3 ppm, about 30 ppm deshielded compared to $(C_6H_5)_2Si(OH)_2$, the literature monomer most like it (Marsmann, 1981) This is at least consistent with our prediction that Si will be strongly deshielded in the "two-ring" compared to ring structures with n≥3.. It is also worth noting that for the $[(CH_3)_2SiO]_n$, n=3-6 series, the experimental Si shift values are -9, -20, -23 and -23 ppm, respectively (Marsmann, 1981). Thus, the n=3 ring is indeed deshielded compared to thehigher order ones by about 11 ppm, similar to our calculated difference of 15 ppm.

From one point of view the use of simple models such as $(SiH_2O)_n$ is certainly less desirable than the use of much larger, more complicated models correct through the second nearest-neighbour atoms, e. g. ones in which the H's are replaced by OH's Certainly, such larger cluster models will be used as computational capabilities improve. However, for understanding the common features of shielding trends in related materials and for building a bridge of understanding betwen solids and chemically related gas phase molecules studies on simpler molecules can be very valuable. Such molecules are also simple enough to be understood using qualitative MO models, which may provide a deeper understanding of shielding trends. For example, in Lindsay and Tossell (1991) we showed that the same qualitative MO theory which allows us to at least rationalize orbital energies as a function of angle and equilibrium geometries as a function of electron count in AH_2 type molecules can be used to understand why the Si NMR shielding decreases with <Si-O-Si in $(SiH_3)_2O$ (an analog of an AH_2 molecule with A=O and H replaced by the sigma bonding orbital of - SiH_3). Briefly, as the <Si-O-Si decreases the Si-O sigma-bonding orbital of b_2 symmetry is destabilized and its contribution to the paramagnetic shielding increases in magnitude. At the same time, one component of the O2p pi non-bonding orbital becomes a slightly bonding a_1 symmetry orbital. As it is stabilized its shielding contribution decreases in magnitude. These two orbitals contribute more than half the total magnitude of the paramagnetic shielding, so it is the balance of their contributions which determines the change in total shielding with angle, as shown in Table 4 .

Table 4. Values of orbital eigenvalues for the analogs of the $1b_1$, $2a_1$ and $1b_2$ canonical MO's of $(SiH_3)_2O$ at <Si-O-Si=140° and 120°, along with their contributions , $\sigma P(i)$, to the paramagnetic shielding and the total paramagnetic shielding over all orbitals, σP. Eigenvalues in atomic units and shieldings in ppm.

	<Si-O-Si	
	140	120
$\varepsilon(1b_1)$	-0.446	-0.448
$\varepsilon(2a_1)$	-0.460	-0.484
$\varepsilon(1b_2)$	-0.483	-0.467
$\sigma P(2a_1)$	-70.5	-65.9
$\sigma P(1b_2)$	-126.0	-140.2
$\sigma P(tot)$	-460.8	-471.6

Of course, this analysis is somewhat incomplete since we have the decomposition of the paramagnetic shielding over occupied orbitals but no specific information (other than restrictions imposed by symmetry) on the occupied-unoccupied orbital mixings determining this term. Nonetheless, the analysis may be useful for rationalizing trends in shieldings in related systems, e. g. at bridging S in $(SiH_3)_2S$.

P AND SI SHIELDING ANISOTROPIES

Another quantity potentially more useful than the isotropic NMR shielding itself is the full shielding tensor, which can give very detailed information on three dimensional structure and bonding. We have in general found that shielding tensor components, relative to the isotropic average, are actually somewhat easier to predict than are absolute values of the isotropic shielding, presumably because the chemical anisotropy resides in the frontier orbitals, which are fairly well treated in the MO calculation. For example, in early studies on P fluorides and oxyfluorides (Tossell and Lazzeretti, 1987), using experimental geometries and polarized triple zeta bases (with two P3d functions), we found reasonably accurate shielding anisotropies, as shown in Table 5.

Table 5. Comparison of calculated shilelding constant anisotropies and average values with experiment (in ppm) for P fluorides and oxyfluorides

molecule	$\Delta\sigma(exp)$[a]	$\Delta\sigma(calc)$	$\sigma(exp)$	$\sigma(calc)$[b]
PF_3	182	243	223	373
PF_3O	278	358	355	507
HPO_4^{-2}	-102	-108	313	474
PO_3F^{2-}	-145	-117	322	478

a. $\Delta\sigma$ is the difference of shieldings parallel and perpendicular to the three fold axis
b. shifts have been converted to absolute shieldings using Jameson et al., (1990)

One obvious conclusion from the data in Table 5 is that the conventional CHF method gives absolute shielding much too large for these molecules, due to the imperfect cancellation of core electron contributions to the diamagnetic and paramagnetic shieldings. Note, however, that the shifts along this nearly isoelectronic series are given well. Also, the anisotropies have the right sign and magnitude, although they are generally exaggerated. One interesting point emerging from the HPO_4^{-2} calculations was that the proton had to be included, i. e. a bare PO_4^{-3} group with the HPO_4^{-2} geometry gave an anisotropy only about 25 % as large as HPO_4^{-2} itself. We obtained similarly good results for the species $P_3O_9^{-3}$ (Lindsay and Tossell, 1991), for which our calculated shielding anisotropy was (fortuitously) within a ppm of the experimental value of 249. For the above "three-ring" we were of course able to include in the calculation all the first neighbour O's and second neighbour P's which generated the anisotropy.

We found the Si shielding anisotropy in forsterite, Mg_2SiO_4, much more difficult to accurately describe (Tossell, 1992b). In this mineral the SiO_4 tetrahedra are unpolymerized, being held together by Mg^{2+} counterions,. and they are only slighly distorted from T_d symmetry. The shielding anisotropy is, however, fairly substantial. To explain this result we tried a number of different models, as shown in Table 6. In particular we have used the conventional CHF method and the LORG method (both as implemented in the program RPAC), several different basis sets, and both bare silicate anions and anions stabilized by point charges, designated PC.

Table 6. Experimental and calculated values for the ^{29}Si NMR shielding tensor in forsterite (calculated values from the RPAC program unless otherwise noted)

		σ	11	Δσ(relative to average) 22	33
exp[a]		431.7	-24.4	-7.9	+32.2
free SiO4^{-4}, forsterite geom., 6-31G*	CHF	515.9	-13.3	-5.9	+19.2
	LORG	449.0	-18.7	-8.5	+27.2
6-31G*(Si triple 3d)	CHF	540.7	-6.6	-3.0	+9.6
	LORG	476.3	-8.8	-4.6	+13.4
[6s5p1d/4s3p1d]	CHF	528.3	-13.1	-5.7	+18.8
SiO4PC4, all <Si-O-PC=101°, 6-31G*	CHF	502.1	-34.3	+5.8	+28.6
	LORG	433.7	-30.8	+0.2	+30.5
SiO4PC4, forsterite values for <Si-O-PC, 6-31G*	CHF	506.3	-32.4	+7.9	+24.5
	LORG	437.7	-28.9	+2.3	+26.6
a. Weiden and Rager	(1985)				

As expected, the LORG values for the isotropic shielding are considerably smaller than the CHF and much closer to experiment. The effect of the point charge stabilization is to somewhat reduce the isotropic shielding but to substantially increase the anisotropy. Using accurate second-nearest-neighbour angles improves the shielding anisotropy slightly. A more flexible basis set actually gives poorer results for the unstabilized cluster. Forsterite is an important test case since it establishes what level of description in the second-neighbour sphere is necessary for describing the anisotropy, which is potentially measureable through its effect upon relaxation times in silicate glasses and liquids and may help to establish theSi species existing within such phases. General trends in Si shielding anisotropies are also discussed in the paper by Wolff in this volume.

SYMMETRY AND POINT CHARGE EFFECTS ON SI SHIELDINGS

We first used a SiO4PC4 model in describing the shielding in Si(OH)4 and various ions derived from its deprotonation (Tossell, 1991a). In that study we found that the series Si(OH)4 to SiO4^{-4} showed a very slightly "sagging" tendency, with Si(OH)3O$^-$ having the smallest shielding. Tetrahedral symmetry Si(OH)4 was also found to have a shielding higher than that of the equilibrium S4 symmetry conformer by almost 100 ppm and we showed that replacing the H's with point charges (i. e. the bare proton with no basis functions) changed the shielding by only a few ppm. We were able to relate the trends in shielding (which were almost entirely determined by the magnitude of the paramagnetic contribution) to the relative eigenvalues of the Si-O sigma bonding CMO's in these species.

For example, the t_2 bonding MO was about 4.7-5.0 eV more stable than the (nonbonding) HOMO in both SiO_4^{-4} and in S_4 symmetry $Si(OH)_4$ or SiO_4PC_4 but was 6.6-7.0 eV more stable for T_d symmetry $Si(OH)_4$ or SiO_4PC_4. The stabilization of this sigma-bonding MO reduced the magnitude of its paramagnetic deshielding, giving larger overall shielding values for the protonated T_d symmetry species. Although this interpretation is clearly related to the semiempirical MO energy scheme developed in Tossell (1984) it is also related to the second-neighbour bond polarization approach developed by Sternberg (1988). Basically, any second-neighbour influence which stabilizes the Si-O bonding orbital will tend to reduce the magnitude of its paramagnetic shielding contribution and thus to shield the Si nucleus.

ZN AND CD CHLORIDES AND BISULFIDES

In addition to our studies of the NMR properties of anions and network-forming cations in minerals and glasses, we have begun a series of calculations on the shieldings of Zn and Cd complexes in aqueous solution. A long-standing problem in transition metal geochemisty has been understanding how metals such as Zn and Cd, whose predominant sulfide ores have very low solubility constants, can reach appreciable concentrations in natural waters. Chloride and bisulfide complexes have been implicated in the solubilization of these metals but identification of the complexes is difficult since they are transparent in the visible and their vibrational spectra are often complex and poorly resolved. In previous experimental studies, average shieldings for a given solution composition along with assumed speciation models and equilibrium constants were used to determine the shieldings for individual complexes (e. g. Maciel et al., 1977).

We are engaged in systematically studying the Zn and Cd chloride and bisulfide complexes to determine trends in complex strucure, stabilities and properties. Some of the simpler chloride species have previously been studied by Nakatsuji et al. (1984), who also advanced a qualitative model based on metal valence p orbital populations to interpret trends in the paramagnetic and total shieldings. Initial studies of ZnX_4^{-2} complexes, where X=Cl,OH,SH and CN, established that polarized double zeta basis set SCF calculations could reproduce structures, vibrational spectra and NMR shifts, if the shieldings were corrected for distant core electron contributions (Tossell, 1990b). We show in Table 7 calculated bond distances, totally symmetric stretching frequencies and NMR shifts, compared to $Zn(OH_2)_6^{2+}$, for the various species.

Table 7. Calculated bond distances (in Å), totally symmetric stretching frequences (in cm^{-1}) and NMR shifts (in ppm compared to $Zn(OH_2)_6^{2+}$) for ZnX_4^{-2} species. Experimental values are given in parentheses

X	R	ν	$\Delta\sigma$
Cl-	2.31 (2.28, 2.30)	297 (275-282)	-197 (-253)
CN-	2.11 (2.02)	3.0×10^2 (339)	-249 (-284)
OH-	1.96	4.9×10^2 (465)	-195 (-220)
SH-	2.40 (2.35)	2.4×10^2, 246	-317 (-350 to -360)

There is perhaps a slight tendency for the more ionic interaction with Cl^- to be described somewhat better than the more covalent interactions. Note that all these calculations are on bare gas-phase species with no solvent interactions. For $X=CN^-$ and OH^- we have so far only estimated frequencies, with the ligand group held rigid. Trends in shielding are well reproduced, showing in particular the deshielding expected for SH^- compared to OH^-

A more difficult task is to evaluate the shieldings for species such as "aqueous ZnCl", which must actually be modeled as at least the species $ZnCl(OH_2)_a$, with a the variable number of waters of hydration. A more complete study would of course incorporate the effect of the solvent, at least at the level of a polarizable continuum, and would probably need to consider H-bonding in the second coordination shell. We have previously published results for the $ZnCl_n(OH_2)_a^{2-n}$, n=1-4, series (Tossell, 1991b), obtaining good agreement with available x-ray structural data, some understanding of the Raman data and a preliminary match of some of the observed NMR shieldings to the various species present. Later studies on the Cd chlorides and bromides (Butterworth et al, 1992) and the Zn and Cd bisulfides (Tossell and Vaughan, 1992) have clarified some of the systematic trends in the shielding. Results for Cd chlorides and their hydrated analogs are shown in Table 8, obtained at minimum energy geometries (unless otherwise noted) using essentially polarized double zeta basis sets and the conventional CHF method with core correction.

Table 8. Calculated isotropic NMR shieldings for Cd chloride complexes. Core corrected absolute shieldings and shifts compared to $Cd(OH_2)_6^{2+}$ are given and compared with experiment

	σ_{corr}	$\Delta\sigma_{corr}$	
		calc.	exp.
Cd	4781	-722	-1106[a]
Cd^{+2}	4769	-710	
$Cd(OH_2)_6^{2+}$	4059	0	0
$CdCl^+$	4370	-311	
$Cd(OH_2)_5Cl+$	3955	104	89[b]
$CdCl_2$	3880	179	
$CdCl_2(OH_2)_2$	3804	255	
$CdCl_2(OH_2)_4$	3848	211	114[b]
$CdCl_3^-$	3724	335	
$CdCl_3(OH_2)^-$	3755	304	296[b]
$CdCl_3(OH_2)_2^-$	3882	177	

CdCl$_4{}^{-2}$, R=2.453Å(exp.)	3608	451	451-474
R=2.574Å(calc)	3802	257	
CdCl$_5{}^{-3}$ R=2.53,2.56Å (exp.)	3741	318	
R=2.72,2.66Å (calc.)	3955	104	247-329 (solution), 188,172 (solids)

a. Kruger et al., (1974)
b. Ackerman et al. (1979)

Our results show first that for bare chloride species, CdCl$_n{}^{2-n}$, the shielding is actually a minimum for n=3 if one uses comparable geometries (i. e. calculated geometries for all species). The n=4 and 5 species have higher shieldings, because the shielding depends on the Cd-Cl bond distance as well as the number of Cl's and Cd-Cl bond distances are considerably larger in the n=4,5 species. Considering the substitution of Cl$^-$ for H$_2$O , starting from the Cd(OH$_2$)$_6{}^{2+}$ species dominant in the solutions of non-complexing ligands, we find a deshielding of about 100 ppm per substitution. For the n=2 case, removal of two H$_2$O molecules allows closer approach of the Cl atoms and therefore a lower shielding. None of the n=2 species considered match well against the experimental data obtained using assumed equilibrium constants. For n=3 a similar trend is observed; addition of one H$_2$O molecule slightly increases the Cd-Cl distance giving a slight shielding, while addition of two waters substantially increases R(Cd-Cl) and gives a much increased shielding. Results for the CdCl$_3$(OH$_2$)$^-$ species are in reasonable agreement with experiment. For CdCl$_4{}^{-2}$ the relative shielding at the experimental distance is in good agreement with the experimental value, which was actually fixed by comparison with solid state data. At the calculated distance, which is 0.12Å larger, the shielding is also considerably larger (due to reduction in the magnitude of the paramagnetic term). The CdCl$_4$ geometry optimization was performed with effective core potentials and an unpolarized double zeta basis, which may account for the exaggerated bond length. However, for CdCl$_5{}^{-3}$ we used our usual all-electron polarized double zeta bases and still got apparently exaggerated distances (and the wrong order of apical and equatorial distances) and rather erratic shielding results. Note that the experimental values for the shielding of apparent CdCl$_5{}^{-3}$ species are also quite variable.

Although our main interest in the bisulfide complexes arises from their geochemical significance, they are also important biochemically. Recently, a Cd(SR)$_3{}^-$ species has been synthesized and characterized by x-ray diffraction and solid state NMR and its possible biological significance discussed (Santos et al, 1991). Modeling this species as Cd(SH)$_3{}^-$, we calculate a minimum energy geometry of C$_{3h}$ symmetry (as observed) and obtain a Cd-S bond distance of 2.485Å, compared to 2.42Å observed. At the experimental geometry, the calculated shielding perpendicular to the three-fold axis is larger than that parallel by

393 ppm, compared to an experimental value of 483. Correcting the shielding components for the presence of the (30) S core electrons gives an anisotropy of 481 ppm, in good agreement with experiment. Perhaps more important to geochemists and surface chemists, the shielding anisotropy for the solution or surface species $Cd(SH)_3(OH_2)^-$, is calculated to be 429 ppm. Such species may thus be better characterized by their shielding anisotropy than by their isotropic shielding values.

Conclusion

Although substantial questions still remain about the most accurate and efficient ways to calculate both the structures and NMR shieldings of the chemical species present in minerals, glasses and solutions, it is clear that conventional *ab initio* SCF MO theory and conventional coupled Hartree-Fock perturbation theory (with somple core electron corrections) can reproduce both many of the experimental trends and some of the specific experimental results. Higher level correlated methods (such as those described in other chapters in this volume) and consideration of larger clusters will certainly eventually lead to improved agreement with experiment, but there is also a great need for the increased use of qualitative concepts, soundly based upon the conclusions from *ab initio* calculations, which can help to order the data and to predict at least the overall trends. At present, it is also quite valuable to examine other properties, e. g. EFG's or vibrational spectra, since NMR data is often not sufficient to completely characterize a condensed phase species. Recent successful applications of *ab initio* Hartree-Fock band theory (Pisani, 1987) raise the possiblity of going beyound the cluster model in the calculation of solid-state shieldings. For example, the Hartree-Fock crystal orbitals from the band calculation might be localized as Wannier-like functions and shieldings then evaulated from those orbitals localized near a given magnetic nucleus.

Acknowledgements

We gratefully acknowledge support from the US National Science Foundation through grants EAR8603499 and EAR9000654, from NATO through CRG 90015 to Tossell and D. J. Vaughan and for NATO support of this Advanced Research Workshop through grant ARW 910712.

References

Ackerman, J. J. H., Orr. T. V., Bartuska, V. J. and Maciel, G. E. (1979), Effect of halide complexation of Cadmium (II) on Cadmium-113 chemical shifts, J. Am. Chem. Soc., 101, 341-347

Brinker, C. J., Brow, R. K., Tallant, D.R. and Kirkpatrick, R. J. (1990) Surface structure and chemisty of high surface area silica gels, J. Noncrystal. Solids, 120, 26-33

Butterworth, P., Hillier, I.H., Burton, N.A., Vaughan, D.J., Guest, M.F. and Tossell., J.A. (1992), Calculations of the structures, stabilities, Raman spectra and NMR spectra of

$CdCl_n(OH_2)_a^{2-n}$, $CdBr_n(OH_2)_a^{2-n}$ and $ZnCl_n(OH_2)_a^{2-n}$ species in aqueous solution, submitted to J. Phys. Chem.

Engelhardt, G. and Radeglia, R. (1984), A semi-empirical quantum-chemical rationalization of the correlation between SiOSi angles and ^{29}Si NMR chemical shifts of silica polymorphs and framework aluminosilicates (zeolites), Chem. Phys. Lett., 108, 271-274

Farnan, I., Grandinetti, P.J., Baltisberger, J. H., Stebbins, J. F., Werner, U., Eastman, M. A. and Pines, A. (1992), Quantification of the disorder in network-modified silicate glasses, Nature, 358, 31-35

Flygare, W.H. and Goodisman, J. (1968), Calculation of diamagnetic shielding in molecules, J. Chem. Phys., 49, 3122-3125

Gibbs, G.V. (1982), Molecules as models for bonding in silicates, Amer. Mineral., 67, 421-450

Hansen, A.E. and Bouman, T.D. (1985), Localized orbital local origin method for calculation and analysis of NMR shieldings, Applications to ^{13}C shielding tensors, J. Chem. Phys., 82, 5033-5047

Jameson, C. J., deDios, A. and Jameson, A. K. (1990), Absolute shielding scale for ^{31}P from gas-phase NMR studies, Chem. Phys. Lett., 167, 575-582

Kohn, S.C., Dupree, R., Mortuza, M.G. and Henderson, C.M.B. (1991), NMR evidence for five- and six- coordinated aluminum fluoride complexes in F-bearing aluminosilicate glasses, Amer. Mineral., 76, 309-312

Kruger, H., Lutz, O., Schwenk, A. and Stricker, G. (1974), Nuclear magnetic resonance studies of ^{111}Cd, Z. Phys., 266, 233-237

Lasaga, A. C. and Gibbs, G. V. (1987), Applications of quantum mechanical potential surfaces to mineral physics calculations, Phys. Chem. Minerals, 14, 107-117

Lazzeretti, P. and Zanasi, R. (1980), Theoretical determination of the magnetic properties of HCl, H_2S, PH_3 and SiH_4 molecules, J. Chem. Phys., 72, 6768-6776

Lindsay, C.G. and Tossell, J.A. (1991), Ab initio calculations of ^{17}O and ^nT NMR parameters ($^nT={}^{31}P, {}^{29}Si$) in H_3TOTH_3 dimers and T_3O_9 trimeric rings, Phys. Chem. Minerals, 18, 191-198

Lippmaa, E., Magi, N., Samoson, S., Engelhardt, G. and Grimmer, A.-R. (1980), Structural studes of silicates by solid-state high-resolution ^{29}Si NMR, J. Am. Chem. Soc., 102, 4889-4893

Lipscomb, W. N. (1966), The chemical shift and other second-order magnetic and electric properties of small molecules, pp. 137-176 in Advances in Magnetic Resonance, Vol. 2, ed. J. S. Waugh

Maciel, G. E., Simeral, L. and Ackerman, J. J. H. (1977), Effect of complexation of Zn(II) on Zinc-67 chemical shifts, J. Phys. Chem. 81, 263-267

Malkin, V.G. and Zhidomirov, G.M. (1990), Quantum-chemical calculations of ^{29}Si chemical shifts in zeolites, Zeolites, 10, 207-209

McKillop, K.L., Gillette, G.R., Powell, D.R. and West, R. (1992), 1,2-Disiladioxetanes: structure, rearrangement, and reactivity, J. Am. Chem. Soc., 114, 5203-5208

Marsmann, H. (1981), ^{29}Si-NMR spectroscopic results, pp. 65-235 in NMR Basic Principles and Progress, Vol. 17, Oxygen-17 and Silicon-29, ed. Diehl, P., Fluck, E. and Kosfeld, R.

Nakatsuji, H., Kanda, K. , Endo, K. and Yonezawa, T. (1984) Theoretical study of the metal chemical shift in nuclear magnetic resonance. Ag, Cd, Cu and Zn complexes, J. Am. Chem. Soc, 106, 4653-4660

Pisani, C. (1987) Hartree-Fock *ab initio* approaches to the solution of some solid-state problems: state of the art and prospects, Int. Rev. Phys. Chem., 6, 367-384

Radeglia, R. and Engelhardt, G. (1985), Correlation of Si-O-T (T=Si or Al) angles and ^{29}Si NMR chemical shifts in silicates and aluminosilicates. Interpretation by semi-empirical quantum-chemical considerations, Chem. Phys. Lett., 114, 28-30

Santos, R. A., Gruff, E. S., Koch, S. A. and Harbison, G. S. (1991), Solid-state ^{199}Hg and ^{113}Cd NMR studies of mercury and cadmium thiolate complexes. Spectroscopic models for [Hg(SCys)$_n$] centers in the bacterial mercury resistance proteins, J. Am. Chem. Soc., 113, 469-475

Sauer, J. (1989), Molecular models in ab initio studies of solids and surrfaces: from ionic crystals and semiconductors to catalysts, Chem. Rev. 89, 199-255

Skibsted, J., Hjorth, J. and Jakobsen, H. J. (1990), Correlation between ^{29}Si NMR chemical shifts and mean Si-O bond lengths for calcium silicates, Chem. Phys. Lett., 172, 279-283

Smith, J. V. and Blackwell, C.S. (1983), Nuclear magnetic resonance of silica polymorphs, Nature, 303, 223-225

Thomas, J. M., Klinowski, J., Ramdas, S., Huner, B.K. and Tennakoon, D.T.B. (1983), The evaluation of non-equivalent tetrahedral sites from 29Si NMR chemical shifts in zeolites and related aluminosilicates, Chem. Phys. Lett., 102, 158-162

Tossell, J.A. (1977), A comparison of silicon-oxygen bonding in quartz and magnesian olivine from X-ray spectra and molecular orbital calculations, Amer. Mineral. 62, 136-141

Tossell, J.A. (1984), Correlation of 29Si nuclear magnetic resonance chemical shifts in silicates with orbital energy differences obtained from X-ray spectra, Phys. Chem. Minerals, 10, 137-141

Tossell, J.A. (1990a), Calculation of NMR shieldings and other properties for three and five coordinate Si, three coordinate o and some siloxane and boroxol ring compounds, J. Noncrystal. Solids, 120, 13-19

Tossell, J.A. (1990b), *Ab initio* calculation of the structures, Raman frequences and Zn NMR spectra of tetrahedral complexes of Zn^{2+}, Chem. Phys. Lett., 169, 145-149

Tossell, J.A. (1991a), Calculation of the effect of deprotonation on the Si NMR shielding for the series Si(OH)$_4$ to SiO$_4^{-4}$, Phys. Chem. Minerals, 17, 654-660

Tossell, J. A. (1991b), Calculation of the structures, stabilities, and Raman and Zn NMR spectra of ZnCl$_n$(OH$_2$)$_a^{2-n}$ species in aqueous solution, J. Phys. Chem., 95, 366-371

Tossell, J. A. (1992a), Theoretical studies of the speciation of Al in F-bearing aluminosilicate glasses, Amer Mineral., in press

Tossell, J.A. (1992b), Calculation of the ^{29}Si NMR shielding tensor in forsterite, Phys. Chem. Minerals, in press

Tossell, J. A. and Lazzeretti, P. (1986), *Ab initio* calculations of ^{29}Si NMR chemical shifts for some gas phase and solid state silicon fluorides and oxides, J. Chem. Phys., 84, 369-374

Tossell, J. A. and Lazzeretti, P. (1987), *Ab initio* calculation of the ^{31}P NMR shielding tensor for the series $POaF_b^{-(2a+b-5)}$, a+b=4 and for HPO_4^{-2}, Chem. Phys. Lett., 140, 37-40

Tossell, J. A. and Lazzeretti, P. (1988), Calculation of NMR parameters for briding oxygens in H3T-O-TH3 linkages (T,T=Al, Si,P), for oxygen in SiH_3O^-, SiH_3OH and SiH_3OMg^+ and for bridging fluorine in $H_3SiFSiH_3^+$, Phys. Chem. Minerals, 15, 564-569

Tossell, J. A. and Vaughan, D.J. (1992), Bisulfide complexes of Zn and Cd in aqueous solution: calculation of structure, stability, vibrational and NMR spectra and of speciation on sulfide mineral surfaces, submitted to Geochim. Cosmochim. Acta

Weiden, N. and Rager, H. (1985), The chemical shift of the ^{29}Si nuclear magnetic resonance in a synthetic single crystal of Mg_2SiO_4, Z. Naturforsch., 40a, 126-130

APPLICATIONS OF DIPOLAR NMR SPECTROSCOPY IN CHARACTERIZING NITROGEN AND PHOSPHORUS SHIELDING TENSORS

R.E. Wasylishen, R.D. Curtis, K. Eichele, M.D. Lumsden,
G.H. Penner, W.P. Power, and G. Wu
Department of Chemistry
Dalhousie University
Halifax, Nova Scotia
Canada B3H 4J3

The application of dipolar NMR techniques to the characterization of nitrogen and phosphorus shielding tensors is discussed. After a brief outline of the theoretical basis of the technique, results from the authors' laboratory are presented. Nitrogen shielding tensors in several compounds containing two-coordinate nitrogen functional groups are characterized and compared to those calculated for related model compounds using the localized orbital-local origin (LORG) method.

1. Introduction

NMR studies of solids are potentially more informative than solution NMR studies since they provide experimentalists with an opportunity to characterize the principal components of shielding tensors as opposed to their traces [1,2,3,4,5,6,7,8,9]. The nuclear magnetic shielding tensor of a particular nucleus can be described by a second-rank Cartesian tensor which is comprised of up to nine independent components [10]. This tensor can be written as a sum of a symmetric part, characterized by up to six unique components, and an antisymmetric part with three unique components [1,11]. To first order, only the symmetric part of the magnetic shielding tensor affects the NMR spectrum, and, for the purposes of the discussion to follow, the antisymmetric portion of the shielding tensor can safely be ignored [1,12]. In one particular axis system, known as the principal axis system (PAS), only the three diagonal elements of the symmetric portion of the nuclear magnetic shielding tensor are non-zero. The NMR line shape of a static powder sample provides the three principal components (the diagonal elements) of the tensor; however, up to three angles are required to specify the orientation of the principal axis system with respect to the molecular frame. In order to obtain unambiguous information concerning the orientation of shielding tensors, single crystal NMR studies are necessary [7,13,14]. Unfortunately, the large single crystals necessary for such measurements of insensitive NMR isotopes (e.g., nitrogen-15) are often not readily available. Also, it is essential to know the orientation of the molecules with respect to the crystal axis system in order to properly analyze the single crystal NMR results. This information is only available from diffraction techniques such as X-ray crystallography. One alternative to single crystal NMR studies that we have found very useful is dipolar NMR [15,16,17]. The technique involves the analysis of NMR spectra resulting from a spin-½ nucleus that is dipolar-coupled to a neighbouring spin. Such an analysis is capable of providing the dipolar coupling constant associated with the spin-pair, the three principal components of the shielding tensor, and two of the angles required to specify the orientation of the shielding tensor. In the discussion which follows we first outline the basic theory required to understand the analysis of NMR line shapes arising from an isolated spin-pair. This discussion is followed by several

J. A. Tossell (ed.), Nuclear Magnetic Shieldings and Molecular Structure, 297–314.
© 1993 *Kluwer Academic Publishers.*

examples from our laboratory that demonstrate the utility of the dipolar NMR technique in characterizing nitrogen and phosphorus chemical shielding tensors. The experiments described here are relatively simple as well as practical and informative. Over the past ten years several groups have made important contributions to the development of dipolar NMR, and novel applications of the technique to a variety of problems. Literature references to this work can be found in a recent review dealing with NMR studies of isolated spin-pairs in the solid state [18]. More recently, Clayden has presented an excellent summary of developments in solid state NMR [19].

2. Theoretical Background

2.1 HETERONUCLEAR SPIN SYSTEMS

For an "isolated" pair of unlike spins, AX, such as ^{15}N-^{2}H, ^{15}N-^{13}C, ^{31}P-^{14}N, or ^{31}P-^{15}N, one has to consider the Zeeman, chemical shielding, direct dipolar and indirect spin-spin interactions. The NMR transition frequency for the A spins (spin-½) depends on the principal components of the A-spin chemical shielding tensor and the orientation of the principal axis system of this tensor in the laboratory frame, defined by the angles θ and ϕ as illustrated in figure 1. The

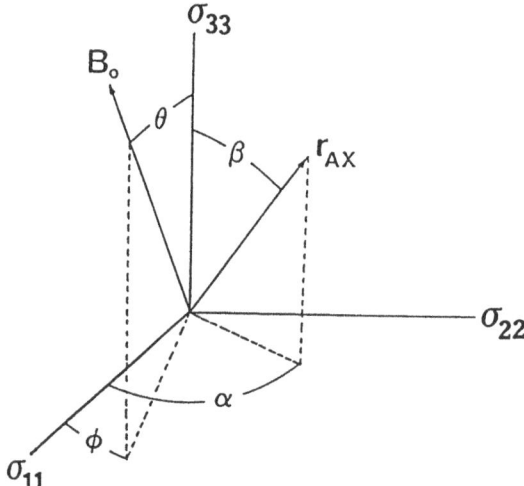

Figure 1. The angles α and β define the orientation of r_{AX} in the PAS of the chemical shielding tensor. θ and ϕ define the orientation of the molecule in the applied magnetic field, B_o.

transition frequency is also dependent on the orientation of the dipolar vector, r_{AX}, in the magnetic field and in the PAS of the shielding tensor. The orientation of r_{AX} in the PAS of the shielding tensor of nucleus A is defined by the angles α and β. Specifically the frequency of the spin-A is given by equation 1.

$$\nu_m(\theta,\phi) = \nu_L - \nu_{CS} - m\nu_D - mJ_{iso} \tag{1}$$

where

$$\nu_L = \frac{\gamma_A}{2\pi} B_o \tag{2}$$

$$\nu_{CS} = (\frac{\gamma_A B_o}{2\pi}) (\sigma_{11}\sin^2\theta\cos^2\phi + \sigma_{22}\sin^2\theta\sin^2\phi + \sigma_{33}\cos^2\theta) \tag{3}$$

$$\nu_D = R_{eff} [1 - 3(\sin\beta\sin\theta\cos(\phi-\alpha) + \cos\beta\cos\theta)^2] \tag{4}$$

$$R_{eff} = R - \frac{\Delta J}{3} \tag{5}$$

$$R = \frac{\gamma_A\gamma_X\hbar}{2\pi} \frac{\mu_o}{4\pi} \langle r_{AX}^{-3} \rangle \tag{6}$$

$$\Delta J = J_\parallel - J_\perp \tag{7}$$

In equation 1, the allowed values of the quantum number m depend upon the spin of nucleus X. For $I=\frac{1}{2}$, $m=+\frac{1}{2}$ or $-\frac{1}{2}$; for $I=1$, m may take on values of 1, 0, or -1. Equation 3 describes the orientation dependence of the chemical shielding for spin-A in the magnetic field. Similarly, equation 4 describes the orientation dependence of the dipolar interaction in the magnetic field relative to the PAS of the shielding tensor. The effective dipolar coupling constant, R_{eff}, is dependent on the direct dipolar coupling constant, R, and the anisotropy of the indirect spin-spin coupling tensor, ΔJ. For the spin-systems considered here we will assume that $(\Delta J/3) \ll R$, hence $R_{eff} \approx R$. The angular brackets in equation 6 indicate that the AX internuclear separation is a motionally averaged value. J_{iso} is the isotropic indirect spin-spin coupling constant. The principal components of the chemical shielding tensor, σ_{11}, σ_{22}, and σ_{33}, are defined such that: $\sigma_{33} > \sigma_{22} > \sigma_{11}$. β is the angle between the dipolar vector r_{AX} and σ_{33}, and α is the angle between the projection of r_{AX} onto the σ_{11}, σ_{22} plane and the σ_{11} axis. In practice, NMR spectroscopists generally measure chemical shifts, or differences in shielding constants between the nuclei of interest and a reference nucleus, as opposed to chemical shielding constants [20]. The components of the chemical shift tensor will be defined using the convention $\delta_{11} > \delta_{22} > \delta_{33}$ (note that δ_{33} corresponds to the most shielded component of the chemical shift tensor). The chemical shift anisotropy (CSA), $\Delta\delta$, is defined as $\delta_{11} - \delta_{33}$ and simply represents the width of the A spin spectrum in the absence of other interactions.

For a given spin-pair, the NMR line shape can be calculated by summing $\nu_m(\theta,\phi)$ over all values of m, $\theta(0° \leq \theta \leq 180°)$ and $\phi(0° \leq \phi \leq 360°)$. We have found the POWDER interpolation routine of Alderman, Solum and Grant [21] most convenient for this purpose. In figure 2, a typical non-axially symmetric chemical shielding powder pattern for an isolated spin-$\frac{1}{2}$ nucleus is shown. In the presence of an adjacent spin-$\frac{1}{2}$ nucleus, the discontinuities at δ_{11}, δ_{22}, and δ_{33} are split by the dipolar interaction (see figure 3). Assuming $J_{iso}= 0$, the angles α and β can be estimated from the splittings at δ_{11}, δ_{22}, and δ_{33} by solving the following equations:

$$\Delta\nu_{11} = | R (1-3\cos^2\alpha \sin^2\beta) | \tag{8}$$

$$\Delta v_{22} = \mid R \; (1-3\sin^2\alpha \; \sin^2\beta) \mid \tag{9}$$

$$\Delta v_{33} = \mid R \; (1-3\cos^2\beta) \mid \tag{10}$$

Note, these expressions will only be valid if $\mid v_{ii} - v_{jj} \mid \gg R$, where v_{ii} and v_{jj} correspond to the frequencies (in Hz) of the principal components of the chemical shift tensor. Three

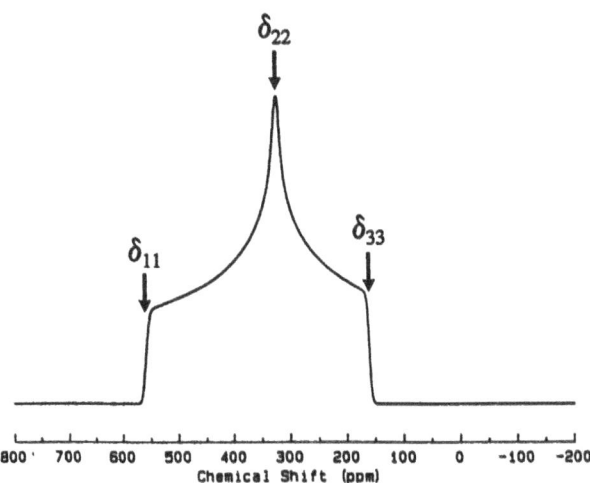

Figure 2. A typical spin-½ NMR powder pattern resulting from anisotropic chemical shielding. δ_{11}=560 ppm, δ_{22}=328 ppm, and δ_{33}=160 ppm.

situations which are pertinent to the present discussion follow from figure 1 and equations 1-10:

1.   If $\beta = 0°$, then r_{AX} is parallel to σ_{33} and the angle α is undetermined.
2.   If $\beta = 90°$, then r_{AX} lies in the σ_{11},σ_{22} plane.
3.   If $\alpha = 0°$, then r_{AX} lies in the σ_{11},σ_{33} plane.

Since the dipolar interaction is axially symmetric, the overall NMR line shape for a powder sample will be insensitive to any rotation of the chemical shift tensor about the dipolar vector. Thus, the orientations of the chemical shift tensors of a spin-pair can not be determined unambiguously from only the powder NMR line shape. In practice, however, this uncertainty can usually be overcome by considering the local symmetry of the molecule and/or results from molecular orbital shielding calculations (vide infra).

Equations analogous to equations 8-10 can be derived if the X-spin is greater than ½. The A-spin NMR line shape for a static powder sample will consist of $2I+1$ overlapping powder patterns corresponding to each of the allowed values of m_X. In general, each of these powder patterns will have a shape analogous to that arising from anisotropic shielding (see figure 2). An example of the powder pattern resulting from the NMR of a spin-½ nucleus dipolar coupled to a spin-3/2 nucleus is shown in figure 4. Note that in this case the shielding anisotropy and direct dipolar coupling constant are comparable. If the X-nucleus quadrupolar coupling constant is comparable to, or greater than its Larmor frequency, then the high-field approximation will not be valid and equation 1 will not properly describe the A-spin line shape [22,23,24,25].

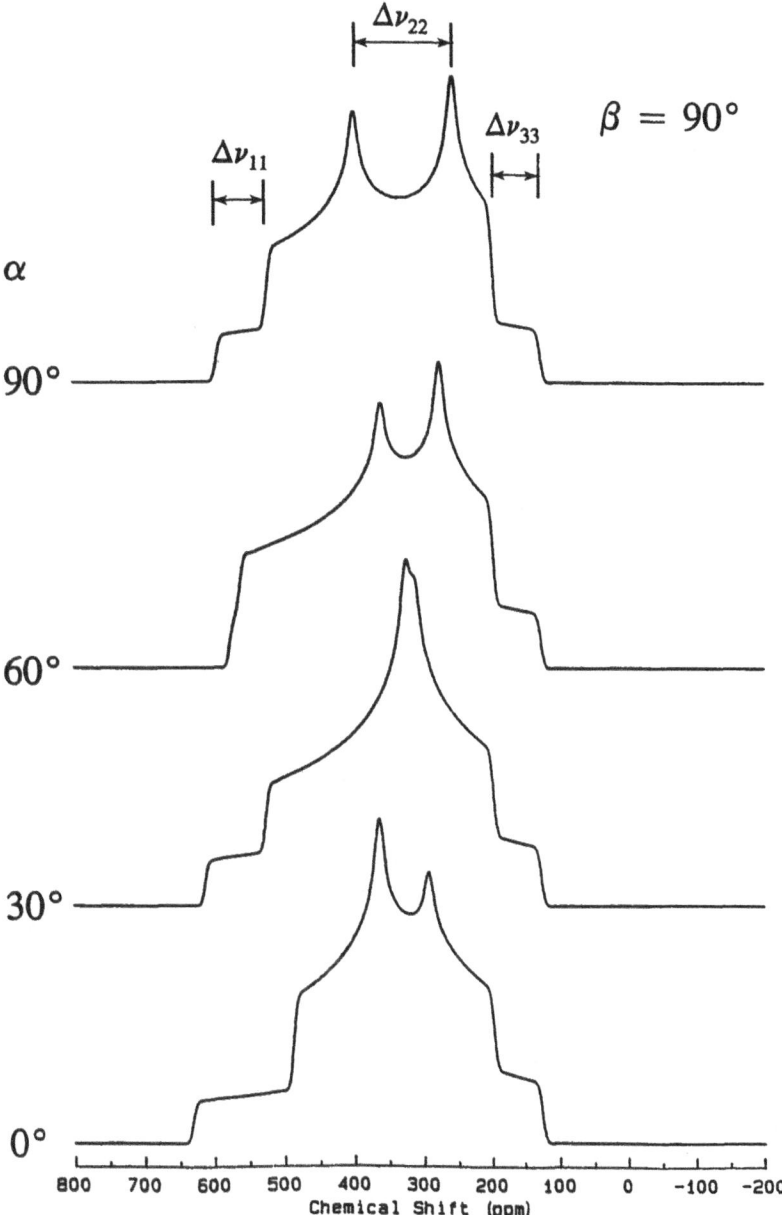

Figure 3. Calculated NMR powder line shape for a spin-½ nucleus (A) dipolar coupled to a heteronuclear spin-½ nucleus (X). The principal components of the A-spin shift tensor are those given in figure 2. The Larmor frequency for the A-spin is 20.3 MHz, R = 1460 Hz, J_{iso} = 0, β = 90° and α = 0°, 30°, 60°, and 90°.

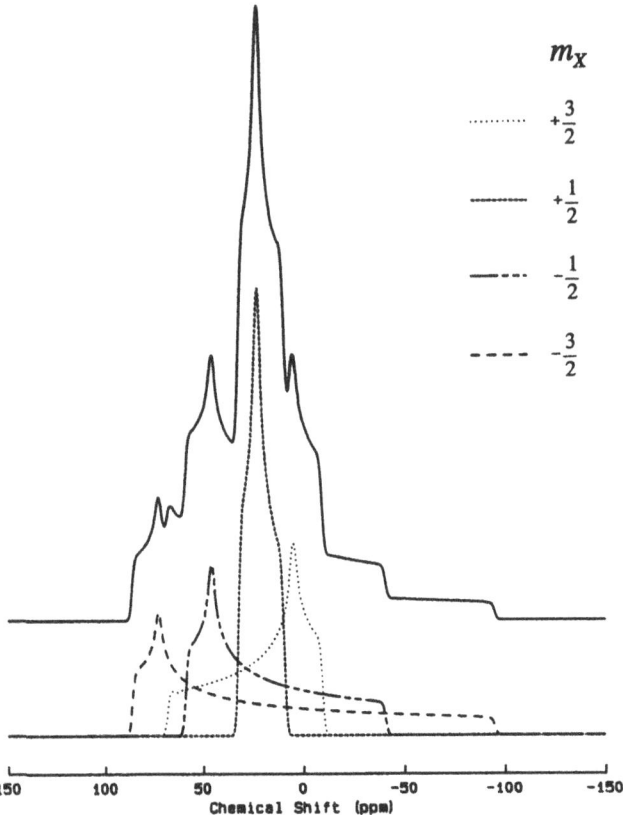

Figure 4. Calculated NMR powder line shape for a spin-½ nucleus(A) dipolar coupled to a spin 3/2 nucleus. Each of the four sub-spectra corresponding to the allowed values of m_x is indicated by the dashed lines. The parameters used to calculate the line shape are: δ_{11} = 46 ppm, δ_{22} = 33 ppm, δ_{33} = -14 ppm, R = 2150 Hz, α = 80°, β = 8°. The Larmor frequency of the A-spin is 81.0 MHz.

2.2 HOMONUCLEAR SPIN-SYSTEMS

If two homonuclear spins have quite different chemical shifts at all orientations of the molecule in the magnetic field, the spin-system can be treated as an AX spin-system and equation 1 is applicable [15,17]. Under these conditions the powder line shape of each member of the spin-pair will consist of two overlapping sub-spectra corresponding to $m = \pm$ ½.

If the two spins have identical chemical shifts at all orientations of the molecule in the magnetic field, then the spin-system may be treated as an A_2 spin-system. For example, two nuclei that are related by a center of inversion will constitute an equivalent spin-pair. Under these conditions the total line shape consists of two sub-spectra which can be described by equation 11 where all terms are equivalent to those given in equations 2-7.

If the chemical shift difference is comparable to the dipolar coupling at any orientation

$$v_{\pm}(\theta,\phi) = v_L - v_{CS} \mp \tfrac{3}{4} v_D \tag{11}$$

of the molecule in the magnetic field, then the spin-system must be treated as an AB spin-system at that orientation. It is important to recognize that in any particular molecule a spin-pair may exhibit A_2, AB and AX behaviour depending on the orientation of the molecule in the magnetic field. In order to describe the powder NMR line shape for a general homonuclear spin-½ pair, one has to use exact expressions for the four transitions, v_i, and their relative intensities, P_i, as indicated in equations 12 [17,26,27].

$$v_1 = \tfrac{1}{2} (v_A + v_B + D + A); \qquad P_1 = 1 - B/D \tag{12a}$$

$$v_2 = \tfrac{1}{2} (v_A + v_B + D - A); \qquad P_2 = 1 + B/D \tag{12b}$$

$$v_3 = \tfrac{1}{2} (v_A + v_B - D + A); \qquad P_3 = 1 + B/D \tag{12c}$$

$$v_4 = \tfrac{1}{2} (v_A + v_B - D - A); \qquad P_4 = 1 - B/D \tag{12d}$$

where $D = [(v_A - v_B)^2 + B^2]^{1/2}$, $A = J_{iso} + v_D$ and $B = J_{iso} - \tfrac{1}{2} v_D$. In equations 12, v_A and v_B describe the shielding contributions to the NMR transition frequencies of spin A and spin B, respectively. For an AB spin-system it is most convenient to express the orientations of the A and B chemical shielding tensors in the PAS of the dipolar tensor. Again information concerning the orientation of the principal components of the A and B shielding tensors in the molecular frame can be obtained from an analysis of the static NMR line shape. A complete discussion of this problem will be presented elsewhere.

2.3 MANIFESTATION OF DIPOLAR INTERACTIONS IN MAS SPECTRA

For a dilute spin-½ system, magic-angle spinning (MAS) causes the powder line shape due to anisotropic chemical shielding to collapse into a sharp peak at the isotropic position, $\delta_{iso} = (\delta_{11} + \delta_{22} + \delta_{33})/3$. If the spinning frequency is less than the total breadth of the line shape in the absence of spinning, the isotropic peak will be flanked by spinning sidebands at integral multiples of the spinning frequency (figure 5). Maricq and Waugh [31] and Herzfeld and Berger [28] have developed convenient methods of recovering the principal components of a shielding tensor from MAS spectra; of course no information about the orientation of the shielding tensor can be obtained from such an analysis. If a pair of dipolar coupled nuclei are also J coupled so that J_{iso} exceeds the MAS line width, then it is possible to carry out a Herzfeld-Berger analysis of each of the sub-spectra corresponding to the allowed values of m, provided the AX approximation is valid [29,30]. In this way the static dipolar coupled line shape can be reconstructed and orientational information can be obtained. We have recently applied this technique to analyze a system containing a non-equivalent homonuclear phosphorus spin-pair (*vide infra*).

The influence of magic-angle spinning on the NMR spectra of homonuclear spin-pairs, in which both spins have the same isotropic chemical shift, has been a subject of much recent interest [31,32,33,34,35]. If the two "equivalent" spins are related by a two fold axis but do not have a center of symmetry, the three principal components of the two chemical shift tensors will have the same magnitude but different orientations in the molecular frame of reference. Therefore the two "equivalent" spins will in fact have different chemical shifts at

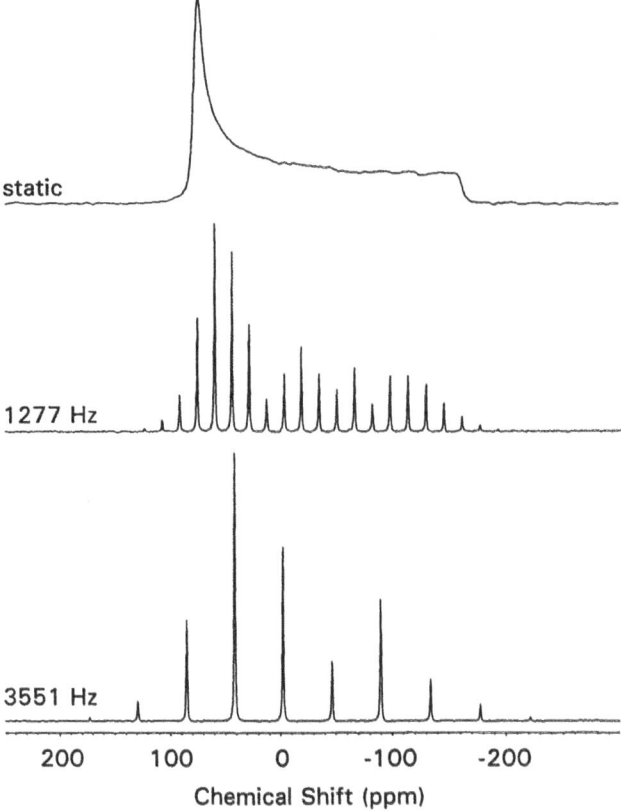

Figure 5. Static and MAS ^{31}P NMR spectra of a phosphine oxide.

certain orientations. In such cases, the presence of the homonuclear dipolar interaction results in non-Gaussian MAS line shapes. Such MAS line shapes are sensitive to chemical shift tensors, the dipolar interaction and their relative orientation; they are also sensitive to the spinning speed. By analyzing the detailed MAS line shape as a function of sample spinning speed, information about the chemical shift tensors and their orientation in the molecular frame may be obtained.

3. Experimental Results Demonstrating the Utility of Dipolar NMR

3.1 NITROGEN SHIFT TENSORS

Nitrogen chemical shifts cover a range of approximately 1000 ppm [36]. See table 1 for typical ranges of isotropic nitrogen chemical shifts for some common functional groups. Here we reference all nitrogen chemical shifts with respect to liquid ammonia at 20°C. The other common reference for nitrogen chemical shifts is nitromethane which is deshielded with respect to liquid ammonia by approximately 380 ppm. In nitrogen-15 NMR studies of solids we use the peak due to the ammonium ion of solid ammonium nitrate as a secondary reference. The

TABLE 1. Typical nitrogen chemical shifts for some functional groups

Compounds	δ_{iso} (ppm)
Alkylamines	90 to -10
Nitriles	240 to 270
Imines	290 to 380
Oximes	320 to 410
Azines	230 to 470
Nitro Compounds	350 to 410
Azo Compounds	450 to 550
Nitrites	ca. 580
Nitroso Compounds	810 to 980

ammonium ion of solid NH_4NO_3 is deshielded by 23.8 ± 0.3 ppm with respect to liquid ammonia at 20°C. The absolute nitrogen shielding constant in liquid ammonia at 20°C is approximately 244.6 ppm [37].

Although isotropic nitrogen chemical shifts are known for hundreds of compounds, much less experimental information is available on nitrogen shielding tensors [9]. Most of the data on nitrogen shielding tensors involve studies of the amide group in amino acids and peptides. Below we will summarize some of the information that we have obtained on compounds containing functional groups where nitrogen is bonded to two other atoms. Specifically, information concerning the orientation of the nitrogen shielding tensor in the oxime [38], imine [24], azo [39,40] and nitroso [41] fragments has been obtained from the dipolar NMR technique. The experimental results are compared to those calculated for simple related model compounds using *ab initio* molecular orbital calculations [42].

The ^{15}N NMR spectrum of a static powder sample of (E)-acetophenone oxime, 1, obtained at 20.3 MHz is shown in figure 6a [38]. The oxime carbon was ^{13}C labelled and the ^1H nuclei were decoupled; thus, the ^{15}N NMR spectrum may be considered to be that of an "isolated" ^{15}N-^{13}C spin-pair. The line shape shown in figure 6b was simulated using $R = 1460$ Hz; this value was calculated from the $C=N$ bond length determined by X-ray diffraction, 127.8 pm. Since $^1J(^{15}N,^{13}C) < 5$ Hz in these systems, indirect spin-spin coupling was not considered in calculating the spectrum in figure 6b. The angle β was estimated from the splitting at δ_{33} using equation 10; in this case $\beta = 90°$. Because the dipolar splitting at the δ_{11} discontinuity is approximately zero, it is obvious from equation 8 that α is near the magic angle. A best fit between the experimental and calculated spectrum was obtained with $\alpha = 52°$ and $\beta = 90°$. Since $\beta = 90°$, the nitrogen nucleus experiences greatest shielding when the planar oxime moiety is perpendicular to the applied magnetic field. This information together with the observation that $\alpha = 52°$ results in the two possible orientations of δ_{11} and δ_{22} shown in figure 7. Only the orientation shown in figure 7a is consistent with molecular orbital calculations (*vide infra*).

Nitrogen-15 NMR experiments similar to those outlined above for the oxime group have

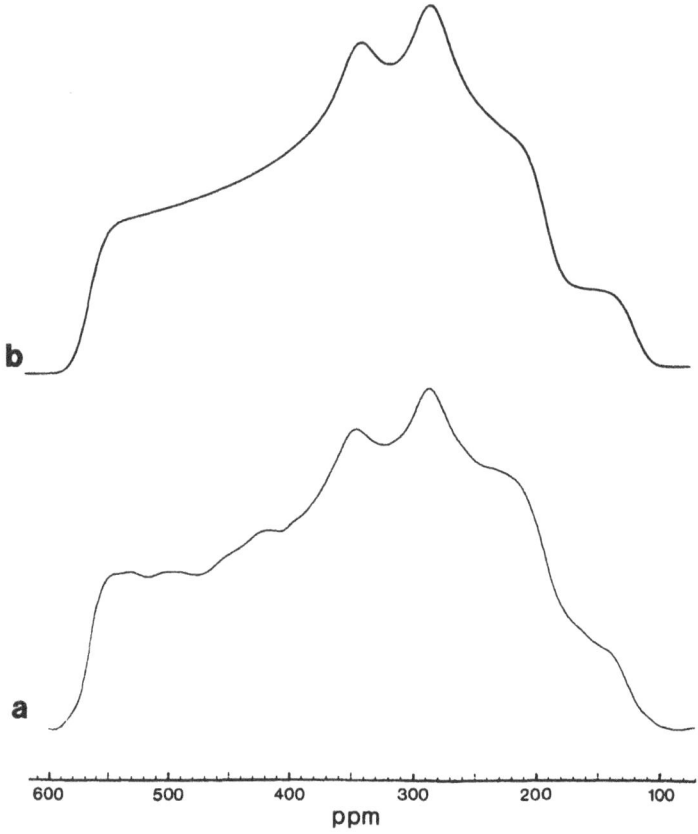

Figure 6. Observed (a) and calculated (b) ^{15}N NMR of a static powder sample of 1 where the oxime carbon and nitrogen are ^{13}C and ^{15}N labelled.

also been carried out for the imine group in benzylideneaniline-^{15}N,^{13}C, 2, and the azo group of *trans*-azobenzene-^{15}N$_2$, 3, and *cis*-azobenzene-^{15}N$_2$, 4, [24,39,40]. Because the two nitrogen atoms of *trans*-azobenzene are related by a center of inversion, the spin system was treated as an A$_2$ spin system. In the case of *cis*-azobenzene, the two nitrogens have the same isotropic chemical shift in the solid state, but the orientations of their respective shielding tensors differ. The MAS ^{15}N NMR spectrum of 4 shows small spinning rate dependent splittings. The detailed MAS line shape is intricately related to the magnitude of the homonuclear dipolar coupling, the principal components of the nitrogen shielding tensor and their orientation with respect to the dipolar vector. The static powder line shape shows second-order features at some orientations. The experimental results for compounds 2-4 are summarized in table 2. In addition, the principal components of the nitrogen shift tensors for the nitrite ion in sodium nitrite, 5, and for the nitroso moiety of a nitrosobenzene derivative, 6, are given [41]. The orientation of the nitrogen shift tensor in compounds 2-4 is analogous to that found for the oxime group in 1. In each case the most shielded direction is approximately perpendicular to the X=N-Y plane and the intermediate component, δ_{22}, is in the same general direction as the lone-pair on nitrogen. However, notice that there are significant variations in α, particularly in *cis*-azobenzene.

In order to compare the experimental results in table 2 with theory, we have carried out

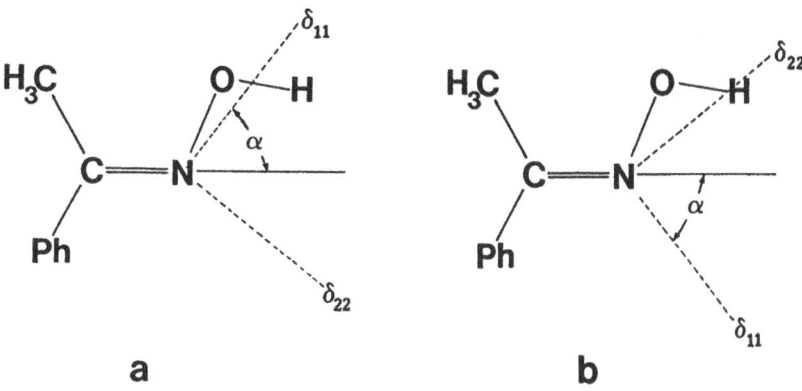

Figure 7. Possible orientations of the principal axis system for the nitrogen shielding tensor in 1 from dipolar NMR.

TABLE 2. Principal components (in ppm) of nitrogen chemical shift tensors in some compounds containing two coordinate nitrogen. The angles α and β were determined using dipolar NMR.

	δ_{11}	δ_{22}	δ_{33}	δ_{iso}	α	β
1, (E)-acetophenone oxime	560	328	160	349	52°	90°
2, benzylideneaniline	610	321	65	332	42°	90°
3, *trans*-azobenzene	1034	391	109	511	37°	83°
4, *cis*-azobenzene	1005	469	112	529	15°	90°
5, sodium nitrite	1160	515	205	627	-	-
6, 4-dimethylamino-nitrosobenzene	1702	554	223	826	-	-

ab initio molecular orbital calculations on very simple model compounds containing these same functional groups. The nitrogen shielding tensors calculated using the program RPAC are summarized in table 3 together with the experimental results. The experimental values of δ_{ii} in table 2 were converted to σ_{ii} values by taking the isotropic absolute value of the nitrogen shielding for liquid ammonia as 245 ppm at 20°C. A 6-311G(2D,1P) basis set was used for all MO calculations; geometries were obtained either from microwave spectroscopic studies or theoretically by geometry optimization. In comparing the results in table 3, the first point to note is that the orientation of the nitrogen shielding tensors in compounds **1-4** is well reproduced by the MO calculations. For example, α in *cis*-azobenzene, **4**, is approximately 22° less than the corresponding value in *trans*-azobenzene, **5**. The calculations predict α to decrease by 16° on going from *trans*-diazene to *cis*-diazene. Secondly, considering the structural differences between the model compounds and compounds **1-6**, the agreement between the observed and

TABLE 3. Comparison of calculated and observed nitrogen shielding tensors. Calculated orientations are given with respect to the C=N, N=N or N=O bond vector.

	σ_{11}	σ_{22}	σ_{33}	α	β
CH$_2$=NOH	-517	-121	119	62°	90°
1	-315	-83	85	52°	90°
CH$_2$=NH	-514	-124	252	48°	90°
2	-365	-76	180	42°	90°
trans-N$_2$H$_2$	-1176	-205	193	40°	90°
3	-789	-146	136	37°	83°
cis-N$_2$H$_2$	-875	-311	181	24°	90°
4	-760	-224	133	15°	90°
NO$_2^-$	-1341	-344	147	33°	90°
5	-915	-270	40	-	-
CH$_3$N=O	-2862	-342	172	46°	90°
6	-1457	-309	22	-	-

calculated principal components is fair. The notable exception is σ_{11} for the nitroso group where the paramagnetic contribution appears to be overestimated by more than 1000 ppm. This component of the shielding tensor depends critically on the n→π* energy gap. In fact, the value of σ_{11} in compounds 1-6 is qualitatively related to the energy associated with the n→π* transition in these compounds. Inclusion of electron correlation effects would certainly bring the molecular orbital calculations in closer agreement with experiment [43].

Often nitrogen is bonded to one or more hydrogens. Such hydrogens are readily exchanged by deuterium. Information about the orientation of the nitrogen shift tensor with respect to the ^{15}N,^{2}H dipolar vector can be obtained from an analysis of the static powder line shape. The ^{15}N NMR spectrum of phthalimide-^{15}N, 7, and phthalimide-^{15}N,^{2}H, 8, is shown in figure 8. Analyses of these spectra indicate: δ_{11} = 202 ppm, δ_{22} = 171 ppm , and δ_{33} = 98 ppm; the least shielded component lies along the N-D bond (α = 0°) while the most shielded is perpendicular to the molecular plane. It should be mentioned that one can also orient the nitrogen shift tensor with respect to the N-H bond vector. However, one must use special pulse sequences designed to suppress the strong ^{1}H,^{1}H homonuclear dipolar interaction.

It is sometimes possible to carry out two independent dipolar NMR experiments to unambiguously fix the orientation of a chemical shielding tensor. For example, in the case of acetanilide we have prepared two different samples, one containing the ^{15}N,^{2}H spin-pair and the

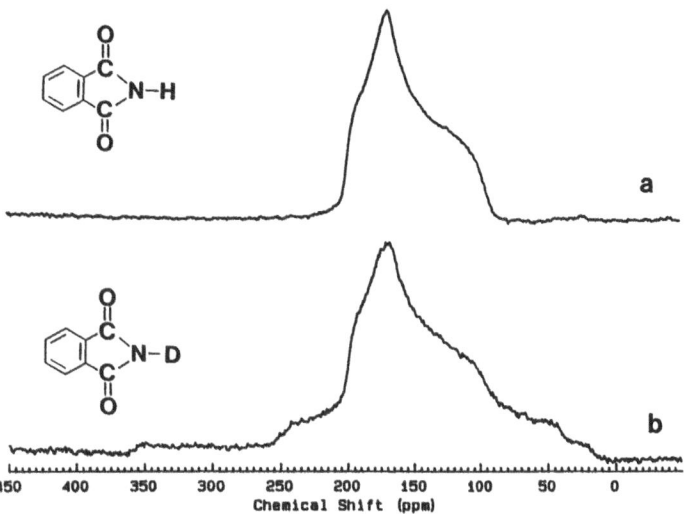

Figure 8. Nitrogen-15 NMR spectrum of a static sample of (a) phthalimide-^{15}N and (b) phthalimide-^{15}N,^{2}H at 20.3 MHz.

other containing the ^{15}N,^{13}C spin-pair. Analysis of the two ^{15}N NMR powder line shapes has allowed us to accurately fix the orientation of the nitrogen shielding with respect to the ^{15}N,^{2}H and ^{15}N,^{13}C bond vectors.

3.2 PHOSPHORUS SHIFT TENSORS

In this section we present two examples from our laboratory which demonstrate the utility of dipolar NMR in characterizing phosphorus shielding tensors. The first system is a new heterocyclic compound which contains a P(III)-P(III) bond, **9** ; the structure is indicated in figure 9. This compound was of particular interest since it is known to have a "butterfly" arrangement in which the phosphorus lone pair electrons adopt an eclipsed conformation. The CP/MAS ^{31}P NMR spectrum consists of a pair of doublets, one centered at 75.8 ppm (P$_1$) and the other at 52.0 ppm (P$_2$) [44]. Each of these doublets is flanked by a number of spinning sidebands that indicate substantial anisotropy in the ^{31}P shielding tensors. The doublets arise from indirect ^{31}P,^{31}P spin-spin coupling, $^1J(^{31}P,^{31}P)$ = 160 Hz. On the basis of the P,P bond length from X-ray diffraction studies, the calculated direct dipolar coupling constant, R, is 1770 Hz. The ^{31}P NMR line shape of a static powder sample is made up of four overlapping powder patterns each characterized by three principal components. The calculated total line shape, figure 9, is in excellent agreement with the observed spectrum. The ^{31}P NMR powder line shape in figure 9 was simulated by first analyzing the MAS spinning sidebands. The intensity distribution of the spinning sidebands associated with each of the four isotropic peaks of the MAS spectrum is characterized by three principal values. Anaylsis of the MAS spectra thus leads to 12 components which define the four sub-spectra shown in figures 9c and 9d. The final fit between the observed and calculated static line shape was obtained by trial and error. The results obtained for the principal components of the chemical shift tensor for P$_1$ are: δ_{11}=129 ppm,

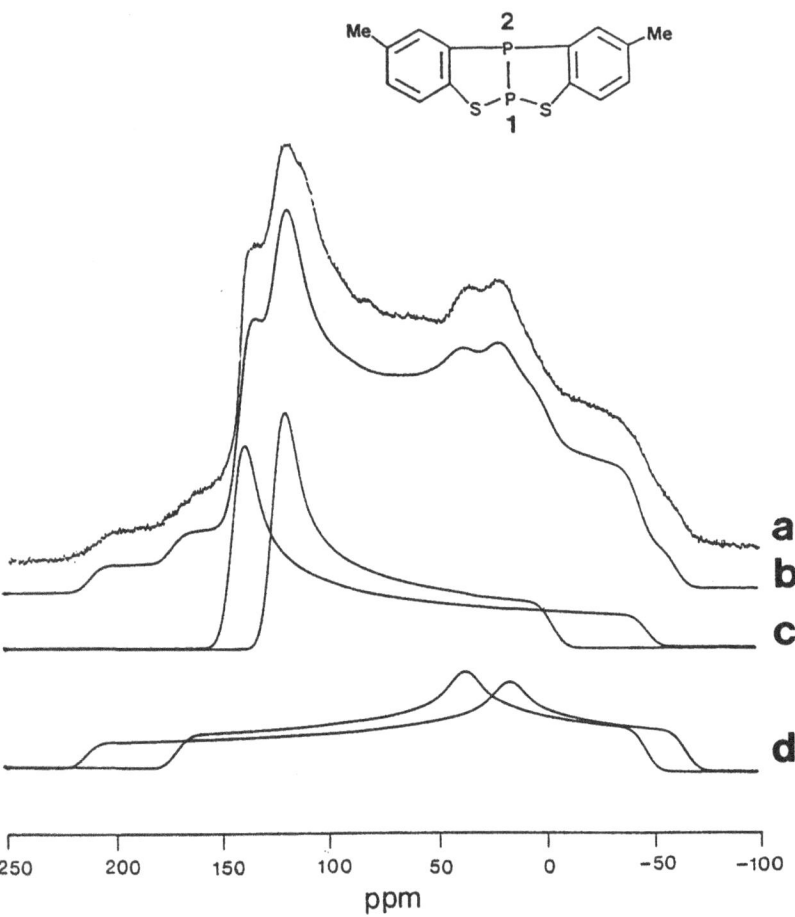

Figure 9. (a) Experimental ^{31}P CP NMR spectrum of a static sample of **9** ; (b) calculated ^{31}P NMR spectrum of **9** using the parameters in the text: (c) and (d) calculated dipolar subspectra for P1 and P2, respectively, using the same parameters as in (b).

δ_{22}=122 ppm, δ_{33}=-25 ppm and for P$_2$: δ_{11}=190 ppm, δ_{22}=25 ppm, δ_{33}=-54 ppm. The orientation of the two phosphorus shift tensors is illustrated in figure 10. The least shielded component, δ_{11}, of P$_1$ and the intermediate component, δ_{22}, of P$_2$ are perpendicular to the approximate mirror plane of the molecule. The observed spectrum is also consistent with δ_{33} (P$_1$) lying at -10° with respect to the P$_1$-P$_2$ dipolar vector and δ_{33} (P$_2$) making an angle of 100° with respect to the P$_1$-P$_2$ dipolar vector. One interesting feature of the spectrum in figure 9 is that it appears to be first order (AX) across the entire line shape. Close inspection of the relative orientation of the two phosphorus shift tensors makes it clear that at almost all orientations of the molecule in the magnetic field the spectrum indeed is first order. The shielding anisotropy for both P$_1$ (154 ppm) and P$_2$ (244 ppm) is exceptionally large. For example, in tetraphenyldiphosphine the shielding anisotropy is only 46 ppm. We are not in a

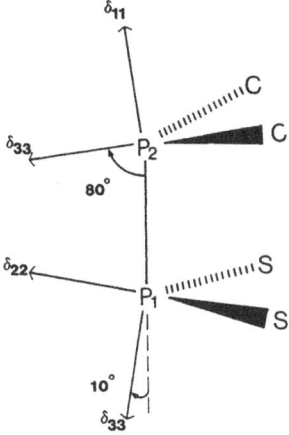

Figure 10. A schematic diagram of the orientation of the two ³¹P chemical shift tensors in **9**.

position to explain the large anisotropies observed for the thiadiphosphole at this time.

Finally, we briefly mention the results of a dipolar NMR study for a compound which contains two dipolar-coupled phosphorus nuclei which are not directly bonded to one another. The ³¹P NMR powder spectrum of a static sample of Lawesson's Reagent, 2,4-bis(4-methoxyphenyl)-1,3-dithia-2,4-diphosphetane-2,4-disulfide, **10**, is shown in figure 11. Analysis of the line shape indicates a direct dipolar coupling constant of 900 Hz which implies a P,P separation of 280 pm. In addition to this structural information, analysis of the ³¹P NMR line

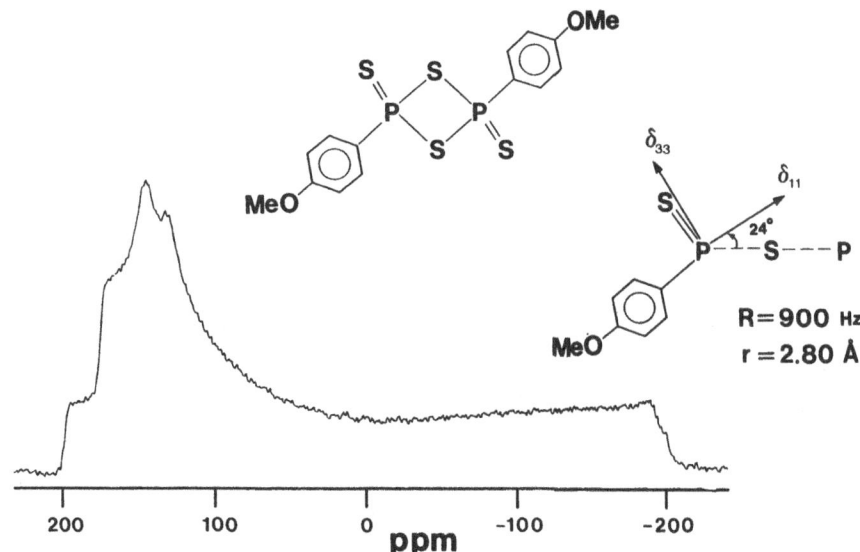

Figure 11. 81.0 MHz CP ³¹P NMR line shape of Lawesson's Reagent, **10**. A preliminary assignment of the orientation of the phosphorus shielding tensor is also indicated.

shape allows one to obtain information concerning the orientation of the phosphorus shielding tensor in the molecular axis system. To date, we have not attempted molecular orbital calculations of the phosphorus shielding tensor in this system. Clearly much more work will be necessary to characterize and understand phosphorus chemical shifts in this and related systems. There are still many open questions concerning the relationship between molecular electronic structure and the nuclear magnetic shielding tensors.

4. Conclusions

In this short review, we have summarized some information that we have obtained regarding nitrogen and phosphorus chemical shielding tensors using solid state NMR. The dipolar NMR technique has been an extremely useful tool in our laboratory. The experiments are relatively easy to perform, and the analysis of the results is generally straightforward, particularly if the molecule possesses one or more elements of symmetry. Measurements at two or more applied magnetic fields can be very helpful in improving the reliability of an analysis. In homonuclear spin-systems where the J coupling constants are too small to be well resolved, variable field studies are often essential. Also, it has been our experience that molecular orbital calculations can be helpful in distinguishing between two alternate orientations of a shielding tensor which are both consistent with dipolar NMR experiments. Of course, the MO results are essential in improving our understanding of the relationship between structure and chemical shifts.

In some cases molecular motion in the solid state can lead to erroneous values of R_{eff} (invariably R_{eff} values are smaller than expected). Fortunately low temperature measurements can largely alleviate this problem; however, it is important to recognize that even at low temperatures librational motions may cause small reductions in R_{eff} [45]. The ^{15}N dipolar NMR studies generally require the synthesis of isotopically labelled compounds. Although this is sometimes tedious and expensive, efficient synthetic procedures are now available for introducing isotopic labels. In spite of these latter shortcomings, we anticipate that the dipolar interaction will continue to provide us with information concerning the orientation of shielding tensors in powder samples.

Acknowledgements:

We wish to thank Professor C.J. Jameson and Dr. A. DeDios for a LORG calculation on formaldoxime, Professors T.D. Bouman and A.E. Hansen for making the RPAC program available. We are grateful to Professor R.J. Boyd for making computer time available to our group, and Dr. Z. Shi for helping us to implement the RPAC program in conjunction with GAUSSIAN 90 on the Stellar GS2500 vector processing computer. We would also like to acknowledge Professor D.W. Alderman for a listing of the POWDER routine, and Professor G. Baccolini for providing us with a sample of compound 9. The research described here was supported by NSERC of Canada operating and equipment grants.

References:

1.   U. Haeberlen, "High Resolution NMR in Solids. Selective Averaging", Academic Press, New York, 1976.
2.   M. Mehring, "Principles of High Resolution NMR in Solids", 2nd ed., Springer-Verlag, New York, 1983.
3.   H.W. Spiess, NMR Basic Principles and Progress 15, 55 (1978).
4.   C.A. Fyfe, "Solid State NMR for Chemists", C.F.C. Press, Guelph, Ontario, 1983.
5.   B.C. Gerstein and C.R. Dybowski, "Transient Techniques in NMR of Solids", Academic Press, New York, 1985.
6.   J.C. Facelli and D.M. Grant, Topics in Stereochem. 19, 1 (1989).
7.   W.S. Veeman, Prog. Nucl. Magn. Reson. Spectrosc. 16, 193 (1984).
8.   J.B. Robert and L. Wiesenfeld, in "Phosphorus-31 NMR Spectroscopy in Stereochemical Analysis", edited by J. G. Verkade and L.D. Quin, VCH Publishers, Inc., 1987, p.151.
9.   T.M. Duncan, "A Compilation of Chemical Shift Anisotropies", The Farragut Press, Madison, Wisconsin, 1990.
10. A.D. Buckingham and S.M. Malm, Mol. Phys. 22, 1127 (1971).
11. F.A.L. Anet and D.J. O'Leary, Concepts in Magn. Reson. 3, 193 (1991); ibid. 4, 35 (1992).
12. S. Ding and C. Ye, Chem. Phys. Lett. 170, 277 (1990).
13. M.A. Kennedy and P.D. Ellis, Concepts in Magn. Reson. 1, 35 and 109 (1989).
14. D.W.Alderman, M.H. Sherwood, and D.M. Grant, J. Magn. Reson. 86, 60 (1990).
15. a.) D.L. VanderHart and H.S. Gutowsky, J. Chem. Phys. 49, 261 (1968).
     b.) D.L. VanderHart, H.S. Gutowsky, and T.C. Farrar, J. Chem. Phys. 50, 1058 (1969).
16. M. Linder, A. Höhener, and R.R. Ernst, J. Chem. Phys. 73, 4959 (1980).
17. K.W. Zilm and D.M. Grant, J. Am. Chem. Soc. 103, 2913 (1981).
18. W.P. Power and R.E. Wasylishen, Ann. Rep. NMR Spectrosc. 23, 1 (1991).
19. N.J. Clayden, Ann. Rep. NMR Spectrosc. 24, 1 (1992).
20. C.J. Jameson and J. Mason, in "Multinuclear NMR", edited by J. Mason, Plenum Press, New York, 1987, p. 51, Chapt. 3.
21. D.W. Alderman, M.S. Solum and D.M. Grant, J. Chem. Phys. 84, 3717 (1986).
22. D.L. VanderHart, H.S. Gutowsky, and T.C. Farrar, J. Am. Chem. Soc. 89, 5056 (1967).
23. M.E. Stoll, R.W. Vaughan, R.B. Saillant, and T. Cole, J. Chem. Phys. 61, 2896 (1974).
24. R.D. Curtis, G.H. Penner, W.P. Power, and R.E. Wasylishen, J. Phys. Chem. 94, 4000 (1990).
25. Q. Teng, M. Iqbal, and T.A. Cross, J. Am. Chem. Soc. 114, 5312 (1992).
26. H. van Willigen, R.G. Griffin and R.A. Haberkorn, J. Chem. Phys. 67, 5855 (1977).
27. K.W. Zilm, A.J. Beeler, D.M. Grant, J. Michl, T.-C. Chou, and E.L. Allred, J. Am. Chem. Soc. 103, 2119 (1981).
28. J. Herzfeld and A.E. Berger, J. Chem. Phys. 73, 6021 (1980).
29. R.K. Harris, K.J. Packer and A.M. Thayer, J. Magn. Reson. 62, 284 (1985).
30. U. Haubenreisser, U. Sternberg and A.-R. Grimmer, Mol. Phys. 60, 151 (1987).
31. M.M. Maricq and J.S. Waugh, J. Chem. Phys. 70, 3300 (1979).

32. S. Hayashi and K. Hayamizu, Chem. Phys. Lett. **161**, 158 (1989); Chem. Phys. **157**, 381 (1991).

33. A. Kubo and C.A. McDowell, J. Chem. Phys. **92**, 7156 (1990).

34. A. Schmidt and S. Vega, J. Chem. Phys. **96**, 2655 (1992).

35. K. Eichele, G. Wu and R.E. Wasylishen, J. Magn. Reson., in press.

36. J. Mason, in "Multinuclear NMR", edited by J. Mason, Plenum Press, New York, 1987, Chapt. 12, p. 335.

37. C.J. Jameson, A.K. Jameson, D. Oppusunggu, S. Wille, P.M. Burrell, and J. Mason, J. Chem. Phys. **74**, 81 (1981).

38. R.E. Wasylishen, G.H. Penner, W.P. Power, and R.D. Curtis, J. Am. Chem. Soc. **111**, 6082 (1989).

39. R.E. Wasylishen, W.P. Power, G.H. Penner, and R.D. Curtis, Can. J. Chem. **67**, 1219 (1989).

40. R.D. Curtis, J.W. Hilborn, G. Wu, M.D. Lumsden, R.E. Wasylishen and J.A. Pincock, submitted for publication.

41. M.D. Lumsden, G. Wu, R.E. Wasylishen and R.D. Curtis, submitted for publication.

42. T.D. Bouman and A.E. Hansen, Int. J. Quantum Chem.: Quantum Chem. Symp. **23**, 381 (1989).

43. a.) T.D. Bouman and A.E. Hansen, Chem. Phys. Lett., in press.
    b.) M. Schindler, J. Am. Chem. Soc. **109**, 5950 (1987).

44. G. Wu, R.E. Wasylishen, W.P. Power and G. Baccolini, Can. J. Chem. **70**, 1229 (1992).

45. J.D. Dunitz, E.F. Maverick and K.N. Trueblood, Angew. Chem. Int. Ed. Engl. **27**, 880 (1988).

COMPARISONS OF SHIELDING ANISOTROPIES FOR DIFFERENT NUCLEI AND OTHER
INSIGHTS INTO SHIELDING FROM AN EXPERIMENTALIST'S VIEWPOINT

K. W. ZILM and J. C. DUCHAMP
Department of Chemistry
Yale University
225 Prospect Street
New Haven, CT 06511

ABSTRACT. In pursuing studies of multiple bonds and strained rings
containing ^{29}Si, ^{31}P and ^{119}Sn, we have needed to consider systematic
methods for comparing the shielding anisotropies of these nuclei with
those for ^{13}C in their wholly organic counterparts. A simple shift
scaling method has been empirically found to be quite useful in this
regard. This work has also lead us to consider semi-empirical
calculations of shielding to develop insight into the factors governing
gross trends in shielding anisotropies.

1. Introduction

The widespread popularity NMR enjoys as an analytical technique stems
from the high degree of correlation between chemical shifts and
molecular structure. Chemists have been quite successful in relating
trends in isotropic chemical shifts to classical chemical descriptions
of structure, bonding and reactivity. This is especially true in organic
chemistry. Upon a cursory inspection of any undergraduate text on the
subject one will find a number of passages relating ^{1}H or ^{13}C chemical
shifts to concepts of aromaticity, resonance, hydridization, etc. Over
the last two decades the availability of solid state NMR methods for
measuring shielding anisotropies has made it possible to experimentally
probe how structural factors affect anisotropic ^{13}C chemical shifts. Ab
initio calculations now reproduce ^{13}C shielding anisotropies quite
accurately, however they have been somewhat less successful in relating
shielding tensor elements to specific aspects of electronic structure.

Advances in synthetic chemistry have provided a fairly large number
of compounds in which a heavier main group element has replaced a carbon
center [1]. Examples include phosphaalkenes, phosphaalkynes, silenes and
siloxiranes. In an effort to compare the bonding in such molecules to
their more familiar organic counterparts we have undertaken a study of
the shielding anisotropies for the heteroatoms. As such we have been
lead to consider methods for sensibly comparing shielding anisotropies
for different NMR active nuclei [2,3]. One of the principal results
coming from this work is that the empirical linear correlations found

315

J. A. Tossell (ed.), Nuclear Magnetic Shieldings and Molecular Structure, 315–334.
© 1993 Kluwer Academic Publishers.

between isotropic chemical shifts for different nuclei in isostructural environments also hold for the anisotropic shifts.

This latter observation has lead us to re-examine the physical basis of these linear relationships. Linear isotropic chemical shift correlations were first rationalized on the basis of early semiempirical theories of chemical shielding [4]. The success of these linear shift correlations in relating anisotropic chemical shifts for different nuclei suggested that we reinvestigate the application of some of these methods for calculating shift tensors. It has been found that Pople's LCAO approach for calculating paramagnetic shielding [5] in combination with Flygare's diamagnetic term [6] produces reasonable results for a number of small molecules. Gross trends in shielding tensor elements are reproduced and the formalism is useful for identifying the major terms that give rise to changes in shielding.

This approach to shielding calculations and the shift scaling technique have provided us with a simple framework in which we can understand many of our experimental observations in studies of anisotropic shielding. Trends in tensor elements can often be attributed to a particular shielding term or changes in hybridization. Differences in shielding tensors for different elements are often accounted for by the Z dependence of the chemical shift.

2. Shift Tensor Comparisons

The synthesis of the first compound containing a Si=Si double bond by West and co-workers [7] marked the beginning of what could be called modern main group chemistry. Prior to this event conventional wisdom argued against the existence of a chemistry much beyond C in the main group that would be as rich in multiply bonded functional groups. Many disilenes (Si=Si) and silenes (Si=C) have since been synthesized as well as diphosphenes (P=P), phosphaalkenes (P=C) and phosphaallenes (C=C=P) [1]. Main group analogues of alkenes, alkynes, cyclopropanes, ketones, etc. containing a variety of combinations of Si, P, Sn and B have been successfully attained. New examples of such main group functionalities are being continuously reported in the chemical literature.

Whenever a new main group compound is successfully synthesized, the question of how it compares structurally and chemically to its wholly organic counterpart arises. Since NMR chemical shifts are known to correlate so well with electronic structure, one is naturally drawn to NMR to provide a spectroscopic basis for this comparison. For multiply bonded systems, shift anisotropies are expected to be especially useful in this regard as the anisotropy in the shielding should mirror the electronic anisotropy inherent in a multiple bond. Such a comparison of the ^{29}Si shielding anisotropy in the first disilene, with the ^{13}C shift anisotropy in typical alkenes, provided some of the strongest evidence to support the assertion that the Si=Si double bond was quite similar to a C=C double bond [8].

Since that time a number of additional shift anisotropies have been measured for systems with multiple bonds between C, Si, P and N. While ^{13}C and ^{29}Si have similar shift ranges and are thus straightforward to compare, ^{31}P and ^{15}N have much larger shift ranges making comparison more difficult. An approach we have found useful is based on a well known observation by Jameson and Gutowsky [4]. These workers found that the shift ranges for different nuclei depend in a periodic fashion upon the inverse cube of the radius of the np or nd valence electrons for a given element. This behavior was rationalized on the basis of their LCAO formulation in which such a term was found to factor out of the expression for the paramagnetic term of the shielding. A number of authors have derived similar approximate expressions for the paramagnetic shielding, all of which have the following basic form:

$$\sigma_p = - \frac{e^2\hbar^2}{2m^2c^2(\Delta E)} <r^{-3}>_{np,nd} \sum_i Q_i \quad (1)$$

In this equation changes in the shielding due to differing bonding arrangements are largely included in the sum over the charge-density bond-order matrix elements collected in the Q_i. As long as the average excitation energy ΔE does not vary, or at most scales with the $<r^{-3}>$ term, this expression suggests that the shifts for differing nuclei in series of isoelectronic and isostructural compounds should be linearly correlated. In fact a number of such correlations for pairs of nuclei X and Y have been experimentally observed [9]. Furthermore the slopes often agree with the prediction that $\delta X/\delta Y = <r_X^{-3}>_{np,nd}/<r_Y^{-3}>_{np,nd}$.

These observations on *isotropic* shifts lead us to surmise that comparisons of shielding *anisotropies* would be more meaningful if this intrinsic Z dependence on shielding ranges was first factored out. The working hypothesis is that if the elements of the shielding tensor for differing nuclei in isostructural environments scale according to the ratio of their respective $<r^{-3}>_{np,nd}$ factors, that the ΣQ_i terms must be comparable. This in turn would imply that the bonding arrangement experienced by the two nuclei is similar. A poor comparison conversely would imply that the bonding is dissimilar.

In order to compare ^{31}P and ^{15}N shielding anisotropies to those for ^{13}C, we have developed isotropic shift correlations [2,3] for as wide a range of functional groups as possible. While comparisons for nuclei in the same column of the periodic table are straightforward, comparisons that cross rows are less readily implemented. In a ^{13}C/^{15}N correlation one must either compare molecules wherein the N and its lone pair are replaced by a C-H, or alternatively a neutral organic molecule must be compared with the charged protonated analogue. In both cases the molecules are isoelectronic and one can argue convincingly in favor of either type of comparison. We have chosen to compare the neutral species since the protonated P containing multiply bonded compounds are not readily available. For precedence we note that Lichter and Roberts [10] have previously demonstrated that the isotropic ^{13}C shifts in hydrocarbons correlate well with the ^{15}N shifts in analogous amines. In their series the nitrogen with its lone pair was substituted for a C-H. We have expanded the range of compounds in this correlation and also made a similar correlation for ^{13}C and ^{31}P shifts. In order to ensure that the changes in shielding being compared are due to local differences in electronic structure for these various functional groups, only compounds with hard ligands have been included insofar as possible. Therefore compounds with alkyl or aryl ligands on the centers in question have been used. Systems with halogen or alkoxy substituents have been excluded. For ^{31}P we find $\delta^{13}C = 0.261\delta^{31}P + 64$ where the ^{13}C shift reference is TMS and that for ^{31}P is 85% H_3PO_4. In the case of ^{15}N the scaling relation is $\delta^{13}C = 0.349\delta^{15}N + 133$ (^{15}N reference CH_3NO_2). In the ^{31}P correlation the points in the linear regression range from those for CH_4/PH_3 to $C=P/P=P$. For the ^{15}N correlation the points start at CH_4/NH_3 and end at benzene/pyridine. Since a large number of compounds were used in these correlations, all of the data represented will not be reproduced here. Details of the isotropic shifts used in these correlations can be found in reference [11].

Using these slopes we can then contrast the shielding anisotropies for these systems. Figure 1 depicts in diagrammatic form how the shielding for each of the principal components of several ^{13}C, ^{15}N and ^{31}P tensors varies in a number of functional groups [11]. In the systems being compared the local symmetry and structural similarity ensure that the principal axis of these shielding tensors are oriented [2,3] in close to the same way in the molecular frame enabling a meaningful comparison. On the left hand side the principal components are indicated relative to the isotropic shielding on the shift scale native to the nucleus in question. The ^{15}N and ^{31}P shielding tensors typically have a spread of 2 to 4 times that for the analogous ^{13}C shielding tensor. When each of the components is converted using the isotropic shift correlations to the ^{13}C shielding scale, the reduced tensors on the right hand side of Figure 1 are obtained. With the exception of the phosphaalkyne and the isonitrile data, the reduced tensors compare very well with their ^{13}C counterparts.

The doubly bonded systems provide a striking comparison. The elements of these reduced shielding tensors for the P containing systems are given in Table 1. In our initial work on the P=P double bond [2] we noted that the σ_{11} component of the ^{31}P shielding tensor came at a much lower field position relative to that expected on the basis of comparison to the ^{13}C shielding in an alkene. At that time we suggested that this result could possibly be due to a low lying electronic transition involving the P lone pair giving an enhanced paramagnetic shift in this component. However the comparison of the ^{31}P shielding tensor in the C=P double bond [3] to the ^{13}C tensor for a typical alkene shows that the large deshielding in the P=P double bond cannot be rationalized so easily. Even though the P in the P=C double bond has a lone pair, its shielding tensor compares quite well to the ^{13}C tensor for the alkene carbon in trans-2-butene. When the ^{31}P shielding tensor in the P=P double bond is reduced to the ^{13}C shift scale, it is also found to now compare quite well to the ^{13}C shielding for the C in the P=C double bond. In both of these latter cases there is a significant deshielding in the σ_{11} component. Taken together, these data then suggest that the effect of the P lone pair is to produce an exceptional paramagnetic shielding for its multiply bonded neighbor. The inference then is that the deshielding observed for both σ_{11} in the ^{13}C spectrum of the C=P bond and σ_{11} in the ^{31}P spectrum of the P=P bond is a result of involvement of the lone pair on the neighboring P.

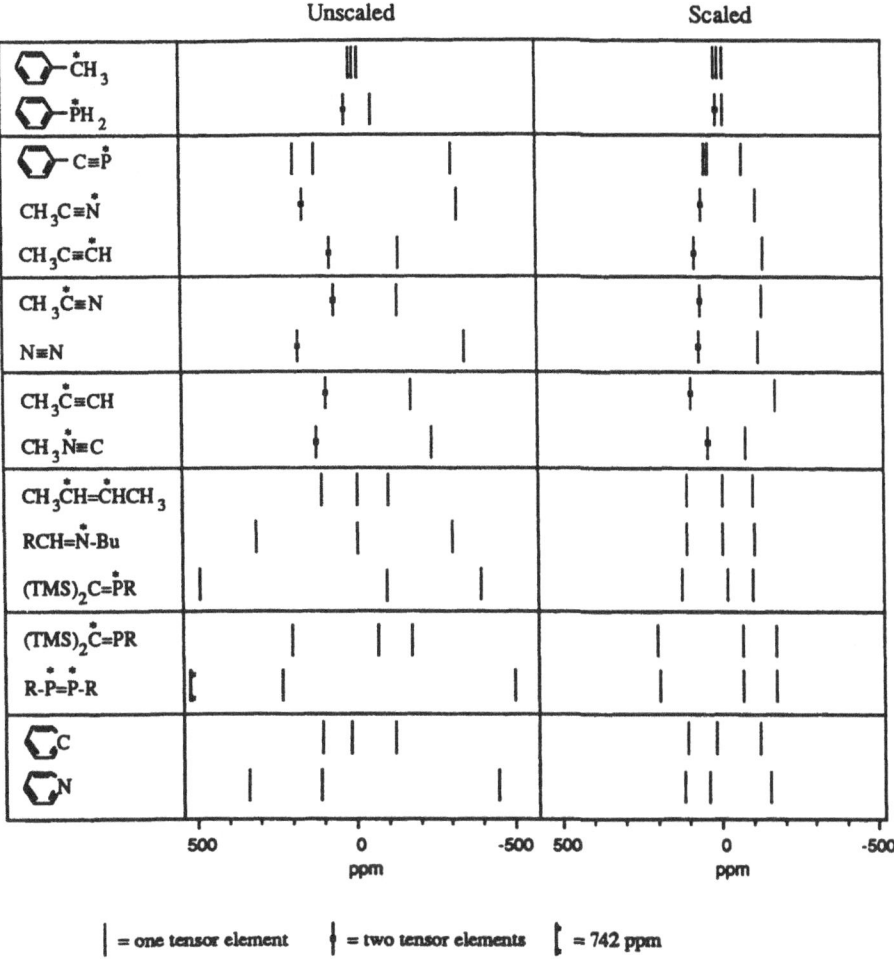

Figure 1. A comparision of scaled and unscaled shift anisotropies for ^{13}C, ^{31}P and ^{15}N in a variety of bonding situations. All shifts are referred to the isotropic on the δ scale. On the left are the unscaled shifts using the natural shielding scale for the nucleus (*) indicated. The right hand side compares the shift tensors scaled to the ^{13}C scale. RCH=N-Bu is t-retinylidene-butylimine [12], (TMS)$_2$C=PR is 2,4,6-t-Bu$_3$C$_6$H$_2$P=C(SiMe$_3$)$_2$.

Table 1. Comparison of Scaled Tensor Elements[a]

observed ^{13}C					scaled ^{31}P				
type	δ_{11}	δ_{22}	δ_{33}	$\Delta\delta$	type	δ_{11}	δ_{22}	δ_{33}	$\Delta\delta$
^{13}C=C[b]	108	-5	102	210	^{31}P=C[c]	124	-25	-98	222
^{13}C=P[c]	188	-40	-147	335	^{31}P=P[d]	194	-64	-130	324
^{13}C≡C-Me[e]	71	71	-143	214	^{31}P≡C-Ar[f]	52	28	-80	132
^{13}C≡C-Me[g]	68	68	-134	202					
^{13}C≡C-Ar[h]	56	47	-103	159					

[a]all shifts relative to isotropic [b]trans-2-butene [c]2,4,6-t-Bu$_3$C$_6$H$_2$P=C(SiMe$_3$)$_2$ [d]1,2-bis(2,4,6-tri-tert-butylphenyl)diphosphene [e]methylacetylene [f]2,4,6-t-Bu$_3$C$_6$H$_2$C≡P [g]IGLO calculation for e [h]IGLO calculation for phenylacetylene

The shielding tensors for a number of other functional groups where a ^{13}C has been replaced by either ^{31}P or ^{15}N also are found to compare quite well when adjusted to the ^{13}C scale (Figure 1). The C=N double bond is seen to be similar to the C=C and C=P bonds just discussed, as are pyridine and benzene. The scaled ^{15}N shielding in nitriles match the ^{13}C shielding in alkynes and the ^{15}N shielding in N$_2$ compares well with the ^{13}C shielding in acetonitrile. Since the shielding tensors compared in this way match well, the bonding in the functional groups being compared are implied to be qualitatively quite similar.

Two notable cases where the comparisons are poor after scaling to the ^{13}C shift scale involve the isonitriles and phosphaalkynes when compared to alkynes. For isonitriles this is to be expected. Since charge separated resonance structures are important for isonitriles and are not for alkynes, the bonding is dissimilar as mirrored in the shielding tensors. The discrepancy between the scaled ^{31}P shielding in the phosphaalkyne and the ^{13}C shielding in methylacetylene is not due to a difference in the bonding, but rather a result of the fact that the molecules compared are not isostructural. The problem arises in that a methyl group has been replaced by an aryl group in the P=C system. In a triply bonded molecule the shielding along the bond is quite sensitive to how the C$_\infty$ symmetry is broken by the other ligands. This symmetry breaking is quite different for the methyl and aryl groups and gives rise to the poor agreement in the σ_{33} components upon scaling. Unfortunately this conjecture has not been experimentally tested at present. Attempts by us to measure the ^{13}C shielding tensor in phenylacetylene failed due to the molecule crystallizing in such a way as to produce a forest of peaks in the low temperature CPMAS spectrum making analysis of sideband intensities impractical. However, calculations using the IGLO method do verify this hypothesis. In Table 1 the IGLO results [11] for methylacetylene and phenylacetylene are compared to the experimental ^{13}C shielding in methylacetylene and the scaled ^{31}P shielding in the Ar-C≡P system. The previous discrepancy in the σ_{33} component is seen to be largely accounted for by the replacement of the methyl group by a phenyl group. In addition the σ_{11} and σ_{22} components are brought into closer agreement. Thus the similarity in the scaled ^{31}P shielding anisotropy with that calculated for phenylacetylene implies the P≡C and C≡C triple bonds are electronically similar.

The principal empirical observation made here is that linear shift correlations between different nuclei in similar bonding arrangements hold for the individual components of the shielding tensor as well as for the isotropic shifts. This provides a basis for a more systematic comparison of the electronic structure of functional groups containing different main group nuclei. Further experimental work is needed to verify whether this observation will hold in general. As ab initio calculations of shielding for elements beyond the second row become more reliable in reproducing experimental data, it will be possible to test this method more exhaustively. Hopefully such calculations will be able to delineate the circumstances in which such correlations will hold, and when they fail for what reasons.

It is also hoped that these observations will provide some insight into the current problems encountered in calculating shielding tensors for heavier elements such as P. The data for the ^{31}P shieldings observed experimentally seem to quite simply follow from a scaling law with respect to ^{13}C shielding tensors. Since ^{13}C shielding tensors can be so accurately calculated it then seems somewhat curious that those calculated for ^{31}P in multiply bonded systems match experiment so poorly. It may prove useful in this regard to computationally investigate why the current ab initio results do not follow the empirical scaling law experimentally observed.

3. Characteristic Shielding Anisotropies

The observation that the trends in anisotropic shieldings presented above are transferable from one nucleus to another reinforces the notion that chemical shielding in largely determined by local electronic structure. While this is mirrored in isotropic shifts, it is even more dramatically driven home by considering characteristic shielding anisotropies. Over the last two decades a wide variety of ^{13}C shielding tensors have been measured. The data collected to date demonstrate that different organic functional groups can be classified according to sets of characteristic principal shielding components. Thus a given functional group is characterized not only by a specific isotropic shift range, but also by a characteristic spread, anisotropy and asymmetry in its ^{13}C shielding tensor. Table 2 constructed from data in Duncan's compilation [12] of shielding tensors illustrates this point. The average δ_{11}, δ_{22}, δ_{33} and δ_{iso} values for aromatic C-H groups, alkynes, alkenes, ketones, aldehydes and methyl groups are shown along with the standard deviations of the data from these values. These average values for the principal components and their separations from one another are quite different for these various functional groups. The standard deviations in the individual tensor components are nearly as small as those for the isotropic shifts clearly demonstrating that each tensor component for each functional group has a characteristic range.

Table 2. Characteristic ^{13}C Shielding Anisotropies[a]

type	δ_{11}	δ_{22}	δ_{33}	δ_{iso}	# of spectra
aromatic C-H	207(22)	146(19)	18(12)	124(11)	75
alkynes	151(9)	151(9)	-79(17)	75(8)	8
alkenes	224(16)	134(21)	37(15)	131(10)	53
ketones	279(5)	245(18)	85(10)	203(5)	4
aldehydes	272(10)	226(13)	91(8)	196(5)	5
R-CH$_3$	33(10)	24(10)	4(2)	20(6)	26

[a] shifts relative to TMS. Average values (standard deviation) for number of spectra reported

These values must then be largely determined by the local electronic anisotropy, i.e. hybridization or bond order, of the center in question and the identity of the atoms to which it is bonded. That is not to say that the second substituent shell does not give rise to important contributions to the total shifts. Rather, such effects are small in comparison to the characteristic differences between tensor elements which are often a substantial fraction of the full isotropic ^{13}C chemical shift range.

From these observations one would suspect that the gross changes in shift tensors upon moving from one functional group to another should be amenable to treatment by simpler semiempirical methods. The development of such methods for shielding tensors could be useful in disentangling how changes in electronic structure or geometry affect different shielding tensor elements. They would also be useful in helping to assign the orientation of shielding tensors in the molecular frame without resorting to more difficult experiments or higher level calculations.

A number of simple approaches to calculating isotropic chemical shifts have been developed since the first observation of the chemical shielding phenomenon. Many of these methods reproduced observed trends in isotropic shift patterns and they still are in wide use by chemists for qualitative discussions of shielding. The concepts developed in these treatments are even used in discussing the results of high level ab initio shielding calculations. However relatively little work has been done on interpretation of shielding anisotropies using such methods. This is not surprising as experimental determinations of most shielding anisotropies came long after the theoretical community had finished with semi-empirical approaches and had moved on to refinement of more accurate but computationally intensive methods for calculating shielding parameters.

4. Approximate Calculations of Shielding Tensors

Our desire to develop a better qualitative understanding of shielding anisotropies has lead us to reconsider some of the previous methods proposed to rationalize isotropic shifts. It is our intention to identify a method suitable for explaining shift anisotropies only at the level of reproducing the characteristic shift tensors observed for different types of functional groups as previously discussed.

Many qualitative approaches to shielding consider the total shielding for an atom A to be subdivided into terms as follows:

$$\sigma^A = \sigma^{AA}_d + \sigma^{AA}_p + \sum_{B \neq A} \sigma^{AB} + \sigma^{A,ring} \tag{2}$$

The first term is just the diamagnetic Langevin term for the atom A. σ^{AA}_p is the leading term for the paramagnetic shielding produced by currents on A. The rest of the diamagnetic and paramagnetic contributions are lumped together in the σ^{AB} terms and the final term which is pertinent to cyclic systems. For the case of ^{13}C in organic molecules, shielding differences have often been taken to be largely due to σ^{AA}_p. Using a GIAO basis Pople [5] derived a simple expression for this term in an LCAO formalism using an average energy approximation and neglecting overlap. This expression has been often used in discussions of trends in isotropic ^{13}C chemical shifts. For systems involving primarily second row elements, the zz component of this term is given by:

$$(\sigma^{AA}_p)_{zz} = - \frac{e^2 \hbar^2}{2m^2 c^2 (\Delta E)} <r^{-3}>_{2p} \left\{ (Q_{AA})_{zz} + \sum_{B \neq A} (Q_{AB})_{zz} \right\} \tag{3}$$

The $(Q_{AA})_{zz}$ terms expressed using charge-density bond-order factors are:

$$(Q_{AA})_{zz} = 2 - 2(P_{x_A x_A} - 1)(P_{y_A y_A} - 1) + 2P^2_{x_A y_A} \tag{4}$$

Since the electron density in each 2p function for most hydrocarbons is approximately unity, the $(Q_{AA})_{ii}$ will all be very close to 2. Therefore this portion of σ^{AA}_p will be isotropic. On the other hand the portion of the paramagnetic term that arises from the influence of 2p electrons on neighboring atoms B will be anisotropic as can be seen from considering the Q_{AB} terms:

$$(Q_{AB})_{zz} = -2P_{x_A x_B} P_{y_A y_B} + 2P_{x_A y_B} P_{y_A x_B} \tag{5}$$

Thus the local paramagnetic term in this treatment consists of an isotropic part for the atomic orbitals centered on the center A, and an anisotropic portion arising from the influence of the 2p electrons on all of the centers B which participate in the bonding to the center A. Application of equation 3 to the calculation of shielding differences does reproduce some of the trends in isotropic ^{13}C shifts that have been observed [13]. However, closer inspection shows that this approach does poorly with respect to absolute shifts. Comparison with measured ^{13}C shielding anisotropies is also not particularly good [12,13].

Part of the problem with this type of approach is the assumption that the diamagnetic terms are isotropic and do not vary much from one molecule to the next. Of course the division into diamagnetic and paramagnetic terms is somewhat artificial as the partitioning depends upon how one has chosen to deal with the gauge problem. If a single origin fixed at the nucleus of interest is chosen as the gauge origin, the diamagnetic term is that originally defined by Ramsey [14]. We will denote this term $\sigma^{(1)}$ to avoid confusion with other expressions for the diamagnetic shielding appropriate to other choices of gauge. Several workers have noted that in this choice of gauge the diamagnetic term does vary significantly from molecule to molecule and that it can be quite anisotropic. Baird and Teo suggested [15] that diamagnetic corrections should be considered in addition to the σ_p^{AA} shielding term in interpretations of ^{13}C isotropic shifts. These workers found the excitation energies that they back calculated from isotropic shifts were then were more readily accommodated with other spectroscopic data relevant to ΔE. However, it is not obvious that such an approach is theoretically justified as the Pople equation was developed using a GIAO basis and thus in principle employed a different choice of gauge.

Since it is well known that the Ramsey diamagnetic term can be calculated quite accurately (vide infra), it would then be useful to have an approximate theory for the paramagnetic term which we shall denote $\sigma^{(2)}$ in this gauge. Using an LCAO approach Karplus and Das [16] developed an expression for $\sigma^{(2)}$ from the Ramsey equations. In the average energy approximation they derived the following expression for the zz component of the Ramsey paramagnetic term $\sigma^{(2)}$:

$$\sigma^{(2)}_{zz} = - \frac{e^2}{m^2 c^2 (\Delta E)} \sum_{\nu,\mu} P_{\mu\nu} \left\{ T_{\nu\mu} - \frac{1}{2} \sum_{\lambda,\rho} P_{\rho\lambda} \, \Xi_{\nu\rho\lambda\mu} \right\} \qquad (6)$$

In equation 6 the $P_{\mu\nu}$ are the usual charge-density bond-order matrix elements and the $T_{\nu\mu}$ and $\Xi_{\nu\rho\lambda\mu}$ terms are defined as:

$$T_{\nu\mu} = \langle\phi_\nu| \frac{1_z 1_z}{r^3} |\phi_\mu\rangle \quad \text{and} \quad \Xi_{\nu\rho\lambda\mu} = \langle\phi_\nu| \frac{1_z}{r^3} |\phi_\rho\rangle\langle\phi_\lambda| 1_z |\phi_\mu\rangle \qquad (7)$$

The functions ϕ_i are the AO's used to form the LCAO ground state wave MO functions and a summation over all occupied MO's is implied in equation 6. The ϕ_i are conveniently classified as those on the center A, ϕ_i^A, and those on the directly bonded centers B, ϕ_i^B. Because of the particular gauge chosen, all the orbital angular momentum operators 1_z are taken about the origin centered at the nucleus A. Matrix elements which only involve the ϕ_i^A are straight forward to calculate. The contribution of these terms to equation 6 as shown by Karplus and Das results in a portion of $\sigma^{(2)}$ which is identical to that given by the Q_{AA} term in equation 3. Analytic expressions also exist for the two center integrals that contribute to equation 7 and have been tabulated by Barfield and Grant [17] and by Kato [18]. The most important integrals of this type have both ϕ_λ and ϕ_μ on center B with the other functions ϕ_i centered on A. These terms are very easily shown to combine to give a contribution identical to that specified by the Q_{AB} terms in equation 3. All of the other integrals which are not identically zero either involve overlap integrals or are smaller by the ratio of $\langle R^{-3}\rangle/\langle r^{-3}\rangle_{2p}$ where R is the A-B bond length. For typical bond lengths these latter terms are ~25 times smaller than those already included and can therefore safely be dropped. Since the LCAO approach being used [5] consistently neglects overlap contributions, the two center terms involving overlap integrals are also dropped.

The resulting LCAO expression for $\sigma^{(2)}$ derived as just indicated then is identical to Pople's GIAO LCAO expression for σ_p^{AA} in equation 3. At first sight this seems somewhat perplexing as one method uses a gauge centered at the origin while the other employs a different gauge choice through the GIAO's. The fact that the two results are the same is a consequence of the complete neglect of overlap terms in both calculations. A theorem to this effect has been previously proven by Slichter [19] who noted that one is free to choose what he terms the "natural" gauge origin as long as overlap terms can be neglected. This result demonstrates that the remainder of the shielding, i.e. that not due to the equivalent $\sigma^{(2)}$ or σ_p^{AA} terms, is necessarily the same at this level of approximation. The realization that equation 3 gives the Ramsey paramagnetic term approximated at the LCAO level then shows that its deficiency in treating shielding anisotropies can be attributed to the neglect of the anisotropy in $\sigma^{(1)}$.

5. Diamagnetic Anisotropies

In order to use the Pople expression for a qualitative shielding anisotropy calculation we should then add on a diamagnetic correction. The work of Flygare [6,20] in calculating Ramsey diamagnetic shielding tensors is especially helpful in this regard. Flygare demonstrated over 20 years ago that this term can be quite accurately calculated using an atom-dipole approach [20]. His diamagnetic shielding anisotropies calculated in this way reproduce the results of ab initio calculations often to better than 1 ppm. These diamagnetic anisotropies were found to be in excellent agreement with the experimental data derived from measurements of shielding anisotropies and spin rotation tensors in combination with known molecular geometries. Flygare also predicted NMR shielding tensors [6] from spin rotation tensors and his calculated diamagnetic shieldings. Many of these NMR shielding tensors have since been measured and they largely do agree with his predictions.

In the qualitative approach taken thus far, only electrons on the center A and its directly bonded neighbors B are included. Therefore our calculation of $\sigma^{(1)}$ will only include this same set of electrons. Table 3 contains this restricted portion of the diamagnetic shielding anisotropies for a number of small organic molecules using Gierke and Flygare's approach. These numbers have been calculated from the full atom-dipole expansion:

$$\sigma_{xx}^{(1)} = \sigma_d^{AA} + \frac{e^2}{2mc^2} \sum_n' \frac{Z_n}{r_n^3} (y_n^2 + z_n^2) + \sigma_{III}^d + \sigma_{IV}^d \qquad (8)$$

The first term in equation 8 is that for the isolated atom A. The next three terms comprise an expansion of the diamagnetic shielding from the electrons on other centers into a point charge term, a dipole term (σ_{III}^d) and a quadrupole term (σ_{IV}^d). The full expressions for these terms can be found in reference [20]. While the point charge term produces good values for $\sigma_{iso}^{(1)}$, the quadrupole term is needed for accurate diamagnetic shielding anisotropies.

Table 3. Diamagnetic Shielding Anisotropies

molecule	$\sigma^{(1)}$			
y↑ └→ x	xx	yy	zz	iso
CH₃-CH₃	290	340	340	323
H₂C=CH₂	280	331	337	316
H-C≡C-H	273	331	331	312
H₂C=C=CH₂ (B)	279	331	338	316
H₂C=C=CH₂ (A)	290	374	374	346
O=C=O	296	436	436	389
O=C=S	298	483	483	421
S=C=S	301	531	531	454
C≡O	280	350	350	327
CH₃-F	301	369	369	346
CH₃-Br	305	534	534	458
CH₃-I	304	630	630	521

For sets of closely related molecules the isotropic diamagnetic shieldings can be seen to not vary significantly in line with conventional expectations. However, as heteroatoms are added the isotropic diamagnetic shift can change by over 100 ppm. Such behavior was previously noted by Barfield and Grant [17] who previously questioned the idea that variations in the diamagnetic component need not be considered in qualitative calculations of differences in isotropic shielding constants. Moreover the diamagnetic anisotropies themselves are not small and they are quite variable.

As a point of clarification it should be emphasized that these variations noted for the diamagnetic term have been calculated in a gauge centered at the nucleus of interest. Thus the discussion is not transferable to other methods employing different choices of gauge origin. In modern high level calculations the majority of the shielding being accounted for in the present fashion is usually contained in a "paramagnetic" term. Therefore is is not inconsistent that an IGLO [21] or LORG [22] calculation will produce fairly constant and isotropic diamagnetic terms. These very accurate methods have simply handled the partitioning of the shielding differently than the qualitative approach being presently considered.

6. Results for Small Molecules

Table 4 lists the terms used in equation 3 for calculating $\sigma^{(2)}$ for the molecules in Table 3. The principal values of the full shielding tensors so calculated are listed in Table 5. Experimental results are included for comparison. The latter were placed on the absolute ^{13}C shielding

Table 4. Q_{AB} Terms and ΔE Values

molecule	$\sum Q_{AB}$			$-\sigma^{(2)}$			$\Delta E(ev)$
	xx	yy	zz	xx	yy	zz	
CH_3-CH_3	0	0	0	149	149	149	13.93
$CH_2=CH_2$	0	1.33	0	211	353	211	9.80
$HC\equiv CH$	-2	1	1	0	298	298	10.45
$H_2C_B=C=CH_2$	0	1.15	0	177	279	177	11.73
$H_2C=C_A=CH_2$	0	1.15	1.15	272	429	429	7.63
$O=C=O$	-2	2	2	0	509	509	8.16
$O=C=S$	-2	2	2	0	588	588	7.06
$S=C=S$	-2	2	2	0	694	694	5.99
$C\equiv O$	-2	1	1	0	487	487	6.39

Table 5. Comparison of Experimental and Calculated Shielding
Anisotropies

molecule	$\sigma_{abs} = 183 - \delta_{TMS}$					σ calculated		
	xx	yy	zz	iso[a]	iso[b]	xx	yy	zz
CH_3-CH_3	179	172	172	174		141	191	191
$CH_2=CH_2$	63	-51	159	57	64.5	69	-22	126
$HC\equiv CH$	273	33	33	113	117.2	273	33	33
$H_2C_B=C=CH_2$	129	25	160	105	115.2	102	52	161
$H_2C=C_A=CH_2$	8	-50	-50	-31	-29.3	18	-55	-55
$O=C=O$	273	-62	-62	50	58.8	296	-73	-73
$O=C=S$	273	-92	-92	29	30.0	298	-105	-105
$S=C=S$	273	-149	-149	-8	-8.0	301	-163	-163
$C\equiv O$	245	-120	-120	2	1.0	280	-137	-137

[a] isotropic calculated from solid state values [12] [b] absolute
isotropic shift taken from reference [23]

scale following the work of Jameson [23]. The majority of the
experimental values quoted were measured in an electromagnet and
referenced to a cylindrical sample of tetramethylsilane (TMS) placed $\perp$
to B_0. Therefore absolute shielding values were calculated using
$\sigma_{abs} = 183 - \delta_{TMS}$. In all cases the MO's used in the approximate
shielding calculations were comprised of simple hybrid orbitals as
constructed in reference [13]. Comparison with the experimental values
is good given the crude approximations employed at this level.

One of the first points to consider is the choice of excitation
energy. This is the crudest approximation of the method and the one
least amenable to sensible interpretation. While one might expect ΔE to
be anisotropic, the level of theory used does not justify such detail
and therefore a single value of ΔE was used. The values in Table 4 were
chosen to reproduce the absolute isotropic shifts. At the very least one
should however expect the ΔE to be somewhat lower than the first
ionization potential and higher than the HOMO-LUMO gap. These numbers
found do in fact generally satisfy this criterion. As such the addition
of the Flygare diamagnetic correction represents a substantial
improvement over numbers obtained by considering just the paramagnetic
portion and the isolated atom diamagnetic term alone. For instance in
acetylene the absolute isotropic diamagnetic shielding increases from
the free atom value of 260 ppm to 312 ppm. If the value of 260 ppm was

used alone the required paramagnetic term would be -155 ppm instead of -211 ppm. The ΔE implied then would be ~ 16 ev which is unreasonably high. Inclusion of the neighboring atom contribution to $\sigma^{(1)}$ in general reduces the ΔE required to reproduce the absolute isotropic shift on the order of 50%.

The differences in the shielding anisotropies in ethylene, acetylene and allene are readily understood on the basis of the decomposition of the shielding into a diamagnetic anisotropy from the neighboring atoms and a paramagnetic anisotropy arising from the multiple bonds. In the following discussion the coordinate system chosen places the x direction along the CC bond, and the y direction in the molecular plane and perpendicular to the CC bond as indicated in Table 3. In ethylene the diamagnetic shielding tensor is approximately axially symmetric with the smallest shielding along the C=C bond in the x direction. Inspection of the table of Q_{AB} values shows that the paramagnetic shielding will also be axially symmetric, however the unique axis, which is the most deshielded, is along y. When these two portions are combined the resulting tensor has three well spaced ($\eta \sim 1$) components as observed experimentally.

In acetylene both the paramagnetic and diamagnetic tensors are axially symmetric as required by the C_∞ symmetry. Since the C centers are sp hybridized, the sigma bond has less p character lowering the Q_{AB} terms for the y and z directions perpendicular to the C≡C axis. Thus the paramagnetic shielding is expected to be smaller along the y axis and larger along the z axis in acetylene than in ethylene as observed. Due to symmetry the paramagnetic contribution along x is identically zero.

In allene the terminal carbon is shielded substantially with respect to ethylene and has a spread in its anisotropy only half as large. The diamagnetic portion for this carbon in allene is essentially identical with that in ethylene. Thus the difference comes from the paramagnetic term. The lesser paramagnetic contribution arises again from a lowering in the p character in the CC sigma bond as well as from an increase in the ΔE. The central carbon in allene is especially unique in that it is very deshielded and at the same time has a rather small spread in its anisotropy. As in acetylene both the y and z components of the paramagnetic shielding are large, and these counterbalance the larger diamagnetic contribution arising from having the two C atoms bonded along the x axis. The paramagnetic portion along the x direction is ~60% less. To match the isotropic shift a lower excitation energy is required, and the resulting anisotropy in the paramagnetic portion almost cancels out that in the diamagnetic portion of the shielding.

The shielding patterns in hydrocarbons are thus seen to be qualitatively accounted for by this simple theory of shielding. In this framework the changes in the anisotropies can be rationalized in terms of changes in hybridization producing differing paramagnetic anisotropies which are balanced by diamagnetic anisotropies arising from the relative geometric placement of the atoms and their associated electron densities.

Other qualitative trends can also be reproduced in simple related series. In the halomethanes the shielding along the C-X bond varies only 16 ppm upon going from CH_3F to CH_3I. On the other hand the shielding perpendicular to the bond varies greatly. This is primarily due to changes in the diamagnetic term (Table 3) as the halogen becomes heavier. This is balanced by a more slowly increasing paramagnetic term which increases due to a lowering of ΔE as one moves from F to I in the series. An opposite trend is noted for CO_2, OCS and CS_2. As the O atoms are replaced successively by S atoms the ΔE lowers giving an increased paramagnetic shielding which is greater than the increase in the diamagnetic term.

At this qualitative level of discussion the method is seen to provide some insight into gross changes in shielding tensor elements, especially for hydrocarbons. Many important trends in ^{13}C shieldings though cannot be reproduced with this simple model as would be expected given the simplistic treatment of the molecular orbitals. For molecules such as ketene, which cannot be so easily described in these simple MO pictures, the method is insufficient. In the halocarbons the anisotropy of the paramagnetic portion is not reproduced leading to a large overestimate in the shielding anisotropy. However the results for hydrocarbons are encouraging and a more sophisticated semiempirical treatment of $\sigma^{(2)}$ should be expected to produce better results. Further development of such methods would be useful as the description of tensorial shifts in terms of a balance between the Flygare diamagnetic term and a Ramsey paramagnetic term lends itself to qualitative interpretation. The former is determined primarily by structure and the identity of the elements bound to a given center, while the latter is affected by the form of the MO's and thus hybridization, conjugation, excitation energies etc.

7. Motional Averaging and Absolute Shielding

For molecules with C_∞ symmetry $\sigma^{(2)}$ is identically zero along the molecular axis and the parallel component of the shielding is determined solely by $\sigma_\parallel^d$, the parallel component of $\sigma^{(1)}$. The value of the isotropic shift then fixes the value of $\sigma_\perp$ since $\sigma_\perp = 0.5(3\sigma_{iso} - \sigma_\parallel^d)$. Any apparent discrepancy between the calculated and observed shielding along the symmetry axis then translates into a discrepancy between the calculated and observed spread in the shielding tensor.

The linear molecules C_2H_2, CO, CO_2, OCS, CS_2, and carbon suboxide are interesting to discuss in this regard. With the exception of CO, all of these molecules are found to have their parallel shielding at 273 ppm (δ_{TMS} = -90 ppm). This constancy is reproduced in the calculation of the Flygare diamagnetic terms, however the experimental values all display less shielding along the parallel component. For the triatomic species the calculated $\sigma_\parallel$ is ~27 ppm more shielded and in CO the difference is 35 ppm. In the case of OCS, CS_2 and CO the isotropic shielding from the solid state measurements (see Table 5) agree well with Jameson's absolute shielding values [23], so an error in referencing the shifts is not likely to be the cause of this disagreement. One could attribute these differences to the approximate nature of the Flygare diamagnetic shielding calculation, however another possibility is that the experimental shielding tensors are motionally averaged.

The shielding tensors for CO_2, OCS and CS_2 were measured in Ar matrix environments [24]. It has been noted previously that the dipolar couplings for small molecules in the Ar matrix environment are subject to some motional averaging [25]. The spread in the shielding tensor for CO has also been found to be a strong function of temperature increasing from 335 ppm at 46 K to 365 ppm at 4.2 K [26]. Beeler et al recorded a spread of 353 ppm at ~20 K in an Ar matrix with a somewhat different isotropic shift [24]. The observation that the spread in the anisotropy is temperature dependent at such low temperatures implies averaging involving very low frequency librations. Such low frequency lattice motions may also have a fairly large amplitude zero point component so that significant averaging will persist even at liquid helium temperatures. Therefore it is important to consider motional averaging when the experimental shielding tensors are to be compared to the values calculated for the rigid molecule in the computer.

Ishol and Scott [27] found that such corrections increased the spread in the ^{15}N shielding anisotropy for N_2 from 520 ppm at 4.2 K to 603 ppm for a "static" molecule. A similar analysis for CO [26] gives 406 ppm for the spread in the ^{13}C shift tensor for a static CO molecule. This compares very well to the calculated spread of 417 ppm and is about 1.15 times larger than that measured by Beeler et al at 20 K. The motional averaging in the triatomic series is expected to be somewhat less. Inspection of the numbers in Table 5 shows that the ratio of the calculated to measured shielding spread is ~ 1.10 implying a third as much motional averaging. This ratio is also in line with the ratio of the calculated to measured ^{13}C-^{13}C or ^{13}C-^{19}F dipolar coupling for a number of similar size molecules studied in Ar matrices [25]. These observations taken together provide strong circumstantial support for the need to apply significant motional averaging corrections to the shielding tensors measured for many of these small molecules even at cryogenic temperatures.

One should not however read too much into this analysis as regards the accuracy of the approximate shielding calculations used here. A weakness of the foregoing treatment is that the anisotropies calculated here are rather sensitive to the value for the absolute isotropic shift. Thus in the cases of CO, OCS and CS_2 where the solid state isotropic shifts agree with those measured by Jameson, the approach is at least self consistent. However the solid state absolute isotropic shielding values often differ from those obtained in gas phase measurements by as much as 10 ppm. If the reported isotropic shifts are incorrect, the calculated paramagnetic portion of the anisotropy will then be in error (assuming the diamagnetic portion is correct). The case of acetylene is a good example. The solid state absolute isotropic shift is 113 ppm whereas the accepted value is 117.2 ppm. Since the calculated diamagnetic shielding along the symmetry axis is the same as that reported, the calculated spread in the anisotropy is also the same as the experimental value. This would imply that the motional averaging is insignificant. However this is known not to be the case from the ^{13}C-^{13}C dipolar coupling measured in this particular experiment [25]. Therefore the combination of motional averaging, errors in referencing the experimental data, and errors in the calculation of the shielding have combined here to provide a wholly fortuitous agreement between theory and experiment.

Regardless of the shortcomings of the approximate shielding calculations used here, the observed temperature dependence of the shielding anisotropies for small molecules demonstrates that motional averaging corrections should be applied to experimental numbers before comparisons with theoretical values are made. When these corrections can be accounted for as done here, the agreement with the approximate shielding tensors calculated generally improves. Without these corrections the calculated shielding anisotropies are usually larger than measured.

8. Conclusions

Our studies of shielding anisotropies for a variety of main group multiply bonded species have lead us to consider methods for comparing shielding tensors for different magnetic nuclei. A scaling law has been observed which provides the experimentalist with a means for comparing the electronic structures of multiple bonds involving different elements. This had lead us to also investigate simple pictures of shielding for the purpose of understanding the characteristic shielding anisotropies observed for different functional groups. A combination of Flygare's approach to the Ramsey diamagnetic term and the use of Pople's LCAO formulation of the Ramsey paramagnetic term reproduces many of the trends observed in ^{13}C shielding anisotropies for hydrocarbons and other small molecules. The methods developed in this work provide a simple framework for understanding trends in anisotropic shielding and with further developments may prove useful for developing further insight into the results coming from high level calculations of shielding tensors. It has been noted that motional averaging corrections should be applied to experimental shielding tensors before comparison is made to calculated values. Furthermore it is important for experimentalists to reference their experimental data as carefully as possible so that accurate conversions to absolute shielding scales can be carried out. Without such precautions the comparisons of ab initio and experimental shielding tensors becomes difficult if high accuracy is the goal.

Acknowledgements: JCD would like to acknowledge the support of his graduate studies by a fellowship from the Heyl Foundation. This research was supported by a grant from the National Science Foundation, CHE-8517584. The authors would also like to thank Prof. C. J. Jameson and Prof. M. Barfield for many useful discussions on the theory of chemical shielding.

References:
[1] a) Raabe, G. and Michl, J. (1985) "Chem. Rev. 85, 419-509.
     b) Cowley, A.H. (1984) "Double Bonding Between the Heavier Main Group Elements: From Reactive Intermediates to Isolable Molecules" Polyhedron 3, 389-432.
[2] Zilm, K. W., Webb, G. G., Cowley, A. H., Pakulski, M., Orendt, A. (1988) "The Nature of the Phosphorus-Phosphorus Double Bond as Studied by Solid-State NMR" J. Am. Chem. Soc. 110, 2032-2038.
[3] Duchamp, J. C., Pakulski, M., Cowley, A. H., Zilm, K. W. (1990) "Nature of the Carbon-Phosphorus Double Bond and the Carbon-Phosphorus Triple Bond as Studied by Solid-State NMR" J. Am. Chem. Soc. 112, 6803-6809.
[4] Jameson, C. J., Gutowsky, H. S. (1964) "Calculation of Chemical Shifts I. General Formulation and the Z Dependence" J. Chem. Phys. 40, 1714-1724.
[5] Pople, J. A. (1962) "Molecular-Orbital Theory of Diamagnetism. I. An Approximate LCAO Scheme" J. Chem. Phys. 37, 53-59.
[6] Flygare, W. H. (1974) "Magnetic Interactions and an Analysis of Molecular Electronic Charge Distribution from Magnetic Parameters" Chem. Rev. 74, 653-687.
[7] West, R., Fink, M. J., Michl, J. (1981) "Tetramesityldisilene, A Stable Compound Containing a Silicon-Silicon Double Bond" Science 214, 1343-1344.
[8] Zilm, K. W., Grant, D. M., Michl, J., Fink, M. J., West, R. (1984) "Electronic Structure of the Silicon-Silicon Double Bond. Silicon-29 Shielding Anisotropy in Tetramesityldisilene" Organometallics 2, 193-194.
[9] see references [2] and [3] and references therein.
[10] Lichter, R. L., Roberts, J. D. (1972) "Nitrogen-15 Magnetic Resonance Spectroscopy. XIV. Natural-Abundance Nitrogen-15 Chemical Shifts of Amines" J. Am. Chem. Soc. 94, 2495-2500.
[11] Duchamp, J. C. (1992) "Characterization of Main Group Functionalities Using Solids NMR: Chemical Shift Anisotropies and 2D NMR in the Solid State" Ph. D. thesis, Yale University.
[12] Duncan, T. M. (1990) "A Compilation of Chemical Shift Anisotropies" Farragut Press, Madison, WI.
[13] Pople, J. A. (1963) "The Theory of Carbon Chemical Shifts in NMR" Mol. Phys. 7, 301-306.
[14] Ramsey, N. F. (1952) "Chemical Effects in Nuclear Magnetic Resonance and in Diamagnetic Susceptibility" Phys. Rev. 86, 243-246.
[15] Baird, N. C., Teo, K. C. (1976) "Effective Excitation Energies in ^{13}C NMR Chemical Shift Calculations" J. Magn. Reson. 24, 87-94.
[16] Karplus, M. and Das, T. P. (1961) "Theory of Localized Contributions to the Chemical Shift. Application to Fluorobenzenes" J. Chem. Phys. 34, 1683-1692.
[17] Barfield, M. and Grant, D. M. (1977) "Magnetic Shielding and Shielding Anisotropies by Semiempirical Molecular Orbital Theory. I. Inclusion of Two-center Integrals of the Type $\langle\phi_B|0_A|\phi_B\rangle$" J. Chem. Phys. 67, 3322-3328.

334

[18] Kato, H. (1970) "Calculation of Proton Chemical Shifts" J. Chem. Phys. 52, 3723-3726.
[19] Slichter, C. P. (1978) "Principles of Magnetic Resonance" Springer-Verlag, New York, 325-327.
[20] Gierke, T. D., Flygare, W. H. (1972) "An Empirical Evaluation of the Individual Elements in the Nuclear Diamagnetic Shielding Tensor by the Atom Dipole Method" J. Am. Chem. Soc. 94, 7277-7283.
[21] Kutzelnigg, W. (1979) "Theory of Magnetic Susceptibilities and NMR Chemical Shifts in Terms of Localized Quantities" Isr. J. Chem. 19, 193-200.
[22] Hansen, A. E., Bouman, T. D. (1985) "Localized Orbital/Local Origin Method for Calculation and Analysis of NMR Shieldings. Applications to ^{13}C Shielding Tensors" J. Chem. Phys. 82, 5035-5047.
[23] Jameson, C. J. and Jameson, A. K. (1982) "Gas-Phase ^{13}C Chemical Shifts in the Zero-Pressure Limit: Refinements to the Absolute Shielding Scale for ^{13}C" Chem. Phys. Lett. 134, 461-466.
[24] Beeler, A. J. et al (1984) "Low Temperature ^{13}C Magnetic Resonance in Solids. 3. Linear and Pseudolinear Molecules" J. Am. Chem. Soc. 106, 7672-7676.
[25] Zilm, K. W., and Grant, D. M. (1981) "Carbon-13 Dipolar Spectroscopy of Small Organic Molecules in Argon Matrices" J. Am. Chem. Soc. 103, 2913-2922.
[26] Gibson, A. A. V., Scott, T. A., Fukushima, E. (1977) "Anisotropy of the Chemical Shift Tensor for Solid Carbon Monoxide" J. Magn. Reson. 27, 29-33.
[27] Ishol, L. M. and Scott, T. A. (1977) "Anisotropy of the Chemical Shift Tensor for Solid Nitrogen" J. Magn. Reson. 27, 23-28.

AN ORIGIN-INDEPENDENT THEORY FOR CALCULATION OF NMR SHIELDING CONSTANTS

JAN GEERTSEN
Blaaklokkevej 81
5250 Odense SV
Denmark

A recently developed origin-independent theory for calculation of magnetic shielding constants is reviewed and applied in direct calculations for the first time. Comparisons are made with the usual (origin-dependent) approach in terms of basis set and correlation effects.

1. Introduction

As is well known, the magnetic field **B** does not appear directly in the electronic Hamiltonian but rather it is represented by the vector potential **A** which is related to **B** according to

$$\mathbf{B} = \nabla \times \mathbf{A} \tag{1}$$

However, the vector potential is to some extent arbitrary since if Λ is a real, differentiable scalar field then evidently both **A** and

$$\mathbf{A'} = \mathbf{A} + \nabla\Lambda \tag{2}$$

yield the same **B**. In spite of this "gauge" dependence of operators the choice of gauge should clearly have no physical consequences if the exact solutions of the Schrödinger equation are available. All predictions of physical observables must be gauge independent (Cohen-Tannoudji *et al.* 1977) - otherwise their values would be arbitrary.

But gauge invariant results are not guaranteed for a particular approximation scheme and because all realistic calculations are based on approximations, for example perturbation theory, gauge problems can actually be a source of major difficulties in the proper interpretation of the derived results, as well as a source of many misconceptions.

The usual description of the nuclear magnetic shielding problem is based upon conventional perturbation theory where the separation of the Hamiltonian into a field-free (or "dark") Hamiltonian, H_0 and a field-dependent

335

J. A. Tossell (ed.), Nuclear Magnetic Shieldings and Molecular Structure, 335–349.
© 1993 *Kluwer Academic Publishers.*

part is used in all gauges. Conventional perturbation theory is a socalled "hybrid procedure" where only the potentials - and not the wave functions - are modified under a given gauge transformation. Hence one should in general be extremely careful when interpreting the computational outcome of this method since gauge invariant results are not necessarily obtained.

In the following it will be reviewed how it is possible to formulate an alternative (conventional) theory so that the computed results for the magnetic shielding constant become truly invariant. This method was originally developed using polarization propagator theory (Geertsen, 1991) but - as discussed in this work - it may be used in *any* sum-over-states formalism. Illustrative applications are given for the HF molecule.

2. Theory

2.1. CONVENTIONAL FORMULATION

The usual (conventional) formulation of the susceptibility and the magnetic shielding problems is based upon conventional perturbation theory, i.e. :

$$H = H(\mathbf{A}, A_0) = H_0 + \Delta H \tag{3}$$

where

$$H_0 = H(\mathbf{0}, 0) = \sum_{i=1}^{N} \frac{\mathbf{p}_i^2}{2m} + V \tag{4}$$

$$\Delta H = \sum_{i=1}^{N} \left\{ -\frac{e}{2mc}(\mathbf{A}_i \cdot \mathbf{p}_i + \mathbf{p}_i \cdot \mathbf{A}_i) + \frac{e^2}{2mc^2}\mathbf{A}_i^2 \right\} + eA_0 \tag{5}$$

and N is the number of electrons.
For a homogenous magnetic field **B** the vector potential is (see, e.g., Reitz *et al.* 1979)

$$\mathbf{A}_i(\mathbf{r}_i') = \frac{1}{2}\mathbf{B} \times (\mathbf{r}_i - \mathbf{r}_i') + \sum_I \frac{\mathbf{\mu}^I \times (\mathbf{r}_i - \mathbf{R}_I)}{|\mathbf{r}_i - \mathbf{R}_I|^3}$$

$$= \mathbf{A}_i^{(1)}(\mathbf{r}_i') + \mathbf{A}_i^{(2)} \tag{6}$$

where $\mathbf{r}_i'$ is a constant vector. The vector potential in the above equation satisfies the Coulomb gauge condition, $\nabla \cdot \mathbf{A} = 0$.
Using the two empirical expressions for the interaction energies related to the susceptibility (χ) and the magnetic shielding (σ)

$$\Delta E_\chi = -\frac{1}{2}\mathbf{B}\cdot\underline{\chi}\cdot\mathbf{B} = -\frac{1}{2}\mathbf{B}\cdot(\underline{\chi}^d + \underline{\chi}^p)\cdot\mathbf{B}$$

$$= \Delta E_\chi^d + \Delta E_\chi^p \tag{7}$$

$$\Delta E_\sigma = \sum_I \boldsymbol{\mu}^I\cdot\underline{\sigma}\cdot\mathbf{B} = \sum_I \boldsymbol{\mu}^I\cdot(\underline{\sigma}^{dI} + \underline{\sigma}^{pI})\cdot\mathbf{B}$$

$$= \Delta E_\sigma^d + \Delta E_\sigma^p \tag{8}$$

it is a well known result that perturbation theory gives the following diamagnetic and paramagnetic contributions :

$$\Delta E_{\chi,\sigma}^d(\mathbf{r}') = \frac{e^2}{mc^2}\, 2^{l-2}\, <o|\mathbf{A}^{(k)}\cdot\mathbf{A}^{(l)}|o> \tag{9}$$

$$\Delta E_{\chi,\sigma}^p(\mathbf{r}') = \left(\frac{e}{mc}\right)^2 2^{l-1}\sum_{n\neq0}\frac{<o|\mathbf{A}^{(k)}\cdot\mathbf{p}|n><n|\mathbf{A}^{(l)}\cdot\mathbf{p}|o>}{E_o - E_n}$$

$$= \left(\frac{e}{mc}\right)^2 2^{l-2} \ll \mathbf{A}^{(k)}\cdot\mathbf{p}\,;\,\mathbf{A}^{(l)}\cdot\mathbf{p} \gg_{E_o} \tag{10}$$

The above equations are written in terms of second quantized operators (see, e.g., Jørgensen and Simons 1981) so that the summations over electrons have disappeared; $(k,l) = (1,1)$ is used with respect to χ and $(k,l) = (1,2)$ for σ. Furthermore, the polarization propagator has been introduced (Oddershede 1978, 1987; Jørgensen and Simons 1981; Geertsen et al. 1991).
Using Eqs.(7)-(10) we see that the diamagnetic contributions to the susceptibility and shielding tensors are obtained as ground state average values whereas the paramagnetic components may be written in terms of the polarization propagator (Geertsen 1989, 1991). That is,

$$\underline{\chi}^d(\mathbf{r}') = -\frac{e^2}{4mc^2}<o|(\mathbf{r}-\mathbf{r}')^2\underline{1} - (\mathbf{r}-\mathbf{r}')(\mathbf{r}-\mathbf{r}')|o> \tag{11}$$

$$\underline{\chi}^p(\mathbf{r}') = -\left(\frac{e}{2mc}\right)^2 \ll \mathbf{l}(\mathbf{r}')\,;\,\mathbf{l}(\mathbf{r}')\gg_{E_o} \tag{12}$$

$$\underline{\sigma}^{dI}(\mathbf{r}') = \frac{e^2}{2mc^2}<o|(\mathbf{r}-\mathbf{r}')\cdot\frac{\mathbf{r}-\mathbf{R}_I}{|\mathbf{r}-\mathbf{R}_I|^3}\underline{1} - (\mathbf{r}-\mathbf{r}')\frac{\mathbf{r}-\mathbf{R}_I}{|\mathbf{r}-\mathbf{R}_I|^3}|o> \tag{13}$$

$$\underline{\sigma}^{\text{pI}}(\mathbf{r}') = \frac{1}{2}\left(\frac{e}{mc}\right)^{2} \ll \mathbf{l}(\mathbf{r}') ; \mathbf{M}^{\text{I}} \gg_{E=0} \tag{14}$$

where

$$\mathbf{l}(\mathbf{r}') = (\mathbf{r} - \mathbf{r}') \times \mathbf{p} \tag{15}$$

and

$$\mathbf{M}^{\text{I}} = \frac{\mathbf{l}(\mathbf{R}_{\text{I}})}{|\mathbf{r} - \mathbf{R}_{\text{I}}|^{3}} \tag{16}$$

Clearly the sum of dia- and paramagnetic contributions to each property should be unaffected by the (arbitrary) choice of gauge. Based on propagator theory Geertsen (1989) showed that this requirement is in fact fulfilled for the susceptibility provided that the basis set is complete; - even though conventional perturbation theory is a hybrid procedure where only the potentials (and not the wave functions) are modified under a given gauge transformation. As we shall now demonstrate, we can easily make a "combined proof" which holds also for the shielding.

Let us consider a general gauge transformation of the vector potential

$$\mathbf{A} \rightarrow \mathbf{A}' = \mathbf{A} + \nabla\Lambda \tag{17}$$

where $\Lambda = \Lambda(\mathbf{r})$ is a real scalar function. In the new (primed) gauge Eq.(10) gives

$$(\Delta E_{\chi,\sigma}^{p})' = \left(\frac{e}{mc}\right)^{2} 2^{l-2} \ll (\mathbf{A}^{(k)} + \nabla\Lambda)\cdot\mathbf{p} ; (\mathbf{A}^{(l)} + \nabla\Lambda)\cdot\mathbf{p} \gg_{E=0} \tag{18}$$

Assuming that the Coulomb gauge condition is preserved under the transformation we have

$$(\nabla\Lambda)\cdot\mathbf{p} = (im/\hbar)[H_{0},\Lambda] \tag{19}$$

which allows us to write Eq.(18) in terms of a ground state average value (Geertsen 1989) :

$$\begin{aligned}
(\Delta E_{\chi,\sigma}^{p})' = {}& \Delta E_{\chi,\sigma}^{p} \\
& + (e^{2}/mc^{2}\hbar) \, 2^{l-2} <\text{o}| \, i[\Lambda, (\mathbf{A}^{(k)} + \mathbf{A}^{(l)})\cdot\mathbf{p}] \\
& \qquad\qquad + (m/\hbar)[\Lambda,[\Lambda,H_{0}]]|\text{o}>
\end{aligned} \tag{20}$$

Evaluation of the single and double commutators gives

$$[\Lambda , (\mathbf{A}^{(k)} + \mathbf{A}^{(l)})\cdot\mathbf{p}] = i\hbar\ (\mathbf{A}^{(k)} + \mathbf{A}^{(l)})\cdot(\nabla\Lambda) \tag{21}$$

$$[\Lambda ,[\Lambda ,H_0]] = -(\hbar^2/m)(\nabla\Lambda)^2 \tag{22}$$

- and therefore :

$$(\Delta E^p_{\chi,\sigma})' = \Delta E^p_{\chi,\sigma} - \frac{e^2}{mc^2}\ 2^{l-2} <0|(\mathbf{A}^{(k)} + \mathbf{A}^{(l)})\cdot\nabla\Lambda + (\nabla\Lambda)^2|0> \tag{23}$$

Similarly the diamagnetic part gives

$$(\Delta E^d_{\chi,\sigma})' = \frac{e^2}{mc^2}\ 2^{l-2} <0|(\mathbf{A}^{(k)} + \nabla\Lambda)\cdot(\mathbf{A}^{(l)} + \nabla\Lambda)|0>$$

$$= \Delta E^d_{\chi,\sigma} + \frac{e^2}{mc^2}\ 2^{l-2} <0|(\mathbf{A}^{(k)} + \mathbf{A}^{(l)})\cdot\nabla\Lambda + (\nabla\Lambda)^2|0> \tag{24}$$

which means that $\Delta E_{\chi,\sigma}$ (and hence χ, σ) is gauge invariant

$$(\Delta E^d_{\chi,\sigma})' + (\Delta E^p_{\chi,\sigma})' = \Delta E^d_{\chi,\sigma} + \Delta E^p_{\chi,\sigma} \tag{25}$$

provided that the basis set is complete. However, in actual calculations the basis set is not complete and consequently *any* results for the two properties may be obtained.

Experimentally the susceptibility and shielding *constants* are often the only accessible quantities

$$\chi = \frac{1}{3}\,\mathrm{Tr}\,\underset{\sim}{\chi} \tag{26}$$

$$\sigma^J = \frac{1}{3}\,\mathrm{Tr}\,\underset{\sim}{\sigma}^J \tag{27}$$

and are given by :

$$\chi^d(\mathbf{r}') = -\frac{e^2}{6mc^2} <0|(\mathbf{r} - \mathbf{r}')^2|0> \tag{28}$$

$$\chi^p(\mathbf{r}') = -\frac{1}{3}\left(\frac{e}{2mc}\right)^2 \mathrm{Tr} \ll \mathbf{l}(\mathbf{r}') ; \mathbf{l}(\mathbf{r}') \gg_{E=0} \tag{29}$$

$$\sigma^{dI}(\mathbf{r'}) = \frac{e^2}{3mc^2} <0|(\mathbf{r} - \mathbf{r'}) \cdot \frac{\mathbf{r} - \mathbf{R}_I}{|\mathbf{r} - \mathbf{R}_I|^3}|0> \tag{30}$$

$$\sigma^{pI}(\mathbf{r'}) = \frac{1}{6}\left(\frac{e}{mc}\right)^2 \mathrm{Tr} \ll \mathbf{l}(\mathbf{r'}) \, ; \, \mathbf{M}^I \gg_{E=0} \tag{31}$$

Let us now consider a special class of gauge transformations which corresponds to a shift of origin for the vector potential

$$\mathbf{r'} \rightarrow \mathbf{r''} = \mathbf{r'} + \mathbf{d} \tag{32}$$

where $\mathbf{d}$ specifies the displacement. Such a transformation is described by the following type of gauge function (Arrighini *et al.* 1968; Lazzeretti and Zanasi 1985) :

$$\Lambda = -\frac{1}{2}(\mathbf{B} \times \mathbf{d}) \cdot (\mathbf{r} - \mathbf{r'}) = \mathbf{d} \cdot \mathbf{A}^{(1)}(\mathbf{r'}) \tag{33}$$

Since

$$\nabla \Lambda = -\frac{1}{2}\mathbf{B} \times \mathbf{d} \tag{34}$$

the new vector potential is given by

$$\begin{aligned}
\mathbf{A}_{new} &= \mathbf{A}^{(1)}(\mathbf{r'}) + \mathbf{A}^{(2)} + \nabla \Lambda \\
&= \frac{1}{2}\mathbf{B} \times (\mathbf{r} - \mathbf{r''}) + \mathbf{A}^{(2)} \\
&= \mathbf{A}(\mathbf{r''})
\end{aligned} \tag{35}$$

By simply changing $\mathbf{r'}$ to $\mathbf{r''}$ in Eqs.(28)-(31) we can thus write the dia- and paramagnetic contributions relative to the new origin, $\mathbf{r''}$ as functions of the variables $\mathbf{r'}$ and $\mathbf{d}$ (Arrighini *et al.* 1968; Lazzeretti and Zanasi 1985; Geertsen 1989, 1991, 1992) :

$$\chi^d(\mathbf{r''}) = \chi^d(\mathbf{r'}) + \Delta \chi^d \tag{36}$$

$$\chi^p(\mathbf{r''}) = \chi^p(\mathbf{r'}) + \Delta \chi^p \tag{37}$$

$$\sigma^{dI}(\mathbf{r''}) = \sigma^{dI}(\mathbf{r'}) + \Delta \sigma^{dI} \tag{38}$$

$$\sigma^{pl}(\mathbf{r}'') = \sigma^{pl}(\mathbf{r}') + \Delta\sigma^{pl} \tag{39}$$

where

$$\Delta\chi^{d} = -\frac{e^2}{6mc^2}\left\{\frac{2}{e}<0|\boldsymbol{\mu}^{e}(\mathbf{r}')|0>\cdot\mathbf{d} + N\,\mathbf{d}^2\right\} \tag{40}$$

$$\Delta\chi^{p} = -\frac{1}{3}\left(\frac{e}{2mc}\right)^2\sum_{\alpha}\varepsilon_{\alpha\beta\gamma}\ll\mathbf{p}_{\alpha}\,;\,2\mathbf{l}_{\gamma} - \varepsilon_{\gamma\delta\lambda}\mathbf{p}_{\lambda}\mathbf{d}_{\delta}\gg_{E_0}\mathbf{d}_{\beta} \tag{41}$$

$$\Delta\sigma^{dl} = -\frac{e}{3mc^2}<0|\mathbf{E}^{I}|0>\cdot\mathbf{d} \tag{42}$$

$$\Delta\sigma^{pl} = -\frac{1}{6}\left(\frac{e}{mc}\right)^2\sum_{\alpha}\varepsilon_{\alpha\beta\gamma}\ll\mathbf{p}_{\beta}\,;\,\mathbf{M}_{\gamma}^{I}\gg_{E_0}\mathbf{d}_{\alpha} \tag{43}$$

The operators for the electronic contribution to the dipole moment and for the electric field ar $\mathbf{R}_I$ due to all electrons are given by

$$\boldsymbol{\mu}^{e}(\mathbf{r}') = -e\,(\mathbf{r} - \mathbf{r}') \tag{44}$$

and

$$\mathbf{E}^{I} = \frac{e\,(\mathbf{r} - \mathbf{R}_I)}{|\mathbf{r} - \mathbf{R}_I|^3} \tag{45}$$

respectively, and the Levi-Civita symbol has been introduced. More explicit versions of Eqs.(41) and (43) obtained by carrying out the summations are given by Geertsen (1991, 1992).
From these equations it follows that within the present formulation exact origin independence of the susceptibility

$$\chi^{d}(\mathbf{r}') + \chi^{p}(\mathbf{r}') = \chi^{d}(\mathbf{r}'') + \chi^{p}(\mathbf{r}'') \tag{46}$$

$$\Downarrow$$

$$\Delta\chi^{d} = -\Delta\chi^{p} \tag{47}$$

is obtained only if the following constraints are exactly fulfilled :

$$<0|\mu_\alpha^e(\mathbf{r'})|0> = \frac{e}{2m}\,\varepsilon_{\alpha\beta\gamma}\ll \mathbf{p}_\beta\,;\mathbf{l}_\gamma(\mathbf{r'})\gg_{E=0} \tag{48}$$

$$mN\delta_{\alpha\beta} = -\ll \mathbf{p}_\alpha\,;\mathbf{p}_\beta\gg_{E=0} \tag{49}$$

From the latter equation the Thomas-Reiche-Kuhn (TRK) sum-rule in the dipole velocity approximation may be derived :

$$N = -\frac{1}{3m}\,\mathrm{Tr}\ll \mathbf{p}\,;\mathbf{p}\gg_{E=0} \tag{50}$$

For the shielding it is similarly required that

$$\sigma^{dl}(\mathbf{r'}) + \sigma^{pl}(\mathbf{r'}) = \sigma^{dl}(\mathbf{r''}) + \sigma^{pl}(\mathbf{r''}) \tag{51}$$

$$\Downarrow$$

$$\Delta\sigma^{dl} = -\Delta\sigma^{pl} \tag{52}$$

which gives a sum-rule for the electric field :

$$<0|\mathbf{E}_\alpha^l|0> = \frac{e}{2m}\,\varepsilon_{\alpha\beta\gamma}\ll \mathbf{p}_\gamma\,;\mathbf{M}_\beta^l\gg_{E=0} \tag{53}$$

However, in actual calculations employing a finite basis set the above sum-rules are *not* fulfilled which means that neither the susceptibility nor the shielding will be unchanged when the origin is shifted.

The requirement of a complete basis is thus a necessary - but not a sufficient - condition; the hypervirial theorem for the operators should also be satisfied. This is the case for the random phase approximation (RPA), for example, provided that the basis is complete. On the background of these considerations it follows that *if* the diamagnetic contributions are computed as SCF average values the paramagnetic terms *should* be calculated within the RPA. Anyway, it is often argued that χ^d, σ^d do not change much when introducing more correlation and hence only χ^p, σ^p are studied at correlated levels beyond RPA. Naturally, in such cases one will observe a stronger origin dependence (relative to the SCF-RPA case); moreover, even in a complete basis one will not get an invariant answer.

Often the gauge origin is chosen as the center of mass. In the case of the susceptibility this choice has to some extent an *experimental* justification since the paramagnetic term is related to the rotational g-factor which is separately observable (Gordy and Cook 1970; Geertsen and Scuseria 1989; Sauer *et al*. 1992). But in calculations of χ and σ there is no *theoretical* justification for this particular choice.

2.2 INVARIANT FORMULATION

Using the same philosophy as for the susceptibility problem (Geertsen 1989) also the diamagnetic contribution to the magnetic shielding constant can be expressed in terms of the propagator. The theory has been described in detail elsewhere (Geertsen 1991) and here we will simply state the result that

$$\sigma^{dI}(\mathbf{r}') = \frac{e}{6m^2c^2} \, \mathrm{Tr} \ll \mathbf{p} \, ; \mathbf{J}^I(\mathbf{r}') \gg_{E=0} \tag{54a}$$

or

$$\sigma^{dI}(\mathbf{r}') = \frac{1}{3} \, \mathrm{Tr} \left[\underline{\Omega}^I(\mathbf{r}') \right] \tag{54b}$$

where the tensor Ω has been defined as

$$\underline{\Omega}^I(\mathbf{r}') = \frac{e}{2m^2c^2} \ll \mathbf{p} \, ; \mathbf{J}^I(\mathbf{r}') \gg_{E=0} \tag{55}$$

and where

$$\mathbf{J}^I(\mathbf{r}') = \mathbf{E}^I \times \mathbf{l}_0 - e\mathbf{r}' \times \mathbf{M}^I - ieh\Delta\mathbf{J}^I \tag{56a}$$

$$\Delta\mathbf{J}^I = \frac{\mathbf{r}}{|\mathbf{r} - \mathbf{R}_I|^3} + \frac{3}{2} \frac{\mathbf{r} - \mathbf{R}_I}{|\mathbf{r} - \mathbf{R}_I|^5} \times (\mathbf{r} \times \mathbf{R}_I) + \frac{4\pi}{3} \delta(\mathbf{r} - \mathbf{R}_I)\mathbf{r} \tag{56b}$$

$$\mathbf{l}_0 = \mathbf{r} \times \mathbf{p} \tag{56c}$$

The last term in Eq.(56a) is included in order to make the total operator Hermitian and $\mathbf{J}^I(\mathbf{r}')$ has been designed in such a way that an origin independent result for the total shielding is ensured. Let us demonstrate that σ^I has this property.

An origin transformation from $\mathbf{r}'$ to $\mathbf{r}''$ (see Eq.(32)) gives

$$\sigma^{dI}(\mathbf{r}'') = \sigma^{dI}(\mathbf{r}') + \Delta\sigma^{dI} \tag{57}$$

$$\Delta\sigma^{dI} = -\frac{e^2}{6m^2c^2} \, \mathrm{Tr} \ll \mathbf{p} \, ; \mathbf{d} \times \mathbf{M}^I \gg_{E=0}$$

$$= \frac{e^2}{6m^2c^2} \, \mathrm{Tr} \ll \mathbf{d} \times \mathbf{p} \, ; \mathbf{M}^I \gg_{E=0} \tag{58}$$

For the paramagnetic term we have

$$\sigma^{pI}(\mathbf{r}'') = \sigma^{pI}(\mathbf{r}') + \Delta\sigma^{pI} \tag{59}$$

where

$$\Delta\sigma^{pI} = -\frac{e^2}{6m^2c^2} \text{Tr} \ll \mathbf{d} \times \mathbf{p} ; \mathbf{M}^I \gg_{E=0} \tag{60}$$

Comparing Eqs.(58) and (60) it is seen that

$$\Delta\sigma^{dI} = -\Delta\sigma^{pI} \tag{61}$$

Hence, within this formulation the shielding constant is invariant :

$$\sigma^I(\mathbf{r}'') = \sigma^I(\mathbf{r}') \tag{62}$$

As previously pointed out (Geertsen, 1991) the tensor Ω appearing in the expression for σ^{dI} in Eq.(54b) is not equal to the diamagnetic shielding tensor. For any component of Ω it holds that

$$\Omega^I_{\alpha\beta} \neq \sigma^{dI}_{\alpha\beta} \tag{63}$$

but when the *trace* in Eq.(54b) is formed, however, we get the *trace* of the σ^{dI} tensor, i.e. the diamagnetic shielding constant.

Hence, the physical significance of Ω primarily lies in its trace and one would therefore not necessarily expect the individual components to be of separate interest. It turns out, however, that certain sums of Ω-components other than the trace are related to equivalent sums with respect to σ^{dI} . As an example it may be shown that

$$\Omega^I_{\alpha\beta}(\mathbf{r}') + \Omega^I_{\beta\alpha}(\mathbf{r}') = \sigma^{dI}_{\alpha\beta}(\mathbf{r}') + \sigma^{dI}_{\beta\alpha}(\mathbf{r}') \qquad ; \qquad (\alpha \neq \beta) \tag{64}$$

- but contrary to the trace the above sum is not origin independent.

Let us finally mention that although Eq.(54a) has been written in terms of the polarization propagator the theory may be used in *any* sum-over-states method without destroying the property of origin independence. This is trivially seen by noting that (see, e.g., Jørgensen and Simons 1981) :

$$\ll \mathbf{p} ; \mathbf{J}^I(\mathbf{r}') \gg_{E=0} = 2 \sum_{n \neq 0} \frac{<o|\mathbf{p}|n><n|\mathbf{J}^I(\mathbf{r}')|o>}{E_o - E_n} \tag{65}$$

3. Results and discussion

The problems associated with the conventional method will now be illustrated by calculations of the hydrogen magnetic shielding constant in the HF molecule using three basis sets I, II, and III consisting of 72, 91, and 119 Gaussian-type-orbitals (GTO's), respectively (see Table 1).

TABLE 1. Basis sets used in calculations for hydrogen fluoride

Basis	GTO's	Description
I	72	Lazzeretti *et al.* (1983); Geertsen *et al.* (1987)
II	91	F : same as in basis I but uncontracted
		$+ \xi^s = 0.0875 ; 0.0300$
		H : Geertsen *et al.* (1986), Table VIII, basis 101
		$+ \xi^p = 0.02 ; \xi^d = 1.44$
III	119	Basis II [a] $+$
		F : $\xi^p = 0.059326 ; \xi^d = 0.022 ; \xi^f = 1.0$
		H : $\xi^p = 0.004 ; \xi^d = 1.6 ; 0.5$

[a] The d-exponent of 1.44 for H used in basis II
is not used in basis III.

In Fig. 1 the magnetic shielding constant of hydrogen has been plotted as a function of gauge origin along the internuclear axis (relative to the center of mass) using the three basis sets. The diamagnetic terms are computed as SCF average values whereas the paramagnetic contributions are evaluated in RPA.

As expected the curves become closer to horizontal lines when enlarging the basis set size; - but even for the largest basis there is a pronounced origin dependence. It is important to notice that the basis set effect is extremely origin dependent. It is obvious that a range of origins exist where the basis set effect is positive - in another interval a negative effect is obtained. On the other hand, an "observer" standing - 0.3 a.u. away from the center of mass will claim that when going from basis I to basis II there is no basis set effect at all. It is thus clear that within the conventional method basis set extensions can not be studied using a single arbitrarily chosen origin (like e.g. the center of mass). Such calculations could in fact be quite misleading since they may suggest convergence when it has not been achieved.

Corresponding curves using basis I are given in Fig. 2 but now the paramagnetic terms are studied at various correlated levels (Oddershede 1978,1987; Geertsen *et al.* 1991). As expected from our above considerations the origin dependence increases when correlation beyond RPA is introduced. It also becomes clear that not only the basis set effect is origin dependent but that a similar behaviour is found for the dependence of correlation. Both positive and negative (and arbitrarily large) effects may be obtained - and

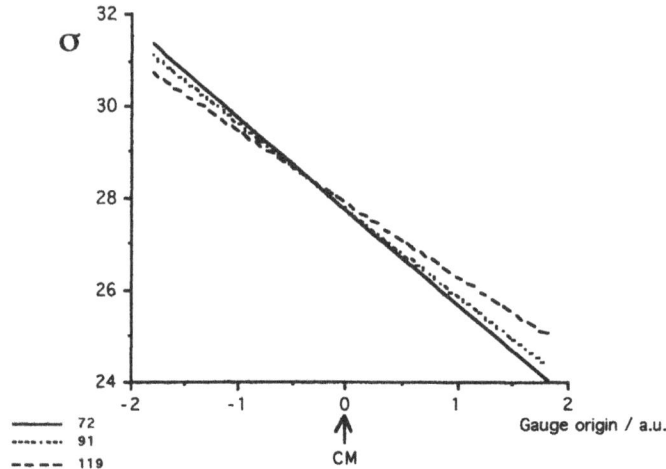

Fig. 1   The magnetic shielding constant of hydrogen in HF (ppm) calculated
         in SCF/RPA as a function of gauge origin using three basis sets.
         Conventional method

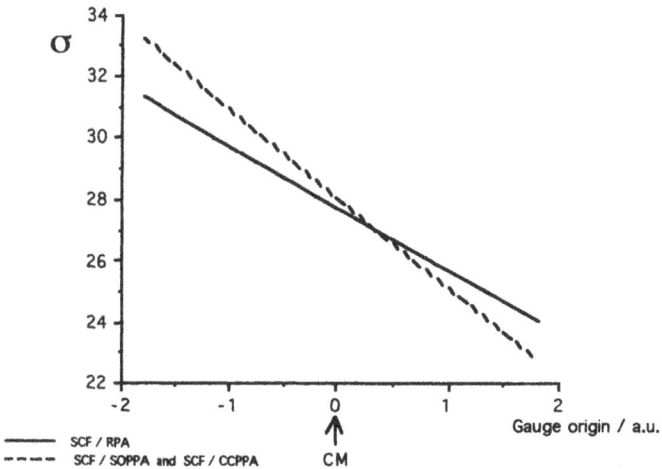

Fig. 2   The magnetic shielding constant of hydrogen in HF (ppm) using basis set I.
         The diamagnetic term is obtained as an SCF average value and the
         paramagnetic contribution is calculated at three levels of correlation.
         Conventional method

again, origins exist where no change is observed. In addition, for some origins one may find that the correlation contributions determined in SOPPA (second-order polarization propagator approximation) and CCPPA (coupled cluster polarization propagator approximation) have opposite signs, although this is not seen in the present example.

As we have demonstrated one should be very careful when applying conventional perturbation theory in the calculation of shielding constants. Obtaining this quantity with an accuracy appropriate for comparison with experimental data could be a formidable task.

Therefore various criteria for choosing the "best" origin have been proposed (Chan and Das 1962; Sadlej 1975; Yaris 1976). Approaches using so-called gauge invariant atomic orbitals (GIAO) have been developed (Ditchfield 1972; Helgaker and Jørgensen 1991) where the vector potential is incorporated into the basis functions; Kutzelnigg (1980) and Schindler and Kutzelnigg (1982) introduced an individual gauge for each localized orbital (IGLO) and Hansen and Bouman (1985) proposed a conceptually similar local orbital / local origin (LORG) approach. A "distributed origin gauge with origin at the nuclei" (DOGON) method was introduced quite recently by Stephens et al. (1989), Stephens and Jalkanen (1989), and Lazzeretti et al. (1991). Finally use of polarized basis sets (Jaszunski and Roos 1984) and locally dense basis sets (Chesnut and Moore 1989) has been made.

In the invariant method presented in the previous section origin independent results are guaranteed even in an incomplete - and arbitrarily small - basis set. Computational results using this theory have not previously appeared. Fig. 3 illustrates an application for hydrogen in the HF molecule

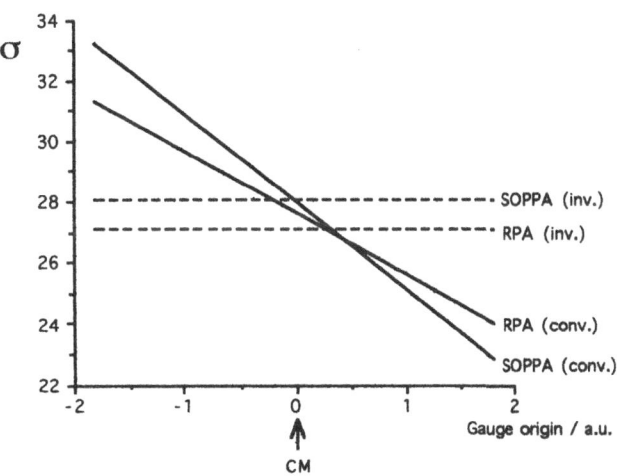

Fig. 3   The conventional and gauge origin independent (invariant) magnetic shielding constant of hydrogen in HF using basis set I (ppm). Experiment : 28.8 ± 0.5 (quoted by Lazzeretti and Zanasi (1985))

348

based on the earlier used 72 basis functions (basis I). As expected horizontal lines are obtained when σ^H is plotted as a function of gauge origin and, as for the susceptibility (Geertsen 1992), the extra correlation contribution included when going from RPA to SOPPA is not insignificant. In the latter approach a result of 28.1 ppm is predicted which is acceptable within a basis set of the considered size (experimental data are given in the figure).

In this particular example the intersection between the SOPPA curves is very close to the center of mass (- 0.01 a.u.). Fortunately this origin was chosen in our previous conventional calculations of the hydrogen shielding constant in HF (Oddershede and Geertsen 1990) where the same basis was employed.

Acknowledgement

This work was supported in part by the Danish Natural Science Research Council (grant No. 11-9004). All computed results in this paper were obtained using a modified version of the set of programs written by Scuseria (1986) and Scuseria et al. (1987,1988).

References

Arrighini, G.P., Maestro, M., and Moccia, R. (1968); J. Chem. Phys. **49**, 882

Chan, S.I., and Das, T.P. (1962); J. Chem. Phys. **37**, 1527

Chesnut, D.B., and Moore, K.D. (1989); J. Comput. Chem. **10**, 648

Cohen-Tannoudji, C., Diu, B., and Laloë, F. (1977); *Quantum Mechanics* (Hermann/Wiley, Paris)

Ditchfield, R. (1972); J. Chem. Phys. **56**, 5688

Geertsen, J., Oddershede, J., and Sabin, J.R. (1986); Phys. Rev. **A34**, 1104

Geertsen, J., Oddershede, J., and Scuseria, G.E. (1987); Intern. J. Quantum Chem. Symp. **21**, 475

Geertsen, J. (1989); J. Chem. Phys. **90**, 4892

Geertsen, J., and Scuseria, G.E. (1989); J. Chem. Phys. **90**, 6486

Geertsen, J. (1991); Chem. Phys. Lett. **179**, 479

Geertsen, J., Eriksen, S., and Oddershede, J. (1991); Advan. Quantum Chem. **22**, 167

Geertsen, J. (1992); Chem. Phys. Lett. **188**, 326

Gordy, W., and Cook, R.L. (1970); *Microwave Molecular Spectra* (Interscience, New York)

Hansen, A.E., and Bouman, T.D. (1985); J. Chem. Phys. **82**, 5035

Helgaker, T., and Jørgensen, P. (1991); J. Chem. Phys. **95**, 2595

Jaszunski, M., and Roos, B.O. (1984); Mol. Phys. **52**, 1209

Jørgensen, P., and Simons, J. (1981); *Second Quantization-based Methods in Quantum Chemistry* (Academic Press, New York)

Kutzelnigg, W. (1980); Israel J. Chem. **19**, 193

Lazzeretti, P., Rossi, E., and Zanasi, R. (1983); Phys. Rev. **A27**, 1301

Lazzeretti, P., and Zanasi, R. (1985); Phys. Rev. **A32**, 2607

Lazzeretti, P., Malagoli, M., and Zanasi, R. (1991); Chem. Phys. **150**, 173

Oddershede, J. (1978); Advan. Quantum Chem. **11**, 275

Oddershede, J. (1987); Advan. Chem. Phys. **69**, 201

Oddershede, J., and Geertsen, J. (1990); J. Chem. Phys. **92**, 6036

Reitz, J.R., Milford, F.J., and Christy, R.W. (1979); *Foundations of Electro - magnetic Theory*, (Addison-Wesley, Massachusetts)

Sadlej, A.J. (1975); Chem. Phys. Lett. **36**, 129

Sauer, S.P.A., Oddershede, J., and Geertsen, J. (1992); Mol. Phys. (in press)

Schindler, M., and Kutzelnigg, W. (1982); J. Chem. Phys. **76**, 1919

Scuseria, G.E. (1986); Chem. Phys. Lett. **127**, 236

Scuseria, G.E., Scheiner, A.C., Rice, J.E., Lee, T.J., and Schaefer, H.F. (1987); J. Chem. Phys. **86**, 2881

Scuseria, G.E., Janssen, C.L.,and Schaefer, H.F. (1988); J. Chem. Phys. **89**, 7382

Stephens, P.J., and Jalkanen, K.J. (1989); J. Chem. Phys. **91**, 1379

Stephens, P.J., Jalkanen, K.J., Lazzeretti, P., and Zanasi, R. (1989); Chem. Phys. Lett. **156**, 509

Yaris, R. (1976); Chem. Phys. Lett. **38**, 460

CORRELATED AND GAUGE INVARIANT CALCULATIONS OF NUCLEAR SHIELDING CONSTANTS

STEPHAN P.A. SAUER AND J. ODDERSHEDE
Department of Chemistry
Odense University
DK - 5230 Odense M
Denmark

ABSTRACT. We discuss applications of a method proposed by Geertsen for gauge invariant calculations of NMR shielding constants. We apply it to small molecules where it is possible to use high quality basis sets. The method is derived using sum-over-states arguments without any reference to polarization propagator theory. We demonstrate the necessity of having basis set converged CHF/RPA results before firm conclusions about the importance of inclusion of higher order electron correlation in the shielding calculations can be made. It is shown how the shielding constants are modified when correlation is computed within the second order polarization propagator approach (SOPPA).

1. Introduction

From a theoretical point of view there are three main problems in the calculations of nuclear magnetic resonance (NMR) shielding constants:
(1) the gauge origin problem
(2) the effect of electron correlation, and
(3) the basis set problem.

Unfortunately, the problems are related, so it often makes little sense to solve one problem and not the other two. For instance, inclusion of higher order correlation in a small basis set calculation may give results of poorer quality than the coupled-perturbed Hartree-Fock /random phase approximation (CHF/ RPA) calculations in converged basis sets. It is also clear that the gauge origin problem is coupled to the basis set problem. In a complete basis set there is no gauge origin problem.

Any observable quantity, which describes the coupling between an atomic or molecular system and external fields, must of course be independent of the particular choice of the mathematical representation of the external field. In the case of an NMR experiment one observes the coupling between an external static magnetic field (with the magnetic induction B) and the magnetic moments M_N of the nuclei. The change in the molecular energy to first order in B resulting from this coupling can be written as :

$$\Delta E(B,M) = -\sum_N B \, E_N^{(1,1)} \, M_N \tag{1}$$

351

J. A. Tossell (ed.), Nuclear Magnetic Shieldings and Molecular Structure, 351–365.
© 1993 *Kluwer Academic Publishers.*

Because of the presence of the electrons surrounding the nuclei, the coupling term $E_N^{(1,1)}$ deviates slightly from its value for vacuum. The difference is called the nuclear shielding tensor σ_N

$$E_N^{(1,1)} = 1 - \sigma_N \tag{2}$$

where 1 is a unit matrix. The external magnetic field and the magnetic field created by the magnetic nuclei at position R_N enter the Hamiltonian in form of their vector potentials (in SI units).

$$A(R) = \frac{1}{2}B \times (r-R) + \frac{\mu_0}{4\pi} \sum_N \frac{M_N \times (r-R_N)}{|r-R_N|^3} \tag{3}$$

Thus the vector potential originating from an external static magnetic field is not unique, because it depends on the arbitrary parameter R, called the gauge origin. This has the consequence that the total shielding tensor

$$\sigma_N = \sigma_N^d + \sigma_N^p \tag{4}$$

in finite basis set calculations becomes a *linear* function of the position of the gauge origin R (Lazzeretti and Zanasi 1985)

$$\sigma_N(R+d) = \sigma_N(R) + dC_N^{gauge} \tag{5}$$

and that the relative magnitude of the diamagnetic, σ_N^d, and the paramagnetic, σ_N^p, parts of the nuclear shielding tensor is a function of the displacement of the gauge origin, d, and thus arbitrary.

Historically, the use of gauge including atomic orbitals (GIAO) (Ditchfield 1974) or London orbitals (London 1937) represents the first solution to the gauge origin problem. Lately, this approach has been made computationally very feasible utilizing the techniques and experiences from analytic derivative methods (Wolinski et al. 1990, Helgaker and Jørgensen 1991). Another way of circumventing the gauge origin problem is the use of local gauge origins for each localized molecular orbital or groups of orbitals. In doing so the main problem with the common gauge origin methods - the non-cancellation of long range contributions - is eliminated. The two methods applying this technique, the individual gauge localized orbital (IGLO) method (Kutzelnigg 1980, Schindler and Kutzellnigg 1982) and the localized orbital/local origin (LORG) method (Hansen and Bouman 1985 and 1989) differ only in details.

In all of the above methods there is still kept a reference to a particular gauge origin (the localized molecular orbital or the atomic positions at which the GIAO's are centered). On the other hand, the method proposed by Geertsen (1991) is truly gauge origin independent as there is no reference to any origin in this method. The basic step in this approach is the reformulation of the diamagnetic term such that it becomes similar to the paramagnetic term, i.e. they are both expressed as a sum-over-states. It is then possible to add a term to the sum-over-states expression for σ^d, a term that makes no contribution to σ^d in the classical, average-value-over-the-ground-state-form. However, this term will cancel the gauge term from σ^p and it hence makes the total shielding origin independent. In a way one may say that the philosophy of the Geertsen method is the opposite of that of the local origin methods where the long range cancellation is obtained by re-expressing the paramagnetic term. The original derivation of the gauge invariant theory by Geertsen (1991) was formulated within the polarization propagator formalism. However, it is equally easy to derive the relations without any reference to propagator theory as we shall show in the next section.

The rest of this communication is concerned with accurate calculations on small molecules. In sec. 3 we give numerical examples illustrating the differences between gauge invariant and conventional common gauge origin dependent calculations as a function of the size of the basis set. Finally in sec. 4, we discuss the importance of including higher order electron correlation in the calculation of shielding constants.

2. Sum-over-states derivation of gauge invariant nuclear shieldings

Conventionally, the symmetrized diamagnetic part of the nuclear shielding tensor is expressed as a ground state average value

$$\sigma_N^d(R) = \frac{e^2}{4m}\frac{\mu_0}{4\pi}\langle 0|\sum_j \frac{(r_{jN}\cdot r_{jO}1-r_{jN}r_{jO}) + (r_{jN}\cdot r_{jO}1-r_{jN}r_{jO})^T}{r_{jN}^3}|0\rangle \tag{6}$$

Here T indicates the transposed matrix, $r_{jN} = r_j - R_N$, that is, the position of electron j relative to nucleus N, and $r_{jO} = r_j - R$, that is, the position of electron j relative to the gauge origin. Thus, changing the gauge origin R by a distance d will not affect r_{jN} whereas $r_{jO} \to r_{jO} - d$ and hence σ_N^d is a linear function of d (see Eq. (5)). The first step in the derivation of a gauge invariant form for total shielding is to express σ_N^d as a sum-over-states. This is accomplished by noticing that

$$(\mathbf{v}\mathbf{r}_O 1-\mathbf{v}\mathbf{r}_O) + (\mathbf{v}\mathbf{r}_O 1-\mathbf{v}\mathbf{r}_O)^T = \frac{1}{i\hbar}\{[\mathbf{v}\times l_O \mathbf{r}] + [\mathbf{v}\times l_O \mathbf{r}]^T\} \tag{7}$$

where $\mathbf{v}$ is any vector and $l_O = \mathbf{r}_O \times \mathbf{p}$. Inserting Eq. (7) with $\mathbf{v} = r_{jN} / r_{jN}^3$ and introducing the resolution of the identity in Eq.(6) we find

$$\sigma_N^d(R) = \frac{e^2}{4m}\frac{\mu_0}{4\pi}\frac{1}{i\hbar}\sum_n \{\langle 0|\frac{r_N\times l_O}{r_N^3}|n\rangle\langle n|r|0\rangle - \langle 0|r|n\rangle\langle n|\frac{r_N\times l_O}{r_N^3}|0\rangle$$

$$+ (\langle 0|\frac{r_N\times l_O}{r_N^3}|n\rangle\langle n|r|0\rangle)^T - (\langle 0|r|n\rangle\langle n|\frac{r_N\times l_O}{r_N^3}|0\rangle)^T\} \tag{8}$$

where we have omitted the summation over electrons, which is equivalent to writing the operators in their second quantized from. Using that $\{ |n\rangle \}$ are eigenstates of the Hamiltonian we obtain

$$\sigma_N^d(R) = \frac{e^2}{4m}\frac{\mu_0}{4\pi}\frac{1}{i\hbar}\sum_n \{ \frac{\langle 0|\frac{r_N \times l_O}{r_N^3}|n\rangle\langle n|[r,H]|0\rangle + \langle 0|[r,H]|n\rangle\langle n|\frac{r_N \times l_O}{r_N^3}|0\rangle}{E_0 - E_n}$$
$$+ \frac{(\langle 0|\frac{r_N \times l_O}{r_N^3}|n\rangle\langle n|[r,H]|0\rangle)^T + (\langle 0|[r,H]|n\rangle\langle n|\frac{r_N \times l_O}{r_N^3}|0\rangle)^T}{E_0 - E_n} \} \tag{9}$$

Application of the operator relationship

$$[r,H] = \frac{i\hbar}{m}p \tag{10}$$

leads to the form

$$\sigma_N^d(R) = \frac{e^2}{4m^2}\frac{\mu_0}{4\pi}\sum_{n\neq 0} \{ \frac{\langle 0|\frac{r_N \times l_O}{r_N^3}|n\rangle\langle n|p|0\rangle + \langle 0|p|n\rangle\langle n|\frac{r_N \times l_O}{r_N^3}|0\rangle}{E_0 - E_n}$$
$$+ \frac{(\langle 0|\frac{r_N \times l_O}{r_N^3}|n\rangle\langle n|p|0\rangle)^T + (\langle 0|p|n\rangle\langle n|\frac{r_N \times l_O}{r_N^3}|0\rangle)^T}{E_0 - E_n} \} \tag{11}$$

Comparing this form of σ_N^d tensor to the symmetrized paramagnetic shielding tensor

$$\sigma_N^p(R) = \frac{e^2}{4m^2}\frac{\mu_0}{4\pi}\sum_{n\neq 0} \{ \frac{\langle 0|\frac{l_N}{r_N^3}|n\rangle\langle n|l_O|0\rangle + \langle 0|l_O|n\rangle\langle n|\frac{l_N}{r_N^3}|0\rangle}{E_0 - E_n}$$
$$+ \frac{(\langle 0|\frac{l_N}{r_N^3}|n\rangle\langle n|l_O|0\rangle)^T + (\langle 0|l_O|n\rangle\langle n|\frac{l_N}{r_N^3}|0\rangle)^T}{E_0 - E_n} \} \tag{12}$$

we see a great deal of symmetry in the two expressions. Loosely speaking, the diamagnetic tensor is obtained from σ_N^p by "moving $r_O \times$ from one of the two operators to the other".

We note (1) that this reformulation holds for the full, symmetrized shielding tensor but not for the nonsymmetrized tensor, (2) that operator relationships are used to derive Eq. (11) from Eq. (6) so that the two expressions only give the same numerical answer in complete basis sets, and (3) that the reformulation does not remove the gauge origin dependence of σ_N. Point (3) can be achieved by observing that if one can find an operator $\hat{O}$ for which $[\hat{O}, r] + [\hat{O}, r]^T = 0$, then from Eq. (7)

$$(\nabla \tau_o 1 - \nabla r_o) + (\nabla \tau_o 1 - \nabla r_o)^T = \frac{1}{i\hbar}\{[\hat{O} + \mathbf{v} \times l_o \mathbf{r}] + [\hat{O} + \mathbf{v} \times l_o \mathbf{r}]^T\} \tag{13}$$

and the symmetrized diamagnetic shielding tensor remains unchanged, if $\hat{O}$ is added to $(r_{jN}/r_{jN}^3) \times l_O$ in Eq.(11). It is straightforward to show (Geertsen 1991) that

$$[O_x, x] = [O_y, y] = [O_z, z] = 0 \tag{14}$$

if we define

$$O = (\frac{r_N}{r_N^3} \times R) \times p \tag{15}$$

where R is the gauge origin. Geertsen's expression, however, is only valid for the diagonal elements of the shielding tensor, because the mixed commutators like e.g. $[\hat{O}_x, y]$ are in general non-zero. But realizing that $[\hat{O}_x, y] = - [\hat{O}_y, x]$, we can see that Eq.(13) is valid for all components of the symmetrized tensor. When $\hat{O}$ is added to $(r_{jN}/r_{jN}^3) \times l_O$ in Eq.(11) it is again a simple, but tedious exercise to show that the change in the trace of the sum-over-states expression for σ_N^d in Eq.(11) due to a displacement d of the gauge origin R is *exactly* the same term (with opposite sign) as the corresponding change in the trace of the paramagnetic shielding tensor σ_N^p in Eq.(12). The total shielding constant, the trace of the total shielding tensor, is thus independent of the arbitrary displacement d of the gauge origin R and hence gauge origin independent for any basis set. This exact algebraic cancellation is, however, not fulfilled for the individual components of the total shielding tensor, and the tensor itself is therefore only gauge origin independent in the limit of a complete basis set. It should also be noted that it is computationally convenient to work with Hermitian operators. The operator defined in Eq.(15) is not Hermitian. However, if we add to $\hat{O}$ another term

$$\hat{O}_H = -i\hbar\{\frac{r}{r_N^3} + \frac{3}{2}\frac{r_N}{r_N^5} \times (r \times R_N) + \frac{4}{3}\pi\delta(r_N)r\} \tag{16}$$

which also commutes with r and thus can be added to $(r_{jN}/r_{jN}^3) \times l_O$ in Eq.(11) without changing σ_N^d, then $\hat{O} + \hat{O}_H$ is Hermitian. In conclusion, the gauge invariant calculation of the shielding constant employs the standard expression for σ_N^p in Eq.(12) and the expression for σ_N^d in Eq.(11) with $(r_{jN}/r_{jN}^3) \times l_O$ replaced by $(r_{jN}/r_{jN}^3) \times l_O + \hat{O} + \hat{O}_H$.

As a final comment we point out that Eqs. (11) and (12) are examples of polarization propagators in the limit of E=0. A polarization propagator is defined as (see e.g. Oddershede 1987)

$$\langle\langle A; B\rangle\rangle_E = \sum_{n \neq 0} \{\frac{\langle 0|A|n\rangle\langle n|B|0\rangle}{E + E_0 - E_n} - \frac{\langle 0|B|n\rangle\langle n|A|0\rangle}{E + E_n - E_0}\} \tag{17}$$

for any pair of one-electron operators A and B. Hence,

$$\sigma_N^d(R) = \frac{e^2}{4m^2}\frac{\mu_0}{4\pi}\{ \langle\!\langle \frac{r_N \times l_O}{r_N^3} + \hat{O} + \hat{O}_H ; p \rangle\!\rangle_{E=0} + \langle\!\langle \frac{r_N \times l_O}{r_N^3} + \hat{O} + \hat{O}_H ; p \rangle\!\rangle_{E=0}^T\} \quad (18)$$

and

$$\sigma_N^P(R) = \frac{e^2}{4m^2}\frac{\mu_0}{4\pi}\{ \langle\!\langle \frac{l_N}{r_N^3} ; l_O \rangle\!\rangle_{E=0} + \langle\!\langle \frac{l_N}{r_N^3} ; l_O \rangle\!\rangle_{E=0}^T\} \quad (19)$$

and this is the way we actually evaluate the shieldings (Oddershede and Geertsen 1990b). However, any sum-over-states, coupled perturbed or finite field method can be used to calculate σ_N gauge invariantly by means of Eqs.(11) and (12).

3. Characteristics of gauge invariant shielding calculations

We have programmed the gauge invariant expressions for σ_N^d Eq.(18), as part of the RPAC molecular properties package (Bouman and Hansen 1990b) and all results reported here are computed with this version of RPAC. The calculations are performed using the first and second order polarization propagator approach. The latter approximation is normally referred to by the acronym SOPPA (Nielsen et al. 1980) while the first order, or random phase approximation (RPA), is the same as the coupled perturbed Hartree-Fock (CHF) method. The question of whether or not CHF/RPA contains electron correlation is a matter of the choice made for the zeroth order Hamiltonian (Sabin and Oddershede 1993 and Paidarova et al. 1991). If the perturbing field is included in the zeroth order Hamiltonian then CHF/RPA is the uncorrelated limit and since this is the common convention in theory of NMR shieldings we shall also adopt this notation here. Within the polarization propagator formalism, the SOPPA represents the next order in perturbation theory, the perturbation being the fluctuation potential, that is, the difference between the unperturbed Hamiltonian (no field) and the Hartree-Fock Hamiltonian.

From the point of view of polarization propagator theory it is optimal to have the expressions for molecular properties as sum-over-states, because (see Eq.(17)) this is what we determine directly (Oddershede and Sabin 1991). By directly we mean that we have equations for $<<A;B>>_E$ that permits us calculate the propagator *without* actually performing the sum-over-states. We cannot calculate ground state averages directly. It is therefore necessary to be able to reexpress σ_N^d from Eq.(6) as a sum-over-states as in Eq. (11) in order to compute both σ_N^d and σ_N^P at the same level of theory. That was not the case in our first polarization propagator calculations of shielding constants (Oddershede and Geertsen 1990 and Paidarova et al. 1991) where σ_N^P was calculated in SOPPA while σ_N^d was obtained as a Hartree-Fock ground state average. The same holds also for the second order extension of LORG called SOLO (Bouman and Hansen 1990a, 1992). This ambiguity is now removed, in addition to being able to perform the calculations in a gauge origin independent fashion.

Figure 1 shows a typical comparison of a conventional common gauge origin, and a gauge origin independent calculation. For a linear molecule, Eq. (5) reduces to a scalar relationship since only displacements along the internuclear axis causes an origin dependence. Hence, we see that the shieldings are tilted lines as a function of the origin displacement in the common origin calculations with the slopes diminishing as the basis set is improved. The best calculation is performed with a [9s9p5d2f] basis set consisting of 75 CGTO's/atom. With this

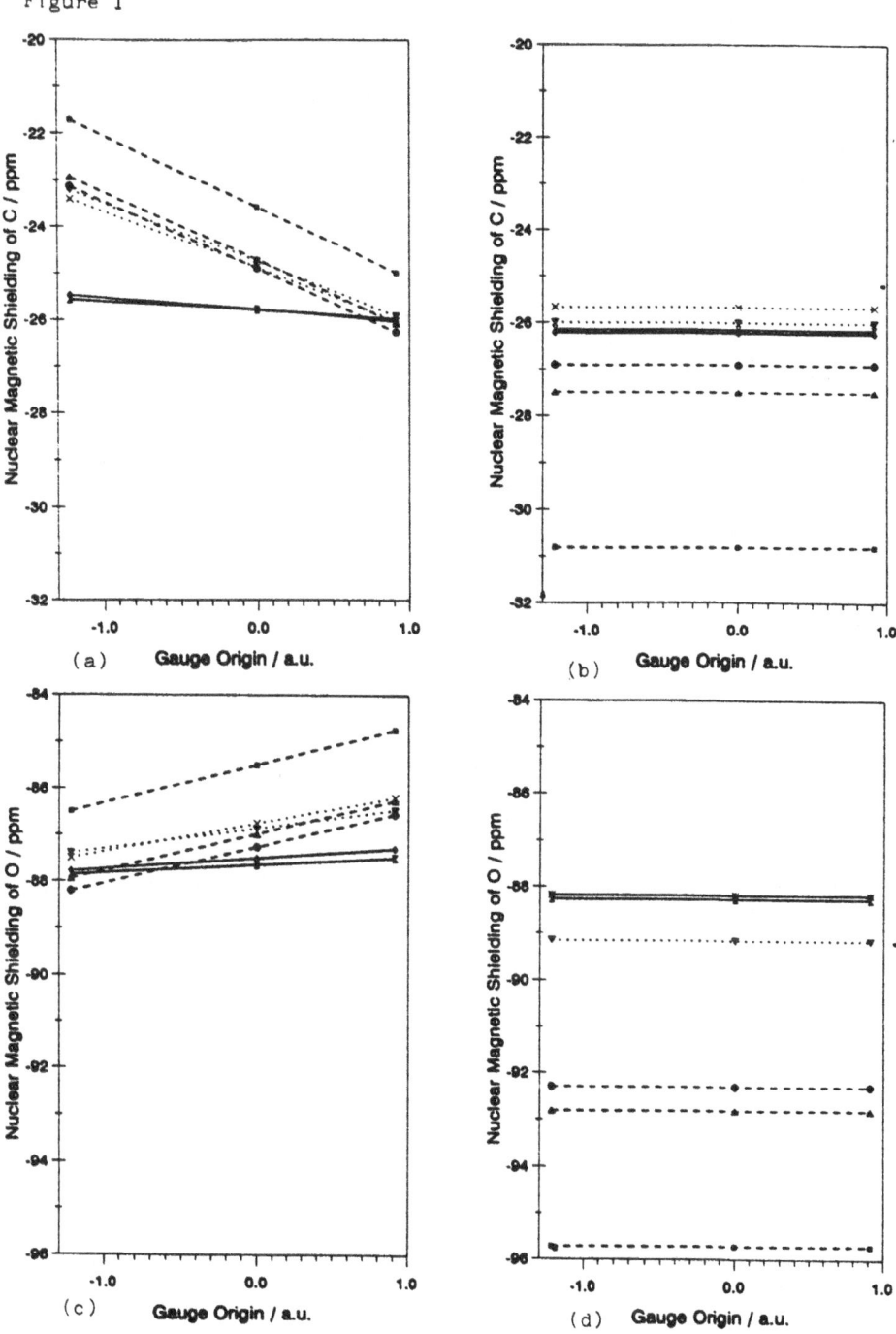

Figure 1

Figure 1. The total nuclear magnetic shieldings (in CHF) as a function of the gauge origin (relative to the center-of-mass) for CO. Figs. 1 (a) and (c) display results obtained using the common origin method while Figs. 1 (b) and (d) give results from gauge origin independent calculations. The figures show the basis set dependencies of σ_C and σ_O. The legends for the various lines are: ■ basis set [9s7p3d], level (3) described below on this page, labelled spd in Fig. 2, ▲ basis set spd with an extra tight p-function, labelled sptd in Fig. 2, ● one extra tight p-function added to the previous set; this is labelled spttd in Fig. 2, ▼ one tight and one diffuse d-funcion are added to spttd (see level (5) described below) obtaining the spttdtd basis set, and finally one f-function (△) and two f-functions (✕) are added to the previous basis set obtaining spttdtdf and spttdtdfs, respectively.

basis set both the origin dependent and independent calculations give nearly the same answer for any "reasonable" choice of gauge origin.

The convergence of the calculations with increasing basis set size is not monotonic, even though the general trend is that the common gauge origin and the gauge origin independent calculations approach the converged result from opposite directions. Fig. 1 also illustrates the necessity of including tight basis functions, i.e. functions with large exponents, in the basis set. In particular, tight p-functions are important. It is mainly the diamagnetic part of σ_N that is affected by addition of large exponent functions. The reason for that is the appearance of the p-operator in Eq. (10). This means that derivatives of s-functions, i.e. p-functions, with exponents similar to those of the s-functions ought to be included in the basis set in order to obtain a complete expansion of the velocity matrix element. Since we have s-functions with large exponents we must also include tight p-functions in the basis set. This is the same kind of argument used by Jaszunski and Roos (1984) and Roos and Sadlej (1985) in their construction of "polarized" basis sets.

We have designed a general procedure for choosing basis sets which fulfil the criterion of given very nearly the same answer for σ in both the gauge origin independent and conventional common gauge origin calculations:

(1)   An (11s7p) set from Van Duijneveldt (1971) tables is contracted to [8s7p].

(2)   The three most diffuse s-orbitals are replaced by four diffuse orbitals which gives a [9s7p] set.

(3)   Three d-orbitals (5 components) taken from Dunning (1989) are added resulting in a [9s7p3d] set.

(4)   Two tight p-functions are added. The exponents are obtained in an even-tempered fashion using the exponent ratio that corresponds to the ratio between the two most tight p-functions in the [9s7p3d] set.

(5)   One tight and one diffuse d-function are added, again using the same ratios as the ones in the Dunning d-functions and the even-tempered prescription.

(6)   The final [9s9p5d2f] set is obtained by adding two f-functions, taken from Dunning's (1989) tabulation, to the previous set.

Figure 2

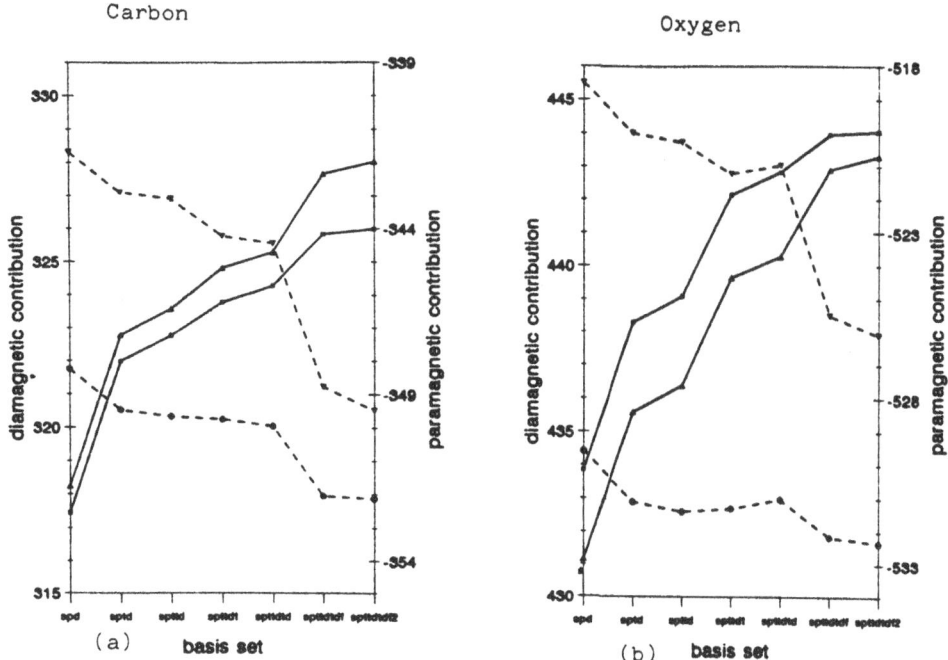

Figure 2. The diamagnetic (solid lines) and paramagnetic (dashed lines) contributions to the carbon, Fig. 2 (a), and oxygen, Fig. 2 (b), shieldings in CO as a function of the basis set, described in the caption of Fig. 1. RPA/CHF curves are labelled by ● and ■ while the SOPPA curves carry the legends ▲ and ▼.

This basis set represents a good balance between the various angular moment quantum numbers and between tight and diffuse basis functions. It has been used to compute both shieldings and magnetizabilities for molecules consisting of first and second row atoms (Sauer *et al.* 1993). Point (3) is necessary because it turned out that the van Duijneveldt basis set was incomplete in the diffuse s-space. We discovered this effect from changes incurred when using a six component Cartesian gaussian d-orbital in conjunction with original van Duijneveldt sp basis set. However, when spanning the space as discussed in point (3) with four s-orbitals, it did not matter whether or not we used 5 or 6 components for the d-orbitals.

It is a well established fact (Amos 1979, Feller *et al.* 1987) that σ_N^d expressed as a ground state average, Eq. (6), is little basis set dependent. It is also clear from the proceeding discussion that one should expect a more pronounced basis set dependence of σ_N^d computed according to Eqs. (11) or (18). We have illustrated the basis set dependence of σ^p and σ^d for both nuclei of CO in Fig. 2. We see that σ_O^d and σ_C^d show the expected behaviour. However, when σ^p has converged the same holds for σ^d. Since basis set converged shieldings cannot be obtained without σ^p being converged, the basis set convergence of σ^d thus poses little extra requirement on the basis set size. We see that the same conclusions hold at both the RPA and

TABLE 1 NMR Shielding Constants of the Heteroatom (in ppm) in H_2O, CH_4, NH_3 and HF

Basis set	H_2O	CH_4	NH_3	HF
CHF[a] 6-31G [3s2p/2s]	327	206	267	411
CHF[a] 6-311G* [4s3p1d/3s1p]	348	195	271	415
CHF[b] [9s9p5d1f/5s3p]	324	193	260	411
MP2[a] 6-31G [3s2p/2s]	344	212	279	422
MP2[a] 6-311G* [4s3p1d/3s1p]	359	202	286	427
SOPPA[b] [9s9p5d1f/5s3p]	329	194	265	410
Experiment	334[c]	197.35[d]	264.5[e]	410±6[f]

[a]Fukui et al. (1992).
[b]This calculation using the basis sets (with one f-function only) discussed in sec. 3 of the text.
[c]Appleman et al. (1974).
[d]Jackowski and Raynes (1977).
[e]Kukolich (1975).
[f]Hinderman and Cornwall (1968).

SOPPA level of theory. However, there is an interesting effect of the f-functions at the correlated level that is *not* seen at the uncorrelated level. If basis set convergence studies were performed at the CHF level only one might have arrived at the erroneous conclusion that f-functions were not needed in the basis set. We saw a similar effect in the calculation of the magnetizability of BH at the correlated and uncorrelated level (Sauer et al. 1992).

4. Examples of Correlated Gauge Origin Calculatives

Since we are preparing a more detailed account of correlated, gauge origin independent calculations for publication elsewhere (Sauer et al. 1993), we shall only give a brief summary of some of our findings here.

We first address the importance of having basis set converged CHF results before correlation is added. In Table 1 we have compared our calculated shieldings for few second row hydrides with those obtained by Fukui et al. (1992) in much smaller basis sets. We see that already at the CHF level of approximation the results of Fukui et al. (1992) are larger than the experimental shieldings and since correlation in all cases increase σ, this tendency is even more pronounced at the correlated, MP2, level of approximation. However, when converged basis sets are used the CHF shieldings are reduced and there is now "room" for the correlation contribu-

TABLE 2 NMR Shielding Constant of N in the N_2 Molecule (in ppm)

Basis set	Method	Reference	σ_N
[8s4p3d], 76 CGTO's	CHF	Oddershede and Geertsen (1990)	-106.5
	SOPPA	Oddershede and Geertsen (1990)	-72.2
[9s9p5d2f], 150 CGTO's	CHF	Sauer et al. (1993)	-109.2
	SOPPA	Sauer et al. (1993)	-82.2
-	Experiment	Jameson et al. (1981)	-61.6

TABLE 3 NMR Shielding Constant of F in the HF Molecule (in ppm)[a].

Shielding	CHF	SOPPA
σ_F^d	481.19	479.67
σ_F^p	-70.15	-69.38
σ_F^{total}	411.03	410.29

[a]Gauge origin invariant calculation using a [9s9p5d1f/5s3p]
basis set. CM is used as the origin of the coordinate system.

tions, which when added in SOPPA give shieldings in good agreement with experiments. This effect of basis set is particularly pronounced in the present examples where the CHF shieldings deviate very little from the experiments. However, the effect is always there but may not be so evident if the CHF shieldings are much smaller than the experimental ones.

The second point we wish to make in relation to Table 1 is that MP2 correlation is larger than SOPPA correlation. It has been pointed out several times that MP2 tends to overestimate correlation which is also what we see here. In the present examples the SOPPA results are in excellent agreement with experiment. However, when the correlation effects are larger, like in CO and N_2 (Sauer et al. 1993), SOPPA will underestimate the correlation effects. This underestimation is especially pronounced in large basis sets. One thus has the interesting effect illustrated in Table 2 for N_2; the best agreement with experiment is obtained in the smaller basis set. This again illustrates our first point. It is our conclusion from these studies that second

TABLE 4 NMR Shieldings (in ppm)[a].

Molecule	Atom	CHF	SOPPA	Experiment	$\dfrac{\text{CHF - SOPPA}}{\text{CHF - Exp.}}$ %
CO	σ_C	-26.1	-21.4	0.6 ± 0.9[b]	18
	σ_O	-88.3	-82.8	-40.1 ± 17.2[c]	11
N_2	σ_N	-109.2	-82.2	-61.6[d]	57
HCN	σ_C	69.2	72.1	82.1[e]	21
	σ_N	-53.2	-35.7	-20.4[d]	53

[a]Gauge origin independent calculations with the [9s9p5d2f] basis set for the heavy atom   and [5s3p] for H.
[b]Raynes et al. (1989).
[c]Frerking and Langer (1981) and Wasylishen et al. (1984).
[d]Jameson et al. (1981).
[e]Jameson and Jameson (1987).

order correlation introduced via polarization propagator method will not recover the full difference between CHF and experiments in cases where the correlation contributions are large.

The third and final observation made from Table 1 refers to the sign of the correlation contribution to σ. We see that it is positive in most cases, an observation also made by other authors at the meeting (Hansen 1992). Traditionally, this is explained by the following series of observations:

- the paramagnetic term is most affected by inclusion of correlation

- σ^p is inversely proportional with the excitation $E_n - E_0$

- $(E_n - E_0)_{RPA} < (E_n - E_0)_{correlated}$ in most cases (but not always)

Hence $\sigma^p_{corr} > \sigma^p_{CHF}$ and the same should hold for the total shieldings. We see from Table 1 that σ_F in HF is a notable exception from this rule. The origin of this exception is further elaborated on in Table 3 where we see that σ^p shows the expected behaviour but that the change is so small that it is offset by an opposite and numerically larger effect in σ_F^d. It is interesting to note that flourine and carbon in CH_3F are examples of other cases where $\sigma_{corr} < \sigma_{CHF}$ (Facelli et al. 1990, Hansen 1992).

It is relatively well established that correlation contributions to shieldings are important for atoms involved in multiple bonds, especially in cases where the atoms have lone pairs as well (Oddershede and Geertsen 1990, Paidarova et al. 1991, Bouman and Hansen 1990a and 1992). Our currently best calculations for molecules of this nature are given in Table 4. We see the tendency discussed earlier of an underestimation of the correlation contribution calculated by

SOPPA. However, it is rather puzzling that the per cent difference between SOPPA and experiment is so much larger for N than for C. It is not possible to make a similar statement about O due to the large error bars on the experiment (Frerking and Langer 1981 and Wasylishen *et al.* 1984). The origin of the difference between the C and N shieldings (in the very limited sampling currently available) is not clear to us, but it may be interesting to take another look at some of the absolute shielding scales.

5. Conclusions

We have discussed the calculation of NMR shieldings using a combination of the polarization propagator method and a newly proposed (Geertsen 1991) gauge origin invariant formulation. In particular, we have addressed the following points:

- Both the paramagnetic *and* the diamagnetic contribution to the NMR shielding tensor are expressed as sum-over-states in the formulation leading to the gauge invariant theory of Geertsen (1991, 1992).

- The gauge invariant theory is derived without any reference to polarization propagator theory; the basic ingrediences are the operator relationships in Eqs. (7) and (13).

- We have constructed a 75 GTO basis set / non-H atom which is converged with respect to the calculation of both σ^d and σ^p, at the uncorrelated as well as at the correlated level.

- There is a similar basis set dependence of σ^d and σ^p. That means, if one is converged the other one is also.

- We find no monotonic convergence of σ towards the basis set limit.

- It is essential that correlation is added to the basis set *converged* CHF results since the correlation contributions are basis set dependent.

- The SOPPA method does not give the full difference between CHF and experiment. Higher than second order terms are needed in the polarization propagator formalism for shielding calculations in molecules with multiple bonds.

- MP2 calculations (Fukui *et al.* 1992, Gauss 1992) and multi-configuration IGLO (Kutzelnigg 1992) give much larger correlation contributions than SOPPA. In some cases the correlation will be overestimated by these methods.

In summary, we have shown how the truly gauge origin independent method proposed by Geertsen (1991) performs in large basis set calculations.

364

Acknowledgements

This work is supported by grants from the Danish Natural Science Research Council (Grants No. 11-9004 and 11-9678). One of the authors (SPAS) would like to thank the Danish Research Academy for financial support.

References

Amos, R.D. (1979) Chem. Phys. Lett. **68**, 536.
Appleman, B., Tokuhiro, T., Fraenkel, G. and Kern, C. (1974) J. Chem. Phys. **60**, 2574.
Bouman, T.D. and Hansen, Aa. E. (1990a) Chem. Phys. Lett. **175**, 292.
Bouman, T.D. and Hansen, Aa.E. (1990b) RPAC Molecular Properties Package, Version 9.0 (private communication).
Bouman, T.D. and Hansen, Aa.E. (1992) Chem. Phys. Lett. **197**, 59.
Ditchfield, R. (1974) Mol. Phys. **27**, 789.
Dunning, T.H. (1989) J. Chem. Phys. **90**, 1007.
Facelli, J.C., Grant, D.M., Bouman, T.D. and Hansen, Aa. E. (1990) J. Comput. Chem. **11**, 32.
Feller, D., Boyle, C.M. and Davidson, E.R. (1987) J. Chem. Phys. **86**, 3424.
Frerking, M.A. and Langer, W.D. (1981) J. Chem. Phys. **74**, 6990.
Fukui, H., Miura, K. and Matsuda, H. (1992) J. Chem. Phys. **96**, 2039.
Gauss, J. (1992) Chem. Phys. Lett. **191**, 614.
Geertsen, J. (1991) Chem. Phys. Lett. **179**, 479.
Geertsen, J. (1992) This proceeding.
Hansen, Aa. E. (1992) This proceeding.
Hansen, Aa.E. and Bouman, T.D. (1985) J. Chem. Phys. **82**, 5035.
Hansen, Aa.E. and Bouman, T.D. (1989) J. Chem. Phys. **91**, 3552.
Helgaker, T. and Jørgensen, P. (1991) J. Chem. Phys. **95**, 2595.
Hinderman, D. and Cornwall, C. (1968) J. Chem. Phys. **48**, 2017 and 4148.
Jackowski, K. and Raynes, W.T. (1977) Mol. Phys. **34**, 465.
Jameson, A.K. and Jameson, C.J. (1987) Chem. Phys. Lett. **134**, 461.
Jameson, C.J., Jameson, A.K., Opposunggu, D., Willie, S., Burrell, P.M. and Mason, J. (1981) J. Chem. Phys. **74**, 81.
Jaszunski, M. and Roos, B.O. (1984) Mol. Phys. **52**, 1209.
Kukolich, S.G. (1975) J. Am. Chem. Soc. **97**, 5704.
Kutzelnigg, W. (1992) This proceeding.
Kutzelnigg, W. (1980) Israel J. Chem. **19**, 193.
Lazzeretti, P. and Zanasi, R. (1985) Phys. Rev. A **32**, 2607.
London, F. (1937) J. Phys. Radium **8**, 397.
Nielsen, E.S., Jørgensen, P. and Oddershede, J. (1980) J. Chem. Phys. **73**, 6238.
Oddershede, J. and Sabin, J.R. (1991) Intern. J. Quantum Chem. **39**, 371.
Oddershede, J. and Geertsen, J. (1990) J. Chem. Phys. **92**, 6036.
Oddershede, J. (1987) Adv. Chem. Phys. **69**, 201.

Paidarova, I., Komasa, J. and Oddershede, J. (1991) Mol. Phys. **72**, 559.

Raynes, W.T., McVay, R. and Wright, S.J. (1989) J. Chem. Soc., Faraday Trans. 2 **85**, 759.

Roos, B.O. and Sadlej, A.J. (1985) Chem. Phys. **94**, 43.

Sabin, J.R. and Oddershede, J. (1993) Theoret. Chem. Acta, in press.

Sauer, S.P.A., Oddershede, J. and Geertsen, J. (1992) Mol. Phys. **76**, 445.

Sauer, S.P.A., Paidorova, I. and Oddershede, J (1993) In preparation.

Schindler, M. and Kutzelnigg, W. (1982) J. Chem. Phys. **76**, 1919.

Van Duijneveldt, F.B. (1971) IBM Technical Report RJ 945.

Wasylishen, R.E., Mooibroek, S. and MacDonald, J.B. (1984) J. Chem. Phys. **81**, 1057.

Wolinski, K., Hinton, J.F. and Pulay, P. (1990) J. Am. Chem. Soc. **112**, 8251.

CARBON-13 CHEMICAL SHIELDING TENSORS IN SUGARS: SUCROSE AND METHYL-α-D-GLUCOPYRANOSIDE.

David M. Grant,
Julio C. Facelli,
D. W. Alderman and
Mark H. Sherwood

Department of Chemistry and
Utah Supercomputing Institute
University of Utah
Salt Lake City, Utah 84112, USA

ABSTRACT. This work discusses the benefits of measuring complete ^{13}C chemical shift tensors in single crystals. The comparison of these shift tensors with theoretical chemical shielding tensors provides a way to understand the electronic and the molecular structure of reasonably large molecules. Complete symmetrical chemical shift tensors, with six components each, are used to obtain information on conformational features important in the two sugars, sucrose and methyl-α-D-glucopyranoside. The theoretical calculations are sufficiently accurate to assist in making assignments of resonances arising from very similar carbon atoms in each sugar. It is shown that the best quantum chemical calculations of these ^{13}C chemical shielding tensors correlate the experimental tensor values in sucrose and in a revised assignment scheme for methyl-α-D-glucopyranoside with a standard deviation of about 3 ppm. The previous tentative assignment of the tensors in methyl-α-D-glucopyranoside gives a standard deviation of 4.8 ppm when theory is compared with experiment. These results provide a strong statistical argument for permuting the initial C_4 and C_5 tensor assignments in this molecule. The six parameters of the shift tensor provide not only three principal values but also the orientational information contained within the shift tensor. Thus, there is an advantage in working with all six parameters of a complete chemical shift tensor over either the three principal shift components obtained on a powder sample or the single isotropic shift recorded in solids by magic angle spinning. These results also indicate that theoretical chemical shielding calculations correlate with experimental shift values and can address subtle structural features with an accuracy comparable to that recorded for x-ray and neutron diffraction studies.

J. A. Tossell (ed.), Nuclear Magnetic Shieldings and Molecular Structure, 367–384.
© 1993 *Kluwer Academic Publishers.*

1. INTRODUCTION.

From the first application of NMR to chemistry, the chemical shift has been the premier NMR structural correlation parameter. This is because of its intimate dependence on the electronic foundations of molecular structure and its sensitivity to chemical factors, as its very name implies. Only in the past few years have theoretical calculations become sufficiently accurate, due to the availability of adequate computer resources, to predict and to correlate experimental chemical shift variations of less than 5 ppm in reasonably large molecules. Thus, our basic understanding of chemical shielding[1,2] is just now beginning to provide an adequate foundation for interpreting shift values in more than an empirical way. Theoretical tools and expertise for dealing with moderate size molecules now exist to interpret shift tensor data obtained on a variety of solid powders and single crystal samples. Furthermore, only in the past decade has a sufficient body of experimental information on chemical shift tensors become available upon which structural correlations may be based. The availability of good chemical shift tensor data on a variety of compounds has been essential for the flourishing of chemical shielding theory in recent years. Recognition that the chemical shift interaction is a tensor quantity with six parameters per nucleus is necessary to appreciate fully the conceptual foundations of this NMR property. Compared to typical liquid chemical shifts, these six tensor parameters provide considerably more information for characterizing the three dimensional (3D) electronic structure of molecules. These six shift parameters may be expressed as three principal values in a principal axes frame plus three orientational parameters relating this frame to a designated reference frame; the relative orientations of two frames can be described by three Euler angles. Such data provide a rich source of structural information not available from single isotropic chemical shifts obtained in liquid samples. To a large extent, the absence of these 3D data in isotropic liquids accounts for our limited understanding of liquid chemical shifts and their origin.

Some of the earliest work[3,4] on liquid ^{13}C chemical shifts identified important correlations between isotropic chemical shifts and molecular conformations. Attention is now directed to a consideration of conformational effects on the complete chemical shift tensor obtainable only in solid state samples, but more especially in single crystal samples where the shift tensors of many atoms can be studied simultaneously. Using one dimensional, 1D, goniometer methods, one is generally limited to crystals of 20 magnetically nonequivalent carbon atoms or less. The broader lines found in solids prevent the identification of many overlapping peaks in such 1D rotational patterns. The single axis goniometer probe is now superseded with a flipper probe[5,6] capable of obtaining spatially correlated two dimensional, 2D, NMR spectra. This flipping method produces, for the first time, useful spectra on relatively large and important molecules in single crystals and has the potential for indexing approximately an order of magnitude more resonances (i.e., 100-200 nuclei).

Carbon-13 chemical shift tensors in sugars are a rich source of information because various sugars provide a great variety of not only molecular configurations and conformations, but they also manifest hydrogen bonding and other long range effects that have an impact on chemical shifts in biological systems. An extensive body of x-ray and neutron diffraction data on sugars is also available to assist in the structural correlations.[7] Sugar ^{13}C shifts have been studied extensively in solution[8] and in powder samples by CP/MAS,[9] but only recently have single crystal data on sugars been reported.[6,10]

In this arena of six tensor parameters per nucleus, it is essential that molecular systems selected for study have a rationale for assigning the shift tensors. In sugars local symmetry, defined by the CO bond vector in both the alcohol and the ether molecular fragments, assists in surmounting this assignment hurdle. Diffraction data, from either x-ray or neutron sources, are used to specify the orientations of dominant local symmetry features which govern the placement of the tensor's principal axes within a unit cell. In addition, theoretical calculations of chemical shielding, to be discussed below, have improved to the point where they can be used to make tensor assignments that are impossible from experimental structural data alone. Interestingly, much of the scatter in the theoretical shielding calculations is comparable to the uncertainties in diffraction parameters used to obtain the molecular wave functions. Thus, the intrinsic accuracy of chemical shift data exceeds that of the theoretical calculations of shielding parameters limited by quantum mechanical approximations employed in larger molecules and by possible uncertainties in the structural data used to obtain the wave functions.

2. EXPERIMENTAL ^{13}C CHEMICAL SHIFT TENSORS IN SUGARS.

2.1. ISOTROPIC ^{13}C CHEMICAL SHIFTS IN SUGARS.

Solid and solution spectra on sucrose have shown that isotropic shifts in solids differ significantly from those in solution (see Figure 1); e.g., a difference of more than 5 ppm is observed for C'$_3$. The solution and solid experimental isotropic shifts for sucrose are given in Table 1 along with the principal values of the solid shifts.[6] The resonances in sucrose solutions eluded proper assignment until relatively recently,[8] and the most favored interpretation of these solution shifts now requires kinetic averaging of the sucrose structure.[11] Kinetic changes in conformation, hydrogen bonds, etc., are reasonable explanations for these variations, but the rapid conformational fluctuations of molecules in solution generally preclude a detailed structural analysis of chemical shifts. In contrast, single crystal data, where the molecules are conformationally locked, provide specific structural information for characterizing chemical shift tensors.

It is apparent from Figure 1 that the isotropic shifts in sucrose vary significantly between the solution and solid states. Note the relatively sizable changes (> 2 ppm) in the isotropic shifts in C$_4$, C'$_1$, C'$_3$, C'$_4$ and C'$_6$. The larger shift changes between the solution and solid samples are primarily in the furanyl carbons, suggesting that the structural variations within the furanyl five member ring are more sensitive to a change in state than are the variations

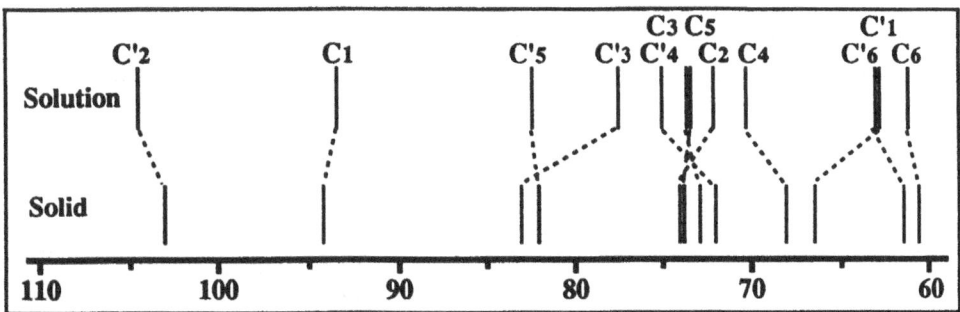

Figure 1. Comparison of Solution and Solid Isotropic Sucrose Chemical Shifts in ppm.

TABLE 1. The Experimental ^{13}C Principal Shift Values in ppm from TMS in Sucrose (See Refs. 6 and 8).

Carbon	δ_{11}	δ_{22}	δ_{33}	δ_{iso}	δ_{soln}
C_1	123.8	89.1	69.1	94.0	92.9
C_2	94.2	74.5	53.6	74.1	72.0
C_3	91.8	73.3	53.6	72.9	73.6
C_4	86.0	77.1	40.7	67.9	70.2
C_5	93.7	82.8	44.2	73.6	73.3
C_6	86.3	67.2	28.1	60.5	61.1
C'_1	87.1	74.4	38.0	66.5	63.3
C'_2	120.6	97.7	89.9	102.7	104.4
C'_3	104.1	77.4	68.5	83.3	77.4
C'_4	103.6	66.1	46.7	72.1	75.0
C'_5	105.5	92.2	48.4	82.0	82.2
C'_6	84.3	69.8	29.8	61.3	63.4

within the pyranyl six member ring. The fairly large change for C_4, noted in Figure 1, could be due to the absence of a hydrogen bond in the solid sample, and its inevitable presence in solution. The four resonances, C_2, C_3, C_5 and C'_4, invert their order between the solution and solid samples, and this illustrates the problem of using liquid chemical shifts to assign isotropic solid shifts.

The solution and isotropic solid shift values given in Table 2 for methyl-α-D-glucopyranoside,[10] also exhibit major changes between solid and solution states. Similar to

TABLE 2. The Experimental ^{13}C Principal Shift Values in ppm from TMS in Methyl-α-D-Glucopyranoside (Ref. 10) Permuted According to the GIAO/6-31G Reassignment.

Carbon	δ_{11}	δ_{22}	δ_{33}	δ_{iso}	δ_{soln}
C_1	119.7	95.1	90.0	101.6	97.8
C_2	89.8	75.6	51.0	72.1	72.0
C_3	94.5	68.7	61.2	74.8	69.2
C_4	86.8	79.3	53.7	73.3	67.8
C_5	95.8	73.9	59.4	76.4	67.2
C_6	88.8	69.2	32.6	63.5	62.4
methyl	87.1	72.3	11.5	57.0	56.2

sucrose, only the C_2 and C_6 carbons in this sample have shifts that are within 2 ppm of the solution values. Such isotropic shifts, unfortunately, are insufficient to provide extensive molecular structural information.

2.2. EXPERIMENTAL ^{13}C CHEMICAL SHIFT TENSORS IN SUGARS.

Figure 2 exhibits schematically the principal values of the chemical shift tensor data on sucrose[6] (see Table 1) and on methyl-α-D-glucopyranoside[10] (see Table 2). The data in the latter molecule involve a reassignment of the C_4 and C_5 carbons as a result of theoretical considerations to be discussed below. All the principal sugar shifts are interleaved in Figure 2, so that the similarities and variations between corresponding carbons in the two glucose moieties can be compared directly. The primary alcohols, which tend to appear upfield, have quite similar principal values. This likely reflects the similarity of structural

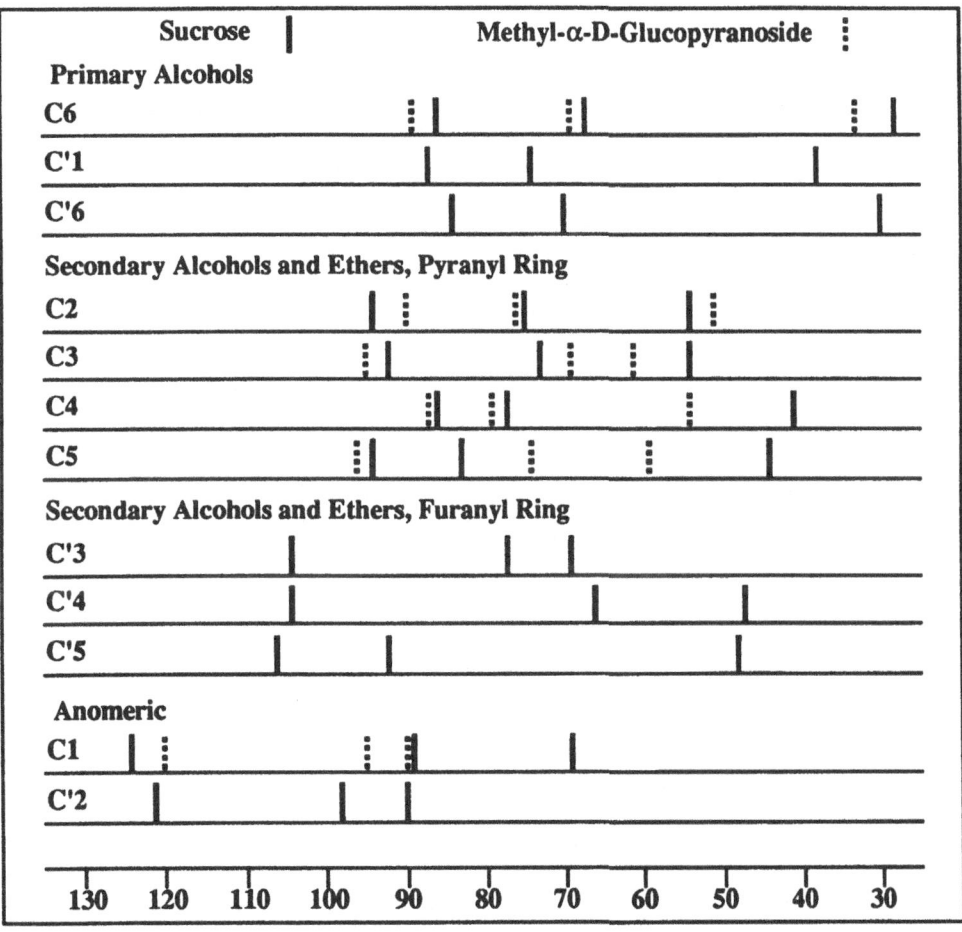

Figure 2. Principal Values of Methyl-α-D-Glucopyranoside and Sucrose Shift Tensors.

features among the primary alcohol carbons. The principal shift components in the anomeric carbons, which appear at lower fields, also have unique characteristic patterns that distinguish them from each other and from the remaining carbons. The scatter among the anomeric carbons exists primarily in their δ_{22} and δ_{33} values. In methyl-α-D-glucopyranoside the methyl carbon, not presented in Figure 2 but given in Table 2, has characteristic upfield values that may be readily identified. Finally, it is noted that the δ_{11} values of the furanyl carbons are quite similar. Further recognition of characteristic spectral patterns in the remaining diverse tensors is not obvious. Characteristic principal shifts for the primary alcohols and the anomeric carbons allow these carbons to be separated from the other carbons, thereby simplifying the assignment of the remaining sugar tensors. This is especially helpful in sucrose where the larger number of similar carbons increases the general difficulty of the assignment problem.

In single crystal NMR, the assignment problem often is complicated further by the increase of data due to a proliferation of spectral lines resulting from multiple molecules within each unit cell. For example, there are two sucrose molecules per unit cell and twenty-four lines are found in the sucrose spectrum, except when the magnetic field lies along or in the plane perpendicular to the crystal's C_2 screw symmetry axis. In these orientations the number of carbon peaks collapses to twelve due to symmetry. The NMR spectrum is unaffected by the translational part of a C_2 screw axis operation. In methyl-α-D-glucopyranoside there are four molecules per unit cell with three mutually perpendicular C_2 screw symmetry axes. A maximum of twenty-eight lines can be observed except when the magnetic field is along or in one of the planes perpendicular to these C_2 screw axes. Thus, the number of peaks collapses to either fourteen or seven when the magnetic field explores the various symmetries of the unit cell. In an elegant experiment, Sastry, Takegoshi and McDowell[10] capitalized on the methyl-α-D-glucopyranoside crystal's symmetry to simplify their spectrum. By placing their goniometer rotation axes successively along the three C_2 screw axes, the number of observable peaks never exceeded fourteen.

The lack of a consistent pattern among the principal shifts in Figure 2, especially for the secondary alcohols and ethers, introduces a serious challenge for assigning and interpreting these tensors. However, the assignment process is aided significantly by using all six tensor parameters because both the magnitudes and the orientation of principal shift values contribute information for identifying the tensors with the molecular geometry of a specific nucleus. The tensor assignments in Figure 2 depended primarily upon the orientation of the highest field principal axis with respect to the CO bond directions. This results from the high shift anisotropy associated with CO bonds.

After identifying sets of tensors related to each other using the symmetry of the unit cell and after dividing the ^{13}C nuclei of these compounds into various chemical types, the data are then interpreted with linear fitting techniques. To illustrate, anomeric carbons are identified by their unique shift components. In sucrose, the different orientation of the two OCO planes for C_1 and C'_2 is then used to complete the assignment of the anomeric carbons. For the remaining carbons, the initial assignments were made by locking the tensor orientation to the CO bond and to an effective local mirror plane in the various molecular fragments. When considering only atoms directly bonded to the carbon of interest, there is a local mirror plane in the $C^{13}CH_2O$ or $C_2^{13}CHO$ moieties appearing, respectively, in the primary alcohols or secondary alcohol and ether groups. This rather simple model, referred to as the local symmetry model, allows all the sucrose shift components, spread over a range of about 75 ppm, to be fitted and assigned using linear regression techniques with a standard deviation, σ, of about 8 ppm.[6] This assumption neglects the importance of the dihedral angle defined by atoms attached to the various CO bonds. The local symmetry fit of the sucrose data was sufficient to eliminate from consideration all but a few of the very large number of permutational assignments possible for the tensors of the two sucrose molecules in a unit cell.

In sucrose the assignment of six primary alcohol tensors to three pairs of carbons has 48 permutational possibilities (i.e., 2^3x3!) while the fourteen tensors of the seven pairs of unique secondary alcohols and ethers have 645,120 possible permutational assignments (i.e., 2^7x7!). In spite of the very large number of permutational possibilities, the local frame assignments were made with reasonably high reliability except only for the permutation of the C_4 and C_5 carbon tensors in the glucose ring. Because the directional features of the local environments associated with the C_4 and C_5 carbons, including the CO bonds, have a near inversion symmetry relationship between them, the respective orientations of these shift tensors are nearly identical. Therefore, the local frame model can not clearly resolve the two alternative assignments[6] for the C_4 and C_5 sucrose tensors. Use of experimental ^{13}C-H dipolar information helped to resolve some of these ambiguities in the initial tentative tensor assignments (i.e., between C_4 and C_5) in sucrose, but such data does not yet exist for dealing with similar assignment difficulties existing in methyl-α-D-glucopyranoside, where the near symmetry condition also exists between the C_4 and C_5 carbon nuclei.

In methyl-α-D-glucopyranoside the assignment of the chemically equivalent but magnetically different C_2 tensors to the four molecules per unit cell is also challenging. Two different pairs of C_2O_2 bonds, which dominate these chemically identical tensors, lie almost in the plane perpendicular to the C_{2X} screw axis and thus two pairs of C_2O_2 bond directions are left nearly inverted with a C_{2X} screw axis rotation. This type of near inversion symmetry makes the assignment of the two related tensors difficult using local symmetry arguments. Unlike the uncertainties in the resolution of the C_4 and C_5 tensors, which have similar orientations but different principal values, the resolution of two very similar C_2 tensors with identical principal values depends upon only orientational features. The diagonal tensor elements for chemically identical nuclei are the same in a Cartesian representation of the unit cell whenever the three axes lie along C_2 screw axes. Conversely, the δ_{xy} and δ_{zx} off-diagonal elements change sign in symmetry related tensors under a C_{2X} screw axis rotation, but the δ_{yz} off-diagonal element is left unchanged. Corresponding relationships obtain also among the four remaining tensors for the C_{2Y} and C_{2Z} screw axis operations.

2.3. ICOSAHEDRAL TENSOR REPRESENTATION OF CHEMICAL SHIFTS AND SHIELDINGS.

The manner in which one should compare chemical shift tensors (e.g., experimental versus theoretical) has received a considerable amount of our attention.[12] Sets of principal values, such as presented in Figure 2, neglect the orientational relationships and therefore are deficient for use in the comparison of two complete tensors. It is common when reporting principal values to include along with the ith principal value the directional cosines (e.g., l_i, m_i and n_i) of the ith principal axis relative to the reference frame of choice. In this representation it is difficult to compare two tensors as the principal values and the directional cosines are dissimilar quantities. Comparing tensors expressed in the common Cartesian representation is also a problem as the three diagonal terms have a different character from those of the three off-diagonal terms, and compensation for this disparity with different metric weighting factors is required to compare statistically any two Cartesian tensors. In theoretical derivations associated with chemical shifts, it is convenient to use irreducible spherical tensors with a scalar of zero rank and five second rank terms. While this representation capitalizes upon the symmetry power of irreducible spherical tensors, it still suffers from two different metric values, one for the scalar and another for the second rank components.

The icosahedral representation[12] of a shift tensor, as used in this paper, consists of six shifts obtained when the field is along the six unique directions associated with the twelve vertices of an icosahedron. The relevant geometrical constructs in Figure 3 show both the Cartesian frame and the corresponding icosahedron properly oriented in this frame. The six icosahedral directions are numbered with Arabic numerals for the selected Cartesian reference frame and may be related to two standard representations: either the Cartesian frame or the principal value frame with relevant direction cosines. These expressions are as follows:

$$\delta_1 = a^2\delta_{xx} + b^2\delta_{yy} - 2ab\delta_{xy} = (al_1 - bm_1)^2\delta_{11} + (al_2 - bm_2)^2\delta_{22} + (al_3 - bm_3)^2\delta_{33}$$

$$\delta_2 = a^2\delta_{xx} + b^2\delta_{yy} + 2ab\delta_{xy} = (al_1 + bm_1)^2\delta_{11} + (al_2 + bm_2)^2\delta_{22} + (al_3 + bm_3)^2\delta_{33}$$

$$\delta_3 = a^2\delta_{yy} + b^2\delta_{zz} - 2ab\delta_{yz} = (am_1 - bn_1)^2\delta_{11} + (am_2 - bn_2)^2\delta_{22} + (am_3 - bn_3)^2\delta_{33}$$

$$\delta_4 = a^2\delta_{yy} + b^2\delta_{zz} + 2ab\delta_{yz} = (am_1 + bn_1)^2\delta_{11} + (am_2 + bn_2)^2\delta_{22} + (am_3 + bn_3)^2\delta_{33}$$

$$\delta_5 = a^2\delta_{zz} + b^2\delta_{xx} - 2ab\delta_{zx} = (an_1 - bl_1)^2\delta_{11} + (an_2 - bl_2)^2\delta_{22} + (an_3 - bl_3)^2\delta_{33}$$

$$\delta_6 = a^2\delta_{zz} + b^2\delta_{xx} + 2ab\delta_{zx} = (an_1 + bl_1)^2\delta_{11} + (an_2 + bl_2)^2\delta_{22} + (an_3 + bl_3)^2\delta_{33}$$

where the standard Cartesian components are given by δ_{xx}, δ_{yy}, δ_{zz}, δ_{xy}, δ_{yz} and δ_{zx}. The principal values are δ_{11}, δ_{22} and δ_{33}, while the direction cosines of the ith principal axis are l_i, m_i and n_i, and the geometrical constants are $a = \cos \varphi = 0.8507$ and $b = \sin \varphi = 0.5257$. The $\varphi (= 31.72°)$ is one half of the angle subtended by two adjacent icosahedral vertices.

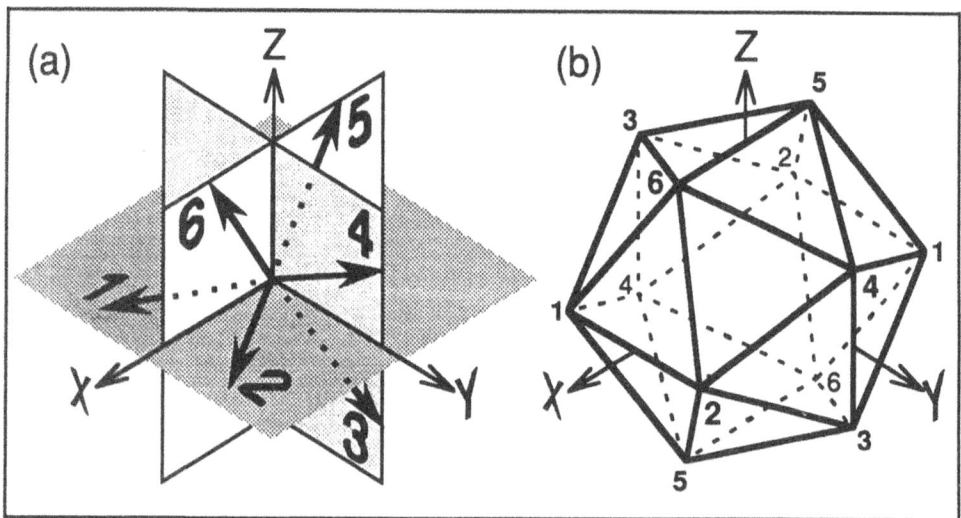

Figure 3. (a) Geometrical Relationships of the Six Icosahedral Directions to the Cartesian Frame. (b) The Six Directions are Identified with the Twelve Icosahedral Vertices.

The icosahedral representation enjoys the advantage that all six components have the same metric and work equally well in any coordinate frame (e.g., laboratory, unit cell or molecular frames). The representation also exhibits the orthogonality properties of irreducible sets, but one of its principal benefits in this study is that the statistical comparison of theoretical and experimental tensors takes place naturally, as all six terms are on the same basis and share a common metric. The six components also may be compared visually more readily in the icosahedral representation This will be obvious in the remaining Figures where the theoretical and experimental results are plotted and where the correlation between these quantities are analyzed statistically. The uniform metric of the icosahedral representation makes linear fitting of such plots valid and places the data analysis on an equal foundation. Moreover, as the vertices in the icosahedral geometry are at a maximum angular separation from each other, the structure optimizes the accuracy with which the tensor can be specified.[13]

3. THEORETICAL CALCULATIONS.

A new implementation of the GIAO method[14,15] and standard Pople basis sets[16] were used in these calculations. The smaller methyl-α-D-glucopyranoside molecule is more amenable for extensive analysis using standard quantum mechanical methods, but sucrose still may be treated at reasonable levels of sophistication. The experimental neutron diffraction results on sucrose[17] and methyl-α-D-glucopyranoside[18] provided the structural data required in the calculations. Theoretical chemical shielding tensors are calculated for the molecule nearest the origin of the unit cell and are reported in the following figures in the icosahedral representation.[12] Standard conventions used in this work are as follows: experimental shifts are in ppm downfield from TMS and theoretical shieldings are in ppm upfield from a bare nucleus.

3.1. METHYL-α-D-GLUCOPYRANOSIDE.

The theoretical chemical shieldings in methyl-α-D-glucopyranoside are compared in Figure 4 with the experimental chemical shifts for three different basis sets: STO-3G, 3-21G and 6-31G and the corresponding statistical data for the comparison are found in Table 3. It

TABLE 3. Statistical Parameters - Theoretical Correlation of Experimental Sugar Tensors.

Compound Basis Set	Slope	Intercept ppm	σ of Fit ppm	R^2
Methyl-α-D-Glucopyranoside STO-3G	- 0.851	253.3 ppm	5.3 ppm	0.898
3-21G	- 0.936	220.8 ppm	3.2 ppm	0.966
6-31G	- 0.994	214.8 ppm	3.4 ppm	0.966
Sucrose STO-3G	- 0.829	249.6 ppm	4.6 ppm	0.913
3-21G	- 0.857	214.0 ppm	2.7 ppm	0.969

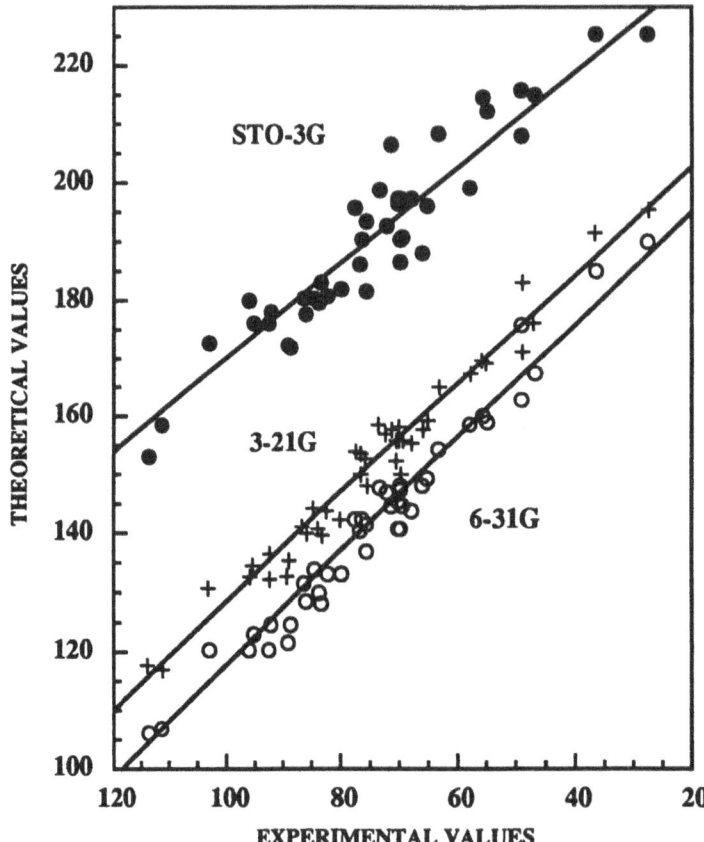

Figure 4. Icosahedral Theoretical Shieldings vs. Icosahedral Experimental Shifts for the Basis Sets: STO-3G, •; 3-21G, +; and 6-31G, ο; in Methyl-α-D-Glucopyranoside.

is noted that the scatter in Figure 4 is considerably reduced for the two largest basis sets, i.e., 3-21G and 6-31G, even though the simpler STO-3G basis set exhibits the same general correlation between theory and experiment. Also, the statistical parameters given in Table 3 improve as the size of the basis set increases. However, to achieve the best statistical measures given in Table 3, the shift assignments of the C_4 and C_5 carbons in methyl-α-D-glucopyranoside must be permuted from those given in the original literature.[10]

The four theoretical C_2 values are compared with the experimental tensor reported in Ref. 10, all properly transformed into the icosahedral representation. These calculations test the consistency of the theoretical and experimental assignments of all four chemically identical but magnetically different C_2 tensors. Using the 6-31G basis set for the four different molecular orientations in the unit cell, the different locations permute the icosahedral shifts depending upon which symmetry related molecule is considered. The A molecule of methyl-α-D-glucopyranoside is the closest to the origin of the unit cell and designated with the identity operation, E. The tensors for the remaining three molecules (i.e., the B, C and D molecules) derive, respectively, from the tensor of the A molecule with a C_{2X},

Molecule	δ_1	δ_2	δ_3	δ_4	δ_5	δ_6	rms
TABLE 4. Icosahedral Shift Tensors in Methyl-α-D-Glucopyranoside for the C_2 Carbons.[†]							
Experiment[10]	69.0	69.9	55.4	69.4	82.9	86.2	distance ppm
Theory A (E)	71.0	70.4	55.8	74.9	87.6	84.4	3.16
B (C_{2X})	70.4	71.0	55.8	74.9	84.4	87.6	2.51
C (C_{2Y})	70.4	71.0	74.9	55.8	87.6	84.4	9.95
D (C_{2Z})	71.0	70.4	74.9	55.8	84.4	87.6	9.78

[†] The experimental tensor reported in Ref. 10 was presumed to be closest to the origin. Icosahedral shifts for the four molecules in the unit cell were obtained from theoretical shielding values using the information contained in Table 3. The rms distance indicates that the theoretical tensor for molecule B is closest to the reported experimental tensor.

C_{2Y} and C_{2Z} screw axis operation as indicated in Table 4. The theoretical chemical shielding values may be converted to the experimental shift scale using the - 0.994 slope and the 214.8 ppm intercept found in Table 3. In this form, a direct comparison of the theoretical tensors with the experimental shielding data is possible.

The *rms distance,* given in Table 4, measures an effective distance between the various theoretical tensors and the reported experimental tensor. This distance provides a statistical assessment of the agreement between theory and experiment for the several C_2 orientations. The rms distances between the experimental tensor and those of molecules A and B are almost the same, but the theoretical tensor for the B molecule agrees slightly better with the experimental tensor even though in Ref. 10 the experimental tensor was tentatively assigned to the molecule closest to the origin. In this instance a slight improvement in the statistics of Table 3 could be realized if the structural data for the B molecule were used in the fit of the experimental tensors. The effect of such an assignment change on the 6-31G statistical data in Table 3 is to increase R^2 to 0.968 and to decrease σ to 3.3 ppm. The slope and the intercept remain essentially unchanged. Nevertheless, the statistical improvement is sufficiently small that it fails to support a definitive assignment of the C_2 carbons at present levels of theoretical accuracy. There is, of course, no ambiguity between the reported experimental tensor and the two theoretical C_2 tensors given for the C and D molecules.

The orientation of the C_2O_2 bond, used in a local symmetry model to make assignments, fails to discriminate between the very similar A and B molecular pair (and for that matter also between the C and D pair) of C_2 tensors. The orientations of the principal axis of the high field principal value, δ_{33}, of the experimental tensor and the corresponding tensor related by the C_{2X} screw axis operation both lie almost equidistantly from the pair of C_2O_2 vectors given in the diffraction structure of methyl-α-D-glucopyranoside.[18] Consequently, any C_2 assignments based on a local symmetry criterion can only be ambiguous. The theoretical calculation of the δ_{33} direction also exhibits deviations from the two C_2O_2 directions in A and B, confirming the problem facing the local symmetry model. The slight improvement in the experimental and theoretical fit for molecule B depends upon a better orientational match of the two principal components perpendicular to the C_2O_2 bond direction. Such dihedral angle information, related to molecular conformations, makes the use of theoretical calculations superior to the simple local symmetry model in its present form.

As indicated above, the statistical measures for assigning pairs of C_2 tensors vary by such small relative amounts that the only purpose for this analysis is merely to identify the assignment's inadequacies not to suggest that it be altered. Ambiguities of this type are rather common in single crystal work when chemically identical nuclei in symmetry related molecules are present in the unit cell. This is an appealing case to illustrate such problems. Fortunately, the values of the principal values given in Figure 2 and in Table 2 are not affected by uncertainties in the unresolved C_2 assignments. Only a modest difference in the directional orientations of these tensors is implied by this type of ambiguity in the assignment. When only slight differences exist in the spatial orientation of two chemically identical tensors, the level of theoretical accuracy must be sufficient to resolve such uncertainties. For example, the rms distance between the C_{2A} and C_{2B} molecules is 1.88 ppm for the two theoretical shieldings. This rms distance is 1.97 ppm between the reported experimental tensor and the corresponding tensor obtained from a C_{2X} screw axis operation. To choose the assignments for these two molecules, it is obvious that the standard deviation of the overall agreement must be less than this 1.97 ppm rms distance.

Two chemically equivalent tensors with only slightly different orientations constitute an excellent example of a typical assignment problem facing single crystal work. Normally, experimental methods are unavailable to address minor discrepancies in orientation between chemically identical nuclei, as found in C_2, and only theoretical methods can even address such assignment problems. Thus, as structural models and theoretical basis sets improve, we will continue to explore this C_2 tensor example. Should ambiguities exist between two chemically different tensors, then spin labeling methods always exist as a way to solve such problems, presuming the tensors are not coincidentally identical.

A detailed discussion of the assignments of the C_4 and C_5 tensors is more pertinent to this work because the principal shift values for C_4 and C_5 differ by 5-9 ppm even though the orientations of the principal axes are very similar and may lead to a false assignment when the criterion is based solely on the orientation of one or more principal shift axes. The 6-31G theoretical correlation of the experimental data using the original literature assignments yields a standard deviation of 4.8 ppm, compared with the 3.4 ppm given in Table 3. Thus, the original assignments degrade considerably the statistical measures even though their principal axes have very similar orientations. For the largest 6-31G basis set, Figure 5 compares the data for the two alternative assignments. The scatter associated with the initial assignment has at least one point that deviates from the correlation line by more than 10 ppm, and other filled points also show significant deviations from the line. The exchange of the C_4 and C_5 assignments, shown on the right side of Figure 5, improves in a dramatic manner the visual correlation between theory and experiment and supports the statistical results. It is observed, for the STO-3G approximation and the simple local frame model, that the two alternative permutations cannot be distinguished with the same high statistical confidence due to the increased scatter in the data. Fortunately, at improved levels of quantum mechanical sophistication, the distinction between C_4 and C_5 can be made with a higher level of confidence. These results illustrate the importance of combining theory and experiment as a way to optimize the interpretation of the very large amount of information that is obtainable from single crystal NMR data.

Using local symmetry arguments, the diffraction and tensor orientational data fail to provide an unequivocal assignment of the C_4 and C_5 tensor sets, to say nothing of the subtle

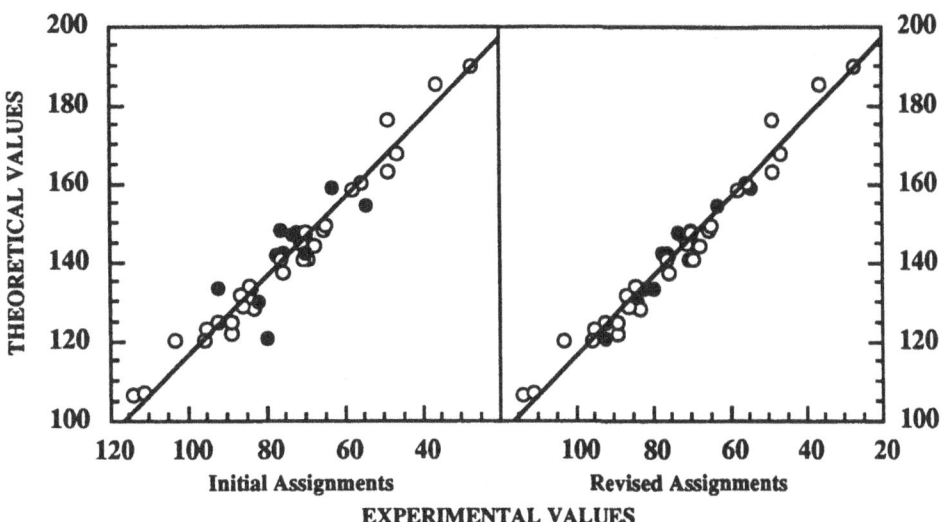

Figure 5. A Comparison of Icosahedral 6-31G Theoretical Shieldings vs. Icosahedral Experimental Shifts for Alternative Tensor Assignments in Methyl-α-D-Glucopyranoside. (Filled circles indicate the C_4 and C_5 tensor values.)

C_{2A} and C_{2B} tensor distinctions discussed above. Only by using larger quantum mechanical basis sets was it possible to sort out the C_4 and C_5 assignments. Thus, limitations in the theory for assigning and explaining experimental chemical shift tensors also identify the present reliability of theoretical methods in use for molecular sizes found in this study. As standard deviations move toward zero and as the correlation slope approaches negative unity in a theoretical and experimental comparison, the basis set may be considered to represent more faithfully the critical shielding factors. It should be noted for the largest 6-31G basis set that the correspondence between theory and experiment has a slope of almost negative unity, i.e., - 0.994, even though the R^2 and σ values are comparable to those found for the 3-21G basis. Perhaps this indicates that the scatter due to intermolecular effects (e.g., possible intermolecular hydrogen bonding or other intermolecular shielding terms) is at the ± 3 ppm level and that such intermolecular omissions in the computational model have obscured much of the benefit of the larger 6-31G basis set. Intermolecular effects of the lattice on shielding, though of great importance in solid state studies, is still largely unresolved theoretically.

3.2. SUCROSE.

The theoretical results on sucrose at the STO-3G and 3-21G levels are compared with the experimental data in Figure 6 where the reduction in scatter for the larger 3-21G basis set is easily observed. The statistical parameters for these two theoretical calculations on sucrose are also given in Table 3. The correlation for the STO-3G quantum mechanical basis is considerably better than that found in the simple local symmetry model and reduces the standard deviation of the fit from about 8 ppm to 4.6 ppm. The theoretical calculations for sucrose at the 3-21G level reduce the value of σ even further to 2.7 ppm, and the corresponding slope, intercept and R^2 also indicate improvement. Until the scatter can be

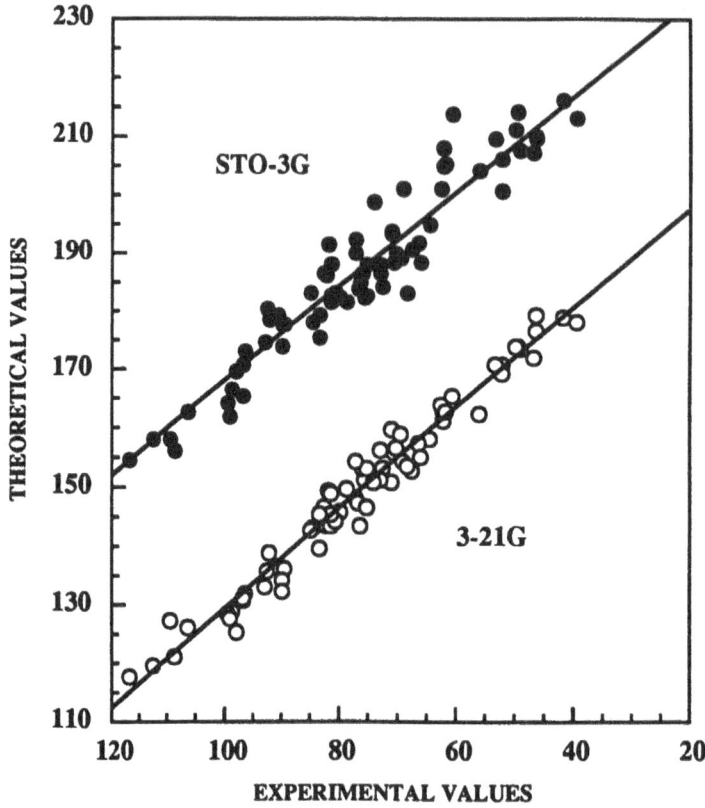

Figure 6. Icosahedral Theoretical Shieldings vs. Icosahedral Experimental Shifts for Sucrose Tensors with the STO-3G and 3-21G Basis Sets.

reduced in the 6-31G treatments of methyl-α-D-glucopyranoside, it would seem for sucrose that the 3-21G level is sufficient for testing assignments when other modeling errors of ± 3 ppm exist. At the 3-21G level, the theoretical calculations have been used to confirm the problematical assignments of C_4 and C_5 given[6] for sucrose and reversing the reported assignments[6] also degrades the fit (σ increases from 2.7 ppm to 3.8 ppm). Thus, theory at the 3-21G level supports the initial, tentative assignments of the sucrose shift tensors.

Other than the large anisotropy associated with CO bonds, no other single structural feature appears to dominate the diverse sugar data. Variations in some of the components of about ± 10 ppm would appear to be associated with changes in various CCCC and CCOH dihedral angles. Many remaining structural features appear to contribute shifts of about 1-3 ppm to the tensor components. Apparently, modest deformations of geminal angles and bond distances, are sufficiently small in sucrose that the associated changes in the shifts appear to be in the 1-2 ppm range. The effect of hydrogen bonds, except for their indirect influence on the conformation of the CCOH dihedral angles, also is rather small with differences in shift values of about 2 ppm. Unfortunately, the magnitude of these structural effects is below the scatter found in Figure 6. Thus, many substituent and conformational features still can not be discussed at the 3-21G level of treatment. Improvements in the accuracy of these theoretical methods will be required to understand refined structural features that influence the ^{13}C chemical shift tensor.

A final test of the theoretical method's reliability is shown in Figure 7 where the 3-21G results for both sucrose and methyl-α-D-glucopyranoside are overlaid in the same plot. The revised assignments for C4 and C5 are used in this figure for methyl-α-D-glucopyranoside. It is gratifying that these combined results fall along the same correlation line and that there is no apparent degradation in the scatter by combining the data for the two molecules. The largest deviations are for the methyl carbons in methyl-α-D-glucopyranoside. Thus, it may be concluded that molecular specific dependencies are not evident in these calculations. Furthermore, subtle differences between the glucose moieties common in both of these two molecules have been reproduced faithfully by these calculations even at the 3-21G approximation. These results contribute to the strong suspicion that the nearly identical scatter in the 3-21G and 6-31G treatment of methyl-α-D-glucopyranoside is due to unspecified contributions of about ±3 ppm that obscures the greater accuracy expected for the 6-31G basis set over that of the 3-21G calculation. The relatively high correlation, exhibited in the combined data of Figure 7, provides evidence that the theoretical methods have progressed to the point where greater confidence in these techniques is justified.

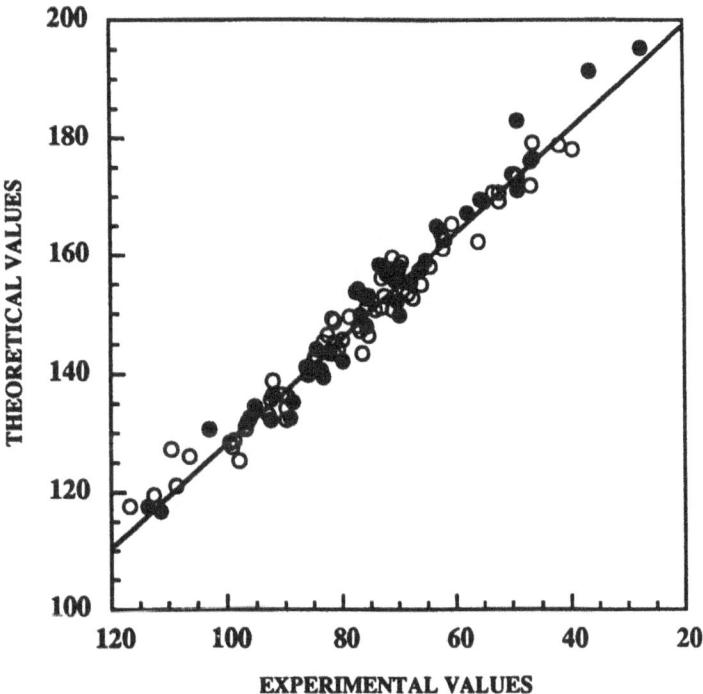

Figure 7. The 3-21G Sucrose, o, and Methyl-α-D-Glucopyranoside, •, Icosahedral Tensors Overlaid to Show the Overall General Agreement between the Two Molecules.

4. INTERMOLECULAR SHIELDING CONTRIBUTIONS.

To estimate the importance of possible intermolecular effects, two additional exploratory calculations were executed for methyl-α-D-glucopyranoside. In the first calculation, this molecule was surrounded by a layer (the equivalent of a first solvation shell) of methanol molecules positioned to mimic the first neighbor glucopyranoside molecules in the unit cell. Second, the recent proposal of Ausburger, Dykstra and Oldfield[19] on the importance of electric charges on ^{13}C chemical shieldings was studied with calculations performed by positioning a unit charge at different positions in a cubic box. The box was approximately 6 Å on edge. The methyl-α-D-glucopyranoside was located at the center of a box, and chemical shielding maps were obtained for each of the seven unique ^{13}C shielding tensors as a function of the unit charge's position in the box. The STO-3G basis was used in the methanol solvation calculation while the 3-21G was used for the electrical charge calculation. Corresponding calculations on sucrose presently are not feasible due to limitations in our computer resources. This work, while supportive of both proposals, exhibited extensive scatter that leaves the conclusions somewhat tentative.

It is difficult to obtain reliable estimates of intermolecular effects on chemical shieldings in larger molecules. Though estimates of the intermolecular effects on shielding have been attempted,[20] a comprehensive and a convenient resolution to this problem still must be found. In this study the well-known difficulties of including intermolecular interactions were compounded by the size of the molecular systems. Using the STO-3G approximation in methyl-α-D-glucopyranoside, calculations for both an isolated molecule and one surrounded by methanol molecules exhibit only minor shifts, less than ±2 ppm, from such intermolecular effects. However, the inclusion in the calculation of attached methanol molecules to represent possible hydrogen bonding effects does not improve, in any statistically significant way, the overall fit of the experimental data.

The intermolecular electrostatic charge model produced shifts in the 3-4 ppm range for a unit electron charge. For more realistic charges of fractional electrons the effect again falls below the ±3 ppm limitations inherent in the present theory. A detailed discussion of these results will be published later.

5. CONCLUSIONS.

The results presented here show that present quantum chemical calculations of the chemical shielding tensor are sufficiently reliable to assist spectroscopists in the difficult problem of assigning ^{13}C shielding tensors in complicated molecules, especially in the typical case where very similar tensors are found within a molecule or between chemically identical nuclei in different molecules in the unit cell.

The current computational methods allow us to calculate the ^{13}C shieldings in sugars with standard deviations of about 3 ppm. The lack of a significant improvement of the standard deviations of our calculation when the basis set is enlarged from 3-21G to 6-31G would indicate that further developments in the calculations may require proper modeling of a variety of intermolecular shielding effects and/or the use of more accurate structural data to obtain better quantum mechanical wave functions. Theory, capable of providing detailed electronic and structural explanations of the foundations of chemical shielding, is now correlating experimental shift data sufficiently well that even greater benefits portend for chemical shift and structural studies in the future. Single crystal techniques in NMR, using 2D flipper probes,[5,6] also have developed to the point where complete chemical shift tensors are obtainable in molecules of several hundred Daltons even in compounds with only natural levels of ^{13}C isotopic abundance.

6. ACKNOWLEDGMENTS.

We acknowledge funding from NIH grant GM08521 from the Institute of General Medical Sciences and computer time from the Utah Supercomputing Institute, funded by the State of Utah and the IBM corporation. The authors are grateful for a copy of the GIAO program obtained from Wolinski, Hinton and Pulay.[15]

7. REFERENCES.

1. Facelli, J.C., Grant, D.M. and Michl, J. '^{13}C Shielding Tensors: Experimental and Theoretical Determination', *Acc. Chem. Research*, **1987**, *20*, 152-158.

2. Facelli, J.C., Grant, D.M. 'Molecular Structure and Carbon-13 Chemical Shielding Tensors Obtained from Nuclear Magnetic Resonance', in *Topics in Stereochemistry*, **1989**, *19*, 1-61, E.L. Eliel and S.H. Wilen (eds.).

3. Grant, D.M. and Paul, E.G. 'Carbon-13 Magnetic Resonance II. Chemical Shift Data for the Alkanes', *J. Am. Chem. Soc.*, **1964**, *86*, 2984-2990. A Citation Classic, see Grant, D.M. *Current Contents/Eng. Techn. & Appl. Sci.*, **1984**, *15 (9)*, 16.

4. Dalling, D.K. and Grant, D.M. 'Carbon-13 Magnetic Resonance IX. The Methylcyclohexanes', *J. Am. Chem. Soc.*, **1967**, *89*, 6612-6622. A Citation Classic, see Dalling, D.K. *Current Contents/Phys. Chem. & Earth Sci.*, **1983**, *23 (38)*, 22.

5. Carter, C.M., Alderman, D.W. and Grant, D.M. 'Two-Dimensional Chemical-Shift-Anisotropy Correlation Spectroscopy', *J. Magn. Reson.*, **1985**, *65*, 183-186.

6. Sherwood, M.H., Alderman, D.W. and Grant, D.M. 'Two-Dimensional Chemical-Shift Tensor Correlation Spectroscopy. Multiple-Axis Sample-Reorientation Mechanism', *J. Magn. Reson.*, **1989**, *84*, 466-489.

7. Jeffrey, G.A. 'Crystallographic Studies of Carbohydrates', *Acta. Cryst.*, **1990**, *B46*, 89-103.

8. Bock, K. and Thøgersen, H. 'Nuclear Magnetic Resonance Spectroscopy in the Study of Mono- and Oligosaccharides', *Annual Reports on NMR Spectroscopy*, **1982**, *13*, 2-57, G.A. Webb (ed.).

9. Taylor, M.G., Marchessault, R.H., Perez, S., Stephenson, P.J. and Fyfe, C.A. '^{13}C CP/MAS Nuclear Magnetic Resonance of Crystalline Methylxylopyranosides', *Can. J. Chem.*, **1985**, *63*, 270-273.

384

10. Sastry, D.L., Takegoshi, K., McDowell, C.A. 'Determination of the ^{13}C Chemical-Shift Tensors in a Single Crystal of Methyl-α-D-Glucopyranoside', *Carbohydr. Res.*, **1987**, *165*, 161-171.

11. Poppe, L. and van Halbeek, H. 'The Rigidity of Sucrose: Just an Illusion?' *J. Am. Chem. Soc. 1992,114*, 1092. See also the excellent collection of references in this Communication to the Editor.

12. Alderman, D.W., Sherwood, M.H. and Grant, D.M. 'Comparing, Modeling and Assigning Chemical Shift Tensors in the Cartesian, Irreducible Spherical and Icosahedral Representations.', *J. Magn. Reson,* **1992**, *in press*.

13 Alderman, D.W., Sherwood, M.H. and Grant, D.M. 'Two Dimensional Chemical Shift Tensor Correlation Spectroscopy. Analysis of Sensitivity and Optimal Measurement Directions', *J. Magn. Reson,* **1990**, *86*, 60-69.

14. Ditchfield, R. 'Self-Consistent Perturbation Theory of Diamagnetism. I. A Gauge-Invariant LCAO Method for NMR Chemical Shifts', *Mol. Phys.*, **1974**, *27*, 789-807.

15. Wolinski, K., Hinton, J.F., Pulay, P. 'Efficient Implementation of the Gauge-Independent Atomic Orbital Method for NMR Chemical Shift Calculations', *J. Am. Chem. Soc.*, **1990**, *112*, 8251-8260.

16. Hehre, W.J., Ditchfield, R., Pople, J.A. 'Self-Consistent Molecular Orbital Methods. XII. Further Extensions of Gaussian-Type Basis Sets for Use in Molecular Orbital Studies of Organic Molecules', *J. Chem. Phys.*, **1972**, *56*, 2257-2266.

17. Brown, G.M. and Levy, M.A. 'Further Refinement of the Structure of Sucrose Based on Neutron-Diffraction Data', *Acta. Crystallogr.*, **1973**, *B29*, 790-797.

18. Jeffrey, G.A., McMullan, R.K. and Takagi, S. 'A Neutron Diffraction Study of the Hydrogen Bonding in the Crystal Structures. Methyl-α-D-Mannopyranoside and Methyl-α-D-Glucopyranoside', *Acta. Crystallogr.*, **1977**, *B33*, 728-737.

19. Ausburger, J.D., Dykstra, C.E. and Oldfield, E. 'Correlation of Carbon-13 and Oxygen-17 Chemical Shifts and the Vibrational Frequency of Electrically Perturbed Carbon Monoxide: A Possible Model for Distal Ligand Effects in Carbon-monoxyheme Proteins', *J. Am. Chem. Soc.*, **1991**, *113*, 2447-2459.

20. Jameson, C.J. 'Theoretical and Physical Aspects of Nuclear Shielding' in A Specialist Periodical Report: Nuclear Magnetic Resonance, **1992**, *21*, 36-69, G.A. Webb (ed.).

IGLO CALCULATIONS OF ^{29}Si NMR CHEMICAL SHIFT ANISOTROPIES IN SILICATE MODELS

R. WOLFF[*], C. VOGEL, R. RADEGLIA[*]
Analytical Center Berlin
Rudower Chaussee 5
O-1199 Berlin-Adlershof
Germany

ABSTRACT. For monosilicate anions with threefold symmetry, the silicon shielding tensor components have been calculated in dependence on the SiO bond lengths and the OSiO bond angles. The parallel shielding component linearly correlates with the bond length of the SiO bond in the same direction (symmetry axis direction) as observed by experiments. The analysis of the localized orbital contributions to the change of the shielding by geometry variations shows that this is an indirect effect. The change of the parallel shielding component is mainly due to the three other SiO bonds, whereas the SiO bond in the symmetry axis direction contributes to the change of the perpendicular shielding component.

1. INTRODUCTION

For many years, the structure of silicates has been a field of research using X ray or neutron diffraction, ^{29}Si-NMR spectroscopy, and quantum chemical calculations. In this paper, we will use nonempirical quantum chemical calculations of the ^{29}Si NMR shielding tensor components of the silicate models SiO_4^{4-} in C_{3v} symmetry to understand a well established relation between the parallel component of the shielding tensor $\sigma_{\parallel}$ in certain silicates and the SiO bond

[*]Present address: Federal Institute for Materials Research and Testing (BAM), Branch Office Adlershof,
Rudower Chaussee 5, O-1199 Berlin-Adlershof, Germany

J. A. Tossell (ed.), Nuclear Magnetic Shieldings and Molecular Structure, 385–399.

length r(Si-O) in the same direction. Grimmer et al. [1,2] observed this relation for a number of silicates, the SiO_4 building units of which having C_{3V} symmetry. In Fig. 1

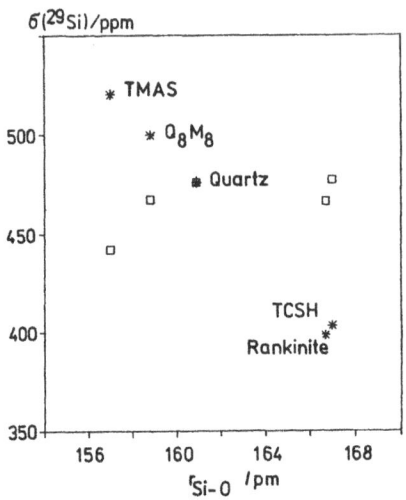

Figure 1. ^{29}Si NMR Shielding tensor components for the SiO_4 building units in silicates by Grimmer et al. [1,2]; parallel components $\sigma_{\parallel}$: * , perpendicular components $\sigma_{\perp}$: □

TABLE 1. Experimental data for the silicates in Fig. 1 from [1,2] and the references cited therein. The shift components $\sigma_{\parallel}$ and $\sigma_{\perp}$ from [1,2] have been related to σ^{TMS} = 368.5 ppm from [3] to get absolute values of the shielding.

Compounds	$r(SiO_a)$ /pm	$r(SiO_b)$ /pm	$\sphericalangle O_a SiO_b$ /grad	$\sigma_{\parallel}$ /ppm	$\sigma_{\perp}$ /ppm	$\Delta\sigma$ /ppm
TMAS	157	160	111	520.5	441.5	79
Q_8M_8	158.8	160	109.9	499.8	467.0	32.8
Quartz	160.9	160.9	109.4	475.9	475.9	0
Rankinite	166.7	160	106.5	398.8	466.3	-67.5
TCSH	167	161	104	403.5	477.5	-74

and Tab. 1 the silicates rankinite $Ca_3Si_2O_7$, tricalciumsi-licatehydrate $Ca_6[Si_2O_7(OH)_6]$ (TCSH), low-quartz SiO_2, and tetramethylammoniumsilicate $[N(CH_3)_4]_8Si_8O_{20}$ (TMAS) have been selected to show this relation. The siloxane $[(CH_3)_3Si]_8Si_8O_{20}$ (Q_8M_8) with the same double four ring structure like TMAS fits into this relation, too. The pertinent perpendicular shielding tensor components $\overline{\sigma}_\perp$ do not correlate as well with the same SiO bond lengths. They are less sensitive to the bond lengths variation than the parallel components.

Though well documented by a number of measurements, these relations are not easy to understand. From the theoretical point of view one should expect, that a bond length variation in a certain direction would be connected with a change of the shielding tensor component perpendicular to this direction. The parallel component should stay unaffected.

To elucidate the relation between the silicon shiel-ding tensor components and the crystal geometry of the silicates we made IGLO (Individual Gauge for Localized Orbitals) calculations introduced by Kutzelnigg et al. [4]. IGLO results are especially convenient in this case. According to the use of localized molecular orbitals the shielding tensor components are calculated as a sum of the shielding contributions of the different SiO bonds, of the lone pairs and of the inner shells. Therefore, the influen-ces of the different SiO bonds of the SiO_4 building unit to the components of the shielding tensor can be studied.

Bringing together the results of crystal structure analysis and the ^{29}Si NMR results using quantum chemical calculations, we hope to get more insight into the back-ground of the empirical relations.

2. MODELS AND METHODS

The SiO_4 building unit of silicates has been modelled by SiO_4^{4-} in C_{3v} symmetry with the SiO_a bond along the symmetry axis and three symmetric SiO_b bonds (Fig.2). For comparison, the $Si(OH)_4$ model was taken into account. The starting geometry has tetrahedral angles at the silicon atoms, $r(Si-O) = 160$ pm, $r(O-H) = 94$ pm, and SiOH $=118.8°$. The distortion of the SiO_4 tetrahedron to get the C_{3v} symmetry of the models has been accomplished by geometry

variations:
1) $r(Si-O_a)$ has been changed,
   $r(Si-O_b) = 160$ pm and $\sphericalangle O_a SiO_b = 109.47°$ remain constant.
2) The three angles $\sphericalangle O_a SiO_b$ have been changed by the same
   value, $r(Si-O_a) = r(Si-O_b) = 160$ pm remain constant.

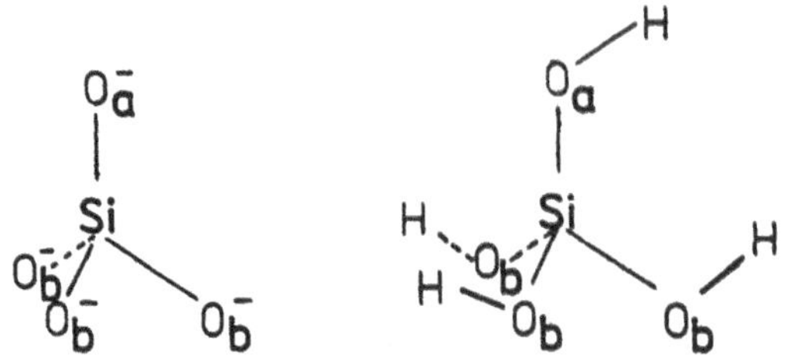

Figure 2. $SiO_4{}^{4-}$ and $Si(OH)_4$ models of the SiO_4 silicate
building units

The components of the shielding tensors have been calcula-
ted using the IGLO program [4] with the original basis set
II which was already successfully used for the calculation
of silicon shielding constants [5]. The basis II is a
Huzinaga basis set with additional polarization functions:
Si : (11s,7p) contracted to [5,6x1;2,5x1] with 2 d sets (η_d
     = 0.35; 1.4)
O : (9s,5p) contracted to [5,4x1;2,3x1] with 1 d set (η_d
     = 1.0)
O⁻ : as O with 2 diffuse functions (η_s = 0.09; η_p = 0.07)
H : (5s) contracted to [3,1,1] with 1 p set (η_p = 0,65)
     The calculations were carried out using a workstation
IBM RISC/6000.

3. RESULTS AND DISCUSSION

3.1. The [29]Si shielding constants of the high symmetric
starting models

For $SiO_4{}^{4-}$ and $Si(OH)_4$ the [29]Si shielding constants with
their LO contributions are collected in Tab. 2.

TABLE 2. The ^{29}Si shielding constants σ in ppm of the starting models and their localized orbitals (LO) contributions

	SiO_4^{4-}	$Si(OH)_4$
K	496.12	496.12
L	268.04	267.98
4 SiO bonds	-342.46	-334.59
all oxygen lone pairs	19.65	21.94
4 OH bonds	-	-4.42
$\sigma(^{29}Si)$	441.35	447.03

The shielding contributions of the Si K shell are practically the same in all cases with fourfold coordinated silicon atoms investigated upto now. The shielding contributions of the Si L shell in the two models do not depend on the termination of the silicate building unit SiO_4, too. The contributions of the inner shells and of the non-bonding oxygen lone pairs are diamagnetic. The shielding contributions of the SiO and OH bonds are paramagnetic. These bonding parts of the valence shell are most sensitive to the models. But as a whole, the results of both models are comparable to each other. $\sigma(^{29}Si)$ for SiO_4^{4-} compares quite well to the experimental value of 439.5 ppm of the same species in the absolute ^{29}Si shielding scale [3].

The starting model SiO_4^{4-} has no ^{29}Si shielding anisotropy (comparably to quartz in Fig.1). In contrast, the starting model $Si(OH)_4$ (cf. Fig. 2) shows a ^{29}Si chemical shielding anisotropy. It compares not as well to the geometric situation we are interested in, and the analysis of its LO contributions is not as clear as in the SiO_4^{4-} model. Therefore, in the following only the features of the shielding tensor components for the geometrically changed SiO_4^{4-} model will be used to discuss.

3.2 The change of the SiO_a bond length in SiO_4^{4-}

To separate the geometric influences to the ^{29}Si shielding tensor components from each other firstly the length of the SiO_a bond in the symmetry axis direction has been changed. At the same time the tetrahedral angles at the silicon atom and the SiO$_b$ bond lengths of 160 pm have been kept constant. The influences of the bond angles at the silicon atom will be considered in the next section. The bond lengths of 160 pm of the non-axial bonds approximately correspond to the structures of the selected silicates (Tab. 1).

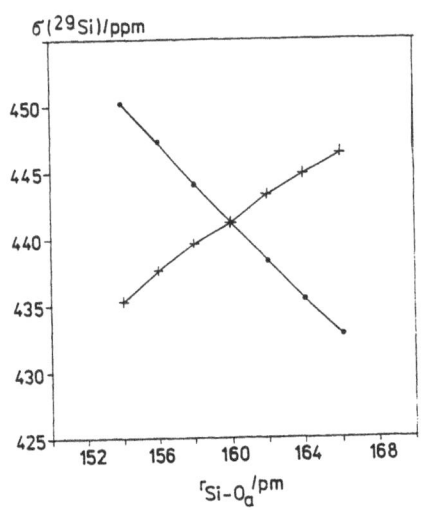

FIGURE 3. The ^{29}Si shielding tensor components $\sigma_\parallel$ ($\bullet$) and $\sigma_\perp$($+$) in dependence on the SiO$_a$ bond length by IGLO calculations

The results of the IGLO calculations in Fig. 3 confirm the experimental results in Fig. 1. There is found a good correlation between $\sigma_\parallel$ and r(Si–O$_a$). With increasing axial bond length the parallel component decreases and the perpendicular component $\sigma_\perp$ increases. The difference of the sensitivities of both is less distinctive compared with the experimental results.

To find an explanation for this result we looked to the LO contributions to the ^{29}Si shielding components. The shielding contributions of the four SiO bonds are decisive for the change of the ^{29}Si shielding anisotropy in dependence on the SiO$_a$ bond length. A smaller part comes from the L shell, whereas the oxygen lone pairs do not contribute to the change of the anisotropy. The influences of the SiO_a bond and of the three SiO_b bonds to $\sigma_\parallel$ and $\sigma_\perp$ have been compared in Fig.4. The changed SiO_a bond does not contribute to the change of $\sigma_\parallel$. It only contributes to the change of $\sigma_\perp$. This is the result we expected (vide supra). The change of $\sigma_\parallel$ only comes from the three SiO_b bonds which remained geometrically unchanged. The SiO_b bonds also contribute to the change of $\sigma_\perp$ diminishing the increase of $\sigma_\perp$ by the SiO_a bond with increasing SiO_a bond length. In summary, the change of $\sigma_\parallel$ by the SiO_a bond length variation is only caused by the SiO_b bonds, whereas all the SiO bonds contribute to the change of the $\sigma_\perp$ component.

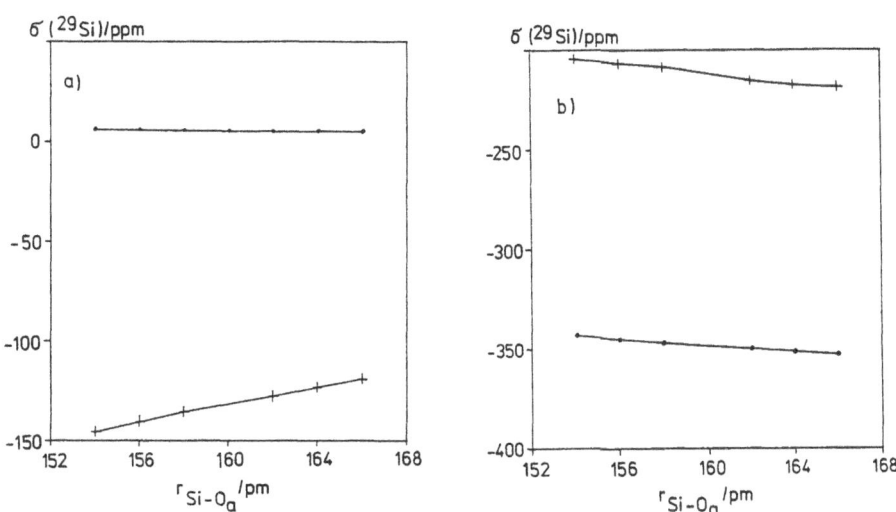

FIGURE 4. Localized shielding contributions of the SiO bonds to $\sigma_\parallel$ (•) and $\sigma_\perp$ (+) in dependence on the SiO_a bond length:
a) contributions of the SiO_a bond
b) contributions of the three SiO_b bonds

How could be understood the decisive role of the rigid part of the model? The IGLO results give the gauge origins of the localized molecular orbitals which are at the same time the centroids of charges of the localized MOs. So, one special feature of the electron distribution in the vicinity of each SiO bond is given. Fig. 5 shows a part of the SiO_4^{4-} model. The points are the coordinates of the atoms ($r(Si-O_a)$ from 154 pm to 166 pm), the crosses show the respective positions of the centroids of charges. In

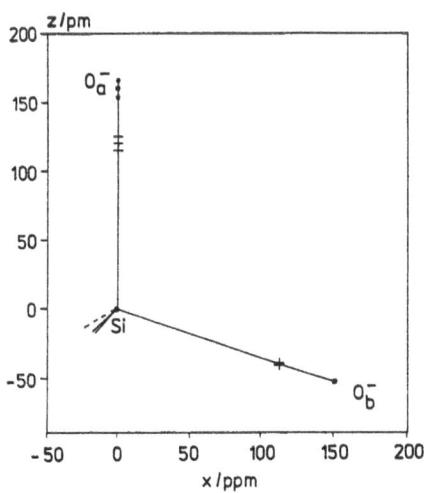

FIGURE 5. Part of the SiO_4^{4-} model which shows the SiO_a bond length variation (• positions of the atoms) and the induced changes of the gauge origins resp. of the centroids of charges (+) of the localized MOs at the SiO_a and SiO_b bonds

the starting model they are situated on the straight bonds near to the oxygen atoms and characterize the polarity of the SiO bond. With increasing SiO_a bond length the position of the centroids of charges of the same bond goes to higher z values. The position of the centroids of charges at the SiO_b bonds are also changed to higher z values but, in this case, not along the bonds. A change of 12 pm in the SiO_a bond length corresponds to a change of 8 pm of the positions of the centroid of charge at the SiO_a bond and of 1.5 pm at each SiO_b bond. This picture should demonstrate that a change of only one bond length influences the electron distribution in the whole model, even in its rigid parts. It should declare why from this parts could come contributions to the silicon shielding components. It shows that, with increasing SiO_a bond lengths, the O_aSiO_b angles have a tendency to decrease by changes of the centroids of charges. The influences of these angles to the [29]Si shielding components will be studied in the next section.

3.3 The change of the O_aSiO_b bond angles in SiO_4^{4-}

In the silicates in Tab. 1 the increasing SiO_a bond lengths are connected with decreasing adjacent O_aSiO_b bond angles and vice versa. In comparison to the influences of the changed SiO_a bond length on the parallel and perpendicular [29]Si shielding tensor components we will look to the influences of the changed O_aSiO_b bond angles to the change of the shielding components using IGLO calculations in

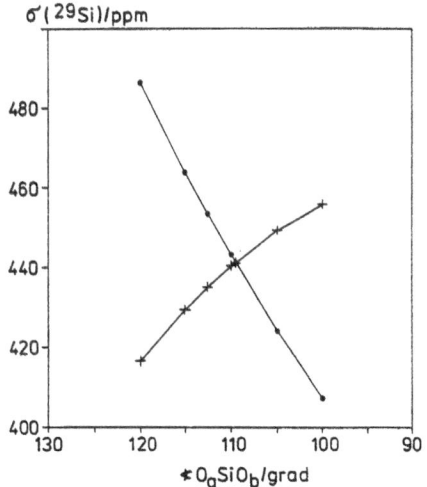

Figure 6. The [29]Si shielding tensor components $\sigma_{\parallel}$ (•) and $\sigma_{\perp}$(+) in dependence on the O_aSiO_b bond angle variations by IGLO calculations

394

Fig. 6. Increasing SiO$_a$ bond lengths at the x axis in Fig.3 correspond to decreasing O$_a$SiO$_b$ bond angles at the x axis in Fig. 6 to compare the results. The changes of $\sigma_{\parallel}$ and $\sigma_{\perp}$ show a similar pattern compared to the results of the bond lengths variation. The sensitivity of the parallel component to the change of the angles is more pronounced and nearer to the experimental pattern. Both influences from the changes of the bond length and of the bond angles complement each other. Both influences contribute to the observed relation of the experimental shielding components and the bond lengths in Fig. 1.

In this case, it is not as easy to express any expectation about the contributions of the different bonds to the shielding components. The change of the O$_a$SiO$_b$ bond angles should affect both adjacent bonds, the bond lengths of which were remained constant. In fact, the IGLO results

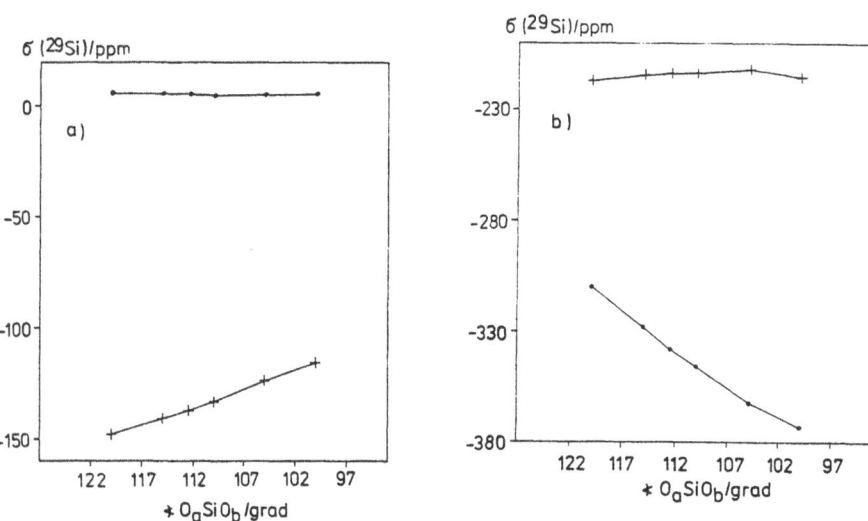

Figure 7. Localized shielding contributions of the SiO bonds to $\sigma_{\parallel}$ ($\bullet$) and $\sigma_{\perp}$ (+) in dependence on the variation of the O$_a$SiO$_b$ angles:
a) contributions of the SiO$_a$ bond
b) contributions of the SiO$_b$ bonds

in Fig. 7 show that the shielding contributions of the
bonds due to the bond angle variation are quite similar to
that due to the bond length variation. The SiO_a bond
contributes to the perpendicular component $\sigma_\perp$, and only the
three SiO_b bonds contribute to the parallel component $\sigma_\parallel$.

The common result of both bond length and bond angle
variations shows that the SiO_a bond in the symmetry axis
direction does not contribute to the parallel component $\sigma_\parallel$,
though the bond length of the same bond correlates very
well with $\sigma_\parallel$. There are rather the three non-axial SiO_b
bonds which alone contribute to $\sigma_\parallel$. The perpendicular
component of the shielding tensor is mainly due to the SiO_a
bond.

Looking to the position of the centroids of charges in
Fig.8 it can be seen that the centroids of charges (cros-

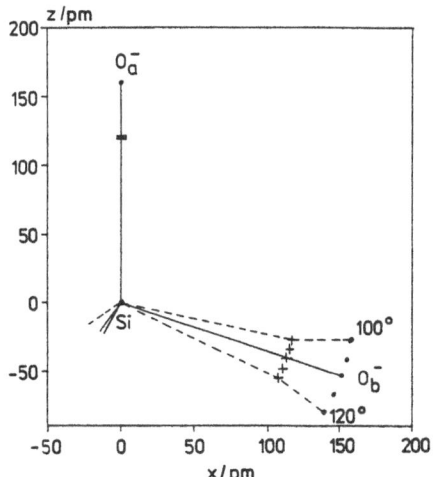

Figure 8. Part of the SiO_4^{4-} model which shows the O_aSiO_b
bond angle variation (● positions of the atoms) and the
induced changes of the gauge origins resp. the centroids of
charges (+) of the localized MOs at the SiO_a and SiO_b bonds

ses) change their position at the SiO_b bonds according to the bond angle variation. They are no more situated on the straight line between the silicon and oxygen atoms except in the case of the starting model. At the SiO_a bond a change of the position of the centroids of charges (change of the polariy) has been induced, too, though this bond was kept constant.

The change of the centroids of charges in geometrically unchanged parts of the model should be regarded as a hint upon the variation of the charge distribution in the whole model even if only one part of it is changed. Comparing the changes of the centroids of charges at the SiO_a bonds in Fig. 5 and Fig. 8 and taking into account that the changes of the perpendicular components $\sigma_\perp$ due to the SiO_a bond is even stronger in the case of the bond angle variation (Fig. 7) than in the case of bond length variation (Fig. 4) it is evident that the change of the bond polarity cannot be the only reason for the change of the shielding contribution. There must be other factors which influence the chemical shielding, e.g. a variation of the relation of s, p and d orbitals occupation or of the excitation energies. Unfortunately, these influences cannot be easily extracted from the results of the IGLO calculations.

3.4 Comparison of theoretical and experimental ^{29}Si NMR chemical shielding anisotropies

A direct comparison of experimental and theoretical results is possible using the anisotropy $\Delta\sigma = \sigma_\parallel - \sigma_\perp$. Fig. 9 shows the anisotropy of the regarded silicates of Tab.1 first over the SiO_a bond length and second over the O_aSiO_b angle. Additionally, the theoretical results are shown. For the comparison one has to take into account that the theoretical values contain the dependence of only one geometric parameter while the experimental values contain the complex geometry. The sign of the theoretical anisotropies has been correctly described but the actual values are too small, also, if one takes into account that the bond length and bond angle influences intensify each other. Probably, this may be a shortcoming of the very restricted model. Rankinite and TCSH have one bridging oxygen atom in the SiO_4 building unit to another silicon atom. In this case the theoretical values are nearer to the experimental ones compared with Q_8M_8 and TMAS with four resp. three bridging oxygen atoms to other silicon atoms. Therefore, a larger geometric fragment around the silicon atom in question has to be taken into account, if one has an interest in a more precisely calculation of ^{29}Si chemical shielding anisotropies of certain silicates. Tossell [6] pointed out to this shortcoming of the SiO_4^{4-} model, too. Nevertheless, the model SiO_4^{4-} has been shown to be very useful for the elucidation of the role of the different SiO bonds in the silicate building unit and their influences to the ^{29}Si shielding tensor components.

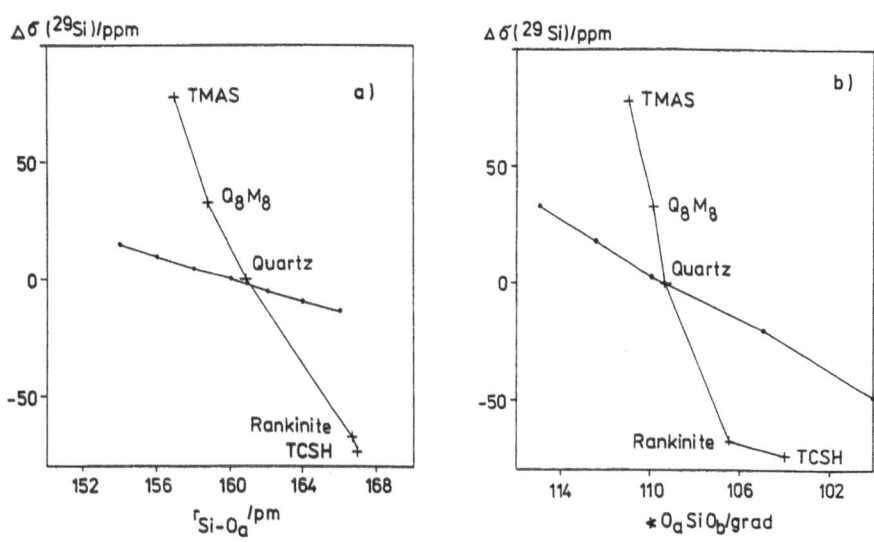

Figure 9. The shielding anisotropy $\Delta\sigma = \sigma_{\parallel} - \sigma_{\perp}$ by experiments (+) and by the IGLO calculations (•)
a) in dependence on SiO_a bond length variation
b) in dependence on the O_aSiO_b bond angle variations

4. CONCLUSIONS

The results of the nonempirical IGLO calculations of the components of the ^{29}Si nuclear shielding tensor of the SiO_4^{4-} model in C3V symmetry agree with the observed correlation between the parallel component of the shielding tensor and the bond length of the axial SiO bond.

The quantum chemical consideration gives the possibility to separate different influences: the bond length and bond angle influences by the model and the molecular orbital contributions to the shielding by the IGLO method. Therefore, the following conclusions can be drawn:

  (i) The axial SiO bond in the symmetry axis direction does not contribute to the change of the parallel component of the shielding tensor though it has different bond lengths in different silicates.

 (ii) The change of the parallel component is only determined by the other three non-axial bonds which remain geometrically unchanged.

(iii) The good correlation between the bond lengths in the symmetry axis direction with the parallel shielding component comes from indirect effects: the bond lengths variation of one bond firstly influences the electron distribution near the other bonds and, secondly, it is connected with a change of the adjacent bond angles. Due to these effects, the other three bonds can contribute to the change of the parallel component. These contributions correlate to the bond length of the axial bond.

5. REFERENCES

[1] Grimmer, A.-R., Peter, R. Fechner, E. and Molgedey, G. (1981) 'High-resolution ^{29}Si NMR in solid silicates. Correlations between shielding tensor and Si-O bond length', Chemical Physics Letters, 77, 331-335

[2] Grimmer, A.-R. (1985) 'Correlation between individual Si-O bond lengths and the principal values of the ^{29}Si chemical shift tensor in solid silicates', Chemical Physics Letters, 119, 416-420

[3] Jameson, C.J. and Jameson, A. K. (1988), 'Absolute shielding scale for ^{29}Si', Chemical Physics Letters, 149,300-305

[4] Kutzelnigg, W., Fleischer, U. and Schindler, M. (1990) 'The IGLO-method: ab-initio calculation and interpretation of NMR chemical shifts and magnetic susceptibilities', in Diehl, P., Fluck, E., Günther, H., Kosfeld, R., Seelig, J. (eds.), NMR Basic Principles

and Progress, Springer-Verlag, Berlin-Heidelberg, 23, 167-262

[5]   Fleischer, U., Schindler, M. and W. Kutzelnigg (1987) 'Magnetic properties in terms of localized quantities. VI. Small hydrides, fluorides, and homonuclear molecules of phosphorus and silicon', The Journal of Chemical Physics, 86, 6337-6347

[6]   Tossell, J. A. (1992) 'Calculation of the ^{29}Si NMR shielding tensor in forsterite', Physics and Chemistry of Minerals, in press

ACKNOWLEDGEMENTS

The authors thank W. Kutzelnigg and U. Fleischer for the release of the IGLO program and for the help in using it. The supports of the Deutsche Forschungsgemeinschaft and the Fonds der Chemischen Industrie are gratefully acknowledged.

EFFECTS OF ISOTOPIC SUBSTITUTION AND TEMPERATURE ON NUCLEAR MAGNETIC SHIELDING

W. T. RAYNES
Dept. of Chemistry
The University of Sheffield
Sheffield   S3 7HF
United Kingdom

ABSTRACT. The effects of isotopic substitution and temperature on nuclear magnetic shielding are related to the intermolecular shielding surface and the intermolecular shielding function which determine them. The pair interaction model is examined in some detail. An improved expression for the electric field contribution is presented as is an up-to-date table of _ab initio_ values of the electric field coefficients (the shielding polarizabilities and hyperpolarizabilities). Calculations of the temperature dependences of the second shielding virial coefficients of the proton and carbon-13 shielding in methane confirm that existing theory is inadequate. The theory of isotope effects and temperature on nuclear shielding at zero density is presented and some recent work is briefly discussed.

1. Introduction

Isotope effects on nuclear shielding are normally very easy to measure. There is now a vast abundance of experimental data available covering many nuclei. Although the physical principles involved are well understood, their application to more than just a few diatomic and small polyatomic molecules faces what at present are insuperable obstacles. Isotope effects originate from a phenomenon which is entirely intramolecular (but see later) - 'the nuclear shielding surface' i.e. the way in which the shielding of a chosen nucleus in a free molecule changes with the internal geometry of the molecule.

The effects of temperature are rather more difficult to measure not least because of the temperature dependence of the NMR reference. Two distinct phenomena contribute to the temperature dependence of nuclear shielding - one is that same nuclear shielding surface which determines the isotope effect whilst the other is entirely intermolecular i.e. the way in which the nuclear shielding changes with the approach of other molecules. In the case of gases, upon which we concentrate sole attention, the two phenomena are easily separated by plotting the observed chemical shift δ against the molar volume V_m of the gas. Thus [1]

$$\delta = \delta_0 + \frac{\sigma_1}{V_m} + \frac{\sigma_2}{V_m^2} + \dots \quad , \tag{1}$$

where $\delta_0 = \sigma_0 - \sigma_{ref}$ is the difference in shielding between the nucleus of interest at zero

J. A. Tossell (ed.), Nuclear Magnetic Shieldings and Molecular Structure, 401–420.
© 1993 Kluwer Academic Publishers.

density and that of the reference nucleus and σ_1, σ_2 etc. are the so-called NMR virial shielding coefficients. δ_0 (through both σ_0 and σ_{ref}) is temperature dependent as are σ_1, σ_2 etc. The procedure one generally follows is to plot δ for a chosen nucleus in a pure gas against V_m^{-1} and then repeat over a range of temperatures. Apart from a few cases [2] one finds that each plot is linear leading to values of δ_0 and σ_1 over the chosen range of temperature. In the case of a mixture of two gases labelled 1 and 2 one has a more general version of eq. (1), viz

$$\delta = \delta_0 + \frac{1}{V_m} \left\{ x\sigma_1^{(11)} + (1-x)\sigma_1^{(12)} \right\} + .. \quad , \tag{2}$$

where x is the mole fraction of gas 1 which contains the nucleus on interest, $\sigma_1^{(11)}$ refers to the effect of intermolecular interactions involving only molecules of type 1 whilst $\sigma_1^{(12)}$ refers to intermolecular interactions involving a molecule of type 1 and a molecule of type 2. Since $\sigma_1^{(11)}$ can be identified with σ_1 in eq. (1), it can be obtained from a study of gas 1 in its pure form leaving $\sigma_1^{(12)}$ to be determined from the gas mixture using eq (2). $\sigma_1^{(12)}$ will also be temperature dependent.

In the following we deal first with the temperature dependence of $\sigma_1^{(12)}$, of which that of $\sigma_1^{(11)}$ is just a special case. After that we treat the temperature dependence of σ_0 together with isotope effects and their temperature dependence since, as stated above, they have a common origin.

2.   Intermolecular Effects

2.1.   GENERAL

The analysis of σ_1 (we drop the superscript '12') proceeds by surrounding the molecule containing the nucleus of interest with a cavity of say 20Å radius, so that molecules outside the cavity are too far removed individually to interact with the molecule. Molecules inside the cavity do interact individually with the molecule of interest. We then write

$$\sigma_1 = (\sigma_1)_b + (\sigma_1)_{loc} \quad , \tag{3}$$

where $(\sigma_1)_{loc}$ is the 'local' contribution due to the molecules in the cavity. $(\sigma_1)_b$ is the 'bulk-susceptibility effect' arising from the molecules outside the cavity which are collectively magnetized by the applied field of the NMR spectrometer and hence contribute to σ_1. The magnitude of $(\sigma_1)_b$ depends on the shape of the sample tube, the direction of the applied field and the molar magnetic susceptibility χ_m of the gas, viz

$$(\sigma_1)_b = \kappa\chi_m \quad , \tag{4}$$

where κ is a shape factor. The only two important cases here are those of a cylindrically shaped sample tube with the applied field (a) perpendicular to the cylinder axis in which case $\kappa = 2\pi/3$ and (b) parallel to the cylinder axis in which case $\kappa = -4\pi/3$. A serious problem arises in the case of proton shielding where $(\sigma_1)_b$ is normally substantially greater than $(\sigma_1)_{loc}$ in numerical terms. As present methods of measuring susceptibilities are not particularly precise large errors can occur in the $(\sigma_1)_{loc}$ values obtained from eq. (3) after measurement of σ_1. Another problem is that for many compounds susceptibility values are unknown and must be estimated by extrapolation from values of similar compounds.

$(\sigma_1)_{loc}$ is related to the interaction between a pair of molecules by the expression [1]

$$(\sigma_1)_{loc} = \tfrac{1}{2} N \int \sigma_{pair} \exp(-u/kT)\, d\tau \qquad , \qquad (5)$$

where N is the Avogadro constant, u is the intermolecular potential function, σ_{pair} is the change in shielding constant for molecule 1 in the presence of molecule 2 at a fixed mutual orientation and distance symbolised by τ (see Fig. 1) where

$$d\tau = r^2 dr\, \sin\theta_1\, \sin\theta_2\, d\theta_1\, d\theta_2\, d\phi \qquad , \qquad (6)$$

and the integration is taken over all distances and mutual orientations of the two molecules. σ_{pair} is the intermolecular shielding function. The standard approach [1,3] is to divide $(\sigma_1)_{loc}$ into three parts viz. $(\sigma_1)_a$, $(\sigma_1)_E$ and $(\sigma_1)_w$ corresponding respectively to three conceptually distinct parts of σ_{pair} :- the magnetic anisotropy contribution, $(\sigma_{pair})_a$,

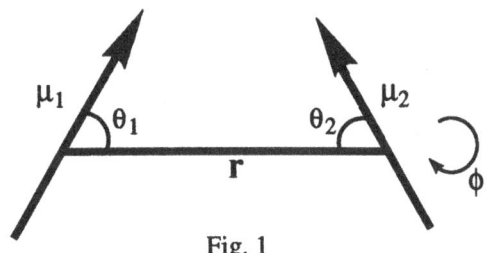

Fig. 1

the polar effect, $(\sigma_{pair})_E$, and the van der Waals contribution, $(\sigma_{pair})_w$. In the following subsections we consider these parts separately. Here we note that the use of Lennard-Jones and Stockmayer potentials in the exponential factor of eq. (5) leads to integrals which involve the functions $H_n(y)$ introduced by Buckingham and Pople [4]. They are defined by

$$H_n(y) = y^{\frac{27-n}{6}} \sum_{k=0}^{\infty} \Gamma\left(\frac{6k+n-3}{12}\right) \frac{y^k}{k!} \qquad , \qquad (7)$$

where

$$y = 2(\varepsilon/kT)^{\frac{1}{2}} \quad , \tag{8}$$

and ε is the well-depth in the Lennard-Jones potential. $\Gamma(x)$ is the gamma function of x. Tables of $H_n(y)$ for $n = 6\text{-}17$ and $y = 0.6$ to 3.2 have been provided by Buckingham and Pople [4]. The model we are using is occasionally referred to as the RBB model.

2.2 THE MAGNETIC ANISOTROPY CONTRIBUTION

Consider an axially symmetric molecule labelled 2 having magnetizability components $\xi_{\|}$ and $\xi_{\perp}$ parallel and perpendicular respectively to the axis of symmetry. Let this axis point along the z-axis (see Fig. 2) and consider the induced magnetic field at the point P at distance r in the yz-plane from the point O where the molecule resides. This field is due to the moment created in the molecule by the applied field B_0 of the NMR spectrometer. When B_0 points along the molecular z-axis, the induced field at P along the negative z-axis, assuming the point dipole approximation, is $-(\mu_0/4\pi)\xi_{\|}B_0(3\cos^2\theta_2-1)r^{-3}$. We use here and subsequently S.I. units. When B_0 points along the y and x-axes the fields along the negative y-axis and x-axis are respectively $-(\mu_0/4\pi)\xi_{\perp}B_0(3\sin^2\theta_2-1)r^{-3}$ and $(\mu_0/4\pi)\xi_{\perp}B_0r^{-3}$. To obtain the mean shielding at P we take an average of these three fields and then divide by B_0 to obtain

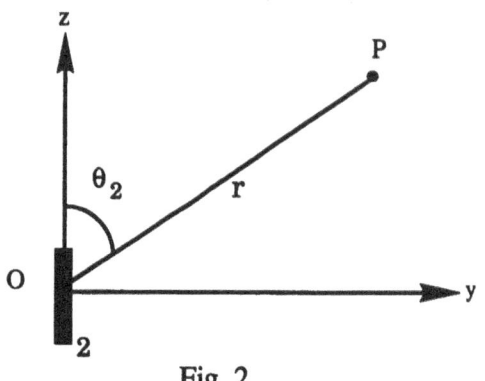

Fig. 2

$$(\sigma_{pair})_a = -\frac{1}{3}(\mu_0/4\pi)(\xi_{\|} - \xi_{\perp})(3\cos^2\theta_2 - 1)r^{-3} \quad . \tag{9}$$

This result is not yet averaged over the molecular motion. As long as molecule 2 can rotate freely $(\sigma_{pair})_a$ is zero. However, as this molecule approaches the molecule of interest certain orientations become more favoured. In the original treatment [1] this was taken into account by the use of a Stockmayer potential

$$u = 4\varepsilon\left\{\left(\frac{r_0}{r}\right)^{12} - \left(\frac{r_0}{r}\right)^6\right\} + (\mu_1\mu_2/4\pi\varepsilon_0)r^{-3}(2\cos\theta_1\cos\theta_2 + \sin\theta_1\sin\theta_2\cos\phi), \tag{10}$$

where μ_1 and μ_2 are the electric dipole moments of the two molecules. This led to the result

$$(\sigma_1)_a = -\frac{N\pi}{1080}\left(\frac{\mu_0}{4\pi}\right)\left(\xi_{\parallel} - \xi_{\perp}\right)\left\{\tau^2 H_9\,(y) + y^4\left(\frac{\tau^4}{70}\right)H_{15}\,(y) +..\right\}, \quad (11)$$

where τ is a dimensionless parameter given by

$$\tau = \mu_1\mu_2/(4\pi\varepsilon_0)\varepsilon r_0^3 \quad . \quad (12)$$

According to eq. (11) for a nonzero magnetic anisotropy contribution to occur it is necessary that both molecules possess electric dipole moments. Thus highly anisotropic perturbers such as CS_2 and C_6H_6 would be predicted to give no $(\sigma_1)_a$. This is clearly unsatisfactory. Some improvement could be achieved by including interactions involving molecular electric quadrupole moments in u. However, it is obvious that it is the Lennard-Jones part of u that needs to be modified to yield more realistic results. Calculations using eq. (11) generally give small values of $(\sigma_1)_a$ so that it is only in proton resonance where other contributions to $(\sigma_1)_{loc}$ are small that $(\sigma_1)_a$ needs to be considered. Even here it is calculated to be small but it seems likely that the use of a more realistic intermolecular potential would lead to somewhat larger values than those calculated hitherto. It is possible that progress along the lines used by Jorgensen [5] could be made.

2.3.   THE ELECTRIC FIELD CONTRIBUTION

When a molecule is placed in a uniform electric field E, a component of the magnetic shielding tensor of a chosen nucleus in the molecule is given by [6]

$$\sigma_{\alpha\beta} = \sigma_{\alpha\beta}^{(0)} + \sigma_{\alpha\beta\gamma}^{(1)}\,E_\gamma + \tfrac{1}{2}\sigma_{\alpha\beta\gamma\delta}^{(2)}\,E_\gamma E_\delta + ... \quad , \quad (13)$$

where standard tensor notation is used. In practice one often wishes to deal with the change in shielding averaged over all directions of the magnetic field but with the electric field held fixed with respect to the molecule. In the original treatment [6], which referred to a proton in an X-H bond, this was given as

$$\Delta\sigma = - AE_z - BE^2 \quad , \quad (14)$$

where the z-axis is taken along the X-H bond from X to H. A and B were regarded as parameters characteristic of the X-H bond. A later treatment [7] showed that the form of the electric field dependence is determined by the nuclear site symmetry. Subsequently, we shall confine our principal attention to nuclei possessing $C_{\infty v}$ and C_{3v} site symmetries. This includes nuclei in a large class of molecules; for example, the proton in the hydrogen halides, all three nuclei in HCN, the carbon, fluorine and nitrogen nuclei in CH_3F, CH_3CN and NH_3 all possess one or other of these site symmetries. Apart from a few very simple molecules (CH_4, C_2H_2 etc.) the protons of X-H bonds do not possess either of these symmetries, although it may be assumed as a first approximation in some cases. For $C_{\infty v}$ and C_{3v} site symmetries the shielding change is given by

$$\Delta\sigma = - A_{\parallel} E_{\parallel} - B_{\parallel} E_{\parallel}^2 - B_{\perp} E_{\perp}^2 - ... , \qquad (15)$$

where the subscripts $\parallel$ and $\perp$ denote directions parallel and perpendicular respectively to the axis defining the C_{3v} site symmetry. Negative signs are used explicitly in eq. (15) to render the parameters $A_{\parallel}$, $B_{\parallel}$ and $B_{\perp}$ positive as, in most cases, this axis is conventionally chosen to point in the direction that leads to a reduction in the shielding. The expressions 'shielding polarizabilities' and 'shielding hyperpolarizabilities' have been introduced for $A_{\parallel}$ and for $B_{\parallel}$ and $B_{\perp}$ respectively [9].

In the early work the parameters A and B had to be treated empirically and obtained by fitting to experimental data. In the near future it is likely that good ab initio values will be available for nuclei in a wide range of simple molecules. This is due to in particular to developments which permit the derivatives of properties to be calculated directly and which employ large basis sets. Results, many of them very recent, are given in Table 1. Many of the results have been obtained by Augspurger and Dykstra [9]. Unfortunately, they have only reported one B value (usually $B_{\parallel}$) for each of the nuclei they considered. In most cases there is also an independent $B_{\perp}$ value as well. For the C and O shieldings in H_2CO, the N shielding in NH_3 and the O shielding in H_2O there are actually three independent B values [7]. For the proton shielding in CH_4 and SiH_4 they use a different axis system from the present one so that with only one B value calculated it is not possible to give their values of $B_{\parallel}$ and $B_{\perp}$. There also appears to be a wrong sign for their A coefficients. In Table 1 the signs of their A values are changed from those they quote after allowing for the different definitions used.

Also given in Table 1 are some previously unpublished results (denoted by the superscript 'g') from the laboratory of the writer [14]. These results were obtained using a program written and developed by P. Lazzeretti; R. Zanasi, P. W. Fowler and E. Steiner. As can be seen they are in generally good agreement with Augspurger and Dykstra's values [9] although there are some differences. The existence of opposite signs for the B values of C in CH_4 and Si in SiH_4 is intriguing. Inspection of the individual components $\sigma_{xxzz}^{(2)}$ and $\sigma_{zzzz}^{(2)}$ shows that the former is positive in both cases (giving increased shielding in an electric field) whilst the latter is negative in both cases; however, the magnitudes involved are such that there is a positive average for CH_4 but a negative average for SiH_4. Upon changing to the axis system of Augspurger and Dykstra [7], excellent agreement is obtained with their B values. Also given in the table are the three independent B values for the C and the O shieldings in H_2CO.

Table 1 lists results in S.I. units. Other authors prefer atomic units (ppm a.u.) or c.g.s. units. The required conversion factors between these units and S.I. units are as follows:-

linear coefficients:

$$1 \text{ ppm a.u.} = 0.019447 \times 10^{-16} \text{ mV}^{-1}$$
$$1 \text{ c.g.s. unit} = 0.3336 \times 10^{-4} \text{ mV}^{-1} \quad ,$$

quadratic coefficients:

$$1 \text{ ppm a.u.} = 0.037789 \times 10^{-28} \text{ m}^2\text{V}^{-2}$$

$$1 \text{ c.g.s. unit} = 0.1113 \times 10^{-8} \text{ m}^2\text{V}^{-2} \quad .$$

In the original application of eq. (14) to gases the field E was regarded as a sum of three fields due to (a) an electric dipole moment in the perturbing molecule, (b) an electric quadrupole moment in the perturbing molecule and (c) an electric dipole moment in the polarizable perturbing molecule induced by the electric dipole moment of the molecule containing the nucleus of interest. For the field due to (c) it was assumed that the electrical polarizability of the perturbing molecule was isotropic. We now relax this assumption and give the perturbing molecule a mean polarizability α_2 and a polarizability anistropy $\Delta\alpha_2 = \alpha_{\parallel} - \alpha_{\perp}$. With this, with eq. (15) instead of eq. (14), and with the Stockmayer potential (eq. (10)) in eq. (5) we obtain [16]

$$(\sigma_1)_E = \sigma_1(E_{\parallel}) + \sigma_1(E_{\parallel}^2) + \sigma_1(E_{\perp}^2) \quad , \tag{16}$$

where

$$
\begin{aligned}
\sigma_1(E_{\parallel}) = -\frac{(\pi N A_{\parallel}/6y^2)}{4\pi\varepsilon_0} \Big[& \mu_2 \left\{ (\tau/3) H_6(y) + (y^4 \tau^3/200) H_{12}(y) + .. \right\} \\
& + (2\mu_1 \Theta_2^2/y^2 kTr_0^5) \left\{ H_8(y) + .. \right\} \\
& + (4\mu_1\alpha_2/r_0^3 y^2) H_6(y) + (\tau^2 y^4/40) H_{12}(y) + .. \right\} \\
& + (8\mu_1\Delta\alpha_2\tau^2 y^2/3r_0^3) \left\{ H_{12}(y) + .. \right\} + \Big] \quad ,
\end{aligned}
\tag{17}
$$

$$
\begin{aligned}
\sigma_1(E_{\parallel}^2) = -\frac{(\pi N B_{\parallel}/y^2 r_0^3)}{(4\pi\varepsilon_0)^2} \Big[& (\mu_2^2/y^2) \left\{ \tfrac{2}{9} H_6(y) + (\tau^2 y^4/100) H_{12}(y) + .. \right\} \\
& + (\Theta_2^2/r_0^2 y^2) \left\{ \tfrac{1}{3} H_8(y) + (103\tau^2 y^4/1400) H_{14}(y) + .. \right\} \\
& + (\mu_1\mu_2/r_0^3) \left\{ \tfrac{4}{15}\tau\alpha_2 + 32\tau\Delta\alpha_2/675) H_{12}(y) + .. \right\} \\
& + .. \Big]
\end{aligned}
\tag{18}
$$

Table 1. Calculated values of the coefficients in eq. (15). For ^{1}H and ^{19}F shielding the sign of $A_\parallel$ assumes that the electric field is directed along the X-H and X-F bonds from left to right. For the remaining nuclei the field direction for the given sign of $A_\parallel$ is indicated in the footnote.

Compound	Nucleus	$A_\parallel$/10^{-16} mV^{-1}	$B_\parallel$/10^{-28} m^{2}V^{-2}	$B_\perp$/10^{-28} m^{2}V^{-2}
H atom	H	—	8.202[f]	8.202[f]
H_2	H	0.974[a], 0.978[b,] 1.12[h]	2.529[a], 3.546[b]	3.350[a],
HF	H	1.50[c,h], 1.39[d]	−6.46[c,] −5.6[d]	0.79[c], 0.5[d]
		1.52[e], 1.58[b]	−5.68[e], −6.21[b]	1.60[e]
HCN	H	1.05[b],1.06[g]	3.27[b], 2.50[g]	1.02[g]
HCCH	H	1.31[b], 1.36[g]	0.26[b], −7.22[g]	2.70[g]
CH_4	H	1.52[b], 1.55[g]	0.29[g]	3.04[g]
SiH_4	H	1.21[b], 1.20[g]	7.51[g]	
HCl	H	2.29[g], 2.67[h]	−6.50[g]	3.37[g]
HBr	H	3.05[h]		
CH_4	C	—	5.10[g], 6.80[b]	5.10[g], 6.80[b]
CH_3F	C	3.59[e]	48.7[e]	59.4[e]
CO	C	−7.28[b]	10.12[b]	
HCN	C	8.22[b], 8.34[g]	−28.40[b], −28.14[g]	10.99[g]
HCCH	C	14.27[b], 14.48[g]	20.91[b], 21.80[g]	4.40[g]
H_2CO	C	−13.56[b], −14.95[g]	−73.21[b], −74.73[g]	2.25[g] ($B_\perp' = -31.84$[g])
HCN	N	37.15[b]	214.19[b]	
NH_3	N	0.99[b]	−30.48[b]	
CO	O	29.69[b]	111.6[b]	
H_2O	O	7.80[b]	81.99[b]	
H_2CO	O	136.50[b], 127.48[g]	1338.5[b], 904.33[g]	229.6[g] ($B_\perp' = 573.4$[g])
HF	F	11.77[e], 12.2[c]	143.2[e], 168.7[c]	20.34[e], 38.5[c]
		13.7[d]	168.0[d]	25.7[d]
CH_3F	F	15.4[e]	287.9[e]	106.9[e]
SiH_4	Si	—	−131.1[b], −124.3[b]	−131.1[b], −124.3[b]

Field directions:- CO: from C to O for the carbon shielding and for the oxygen shielding; HCN: from C to N for both C and N shielding; HCCH: from C_1 to C_2 for the C_2 shielding; H_2CO: from C to O for both C and O shielding, $B_\perp'$ is perp. to molecular plane; NH_3: along C_3 axis from plane of protons to N; H_2O: along C_2 axis from midpoint of protons to O; CH_3F: from C to F for the C shielding.

References: [a]Sadlej and Raynes [8]; [b]Augspurger and Dykstra [9]; [c]Day and Buckingham [10]; [d]Zaucer and Azman [11]; [e]Packer and Raynes [12]; [f]Marshall and Pople [13] - see [8]; [g]Grayson and Raynes [14]; [h]Volodicheva and Rebane [15].

and

$$\sigma_1 (E_\perp^2) = - \frac{(\pi N B_\perp / y^4)}{(4\pi\varepsilon_0)^2} \left[(\mu_2^2/r_0^3) \left\{ \tfrac{4}{9} H_6 (y) + (\tau^2 y^4/150) H_{12} (y) + .. \right\} \right.$$
$$+ (\Theta_2^2/r_0^5) \left\{ \tfrac{2}{3} H_8 (y) + 929 \ \tau^2 y^4/16800) H_{14} (y) + .. \right\}$$
$$\left. + (\mu_1 \mu_2/r_0^6) \left\{ (\tfrac{1}{15} \tau y^2 \alpha_2 + \tfrac{11}{225} \tau y^2 \Delta\alpha_2) H_{12} (y) + .. \right\} + .. \right] \quad . \quad (19)$$

Eq. (16) reverts to the original form of $(\sigma_1)_E$ if one writes $\Delta\alpha_2 = 0$ and $B_\parallel = B_\perp$ in eqs. (17)-(19) and changes back to c.g.s. units by removing the factors involving $4\pi\varepsilon_0$.

Table 2 presents the results of calculations to determine the contributions of the terms in eq. (17) to $\sigma_1(E_\parallel)$. The molecule under study is HCl for which the following parameters were used:- $\mu_1 = 3.443 \times 10^{-30}$ Cm (=1.034D); $\Theta = 12.92 \times 10^{-40}$ Cm^2 [17]; $\alpha = 2.63 \times 10^{-30}$ m^3, $\varepsilon/k = 218.0$K and $r_0 = 3.506 \times 10^{-10}$ m. It was assumed that $\Delta\alpha = \alpha/5$.

The value of A taken from Table 1 was 2.29×10^{-16} mV^{-1}. This was substantially less than the earlier empirical value [1] of 13.34×10^{-16} mV^{-1} (40×10^{-12} c.g.s. units). Values of $(\sigma_1)_E$ were calculated for temperatures of 200K and 300K.

Only the linear terms need be considered as the contributions of quadratic electric fields are negligible for HCl and for proton shielding in other polar molecules. As the table shows the polarizability anisotropy contribution is not negligible being 1.1 in a total 30.0×10^{-6} ppm m^3 mol^{-1} at 300K rising to 3.1 in a total 56.5×10^{-6} ppm m^3 mol^{-1} at 200K. In particular the polarizability anisotropy contributes almost one quarter of the dipole-induced dipole contribution at 300K and almost one half at 200K. For a molecule with similar parameters but with $\Delta\alpha = \alpha/3$ the polarizability anisotropy contribution exceeds that of the mean polarizability at 200K. The experimental value of $(\sigma_1)_{loc}$ for HCl gas at 300K is -104×10^{-6} ppm m^3 mol^{-1} [1] and therefore about 30% of this is contributed by $(\sigma_1)_E$ and 70% by $(\sigma_1)_W$ at this temperature.

The model we use for $(\sigma_{pair})_E$ assumes that the electric field is uniform across the molecule. This cannot be strictly true, of course, and it has been suggested that electric field gradients are important [18]. Coefficients for the dependence of shielding on an electric field gradient have been calculated by Augspurger and Dykstra [9]. The model also does not incorporate the possibility of a site effect even though one has been shown by ^{19}F resonance to occur for polar molecules [19].

Table. 2. Contributions of dipole-dipole, dipole-quadrupole and dipole-induced dipole interactions to $(\sigma_1)_E$ in units of 1×10^{-6} ppm m^3 mol^{-1} for HCl gas at 200K and 300K. Induced contributions due to the mean polarizability α_2 and the polarizability anisotropy $\Delta\alpha_2$ are shown separately.

	200K	300K
dipole-dipole	−23.2	−11.9
dipole-quadrupole	−25.2	−13.3
dipole-induced dipole		
α_2	−5.0	−3.7
$\Delta\alpha_2$	−3.1	−1.1

2.4 THE VAN DER WAALS CONTRIBUTION

The development of a good model for this contribution has proved very much more difficult than for the other two. This is unfortunate since $(\sigma_1)_w$ is always nonzero and is very often the largest or even the sole contributor to $(\sigma_1)_{loc}$. It has been established, however, that $(\sigma_1)_w$ is always negative. In the original treatment [1] only the attractive dispersion forces between molecules were considered. This attraction is attributed to fluctuating electric fields which each of the molecules exerts on the other thereby attracting the electrons of one molecule to the nuclei of the other. Although this field averages to zero, its mean square $\overline{F^2}$ does not. The shielding change of the nucleus of interest due to this field from the perturbing molecule is written

$$(\sigma_{pair})_w = -B \overline{F^2} , \tag{20}$$

where B in eq. (20) is equated with B in eq. (14). A model for $\overline{F^2}$ gives [1]

$$\overline{F^2} = \frac{3\alpha_2 I_2}{\left(4\pi\varepsilon_0\right)^2 r^6} , \tag{21}$$

where I_2 is the ionization energy of the perturbing molecule and r and α_2 are as previously defined. Using the Stockmayer potential one finds

$$(\sigma_1)_w = -\left(\frac{\pi NB}{y^4 r_0^3}\right) \frac{\alpha_2 I_2}{(4\pi\varepsilon_0)^2} \left\{ H_6(y) + (\tau^2/48) y^4 H_{12}(y) + .. \right\} . \quad (22)$$

Early studies of gases involved fitting experimental values of $(\sigma_1)_{loc}$ for a nucleus in a chosen compound in the pure form and in binary mixtures with a variety of gases to the sum of eqs. (16) and (22). In this way values of A and B were obtained empirically. Rummens and Bernstein [20] refined the general model for $(\sigma_1)_w$. They took account of the "site effect" ie. if the chosen nucleus lies close to the periphery of the molecule in which it is contained then the right hand side of eq. (22) must be multiplied by a "site factor" S_6^g given by

$$S_6^g = 1 + 5 q_0^2 H_8(y)/H_6(y) + 14 q_0^4 H_{10}(y)/H_6(y) + .. \quad , \quad (23)$$

where $q_0 = d/r_0$ is dimensionless. d is the distance of the nucleus of interest from the centre of the molecule and r_0 is as defined above. Rummens and Bernstein [20] also tried to allow for the influence of the short-range repulsive forces by postulating that $(\sigma_{pair})_w$ varies as the total interaction energy. Thus

$$(\sigma_{pair})_w = -B \overline{F^2} \left\{ 1 - \left(\frac{r_0}{r}\right)^6 \right\} \quad , \quad (24)$$

so that the dispersion forces give a deshielding term dependent on r^{-6} and the repulsive forces give a shielding term dependent on r^{-12}. They give an involved expression for the inclusion of both the site effect and the repulsive effect which will not be given here.

Over the years a satisfactory test of eq. (22) has been hard to make because good experimental values of $(\sigma_1)_w$ at different temperatures for nonpolar molecules were not available nor were values of B. The position today is rather better and we shall test eq. (22) using some recently reported experimental values for $(\sigma_1)_w$ in methane as a function of temperature together with the new ab initio B values given in Table 1. Methane is a very suitable molecule for study:- it has no magnetic anisotropy nor does it have electric dipole and quadrupole moments so that one may identify $(\sigma_1)_{loc}$ with $(\sigma_1)_w$ and the part of eq. (22) involving τ vanishes. Furthermore, there are two nuclei to be studied. Values of $(\sigma_1)_w$ for these nuclei at a given temperature should differ only because of the difference in B of the two nuclei plus a site factor contribution. The temperature dependence of $(\sigma_1)_w$ for both nuclei resides in the ratio $H_6(y)/y^4$ as the temperature dependence of the site factor is negligible. This means that according to the model the temperature dependences should be the same apart from a numerical factor.

Results are presented in Table 3. Experimental values [21,22] for $(\sigma_1)_w$ for ^1H and ^{13}C resonance over the range 180-380K are given in the second and third columns. The numerical values of $(\sigma_1)_w$ in the columns can be taken to be in the original c.g.s. units of ppm cm^3 mol^{-1}. The ^1H results bear an error of ±3 units but there is a possible constant error due to the uncertainty in the χ_m value used to determine $(\sigma_1)_b$. The ^{13}C results mostly bear an error of ±6 units but for temperatures below 200K the error is ±10 units. In each case the numerical value of $(\sigma_1)_w$ rises as the temperature falls. However, whereas it increases by a factor of about three for the ^1H results over the temperature range, the increase for the ^{13}C results is only about one third. This in itself contradicts eq. (22) which predicts an increase of about one fifth over the range as shown by $H_6(y)/y^4$ in the fourth column of the table.

The theory is also unsatisfactory in reproducing the absolute magnitudes of $(\sigma_1)_w$ for the carbon-13 shielding. In the final columns of Table 3 are given calculated values of $(\sigma_1)_w$ for both ^1H and ^{13}C shielding over the temperature range using the following data: $r_o = 3.817 \times 10^{-10}m$, $\varepsilon/k = 148.2K$, $\alpha_2 = 2.60 \times 10^{-30}m^3$ and $I_2 = 20.8 \times 10^{-5}J$. For the proton shielding a mean B value, $(B_\parallel + 2B_\perp)/3$, of $1.93 \times 10^{-28}m^2V^{-2}$ was used whilst for the carbon-shielding B was taken to be $5.95 \times 10^{-28}m^2V^{-2}$ which is a rotationally averaged mean of the two values given in Table 1. The site factor S_6^g used for the ^1H results is 1.31 and, as stated above, is temperature-independent to this level of accuracy. Although the ^1H results are too small, they are approximately of the right order of magnitude. However, the ^{13}C results are small by much more than an order of magnitude. This stems, of course, from the fact that the calculated B value for the carbon shielding is only three times that for the proton shielding whilst the observed $(\sigma_1)_w$ value are some 20-40 times larger. The unsatisfactory nature of eq. (22) noted here is not new. Jameson and co-workers [2,23] have shown in both ^{129}Xe and ^{19}F resonance that $(\sigma_1)_{loc}$ does not vary with polarizability in the way predicted by this equation.

An examination of the nature of $(\sigma_{pair})_w$ for two interacting xenon atoms was undertaken by Jameson [24] without the assumption of any particular pair interaction model. Here, of course, the intermolecular shielding function can be identified with $(\sigma_{pair})_w$ and there is no angle-dependence. Using a good Xe-Xe intermolecular potential and taking twenty-one experimental values of σ_1 (T) over a range of temperature she obtained a numerical shielding function by inversion for the xenon-xenon interaction. A nonmonotonic dependence on r was found. A rather similar nonmonotonic dependence was found when

Table 3. Observed and calculated data concerning the temperature dependence of $(\sigma_1)_w$ for the ^{1}H and ^{13}C shielding in methane. Values of $(\sigma_1)_w$ are in the units 1×10^{-6} ppm m^3 mol^{-1}.

T/K	$(\sigma_1)_w$ - observed		H$_6$ (y)/y^4	$(\sigma_1)_w$ - calculated	
	[1]H [21]	[13]C[21,22]		[1]H	[13]C
180	-16.6	-314	6.90	-3.0	-6.9
190	-16.6	-312	6.76	-3.0	-6.7
210	-14.0	-275	6.49	-2.8	-6.5
220		(-267)	6.42	-2.7	-6.4
230	-12.1	-265	6.29	-2.7	-6.3
255	-10.3		6.14	-2.7	-6.1
265		-253	6.05	-2.5	-6.0
280	-9.1	-245	6.02	-2.5	-6.0
290	-6.0	-242	5.97	-2.5	-6.0
300	-5.3[a]	(-224)	5.89	-2.5	-5.9
305		-239	5.87	-2.5	-5.9
310	-7.9	-234	5.87	-2.5	-5.9
320	-5.0	-234	5.87	-2.5	-5.9
325			5.86	-2.5	-5.9
345	-4.5	-233	5.77	-2.5	-5.7
370		-230	5.75	-2.5	-5.7
380	-5.9	(-209)	5.74	-2.4	-5.7

[a]Mean of nine independent determinations by other workers - see Table 3 of [21]; [b]results in parentheses are from [22].

experimental values of $d\sigma_1(T)/dT$ were used. It was then attempted to fit this numerical function to several analytical function of the type exp $[-\alpha r^n]$, exp $[-\alpha(r-\beta)^n]$, $[u(r)]^n$ where n = 1 and 2 and α and β are adjustable parameters. Linear combinations of these functions were also tried. Satisfactory agreement was not obtainable with any of them. The best agreement was given when $(\sigma_{pair})_w$ was taken to be proportional to $[u(r)]^2$. However, there is no theoretical reason why $(\sigma_{pair})_w$ should vary as the square of the intermolecular potential.

The existence of a nonmonotonic intermolecular shielding function seems surprising initially but for two atoms which form a chemical bond, as in F_2 and CO, nonmonotonic variations of σ with r have been known for many years [24,25]. It is interesting to note that

Grayce and Harris [26] have recently used electron gas theory to calculate the nuclear magnetic shielding for the $^3\Sigma_u^+$ state of the H_2 molecule. This approximates to interacting closed-shell atoms. They also obtain a nonmonotonic shielding function which resembles that of Jameson discussed above.

Possible direct evidence for an angle-dependent σ_{pair} has been obtained from studies of xenon gas mixtures with nonspherical molecules [27].

A number of ab initio calculations of intermolecular effects on shielding have appeared [28-31] and there are several reviews of the subject [32-35], the most recent of which, however, is in 1980.

3. Intramolecular Effects

3.1. TERMINOLOGY AND NOTATION

We now turn our attention to δ_0 of eq. (1) or more exactly to σ_0 defined below eq. (1). Our interest is in the temperature dependence of σ_0 which, as stated earlier, is due entirely to intramolecular phenomena. Although isotope effects on shielding constants are conceptually quite distinct from those of temperature, both have a common origin - an averaging over the shielding surface. Isotope effects are very much easier to measure than the temperature dependence of the shielding. As a consequence there exists an abundance of experimental data which has been discussed in a number of reviews [36-39].

The word 'isotopomers' is widely used nowadays to denote molecules which differ in isotopic substitution in one or more positions. This is both a more general and a more useful definition than the original [40] which defined isotopomers as molecules which differ in isotopic substitution at one place only. (However, there are still some authors who, erroneously in the writer's opinion, use the word "isotopes" instead of isotopomers or who use specially concocted phrases such as "isotopic molecules" or "isotopic variants"). An isotope shift is then the difference in shielding between corresponding nuclei in two isotopomers of the same compound. These shifts are sometimes referred to as "intrinsic isotope shifts". A primary isotope shift is one involving the shielding difference between the newly substituted nucleus and the one which it replaced - such as, for example, the shielding difference between the deuteron in CH_3D and a proton in CH_4. A secondary isotope shift refers to the shielding difference between corresponding nuclei not involved in the substitution - such as, for example, a proton of CH_3D and a proton of CH_4. This separation into primary and secondary isotope shifts is very useful for classifying and discussing data. However, in the final analysis it is of no special importance since all isotopomers of a compound have the same shielding surface and it is just a question of whether the site of isotopic substitution is at the nucleus of interest or somewhere else.

It is to be noted that since we have defined isotope shifts as shielding differences between corresponding positions in two isotopomers, shielding difference such as that between the protons in CH_3CH_2D and those between the deuteron and the protons in this compound are not strictly isotope shifts even though they are due to isotopic substitution. They could be called 'internal isotope shifts' although this is, perhaps, a rather vague expression. Other classes of isotope shifts are (a) solvent isotope shifts where the resonance signal of a solute is shifted when a solvent is changed from one isotopomer to another and (b) equilibrium isotope shifts where resonance signals are shifted upon isotopic substitution due to an equilibration change. We shall not be concerned with these shifts in this review, but with those referred to above as intrinsic isotope shifts.

The formal definition and notation of isotope shifts used most frequently today was suggested by Gombler [41]. This employs the symbol ${}^{n}\Delta X({}^{M/m}Y)$ which is defined thus:

$$
{}^{n}\Delta X({}^{M/m}Y) = \sigma_x \text{ (for } {}^{m}Y) - \sigma_x \text{ (for } {}^{M}Y) \quad , \tag{25}
$$

where n denotes the number of bonds separating the observed nucleus X from the site of isotopic substitution in the compound Y. M and m refer to the nuclear masses with M>m. Since heavy isotopic substitution in a free molecule normally leads to greater shielding for all nuclei (including the heavy nucleus) than in the original isotopomer, it is clear from eq. (25) that the more positive the isotope shift the less positive the shielding. This definition was chosen to run parallel to the IUPAC recommendations [42] for chemical shifts. In the writer's opinion both definitions are unfortunate since it seems inherently more sensible [43] for positive chemical shifts to accompany increased shielding.

3.2. THE THEORY OF ISOTOPE AND TEMPERATURE EFFECTS

For an isolated molecule with an arbitary geometry not far removed from equilibrium we can expand the shielding as a power series in symmetry coordinates, S_i. Thus, adopting the Born-Oppenheimer approximation,

$$
\sigma = \sigma_e + \sum_i \left(\frac{\partial \sigma}{\partial S_i}\right)_e S_i + \frac{1}{2}\sum_{i,j} \left(\frac{\partial^2 \sigma}{\partial S_i \partial S_j}\right)_e S_i S_j + .. \tag{26}
$$

Each coordinate S_i is a linear combination of either bond length changes or bond angle changes with respect to equilibrium geometry. Standard techniques are available [44] for constructing these coordinates. The essential feature of a symmetry coordinate is that it transforms as one of the symmetry species of the point group of the molecule. Eq. (26) describes the shielding surface in the region near equilibrium and is valid provided that S_i, S_j etc. are not too large. σ_e, the shielding at equilibrium geometry, and the derivatives $(\partial\sigma/\partial S_i)_e$, $(\partial^2\sigma/\partial S_i\partial S_j)_e$ etc. are to be regarded as parameters to be obtained from experimental measurement as described below or from ab initio calculation.

The advantage of using symmetry coordinates in eq. (26) rather than using an alternative expansion involving valence coordinates (bond length changes and bond angle changes) is that the number of nonzero coefficients $(\partial\sigma/\partial S_i)_e$ etc. is reduced to a minimum. Eq. (26) for the shielding surface near equilibrium can be compared with the potential energy surface near equilibrium : the former is obviously more complicated since σ_e will never be zero and at least one of the $(\partial\sigma/\partial S_i)_e$ must be nonzero; in addition if the nuclear site symmetry is lower than that of the point group of the molecule there will be more independent second derivatives $(\partial^2\sigma/\partial S_i\partial S_j)_e$ than there are harmonic force constants. Fowler [45] has given a table specifying the number of independent coefficients for property surfaces involving the various site symmetries.

To proceed further one relates each symmetry coordinate to the various reduced normal coordinates q_r [46]. Thus

$$S_i = \sum_r \overline{L}_i^r q_r + \frac{1}{2}\sum_{r,s}\overline{L}_i^{rs} q_r q_s + .. \quad , \tag{27}$$

The relationship is nonlinear and the coefficients $\overline{L}_i^r$, $\overline{L}_i^{rs}$ etc. are referred to as elements of the $\overline{L}$ tensor. These elements can be calculated from the molecular geometry and the harmonic force constants. It will be noted that since the S_i are mass-independent and the q_i are mass-dependent, the $\overline{L}_i^r$, $\overline{L}_i^{rs}$ etc. must also be mass-dependent. Substituting from eq. (27) into eq. (26) gives the shielding surface in terms of reduced normal coordinates viz

$$\sigma = \sigma_e + \sum_{i,r}\left(\frac{\partial\sigma}{\partial S_i}\right)_e \overline{L}_i^r q_r$$
$$+ \frac{1}{2}\sum_{i,r,s}\left[\left(\frac{\partial\sigma}{\partial S_i}\right)_e\overline{L}_i^{rs} + \sum_j\left(\frac{\partial^2\sigma}{\partial S_i\partial S_j}\right)_e \overline{L}_i^r\overline{L}_j^s\right]q_r q_s +.. \quad , \tag{28}$$

where the respective coefficients multiplying q_r, $q_r q_s$ etc. are obviously the derivatives $(\partial\sigma/\partial q_r)_e$, $(\partial^2\sigma/\partial q_r\partial q_s)_e$ etc. and are isotope-dependent. It will be noted that the first derivative $(\partial\sigma/\partial S_i)_e$ contributes to the second derivative $(\partial^2\sigma/\partial q_r\partial q_s)_e$.

If one now obtains the average for a particular vibrational and rotational state $\langle q_r\rangle_{vJ}$, $\langle q_r q_s\rangle_{vJ}$ etc. one can obtain the value of σ in that state. However, it is normal to proceed directly to the thermal averages $\langle q_r\rangle^T$ and $\langle q_r q_s\rangle^T$ and obtain a value $\langle\sigma\rangle^T$ for the shielding of the chosen isotopomer at temperature T. The thermal average can be calculated from the formulas of Toyama, Oka and Morino [47] viz

$$\langle q_r\rangle^T = -\frac{1}{2\omega_r}\left[\, 3k_{rrr}\coth(hc\omega_r/2kT) + \sum_s{}' d_r k_{rss}(hc\omega_s/2kT)\right.$$
$$\left. -\frac{kT}{2\pi c}\left(\frac{1}{hc\omega_r}\right)^{\frac{1}{2}}\sum_\alpha \frac{a_r^{(\alpha\alpha)}}{I_{\alpha\alpha}^{(e)}}\right] , \tag{29}$$

and

$$<q_r q_s>^T \ = \ <q_r^2>^T \ = \ \frac{d_r}{2} \coth (hc\omega_r /2kT) \qquad . \qquad (30)$$

In eqs. (29) and (30) ω_r and ω_s are frequencies of the vibrational modes r and s, k_{rss} is a cubic anharmonic force constant, $a_r^{(\alpha\alpha)}$ is the derivative of the moment of inertia about the α-axis with respect to the r-th reduced normal coordinate at equilibrium geometry, $I_{\alpha\alpha}^{(e)}$ is the equilibrium moment of inertia about the α-axis and d_r is the degeneracy of mode r. In the summation of the second term of eq. (29) s≠r. The part of $<q_r>^T$ involving the moment of inertia gives the contribution of rotational distortion whilst the remaining part is due to vibrational anharmonicity.

Summarising we can write

$$<\sigma>^T = \sigma_e + \sum_r \left(\frac{\partial\sigma}{\partial q_r}\right)_e <q_r>^T + \frac{1}{2}\sum_{r,s}\left(\frac{\partial^2\sigma}{\partial q_r \partial q_s}\right)_e <q_r q_s>^T + ... \qquad , \qquad (31)$$

where the derivatives $(\partial\sigma/\partial q_r)_e$ and $(\partial^2\sigma/\partial q_r \partial q_s)_e$ are given explicitly in eq. (28). For a particular isotopomer eq. (31) gives the temperature dependence of the shielding. The difference between $<\sigma>^T$ for one isotopomer and $<\sigma'>^T$ for a second isotopomer will then give the isotope shift at T. Since σ_e will then cancel out, the isotope shift is determined primarily by $(\partial\sigma/\partial S_i)_e$. A very detailed treatment of the theory of isotope shifts has been presented by Jameson and Osten [48]. Jameson has also given a general account of the rovibrational-averaging of molecular properties [49].

3.3   PRINCIPAL FEATURES

There are several generalisations concerning isotope shifts, some of which have already been noted above, that have been known for many years [36, 48, 50, 51]:-

(1)   Upon substitution with a heavier isotope the shielding of all nuclei increases.

(2)   The magnitude of the shift diminishes as the number of bonds separating the site of isotopic substitution from the nucleus of interest increases. One-bond isotope shifts are particularly large relative to others.

(3)   The magnitude of the shift depends on the nucleus involved and reflects the chemical shift range of that nucleus.

(4)   The magnitude of the shift is related to the fractional change in mass upon isotopic substitution. Shifts due to deuterium substitution are by far the largest for a given

resonance.

(5)   The shifts are additive to a very good approximation.

3.4.   PRESENT STUDIES

We now mention a number of laboratories where studies relevant to isotopic and temperature effects are currently being undertaken and highlight some of their recent work.

Ab initio calculations of the bond length dependence of shielding in a wide range of molecules are being carried out by Chestnut and coworkers. Of particular interest is their recent result [52] which shows that the stretching of an AB bond in the chain of atoms A-B-C can have a greater effect on the C nuclear shielding than on either the A or B nuclear shielding. Isotope shifts for CO obtained from correlated wave functions have been obtained by Oddershede's group [53].

In the writer's laboratory recent work has included the ab initio calculation of the carbon-13 shielding surface of the methane molecule and the temperature dependence and isotope shifts of its deuterated isotopomers [54]. Parallel to this was an experimental investigation into the temperature dependence of the carbon-13 and proton isotope shifts of methane and some of its isotopomers [55].

Jameson's group continues to make major contributions as is evident from the number of citations given in this review. Further work not cited involves isotope shifts, and temperature dependence studies of MF_6 and $M(XY)_6$ molecules [56,57]. More recently they have carried out a thorough investigation involving both ab initio calculation and experiment of the temperature dependence of the shielding in NH_3 and PH_3 and their trideuterated isotopomers [58]. A comprehensive review of isotope effects on shielding and spin-spin coupling has recently appeared [59].

Wasylishen's group has made important experimental contributions [60,61]. Finally, the work of the Moscow group led by Sergeyev must be mentioned [62,63]. A recent thorough review of isotope effects on coupling constants contains some important point concerning isotope shifts [64].

4.   Discussion

In this review we have treated intra- and intermolecular effects separately. This is justified to a high degree not not absolutely. Beckett and Carr [65] found a density-dependence for the isotope shift $\sigma(D_2)$ - $\sigma(HD)$ - see also [19]. Very recent work [66] has shown that the 'internal proton isotope shift' in gaseous CH_3CHD_2 measured at 600MHz is 11.271Hz at 17 atm. pressure and 11.361Hz at 31 atm. pressure.

The main difficulty in predicting isotope shifts from theory or in calculating shielding derivatives from experimental isotope shifts arises from the limited knowledge of anharmonic force constants and other vibration-rotation data. This may be overcome, however, to some extent by ab initio calculations. For intermolecular effects the most important requirement for substantial future progress is a flexible form for $(\sigma_{pair})_w$. Again correlated ab initio calculations on simple systems will be very helpful in this when they become feasible.

References

[1] Raynes, W. T., Buckingham, A.D., and Bernstein, H. J. (1962), J. chem. Phys. 36, 3481.

[2] Jameson, A. K., Jameson, C. J., and Gutowsky, H. S. (1970), J. chem. Phys. 53, 2310.

[3] Buckingham, A.D., Schaefer, T. and Schneider, W. G. (1960), J. chem. Phys. 32, 1227.

[4] Buckingham, A.D., and Pople, J. A. (1955), Trans. Farad. Soc. 51, 1173.

[5] Jorgensen, W. L. Binning, R. C. and Bigot, B. (1981), J.A.C.S. 103, 4393.

[6] Buckingham, A.D. (1960), Canad. J. Chem. 38, 300.

[7] Raynes, W. T., and Ratcliffe, R. (1979), Molec. Phys. 37, 571.

[8] Sadlej A.J., and Raynes, W. T., (1978), Molec. Phys. 35, 101.

[9] Augspurger, J.D., and Dykstra, C.E. (1991), J. Phys. Chem. 95, 9230.

[10] Day B., and Buckingham, A.D. (1976), Molec. Phys. 32, 343.

[11] Zaucer, M., and Azman, A. (1979), Z. Naturf. 349, 1279.

[12] Packer M.J., and Raynes, W. T. (1990), Molec. Phys. 69, 391.

[13] Marshall T., and Pople, J.A. (1958), Molec. Phys. 1, 199.

[14] Grayson, M., and Raynes, W. T. (unpublished work).

[15] Volodicheva, M. I. and Rebane, T. (1985), Teor. Eksp. Khim 21, 391 (p. 357), Engl. translation.

[16] McGrother, S. and Raynes, W. T. (to be published).

[17] Stone, A. J., and Alderton, M. A. (1985), Molec. Phys. 56, 1047.

[18] Batchelor, J. G. (1975), J. Am. Chem. Soc. 97, 3410.

[19] Jameson, C. J., Jameson, A. K. and Oppusungu, D. (1984), J. Chem. Phys. 81, 2313.

[20] Rummens, F. H. A. and Bernstein, H. J. (1965), J. chem. Phys. 43, 2971.

[21] Bennett, B. B., and Raynes, W. T. (1991), Mag. Res. in Chem. 29,. 946.

[22] Jameson, A. K., Moyer, J. and Jameson, C. J. (1978), J. chem. Phys. 68, 2873.

[23] Jameson, C. J., and Jameson, A. K. (1985), J. magn. Res. 62, 209.

[24] Jameson, C. J. (1975), J. chem. Phys. 63, 5296.

[25] Stevens, R. M. and Karplus, M. (1968), J. chem. Phys. 49, 1094.

[26] Grayce, C. J. and Harris, R. A. (1991), Molec. Phys. 72, 523.

[27] Jameson, C. J., Jameson, A. K. and Cohen, S. M. (1977), J. chem. Phys. 66, 5226.

[28] Ditchfield, R. (1976), J. chem. Phys. 65, 3123.

[29] Jackowski, K., Raynes, W. T. and Sadlej, A. J. (1978), Chem. Phys. Lett, 54, 128.

[30] Riley, J. P., Hillier, I. H. and Raynes, W. T. (1979), Molec. Phys. 38, 353.

420

[31]   Giessner-Prettre, C. and Ferchiou, S. (1983), J. magn. Res. 55, 64.
[32]   Govil, G. (1973), Appl. Spectroscopy Rev. 7, 47.
[33]   Homer, J. (1975), Appl. Spectroscopy Rev. 9, 1.
[34]   Rummens, F. H. A. (1975), NMR Basic Principles and Progress, 10, 5.
[35]   Jameson, C. J. (1980), Bull. Magn. Res. 3, 3.
[36]   Batiz-Hernandez, H. and Bernheim, R. A. (1967), Progr. NMR Spectros., 3, 63.
[37]   Hansen P. E. (1983), Ann Rep. NMR Spectroscopy, 15, 105.
[38]   Forsyth, D. A. (1984), Isotopes in Org. Chem. 6, 1.
[39]   Hansen, P. E. (1988), Progr. NMR Spectroscopy, 20, 207.
[40]   Randall, E.W., Ellner, J. J. and Zuckermann, J. J. (1966), J.A.C.S., 88, 622.
[41]   Gombler W. (1982), J. Am. Chem. Soc., 104, 6616.
[42]   IUPAC recommendations (1972), Pure and Applied Chem. 29, 627 and (1976), 45, 219.
[43]   Raynes, W. T. (1971), Specialist Periodical Report on NMR, The Chem. Soc., 1, 1.
[44]   Wilson, E.B., Decius, J. C. and Cross, P. (1955), Molecular Vibrations, McGraw Hill.
[45]   Fowler, P. W. (1982), Chem. Phys. Lett, 85, 313.
[46]   Hoy, A. R., Mills, I. M. and Strey, W. (1972), Molec. Phys. 24, 1265.
[47]   Toyama, M., Oka, T. and Morino, Y. (1964), J. mol. Spec. 13, 193.
[48]   Jameson, C. J. and Osten, H. J. (1986), Ann. Reports on NMR Spectros., 17, 1.
[49]   Jameson, C. J. (1991) in Theoretical Models of Chemical Bonding. Part 3, edited by Z. B. Maksic, Springer-Verlag, 457.
[50]   Jameson, C. J. (1977), J. chem. Phys. 66, 4983.
[51]   Jameson, C. J. and Osten, H. J. (1984), J. chem. Phys. 81, 4293.
[52]   Chestnut, D. B. and Wright, D. W. (1991), J. Comput. Chem. 12, 546.
[53]   Paidarova, I., Komasa, J. and Oddershede, J. (1991), Molec. Phys. 72, 559.
[54]   Raynes, W. T., Fowler, P. W., Lazzeretti, P., Zanasi, R. and Grayson, M. (1988), Molec. Phys. 64, 143.
[55]   Bennett, B. and Raynes, W. T. (1989), Spec. Acta, 45A, 1267.
[56]   Jameson, C. J., Jameson, A. K. and Oppusunggu, D. (1986), J. chem. Phys. 85, 5480.
[57]   Jameson, C. J., Rehder, D. and Hoch, M. (1987), J.A.C.S. 109, 2589.
[58]   Jameson, C. J., de Dios, A. C. and Jameson, A. K. (1991), J. chem. Phys. 95, 1069, 95, 9042.
[59]   Jameson, C. J., (1991) "Isotopes in the Physical and Biomedical Sciences" (Edit. E. Buncel and J. R. Jones), 2, 1.
[60]   Wasylishen, R. E. and Friedrich, J. O. (1987), Canad. J. Chem. 65, 2238
[61]   Leighten, K. L. and Wasylishen, R. E. (1987), Canad. J. Chem. 65, 1469.
[62]   Strelenko, Y. A., Torocheshnikov, V. N. and Sergeyev, N. M. (1990), J. mag. Res. 89, 123.
[63]   Roznyatovsky, V. A., Sergeyev, N. M. and Chertkov, V. A. (1991), Mag. Res. in Chemistry, 29, 304.
[64]   Sergeyev, N. M. (1990), NMR Basic Principles and Progress. Springer Verlag, 22, 33.
[65]   Beckett, J. R. and Carr, H. Y. (1981), Phys. Rev. 24A, 144.
[66]   Jackowski, K., Prime, A. and Raynes, W. T. (unpublished work).

THE ROLE OF NMR SHIFTS IN STRUCTURAL STUDIES OF GLASSES, CERAMICS AND MINERALS.

R. DUPREE, S.C. KOHN, C.M.B. HENDERSON, A.M.T BELL

University of Warwick *University of Manchester* SERC
Physics Department *Geology Department* *Daresbury Laboratory*
Coventry CV4 7AL *Manchester M13 9PL* *Warrington WA4 4AD*
U.K. *U.K.* *U.K.*

ABSTRACT. Experimental data on chemical shifts in a variety of crystalline and amorphous silicates and ceramics are reviewed and the published correlations between ^{29}Si chemical shifts and structural parameters are critically examined. None of the correlations accurately predict the ^{29}Si spectra of tridymite or $K_2MgSi_5O_{12}$ leucite, but those of Sherriff and Grundy (1986) and Sternberg (1988) are superior to the simpler geometric functions. However all except the Sherriff and Grundy correlation can be used to provide some information on bond angle distributions in simple glasses whose medium range order is poorly understood. ^{15}N, ^{29}Si and ^{13}C shifts in ceramics are potentially very informative, but no correlations between shift and structure in these materials are currently available. Much further theoretical work on the calculation of chemical shifts is required before quantitative structural information can be reliably deduced from NMR shifts in the materials discussed here.

1   Introduction

An experimentalist uses the chemical shift as a probe of structure. Whilst appreciating that it is not possible at present, to calculate the shift from first principles let alone characterise a structure by one number (or three from the shift tensor) he would like some simple functional form for the variation of shift with some simple structural parameter. In this paper we will attempt to show how even limited knowledge of the connection between shift and structure can be of value and to point out how the limitations in our current knowledge restrict the information content of an NMR spectrum. Some suggestions for future calculations will be made. We shall restrict our discussion to three classes of materials, glasses, framework silicate minerals and analogues, and ceramics, and will concentrate (mostly) on ^{29}Si NMR.

2   Background

Since the first reasonably high resolution NMR measurements on silicates correlations between the shift and various structural parameters have been suggested. The large majority of these have been concerned with ^{29}Si tetrahedrally coordinated to oxygen since this is by far the most common structural unit in these materials. Some work has been done on six-coordinated silicon (Stebbins and Kanzaki (1991)) and on ^{27}Al (e.g. Phillips et al. (1989)) but the complication of the quadrupolar contribution to the position and the broader lines has limited the amount of

J. A. Tossell (ed.), Nuclear Magnetic Shieldings and Molecular Structure, 421–434.
© 1993 *Kluwer Academic Publishers.*

data. There has been very little attempt to quantitatively correlate the shifts of other readily accessible nuclei such as ^{23}Na or ^{11}B with details of structure. The situation up to 1986 has been reviewed by Engelhardt and Michel (1987) and a more complete discussion is given there. The simplest correlations are with mean Si-O-Si bond angle, $\bar{\alpha}$, or mean Si-O bond length. The first of these has apparently given a reasonable agreement with experiment for instance the chemical shift, δ, being given by

$$\delta = -0.6192\bar{\alpha} - 18.68 \tag{1}$$

for a number of SiO$_2$ polymorphs and Q^4(0Al) zeolites (Engelhardt and Radeglia (1984)). A less good but still significant correlation was found between the bond length and the shift (e.g. Smith and Blackwell (1983)). These correlations are entirely empirical but show that both these parameters are related to the shift. More physically based correlations have been suggested by a number of authors in particular Engelhardt and Radeglia pointed out that for the typical net charges for tetrahedral SiO$_4$ units the chemical shift varies approximately linearly with the orbital electronegativity of the oxygens in the SiO bonds. This in turn can be related to the degree of s hybridization, ρ, of the oxygen bond orbitals by $\rho = \cos\alpha/(\cos\alpha - 1)$. Thus for tetrahedral units a correlation such as given in eqn 2 might be expected.

$$\delta = -A \overline{\cos\alpha}/(\overline{\cos\alpha - 1}) + B \tag{2}$$

In fact the agreement with experiment is of similar quality to that of eqn 1. Another physically based correlation which gives a slightly better agreement with experiment is based on the mean secant of the four Si-O-Si bond angles (Smith and Blackwell (1983)). More recently the effect of the second coordination sphere on the shift has been calculated by Sternberg (1988) and Hallas and Sternberg (1989) using a point charge approximation. The chemical shift is given by the sum of two contributions, δ^{fs} the first sphere shift of an isolated unit and the shift caused by the Si-O (or Al-O) bond polarization which is dependent on the Si-O-X bond angle and on X (eqn 3).

$$\delta = \delta^{fs} + A_{pol}\sum_{i}^{4} (V_{SiO})_i \tag{3}$$

A_{pol} is dependent on the bond polarity and is determined from experiment. For SiO$_2$ polymorphs and dealuminated zeolites this also gives a nonlinear variation of shift with bond angle and only differs significantly from previous correlations at large ($> 160°$) and small ($< 140°$) angles for which there is little experimental data. A further correlation between the chemical shift and structural parameters for silicate minerals was suggested by Sherriff and Grundy (1988). This was based on the magnetic anisotropy and valence of the bond between the oxygen and the second neighbour cation to the silicon. The expression for the magnetic anisotropy is given by

$$\chi' = \sum s_i ((1-3 \cos^2 \theta_i)/3R_i^3) \tag{4}$$

where R is the distance between each silicon atom and the midpoint of the oxygen-second

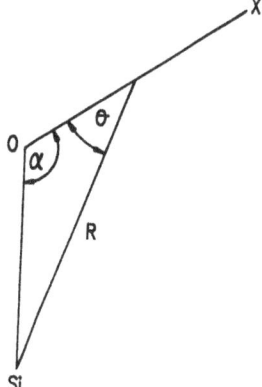

Figure 1 Diagram showing the angles and lengths used in eqn (4). (after Sherriff and Grundy (1988))

neighbour cation bond, r the length of this bond and Θ the angle between them (see fig 1). The bond valence s_i is given by $s_i = \exp[\ (r_o - r_i)/0.37]$ where r_o, the bond valence parameter, was taken from Brown and Altermatt (1985). A total of 76 data points were used to give

$$\delta = 359.11\chi' - 44.37 \tag{5}$$

A somewhat better agreement with experiment was found if the hybridization of the bridging oxygen was included in which case eqn 4 becomes

$$\chi'' = \sum s_i\ ((1 - 3\cos^2 \theta_i)/3R_i^3)\ (\cos \alpha_i /(\cos \alpha_i - 1)) \tag{6}$$

giving the chemical shift as

$$\delta = 650.08\chi'' - 56.06 \tag{7}$$

Over the range of samples used this probably gives the best overall correlation although it is less amenable to direct stuctural interpretation and perhaps is best used to discriminate between different structural models. It includes the hybridization term (as does that of Sternberg) and has a distance parameter R which is related to that of Sternberg. One problem with all the correlations is that the quality of the crystal structures may not be very good, small variations in bond angle/length between one determination and another can easily give a calculated shift difference of ~ 1ppm.

All of the shift correlations discussed above assume that a partial chemical shift per silicon oxygen bond can be defined. The total shift being the sum of these partial values

$$\delta = \sum_{i=1}^{4} \delta_i \tag{8}$$

Thus the bonding and chemical shift contribution at each corner of the tetrahedron is assumed independent of that at adjacent corners; whilst this is a reasonable first order approach there is some evidence (presented later) that it may not always apply.

3   Glasses

It is difficult to obtain much structural information about glasses because the lack of long range order means that scattering techniques such as XRD etc are of limited applicability. In particular information beyond the nearest neighbour distance is limited. The different coordination polyhedra which can be present in some glasses are readily distinguished using NMR since for e.g. ^{29}Si typical shift ranges for SiO_4 units are in the range -60ppm to -120ppm and for SiO_6 between -180ppm and -220ppm. For ^{27}Al the situation is more complicated with AlO_5 polyhedra occurring much more frequently than in crystalline materials. These usually occur with a peak position ~ 30-40ppm compared with 70-40ppm for AlO_4 and < 20ppm for AlO_6. However the shift can be quite strongly affected by the second coordination sphere, e.g. replacement of Al by P causing a negative shift of ~ 20ppm for both tetrahedral and octahedral Al. For aluminoborate glasses this has led to Al peaks at 30ppm being assigned to $Al[OB_4]$ by some authors and to $Al[OB_5]$ by others (Dupree et al. (1985a), Hahnert and Hallas (1987), Hallas and Sternberg (1989) and Bunker et al. (1991)). A central issue in simple binary glasses has been the distribution of non bridging oxygens which is related to the amount of "order" in the glass. Fortunately for Na_2O-SiO_2 and Li_2O-SiO_2 glasses Q^4, Q^3 and Q^2 units are resolvable, however the shift difference between Q^4 and Q^3 etc decreases with electronegativity, see fig 2, so that it becomes more difficult to resolve the different Q species for glasses containing other alkalis or alkaline earths. The influence of the second coordination sphere on the silicon shift can give useful information and examples are

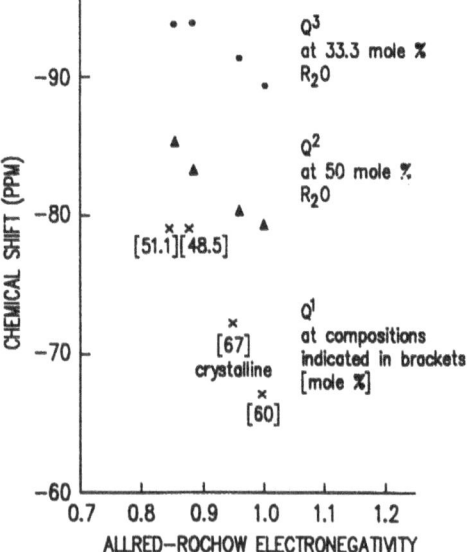

Figure 2 ^{29}Si shifts for different Q^n species v electronegativity for alkali silicate glasses (after Dupree et al.(1986))

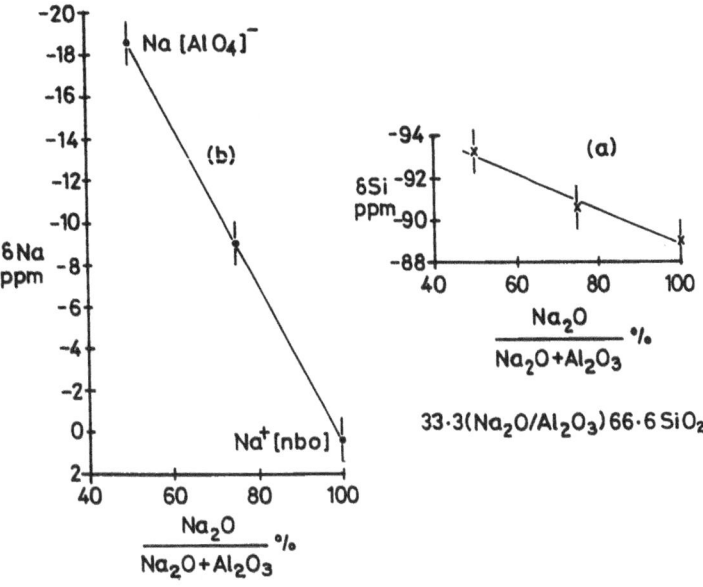

Figure 3 Effect of substitution of Al_2O_3 for Na_2O in a $(Na_2O.Al_2O_3)_{0.33}(SiO_2)_{0.66}$ glass on (a) the ^{29}Si shift and (b) the ^{23}Na shift. (after Dupree et al (1985b)

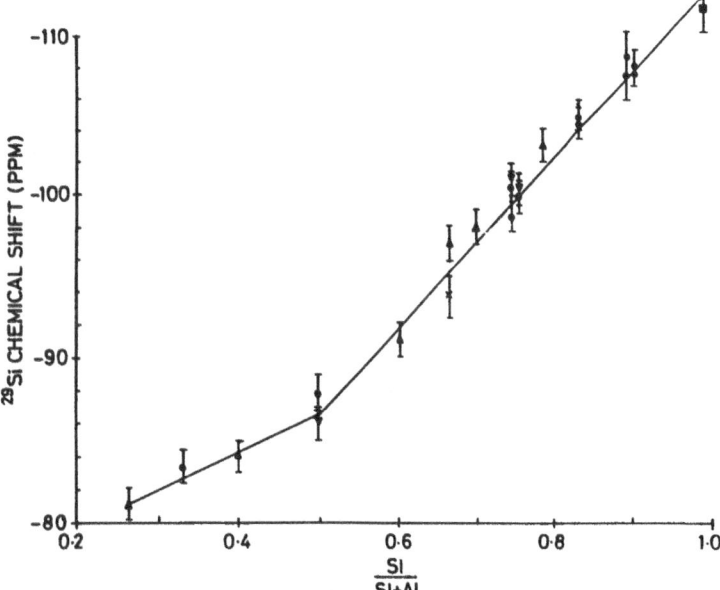

Figure 4 Variation of ^{29}Si shift with $Si/(Si+Al)$ ratio for framework aluminosilicate glasses (after Dupree (1988) and Oestrike et al.(1987))

shown in figs 3 and 4. Figure 3a shows the effect on the ^{29}Si shift as Na$_2$O is replaced by Al$_2$O$_3$ (Dupree et al. (1985b)). The change in the shift ~ 4ppm for 50% Al$_2$O$_3$ is comparable with that observed in zeolites when a silicon in the second coordination sphere is replaced by aluminium. Much stronger effects are seen in the sodium position (fig 3b) which changes by ~ 18ppm for a sodium to aluminium ratio of one. The true chemical shift for ^{23}Na is moved by ~ 2ppm upfield from the peak position in Na$_2$O.2SiO$_2$ glass because of the second order quadrupolar shift. Whilst there may be some change in the quadrupolar interaction as aluminium is added to the glass most of this difference in position must come from a large change in shift. Whilst confirming that [AlO$_4$]$^-$/Na$^+$ association occurs in these glasses, there do not appear to be calculations of ^{23}Na shifts even in simple materials. Figure 4 shows the ^{29}Si shift v Si/(Si+Al) ratio for a number of framework (i.e. all bridging oxygen) glasses (Oestrike et al. (1987), Dupree (1988)). The modifier cation has only a small effect and the dominant cause of the shift changes is the increasing number of Al atoms in the next neighbour shell. For an Si/(Si+Al) ratio of < 0.5 it is no longer possible for aluminium avoidance to be maintained as at a ratio of 0.5 each Si would have 4Al neighbours. The smaller slope indicates that there is only a small increase in the number of Al next nearest neighbours beyond this composition whilst the second neighbour Al content increases. The width of the lines, which would be very broad for a statistical distribution of aluminium over the tetrahedral sites, also gives evidence for a restricted range of environments.

The width of the line can also be used in some systems to derive the Si-O-Si bond angle distribution, an important parameter in describing the intermediate range order in the glasses. We have discussed this several times before (Dupree and Pettifer (1984), Pettifer et al. (1988)) so only brief details will be given here. If the bond angle distribution is V(α), then:-

$$V(\alpha) = X(\delta_i) \left| \frac{d\alpha}{dF} \right| \tag{9}$$

where $X(\delta_i)$ is the distribution function of the partial shifts. The distribution function of the total shift W(δ) is then given by:-

$$FT[W(\delta)] = FT[X(\delta_i)]^4 \tag{10}$$

where FT denotes Fourier Transformation and W(δ) is the measured NMR line. For SiO$_2$ this is approximately Gaussian, thus $X(\delta_i)$ can be readily determined and hence V(α) using one of the correlations described earlier. Figure 5 shows the bond angle distribution obtained for three different correlations together with the distribution determined using X-rays. Whilst there is qualitative agreement in that the most probable bond angle determined from the NMR data lies between 142° and 151° compared with 146° from the X-ray data and all the NMR results give a narrower distribution function than that deduced using X-rays, there are considerable differences in the details. Very recently Farnan et al. (1992) have used 2D DAS on ^{17}O in some potassium silicate glasses to derive information on the Si-O-Si bond angle distribution. This has the advantage of involving only one angle so that it does not rely on the additivity assumed (eqn. (8)) above. The correlation with angle is with the electric field gradient at the oxygen site and with its asymmetry parameter η rather than the chemical shift which seems to vary only slightly. They found that the bond angle distribution in this material was narrower than that deduced from X-ray measurements on SiO$_2$. Preliminary results for SiO$_2$ using ^{17}O

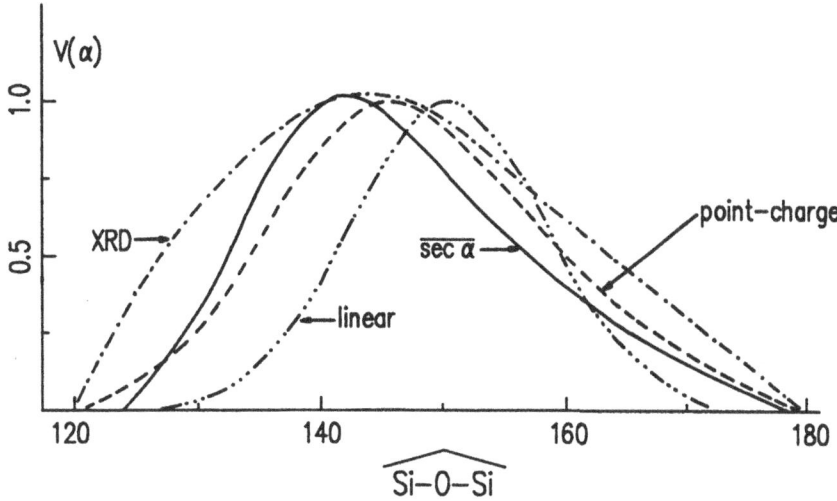

Figure 5 The Si-O-Si bond angle distribution in SiO_2 using :- (i) the linear correlation, (ii) the point charge model, (iii) the secant correlation and (iv) the distribution deduced using X-rays (after Pettifer et al. (1988))

DAS are much closer to the distributions deduced from ^{29}Si NMR than the X-ray $V(\alpha)$ and lie somewhere between a and b on figure 5 (Farnan and Grandinetti pers. comm.).

4   Minerals

It is of interest to compare the predictions of these different correlations with the spectrum observed for tridymite as we did some time ago (Pettifer et al. (1988)) but this time including the Sherriff and Grundy (1988) predictions. This is shown in figure 6(A) and (B). As we noted before none of the models match the spectrum well, the linear model is worst with a peak at −116ppm which is not observed and no peak ~ −109ppm. The secant model predicts too narrow a spectrum and has only a low intensity at −114.5ppm and −109ppm. The point charge model produces three peaks in roughly the correct positions but their relative amplitudes are in poor agreement with experiment. The Sherriff and Grundy model predicts the high frequency part of the spectrum well, but on the low frequency side there is a difference between experiment and the model which increases as the shift becomes more negative. Thus there is a peak predicted at −116.5ppm compared with the most negatively shifted peak of the spectrum at −114.2ppm a difference of 2.3ppm. The peak at −115ppm may correspond to the experimental peak at −113.2ppm a difference of 1.8ppm etc. However some caution is required since the calculated peak positions may not come in the same order as experiment, it is only by knowing which site corresponds to each part of the spectrum that a truly valid comparison can be made.

A case where this is known and where it is of interest to compare the different correlations is in potassium magnesium leucite ($K_2MgSi_5O_{12}$) which is based on the framework topology of the mineral leucite ($KAlSi_2O_6$) but with Mg and Si in tetrahedral sites. Natural

428

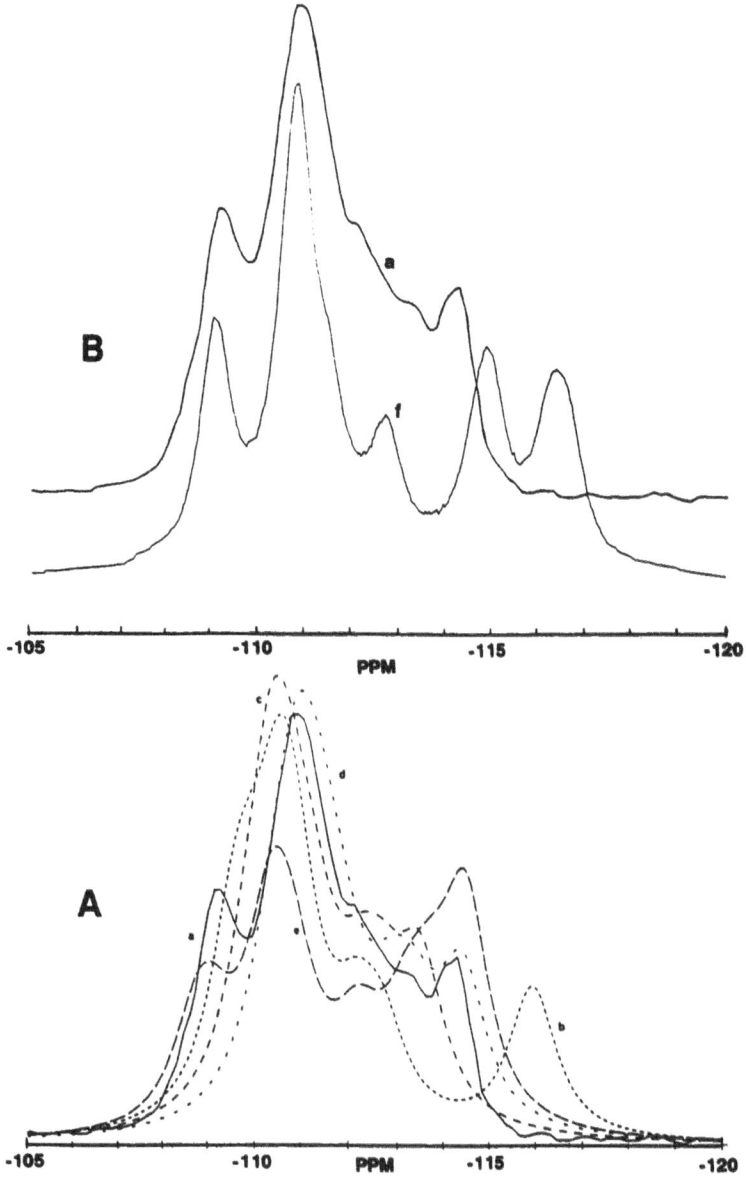

Figure 6 The ^{29}Si spectrum of tridymite. 6(A) lower. Experiment (a) compared with simulations based on (b) the linear correlation, (c) the secant correlation, (d) the s-hybridisation correlation, and (e) the point charge model; 6(B) upper. Experiment compared with the simulation (f) using eqns (6) and (7).

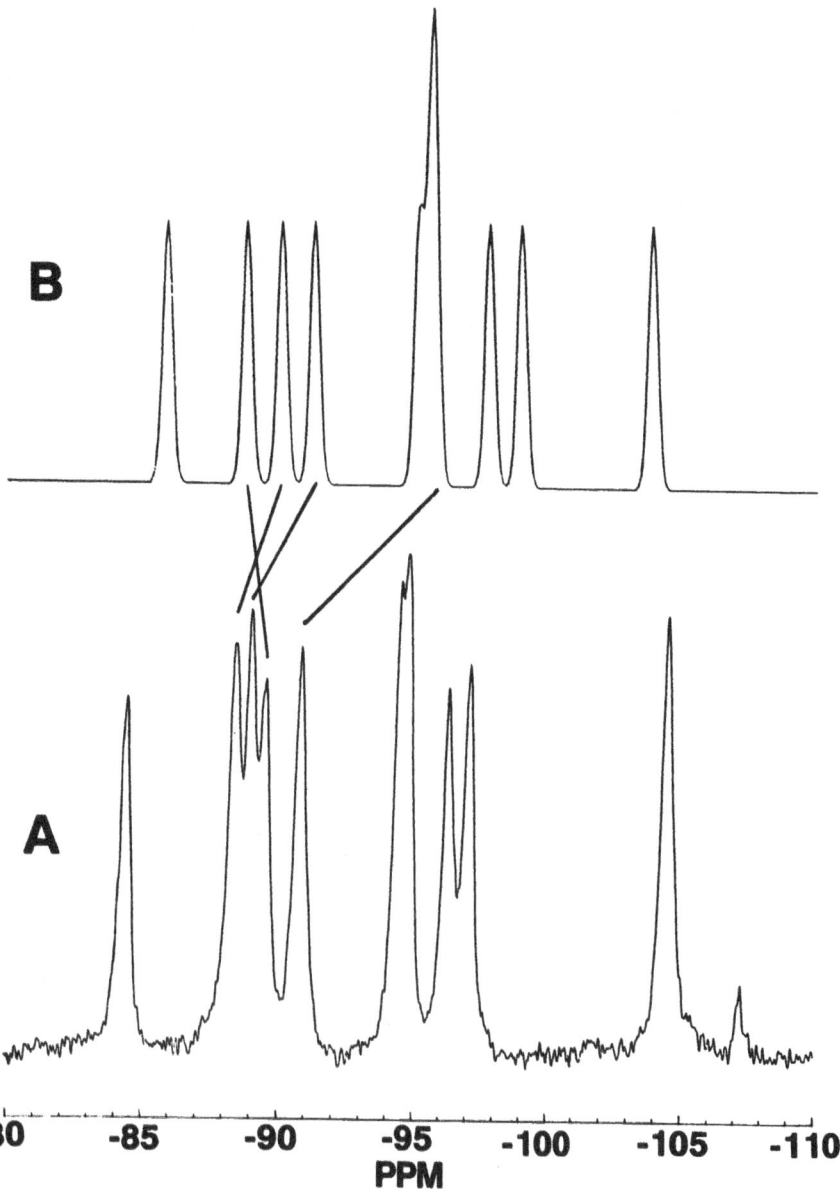

Figure 7 The spectrum for KMg leucite ($K_2MgSi_5O_{12}$) (lower), compared with the simulation using eqns (6) and (7), (upper).

leucite is tetragonal with three distinct tetrahedral sites in the unit cell which contains 48 tetrahedral cations and aluminium and silicon are partially disordered over these sites. KMg leucite prepared hydrothermally is of lower symmetry than natural leucite and the [29]Si spectrum consists of 10 narrow lines spread over 20ppm (Kohn et al. (1991)) indicating that the material is fully ordered with 12 distinct T sites in the unit cell. The chemical shift anisotropies obtained from static and slow spinning MAS experiments show that two of the lines are due to silicon linked to four other silicon sites ($Q^4(4Si)$) and the other 8 were due to Si linked to 3Si and 1Mg ($Q^4(3Si)$). As the structure was unknown and the XRD complex a ^{29}Si COSY spectrum was run to determine the connectivity of the sites. The structure has now been determined by Bell et al. (1992) enabling us, when combined with the connectivity data from the COSY experiment, to make a quantitative comparison with the different correlations. Figure 7 shows the experimental spectrum together with that calculated using equations 6 and 7. Overall there is a considerable degree of similarity between them, the theoretical spectrum being slightly compressed compared with experiment. The worst agreement with experiment is for the peak at -91.0ppm in the experimental spectrum, which is from the $Q^4(4Si)$ site with relatively small bond angles (from $130°$ to $140.6°$), it is predicted to be at -95.6ppm a

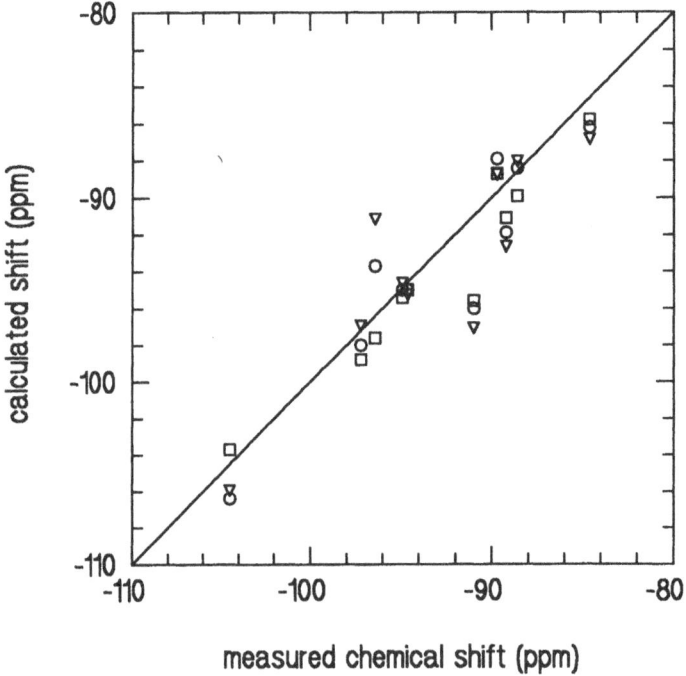

Figure 8 Experimental [29]Si shifts v calculated shifts for KMg leucite. Squares using eqn (6) and (7), circles point charge model, triangles mean secant correlation.

difference of 4.6ppm. The order of the peaks around −89ppm is changed emphasising the need for caution in the assignment of peaks if the connectivity is not known. The other correlations give a somewhat worse agreement with experiment with that of Sternberg being the best provided that A_{pol} in eqn.3 is changed from −284ppm/Hartree to −200ppm/Hartree and δ_{fs} from -53ppm to -59.5ppm. for the sites with one Mg neighbour. This value for A_{pol} is close to that suggested by Sternberg, −190ppm/Hartree, for use with second sphere atoms other than silicon. Peak 5 the $Q^4(4Si)$ peak at −91ppm is predicted to be at −96ppm, the other peaks which deviate most from experiment are peak 8 (at −96.4ppm) which is calculated to be 2.7ppm more positive than is observed and peak 3 (at −89.2ppm) for which the calculation gives −91.9ppm. Figure 8 shows a comparison of the calculated and measured shifts for these two correlations together with a mean secant correlation where the slope and constant for the $Q^4(3Si)$ lines has been adjusted to give a best fit to the data (i.e. to −31 and −135ppm from −55.8 and −176.7 respectively). No correlation is able to fit peak 5. Assuming the crystal structure is correct, this may be because of dynamic effects so that the X-rays see a mean bond angle that is different from that seen by NMR or because the correlations are extended beyond their normal range. All other $Q^4(4Si)$ sites known to us have mean bond angles $\geq 139°$. This site also has the largest variation of Si-O distance amongst the 4 neighbours (0.026Å) so this too may have an effect on the correlations. The other peak which is not fitted very well (except for the Sherriff and Grundy correlation where it is in no worse agreement with experiment than several other peaks) is peak 8 at −96.4ppm. This has the largest range of bond angles from 119.8° to 160.5° and may be an indication that simple additivity, eqn.8, is breaking down. It should be noted that the simple average angle correlation (eqn (1)) predicts completely unrealistic mean bond angle of 116.8° for peak 5 compared with the experimental value of 134.9°. It is therefore recommended that this correlation is not used outside the narrow angular range where it was first formulated, even there caution is required as it predicts −109.6ppm instead of −104.5ppm for the $Q^4(4Si)$ peak 10 of our KMg leucite sample.

5   Ceramics

The fairly large amount of NMR work on ceramics seems not to have produced any successful correlations between the shift and structure. Table 1 lists some ^{29}Si, ^{13}C and ^{15}N shifts in the polymorphs of Si_3N_4 and some polytypes of SiC. In some sense these are good candidates for shift calculations since they contain only two, light, elements and the shifts of both are known to high accuracy. The ^{29}Si shifts are all less negative than for silicon bonded to oxygen and it is clear that the correlations discussed earlier are not applicable since for example eqn.6 and 7 would require Si-X-Si angles of $< 100°$ to change the sign of the $(1-3\cos^2\Theta)$ term which is neccessary for shifts more positive than −56ppm. The ^{15}N shift is more sensitive to its surroundings than either ^{13}C or ^{29}Si as evidenced from the 4 different sites in $\alpha-Si_3N_4$ and also from work in sialon systems (Kruppa et al. (1991)).

Figure 9 shows some ^{29}Si shift ranges that occur in oxynitride ceramics, there is considerable overlap of the different coordination tetrahedra and the shifts for SiN_4 extend from −20ppm (sialon polytypoid 21R) (Smith (1992)) to −65ppm ($LaSi_3N_5$) (Dupree et al. (1989)). It is of interest that the shifts in lanthanum containing phases are more negative by ~ 15ppm than in the yttrium based phases. The ^{29}Si shifts in a number of phases in the $SiO_2-Si_3N_4$-$AlN-Al_2O_3$ systems e.g. β'-sialon ($Si_{6-z}Al_zN_{8-z}O_z$) are remarkably insensitive to z. Two

TABLE 1. ^{29}Si, ^{13}C and ^{15}N shifts
in some simple ceramics

	δ_{Si}/ppm	δ_C/ppm	ref
β-SiC	-17.2	23.7	1,2
4H SiC	-19.7 -22.5	14.7 21.5	3
6H SiC	-14.3 -20.4 -24.9	15.2 20.2 22.7	4,5
15R SiC	-14.6 -20.5 -24.1	16.0 20.7 22.7	1,4,5
		δ_N *	
α-Si_3N_4	-49.0 -47.1	51.6 53.4 64.0 74.6	6,7,8
β-Si_3N_4	-48.5	50.9 68.4	6,7,8
Si_2N_2O	-61.6	40.3	6,7,8

1. G R Finlay et al. (1985), 2. G W Wagner et al. (1989)
3. M Leach (1990), 4. J R Guth et al. (1987)
5. J S Hartman et al. (1987), 6. R K Harris et al. (1990)
7. D Kruppa et al. (1991), 8. R Dupree et al. (1985c)

* ^{15}N shift is relative to $^{15}NH_4NO_3$

explanations for this have been proposed, either the silicon coordinates to four nitrogen atoms at all levels of substitution (the mixed coordinations being restricted to aluminium tetrahedra) (Dupree et al. (1988, 1989)), or the chemical shift remains unaltered for the $SiN_{4-x}O_x$ (x=0-4) tetrahedra when an Al-O group replaces an isoelectronic Si-N unit since the groups have similar electronegativities (Leach (1990)). Theoretical work on this type of system might clarify this.

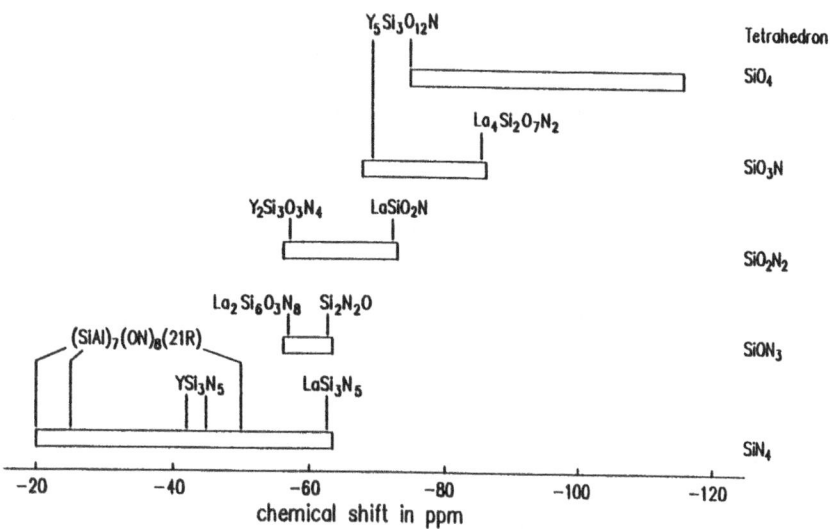

Figure 9 ^{29}Si shift ranges in oxynitrides

6   Conclusion

All the correlations of shift with structure discussed in this work are somewhat imprecise. The formulation of Sherriff and Grundy (1988) seems to give the best overall correlation for the ^{29}Si shift in silicates. Simpler shift correlations based solely on the Si-O-Si bond angle are able to give useful structural information about glasses.   Extreme care is needed if these correlations, particularly the linear shift v angle correlation, are used outside the range of angles/environments for which they were formulated.   There is a need for successful shift/structure correlations in ceramic materials and for calculations of the shift for nuclei other than ^{29}Si e.g. for ^{23}Na, ^{15}N and ^{13}C in the materials discussed above.

7   Acknowledgement

This work was supported by the S.E.R.C. and the N.E.R.C.

8   References

Bell, A.M.T., Fitch, A.N., Cernik, R.J., Champness, P.E., Henderson, C.M.B., Redfern, S.A.T., and Kohn, S.C. (1992) Acta Cryst B (submitted).
Brown, I.D., and Altermatt, D. (1985) Acta Cryst. B41, 244-247.
Bunker. B.C., Kirkpatrick, R.J., Brow, R.K., Turner, G.L., and Nelson, C. (1991)

434

J. Am. Ceram. Soc. 74, 1430-1438.

Dupree, R., and Pettifer, R.F. (1984) Nature 308, 523-525.

Dupree, R., Holland, D., and Williams, D.S. (1985a) Phys. Chem. Glasses 26, 2.

Dupree, R., Holland, D., and Williams, D.S. (1985b) J. Physique Colloque C8, 46, 119.

Dupree, R., Lewis, M.H., Leng-Ward, G., and Williams, D.S. (1985c) J. Mater. Sci. Lett. 4, 393-395.

Dupree, R. (1988) Experimentelle Technik der Physik 36, 315-325.

Dupree, R., Lewis, M.H., and Smith, M.E. (1988) J.Appl.Cryst. 21, 109-116.

Dupree, R., Lewis, M.H., and Smith, M.E. (1989) J. Am. Chem. Soc. 111, 5123.

Engelhardt, G., and Michel, D. (1987) High resolution solid state NMR of silicates and zeolites. J. Wiley & Sons.

Engelhardt, G., and Radeglia, R. (1984) Chem. Phys. Lett. 108, 271-274.

Farnan, I., Grandinetti, P.J., Baltisberger, J.H., Stebbins, J.F., Werner, U., Eastman, M.A., and Pines. A. (1992) Nature 358, 31-35.

Finlay, G.R., Hartman, J.S., Richardson, M.F., and Williams, B.L. (1985) J. Chem. Soc. Chem. Commun. 159-161.

Guth, J.R., and Petuskey, W.T. (1987) J. Phys. Chem. 91, 5361-5364.

Hahnert, M., and Hallas, E. (1987) Rev. Chim. Miner. 29, 221.

Hallas, E., and Sternberg, U. (1989) Mol. Phys. 68, 315-326.

Hartman, J.S., Richardson, M.F., Sherriff, B.L., and Winsborrow, B.G. (1987) J. Am. Chem. Soc. 109, 6059-6067.

Harris, R.K., Leach, M.J., and Thompson, D.P. (1990) Chem. Mater. 2, 320.

Kruppa, D., Dupree, R., and Lewis, M.H. (1991) Materials Letters 11, 195.

Kohn, S.C., Dupree, R., Mortuza, M.G., and Henderson, C.M.B. (1991) Phys. Chem. Minerals 18, 144-152.

Leach, M. (1990) Ph.D. thesis University of Durham.

Oestrike, R., Yang, W.H., Kirkpatrick, R.J., Hervig, R.L., Navrotsky, A., and Montez. B. (1987) Geochim. Cosmochim. Acta, 51 2199.

Pettifer, R.F., Dupree, R., Farnan, I., and Sternberg. U. (1988) J. Non-Cryst. Sol. 106, 408-412.

Phillips, B.L., Kirkpatrick, R.J., and Putnis, A. (1989) Phys. Chem. Minerals 16, 591-598

Sherriff, B., and Grundy, H.D. (1988) Nature 332, 819.

Smith, J.V., and Blackwell, C.S. (1983) Nature 303, 223-225.

Smith, M.E., (1992) J.Phys. Chem. 96, 1444

Stebbins, J.F., and Kanzaki, M. (1991) Science 251, 294

Sternberg, U. (1988) Mol. Phys. 63, 249.

Wagner, G.W., Na, B.K., and Vannice, M.A. (1989) J. Phys. Chem. 93, 5061-5064.

THE INFLUENCE OF STRUCTURE AND GEOMETRY ON THE ³¹P, ²⁹Si, ¹³C AND ¹H CHEMICAL SHIFTS

U. STERNBERG
Friedrich-Schiller-Universität Jena
Institut für Optik und Quantenelektronik
Max-Wien-Platz 1
O-6900 Jena
Germany

ABSTRACT. Starting with a description of the molecular wave function using bond orbitals a general formula was derived describing the change of an expectation value with bond polarization. This formalism is applied to explain the influences of substituents in the second coordination sphere on the chemical shift.

The outlined theory gives a clue to the major influences of an atom (or group) X in the second coordination sphere on the chemical shift of a nucleus a: (i) the bond order of the adjacent bonds, (ii) the charge on the atom x, (ii) the a-b-x bond angles, dihedral angles and the a-x distances.

In the case of ²⁹Si-chemical shifts the theory gives an explanation for the empirical predicted Si-O-Si bond angle dependence of the chemical shift and in the case of ³¹P-chemical shifts in phosphates the influence of d-p-π-bonds is demonstrated.

In the case of hydrogen bonds it could be derived that the ¹H-chemical shifts depends in the first line on $1/r_{O..H}$.

Taking into account the polarization of the C-C-π bond one can explain ¹³C-chemical shifts of many vinyl compounds and only two empirical parameters are needed to account for the large span of observed shifts.

1. Introduction

The chemical shift is considered in most N.M.R. experiments as the parameter of major interest. From chemical shifts details of the structure and electron distributions in molecules and crystal lattices can be derived. In most cases the chemical shift is related to the structure using empirical rules, increment systems or empirical formulas connecting electronegativities and bond angles or distances to chemical shifts.

The aim of this paper is to introduce a semi-empirical concept that allows to derive some of the former empirical rules from quantum chemical starting points. This concept is named bond polarization theory and it should be demonstrated that it allows to predict chemical shifts from molecular structures with a minimum of empirical parameters.

J. A. Tossell (ed.), Nuclear Magnetic Shieldings and Molecular Structure, 435–447.
© 1993 *Kluwer Academic Publishers.*

2. Bond Polarization Theory

Since we are mainly interested in the change of a molecular property and not in its absolute value we divide the molecular system under study into two parts named A and B and investigate only the change in A that is induced by B. The part B can be an ion, an atom or a group. The best suited quantum chemical description are in this case localized orbitals. We introduce therefore two center bond orbitals that are linear combinations of to hybrids χ_a and χ_b at the bonded atoms a and b:

$$\psi_i = c_{ai}\chi_a + c_{bi}\chi_b \qquad \psi_i^* = c_{bi}\chi_a - c_{ai}\chi_b \qquad (1)$$

Additionally to the bond orbital ψ its anti-bond ψ^* is introduced. Orthogonality is forced between different bond orbitals and anti-bond orbitals. In this case only the bond polarity b is has to be determined for every bond:

$$c_{ai} = \sqrt{\frac{1+d_i}{2}} \qquad c_{bi} = \sqrt{\frac{1-d_i}{2}} \qquad (2)$$

The ground state wave function of the molecular system Ψ_0 is constructed as Slater determinant of bond orbitals

$$\Psi_0 = \frac{1}{\sqrt{2n!}} Det|\psi_1^+\psi_1^- \ldots \psi_i^+\psi_i^- \ldots \psi_n^+\psi_n^-| \qquad (3)$$

where + and - designates the possible spin states. Let us now introduce exited configurations Ψ_I by substituting one or two bonds by anti-bonds:

$$\Psi_I = \Psi \binom{j^*}{i} = \frac{1}{\sqrt{2n!}} Det|\ldots \psi_j^{+*}\psi_i^- \ldots| \qquad (4)$$

The wave function including configuration interaction will contain configuration of the type $\Psi(i \rightarrow j^*)$ describing delocalizations and $\Psi(i \rightarrow j^*, k \rightarrow l^*)$ for double excitations. Since we are interested in polarizations only configurations of the type $\Psi(i \rightarrow i^*)$ are needed where an electron from bond i is exited into its own anti-bond i^*. We are interested in the polarization of the bonds within the molecular part A by the charge distribution of part B and write therefore for the molecular wave function including polarization Ψ_p :

$$\Psi_p = \Psi_0 + \sum_{i \in A}^{n_A} \frac{\langle \Psi_0 | \hat{F}_B^0 | \Psi \binom{i^*}{i} \rangle}{E_0 - E\binom{i^*}{i}} \Psi\binom{i^*}{i} + \sum_{i \in B}^{n_B} \ldots \qquad (5)$$

The first sum in (5) runs only over the n_A bonds of the molecular part A and the perturbation by B is expressed using the Fock operator $(F^0)_B$.

Next we calculate the expectation value of an one electron operator $\hat{O}$ inserting the wave function (5) and neglecting all terms that are of higher than the first order. The expectation value of a Slater determinant from bond orbitals gives a sum over bond contributions and we can easily separate the contributions of part A of the molecular system:

$$\langle \psi_P | \hat{\sigma} | \psi_P \rangle_A = 2 \sum_{i \in A}^{n_A} \{ \langle i | \hat{\sigma} | i \rangle + \frac{2}{\Delta E_i} \langle i | \hat{F}_B^o | i^* \rangle \langle i^* | \hat{\sigma} | i \rangle \}$$

$$= \sum_{i \in A}^{n_A} \{ c_i + A_i \langle i | \hat{F}_B^o | i^* \rangle \} \qquad (6)$$

This formula is of central interest throughout the paper and should be interpreted in more detail (see Sternberg (1988)). For an expectation value of an one electron operator we obtain a sum over the n_A bond contributions of the type $\langle i | \hat{O} | i \rangle$ that can be regarded as bond increments C_i. To every bond contribution C_i is added a term that depends linearly on the polarization energy of the bond $\langle i | (F^o)_B | i^* \rangle$ multiplied with an integral of the type $A_i = \langle i | \hat{O} | i^* \rangle$. Instead of calculating the integrals containing the operator $\hat{O}$ we introduce at this point for every type of bond an empirical constant C_i for the expectation value of the unpolarized bond (bond increment) and the empirical slope A_i describing the linear change of the bond contribution with polarization. We now have only to discuss the matrix elements of the Fock operator.

3. Calculation of Integrals

The matrix elements of the Fock operator in equation (6) involve the calculation of complicated coulomb and exchange integrals over Slater atomic orbitals. If we consider the part B of the molecule as distribution of point charges we can replace the Fock operator $(F^o)_B$ by the point potential V_B. This approximation is in most cases justified because the interaction is of long range. With this assumption simple formulas can be derived but they are no necessary prerequisite of the bond polarization theory. For the polarization energy in eq. (6) the following expression is obtained using eq. (1) and (2) (small overlap contributions are neglected):

$$V_{ab} = \langle i | \hat{V}_B | i^* \rangle = \sqrt{1-d^2} \times \qquad (7)$$

$$\sum_{x \in B}^{n_B} q_x \{ \langle \chi_a | \frac{1}{|\vec{R}_{ax} - \vec{r}|} | \chi_a \rangle - \langle \chi_b | \frac{1}{|\vec{R}_{bx} - \vec{r}|} | \chi_b \rangle \}$$

$$\langle nsp^h | \frac{1}{|\vec{R} - \vec{r}|} | nsp^h \rangle =$$

$$\frac{1}{h+1} \{ (h+1) \frac{1}{R} +$$

$$\frac{\sqrt{h}\sqrt{3}}{3} \frac{(2n+1)}{\zeta} \frac{1}{R^2} \cos\theta +$$

$$\frac{h}{5} \frac{(2n+1)(2n+2)}{\zeta^2} \frac{1}{R^3} [3\cos^2\theta - 1]$$

$$(8)$$

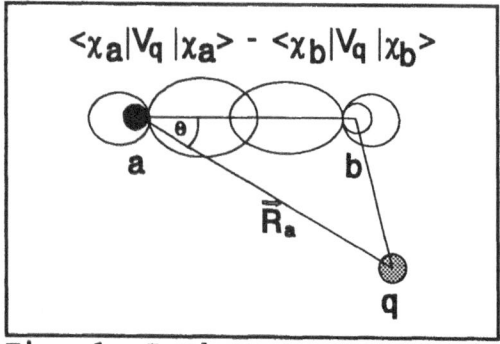

Fig. 1: Bond geometry used in eq. (7) and (8).

According to eq. (7) the bond polarization energy V_{ab} depends linearly on the bond order of the bond between a and b and on the charge q_x. Each term of the sum in eq.(7) is in the case of σ-bonds a difference of two integrals as given in eq. (8) depending on the geometry (see fig. 1) and on atomic parameters (n - principal quantum number, ς - orbital exponent and $h=-1/cos\Theta$ - hybridization). With the exception of the bond order and the charges the formulas can be easily evaluated from the molecular geometry.

4. Bond Polarization and Chemical Shift

Let us now give an interpretation of the bond polarization formula (6) in terms of the chemical shift. At first we introduce an one electron operator $(\hat{O}^{cs})_a$ producing as expectation value the chemical shift (or nuclear shielding) of the nucleus a. Conventionally second order perturbation theory has to be applied to calculate the nuclear shielding but the perturbation can be included into the operator $(\hat{O}^{cs})_a$ using:

$$\hat{H}' = \hat{H}^{(2)} - \sum_{k'} \frac{1}{(E_k^{(0)} - E_0^{(0)})} \hat{H}^{(1)} |k\rangle\langle k| \hat{H}^{(1)} \quad (9)$$

To the nuclear shielding contribute only those terms in $\hat{H}^{(1)}$ and $\hat{H}^{(2)}$ that are bilinear in the magnetic moment of the nucleus a and the external magnetic field. Fortunately the explicit form of this operator is not needed because we introduced empirical parameters for its expectation values.

The operator $(\hat{O}^{cs})_a$ contains in its denominator the distance from the nucleus a to the electron j and therefore this operator will act mainly on bonds that are directly connected to the nucleus of interest a. In this way a very simple definition of the molecular part A is introduced containing only the first bond sphere of atom a. We now write eq. (6) in terms of the chemical shift δ_a:

$$\delta_a = \delta_a^{fs} + \sum_i^{n_a} A_{ab_i}^{pol} V_{ab_i} \quad (10)$$

The first term of eq. (10) is the sum of the n_a (coordination number) bond contributions that add up to the chemical shift of the unpolarized first coordination sphere:

$$\delta_a^{fs} = 2 \sum_i^{n_a} \langle i|\hat{O}_a^{cs}|i\rangle \quad (11)$$

The second term stands for the change of the chemical shift with bond polarization and the slope is defined by:

$$A_{ab_i}^{pol} = \frac{2\sqrt{1-d_i^2}}{\Delta E_i} \langle i^*|\hat{O}_a^{cs}|i\rangle \quad (12)$$

Instead of calculating the integrals in eq. (11) and (12) we calculate only the bond polarization energies V_{ab} and determine the parameters δ^{fs} and A^{pol} by linear regression inserting experimental shifts on the left side of eq.(10). On the first sight we have to determine a large number of empirical parameters but if all bonds of an atom are equivalent (all neighbor atom are of same type) only two empirical parameters are left. The situation for the coordination number one is illustrated in fig. 2.

5. Chemical Shift of Hydrogen Bonded Protons

High resolution solid state N.M.R. investigations of crystalline substances revealed the close connection between proton chemical shift and the geometry of the environment of the hydrogen bonds (see Berglund and Vaughan (1980), Yeffrey and Yeon (1986)). If we limit our considerations on a O-H bond that is polarized by

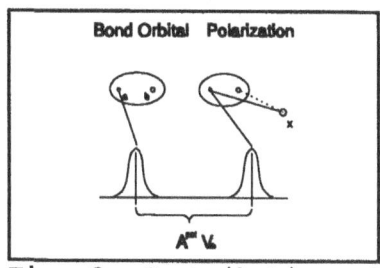

Fig. 2: Contribution of the bond polarization to the chemical shift.

an oxygen we arrive at a situation that is demonstrated in fig. 3. For the chemical shift we obtain:

$$\delta_H = \delta_H^{fs} + A_{HO}^{pol} V_{HO} \qquad (13)$$

$$V_{HO} = \sqrt{1-d^2}\ \{<1s_H|\hat{V}_{O\ominus}|1s_H> - <2sp_O^3|\hat{V}_{O\ominus}|2sp_O^3>\} = \qquad (14)$$

$$\sqrt{1-d^2}\{ -\frac{q_{O\ominus}}{r_{H\cdots O}} + \frac{q_{O\ominus}}{r_{O-O}}[\ 1 + \frac{(5/4)}{\zeta r_{O-O}}Cos\theta + \frac{(9/2)}{(\zeta r_{O-O})^2}(3Cos^2\theta - 1)]\}$$

The leading term in eq. (14) is obviously the contribution with the H⋯O distance in its denominator and we can conclude that to a first approximation the ^1H chemical shift depends reciprocal of the H⋯O distance. The bond angle should be not significant. We plotted therefore chemical shifts of hydrogen bonded protons against $1/r_{H-O}$ including only substances where the position of the proton had be determined using the neutron scattering method (see fig. 4). X ray investigations produce systematically shortened bonds between hydrogen and heavier atoms and had

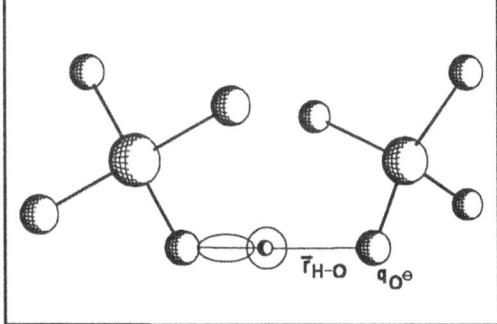

Fig. 3: O-H⋯O hydrogen bond

therefore to be excluded. Keeping in mind the simplicity of the model the scatter within the correlation displayed in fig. 4 is relatively low. If we want to determine the parameters in formula (13) we have to introduce a more comprehensive model. We included into the polarizing potential the neighbor atoms of both oxygen in the form X_β-O-H⋯O-Y_β and calculated the charges on the atoms O,X and Y using simple approximations (see Sternberg and Brunner (1991)). The ^1H chemical shift of an isolated $OH^\bullet$ is calculated to $(\delta^{fs})_H = 4.7$ ppm. For Ca(OH)$_2$ a ^1H shift value of 1.4 ppm was obtained for the protons in $OH^\bullet$ ions. This correspondence

440

indicates that the bond polarization model can be used not only for hydrogen bonds.

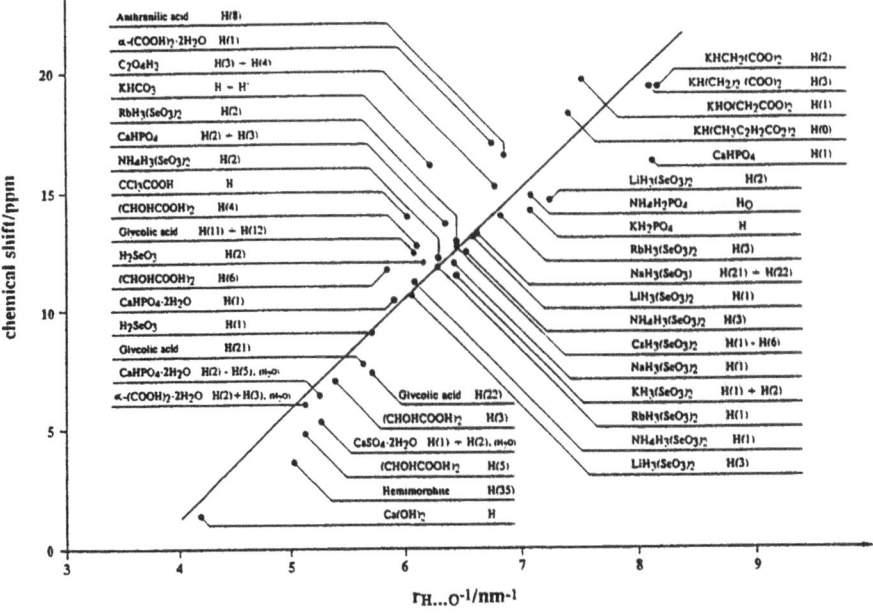

Figure 4: ^{1}H chemical shift plotted against $1/r_{H-O}$

6. ^{29}Si Chemical Shift in SiO$_2$ compounds

From the investigation of quartz polymorphs and dealuminated zeolits empirical correlations were proposed connecting the Si-O-Si bond angles or Si-Si distances with ^{29}Si chemical shifts (see Smith and Blackwell (1984)). In fig. 4 these geometry formulas are compiled and there is an obvious similarity for Si-O-Si bond angles within the range from 135 to 155° (see Pettifer et al. (1988)).

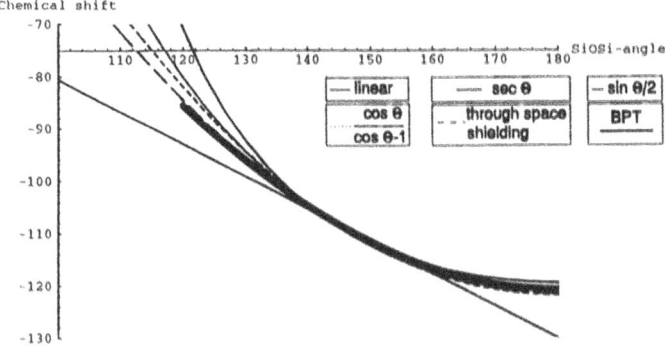

Figure 5: Empirical correlations of ^{29}Si chemical shifts with the Si-O-Si bond angle

In compounds containing SiO_4 tetrahedra the silicon is surrounded by 4 equivalent bonds and we can conclude that only two empirical parameters are needed to predict ^{29}Si chemical shifts:

$$\delta_{Si} = \delta_{Si}^{\beta} + A_{SiO}^{pol} \sum_{i}^{4} V_{SiO_i} \qquad (15)$$

In the next chapter we will show that in some cases an additional parameter is needed to account for π-bond contributions. If only the ß-effect is included the potential that polarizes the 4 Si-O bonds consists of a sum over the 4 next neighbor silicones of the SiO_4 tetrahedron under study (see fig. 6 and eq.(16)). Each second sphere silicon polarizes the for bonds of the SiO_4 tetrahedron. The largest term is in every case the contribution of the adjacent oxygen. From this oxygen contribution one can derive a simple formula for the dependence of the bond angle if we set the Si-O distance fixed to 1.60 Å:

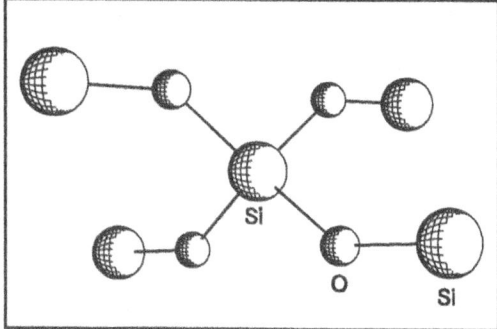

Fig. 6: Second sphere surronding of a SiO_4 tetrahedron

$$V_{SiO} = <i|\hat{V}_x|i^+> = \sqrt{1-d^2} \sum_{x}^{4} q_x \left(<\chi_{Si}|\frac{1}{|\vec{R}_x-\vec{r}|}|\chi_{Si}> - <\chi_O|\frac{1}{|\vec{R}_x-\vec{r}|}|\chi_O>\right) \qquad (16)$$

$$<2sp_O^h|\frac{1}{|\vec{R}_{SiO}-\vec{r}|}|2sp_O^h> = 0.331 - 0.139\frac{\cos\theta\sqrt{-\cos\theta}}{(\cos\theta-1)} - 0.042\frac{(3\cos^2\theta-1)}{(\cos\theta-1)} \qquad (17)$$

The second term of this formula is near to the formulation of Engelhardt and Radeglia (1984). The aim of eq. (17) is not to replace one shift to geometry formula by another but to show their limitations. The bond polarization formula in its general form is valid for all substances containing SiO_4-tetrahedra whereas the simple approximations can only be used for Q^4-networks.

For 20 different Si sites of 11 compounds the bond polarization energies were calculated using eq. (16) and the parameters of eq. (15) were determined by fitting them to experimental shifts (see Sternberg and Prieß (1992)). The correlation showed in the case of the dodecasils large deviations up to 5 ppm. This problem was solved when we included additional π-bond polarization in the same way like in the case of the phosphates (see following chapter 7). The 3d-orbitals of a free silicon atom are unoccupied but they can acquire d-electrons from oxygen due to the π back donation effect. The d-p-π bonds to the oxygens lead to a substantial bond shortening. The results are compiled in the following system of equations:

ΣV_σ	ΣV_π /10^{-3}Hartree			δ/ppm exp.	δ/ppm calculated	
434.39	31.26	1		-107.4	-108.6	α-Quartz
466.83	83.98	1		-109.9	-109.9	α-Cristobalite
477.96	50.25	1		-110.4	-111.5	Chabazite
521.12	144.26	1		-112.2	-112.8	Mordenite Si(1)
562.49	231.79	1		-115.0	-114.0	Si(2)
554.94	145.46	1		-115.0	-115.3	Si(3)
520.70	187.74	1		-113.1	-111.8	Si(4)
438.32	49.72	1	A^{pol}_σ	-107.8	-108.5	Zeolite Y
569.10	182.87	1		-116.2	-115.6	TMA Sodalite
698.55	464.85	1	A^{pol}_π =	-119.8	-119.3	Dodecasil 3C Si(1)
736.12	630.22	1		-117.0	-118.5	Si(2)
730.52	893.05	1	δ^{fs}	-112.8	-112.2	Si(3)
488.86	104.59	1		-113.9	-111.2	Coesite Si(1)
420.82	18.29	1		-108.1	-107.8	Si(2)
701.47	540.40	1		-118.0	-117.8	Dodecasil 1H Si(1)
643.86	407.57	1		-115.6	-116.3	Si(2)
702.67	452.98	1		-121.1	-119.9	Si(3)
723.45	578.28	1		-117.6	-118.7	Si(4)
504.13	128.50	1		-111.0	-111.8	Tridymite (mean value)
566.66	260.67	1		-113.2	-113.7	TPAF Silicalite (mean value)

With the exception of coesite the differences between calculated and observed shifts are not far from the differences between different N.M.R. laboratorys.

The interpretation of ^{27}Al chemical shifts of AlO_4-tetrahedra in terms of the bond polarization is similiar to the case of the SiO_4-tetrahedra with the exception that the influence of π-bonds is negligible (Hallas and Sternberg (1989)).

7. The ^{31}P Chemical Shift of Phosphates

Four σ-bonds are not sufficient for the description of the bond system of phosphorous and we have to include the possibility of the formation of d-p-π bonds. If a bond coordinate system is introduced with the z-axis pointing from the phosphorous to oxygen and from oxygen in the opposite direction the orbitals displayed in fig. 7 have the proper symmetry to form a d-p-π bond. If we adopt the principle of σ-π separation we get consequently two independent bond systems and for their polarization we have to introduce two empirical slopes

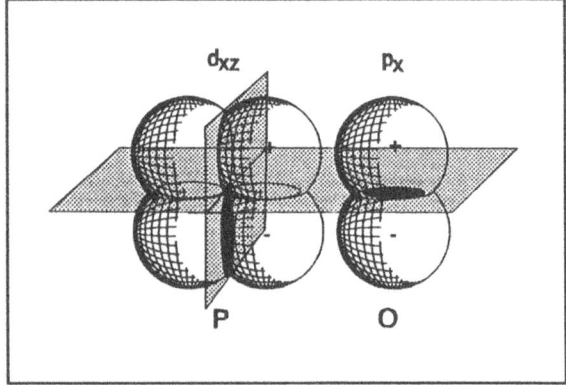

Fig. 7: A d_{xz}-orbital of the phosphorous and a p_x-orbital of the oxygen that can form a d-p-π bond

$(A^{pol})_\sigma$ and $(A^{pol})_\pi$ (see eq. (18)).

$$\delta_P = \delta_P^{fs} + A_{PO\sigma}^{pol} \sum_i^4 V_{PO\sigma_i} + A_{PO\pi}^{pol} \sum_i^4 V_{PO\pi_i} \qquad (18)$$

The formulas for the integrals for the bond polarization energy including d-orbitals are given in Sternberg, Pietrowski and Prieß (1990).

The σ and π bond polarization energies were calculated for 17 phosphates. All positive charges within a radius of 4 Å around a phosphorous were included into the potential for the calculation of the bond polarization. The parameters of eq. (18) were obtained from a least square solution of a system of 20 equations and the experimental ^{31}P shifts were plotted against the calculated values (see fig. 8). For the chemical shift of an isolated $(PO_4)^{3-}$ a value of $\delta^{fs} = 6.12$ ppm was obtained and this value is near to reference substance. The change of the shift with σ-bond polarization was obtained to $(A_{PO\sigma})^{pol} = -81.37$ ppm/Hartree and this value is similar to the value obtained for a Si-O σ-bond with $(A_{SiO\sigma})^{pol} = -77.14$ ppm/Hartree. For the differences between silicon and phosphorous shifts the π bond contribution is found to be responsible. A value of $(A_{PO\pi})^{pol} = 94.58$ ppm/Hartree for P-O and only $(A_{SiO\pi})^{pol} = 22.35$ ppm/Hartree for Si-O was obtained. The π electron polarization is in the case of the ^{29}Si chemical shift only a correction but in the case of the ^{31}P shifts it is the dominating effect. This point leads to another fundamental difference. The largest contribution to the bond polarization depends in the case of the Si-O bond only of the Si-O-Si bond angle (see eq. 17). Because of the nodal plain within the d-p-π-bond (fig. 7) its polarizations depends additionally on the dihedral angle ϕ for the rotation around the

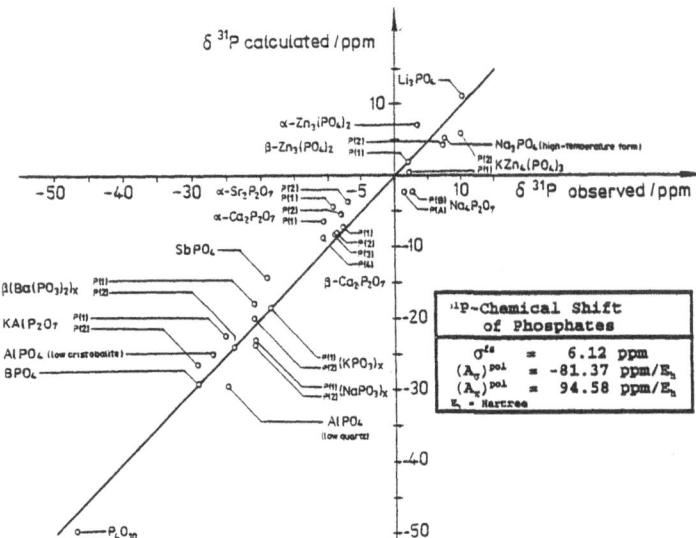

Figure 8: Experimental ^{31}P chemical shifts of phosphates plotted against theoretical values calculated using eq. (18).

P-O bond. Only in the case of a terminal oxygen two π-bonds can be formed resulting again in rotational symmetry. Consequently the ^{31}P shift will depend in a complicated way on the dihedral angle and on the bond angle. Each bond contribution depends additionally on the π-bond order and therefore there are no simple geometry formulas for ^{31}P chemical shifts. The correlation presented in fig. 8 accounts for all these effects and it is now possible to study the influence of the geometry and electron distribution in detail using molecular models. For the phosphate groups in glasses this was carried out by Losso et al. (1992).

8. ^{13}C Chemical Shifts in Vinyl Compounds

Calculating bond polarization energies it can be observed that in the case of σ-bonds the contributions of the bonds pointing into different directions cancel out and only a small difference remains. In the case of an isolated C-C-π bond no such cancellation can occur and large effects can be observed. It can be demonstrated that most shift effects in vinyl compounds can be explained from the π-bond polarization alone. We have then again the fortunate case that the bond polarization formula will contain only two empirical parameters:

$$\delta_C = \delta_C^{\delta} + A_{CC\pi}^{pol} \, V_{cc\pi} \qquad (19)$$

The integrals for the π-bond polarization energy $V_{CC\pi}$ can be calculated from (see Sternberg et al. (1990)):

$$<np_x| \frac{1}{|\vec{R}-\vec{r}|} |np_x> = \qquad (20)$$

$$\frac{1}{2R} + \frac{(2n+1)(2n+2)}{80R^3\zeta^2}[3\sin^2\Theta(\cos^2\Phi-\sin^2\Phi)-(3\cos^2\Theta-1)]$$

Let us now discuss the influence of groups X and Y on the chemical shift of C^* in compounds of the type H_2C^*=CXY. Since the shifts were measured in the liquid state we only used simple standard values for the geometry. If we limit our consideration on the β-effect the angle Φ in eq. (20) can be set to zero. If an electronegative atom is located in γ-position its influence cannot be neglected and it will additionally introduce an influence of the dihedral angle. In the first paper (Sternberg (1988)) the electronegativity differences ΔEN were used to define the charges on the substituents. The formula $d=(0.16+0.035| \, \Delta EN \, |)\Delta EN$ was used to calculate the amount of charge d that was shifted within one bond. These formulas are not necessary for the bond polarization theory and quantum chemical calculations including geometry optimization could have been used as well, but then the calculations could not have been performed using a hand held calculator. Because of the additivity of the point potential the bond polarization energies of the substituents can be regarded as increments and the sum of the X and Y contribution can be inserted into eq. (19) to calculate the chemical shift. In the first paper of Sternberg (1988) a table was presented containing these substituent contributions. In fig. 9 experimental shifts of vinyl compounds are plotted against the bond polarization energies of various substituents X and Y. This plot demonstrates that the bond polarization theory can be regarded as general description for $\beta,\gamma,\delta...$ substituent effects.

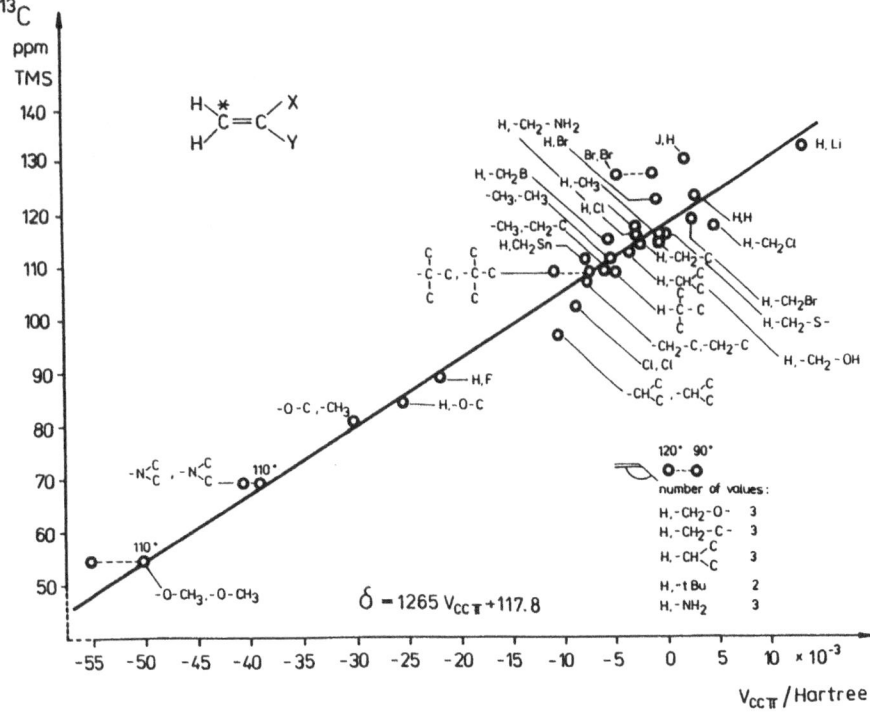

Figure 9: Experimental chemical shifts of various vinyl compounds plotted against the π-bond polarization energy.

9. Summary

The nuclear shealding of an atom is mainly governed by the inner shell electrons of that atoms and the electron distribution within the bonds that make up the first bond sphere. To calculate these contributions first principle quantum mechanical calculations must be performed. The central problem are large paramagnetic contributions to the nuclear shealding that occur when in the vicinity of the nucleus the wave function gets zero on nodal planes.

The chemical shift is a much simpler effect because only the change of the nuclear shealding is of interest and not the absolute value. This change is mainly caused by the polarization of the first sphere bonds by groups, atoms, ions or dipoles within the second or higher coordination spheres. The bond polarization formula (6) gives the connection between the change of an expectation value (in this case the chemical shift) and the bond polarization. This formula opens the possibility to correlate experimental shifts directly to calculated bond polarization energies and to derive formulas that account for the paramount shift effects. For every type of bond within the first bond sphere two empirical parameters must me determined: one parameter for the chemical shift of the unpolarized bond (bond increment) and a slope that accounts for the change of the shielding with bond polarization.

Other molecular properties that are governed by the first bond sphere will depend exactly in the same way from the bond polarization as the chemical shift. This can serve as explanation for large number of correlations of chemical shifts to other molecular properties. The validity of the bond polarization formula has been demonstrated for the deuterium coupling constant and for the atomic charge. The chemical shift will depend linearly on the atomic charge on the same nucleus if only two parameters are needed in the bond polarization formula.

From formulas for the bond polarization energy (7,8) the main factors that influence chemical shifts can be discussed:

(i)   The chemical shift is directly proportional to the bond order of the first sphere bonds. In the case of σ-bonds this value will vary only slightly and it can therefore be contracted in most cases into the empirical slope. If π-bonds are discussed the bond order can vary from zero to one and it must be taken into account explicitly. This can be done using empirical formulas that connect bond shortening and valence.

(ii)   The chemical shift is directly proportional to the charge on polarizing groups. These charges can be determined empirically from electronegativities or from quantum mechanical calculations. The absolute scale is of no concern because only the empirically determined slope will change if another charge scale is introduced.

(iii)   If the other parameters (i) and (ii) kept constant the distances to the polarizing charges and their angles to the first sphere bonds dominate the chemical shifts. In the case of π-bonds large influences of the dihedral angles will show up.

The bond polarization theory does not depend on time consuming computations like conventional quantum chemistry. We decided therefore to integrate the bond polarization into a PC-program named COSMOS (COmputer Simulation of MOlecular Structures) that unites crystallography, molecular modelling and computer visualization of the structures. The calculations within this paper were carried out using the COSMOS program.

References

Berglund, B. and Vaughan, R. W. (1980)
'Correlations Between Proton Chemical Shift Tensors Deuterium Quadrupole Couplings, and Bond Distances for Hydrogen Bonds in Solids', J. Chem. Phys. 73, 2037-2043

Engelhardt, G. and Radeglia, R. (1984)
'A Semi-Empirical Quantum-Chemical Rationalization of the Correlation Between SiOSi Angles and ^{29}Si NMR Chemical Shifts of Silica Polymorphs and Framework Aluminosilicates (Zeolites)', Chem. Phys. Lett. 108, 271-274

Hallas, E. and Sternberg, U. (1989)
'New Aspects of the Semiempirical Theory of ^{27}Al Chemical Shifts and its Application to Glasses', Mol. Phys. 68, 315-326

Jeffrey, G. A. and Yeon, Y. (1986)
'The Correlation Between Hydrogen-Bond Lengths and Proton   Chemical Shifts in Crystals', Acta Cryst. B42, 410-413

Losso, P., Schnabel, B., Jäger, C., Sternberg, U., Stachel, D. and Smith, D. O. (1992)
'^{31}P NMR-Investigations of Binary Alkaline Earth Phosphate Glasses of Ultra Phosphate Composition', J. Non. Cryst. Sol. 143, 265-273

Pettifer, R. F., Dupree, R., Faran, I. and Sternberg, U.(1988)
'Application of the Chemical Shift for the Determination of   Angular Distributions in Glasses', J. Non. Cryst. Sol. 106, 408-412

Smith, J. V. and Blackwell, C. S. (1983)
'Nuclear Magnetic Resonance of Silica Polymorphs',
Nature 303, 223-225

Sternberg, U. (1988)
'Second Sphere Theory for the Interpretation of Chemical Shifts', Molecular Physics 63, 249-267

Sternberg, U., Pietrowski, F. and Priess, W. (1990)
'The Influence of Structure and Geometry on the ^{31}P-Chemical Shift in Phosphates', Z. f. phys. Chem. (Neue Folge), 168, 115-128

Sternberg, U. and Brunner, E. (1991 submitted)
'The Influence of the Short Range Geometry on the Chemical Shift of Protons in Hydrogen Bonds', Chem. Phys. Lett.

Sternberg, U. and Priess, W. (1992 in press)
'The Influence of Structure and Geometry on the ^{29}Si-Chemical Shift',J. Magnetic Resonance

NITROGEN SHIELDING TENSORS

JOAN MASON
Department of Chemistry
The Open University
Milton Keynes MK7 6AA, UK

ABSTRACT.   Nitrogen shielding tensors are compared for singly-bonded nitrogen, and multiply-bonded nitrogen in linear and planar groups, including the peptide link, NO_x groups, and ligands in metal complexes. In isoelectronic (isostructural) locations, nitrogen tensors resemble those for carbon or phosphorus, with scaling by the respective $<r^{-3}>_{np}$ factors. In nitrogen-containing chromophores, however, major deshieldings are observed for axes which mix n_N (lone pair) and π^*orbitals. Thus in biomolecules nitrogen shielding tensors give useful information on protonation and hydrogen bonding, and the shielding tensor affords a criterion for the bending of nitrogen in ligands in metal complexes, as demonstrated for metal nitrosyls. In the dinitrogen oxides N_2O, N_2O_2, N_2O_3, and N_2O_4, there is a substantial contribution to the nitrogen shift from the N-N bond (as shown by LORG calculations), particularly in N_2O_2 in which this bond is very long.

1. Introduction

Nitrogen shielding tensors[1-3] have been measured by a number of methods: by ^{14}N spin-rotation interactions (N_2, NH_3, N_2O, HCN, PN), on oriented single crystals (^{15}N in glycylglycineHCl.H$_2$O, L-histidineHCl.H$_2$O, ^{14}N in nitrates), many in recent years by ^{15}N CP-MAS and powder lineshape analysis, on organic and inorganic molecules, and biomolecules. Concurrent measurement of dipolar interaction between $^{14,15}N$, $^{1,2}H$, ^{13}C, ^{31}P or other nuclei are used to relate the shielding tensor in the molecular frame,[4] two such measurements being needed for unequivocal orientation.[5] Anisotropies have been measured in liquid crystals (MeNC, MeCN, N_2O), and by relaxation methods (PhNO$_2$, pyridine). References to the experimental determinations, including those discussed in the text, are attached to the Tables for ease of retrieval: in very active fields (peptides, for example) only key references are given.

   In the Tables, shifts δ/ppm are relative to neat liquid nitromethane, high frequency positive. Absolute shieldings σ, low frequency positive ($\sigma_{33} > \sigma_{22} > \sigma_{11}$, $\sigma_{iso} = \sigma_{tr}$) are obtained by referring the shifts to an absolute scale for nitrogen[6] based on a spin-rotation measurement for NH$_3$, such that $\sigma(Me^{15}NO_2$, liq.) = -135.8 and $\sigma(NH_3$, liq.) = 244.6 ppm at ambient temperatures (240.6 at -50 °C).[7] Tensor properties given are the span $\Omega = \sigma_{33} - \sigma_{11}$ and the skew $\kappa = 3(\sigma_{22} - \sigma_{iso})/\Omega$; thus for axial tensors, $\Omega = \sigma_\parallel \sim \sigma_\perp$ and $\kappa = \pm 1$.

J. A. Tossell (ed.), Nuclear Magnetic Shieldings and Molecular Structure, 449–471.

1.1 COMPARISONS OF NITROGEN WITH CARBON AND PHOSPHORUS

Following the isoelectronic principle (isoelectronic implying isostructural), similar shielding tensors are expected for nitrogen and carbon in pairs of analogous compounds, as shown in Table 1, with parallelisms of angular momentum relationships and excitation energies, and scaling according to the radial factors $<r^{-3}>_{2p}$ in the paramagnetic term.[8] These factors[9] are approximately in 2 to 1 ratio for N and C (the more electronegative element holds its valence p electrons closer to the nucleus, giving a larger paramagnetic shift). The relationship of nitrogen and phosphorus shieldings is closer since the valencies are the same, and the scaling (the ratio $<r^{-3}>_{2p,N} : <r^{-3}>_{3p,P}$) is now 1 to 0.7.[9]

Table 1   Parallelisms of carbon and nitrogen shielding tensors

Coordn. no.	Symmetry type	Carbon	Nitrogen
4	tetrahedral	CH_4, alkanes	NH_4^+, alkylammonium
3	planar	CO_3^{2-}	NO_3^-
		carboxylate RCO_2^-	nitro, RNO_2
		benzene >CH	pyridinium >NH^+
2	linear	CO_2	NO_2^+
		NCO^-, $MeC{\equiv}N$	NNO, $MeN{\equiv}C$
	ligands:	carbonyl MCO	linear nitrosyl MNO
		cyano MCN, MNC	dinitrogen MNN
1	linear	CO	$N{\equiv}O^+$
		CN^-	NN
		CNO^-	NNO
		$MeN{\equiv}C$	$MeC{\equiv}N$

Note that the comparisons in Table 1 exclude pyramidal or bent nitrogen carrying a lone pair, which is distinctive for nitrogen compared with carbon. A lone pair on nitrogen in delocalised (π) systems plays an important role in the shielding, because of substantial p character of the lone pair electrons and low excitation energies. The relatively low shielding of nitrogen in pyridine and the sizeable increase on protonation (stabilising the lone pair electrons (n_N) with formation of a strong bond to hydrogen) was the first indication of the importance of low-energy n_N - π^* paramagnetic circulations to nitrogen shielding patterns.[1]

2. Singly-bonded nitrogen

Shielding tensors are given in Table 2 for nitrogen in ammonia and some derivatives. The ammonium ions are near-tetrahedral, and anisotropies due to hydrogen bonding were too small to measure, so the tensor elements are equal to the isotropic shielding.

Table 2 Principal components of nitrogen shielding tensors in ammonia and derivatives.

	δ_{iso}	σ_{iso}	σ_{11}	σ_{22}	σ_{33}	Ω	κ	Ref.
$NH_3(g)$	-400.3	263.5	237.7	278.0	278.0	40.3	1	[1]
cis-[Pt($^{15}NH_3$)$_2$(SCN)$_2$][a]	-399.0	263.2	219.9	249.3	320.5	100.6	-0.41	[2]
	-386.0	250.2	212.5	244.1	294.1	81.6	-0.22	
$^{15}NH_4^+$ salts:								[3]
NO_3^-	-355.5	219.7						
SO_4^{2-}	-352.8	217.0						
Cl^-	-338.1	202.3						
Br^-	-336.0	200.2						
I^-	-321.6	185.8						

[a]Two crystal sites.
[1]Kukolich, S.G. (1975), J. Am. Chem. Soc. 97, 5704.
[2]Santos, R.A., Chien, W.J. and Harbison, G.S. (1989), J. Magn. Reson. 84, 357.
[3]Ratcliffe, C.I., Ripmeester, J.A. and Tse, J.S. (1983), Chem. Phys. Lett. 99, 177.

The shielding tensor for ammonia is unusual in having $\sigma_\parallel < \sigma_\perp$ ($\kappa = 1$), a small span, and in the substantial decrease in shielding on protonation (since this lone pair has little p-character, and n_N - σ^* as well as σ - σ^* energies are high). In the ammonium ion the shielding decreases with weakening of the hydrogen bond to the anion, that is, with increase in the positive charge on nitrogen. The deshielding is related to contraction of the $2p_N$ orbitals (via the radial factor $<r^{-3}>_{2p}$), from NH_3 to $NH_3...H^+$ to NH_4^+ with a very weak counterion.

Much smaller deshieldings are observed on the protonation of amines, in which the charge is less localised on the nitrogen. Because of relatively small protonation shifts, there are rather similar shift relationships for amine nitrogen and alkane carbon in corresponding environments (depending on the connectivity), even though the groups are not isostructural. $R-NH_3^+$ tensors (as in histidine, glycine, etc.) have a very small span, about 10 ppm.

There is little change in the NH_3 isotropic shift on formation of ammine ligands in cis-[Pt($^{15}NH_3$)$_2$(SCN)$_2$], with weak bonding to the metal. There is some increase in span of the NH_3 tensor, but considerable decrease in the skew, the complex being planar. The σ_{33} component is particularly sensitive to change in location in the crystal.

Combined shielding tensor and dipolar shift measurements have been used to characterise NH protons and measure the bond distances in polycrystalline and amorphous solids of biological importance, in ^{14}N[10] and in ^{15}N[11] resonance. Protonated imine nitrogen is discussed under the heading of the parent compound.

3. Nitrogen in linear groups

Table 3 Nitrogen tensor components in linear groups; all these tensors have $\kappa = -1$

		δ_{iso}	σ_{iso}	$\sigma_\perp$	$\sigma_\parallel$	Ω	Ref.
MeN≡C		-266	130	10	370	360	1
$NNO(g)$		-240	105	-18	349	367	2,3
$SC^{15}N^-$		˙169	34	-104	311	415	4
MeC≡N		-157	21	-142	346	488	5
$NNO(g)$		-141	5	-174	364	538	2,3
HC≡N(g)		-109	-27	-215	348	563	6
N≡N(g)		-74	-62	-263	340	603	7
$[Mo(dppe)_2(^{15}N_2)_2]$	N_β	-43.4	-92.2	-283	289	572	8
	N_α	-42.0	-93.6	-304	327	631	
P≡N		213	-349	-698	350	1047	9

[1] Yannoni, C.C. (1970), J. Chem. Phys. 52, 2005.
[2] Casleton, K. H. and Kukolich, S. G. (1975), J. Chem. Phys. 62, 2696.
[3] Bhattacharyya P. K. and Dailey B. P. (1973), J. Chem. Phys. 59, 5820.
[4] Dickson, R. M., McKinnon, M. S., Britten J. F. and Wasylishen, R. E. (1987), Can. J. Chem. 65, 941.
[5] Kaplan,S., Pines, A., Griffin, R.G. and Waugh, J.S. (1974), Chem. Phys. Lett. 25, 78; Kennedy, J. D. and McFarlane, W. (1975), Mol. Phys. 29, 593.
[6] Garvey R. M., and De Lucia F.C. (1974), J. Mol. Spectrosc. 50, 38.
[7] Ω from Ishol, L.M. and Scott, T.A. (1977), J. Magn. Reson. 27, 23.
[8] Groombridge, C. J., Mason, J. and Richards, R. L., unpublished work. Isotropic shifts (in thf): Donovan-Mtunzi, S., Richards, R.L. and Mason, J. (1984), J. Chem. Soc., Dalton Trans. 469.
[9] Raymonda, J. and Klemperer, W. (1971), J. Chem. Phys. 55, 232.

Principal components of nitrogen shielding tensors for linear systems are given in Table 3, and Figure 1 is a correlation chart. Maximal shielding is observed for the C_∞ axis, with its largely diamagnetic circulation: most $\sigma_\parallel$ values are in the range 300-370 ppm, resembling the (diamagnetic) shielding of atomic nitrogen, 325 ppm.[12]

The parallel shielding decreases by 60-100 ppm when the π system can delocalise into an aromatic ring, as in the iminophosphenium cation, and benzonitrile with 4-substituents Y (Table 4). In the benzonitriles, some of the largest changes in tensor components (relative to Y = Me), are produced by a positively charged substituent NMe_3^+, which decreases σ_{33} by 11 ppm (and σ_{11} and σ_{22} by 7 and 16 ppm respectively). Similarly the parallel shieldings in dinitrogen are reduced by 51 and 13 ppm on coordination to molybdenum.

The perpendicular tensor components arise from $\sigma \leftrightarrow \pi$ (that is, $\sigma \rightarrow \pi^*$ and $\pi \rightarrow \sigma^*$)

circulations, and for terminal nitrogen, from $n_N \rightarrow \pi^*$ circulations also. The separation of the σ_{11} and σ_{22} components by around 30 ppm in the benzonitriles and iminophosphenium ion reflects conjugation with the aromatic ring. In the benzonitriles the separation is largest (41 ppm) with the π-donor Me_2N as 4-substituent.

For terminal nitrogen in the linear groups the perpendicular shielding $(\sigma_\perp)$ is highest in NNO, and lowest for PN: this has the longest bond (so orbital splittings are smaller), also phosphorus is the least electronegative partner. For N_2O, *ab initio* calculations using the LORG method[13] show comparable (major) contributions to σ^p for the terminal nitrogen from its lone pair and the NN bond, and for the central nitrogen from the NN and NO bonds.

For two-coordinate nitrogen the perpendicular shielding decreases from MeNC to ArNP[+] and to $[Mo(dppe)_2(^{15}N_2)_2]$, with decrease in orbital splittings.

4 Nitrogen in planar systems

Figure 2 is a correlation chart for nitrogen tensors in planar systems, principal values being given in Table 4. Highest shielding and small span are observed for 3-coordinate nitrogen bound only to carbon and hydrogen, in amides or peptides, azinium nitrogen in the 5-membered rings in tryptophan and L-histidine.HCl.H_2O, and iminium groups (Sect. 4.4).

4.1 PEPTIDES AND AMIDES

Shielding tensors have been studied in many peptides.[2,3] Earlier conclusions as to the orientation of the nitrogen tensor were made firm by concurrent observations of $^{13}C^{15}N$ and $^{15}N^1H$ dipolar interactions in L-(1-^{13}C)alanyl-L-(^{15}N)alanine.[5] The axis of lowest shielding σ_{11} is in the amide plane perpendicular to the CN bond, allowing of relatively low-energy CN $\sigma \leftrightarrow \pi$ circulations. Unusually (in contrast to carbon tensors in peptides, and to carbon and nitrogen tensors in aromatic and other delocalised systems) the high-shielding axis σ_{33} is in-plane, in the CN bond direction; the intermediate axis σ_{22} is perpendicular to the plane. These directions are approximate, and vary somewhat in different peptides. In Gly-Gly and related peptide links there is (accidental) near-axial symmetry, with $\kappa \approx +1$, and the σ_{11} axis is at 99° to the CN bond,

σ_{11} is the least-variable component (because of constancy of the CN bond), but is the most sensitive to the geometry of the hydrogen bonding, the σ_{11} axis being nearest to the N-H...O direction. σ_{11} is smaller in an α-helix than in a β-sheet, and sensitive to the handedness of the helix. The out-of-plane component σ_{22} reflects the conformation and secondary structure, being larger in a β-sheet than in an α-helix.[14] σ_{22} is unusually low and σ_{33} unusually high ($\kappa = -0.3$) in the link to proline in which the nitrogen is in a saturated 5-membered ring, although the Ala-Ala isotropic shift is maintained in Ala-Pro. The importance of lattice effects is shown by differences of isotropic shifts in solution.

Table 4 shows similarities of the components in peptides, nylon-6, and asparagine, in which the axes are oriented as in peptides, the σ_{33} axis bisecting the NH_2 group.

Table 4 Principal components of nitrogen shielding tensors in planar groups

	δ_{iso}	σ_{iso}	σ_{11}	σ_{22}	σ_{33}	Ω	κ	Ref.
$^{15}NH_2COCH_2CH(NH_2)COOH.H_2O^a$								
	-276.8	141.0	64.5	107.9	250.6	186	-0.534	1
nylon-6 (-CO^{15}NH-)^b	-268	132	34	154	209	175	0.37	2
glycyl(^{13}C^{15}N)glycineHCl.H$_2$O	-267.4	135.6	30.6	180.8	183.3	153	0.967	3,4
L-alanyl(^{13}C^{15}N)L-alanine	-256.8	121.0	25.1	162.5	175.3	150	0.829	3
alanyl(^{13}C^{15}N)proline	-256	120	22	118	219	197	-0.030	5
RCO^{15}NHR' (peptides)	ca. -260	100-145	10-35	120-180	175-220			3-6
L-tryptophan.HCl		127.8	66.9	121.5	195	128	-0.148	7
tryptophan-26^c		126	70	116	185	115	-0.261	8
L-histidine.HCl.H$_2$O (π-^{15}N)		66.9	-19.8	35.6	184.9	205	-0.459	7
2,4,6-Bu$^t_3$C$_6$H$_2$^{15}N≡P$^+$	-132	-4	-155	-121	266	421	-0.834	9
4-MeC$_6$H$_4$C≡^{15}N	-126	-10	-151	-122	243.8	395	-0.851	10
pyridine-^{15}N	-83	-53	-387	-168	395	782	-0.441	11
Ph^{15}N(O)^{15}N(O)Ph		-66	-220	-43	65	285	0.242	12
PhCH=^{15}NPh	-48	-87	-365	-76	180	545	0.061	13
PhCMe=^{15}NOH	-31.4	-104	-315	-83	85	400	0.157	14
trans-RB^d (-CH=^{15}NH-)	-37.6	-97.8	-406.5	-92.5	205.6	612	0.026	15
RBH$^+$Cl$^-$	-181.2	44.6	-79.2	22.5	189.8	269	-0.246	
RBH$^+$Br$^-$	-186.8	51.6	-69.5	33.8	190.4	260	-0.205	
RBH$^+$I$^-$	-198.5	62.5	-56.7	53.3	190.8	247	-0.111	
bR$_{568}$ (all-trans)^e	-209.3	73.6	-42.9	60.1	203.1	246	-0.165	
bR$_{548}$ (13-cis)	-202.3	66.5	-42.9	39.1	204.1	247	-0.333	
PhNO$_2$(soln)	-2	-134	-399	-32	30	429	0.71	16
KNO$_3$	0^f	-136	-212	-210	14	226	-0.982	17
AgNO$_3$		-142	-212	-206	-7	205	-0.94	18
Pb(NO$_3$)$_2$		-146	-220	-220	2	218	-1.02	
Ba(NO$_3$)$_2$		-151	-227	-227	-0.5	226	-1.01	
ON-NO$_2$(g)^g	63	-199	-390	-218	10	400	-0.14	19
trans-Ph^{15}N=^{15}NPh	130	-267	-789	-146	136	925	0.392	20
Na^{15}NO$_2$	245	-381	-915	-263	36	951	0.372	21
O$_2$N-NO(g)^g	292	-428	-860	-435	10	870	-0.024	19
p-Me$_2$NC$_6$H$_4$^{15}NO	445	-581	-1457	-309	22	1479	0.550	12
ON-NO(g)^g	1835	-1971	-5278	-592	-44	5322	0.78	22

^{a}L-asparagine monohydrate; δ_{iso} assumed +76.1 ppm from NH$_4$Cl reference. ^{b}made from ^{15}N-enriched ε-caprolactam; δ_{iso} assumes glycine zwitterion reference has δ(N) -350 ppm. ^{c}(^{15}N$_{\varepsilon 1}$)-labelled Trp-26 side chain of the coat protein of fd bacteriophage. ^{d}RB is retinylidene-Bu^n-(^{15}N)imine. ^{e}bacteriorhodopsin (568 nm)[ε-^{15}N]Lys-labelled - see text. ^{f}isotropic shift assumed. ^{g}Experimental uncertainties are discussed in the text.

[1]Herzfeld, J., Roberts, J.E. and Griffin, R.G. (1987), J. Chem. Phys. 86, 597.

[2]Powell, D.G. and Mathias, L.J. (1990), J. Am. Chem. Soc. 112, 669.

[3]Oas, T.G., Hartzell, C.J., McMahon, T.J., Drobny, G.P.and Dahlquist, F.W. (1987), J. Am. Chem. Soc. 109, 5956.

[4]cf. Harbison, G.S., Jelinski, L.W., Stark, R.E., Torchia, D.A., Herzfeld, J. and Griffin, R.G. (1984), J. Magn. Reson. 60, 79.

[5]Valentine, K.G., Rockwell, A.L., Gierasch, M. and Opella, S..J. (1987), 73, 519.

[6]Duncan, T.M. (1990), A Compilation of Chemical Shift Anisotropies, Farragut Press, Chicago, p. N-5.

[7]Harbison, G.S., Herzfeld, J. and Griffin, R.G. (1981), J. Am. Chem. Soc. 103, 4752; Roberts, J.E., Harbison, G.S., Munowitz, M.G., Herzfeld, J. and Griffin, R.G. (1987), J. Am. Chem. Soc. 109, 4163.

[8]Cross, T.A. and Opella, S.J. (1983), J. Am. Chem. Soc. 105, 306.

[9]Curtis, R.D., Schriver, M.J. and Wasylishen, R.E. (1991), J. Am. Chem. Soc. 113, 1493.

[10]Sardashti, M. and Maciel, G.E. (1988), J. Phys. Chem. 92, 4620.

[11]Schweitzer, D. and Spiess, H.W. (1974), J. Magn. Reson. 15, 529.

[12]Wasylishen, R.E., communication at the NATO ARW on "The calculation of NMR shielding constants and their use in the determination of the geometric and electronic strucutres of molecules and solids", Maryland, 1992.

[13]Curtis, R.D., Penner, G.H., Power, W.P. and Wasylishen, R.E. (1990), J. Phys. Chem. 94, 4000.

[14]Wasylishen, R.E., Penner, G.H., Power, W.P. and Curtis, R.D. (1989), J. Am. Chem. Soc. 111, 6082.

[15]Harbison, G.S., Herzfeld J. and Griffin, R.G. (1983), Biochem. 22, 1; de Groot, H.J.M., Harbison, G.S., Herzfeld J. and Griffin, R.G. (1989), Biochem. 28, 3346.

[16]Schweitzer, D. and Spiess, H.W. (1974), J. Magn. Reson. 16, 243; Stark, R.E., Vold, R.L. and Vold, R.R. (1977), Chem. Phys. 20, 337.

[17]Bastow, T.J. and Stuart, S.N. (1991), Chem. Phys. Lett. 180, 305.

[18]Santos, R.A., Tang, P., Chien, W.-J., Kwan, S. and Harbison, G.S. (1990), J. Phys. Chem. 94, 2717.

[19]calculated from spin-rotation constants given by Kukolich, S.G. (1982), J. Am. Chem. Soc. 104, 6927 (see text).

[20]Wasylishen, R.E., Power, W. P., Penner, G. H. and Curtis, R. D. (1989), Can. J. Chem. 67, 1219.

[21]Groombridge, C.J., Larkworthy, L.F. and Mason, J., MS in preparation.

[22]calculated from spin-rotation constants given by Western, C.M., Langridge-Smith, P.R.R., Howard, B.J. and Novick, S.E. (1981), Mol. Phys. 44, 145 (see text).

4.2 AZINES, AZINIUM

Few nitrogen tensors have been studied for aromatic azines, despite their chemical importance. Figure 2 gives ranges of principal components observed for protonated nitrogen in five-membered rings of biological interest, and the values for pyridine . It is a general observation for delocalised (pπ) systems in carbon resonance that the out-of-plane axis shows the highest shielding (σ_{33}), depending on relatively high-energy $\sigma \rightarrow \sigma^*$ excitations in the plane. This is borne out in nitrogen resonance, as in the protonated imidazole ring in histidineHCl, in which the σ_{33} axis is (almost) perpendicular to the ring, and the σ_{11} axis along the NH bond. Similar components are observed for tryptophan studied as the hydrochloride, and *in vivo* in a bacteriophage that orients in the magnetic field.

Comparison with pyridine shows the major increases in σ_{11} and σ_{22} on protonation. In pyridine the σ_{11} axis is tangential to the ring, mixing n_N and π^* orbitals, and the σ_{22} axis radial, this shielding being mediated by $\sigma \leftrightarrow \pi$ circulations: the value of σ_{22} for pyridine resembles that of $\sigma_\perp$ in MeCN rather than MeNC (Table 3).

4.3 PLANAR SYSTEMS WITH CN, NN AND NP MULTIPLE BONDS

The ArCN and ArN$^+$P compounds are included in Figure 2 since conjugation with the aromatic system removes the axial symmetry of the CN and PN groups, but these bonds are still effectively triple and form the σ_{33} axis.

In the compounds with bent 2-coordinate nitrogen, (*E*)-acetophenone-oxime, benzylideneaniline, *trans*-azobenzene, and 4-dimethylamino-nitrosobenzene, dipolar-chemical shift measurements support the σ_{22} direction as bisecting the angle at nitrogen (effectively along the lone pair axis). The σ_{11} axis is that of the $n_N \rightarrow \pi^*$ circulation. Figure 2 shows that the σ_{11} component in the planar groups becomes more strongly negative as the π^* LUMOs come down in energy relative to the n_N orbital, with increasing electronegativity of the partner in the double bond, from C=N to N=N and to N=O. Lowest shieldings in Figure 2 are σ_{11} = -789 ppm for azobenzene (which is red-orange, with $n_N \rightarrow \pi^*$ absorption at 448 nm) and -1457 for 1,4-dimethylamino-nitrosobenzene, which is green (cf. Sect. 4.5).

In *trans*-azobenzene there are two molecular sites, differing in the span (925 and 880 ppm) and skew (0.392 and 0.325), while the isotropic shift is maintained. The azo group (CNNC) is effectively planar and centrosymmetric in both sites, with the same N=N bond length , but the torsion angles NNCC are significantly different, 17° and <10°. (The low barrier to twisting of the phenyl group is shown by the torsion angle of 30° observed in the gas phase by eelctron diffraction.)

4.4 ALDIMINE PROTONATION AND THE OPSIN SHIFT

Efects of protonation of nitrogen in a π-system and then of varying the strength of hydrogen-bonding to the counter-ion have been studied in depth in a Schiff base (aldimine) retinylidene-Bu^n-(^{15}N)imine, RB (all-*trans* retinol is vitamin A$_1$). RB models the light-harvesting membrane protein bacteriorhodopsin (bR), which was measured with selective ^{15}N,^{13}C-enrichment.[15] Protonation of RB with a strong acid (HCl) increases σ_{11} by 327 ppm and σ_{22} by 115 ppm (Table 4 and Fig. 2). From HCl to HBr to HI and to weak organic acids (carboxylic acids or phenols) σ_{11} and σ_{22} show small increases with strengthening of the N-H$^+$ bond (σ_{11} and σ_{22} are linearly related) while σ_{33} is little changed. The values for dark-adapted bR resemble those of RBH$^+$ when very weakly hydrogen-bonded.

The ^{15}N and ^{13}C studies demonstrate the major contribution of the counter-ion to the opsin shift in the photocycle. Colour changes in the protein with pH correlate with changes in the ^{15}N shielding, which increases by 16 ppm as neutral purple membrane (568 nm) becomes acid blue membrane (600 nm), reverting with the change to acid purple membrane (565 nm); the (very weak) hydrogen bond in the protein is weaker in the blue form than in the purple forms. The opsin-induced red shift, therefore, is due to interaction of an electronegative group in the protein with the Schiff base end of the chromophore; *cis-trans* isomerisation in the polyene is unlikely to be involved (as shown by comparison with 13-*cis*,15-*syn*-bR$_{548}$, in which only σ_{22} is changed significantly).

4.5 NO$_x$ GROUPS

Figure 3 compares the principal components for planar nitrato, nitro and nitrosyl groups in NO$_x$ molecules and ions (Table 4) with those of the linear oxide N$_2$O. For the N$_2$O$_3$ molecule, spin-rotation hyperfine structure was observed in pulsed beam microwave studies, but with an interaction strength barely larger than experimental errors, so that only one spin-rotation constant could be determined for each nitrogen. Approximate values of the principal components were obtained[16] by use of the solution shifts, 292 ppm (nitroso) and 63 ppm (nitro) relative to nitromethane,[17] to calculate the average spin-rotation shielding terms, and by assuming σ_{33} values (for the out-of-plane axis) comparable with those for the other planar NO$_x$ groups, which are similar among themselves. The values for the nitric oxide dimer N$_2$O$_2$, which is the main component of liquid or solid NO, were calculated[16] from the spin-rotation constants determined by molecular beam electric resonance spectroscopy,[18] with uncertainties of about 200 ppm for σ_{33} and σ_{11} and 40 ppm for σ_{22}.

Nitrate ion is characterised by a small span, the $\sigma_\perp$ value resembling σ_{22} in ON*N*O$_2$. σ_{22} is less negative in nitrobenzene, because of the smaller paramagnetic contribution of the CN bond, and more negative when there is a lone pair on nitrogen in the plane, as in the bent nitroso group in O$_2$N*N*O, Me$_2$NC$_6$H$_4$NO, or N$_2$O$_2$; as expected, nitrite ion is intermediate.

Figure 3 shows the sizeable (σ_{11}) deshielding associated with the n_N - π^* circulation in these bent nitrosyl groups, and to a lesser extent in NO$_2^-$, although LORG calculations[13] suggest that the contribution to σ^p_{iso} of the n_N electrons on the symmetry axis exceeds that of the NO bonds. Nitrite is barely coloured - aqueous nitrite absorbs at 357 nm - but N$_2$O$_3$ is deep blue. In contrast is the high shielding in the nitroso dimer PhN(O)N(O)Ph, which is a diazene (or azo) di-*N*-oxide: the σ_{11} and σ_{22} values match those in nitrobenzene, and σ_{33} matches $\sigma_\perp$ in nitrate ion.

Deshielding in a nitrosyl group is most dramatic in N_2O_2. The σ_{22} and σ_{33} values are slightly more negative than in N_2O_3, but σ_{11} now plummets to a value of -5278 ppm, and σ_{iso} corresponds to an isotropic shift of $\delta = 1835$ which is far beyond the range observed by NMR spectroscopy for diamagnetic compounds. This enormous deshielding is expected to relate to the very low energy of $n_N - \pi^*$ excitation in N_2O_2, and the long and weak NN bond. Remarkably, N_2O_2 is colourless:[19] the long-wavelength electronic absorption is in the near infra red, beyond 1200 nm. The presence of very low-lying magnetically-allowed excited states is shown also by the temperature-independent paramagnetism.

4.6 THE DINITROGEN OXIDES

Properties of the dinitrogen oxides, compared in Table 5, make an interesting sequence. In N_2O both bonds are multiple, but N_2O_4, N_2O_3 and N_2O_2 are all planar, with long, weak N-N bonds, and in N_2O_2 the N-N bond is extraordinarily long, 2.33 Å, in contrast to 1.45Å for N_2H_4. This bond is too long to have any π component: indeed, the delocalisation of oxygen lone pair electrons provides an antibonding (σ^*_{NN}) contribution, the near-rectangular structure being maintained by weak O-O σ-bonding.[19]

Table 5 The dinitrogen oxides

| | O_2N-NO_2 | O_2N-NO | | ON-NO | NNO | |
		N1	N2		N1	N2
r(N-N)/Å[a]	1.78	1.86		2.33	1.13	
∠NNO/°	113	113	105	95(5)	180	
σ_{iso}(N)/ppm	-123	-199	-428	-1971	105	5
$n_N-\pi^*$ absn./nm	340(gas)	665 (ether)		>1200(liq)		

[a]cf. hydrazine H_2N-NH_2 has r(N-N) = 1.45 Å.

LORG calculations[13] for N_2O_2 show the importance to the shielding of a very low-lying π^*_{NO} excited state, which also has π_{NN} and π_{OO} character, and a rather low -lying π^*_{ONNO} state. The N-N bond, very long and very weak, is now the major contributor to the paramagnetic term. The calculated tensor shøws significant antisymmetry, 124 ppm, composed of a negative contribution from the NN bond and a positive one from the nitrogen lone pair. Calculations on diazirine have shown a low-lying excitation and sizeable antisymmetry, 262 ppm.[20]

5.0 Linear and bent nitrosyls in metal complexes

The nitrogen shift is thus a useful criterion for bending at nitrogen, as in the diazenido[21] and nitrosyl ligands.[22] Better still, of course, is the shielding tensor (many of the bent nitrosyls now studied in the solid state are unstable in solution). In five-coordinate {MNO}[8] complexes (that is, with 8 (d+n) electrons) the metal may have a trigonal pyramidal coordination sphere (d^8) with the nitrosyl linear (NO^+), or else a square pyramidal coordination sphere (d^6) with bent apical nitrosyl (NO^-), and similarly for the diazenido ligand ($N=NR^+$ and $N=NR^-$). Our first measurement was of $[RuCl(^{15}NO)_2(PPh_3)_2][BF_4]$,[23] which has one linear and one bent nitrosyl in the solid state. This shows fast fluxionality in solution (between a trigonal pyramidal structure with linear equatorial nitrosyls and the square pyramidal structure observed in the solid), and we observed an equilibrium isotope effect, splitting the averaged ^{15}N resonance, in the semi-^{15}N-enriched compound.[24] The nitrogen shielding tensors extracted from the spinning sideband patterns observed in CPMAS experiments were the first to be observed in metal complexes.

5.1 METAL CARBONYLS AS MODELS FOR THE LINEAR NITROSYL LIGAND

The carbonyl ligand is a useful model for linear nitrosyls , since a number of ^{13}C tensors have been measured: for $[M(CO)_6]$ (M = Cr, Mo, W),[25] $[Fe(CO)_5]$,[26] $[Ni(CO)_4]$,[27] $[M_3(CO)_{12}]$ (M = Fe, Ru, Os), $[Rh_6(CO)_{16}]$ and $[Ir_4(CO)_{12}]$.[28] In the terminal ligand, σ_{33} or $\sigma_{\parallel}$ is little changed from $\sigma_{\parallel}$ in free CO,[29] and the periodicity observed in the isotropic shifts[30] follows the variation in $\sigma_{\perp}$, which increases across the row and down the group of the transition metal. For a given d orbital configuration the ^{13}C shielding increases in parallel with decrease in the back-bonding, as the d orbitals drop below the carbon valence orbitals with increase in nuclear charge of the metal. Deshielding of the ligating carbon correlates with other concomitants of back-bonding which reflect the lengthening and weakening of the CO bond and the shortening and strengthening of the MC bond. The deshielding is enhanced by *trans*-ligands which are good charge donors (particularly polyhapto arenes), by a negative charge on the complex, and (for a given metal) with increase in the number of d electrons, e.g. from d^6 to d^8 to d^{10} complexes of cobalt.

The perpendicular components are slightly more negative in axial bonds, which are usually slightly longer than equatorial bonds in the same complex.

σ_{33} becomes more negative and the other components slightly less so in the face-capping ligand in $[Rh_6(CO)_{16}]$, in which the carbon is 4-coordinate (M_3CO).

5.2 THE LINEAR NITROSYL LIGAND

Table 6 gives nitrogen tensor properties for linear nitrosyl ligands, compared in Figure 4 with those of NNO, and the bent nitrosyl in $[RuCl(^{15}NO)_2(PPh_3)_2]^+$. The $\sigma_{\parallel}$ values are significantly more negative than is likely for the free ligand (NO^+), reflecting greater back-bonding to the nitrosyl compared with the carbonyl ligand. The perpendicular components are somewhat more negative than in NNO, and more negative in a *bis*-nitrosyl than in a mononitrosyl, for the same metal. B*is*-nitrosyls are normally *cis,* and there is delocalisation within the $M(NO)_2$ system, with deviations from linearity of MNO. $[Fe(NO)_2(cystine)]$ is an example of a compound for which the nitrogen shielding tensor has given the structure: the solid is amorphous and insoluble, but the nitrogen tensors clearly show the nitrosyls to be linear and probably *cis*.

Figure 4 shows the sizeable paramagnetic contribution to σ_{11}, and a significant contribution also to σ_{33}, when the nitrosyl bends, in the ruthenium nitrosyl with one bent and one linear ligand. These effects can be seen to some degree in *bis*-nitrosyls.

Table 6 ^{15}N shielding tensors in the linear nitrosyl ligand; all have $\kappa = -1$.

	δ_{iso}	σ_{iso}	$\sigma_\perp$	$\sigma_\parallel$	Ω	Ref.
[Fe(CO)$_3$(^{15}NO)]$^-$	18	-154	-295	127	422	1
[RuCl(^{15}NO)$_2$(PPh$_3$)$_2$][BF$_4$]	26	-162	-309	133	442	2
[Ru(^{15}NO)$_2$(PPh$_3$)$_2$]	37	-173	-284	48	332	
	40	-176	-281	35	316	
[Fe(^{15}NO)$_2$(cystine)]	40	-176	-337	145	482	3

[1]Laska, T.E. and Root, T.W. (1989), quoted by Duncan, T.M. ed. (1990), 'A Compilation of Chemical Shift Anisotropies', Farragut Press, Wisconsin, p. N-4. [PPN][Fe(CO)$_3$(NO)] structure: Pannell, K. H., Chen,Y.-S., Belknap, K., Wu, C.C., Bernal, I., Creswick, M.W. and Huang, H.N. (1983), Inorg. Chem. 22, 418.
[2]Mason, J., Mingos, D. M. P., Schaefer, J., Sherman D. and Stejskal, E. O. (1985), J. Chem. Soc., Chem. Commun., 444. [RuCl(^{15}NO)$_2$(PPh$_3$)$_2$][BF$_4$] structure: Pierpont, C.G and Eisenberg, R. (1972), Inorg. Chem. 11, 1088.
[3]Groombridge, C. J., Larkworthy, L. F. and Mason, J., MS. in preparation.

5.3 THE BENT NITROSYL LIGAND

Table 7 gives tensor properties of bent nitrosyl ligands, for which structures are given in the Chart. In some of these complexes planarity of the basal ligand (with bent apical nitrosyl) is maintained by chemical bonding, as in [Co(NO)(TPP)] and the quadridentate ligands with ethylene bridges; or strong hydrogen bonding as in [Co(NO)(ketox)$_2$]. In some others the nitrogen tensor components are a useful proof of structure, as in *bis*-chelate compounds not constrained to co-planarity, such as the [Co(NO)(rsal)$_2$] complexes (the parent complexes [Co(rsal)$_2$] having tetrahedral coordination). The tensor properties show that all these complexes adopt the pyramidal structure. [CoCl$_2$(^{15}NO)(PPh$_3$)$_2$] shows fast linear-bent fluxionality in solution,and the tensor properties show the nitrosyl to be bent in the solid.

Figure 5 is a correlation chart for a range of bent nitrosyls, in comparison with the bridging nitrosyl in [Ru$_3$(CO)$_{16}$(μ_2-NO)]$^-$, and N$_2$O$_2$. All components show significant deshielding compared with those in linear nitrosyls or the planar dinitrogen oxides, the bent nitrosyl σ_{22} values being comparable with that in N$_2$O$_2$. The σ_{11} values are strongly negative (while not approaching the value for N$_2$O$_2$), and are the most variable. The bridging nitrosyl in [Ru$_3$(CO)$_{16}$(μ_2-NO)]$^-$ shows relatively high σ_{11} shielding as the nitrogen is now 3-

Table 7 Principal components of ^{15}N shielding tensors in bent nitrosyl ligands

Complex	∠MNO/°	δ_{iso}	σ_{iso}	σ_{11}	σ_{22}	σ_{33}	Ω	κ	δ_{soln}	Ref.
[RuCl(^{15}NO)$_2$(PPh$_3$)$_2$]$^+$	136.0	303	-439	-994	-229	-94	900	0.700	flux.	1
[CoCl$_2$(^{15}NO)(PPh$_3$)$_2$]		531	-667	-1149	-550	-301	848	0.414	flux.	2
[Co(^{15}NO)(LL')$_2$]										
LL' =										
(S$_2$CNMe$_2$)$_2$	135.1	499	-535	-877	-672	-357	520	-0.790	501	2
{3,5-NO$_2$)$_2$salox}		679	-815	-1458	-839	-148	1310	-0.055	n.o.	2
(5-NO$_2$salox)		682	-818	-1479	-786	-189	1290	0.074	833.1	2
(naphthen)		709	-875	-1819	-478	-226	1593	-0.898	n.o.	2
(naph-ph)		711	-847	-2069	-357	-113	1956	0.752	n.o.	2
(7-Mesalen)		714	-850	-1968	-291	-291	1677	1.000	710.0	2
(salphen)		717	-853	-1963	-296	-296	1667	1.000	769.7	2
(phsal)$_2$		720	-856	-1348	-740	-480	868	0.401	n.o.	3
(benacen)	123	721	-857	-1536	-707	-327	1209	0.372	723.0	2
(amben)		721	-857	-1681	-576	-315	1366	0.617	734.3	2
(naphth-mph)		721	-857	-1894	-496	-183	1711	0.633	n.o.	2
(bzsal)$_2$		722	-858	-2058	-427	-89	1969	0.657	530.2	3
(acacen)	122	723	-859	-1402	-850	-326	1076	0.025	714.3	2
(bsal)$_2$		728	-864	-1398	-805	-390	1008	0.176	740.5	3
(msal)$_2$		735	-871	-1273	-826	-510	763	0.177	n.o.	3
(esal)$_2$	129	742	-878	-2095	-376	-162	1933	0.779	739.7	3
(TPP) 200 K	127	757	-893	-1853	-626	-198	1655	0.484	770.7	4
(ketox)$_2$	126.3	781	-917	-1260	-1007	-484	776	-0.348	740.3	3
cf. [Ru$_3$(CO)$_{10}$(μ-^{15}NO)]$^{-a}$		424	-560	-1160	-408	-112	1048	0.435	434	5

^{a}bridging NO.

[1]Mason, J., Mingos, D. M. P., Schaefer, J., Sherman D. and Stejskal, E. O. (1985), J. Chem. Soc., Chem. Commun., 444.

[2]Groombridge, C. J., Larkworthy, L. F. and Mason, J., MS. in preparation. Crystal structures, [Co(NO)(S$_2$CNMe$_2$)$_2$] : Enemark, J.H. and Feltham, R.D. (1972), J. Chem. Soc., Dalton Trans., 718; [Co(NO)(acacen)] and [Co(NO)benacen)]: Wiest, R. and Weiss, R. (1971), J. Organomet. Chem. 30, C33; (1972), Rev. Chim. Miner. 9, 655.

[3]Groombridge, C. J., Larkworthy, L. F., Marécaux, A., Povey, D. C., Smith, G. W. and Mason, J., J. Chem. Soc., Dalton Trans. in the press, reporting the crystal structure for [Co(NO)(Ñ-Etsalim)$_2$]. [Co(NO)(ketox)$_2$] crystal structure: Larkworthy, L. F. and Povey, D.C. (1983) J. Cryst. Spectrosc. Res. 13, 413.

[4]Groombridge, C. J., Larkworthy, L. F. and Mason, J., in the press. [Co(NO)(TPP)] crystal structure: Scheidt, W.R. and Hoard, J.L. (1973), J. Am. Chem. Soc. 95, 8281.

[5]Laska, T.E. and Root, T.W. (1989), quoted by T. M. Duncan, ed. (1990), 'A Compilation of Chemical Shift Anisotropies', Farragut Press, Wisconsin, p. N-4; solution spectrum: Stevens, R.E.and Gladfelter, W.L. (1983), Inorg. Chem. 22, 2034.

coordinate, Ru_2NO (the σ_{33} shielding is unexpectedly high).

The apical nitrosyls show a remarkable variation in the span, 520 - 1969 ppm, and the skew, from -0.8 to +1 (with accidental axial symmetry), while maintaining the isotopic shift δ at 680-780 ppm (except that higher shielding is observed with S_4 ligation of cobalt in the basal plane, as in the *bis*-dithiocarbamate). All components reflect significant mixing of the nitrogen and the cobalt paramagnetic circulations (more precisely, the out-of-plane components of the ligand field circulations deshielding cobalt), as observed in solution studies which showed some correlation of ^{15}N and ^{59}Co shifts.[22]

Because of large effects of relatively small changes in small excitation energies, sensitivity may be expected to the internal MNO geometry ($\angle$MNO, length and departure from axiality of the MN bond) and to the conformation of the nitrosyl relative to the ligators in the basal plane. The ligand field is sensitive also to the location of cobalt (usually above the plane), to the ligation in the plane, whether OONN, ONON or N_4, and to the bite of the chelate ligands. As to the conformation, the nitrosyl oxygen avoids good electron donors in the plane, lying over an electron acceptor (such as a phosphine) if one is present; in $[RuCl(^{15}NO)_2(PPh_3)_2]^+$ it lies over the linear nitrosyl. In the square pyramidal nitrosyls of known structure there is eclipsing in $[Co(NO)(S_2CNMe_2)_2]$ (over S), $[Co(NO)(acacen)]$ and $[Co(NO)(benacen)]$ (over N), and staggering in the others: in $[Co(NO)(7\text{-Mesalen})]$ (and $[Co(NO)(salen)]$) the oxygen lies over the ethylene bridge. In $[Co(NO)(ketox)_2]$ and $[Co(NO)(esal)_2]$ it lies over a gap between the two chelate ligands, so this may apply to the other rsal nitrosyls.

The remarkable variation of the tensor components may be explained by sizeable effects of the conformation of the nitrosyl relative to the ligand field of the coligands in the plane, and the small barrier to rotation of the nitrosyl, which is disordered in many of the crystal structures. Such torsional variation could account for compensatory effects in the tensor components, with σ_{22} and σ_{33} increasing as σ_{11} decreases so that the isotropic shift is maintained, Ω and κ increasing in parallel. A low torsional barrier and sensitivity of the projecting nitrosyl to intermolecular forces are shown by the unusual nitrogen tensors we observed for $[Co(NO)(TPP)]$ at ambient temperatures.

6. Nitrosyl swinging in [Co(NO)(TPP)] detected by measurement of the nitrogen tensor

Nitrogen tensor components in $[Co(^{15}NO)(TPP)]$ (Fig. 6) have enabled us to demonstrate swinging or spinning of the nitrosyl in the solid state.[31] From room temperature down to 220 K, unexpectedly, the tensor showed axial symmetry with $\kappa = 1$, a small span (242 ppm), and an isotropic shift 765 ppm (cf. 771 ppm in solution) appropriate to a bent nitrosyl. At 200 K, however, the full span appeared, with $\Omega = 1655$ ppm and $\kappa = 0.484$, but little change in isotropic shift. Since no intermediate rate behaviour was observed (no variation in linewidth), the solid was examined by differential scanning calorimetry. This showed a phase transition at 206.7 K ($\Delta H = 649$ J mol^{-1}); presumably some relaxation in the lattice allows the nitrosyl to revolve about the C_4 axis at higher temperatures. Scheidt and Hoard's X-ray diffraction study[32] at ambient temperatures showed a two-fold disorder of the cobalt relative to the TPP plane, and 4-fold disorder of the nitrosyl (in a given orientation) which is depicted in the 'hydra' diagram in Figure 7 taken from their paper. The oxygen lies in the gap between two porphin nitrogens, pointing towards a pendant phenyl group.

A calculation of limiting slow and fast cases gives Euler angles for the tensor components relative to the C_4 axis. If the σ_{33} axis is set perpendicular to the CoNO plane and the σ_{11} axis along the NO bond direction, transformation of the components measured at 200 K to principal axes defined by C_4 symmetry, averaging of the four positions, and comparison with the values observed at higher temperatures, give a value of 127° for the CoNO angle. In the X-ray study the averaged model gives an NO bond length which is too short, and a rather large CoNO angle, 135.2°. Using indirect evidence as to the displacement of the cobalt from the plane and of the nitrosyl nitrogen from the C_4 axis, the authors estimated a CoNO angle ≤128.5°, which our value of 127° now confirms.

It is likely therefore that Scheidt and Hoard's 'hydra' pictures a swinging of the nitrosyl which gives rise to the averaged tensor above 220 K; also that apparent anisotropies recorded for bent nitrosyls may have been reduced (and asymmetries η increased) by small-angle oscillations of the nitrosyl ligand.

References

[1]Mason, J. (1987) 'Nitrogen', chap. 12 in J. Mason (ed.), Multinuclear NMR, Plenum, New York, pp. 335-367.

[2]Duncan, T.M. (1990), A Compilation of Chemical Shift Anisotropies, Farragut Press, Chicago, N1-9.

[3]Theoretical and physical aspects of nuclear shielding, including shielding tensors, are surveyed annually by Jameson, C.J., chap. 1 in G.A. Webb (ed.), Nuclear Magnetic Resonance: a Specialist Periodical Report, The Royal Society of Chemistry. A later chapter surveys solid state NMR.

[4]Power, W.P and Wasylishen, R.E. (1991), Annu. Rep. NMR Spectrosc. 23, 1.

[5]Hartzell, C.J., Pratum. T.K and Drobny, G. (1987), J. Chem. Phys. 87, 4324.

[6]Jameson, A.K., Jameson C.J., Opposunggu D., Wille W., Burrell P.M., and Mason J. (1981), J. Chem. Phys. 74, 81-88.

[7]Jameson C.J., Jameson, A.K., Cohen, S.M., Parker, H., Opposunggu D., Burrell P.M. and Wille W. (1981), J. Chem. Phys. 74, 1608.

[8]Jameson, C.J. and Mason, J. (1987) 'The Chemical Shift', chap. 3 in J. Mason (ed.), Multinuclear NMR, Plenum, New York; Mason, J. (1979), Adv. Inorg. Chem. Radiochem. 22, 199; (1976), ibid., 18, 197.

[9]Barnes, R.G. and Smith, W.V. (1954), Phys. Rev. 93, 95; Carlson, T.A., Lu, C.C., Tucker, T.S., Nestor, W.W. and Malik, F. B. (1970), 'Eigenvalues, radial expectation values, and potentials for free atoms from Z = 2 to 126 as calculated from relativistic Hartree-Fock-Slater atomic wave functions', Oak Ridge National Laboratory, Oak Ridge, Tennessee.

[10]Griffin, R.G., Bodenhausen, G., Haberkorn, R.A., Huang, T.H., Munowitz, M., Osredkar, R., Ruben, D.J., Stark, R.E. and van Willigen, H. (1981), Phil. Trans. R. Soc. Lond., A299, 475.

[11]Roberts, J.E., Harbison, G.S., Munowitz, M.G., Herzfeld, J. and Griffin, R.G. (1987), J.

Am. Chem. Soc. 109, 4163.

[12]Malli, G. and Froese, C. (1967), Int. J. Quantum. Chem. 1s, 95.

[13]Moore, E.A., LORG=FULL results (MS in preparation). The programs are those of Dupuis, M., Spangler, D., Wendolonski, J.J., Schmidt M.W. and Elbert, St.T., GAMESS, 11/11/1991, Vax (VMS) version; Bouman, T.D. and Hansen, A.E, RPAC 8.6.1. The basis set is Pople's 6-311 G with the addition of 3 d functions on N and O, with the polarisation functions (d) split as written in GAMESS.

[14]e.g. Shoji, A., Ozaki, T., Fujito, T, Deguchi, K., Ando, S., and Ando, I.(1990), J. Am. Chem. Soc. 112, 4693.

[15]de Groot, H.J.M., Smith, S.O., Courtin, J., van den Berg, E., Winkel, C., Lugtenburg, J., Griffin, R.G. and Herzfeld, J., Biochem. (1990), 29, 6873; Lugtenburg, J., Muradin-Szweykowska, M., Heeremans, C., Paardoen, J.A., Harbison, G.S., Herzfeld, J., Griffin, R.G. (1986), J. Am. Chem. Soc., 108, 3104; cf. Table 4.

[16]J. Mason, MS. in preparation.

[17]Andersson, L.-O., Mason, J. (1968), Chem. Commun. 99.

[18]Western, C.M., Langridge-Smith, P.R.R., Howard, B.J. and S.E. Novick, S.E. (1981), Mol. Phys. 44, 145.

[19]Mason, J. (1975), J. Chem. Educ. 52, 445.

[20]Hansen, A.E. and Bouman, T.D. (1989), J. Chem. Phys. 91, 3552.

[21]Haymore, B.L., Hughes, M., Mason, J. and Richards, R.L. (1988), J. Chem. Soc., Dalton Trans., 2935.

[22]Duffin, P.A., Larkworthy, L.F., Mason, J., Stephens, A.N. and Thompson, R.M. Inorg. Chem. (1987), 26, 2034.

[23]Mason, J., Mingos, D.M.P., Schaefer, J., Sherman D. and Stejskal, E.O. (1985) J. Chem. Soc., Chem. Commun., 444.

[24]Mason, J., Mingos, D.M.P., Sherman, D. and Wardle, R.W.M. (1984), J. Chem. Soc., Chem. Commun., 1223.

[25]Oldfield, E., Keniry, M.A., Shinoda, S., Schramm, S., Brown, T.L. and Gutowsky, H.S. (1985), J. Chem. Soc., Chem. Commun., 791-793.

[26]Mahnke, H., Sheline, R.K. and Spiess, H.W. (1974) J. Chem. Phys. 61, 55-60.

[27] Spiess, H.W., Grosescu, R., and Haeberlen, U. (1974), Chem. Phys. 6, 226-234 .

[28]Gleeson, J.W., Vaughan, R.W. (1983), J. Chem. Phys. 78, 5384; Duncan, T.M., Yates, J.T. and Vaughan, R.W. (1980), J. Chem. Phys. 73, 975; Walter, T.H., Reven, L. and Oldfield, E. (1989), J. Phys. Chem. 93, 1320.

[29]Ozier, I., Crapo, M. and Ramsey, N.F. (1968), J. Chem. Phys. 49, 2314; Meerts, W.L., De Leeuw, F.H. and Dymanus, A. (1977), Chem. Phys. 22, 319; Neumann, D.B. and Moskowitz, J.W. (1969), J. Chem. Phys. 50, 2216-2236.

[30]see ref. 8, p. 72ff (coordination shifts).

[31]Groombridge, C.J., Larkworthy, L.F. and Mason, J., Inorg. Chem., in the press.

[32]Scheidt, W.R. and Hoard, J.L. (1973), J. Am. Chem. Soc. 95, 8281

Figure 1 Nitrogen shielding tensors in linear groups

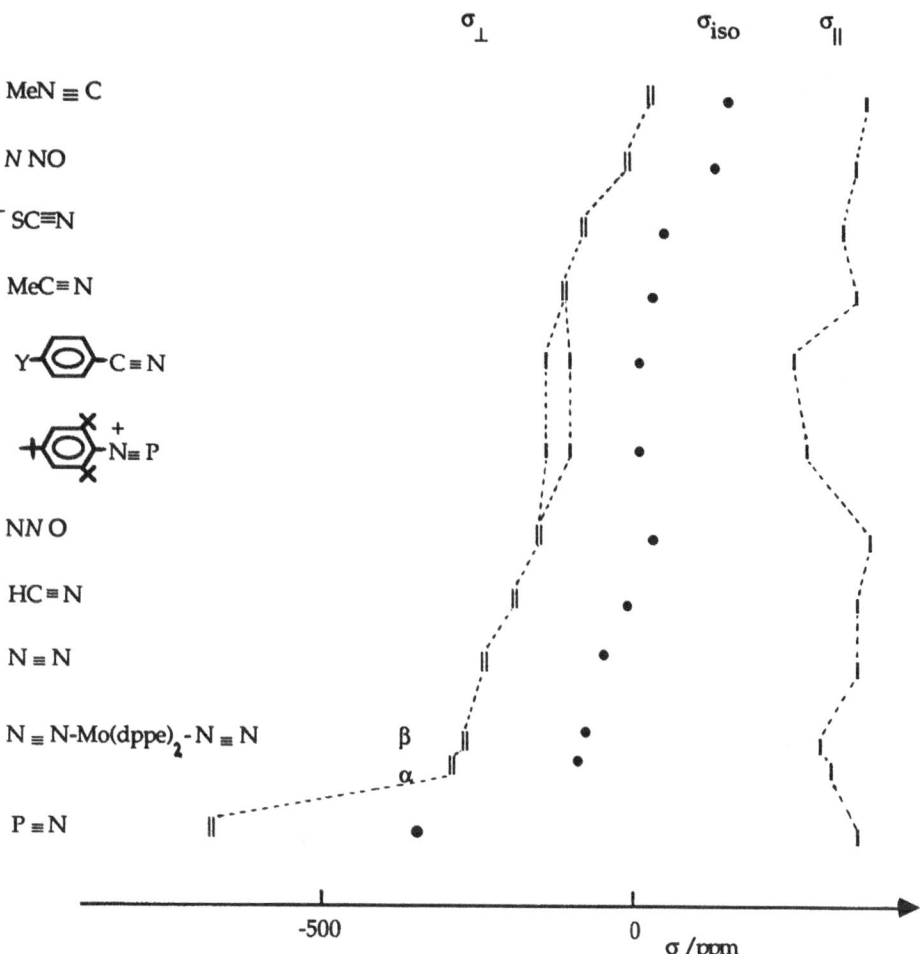

466

Figure 2 Nitrogen tensors in planar CN, NP and NN groups

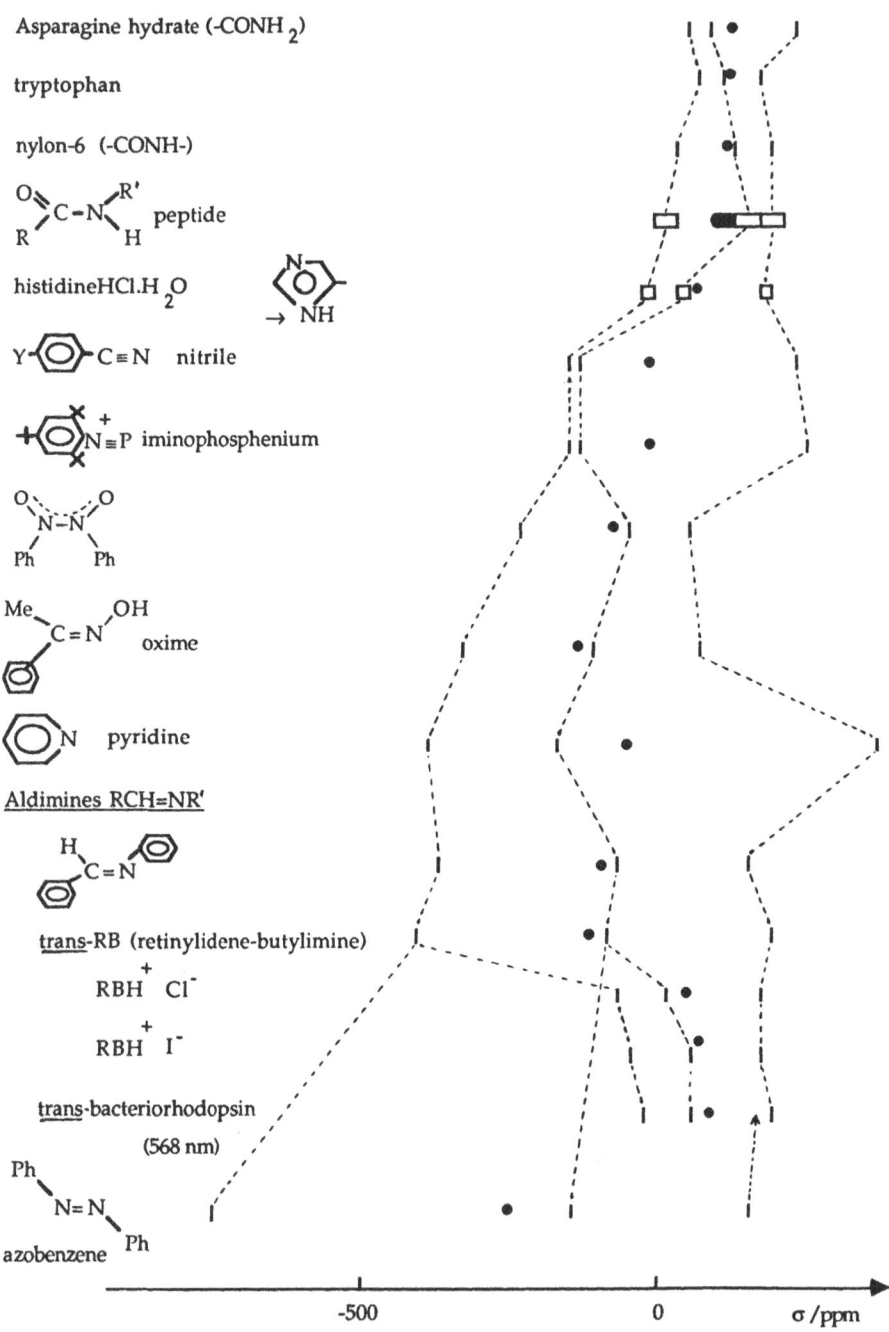

Figure 3 Nitrogen tensors in NO_x groups

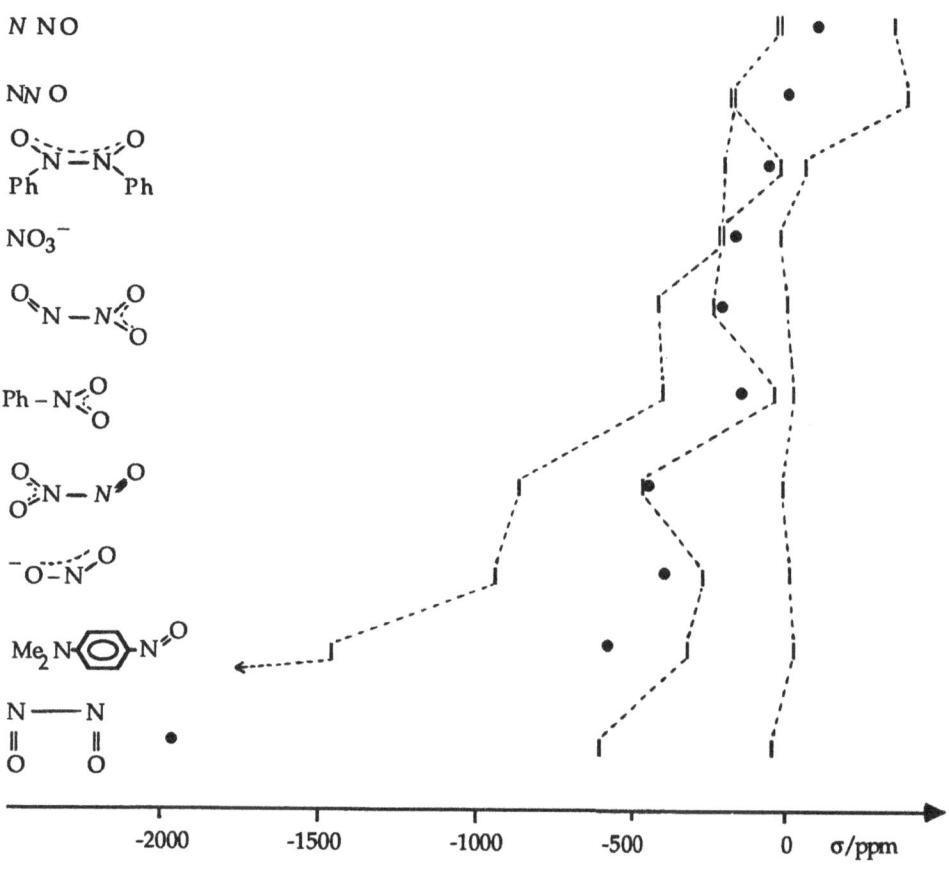

468

Figure 4 Nitrogen shielding tensors in the linear nitrosyl ligand (compounds in Table 6)

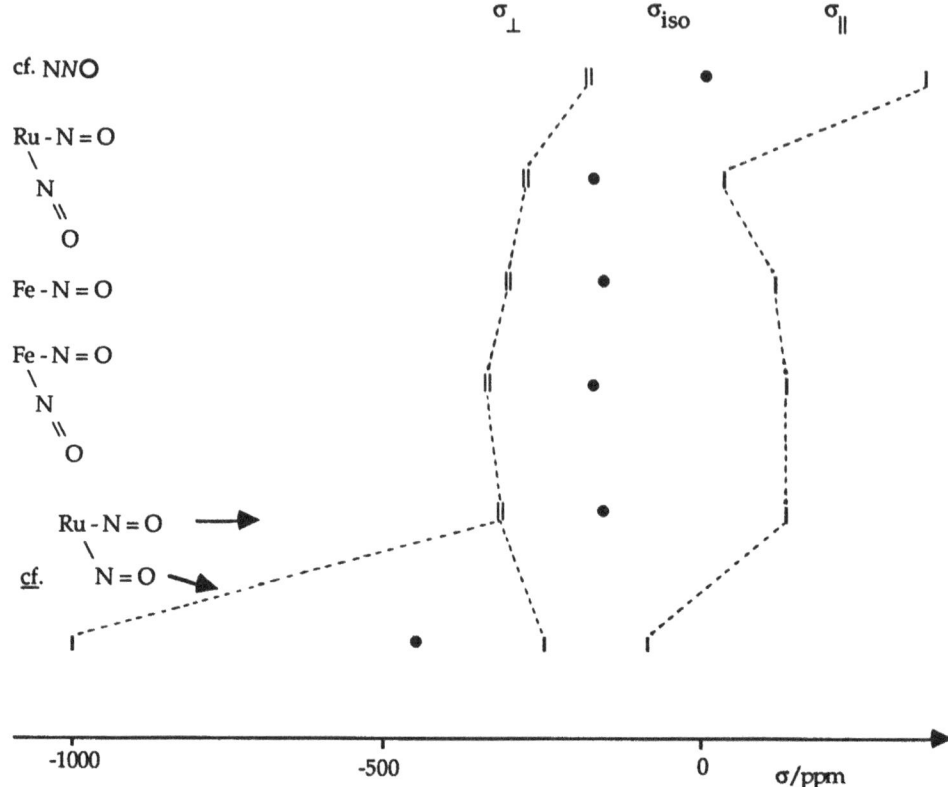

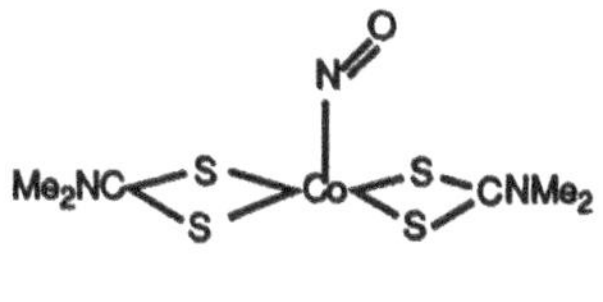

1 [Co(NO)(S₂CNMe₂)₂]

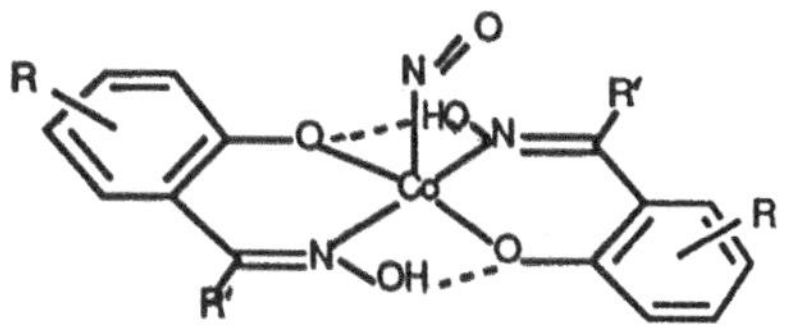

2 [Co(NO)(ketox)₂]: R = H, R' = Me
3 [Co(NO)(5-NO₂salox)₂]: R = 5-NO₂, R' = H

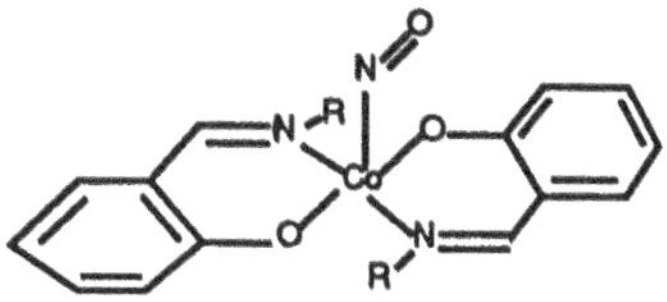

4 [Co(NO)(rsal)₂]: R = Me, Et, Buⁿ, Ph, Bz
written as r = m, e, b, ph, bz

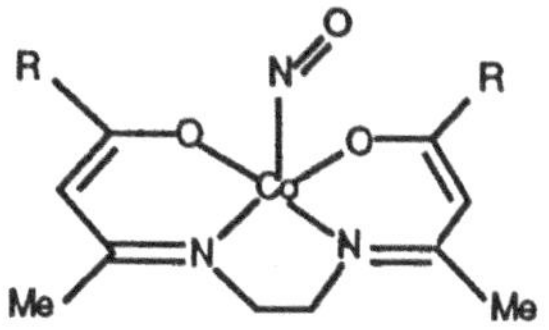

5 [Co(NO)(acacen)]: R = Me
[Co(NO)(benacen)]: R = Ph

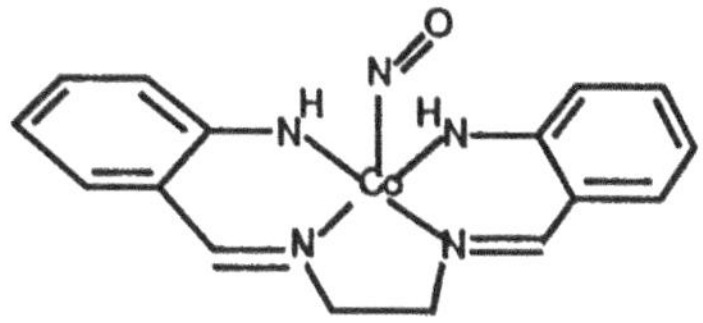

6 [Co(NO)(amben)]

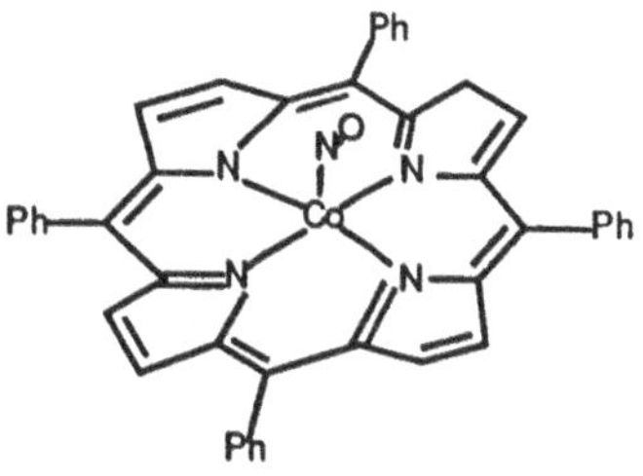

7 [Co(NO)(TPP)]

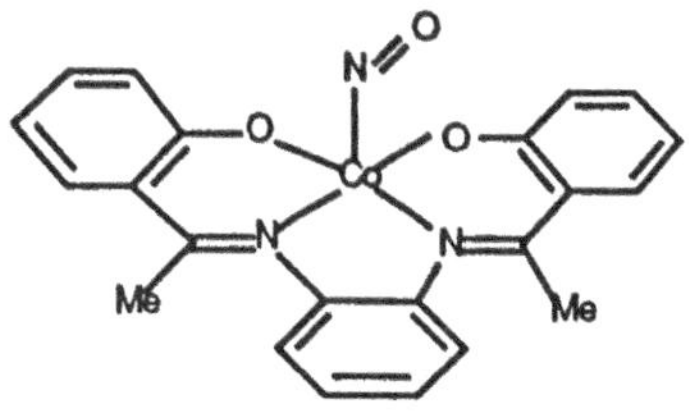

8 [Co(NO)(7-Mesalen)]

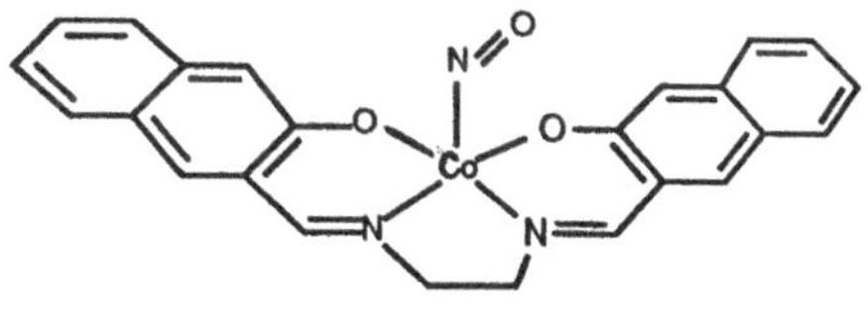

9 [Co(NO)(naphthen)]

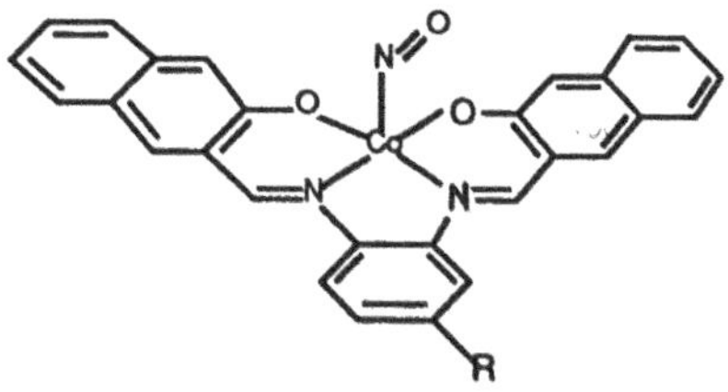

10 [Co(NO)(naphth-ph)]: R = H
[Co(NO)(naphth-mph)]: R = Me

470

Figure 5   Nitrogen tensors in bent nitrosyl ligands

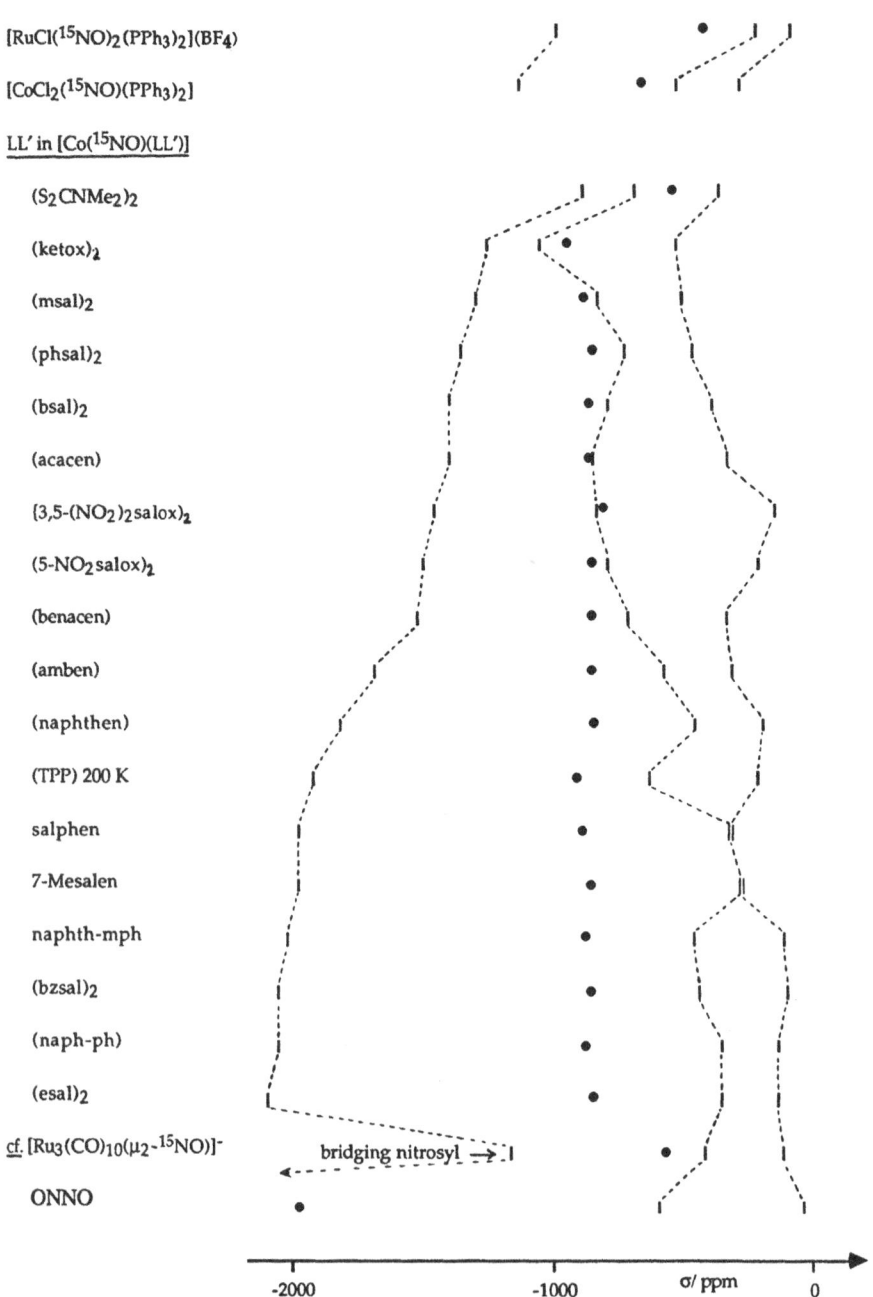

[RuCl(^{15}NO)$_2$(PPh$_3$)$_2$](BF$_4$)

[CoCl$_2$(^{15}NO)(PPh$_3$)$_2$]

LL' in [Co(^{15}NO)(LL')]

   (S$_2$CNMe$_2$)$_2$

   (ketox)$_2$

   (msal)$_2$

   (phsal)$_2$

   (bsal)$_2$

   (acacen)

   {3,5-(NO$_2$)$_2$salox)$_2$

   (5-NO$_2$salox)$_2$

   (benacen)

   (amben)

   (naphthen)

   (TPP) 200 K

   salphen

   7-Mesalen

   naphth-mph

   (bzsal)$_2$

   (naph-ph)

   (esal)$_2$

cf. [Ru$_3$(CO)$_{10}$(μ_2-^{15}NO)]$^-$

   ONNO

bridging nitrosyl →

-2000   -1000   σ/ ppm   0

Figure 6 [Co(^{15}NO)(tetraphenylporphin)] shielding tensor: changes on cooling to 200 K

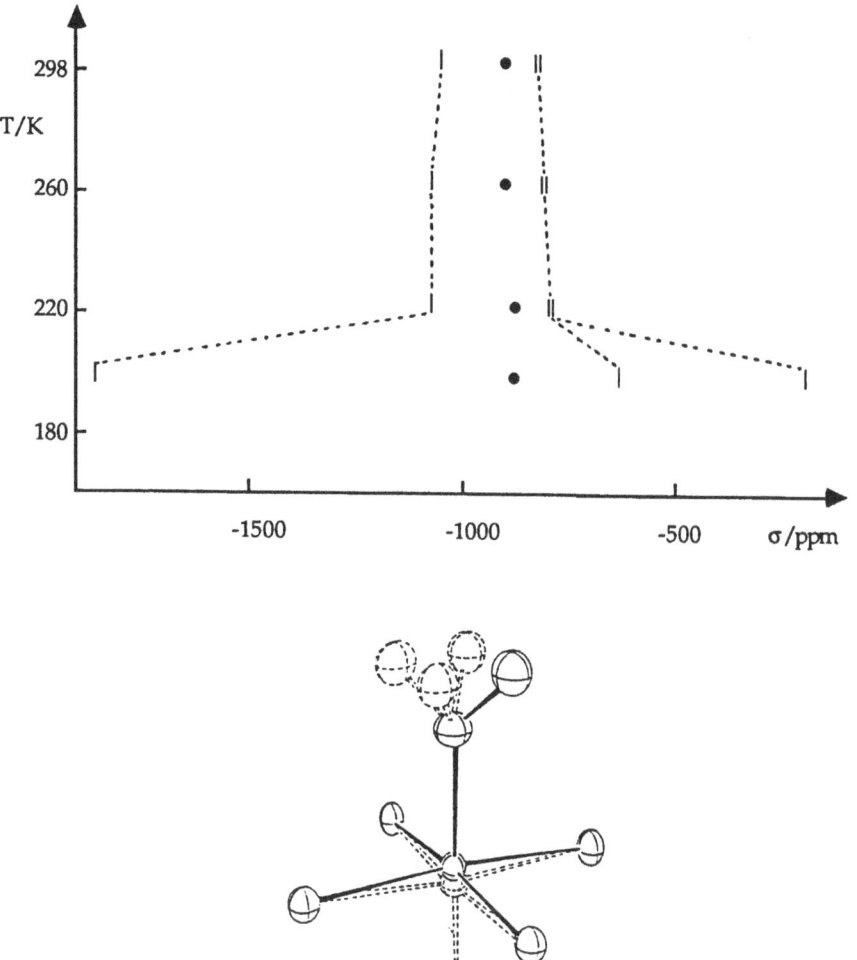

Figure 7 Statistical disorder in [Co(NO)(TPP)]: a single orientation of the Co(NO)N$_4$ group is shown in solid lines (from ref. 32, with permission).

^{1}H and ^{13}C Solid-State NMR Studies of Catalytic Reactions on Molecular Sieves

Michael W. Anderson and Jacek Klinowski

Department of Chemistry, University of Cambridge,
Lensfield Road, Cambridge CB2 1EW, UK.

1. Introduction

Molecular sieves are a class of porous open-framework solids, which includes aluminosilicates (zeolites), aluminophosphates and silicoaluminophosphates of diverse structures. Zeolites, the original molecular sieves, are built from corner-sharing SiO_4^{4-} and AlO_4^{5-} tetrahedra and contain regular systems of intracrystalline cavities and channels of molecular dimensions. The net negative charge of the framework, equal to the number of the constituent aluminium atoms, is balanced by exchangeable cations, M^{n+}, typically sodium, located in the channels which normally also contain water. The name "zeolite" (from the Greek ζεω = to boil and λιθοσ = stone) was coined by Cronstedt [1] in 1756 to describe the behaviour of the newly discovered mineral stilbite which, when heated, rapidly loses water and thus seems to boil.

The general oxide formula of a zeolite is

$$M_{x/n} (AlO_2)_x (SiO_2)_y \cdot m\ H_2O$$

where $y \geq x$. Aluminate tetrahedra cannot be neighbours in the frameworks of hydrothermally prepared zeolites, which means that Al-O-Al linkages are forbidden. This requirement is known as the Loewenstein rule [2]. There are at present around 40 identified species of zeolite minerals (with $1 \leq y/x \leq 5$) and at least 125 synthetic species with a very wide range of aluminium contents.

Zeolites are prepared under mild (60-400°C) hydrothermal conditions in strongly basic media. The type and concentration of the base are important structure-directing factors and a variety of organic bases are now being used in zeolite synthesis. The ZSM series (for Zeolite Socony Mobil) of highly siliceous zeolites is prepared from solutions containing alkylammonium bases. Other elements, such as Ga, Ge, B, Fe and P can substitute for Si and Al in the framework, and there are claims that many other elements can also do so.

Zeolites have a number of interesting physical and chemical properties. The three classes of phenomena which are of greatest practical importance are the ability to sorb organic and inorganic substances, to act as cation exchangers and to catalyse a wide variety of reactions.

The zeolitic channel systems, which may be one-, two- or three-dimensional and may occupy more than 50% of crystal volume, are normally filled with water. When water is removed, other species such as gaseous elements, ammonia, alkali metal vapours, hydrocarbons, alkanols and many other organic and inorganic species may be accommodated in the intracrystalline space. Depending on pore diameter and on

J. A. Tossell (ed.), Nuclear Magnetic Shieldings and Molecular Structure, 473–494.
© 1993 *Kluwer Academic Publishers.*

molecular dimensions, this process is often highly selective, and gives rise to the alternative name for zeolites: molecular sieves. Thus zeolitic sorption is a powerful method for the resolution of mixtures. Commercial applications include thorough drying of organics, separation of hydrocarbons and of N_2 and O_2 in air and the removal of NH_3 and CS_2 from industrial gases.

Cations neutralising the electrical charge of the aluminosilicate framework can be exchanged for other cations from solutions. Zeolites often possess high ion-exchange selectivities for certain cations, and this is used for their isolation and concentration. Molecule sieving properties of zeolites can be further modified by ion exchange. Thus zeolite Na-A sorbs both N_2 and O_2 while Ca-A sorbs nitrogen preferentially to oxygen.

However, it is the ability to catalyse a wide range of reactions, such as cracking, hydrocracking, oxidation and isomerisation of hydrocarbons, which by far overshadows all other applications of zeolites. Rare-earth exchanged and hydrogen forms (prepared indirectly by thermal decomposition of the ammonium form) of some zeolites, such as zeolite Y, mordenite, gmelinite and chabazite, have a cracking activity which is orders of magnitude greater than that of conventional silica/alumina catalysts. Zeolite-based catalysis was first discovered [3] in 1960 and two years later cracking catalysts based on zeolite Y were introduced. They have now almost completely displaced conventional catalysts. The synthetic zeolite ZSM-5, introduced in 1972 [4], is an even more powerful catalyst. Its high silica content (Si/Al ratio is typically 30) gives it high thermal stability, while the channel diameter is very convenient for many applications, particularly in the petroleum industry. The 10-membered channels of ZSM-5 are responsible for the quite striking shape selectivity. Catalytic properties of ZSM-5 include the ability to synthesise gasoline from methanol in a single step (see below). Silicalite, a material which is isostructural with ZSM-5, but contains only small amounts of aluminium is, by contrast to most other zeolites, non-polar (i.e. hydrophobic) and organophilic. Accordingly, silicalite is used in the removal of dissolved organics from water.

Since 1982 several new families of porous solids have been synthesized. The $AlPO_4$ molecular sieves, with structures built from alternating AlO_4 and PO_4 tetrahedra, were the first to be discovered [5]. Some of them have the framework topologies of known zeolites, but many have novel structures. $AlPO_4$ materials are synthesized from gels containing sources of aluminium, phosphorus and at least one organic structure-directing template. Incorporation of a silicon source into an aluminophosphate gel results in the formation of silicoaluminophosphates, SAPO, and the incorporation of a metal, Me (such as Mg, Mn, Fe, Co or Zn), into $AlPO_4$ and SAPO gives the MeAPO and MeAPSO sieves, respectively [6]. Some of these have high Brønsted acidities and thus a considerable potential as heterogeneous catalysts. Specialist monographs and numerous reviews on the structure and properties of molecular sieves are available [7-19].

Synthetic zeolites are usually microcrystalline and furthermore typically contain *four* 10-electron atomic species (Si^{4+}, Al^{3+}, O^{2-} and Na^+) which makes them difficult to study by conventional techniques of structural elucidation. The development of high-resolution solid-state NMR techniques, such as magic-angle spinning (MAS), gave zeolite chemistry a powerful structural tool to monitor all elemental components of such frameworks.

The aim of this review is to survey solid-state NMR results, particularly recent, which are of direct relevance to heterogeneous catalysis on molecular sieves.

2. Chemical Status of Guest Organics in the Intracrystalline Space

The declining oil reserves have stimulated considerable efforts towards the exploration of alternative sources of energy and organic chemicals. One solution is to use the abundant supply of coal as a source of synthesis gas ($CO + H_2$) which is readily converted to methanol (MeOH). MeOH can then be transformed into higher molecular weight hydrocarbons (olefins, aliphatics and aromatics) over shape-selective zeolite catalysts, the most successful of which in this respect is H-ZSM-5, capable of converting MeOH to hydrocarbons up to C_{10}. The selective synthesis of ethylene and propylene, the key intermediates for the production of detergents, plasticizers, lubricants and a variety of chemicals, proceeds over smaller pore zeolites such as chabazite and erionite.

The transformation of methanol to hydrocarbons over zeolite H-ZSM-5 is the basis of several industrially important reactions, such as the MTG or the MTO processes [20,21]. The mechanism of the reaction, particularly as concerns the formation of the first C-C bond and the nature of the interactions between the CH_3OH molecules and the zeolitic framework has been the subject of controversy [22,23]. ^{1}H NMR has been used [24-26] to study the chemistry of methanol adsorbed on H-ZSM-5.

In MAS NMR experiments [25-26] samples were contained inside capsules [27] which could be spun inside the MAS NMR probehead at rates of up to 3 kHz. The design of the capsule allowed the samples to be dehydrated at 400°C under a pressure of 10^{-5} mbar before adsorption of the organic. Capsules were then sealed while keeping the sample at liquid nitrogen temperature in order to prevent the onset of chemical reactions.

High-resolution ^{1}H MAS NMR spectra were recorded using 2 μs (20°) pulses with a repetition time of 5 s. Since the ^{1}H spin-lattice relaxation times of adsorbed alcohols on zeolites were found to be of the order of 0.2 s, such repetition times generate quantitatively reliable spectra.

Hydrogen bonding causes a downfield chemical shift in alcohols because of the deshielding of the proton as a result of electrostatic polarization of the OH bond. In liquid CH_3OH hydrogen bonding causes a downfield shift of 3.1 ppm [Figure 1(b)] relative to CH_3OH in CCl_3D where there is no hydrogen bonding [Figure 1(c)]. The proton MAS NMR spectrum of CH_3OH adsorbed on zeolite H-ZSM-5 [Figure 1(a)] contains a signal at 4.1 ppm corresponding to the methyl protons and another at 9.1 ppm corresponding to the hydroxyl protons. When CD_3OH is adsorbed only the 9.1 ppm signal is observed, which demonstrates that all hydroxyls resonate at the same chemical shift. When 6 molecules of CD_3OH are adsorbed per Brønsted site on zeolite H-ZSM-5 [Figure 2(a)] one signal corresponding to the hydroxyl groups is found at 9.1 ppm. By contrast, when CH_3OD is adsorbed [Figure 2(b)], apart from the signal at 4.1 ppm corresponding to the methyl groups there is a small resonance at ca. 9.4 ppm. Adsorption of CD_3OD demonstrates that this latter signal originates initially from the framework Brønsted acid sites and not from the methyl group [Figure 2(c)]. The low-intensity signals at 0 - 2 ppm are due to the probehead background. This was checked by repeating the experiments under identical conditions with the sample removed.

The large downfield shift of the hydroxyl resonance of the CH_3OH upon adsorption on H-ZMS-5 must be caused by very strong hydrogen bonding and/or direct protonation of the alcohol. Note that all hydroxyl groups in the spectrum given in Figure 2 resonate at the same chemical shift, which indicates that all protons should be equivalent on the timescale of the NMR experiment. Consider the scheme shown in Figure 3, which involves a hydroxyl proton protonating five methanol molecules. It is clear that, in this resonance structure, each molecule of methanol is formally identical to a methoxonium ion. The charged cluster, a supercation, may rotate in the intracrystalline space so that different hydroxyl protons approach the bridging framework oxygen in turn, thus becoming equivalent on the NMR timescale.

At the lower coverage of 2 molecules per Brønsted acid site, the hydroxyl signal moves from 9.1 ppm in Figure 2(a) to 10.5 ppm. This experiment indicates the presence of fast exchange between the proton of the zeolitic acid site and the OD group of the adsorbed methanol.

The downfield shift of the hydroxyl resonance of MeOH is a good measure of the proton donating ability of the solid acid catalyst. In zeolites H-Y and H-L the shifts are considerably smaller than in H-ZSM-5, which is consistent with their lower acidity. The advantage of our MeOH adsorption method for the measurement of catalytic acidity is that it monitors the species to which the proton is donated (the MeOH molecule) rather than the Brønsted acid site.

It is interesting to note that the chemical shift of the hydroxyl resonance is very sensitive to the type of zeolite on which methanol is adsorbed [26]. The change in the hydroxyl chemical shift in H-ZSM-5 is by far the largest (7.8 ppm downfield from MeOH/CCl$_3$D). The position of this resonance also depends on the method of synthesis of ZSM-5. When a zeolite with almost the same Si/Al ratio is prepared using the low-pH "fluoride" route, the corresponding increase of chemical shift is only 6.0 ppm. This result is intriguing, since the overall structure and the Si/Al ratio of the two samples are identical. The major difference is the lack of defect sites in the material synthesised via the fluoride method. This hints that defect sites such as SiOH nests might also be responsible for the extraordinary hydrogen-bonding properties of conventionally prepared H-ZSM-5.

The sodium form of the zeolites shows only very small shifts in the hydroxyl resonance upon adsorption of MeOH. In the case of Na-ZSM-5 the shift is in fact over 1 ppm *upfield* from liquid MeOH (increased shielding of the proton or less hydrogen bonding). This can be explained in terms of coordination of the MeOH to the coordinatively unsaturated sodium cations via the MeOH oxygens, which in effect breaks up the bonding present in the liquid phase. In the series Na-ZSM-5, Na-Y, Na-A there is a progressive increase in the downfield shift of the hydroxyl resonance. This is most easily explained in terms of the increase in Al content of the zeolite framework which increases the total electrostatic charge of the framework oxygen. Consequently, this provides a greater opportunity for hydrogen-bonding of methanol to the zeolite framework adjacent to an aluminium atom in zeolites A or Y.

^{13}C MAS NMR is useful for the elucidation of the nature of the interaction of methanol with zeolitic and silicoaluminophosphate-based molecular sieves prior to the onset of catalytic reactions. *Slow* magic-angle spinning ^{13}C NMR reveals [28] that strongly bound surface CH$_3$-O-Si groups are formed at 250°C when methanol is adsorbed on the molecular sieve SAPO-5. Figure 4(a) shows the ^{13}C MAS NMR spectrum of methanol (MeOH) adsorbed on SAPO-5 at room temperature and not subsequently heated. A single sharp resonance is observed at 50 ppm corresponding to relatively highly mobile adsorbed methanol. After heating the sample to 150°C for 10 minutes 37% of the MeOH has been converted to dimethyl ether (DME) which, judging from its narrow spectral line [Figure 4(b)], is also highly mobile. Further heating to 250°C for an additional 10 minutes results in the spectrum shown in Figure 4(c). Although a number of broad spectral features are in evidence, they all correspond to two chemical species, one at 50 ppm corresponding to methanol and a broader signal at 60 ppm with associated spinning sidebands. The greatly increased linewidth of the latter species together with the presence of sidebands in this typical solid-state spectrum indicate the presence of considerable chemical shift anisotropy resulting from much reduced molecular mobility. The MAS spinning frequency used for these experiments was below 1 kHz. It is clear that the application of normal, much higher, spinning frequencies would average this anisotropy thus concealing important chemical information.

The likely origin of the broad new signal are the strongly bound surface CH_3-O-Si methoxy groups which would have a very similar chemical shift to that of DME. Being anchored at the surface of an aluminium-rich molecular sieve, these methoxy groups undergo both chemical shift and dipole-quadrupole broadening effects brought about by the vicinity of ^{27}Al nuclei. After subsequent heating of the sample to 300°C and above methanol is converted to a mixture of mobile olefins and aliphatics which can be observed with slow MAS and which give narrow spectral lines without spinning sidebands. This demonstrates the progression from a weakly bound methanol molecule to a strongly bound reaction intermediate and finally to a weakly bound hydrocarbon product. By contrast, in zeolite H-ZSM-5 methanol becomes strongly bound to the framework at room temperature without the formation of methoxy groups [28].

3. In Situ Studies of Catalytic Reactions on Molecular Sieves

The catalytic conversion of methanol to hydrocarbons in the gasoline boiling range using zeolite ZSM-5 at ca. 370°C has understandably attracted much attention. ^{13}C MAS NMR can probe directly the role of the active site in shape-selective catalytic reactions on zeolites *in situ*. The kind and quantity of chemical species present inside the particle can now be directly monitored [29-37]. This information, not forthcoming from other techniques, is usefully compared with the composition of the gaseous products to give new insights into reaction pathways on molecular sieves and to assist in the design of new shape-selective catalysts. These experiments have: (i) identified 29 different organic species *in the adsorbed phase* and monitored their fate during the course of the reaction; (ii) observed directly different kinds of shape selectivity in a zeolite; (iii) unequivocally distinguished between mobile and attached species. The results will assist in the design of shape-selective solids and provide a better understanding of catalytic processes in the intracrystalline space.

Shape selectivity of zeolites [3,38-41] arises from the fact that the probabilities of forming various products in the narrow intracrystalline cavities and channels are largely determined by molecular dimension and configuration. Three kinds of shape selectivity have been envisaged [3]. *Reactant selectivity* occurs when only certain molecules can access the intracrystalline space and react there, others being too large to enter the pores. In *product selectivity* only some of the various species formed within the channels and cavities can diffuse out of the crystallite and appear as reaction products. *Restricted transition state selectivity* takes place when certain reactions cannot proceed at all because they would involve transition states requiring more space than is available in the intracrystalline space. The evidence for the existence of the product and transition state selectivities available so far is indirect since it relies on the *absence* of certain species in the products rather than on the *presence* of others in the intracrystalline space, something which has not until now been directly monitored.

^{13}C MAS NMR of sealed H-ZSM-5 samples gives a considerable gain in resolution in comparison with earlier work [42-48]. The spectrum of a sample with adsorbed MeOH and maintained at 20°C [Figure 5(a)], contains a single resonance at 50.8 ppm due to MeOH. After heating the sample to 150°C for 20 mins the spectrum [Figure 5(b)] is composed of two signals, at 50.5 and 60.5 ppm, corresponding to MeOH and DME respectively.

Figure 6 shows the spectrum of a sample treated at 300°C for 35 min. MeOH and DME have been completely converted to a mixture of aliphatics and aromatics. The question arises as to how the various ^{13}C resonances are to be assigned to different hydrocarbon species, although some of them, especially those from methyl groups attached to aromatic rings, overlap. It turns out that all signals can be reliably assigned [49-51]. First, note that most compounds give rise to several NMR peaks. In addition to chemical shift information, and the monitoring of the number and relative intensity of the various ^{13}C signals, two-dimensional ^{13}C spectra have been used [50,51] to determine the connectivity of carbons and the number of protons attached to each carbon atom in the various organics and the details of ^{13}C - ^{1}H couplings, enabling firm assignments for a number of resonances to be made [50].

Table 1. Parameters of the two-dimensional spectrum shown in Figure 7.

Signal	Chemical shift/ppm	Signal multiplicity	J-Coupling /Hz	Tentative assignment [29]	Final assignment
a	24.7	4	135	isobutane	isobutane
b	22.2	4	135	n-hexane isopentane n-heptane	isopentane
c	18.7	4	135	methyl substituted benzenes	methyl substituted benzenes
d	16.7	3	130		
	16.0	4	130	propane	propane
e	-10.7	5	135	methane	methane

A well-resolved two-dimensional J-coupled spectrum [50], measured using no decoupling during part of the evolution period while synchronising the time increment and the rotation period of the MAS spinner, is given in Figure 7. The relative intensities of the N + 1 lines in a spectrum of a spin-1/2 nucleus coupled to N equivalent spin-1/2 nuclei are given by Pascal's triangle as 1 : 1 for N = 1; 1 : 2 : 1 for N = 2; 1 : 3 : 3 : 1 for N = 3 and 1 : 4 : 6 : 4 : 1 for N = 4. Multiplicities of the lines confirm that our assignments, based on conventional one-dimensional spectra, are correct. For example, the resonance at -10.7 ppm in Figure 7 is split into 5 components with a requisite intensity ratio in the 2-D spectrum, which confirms that it must be due to adsorbed methane. Similarly, the 4-fold (methyl) and 3-fold (methylene) signals clearly indicate the presence of propane adsorbed in the intracrystalline space.

The 2D spin diffusion ^{13}C NMR experiment allows us to examine further the spectral assigments obtained from the 1D and the 2D J-resolved experiments [51]. It also provides new details concerning distribution of hydrocarbons in zeolite ZSM-5. Spectral spin diffusion in the solid state involves simultaneous flip-flop transitions of dipolar-coupled spins with different resonance frequencies. The interaction of the X nuclei undergoing spin diffusion with the proton reservoir provides compensation for the energy imbalance. Spin diffusion results in an exchange of magnetization between nuclei responsible for resolved NMR signals, which can be conveniently detected by observing the relevant cross-peaks in the 2D spin-diffusion spectrum. The technique is well established for solids. The rate of spin diffusion is very strongly dependent on the internuclear separation r, being proportional to $1/r^3$ for a rigid crystal lattice and to $1/r^6$ for species undergoing rapid isotropic motion. As a result, for all practical purposes spin diffusion occurs only between nuclei in adjacent functional groups within the same molecule (the intramolecular case) or between nuclei in neighbouring molecules mixed on a microscopic level (the intermolecular case). Both cases are observed in our system.

Figure 8 shows the 2D spin-diffusion spectrum of aliphatic hydrocarbons trapped in the zeolite. The 1D signals, corresponding to the diagonal peaks in the 2D spectrum, have been assigned previously [29,30], but the assignment of signal b has subsequently been questioned [50]. It is clear that n-hexane and n-heptane are present, since signal e comes exclusively from their CH_3 groups. Therefore, considering the chemical shift of CH_2 groups of n-hexane and n-heptane, both hydrocarbons must contribute to signal b. In the J-resolved experiment this signal was not split into a triplet, because no homonuclear proton decoupling was applied during the first half of the evolution period, so that the splitting was obscured by substantial dipolar broadening. By contrast, CH_3 groups of isobutane undergo free rotation, which reduces the dipolar interaction and allows the quartet splitting of signal b to be observed. We note that signals of n-butane in the spin-diffusion spectrum are missing, since cross-polarization has a tendency to underestimate signals of mobile products, so that only those which are as abundant as propane appear.

Zeolite ZSM-5 contains no cages and its channel diameter only allows the hydrocarbon species in the channels to be lined up sequentially. For any two molecules to exchange their positions, access to an unoccupied channel crossing is required, which is difficult to satisfy at high adsorbate loadings. Hydrocarbon molecules are only capable of limited motion along the channels, which does not favour *intermolecular* spin diffusion. Free molecular rotation cannot occur, so that *intramolecular* dipolar interactions are present even for quite mobile functional groups, and make the intramolecular spin diffusion possible. Thus intramolecular spin diffusion in our system is preferred to intermolecular spin diffusion. The assignments of the various signals are given in Table 2.

Munson et al.[35] questioned the suggestion in refs. 29 and 30 that, since CO is observed prior to hydrocarbon formation, it is an intermediate in the reaction. Methanol-^{13}C and formic acid-^{13}C were first coadsorbed on the H-ZSM-5 catalyst, and an *in situ* NMR experiment was performed. It was found that the conversion rate of methanol was not affected by large quantities of CO. The authors then measured the spectra of a sample containing formic acid-^{13}C and unlabelled methanol. [13]CO was not incorporated in the reaction product. The conclusion is that CO is neither an intermediate nor a catalyst in MTG chemistry.

Aronson et al. [52] used ^{13}C NMR to identify the intermediates formed upon the adsorption of 2-methyl-2-propanol on zeolite H-ZSM-5. The adsorbed species can best be described as a highly reactive silyl ether, with the alkyl group covalently bonded to oxygen of the zeolitic framework. They assign the signal at 77 ppm to the carbon bonded to the oxygen, and the signal at 29 ppm as due to aliphatic carbons formed in secondary reactions. Richardson et al. [31] studied the mechanisms by which butadiene oligomerizes in acidic zeolitic catalysts. It was found that oligomerization proceeds primarily by 1,4-addition, and that secondary reactions of the oligomers are strongly dependent upon the properties of the zeolite. Thus the initially formed product undergoes cyclization to form fused rings in zeolite H-Y, but isolated rings are formed in the smaller channels of zeolite H-ZSM-5. These results provide insight into the mechanisms bywhich oxide catalysts are deactivated by pore blockage. The conclusions of Anderson et al. [36] are in agreement with the steric arguments given in refs. [31] and

Table 2. Assignment of the 2D ^{13}C NMR spin-diffusion spectrum in Figure 8.

Diagonal peaks

Signal	Chemical shift (ppm)	Group	Assignment
a	24.7	CH$_3$	isobutane
b	22.3	CH$_3$CH$_2$CH(C̱H$_3$)$_2$	isopentane
		CH$_2$	n-hexane + n-heptane
c	18.7	CH$_3$	methyl-substituted benzenes
d	16.7	CH$_3$ + CH$_2$	propane
e	14.3	CH$_3$	n-hexane + n-heptane
f	11.2	C̱H$_3$CH$_2$CH(CH$_3$)$_2$	isopentane

Cross-peaks

Signals	Assignment	Type of spin diffusion
a - d	isobutane - propane	intermolecular
b - d	isopentane - propane	intermolecular
b - e	n-hexane	intramolecular
	n-heptane	intramolecular
b - f	isopentane	intramolecular

[52]. They used ^{13}C and ^{1}H MAS NMR to monitor the formation of long-chain polymeric hydrocarbons within the intracrystalline space of offretite during the conversion of methanol. The presence of such polymers was attributed to the uninterrupted one-dimensional pore system. In erionite, where the channel system is constricted at every 15 Å, polymers cannot form. It appears that a degree of channel "tortuosity" is preferable in reactions which produce readily oligomerizable molecules, in order to prevent undesirable side reactions.

The oligomerization of propene on zeolite H-Y has been studied [33,37] by variable-temperature ^{13}C MAS NMR. Alkoxy species formed between protonated alkenes and oxygens of the zeolitic framework were found to be important long-lived intermediates in these reactions. Simple secondary or tertiary carbocations are either absent in the zeolite at low temperatures, or are so transient as to be undetectable by NMR even at temperatures as low as 163 K. There was, however, evidence for long-lived alkyl-substituted cyclopentenyl carbocations, which are formed as free ions in the zeolite at room temperature. At 503 K the oligomers crack to form branched butanes, pentanes and other alkanes. The final product was highly aromatic coke. The structure, dynamics and reactivity of an alkoxy intermediate formed from acetylene on zeolite catalysts have been investigated by Lazo et. al. [32].

4. Direct Observation of Shape Selectivity

The *distribution* of adsorbed species in the sample of zeolite with adsorbed methanol treated at 300°C is very different from that observed in the reaction products [29-30]. The principal aromatics expected to be present are *m*- and *p*-xylene, 1,2,4-trimethylbenzene and toluene. However, the main species actually found in the adsorbed phase are *o*- and *p*-xylene and 1,2,4,5-tetramethylbenzene, with smaller amounts of 1,2,4-trimethylbenzene and 1,2,3,5-tetramethylbenzene. The other xylenes, tri- and tetramethylbenzenes are also found but in smaller amounts. The distribution of the three trimethylbenzenes in the adsorbed phases is very different from the thermodynamic equilibrium distribution (see Figure 9). The fact that 1,2,3- or 1,3,5-trimethylbenzenes (with kinetic diameters of 6.4 and 6.7 Å, respectively) are not found among the products, but are present in the adsorbed phase, while the smaller 1,2,4-trimethylbenzene (6.1 Å) is found in both, clearly demonstrates the reality of the concept of product selectivity. The channel dimensions of ZSM-5 are 5.6×5.3 Å, but more space is available at the intersection of the straight and zig-zag channels. While the greater amplitude of thermal vibrations of the framework, which increases the maximum effective size of the channels, allows the smaller isomer to diffuse out of the crystal, the two larger isomers, although formed, are unable to diffuse out at 300°C and must isomerise to 1,2,4-trimethylbenzene.

The distribution of the tetramethylbenzenes in the intracrystalline space is most unexpected (see Figure 10). None of them have ever been reported in the products of the reaction at 300°C and yet all three are clearly present in the adsorbed phase. Because of the restricted intracrystalline space they can only form at channel intersections, but (unlike the trimethylbenzenes) are not generated in the thermodynamic equilibrium distribution. 1,2,3,5-tetramethylbenzene (6.7 Å) should be the dominant species on thermodynamic grounds; in fact it is 1,2,4,5-tetramethylbenzene (6.1 Å) which dominates. The thermodynamically least favoured isomer, 1,2,3,4-tetramethylbenzene (6.4 Å), is found in small quantities. The fact that tetramethylbenzenes are not found in the products again demonstrates product shape selectivity. Their *relative abundance* in the adsorbed phase, on the other hand, shows that an additional kind of shape selectivity occurs within the intracrystalline space. It does not rely on the ability of species to enter or to leave the crystal nor on the size of the transition state: isomerisation is sterically restricted within the crystallite at the active site itself.

It is instructive to compare the shape-selective catalytic conversion of methanol to low-molecular-weight olefins and aliphatics on zeolite ZSM-5 with the results for the molecular sieve SAPO-34 [53]. The sieve, which has the framework topology of the natural zeolite chabazite [54,55] is known to convert MeOH to hydrocarbons with a selectivity for C_2 of 33.8 mol %. In what follows, we shall refer to the gas species leaving the catalyst (and monitored by gas chromatography) as "products" and the compounds monitored by NMR in the intracrystalline space as the "adsorbed phase". Products were analyzed using a high resolution gas chromatograph. A sample of SAPO-34 was prepared with a chemical formula of $SiAl_6P_5O_{24}$.

Figure 11 shows the ^{13}C MAS NMR spectra recorded after reaction at different temperatures. After adsorption of MeOH and no thermal treatment a single signal at 50 ppm from TMS is observed [Figure 11(a)]. After heating to 150°C, a rather poorly resolved spectrum is obtained [Figure 11(b)] consisting of two signals. No further changes are found at 250°C but after heating to 300°C for 15 minutes and 370°C for 10 minutes a multitude of narrow resonances are observed in the aliphatic region (-10 to +40 ppm) [Figures 11(c) and (d)]. No other signals were found. Spectral deconvolution of these resonances, shown in Figure 12, was performed by reference to literature values of ^{13}C NMR chemical shifts [56].

At 300°C the most abundant species in the adsorbed phase are isopentane, propane, isobutane, methane and n-butane with less ethane, neopentane, C_6 and C_7 species. At 370°C the composition of the adsorbed phase is similar except that propane is now more abundant than isopentane. The relative concentrations of adsorbed species were calculated from spectra acquired with high-power proton decoupling but without cross-polarization. Unlike in the case of zeolite H-ZSM-5 (see above) no carbon monoxide intermediate and no aromatics are observed by NMR at any temperature.

The framework structure of SAPO-34 consists of hexagonal prisms (D6R units) linked by four-membered rings. This results in a three-dimensional channel system and a structure containing large ellipsoidal cages 11 Å long and 6.5 Å wide. The cages are stacked in a hexagonal arrangement forming a three-dimensional network of cages linked by the narrower eight-membered windows. The cage and window dimensions for SAPO-34 are almost identical to those in chabazite and therefore similar molecular sieving properties can be expected. Entrance to the chabazite cages may only be gained through the eight-membered ring windows which are nearly circular with a diameter of approximately 3.8 Å. At room temperature chabazite very rapidly sorbs CH_4 and C_2H_6, slowly sorbs n-alkanes but completely excludes branched alkanes.

^{13}C MAS NMR indicates that after heating to 300°C MeOH and DME are completely converted to aliphatic hydrocarbons; no olefins are found. This is understood by reference to the gas chromatography results which show that at lower space velocities the products contain a smaller amount of olefins (Figure 13). By extrapolating the data we predict that under static reactor conditions little or no olefins will be formed. A careful inspection of the chromatography results indicates that the relative amounts of C_1, C_2, C_3, C_4 etc. species (olefins + aliphatics) are roughly similar irrespective of space velocity. In other words, space velocity does not change oligomerization properties, but it does alter the extent of hydrogenation of products. NMR results, which give a fair representation of the total concentration of species formed in the adsorbed phase, can therefore be compared with their concentration in the gaseous products.

The most striking difference between the composition of the adsorbed phase and the products is the preponderance of branched aliphatics up to C_6 in the former. The effect is so clear as to amount to a textbook example of product shape selectivity. Early work has shown that these branched aliphatics are not sorbed by the chabazite crystals from the gas phase at room temperature. The same effect occurs here in reverse: branched-chain hydrocarbons formed *inside* SAPO-34 are not capable of leaving the intracrystalline space even at temperatures as high as 370°C.

Of the C_1, C_2 and C_3 species in the adsorbed phase the concentration of C_3 is always the highest. However, it is C_1 which is the most abundant in the gaseous products, followed by C_2 and C_3 species. The C_1/C_2 concentration ratio in the adsorbed phase is roughly equal to that in the gaseous products at 300°C. This suggests that C_1 and C_2 species have no difficulty in leaving the SAPO-34 crystallite. On the other hand, only 5-10% of the amount of C_3 found in the adsorbed phase is observed in the gas-phase products. Figure 13 illustrates the relative amounts of C_1-C_6 compounds in the products and in the adsorbed phase. The striking differences between the concentrations of C_1 and C_2 on the one hand and C_3 on the other are not easily explained on the basis of diffusion coefficients of aliphatic hydrocarbons in chabazite. At room temperature the diffusion coefficient of propane is ca. 20 times smaller than for ethane and methane. However, at elevated temperatures, such as 150°C, the difference disappears almost completely. It follows that at 370°C the expected diffusion coefficients for methane, ethane and propane should be of the same order of magnitude with a negligible activation barrier. The exclusion of much of the C_3 fraction from the product must therefore be a result of the additional constraints imposed by the presence of branched hydrocarbons which partially block the pore system and significantly alter the diffusional behaviour of other species.

This further illustrates the need for knowing the contents of the intracrystalline space of a shape-selective catalyst during the course of the reaction. Since the available free space is modified by the occlusion of product molecules, it is not sufficient to take into account the crystallographic pore dimensions in order to predict shape-selective action.

The discovery that, despite the composition of the gaseous products, SAPO-34 in fact converts methanol more selectively to C_3 than to C_2 hydrocarbons suggests ways to modify the catalyst so as to enable the C_3 species to escape, thus making it more selective for propylene than ethylene. This might be done by preparing the catalyst with occluded material or partial exchange with cations large enough to prevent the formation of branched hydrocarbons. This would in turn allow the C_2 and C_3 species to diffuse more readily through the channel system without obstruction from the higher hydrocarbons. The fact that such a prediction can be made on the strength of MAS NMR in tandem with gas chromatography illustrates the remarkable potential of this two-pronged approach in the design of novel molecular sieve catalysts.

References

1. Fr. A. Cronstedt, *Kongl. Svenska Vetenskaps Acad. Handlingar,* **17,** 120 (1756).
2. W. Loewenstein, *Am. Mineral.,* **39,** 92 (1954).
3. P. B. Weisz and V. J. Frilette, *J. Phys. Chem.,* **64,** 382 (1960).
4. R. J. Argauer and G. R. Landolt, U.S. Patent, 3, 702, 886 (1972).
5. S. T. Wilson, B. M. Lok, C. A. Messina, T. A. Cannan and E. M. Flanigen, *J. Am. Chem. Soc.,* **104,** 1146 (1982).
6. E. M. Flanigen, B. M. Lok, R. L. Patton and S. T. Wilson, *Pure Appl. Chem.,* **58,** 1351 (1986).
7. W. M. Meier and D. H. Olson, *Atlas of Zeolite Structure Types,* Butterworths, Sevenoaks, Kent (1988).
8. D. W. Breck, *Zeolite Molecular Sieves: Structure, Chemistry and Use,* Wiley, London (1974).
9. R. M. Barrer, *Zeolites and Clay Minerals as Sorbents and Molecular Sieves,* Academic Press, London (1978).
10. R. M. Barrer, *Hydrothermal Chemistry of Zeolites,* Academic Press, London (1982).
11. *Zeolite Chemistry and Catalysis,* ACS Monogr. 171 (1976).
12. *Molecular Sieves - II,* J. R. Katzer (Ed), ACS Symposium Ser. 40 (1977).
13. C. Naccache and Y. Ben Taarit, *Pure Appl. Chem.,* **52,** 2175 (1980).
14. P. A. Jacobs, *Carboniogenic Activity of Zeolites,* Elsevier, Amsterdam (1977).
15. *Catalysis by Zeolites,* B. Imelik, C. Naccache, Y. Ben Taarit, J. C. Védrine, G. Coudurier and H. Praliaud (Eds) Elsevier, Amsterdam (1980).
16. R.E. Wasylishen and C. A. Fyfe, in *Annual Reports on NMR Spectroscopy,* G. A. Webb (Ed) Vol. 12, pp. 1-80, Academic Press, London (1982).
17. J. Klinowski, *Progr. NMR Spectrosc.,* **16,** 237 (1984).
18. J. Klinowski, *Annu. Rev. Mater. Sci.,* **18,** 189 (1988).
19. G. Engelhardt and D. Michel, *High-resolution Solid-State NMR of Silicates and Zeolites,* Wiley, Chichester (1987).
20. S. L. Meisel, J. P. McCullogh, C. H. Lechthaler and P. B. Weisz, *Chemtech,* **6,** 86 (1976).
21. W. W. Kaeding and S. Butter, U.S. Patent No. 3,911,041 (1975).
22. C. D. Chang, *Catal. Rev. Sci. Eng.,* **25,** 1 (1983).

23. G. Winde, A. V. Volkov, A. V. Kiselev and V. I. Lygin, *Russian J. Phys. Chem.*, **49**, 1716 (1975).
24. Z. Luz and A. J. Vega, *J. Phys. Chem.*, **91**, 374 (1987).
25. G. Mirth, J. A. Lercher, M. W. Anderson and J. Klinowski, *J. Chem. Soc., Faraday Trans.*, **86**, 3039 (1990).
26. M. W. Anderson, P. J. Barrie and J. Klinowski, *J. Phys. Chem.*, **95**, 235 (1991).
27. T. A. Carpenter, J. Klinowski, D. T. B. Tennakoon, C. J. Smith and D. C. Edwards, *J. Magn. Reson.*, **68**, 561 (1986).
28. M. W. Anderson and J. Klinowski, *J. Chem. Soc., Chem. Commun.*, 918 (1990).
29. M. W. Anderson and J. Klinowski, *Nature*, **339**, 200 (1989).
30. M. W. Anderson and J. Klinowski, *J. Am. Chem. Soc.*, **112**, 10 (1990).
31. B. R. Richardson, N. D. Lazo, P. D. Schettler, J. L. White and J. F. Haw, *J. Am. Chem. Soc.*, **112**, 2885 (1990).
32. N. D. Lazo, J. L. White, E. J. Munson, M. Lambregts and J. F. Haw, *J. Am. Chem. Soc.*, **112**, 4050 (1990).
33. J. L. White, N. D. Lazo, B. R. Richardson and J. F. Haw, *J. Catal.*, **125**, 260 (1990).
34. E. J. Munson and J. F. Haw, *Anal. Chem.*, **62**, 2532 (1990).
35. E. J. Munson, N. D. Lazo, M. E. Moellenhoff and J. F. Haw, *J. Am. Chem. Soc.*, **113**, 2783 (1991).
36. M. W. Anderson, M. L. Occelli and J. Klinowski, *J. Phys. Chem.*, **96**, 388 (1992).
37. J. F. Haw, B. R. Richardson, I. S. Oshiro, N. D. Lazo and J. A. Speed, *J. Am. Chem. Soc.*, **111**, 2052 (1989).
38. S. M. Csicsery, in *Zeolite Chemistry and Catalysis*, J.A. Rabo (Ed), ACS Monograph, **171**, 680 (1976).
39. N. Y. Chen, W. W. Kaeding and F. G. Dwyer, *J. Am. Chem. Soc.*, **101**, 6783 (1979).
40. W. W. Kaeding, U.S. Patent No. 4,029,716 (1977).
41. W. O. Haag and D. H. Olson, U.S. Patent No. 4,097,543 (1978).
42. E. G. Derouane, J. B. Nagy, P. Dejaifve, J. H. C. van Hooff, B. P. Spekman, J. C. Védrine and C. Naccache, *J. Catal.*, **53**, 40 (1978).
43. E. G. Derouane, P. Dejaifve, J. B. Nagy, J. H. C. van Hooff, B. P. Spekman, C. Naccache and J. C. Védrine, *C. R. Acad. Sc. Paris Ser. C*, **284**, 945 (1977).
44. J. B. Nagy, J. P. Gilson and E. G. Derouane *J. Mol. Catal.*, **5**, 393 (1979).
45. E. G. Derouane, P. Dejaifve and J. B. Nagy, *J. Mol. Catal.*, **3**, 453 (1977).
46. E. G. Derouane and J. B. Nagy, *ACS Symposium Ser.*, **248**, 101 (1984).
47. E. G. Derouane, J. P. Gilson and J. B. Nagy, *Zeolites*, **2**, 42 (1982).
48. C. E. Bronnimann and G. E. Maciel, *J. Am. Chem. Soc.*, **108**, 7154 (1986).
49. J. B. Stothers, *Carbon-13 NMR Spectroscopy*, Academic Press, New York (1972).
50. M. W. Anderson and J. Klinowski, *Chem. Phys. Lett.*, **172**, 275 (1990).
51. W. Kolodziejski and J. Klinowski, *Appl. Catal. A*, **81**, 133 (1992).
52. M. T. Aronson, R. J. Gorte, W. E. Farneth and D. White, D. *J. Am. Chem. Soc.*, **111**, 840 (1989).
53. M. W. Anderson, B. Sulikowski, P. J. Barrie and J. Klinowski, *J. Phys. Chem.*, **94**, 2730 (1990).
54. B. M. Lok, C. A. Messina, R. L. Patton, R. T. Gajek, T. R. Cannan and E. M. Flanigen, *J. Am. Chem. Soc.*, **106**, 6092 (1984).
55. M. Ito, Y. Shimoyama, Y. Saito, Y. Tsurita and M. Otake, *Acta Crystallogr.C*, **41**, 1698 (1985).
56. H. O. Kalinowski, S. Berger and S. Braun, *13C-NMR-Spektroskopie*, Georg Thieme, Stuttgart/New York (1984).

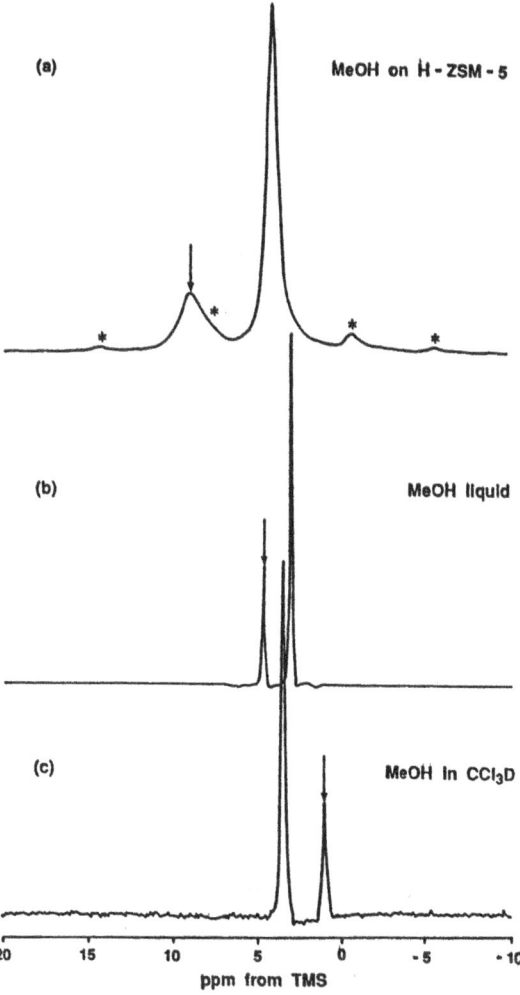

Figure 1. ¹H MAS NMR spectra [26] of (a) CH₃OH adsorbed on zeolite H-ZSM-5; (b) neat liquid CH₃OH; (c) CH₃OH dissolved in CDCl₃. Asterisks denote spinning sidebands, arrows point to the position of the resonance from the methanol hydroxyl.

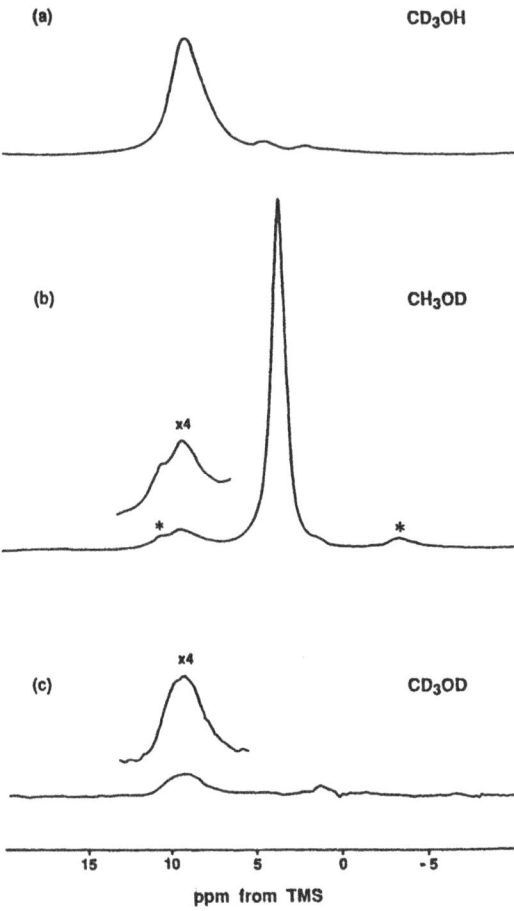

Figure 2. ^{1}H MAS NMR spectra [26] of (a) CD$_3$OH adsorbed on H-ZSM-5; (b) CH$_3$OD adsorbed on H-ZSM-5; (c) CD$_3$OD adsorbed on H-ZSM-5. Asterisks denote spinning sidebands.

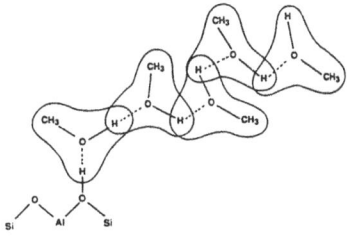

Figure 3. Protonated cluster of hydrogen bonded methanol molecules at the Brønsted acid site [26].

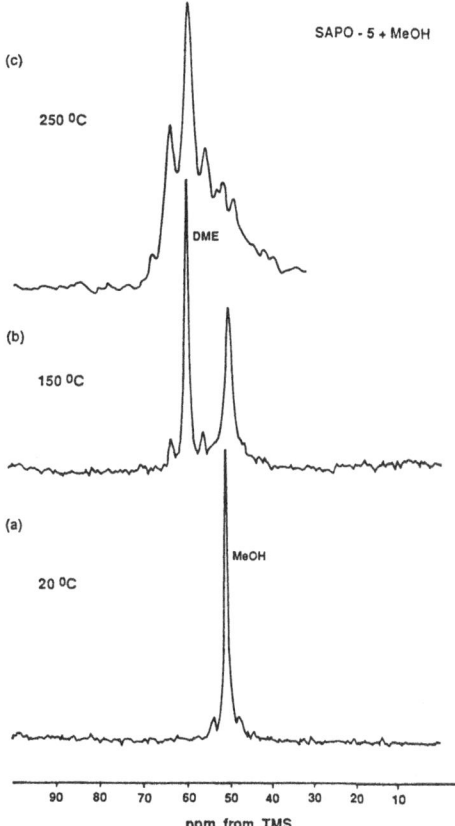

Figure 4. ^{13}C MAS NMR spectra of methanol adsorbed on SAPO-5 [28]. (a) room temperature spectrum; (b) spectrum of a sample heated to 150°C for 10 mins.; (c) sample heated to 250°C for 10 mins.

488

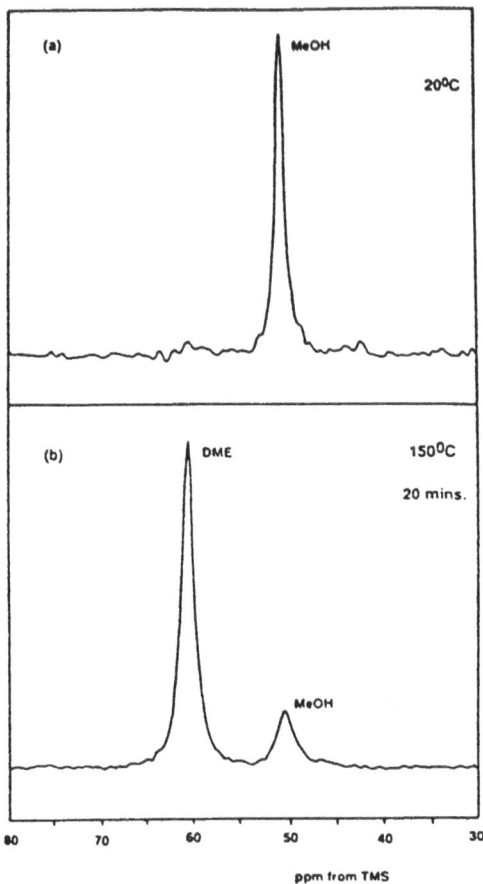

Figure 5. ^{13}C MAS NMR spectra of H-ZSM-5 with 50 torr of adsorbed MeOH and recorded at room temperature [29]. (a) no heating; (b) 150°C for 20 mins. Experiments were performed at room temperature. High-power decoupling (but no cross-polarization) was used with 40° ^{13}C pulses and a 10 s repetition time. Asterisks denote spinning sidebands.

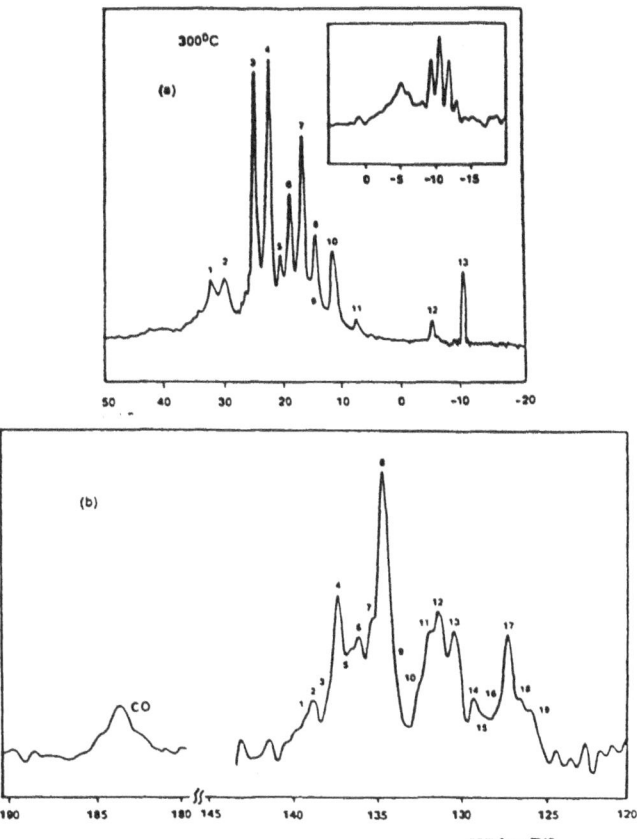

Figure 6. ^{13}C MAS NMR spectra of a sample heated to 300°C for 35 mins and recorded with proton decoupling only (a) aliphatic region; (b) aromatic and CO region [29]. Intensities in (a) and (b) are not on the same scale. The inset shows J-coupling of methane and cyclopropane carbons (recorded without decoupling). Spectral assignments are as follows [peak numbers and intensities (s = strong, m = medium, w = weak) are in brackets, {} for aliphatic and () for aromatic]: i-butane {3,s}; propane {7,s}; n-butane {3,9,m}; n-hexane {1,4,8,m}; i-pentane {1,2,4,10,m}; n-heptane {1,2,4,8,m}; methane {13,m}; ethane {11,w}; cyclopropane {12,w}; o-xylene (4,12,17,s){5}; p-xylene (7,13,s){6}; m-xylene (2,12,15,17,w){5}; toluene (2,13,14,18,w){5}; 1,2,4-trimethylbenzene (4,6,8,11,13,16,m){5,6}; 1,3,5-trimethylbenzene (3,16,w){5}; 1,2,3-trimethylbenzene (5,7,16,19,w){5,8}; 1,2,4,5-tetramethylbenzene (8,11,s){6}; 1,2,3,5-tetramethylbenzene (5,7,10,13,m){5,8}; 1,2,3,4-tetramethylbenzene (7,9,16,w){5,8}. Because of the different Overhauser enhancements of different carbons, peak intensities give only an approximate concentration of the various species.

490

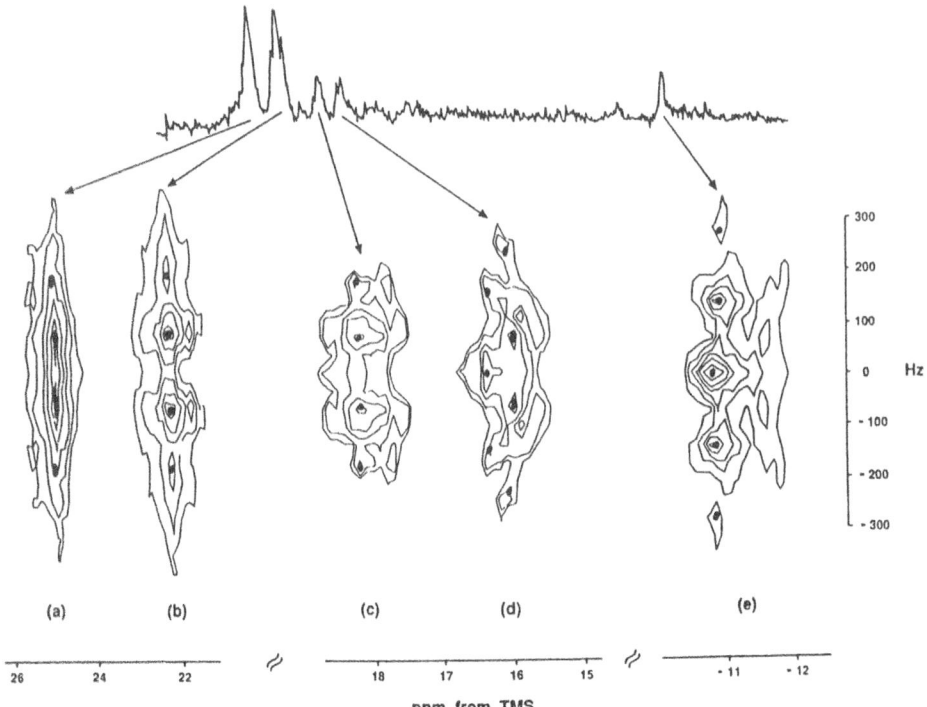

Figure 7. Heteronuclear two-dimensional J-resolved ^{13}C MAS NMR spectrum of zeolite H-ZSM-5 with adsorbed methanol treated at 300°C for 30 mins [50].

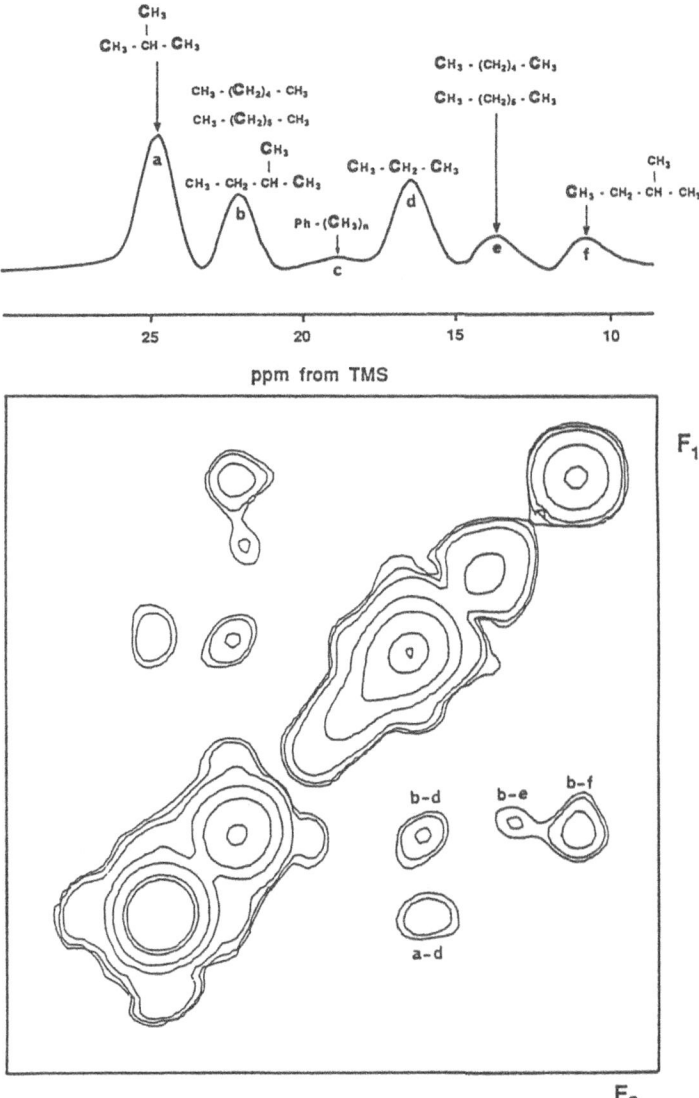

Figure 8. ^{13}C NMR spin diffusion spectrum of products of methanol conversion into gasoline over zeolite ZSM-5 with the projection onto the F_2 axis (corresponding to a conventional spectrum) at the top [51]. Carbon atoms to which individual resonances are assigned are highlighted. For signal assignment see Table 2.

492

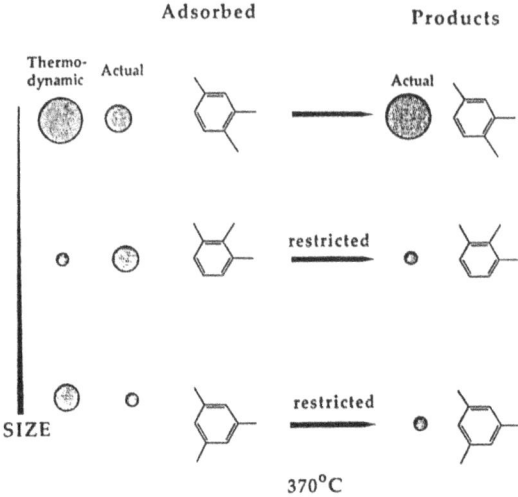

Figure 9. Schematic representation of the expected and actual distribution of the three trimethylbenzenes in the intracrystalline space and in the gaseous products of the reaction. Diameters of the circles are proportional to the intracrystalline content of each compound; the effective molecular size increases from top to bottom of the Figure.

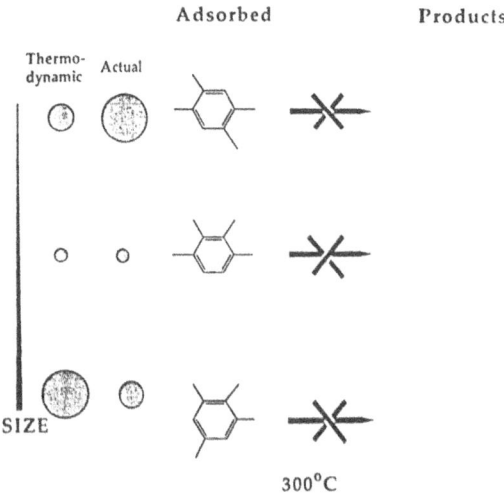

Figure 10. Schematic representation of the expected and actual distribution of the three tetramethylbenzenes in the intracrystalline space and in the gaseous products of the reaction. Diameters of the circles are proportional to the intracrystalline content of each compound; the effective molecular size increases from top to bottom of the Figure.

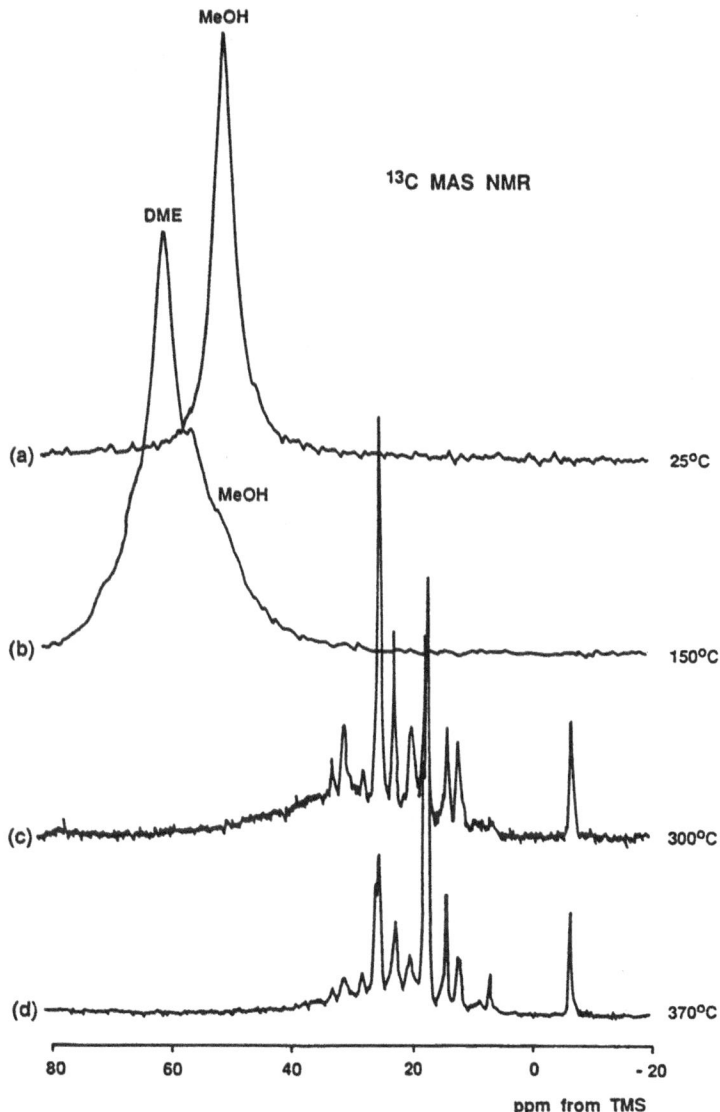

Figure 11. ^{13}C MAS NMR spectra with high-power proton decoupling of the SAPO-34 samples [53]. (a) After adsorption of methanol and no thermal treatment; (b) after heating to 150°C; (c) after heating to 300°C for 15 minutes; (d) after heating to 370°C for 10 minutes.

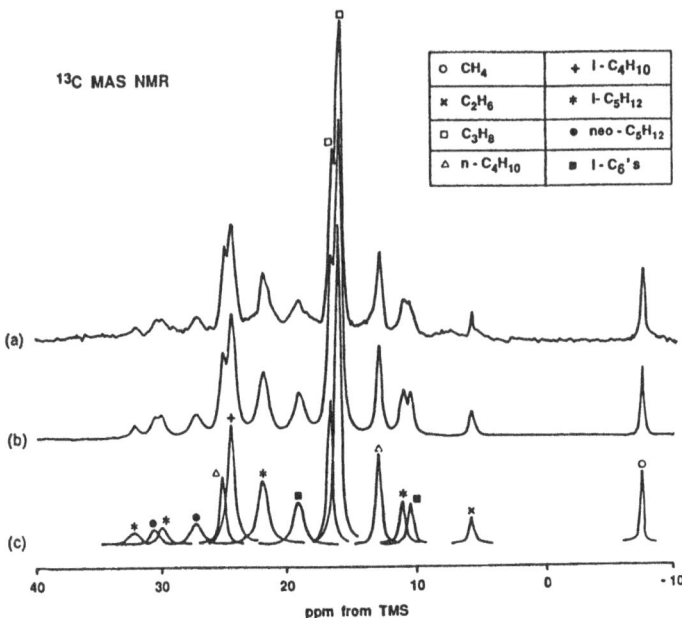

Figure 12. The ^{13}C MAS NMR spectrum [53] of the sample of SAPO-34 with adsorbed methanol treated at 370°C for 10 minutes given in (a) can be simulated (b) using ^{13}C signals corresponding to individual hydrocarbon species (c).

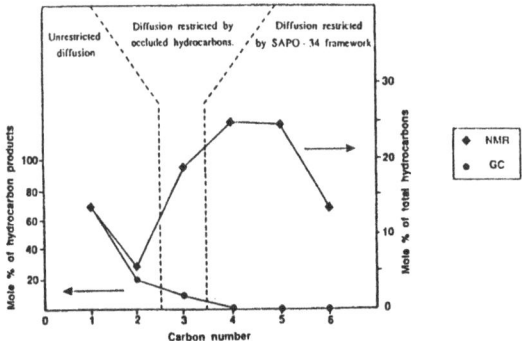

Figure 13. Plot of concentrations of C_1 to C_6 hydrocarbons at 300°C in the gaseous products determined by gas chromatography (GC) and total hydrocarbons determined by NMR [53]. GC data are for LHSV = 0.3 hr^{-1}. NMR and GC data are adjusted so that the concentrations of methane coincide.

Theoretical Study of Chemical Shielding
in Silatrane and Its Derivatives

Joseph H. Iwamiya[†] and Gary E. Maciel*
Department of Chemistry, Colorado State University
Fort Collins, CO 80523-0002

[†] Present address: Lockheed Missiles and Space, Palo Alto Research Laboratories, 3251 Hanover Street, Palo Alto, CA 94304-1191.

* To whom correspondence should be addressed.

J. A. Tossell (ed.), Nuclear Magnetic Shieldings and Molecular Structure, 495–522.

ABSTRACT

The ^{15}N and [29]Si shielding tensors (σ) of silatrane and a set of derivatives are sensitive to substituent-induced structural changes, as reflected in measured chemical shift tensors (δ). As the silicon-nitrogen internuclear distance (r_{Si-N}) decreases, the isotropic ^{15}N chemical shift moves to lower shielding and the isotropic [29]Si chemical shift moves to higher shielding. The magnitude of the ^{15}N chemical shift anisotropy (CSA) decreases, while the [29]Si CSA increases as r_{Si-N} decreases. These changes in the ^{15}N CSA are primarily due to the change in $\sigma_\perp$, which decreases as r_{Si-N} decreases. For [29]Si, all three principal elements change substantially as r_{Si-N} is varied. Variations in the ^{15}N chemical shift elements are interpreted primarily in terms of a direct N---Si transannular interaction. Results of *ab initio* chemical shift calculations on various molecular models are in qualitative agreement with the experimentally observed trends. The calculated quantities indicate that r_{Si-N} variations, i.e., the transannular interaction, not bond angle variations, largely determine the variations in the ^{15}N chemical shift tensor. For the [29]Si chemical shift tensor, the identity of the directly-attached substituent and bond angle distortions also play important roles, especially the former.

INTRODUCTION

Silatranes are a class of organosilicon compounds that nominally feature a pentaco-ordinate silicon atom (structure I, with hydrogens not shown). These compounds are

(I)

interesting from both theoretical and experimental points of view. The interest in silatranes is due to their intriguing molecular structure, physical properties, biological activity and patterns of chemical reactivity [1, 2]. In nearly all of the reported structural studies on silatranes, a degree of pentacoordinate character is shown to exist for the silicon atom, as implied in structure I. Of particular interest in the present study is the existence and influence of a "transannular bond" between the silicon and nitrogen atoms, as indicated by the dashed line in structure I, on the ^{15}N and ^{29}Si chemical shift tensors.

The length of the transannular bond in silatranes is in large part dependent upon the identity of the substituent bound to the silicon atom. Other factors, such as structural modifications on the silatrane framework shown in structure I and the physical environment of the molecule (e.g., solid, liquid, gaseous or solution), also play a role in affecting the length of this "bond" [2-4]. The silicon-nitrogen internuclear distance (r_{Si-N}) has been found to range from 2.02 Å for the chloro derivative (X = Cl) to 2.89 Å for the trans-dimethylphenylphosphine platinum derivative, with the more typical distances falling between 2.02 and 2.22 Å [3]. These distances are considerably smaller than the sum of the Van der Waals radii of 3.5 Å for silicon and nitrogen, yet are larger than a conventional silicon-nitrogen covalent bond length of 1.7 to 1.8 Å found in tetracoordinate silicon compounds [5]. This transannular bond is not like a "normal" silicon-nitrogen bond and one can consider this bond to be a very weak one [2].

To describe the transannular bond, two different models have been proposed [6-8]. The models propose either the involvement of a d_{z^2} silicon atomic orbital (where z is the N---Si internuclear axis) with silicon bonding to oxygen atoms and X via sp^3 hybridization [7] or a three-center, four-electron bond involving the $3p_z$ orbital of silicon and a z-directed X orbital (with silicon bonding to oxygen atoms via sp^2 hybrids [8].

Silatranes have been studied extensively by NMR spectroscopy in solution and numerous discussions of the isotropic chemical shifts of these systems can be found in the literature [1, 2, 9-14]. While the isotropic chemical shift is an enormously useful parameter, it must be remembered that the isotropic chemical shift is the *average* value of the principal elements of the chemical shift tensor. In the principal axis system of the chemical shift tensor, the chemical shift interaction contains *six* pieces of information, the magnitudes of the three principal elements and their orientations [15-17]. Advances in solid-state NMR allows one to consider the orientation dependence, or anisotropy, of the chemical shift interaction [15, 16]. It is now possible to determine the magnitudes of the principal tensor elements conveniently, and the orientation of the principal axes with more effort [16, 18]. *A priori*, it seems likely that, when one attempts to correlate the chemical shift interaction with chemical structure, one will be generally more successful if the correlation is established with an individual element(s) of the chemical shift tensor. Studies that combine the state of the art in NMR, both computationally and experimentally, will invariably lead to greater insight into the chemical bonding within a chemical system.

In the work described here, chemical shift parameters that have been extracted from solid-state ^{15}N and ^{29}Si NMR spectra of a set of silatranes are considered in terms of their correlations with known structural parameters, especially within the framework of *ab initio* chemical shift calculations. *Ab initio* calculations of the ^{15}N and/or ^{29}Si shielding tensor elements of a number of molecular models are presented, with the aim of learning how substituent-induced structural changes of the silatrane framework affect the ^{15}N and [29]Si chemical shift tensors. Ultimately, one would like to be able to learn about transannular bonds by determining chemical shift tensors.

Details of the experimental aspects of this study have been published elsewhere [19]. The previous paper also describes a simple molecular orbital model for describing the effect of the N---Si transannular bond or the ^{15}N chemical shift tensor within the framework of Pople's early approximate MO theory [20, 21].

CALCULATIONAL DETAILS

Ab initio coupled Hartree-Fock (CHF) ^{15}N and [29]Si chemical shielding calculations were performed on a Cyber-205, employing the formalism and approach of Lazzeretti et al. [22, 23]. For silicon, the (12s9p)-[6s5p] basis set of McLean and Chandler was used [24], augmented with d polarization functions having exponents of 0.434 and 0.261 [25]. For nitrogen, a (13s8p)-[5s3p] basis set [26] was used, augmented with d polarization functions having exponents of 1.986 and 0.412 [25]. For fluorine, oxygen and carbon, the (9s5p)-[4s3p] basis set of Dunning was used [27], supplemented with a d polarization function having an exponent of 0.800 [26], 0.750 [28] and 0.600 [25], respectively. For hydrogen bound to oxygen, the (4s)-[2s] basis set of Dunning was used [27]. For hydrogen bound to carbon and nitrogen, the (4s)-[3s] basis set was used [27], supplemented with a p polarization function with an exponent of 0.800 [29] or 1.000 [28], for carbon and nitrogen, respectively.

Due to limitations on the Cyber-205 available to us, the size of the model systems used in the chemical shielding calculations had to be limited. To reduce the size of the computational model and still realistically model the observed changes of the ^{15}N and [29]Si shieldings, it would have been desirable to employ as a model of the silatrane molecule the $XSi(OH)_3$--$N(CH_3)_3$ system, where X is a real substituent. However, this molecular model is too large for a CHF chemical shielding calculation with aforementioned basis sets on the available Cyber-205. We have instead employed for a model of the silatrane molecule the H*Si(OH)$_3$--NH$_3$ system, where H* is a pseudo atom, designed to reflect substituent properties. With this model, the description of the substituent and the chemical bonding around nitrogen atom is simplified and the chemical shielding calculations could be carried out on the available computer. With this simplified model, the best that we can hope for is that the calculated ^{15}N and [29]Si chemical shieldings will give at least a qualitatively correct description of the experimentally observed trends to determine the importance of the substituent effect on the silicon-nitrogen internuclear distance and the identity of the substituent on the ^{15}N and [29]Si chemical shielding parameters. Chemical shielding calculations were also carried out on NH$_3$, N(CH$_3$)$_3$, H*Si(OH)$_3$, SiF$_4$ and XSi(OH)$_3$ (where X = F, H or CH$_3$) to determine the effects of bond angle variations and substituent effects on the ^{15}N or [29]Si shielding tensors.

For the pseudo atom H* bound to the silicon atom, the (4s)-[3s] basis set [27], supplemented with a p polarization function with exponent 0.800, was used [26]. The pseudo atom mimics the identity of the substituent by changing the nuclear charge assigned to the H* atom. For substituents more electronegative than hydrogen, the nuclear charge on H* is increased to a value larger than 1.00. For less electronegative substituents, this parameter is decreased to a value less than 1.00. The choice of the nuclear charge on H* to model the F or CH$_3$ substituents was made by choosing a "nuclear" charge on H* that results in a chemical shielding anisotropy ($\Delta\sigma$) that matches the value *calculated* for FSi(OH)$_3$ or CH$_3$Si(OH)$_3$, respectively. In addition to this charge scaling, the exponents in the (4s)-[3s] basis set used for H* were also scaled to take into account the higher or lower "nuclear" charge, i.e.,

$$\varsigma' = Z^2\varsigma, \tag{1}$$

where ς' is the scaled exponent, Z is the "nuclear" charge on H* and ς is the unscaled, original exponent of the (4s)-[3s] basis set. From the shielding calculations on XSi(OH)$_3$, where X = F or CH$_3$, and H*Si(OH)$_3$, it was determined that a nuclear charge of 1.18 on

H* in H*Si(OH)$_3$ gave roughly the same $\Delta\sigma$ as that found for FSi(OH)$_3$ and a nuclear charge of 0.85 gave roughly the same $\Delta\sigma$ as that found for CH$_3$Si(OH)$_3$. These nuclear charges were used in the H*Si(OH)$_3$--NH$_3$ model to mimic the effects of the F and CH$_3$ on the ^{15}N and ^{29}Si chemical shielding tensors.

For the various molecular models used in the calculations, the following bond lengths and bond angles were used. For the H*Si(OH)$_3$--NH$_3$ model, the H-O, H-N and H*-Si bond lengths were fixed at 0.960, 1.012 and 1.480 Å, respectively. The O-Si bond lengths were chosen to be 1.640 Å when the charge on H* was 1.18, or 1.670 Å when the charge on H* was 1.00 or 0.85. For "computer experiments" in which the N-Si inter-nuclear distance was varied, it was given values of 2.040, 2.120 and 2.175 Å, correspond-ing to the values found experimentally in fluoro-, hydrido- and methylsilatrane [3, 30, 31]. The ∠OSiO values were chosen to be 119.5°, 119.0° or 118.4° and the ∠SiOH values chosen to be 120.5°, 122.5° or 123.1°, depending on the nuclear charge on H* and corre-sponding to the ∠SiOC values reported for the aforementioned silatranes [3, 30, 31]. The ∠HNH angles were fixed at 113° when the nuclear charge on H* was 1.18 or 1.00 or 114° when the charge on H* was 0.85 [2, 3, 30, 31].

In the XSi(OH)$_3$ models (where X = F, H, CH$_3$ or H*), the H-C, F-Si, H-Si, C-Si, O-H and H*-Si bond lengths were chosen to be 1.090, 1.622, 1.480, 1.870, 0.960 and 1.480 Å when these models were fixed at the corresponding fluoro-, hydrido- or methylsilatrane geometries. The O-Si bond lengths were fixed at 1.64 Å for the fluoro-silatrane geometry [30] and 1.67 Å for the hydrido- and methylsilatrane geometries [31]. The ∠OSiO angles were chosen to be 119.5°, 119.0° or 118.4°, and the ∠SiOH angles were given values of 120.5°, 122.5° or 123.1° for the fluoro-, hydrido- or methylsilatrane geometries; the ∠HCH angle was fixed at 109.5°.

For fluoro-, hydrido- and methyltriethoxysilane in their equilibrium geometries, the H-C, F-Si, H-Si, C-Si, O-Si, O-H and H*-Si bond lengths were chosen to be 1.090, 1.540, 1.480, 1.480, 1.700, 1.600, 0.960 and 1.480 Å, respectively. The ∠OSiO values were fixed at 109.9° and the ∠SiOH values were fixed at 119.6°. The ∠HCH values were fixed at 109.5°.

In the NH_3 and $N(CH_3)_3$ models, the H-N, C-N and H-C bond lengths were fixed at 1.012, 1.476 and 1.090Å, respectively [32]. The ∠HNH values in NH_3 were varied between 106.7° and 116.0° and the ∠CNC values in $N(CH_3)_3$ were varied between 111.0° and 116.0°. In the SiF_4 model, the F-Si bond lengths were fixed at 1.556 Å and the ∠FSiF values were fixed at 109.5° [33].

The chemical shielding convention used in the *ab initio* calculations is the σ scale, in which algebraically larger numbers indicate higher shielding, opposite in sense to the δ scale of chemical shifts. The symbol σ is used to label the chemical shielding, and the convention, $\sigma_{33} > \sigma_{22} > \sigma_{11}$, is employed for convenience.

RESULTS AND DISCUSSION

SUMMARY OF EXPERIMENTAL RESULTS

It has been previously shown that the isotropic ^{15}N chemical shift demonstrates a very clear correlation with r_{Si-N} in silatranes [13]. It has been observed that the ^{15}N chemical shift powder patterns have shapes characteristic of axially symmetric chemical shift tensors [17], or very nearly so, and exhibit a substantial variation with the identity of the substituent.

The effect of the substituent on a silatrane structure is clearly seen in r_{Si-N}. More electronegative substituents give geometries with a smaller r_{Si-N} value than those characteristic of silatranes with less electronegative substituents [3]. Since $\delta_{iso} = (\delta_{\parallel} + 2\delta_{\perp})/3$ for an axially symmetric chemical shift tensor and is nothing more than a specific combination (the mean) of the principal elements of the tensor, other combinations or individual elements might be expected to follow analogous trends. Such trends have recently been reported [13], and some are duplicated here. Figure 1 graphically depicts the ^{15}N chemical shift parameters ($\delta_{\parallel}$, $\Delta\delta$, δ_{iso}, and $\delta_{\perp}$) plotted versus r_{Si-N}. One recognizes a very clear correlation for $\Delta\delta$, δ_{iso} and $\delta_{\perp}$, while $\delta_{\parallel}$ appears to be essentially invariant to r_{Si-N}. The chemical shift anisotropy, $\Delta\delta$, defined as $\delta_\| - \delta_\perp$ (where $\delta_\|$ is the unique element), decreases in magnitude as r_{Si-N} decreases, from 44 to 27 ppm for the ethyl and chloro derivatives, respectively. As r_{Si-N} decreases, δ_{iso} increases, as previously observed [13]. It is clear that the observed changes in the ^{15}N chemical shift tensor are driven by changes in $\delta_\perp$, which varies from 7 (ethyl) to 21 (chloro) ppm in the set of samples. Since the local molecular symmetry around the nitrogen atom is approximately C_{3v}, the orientation of the principal axes of the ^{15}N chemical shift tensor can be postulated *a priori* [17]; the $\delta_\|$ element lies along the silicon-nitrogen internuclear axis and the $\delta_\perp$ elements lie in a plane perpendicular to the silicon-nitrogen internuclear axis.

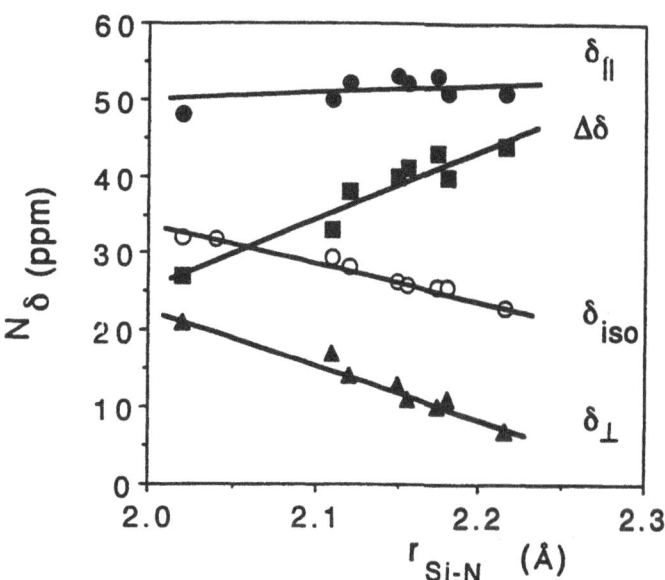

Figure 1. Correlation curves for the experimental ^{15}N chemical shift parameters *vs* r_{Si-N} ($\delta_\|$, $\delta_\| - \delta_\perp = \Delta\delta$, δ_{iso} and $\delta_\perp$).

Examination of the ^{29}Si end of the silatrane system yields even more substantial variations than for the nitrogen end; this, of course, is not surprising, as the substituent is attached directly to silicon. X-ray determined structural data indicate that the deviation of the silicon atom from the plane of the equatorial oxygen atoms increases by more than 0.10 Å, while the deviation of the nitrogen atom from the plane of the equatorial carbon atoms decreases by only 0.06 Å when the substituent is changed from chlorine to an ethyl group [3]. The ^{29}Si chemical shift powder patterns have shapes that correspond to a slightly asymmetric tensor [17], and exhibit substantial variation in their widths.

As in the case of the ^{15}N chemical shift parameters, the [29]Si chemical shift parameters show general trends with variation of r_{Si-N} (i.e., variation of X). Figure 2 is a set of previously reported chemical shift correlation curves for δ_{33}, δ_{11}, δ_{22}, δ_{33} - δ_{11} and δ_{iso} versus r_{Si-N} [19]. The local effect of the directly bound substituent and $\angle OSiO$ variations are expected to play roles in affecting the [29]Si chemical shift tensors. The plots in Figure 2

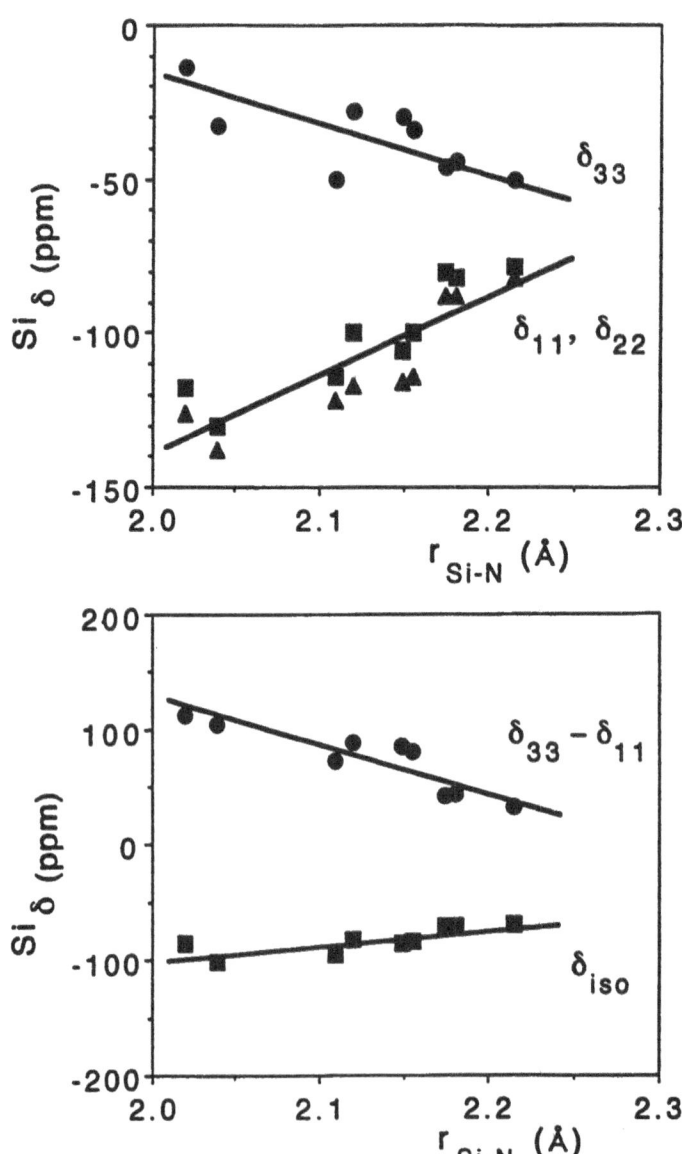

Figure 2.   Correlation curves for some experimental [29]Si chemical shift parameters vs r_{Si-N}. δ_{11} (squares), δ_{22} (triangles).

indicate that, as r_{Si-N} decreases in length, $\delta_{33} - \delta_{11}$ increases, the values of δ_{11} and δ_{22} decrease and δ_{33} increases, with the overall result that δ_{iso} decreases. In contrast to the ^{15}N data, all of the principal elements of the ^{29}Si chemical shift powder pattern show a rough dependence on r_{Si-N}, with the δ_{11} and δ_{22} elements more sensitive to variations in r_{Si-N} than the δ_{33} element.

CALCULATED RESULTS

Figures 1 and 2 suggest very strongly that the ^{15}N and ^{29}Si chemical shift tensor in silatranes can be used to monitor the changes that are induced in the electronic structure of silatranes as the identity of the substituent is varied. It is of interest to determine the relative importance of variations of relevant bond angles, bond lengths or substituent electronic property. "Computer experiments" in which the shielding is calculated can be used to explore, using the *ab initio* formalism of Lazzeretti et al. [22, 23], each of these variations alone; and their affects on the ^{15}N and ^{29}Si shieldings can be determined.

It should be noted that conventional CHF calculations of this type have a strong dependence on the gauge origin, due to the use of Gaussian functions without gauge factors and a common origin [34-37]. CHF calculations of this type are independent of the gauge origin *only in the limit* of a complete basis set [34, 35]. The degree of gauge dependence of the calculated results can be estimated via the Thomas-Reiche-Kuhn sum rule (equation (2)) and the Arrighini-Maestro-Moccia sum rule (equation (3)):

$$(P_\alpha, P_\alpha) = N, \tag{2}$$

$$<r_A/r^3{}_A> = (L_\alpha/r^3, P_\beta)_A = -(L_\beta/r^3, P_\alpha)_A, \tag{3}$$

In equation (2), (P_α, P_α) is a sum of products of the expectation value of the α component of the total electronic linear momentum. In the limit of a complete basis set, the sum that is represented by N is equal to the number of electrons present. In equation (3), the leftmost quantity is the expectation value for the electric field at atom A, the middle and rightmost quantities are expectation values of the first-order density matrix [23, 38-41]. In the limit of a complete basis set, the equalities in equation (3) are upheld. For the shielding calculations presented here, N in equation (2) will not equal the total number of electrons and the equalities in equation (3) will not be upheld. These relations, however, will yield insight into the quality of the wavefunction as well as providing a measure of the quality of the computed shielding quantities for any gauge origin [23].

Figure 3 presents plots of the *calculated* shielding elements for NH_3 and $N(CH_3)_3$

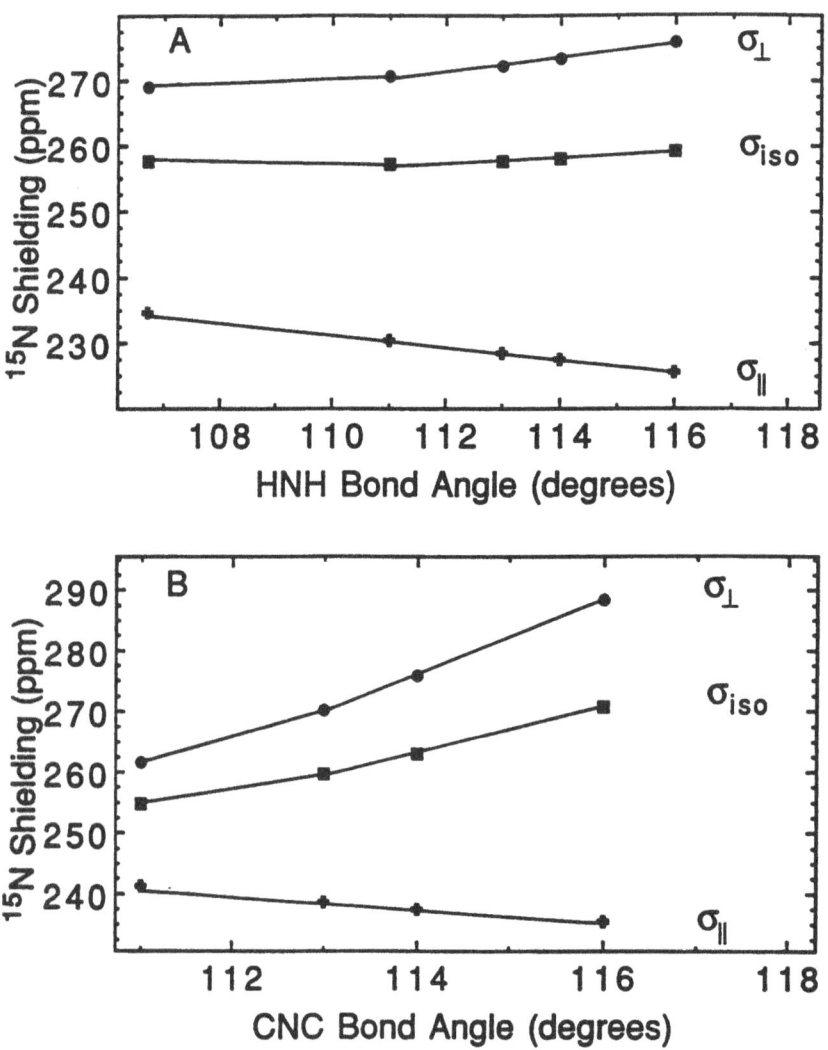

Figure 3. Plot of calculated elements of the ^{15}N chemical shielding tensor in NH_3 (A) and $N(CH_3)_3$ (B) as a function of the HNH and CNC bond angle, respectively.

as a function of the HNH or CNC bond angle, respectively. The ^{15}N shieldings are calculated with the gauge origin located at the center of mass of each molecule. Figure 3A demonstrates that $\angle HNH$ variations, from $106.7°$ to $116.0°$, have a small effect on the ^{15}N chemical shielding in NH_3, in accordance with previously published results [36]. In addition, the calculated shielding results for NH_3 at its equilibrium configuration are in fair agreement with previous calculated and experimental results [34, 42, 43]. The value of $\sigma_\perp$ (representing the σ_{xx} and σ_{yy} elements) shows a small increase to higher shielding and $\sigma_\parallel$ (defined as σ_{zz}) shows a slight decrease to lower shielding as the HNH bond angle increases. The result of these variations is that σ_{iso} is predicted to be almost independent of the HNH bond angle over this range. In particular, over the range of average CNC bond angles measured in silatranes [2, 3] $113°$ to $114°$, the effect of HNH variation over this range is that $\sigma_\perp$ increases by 1.1 ppm, while $\sigma_\parallel$ decreases by 1.0 ppm and σ_{iso} increases by 0.4 ppm. These changes cause $\Delta\sigma$ to increase by 2 ppm, from -44 ppm to -46 ppm.

Figure 3B presents the calculated shielding results for $N(CH_3)_3$, a model which might better represent the local nitrogen atom environment in a silatrane, sans the silicon-nitrogen transannular interaction. The calculated trends are similar to those calculated for NH_3, albeit the angular dependence is somewhat larger. These calculations also predict that CNC angular variations from $113°$ to $114°$ will have a small, but perhaps significant effect on the ^{15}N chemical shielding tensor. The computed $\sigma_\perp$ increases by approximately 6 ppm as the CNC bond angle is varied from $113°$ to $114°$, and $\sigma_\parallel$ decreases by 1 ppm. The result of these slight shifts in the principal elements is that σ_{iso} increases by 4 ppm and $\Delta\sigma$ increases by 7 ppm as the CNC bond angle in $N(CH_3)_3$ is changed from $113°$ to $114°$. In relating the computed shielding trends with the experimental results shown in Figure 1, it should be remembered that the CNC bond angle increases as r_{Si-N} increases. Experimentally, it is observed that $\delta_\perp$ experiences an increase in shielding of 11 ppm, $\delta_\parallel$ appears to be constant, δ_{iso} shifts to higher shielding upfield by 6 ppm and $\Delta\delta$ increases by 16 ppm as the substituent changes from the chloro to the methyl group (r_{Si-N} increases from 2.02 to 2.18 Å). It appears that $\angle CNC$ variations exhibit the same trends as the experimental observations, but the magnitudes of the computed changes are too low to account entirely for the experimental observations. Angular variations apparently do make a significant contribution to the observed ^{15}N chemical shift trends in silatranes, but they are not dominant factors.

Table 1 lists the calculated quantities for the sum rules (equations (2) and (3)) for the NH_3 and $N(CH_3)_3$ calculations. The shielding calculations account for 82% of the total number of electrons present in both the NH_3 and $N(CH_3)_3$ molecules. Equation (3) is nearly satisfied for NH_3, indicating that the ^{15}N shielding quantities should exhibit little gauge dependence. For $N(CH_3)_3$, equation (3) is not satisfied to as high a degree as for NH_3 and the $N(CH_3)_3$ shielding results should exhibit a more substantial dependence on the choice of the gauge origin.

Figure 4 presents the calculated ^{15}N shielding results for the effect of the variation of r_{Si-N} on the ^{15}N chemical shielding in the $H^*Si(OH)_3$--NH_3 model. In the set of computer experiments that yielded these results, the geometry of the model was fixed to that of hydridosilatrane with the exception of r_{Si-N}, which was allowed to vary between 2.04 and 2.175 Å. Figure 4A presents the ^{15}N shielding results calculated with the gauge origin located at the center of mass of the model and Figure 4B presents the results calculated with the gauge origin located at the nitrogen atom. The reason for this dual presentation can be seen in the results shown in Table 2, which presents the results calculated for the ^{15}N shielding sum rules. These results indicate that the ^{15}N shielding results summarized in Figure 4 have a larger gauge dependence than those of Figure 3. Approximately 71% of the total number of electrons is accounted for and equation (3) is satisfied to the same level as in the calculations on the $N(CH_3)_3$ molecule. The fact that these results are so gauge dependent is the reason for showing results obtained with two choices of gauge.

The results shown in Figure 4A (gauge origin located at the center of mass) predict that as r_{Si-N} increases, $\sigma_{\parallel}$ will have a weak dependence and $\sigma_{\perp}$ will have a large dependence (increasing to higher shielding). The result of these dependences is that σ_{iso} moves to higher shielding and the magnitude of $\Delta\sigma$ (defined as $\sigma_{\parallel}$ - $\sigma_{\perp}$) decreases, from 44 ppm to

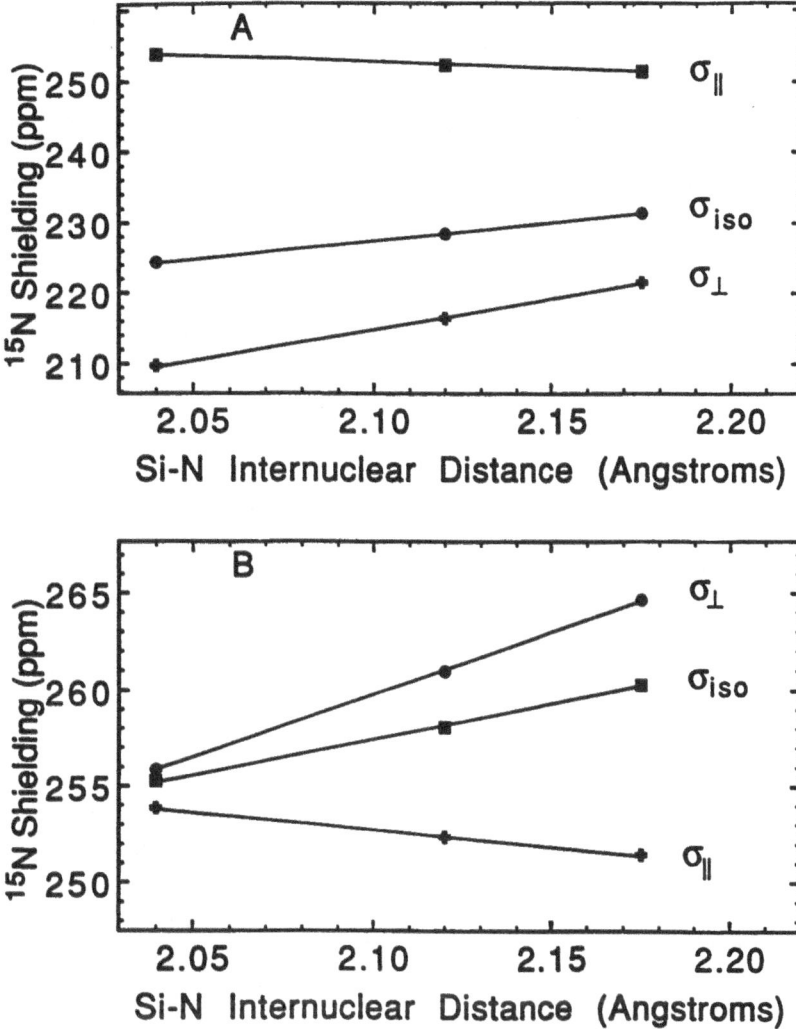

Figure 4. Plot of calculated elements of the ^{15}N chemical shielding tensor ($\sigma_\parallel$, σ_{iso}, $\sigma_\perp$) in $H^*Si(OH)_3$--NH_3 as a function of r_{Si-N}, with the gauge origin located at the center of mass (A) and with the gauge origin located at the nitrogen atom (B).

30 ppm as r_{Si-N} is increased from 2.040 to 2.175 Å. The results shown in Figure 4B (gauge origin located at the nitrogen atom) predict that as r_{Si-N} increases, $\sigma_\perp$ will increase substantially and $\sigma_\parallel$ will have a smaller dependence, decreasing as r_{Si-N} increases. These dependences indicate that σ_{iso} will shift to higher shielding and $\Delta\sigma$ will increase from 2 ppm to 13 ppm as r_{Si-N} is increased from 2.040 to 2.175. Compared to the experimental results, the results of both Figure 4A and Figure 4B show the correct trend for each parame-

Table 1. Calculated quantities for the sum rules corresponding to the ^{15}N shielding results shown in Figure 3.

Angle	$<r_z/r^3>$	$(L_x/r^3, P_y)$	$(L_y/r^3, P_x)$	(P_{ave}, P_{ave})[a]
NH_3				
106.7°	-0.2717	-0.2690	0.2690	8.18 (10)
111.0°	-0.2182	-0.2151	0.2151	8.18 (10)
113.0°	-0.1899	-0.1891	0.1891	8.18 (10)
114.0°	-0.1746	-0.1753	0.1753	8.18 (10)
116.0°	-0.1407	-0.1444	0.1444	8.17 (10)
$N(CH_3)_3$				
111.0°	-0.8251	-0.5326	0.5326	27.93 (34)
113.0°	-0.7219	-0.4643	0.4643	27.95 (34)
114.0°	-0.6623	-0.4253	0.4253	27.95 (34)
116.0°	-0.5356	-0.3429	0.3429	27.97 (34)

[a] The number in parentheses is the total number of electrons in the system.

Table 2. Calculated quantities for the sum rules corresponding to the ^{15}N and ^{29}Si shielding results in the $H*Si(OH)_3$--NH_3 model.[a]

r_{Si-N}	$<r_z/r^3>$	$(L_x/r^3, P_y)$	$(L_y/r^3, P_x)$	(P_{ave}, P_{ave})[b]
^{15}N				
2.040	1.6386	1.0554	-1.0554	36.92 (52)
2.120	1.5484	1.0112	-1.0112	38.88 (52)
2.175	1.4899	0.9814	-0.9814	36.85 (52)
^{29}Si				
2.040	-0.7463	-0.6575	0.6575	36.92 (52)
2.120	-0.7035	-0.6397	0.6397	36.88 (52)
2.175	-0.6765	-0.6271	0.6271	36.85 (52)

[a] With the exception of r_{Si-N}, the bond lengths, angles and charge on $H*$ are fixed to the hydridosilatrane geometry.

[b] The number in parentheses is the total number of electrons in the system.

ter in the figures. The ^{15}N shielding results calculated with the gauge origin located at the

nitrogen nucleus give trends that are closer in magnitude to the experimental observations

than the ^{15}N shielding results obtained with the gauge origin located at the center of mass.

Figure 5 presents the ^{15}N shielding quantities calculated for the $H^*Si(OH)_3$--NH_3

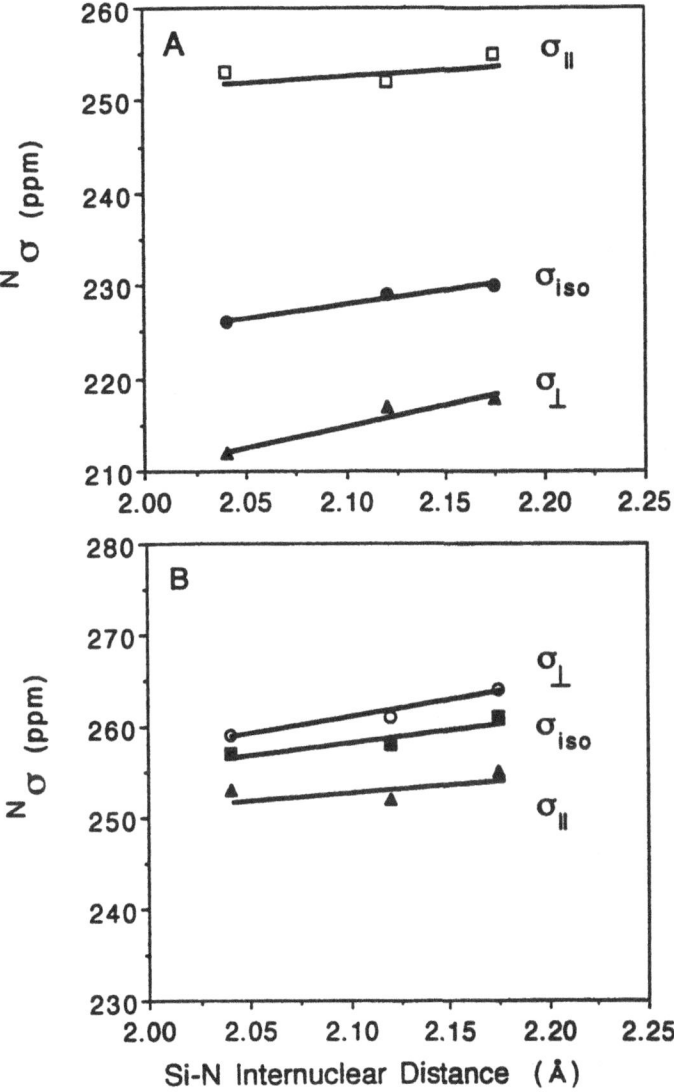

Figure 5.   Plot of calculated elements of the ^{15}N chemical shielding tensor in

H*Si(OH)$_3$--NH$_3$ as a function of r_{Si-N} (with HNH bond angle and H* charge

varied simultaneously, as described in the text) with the gauge origin located at

the center of mass (A) and at the nitrogen nucleus (B).

model in which the geometry and the atomic charge on H* in the model are varied in accordance with the observed geometries of the corresponding silatrane molecules. Figure 5A is a plot of the ^{15}N chemical shielding quantities calculated with the gauge origin located at the center of mass and Figure 5B is a plot of ^{15}N chemical shieldings determined with the gauge origin located at the nitrogen atom. The quantities calculated for evaluating the sum rules are presented in Table 3.

The calculated sum rule quantities in Table 3 indicate that the ^{15}N shielding results for the silatrane model in which the bond lengths, angles and atomic charge on H* are varied to simulate the actual silatrane-type molecule have the same degree of gauge dependence as the shielding results shown in Figure 4. A substantial difference between the characters of Figures 4 and 5 is that the magnitudes of the variations are smaller in Figure 5 than they are in Figure 4. For the same reasons given above for Figure 4, the discussion of the results shown in Figure 5 will be limited to Figure 5B. In this set of calculations, both $\sigma_\perp$ and $\sigma_\parallel$ move slightly to higher shielding as r_{Si-N} increases. The parameter, σ_{iso}, moves by 4 ppm to higher shielding as r_{Si-N} is varied from 2.040 to 2.175 Å. Nevertheless, the results displayed in Figures 4 and 5 at least qualitatively mimic the experimentally observed changes in the ^{15}N chemical shielding of silatranes with variation in r_{Si-N}.

Table 3. Calculated quantities for the sum rules corresponding to the ^{15}N and ^{29}Si shielding results in the H*Si(OH)$_3$--NH$_3$ model.[a]

r_{Si-N}	$<r_z/r^3>$	$(L_x/r^3, P_y)$	$(L_y/r^3, P_x)$	(P_{ave}, P_{ave})[b]
^{15}N				
2.040	1.6430	1.0637	-1.0637	36.98 (52)
2.120	1.5484	1.0112	-1.0112	38.88 (52)
2.175	1.5197	0.9710	-0.9710	35.58 (52)
^{29}Si				
2.040	-0.6652	-0.6022	0.6022	36.98 (52)
2.120	-0.7035	-0.6397	0.6397	36.88 (52)
2.175	-0.7550	-0.6081	0.6081	35.58 (52)

[a] All bond lengths, angles and the charge on H* are varied.

[b] The number in parentheses is the total number of electrons in the system.

The lack of a more quantitative correlation between the experimental solid-state ^{15}N chemical shift observations and the calculated ^{15}N shielding results may be due to a number of different possibilities. The reversal of the principal elements when the gauge origin is moved between the center of mass to the nitrogen atom in the calculated ^{15}N shielding tensor for the silatrane model is probably due to the inadequacies of the basis set used in the *ab initio* calculations. It has been demonstrated previously [34, 22, 43] that if the complexity or quality of the basis set is not high enough, severe gauge dependence and even erroneous results can be obtained, particularly for atoms that have one or more lone electron pairs. The addition of sp-diffuse functions to the basis set used for the nitrogen atom might help in providing a better description of the electronic environment surrounding on the nitrogen atom [34, 43], especially as perturbed by the magnetic field. It is also possible that NH_3 is not a sufficiently good model for the nitrogen atom in silatranes.

Using the same general methodology as discussed above for ^{15}N shielding, the effects of the substituent, the OSiO bond angle and r_{Si-N} on the [29]Si chemical shielding in silatranes were also examined. Table 4 presents the effects of the substituent and OSiO bond angle variations on the calculated [29]Si chemical shielding in $H^*Si(OH)_3$ and $XSi(OH)_3$, a model in which the silicon-nitrogen transannular interaction is absent. The calculated [29]Si shielding results are given with the gauge origin located at the center of mass. As seen in Table 4, the use of H^* in place of X is not detrimental, since the shielding results for $H^*Si(OH)_3$ are qualitatively similar to the results for $XSi(OH)_3$. Table 5 presents the sum rule quantities for the [29]Si shielding calculations; these results indicate that the calculated [29]Si shielding elements have a slightly higher gauge dependence than the calculated ^{15}N results. However, since the location of the center of mass and the silicon atom are relatively close to one another, the gauge dependence will be ignored for this set of calculations.

It is observed in Table 4 that at a fixed OSiO bond angle (109.9°), the [29]Si shielding is predicted to be highly dependent upon the identity of the substituent, although the $H^*Si(OH)_3$ model underestimates the shielding differences. The variation in calculated iso-

Table 4. Effect of the OSiO bond angle and substituent
variations on the calculated ^{29}Si shielding in
$H*Si(OH)_3$.[a]

Model Angle	σ_{iso}	$\sigma_{\parallel}$	$\sigma_{\perp}$	$\Delta\sigma$ [b]
$H* = F$				
109.9°	497	505	494	11
119.5°	500	460	520	-60
$X = F$				
109.9°	529	532	528	4
119.5°	516	478	536	-58
$H* = H, X = H$				
109.9°	486	515	472	43
119.0°	493	481	499	-18
$H* = H_3C$				
109.9°	475	522	452	70
118.4°	483	497	476	21
$X = H_3C$				
109.9°	468	531	436	96
118.4°	484	495	478	17

[a] The unique principal element, $\sigma_{\parallel}$, is aligned along the C_3 rotational axis.

[b] $\Delta\sigma = \sigma_{\parallel} - \sigma_{\perp}$.

tropic ^{29}Si shielding results are qualitatively similar to the experimentally observed chemical shift variation in substituted triethoxysilanes [14, 44, 45], moving to lower shielding as the effective electronegativity of the "substituent" (H*) decreases. The $\sigma_{\perp}$ and $\sigma_{\parallel}$ elements move in opposite directions as the characteristics of H* are varied, with the $\sigma_{\perp}$ elements showing a greater dependence on the H* "substituent". Compared to the trends in experimental ^{29}Si chemical shifts in silatranes, the calculated ^{29}Si shielding elements for the $H*Si(OH)_3$ and $XSi(OH)_3$ models show a discrepancy in the sign of the anisotropies of the shielding tensors. Experimentally, $\sigma_{\parallel} < \sigma_{\perp}$, whereas the opposite is true in the calculated

Table 5. Calculated quantities for the sum rules corresponding to the [29]Si shielding results in the H*Si(OH)$_3$ model.

Model OSiO Angle	$<r_z/r^3>$	$(L_x/r^3, P_y)$	$(L_y/r^3, P_x)$	(P_{ave}, P_{ave})[a]
H* = F				
109.9°	-0.8241	-0.4589	0.4589	28.86 (42)
119.5°	-0.5719	-0.2179	0.2179	28.73 (42)
H* = H				
109.9°	-0.8327	-0.4862	0.4862	28.86 (42)
119.0°	-0.1391	-0.2725	0.2725	28.86 (42)
H* = H$_3$C				
109.9°	-0.8414	-0.4818	0.4814	28.85 (42)
118.4°	-0.2178	-0.2525	0.2525	27.79 (42)

[a] The number in parentheses is the total number of electrons in the system.

results. The experimentally observed ranges in the $\delta_\parallel$ and $\delta_\perp$ elements amount to approximately 8 ppm and 50 ppm, respectively. The calculated results show that the $\sigma_\parallel$ and $\sigma_\perp$ elements vary by about 17 and 42 ppm, respectively, in the H*Si(OH)$_3$ model. Clearly, the identity of the substituent plays a major role in determining changes in the [29]Si shielding of silatranes, but the calculations do not reproduce the correct sign of the shielding anisotropy.

Examination of the effect of $\angle$OSiO variations in the substituted trihydroxysilane models (Table 4) indicates that *distorting* the OSiO bond angle has a substantial effect on the [29]Si chemical shielding. Such an effect had been previously predicted in the case of distorting the SiO$_4^{4-}$ tetrahedron [46]. For two of the three H*Si(OH)$_3$ cases examined, distortion of the OSiO bond angle (above the tetrahedral value) leads to a change in the sign of $\Delta\sigma$. For the particular bond angle distortions examined in this study, the $\Delta\sigma$ values calculated for the H*Si(OH)$_3$ models in which H* = F and H have the same sign as was found experimentally, i.e., $\sigma_\parallel < \sigma_\perp$; for H* = CH$_3$ the calculated $\Delta\sigma$ has the wrong sign. Over the range of about 110° to about 119°, $\sigma_\parallel$ changes by approximately 2.8 to 4.5 ppm/Å, depending upon the substituent; $\sigma_\perp$ changes by approximately 2.6 ppm/Å. More importantly, over the range of typical OSiO bond angles (118° to 120°) measured in silatranes [2, 3], when r_{Si-N} varies from 2.02 to 2.22Å, the variation of $\angle$OSiO in the H*S(OH)$_3$ model will cause the principal elements of the ^{29}Si shielding tensor to shift by approximately 5 to 9 ppm, depending upon the substituent. These calculated variations are much smaller than what is observed experimentally; hence, it is concluded that $\angle$OSiO variations *alone* are not responsible for the observed substituent effects in silatranes.

Figure 6 presents computed results for the dependence of the ^{29}Si chemical shielding (gauge origin located on at the center of mass) for the H*Si(OH)$_3$--NH$_3$ model in which *only* r_{Si-N} is varied. These results are based on the same model as represented in the ^{15}N results presented in Figure 4. There is no variation of the "substituent", H*. Table 2 indicates that the gauge dependence of the ^{29}Si shielding elements calculated for a range of results in r_{Si-N} values for H*Si(OH)$_3$--NH$_3$ are fairly small. Also, since the silicon atom is

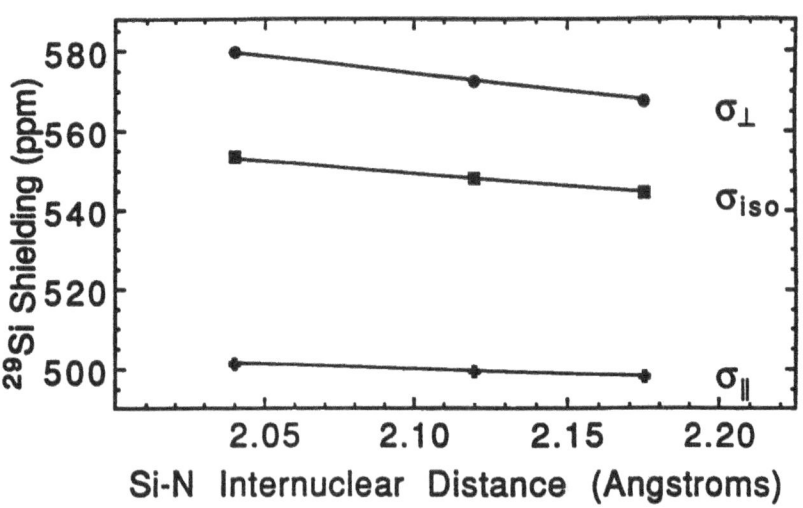

Figure 6. Plot of calculated elements of the ^{29}Si chemical shielding tensor ($\sigma_\perp$, σ_{iso}, $\sigma_\parallel$) in H*Si(OH)$_3$--NH$_3$ as a function of r_{Si-N} with the gauge origin located at the center of mass.

located very close to the center of mass, the differences in the ^{29}Si shielding quantities calculated for these two choices of gauge origin should be small and can be ignored in this discussion. The trend of $\sigma_\perp$ with r_{Si-N} has the correct sense (see Fig. 2), but the magnitudes of the computed changes are too small. The trend of calculated results for $\sigma_\parallel$ has the wrong sense.

Figure 6 indicates that the principal elements of the ^{29}Si shielding tensor all move in the same direction, to lower shielding, as r_{Si-N} increases. Of the principal elements, the $\sigma_\perp$ elements are expected to have a more dramatic dependence upon r_{Si-N} than $\sigma_\parallel$ does. This is borne out in the calculations. In addition, the relative position of the principal elements to one another is correct, as is the dependence of $\sigma_\parallel - \sigma_\perp$ on r_{Si-N}, decreasing as r_{Si-N} increases. However, comparison of the calculated results in Figure 6 with the experimental results in Figure 2 indicates for a fixed, distorted OSiO bond angle at 119° and varying r_{Si-N} underestimates the experimentally observed changes and does not give the correct dependence for the $\sigma_\parallel$ element. Experimentally, $\delta_\parallel$ moves by roughly 8 ppm to higher shielding and $\delta_\perp$ moves roughly 50 ppm to lower shielding as r_{Si-N} increases from 2.04 Å to 2.175 Å. The result is that δ_{iso} moves roughly 30 ppm to lower shielding and $\Delta\delta$ decreases by 58 ppm. The corresponding calculated quantities summarized in Figure 6 indicate that the effects of r_{Si-N} alone will move $\sigma_\parallel$ to lower shielding by only 4 ppm and $\sigma_\perp$ to lower shielding by 12 ppm, with the result that σ_{iso} moves to lower shielding by 9 ppm and $\Delta\sigma$ decreases by 9 ppm. Although, the trends are qualitatively correct in Figure 6, it is apparent that these calculations do not properly account for the magnitudes of the observed changes in the ^{29}Si shielding in silatranes.

Figure 7 presents calculated ^{29}Si chemical shielding parameters obtained in calculations with the gauge origin located at the center of mass for the H*Si(OH)$_3$--NH$_3$ model in which the geometry *and* atomic charge on H* in the model are varied in accordance with the experimentally observed geometries of the correponding silatrane molecules. The results shown in Table 3 indicate that the shielding quantities represented in Figure 7 have a small gauge dependence for the two gauges examined, since the center of mass and the location of the silicon atom are close to one another.

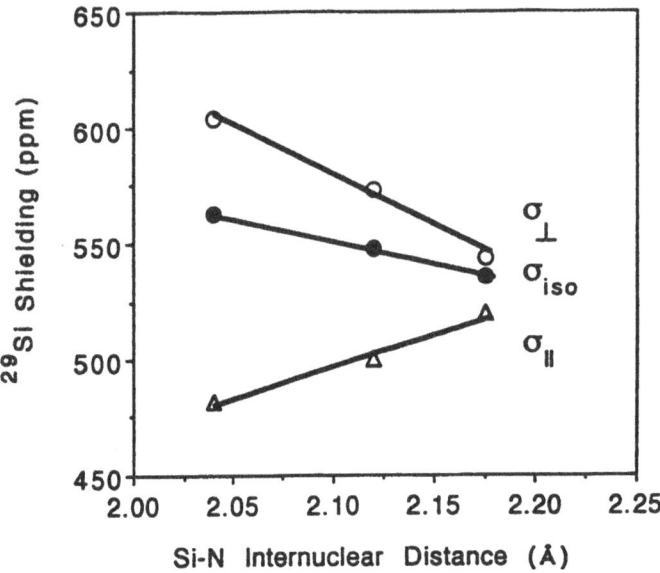

Figure 7. Plot of calculated elements of the ^{29}Si chemical shielding tensor $(\sigma_\parallel, \sigma_{iso}, \sigma_\perp)$ in H*Si(OH)$_3$--NH$_3$ as a function of r_{Si-N} (with OSiO bond angle and H* charge varied simultaneously, as described in the text) with the gauge origin located at the center of mass.

Compared to the experimental results shown in Figure 2, the calculated ^{29}Si chemical shift results appear to be in very good agreement with experiment. All the trends shown in Figure 7 have the same senses as shown for the experimental trends in Figure 2. This might be a reflection of the fact that calculated quantities for the sum rules satisfy equations (2) and (3) quite well.

For this limited data set, the calculated results predict that all of the principal elements of the ^{29}Si chemical shielding tensor have a dependence on r_{Si-N}, with the $\sigma_\perp$ element having the largest dependences, as seen experimentally. From the above discussions on the effects of bond angle distortions, variation of the substituent (H*) and r_{Si-N}, it appears that bond angle and r_{Si-N} variations are *not* the primary causes of the experimentally observed substituent effects on the ^{29}Si chemical shielding in silatranes. To a very large extent, the trends that are observed for the principal elements of the ^{29}Si chemical shielding tensor are apparently due to the direct electronic effects of the attached substituent.

CONCLUSIONS

As substituent-induced variation of ^{29}Si shielding tensor elements appear to be dominated by the "α effect" of the substituent, the ^{15}N chemical shielding tensor appears to be much more promising for providing insight into the silicon-nitrogen transannular interaction in silatranes. The nature of the experimentally observed variation of ^{15}N and ^{29}Si chemical shielding parameters can be largely accounted for by CHF calculations. The results of ^{15}N shielding calculations indicate that the r_{Si-N} dependence plays a major role in the behavior the ^{15}N chemical shift tensor in silatranes. However, due to the gauge dependence of the calculated ^{15}N shielding results and the severe underestimation of the experimental trends, this interpretation is tenuous and improved shielding calculations employing larger basis sets or more realistic models for the nitrogen site in silatranes may be required to get better calculated results [23, 34, 35, 43].

Ab initio ^{29}Si chemical shielding calculations on the silatrane model and of other model indicates that the H*Si(OH)$_3$--NH$_3$ model may be a useful model of the silatrane system. The calculated ^{29}Si shielding trends are in good qualitative agreement with the experimental ^{29}Si results, which reveal that $\sigma_\perp$ is the most sensitive to r_{Si-N}. As one should expect, the shielding calculations imply that the ^{29}Si chemical shielding for the silicon atom in silatranes is dominated by the electronic effect of the directly-attached substituent, although the $\angle$OSiO bond angle distortion and r_{Si-N} variations have significant influences.

ACKNOWLEDGEMENTS

The authors gratefully acknowledge support of this research by NSF Grant No. CHE-9021003. JHI acknowledges the help, support and Cyber-205 time supplied by the

Advanced Technical Computing group at Colorado State University and the financial support of a N.A.S.A. Training Grant.

REFERENCES

1) Voronkov, M. G. (1966) *Pure and Appl. Chem.* **13**, 35.

2) Voronkov, M. G., Dyakov, V. M. and Kirpichenko, S. V. (1982) *J. Organomet. Chem.* **223**, 1.

3) Hencsei, P. and Parkanyi, L. (1985) *Rev. Silicon, Germanium, Tin and Lead Compds.* **8**, 191.

4) Shen, Q. and Hilderbrandt, R. L. (1980) *J. Molec. Struct.* **64**, 257.

5) Turley, J. W. and Boer, F. P. (1968) *J. Am. Chem. Soc.* **90**, 4026.

6) Kutzelnigg, W. (1984) *Angew Chem. Int. Ed. Engl.* **23**, 272.

7) Pauling, L. (1960) *The Nature of the Chemical Bond*, 3rd Ed., Cornell University Press, Ithaca, NY.

8) Rundle, R. E. (1963) *J. Am. Chem. Soc.* **85**, 112.

9) Pestunovich, V. A., Tandura, S. N., Shterenberg, B. Z., Baryshok, V. P. and Voronkov, M. G. (1978) *Izv. Akad. nauk SSSR, Ser. Khim.* 2653.

10) Pestunovich, V. A., Tandura, S. N., Voronkov, M. T., Baryshok, V. P., Zelchan, G. I., Glukhikh, V. I., Engelhardt, G. and Witanowski, M. (1978) *Spectros. Lett.* **11**, 339.

11) Pestunovich, V. A., Tandura, S. N., Shterenberg, B. Z., Baryshok, V. P. and Voronkov, M. G. (1979) *Izv. Akad. nauk SSSR, Ser. Khim.* 2159.

12) Pestunovich, V. A., Tandura, S. N., Shterenberg, B. Z., Baryshok, V. P. and Voronkov, M. G. (1980) *Dokl. Akad. nauk SSSR* **253**, 400.

13) Pestunovich, V. A., Shterenberg, B. Z., Lippmaa, E. T., Myagi, M. Ya., Alla, M. A., Tandura, S. N., Baryshok, V. P., Petukhov, L. P. and Voronkov, M. G. (1981) *Dokl. Akad. nauk SSSR, 258, 1410.*

14) Marsmann, H. (1981) *NMR Basic Principles and Progress* **17**, 65.

15) Haeberlen, U. (1976) *High Resolution NMR in Solids. Selective Averaging*, Academic Press, New York.

16) Ernst, R. R., Bodenhausen, G. and Wokaun, A. (1987) *Principles of Nuclear Magnetic Resonance in One and Two Dimensions*, Oxford University Press, Oxford.

17) Mehring, M. (1983) *Principles of High Resolution NMR in Solids*, 2nd Ed., Springer-Verlag, Berlin-Heidelberg.

18) Fyfe, C. A. (1983) *Solid State NMR for Chemists*, CFC Press, Guelph.

19) Iwamiya, J. H. and Maciel, G. E., in preparation.

20) Pople, J. A. (1962) *J. Chem. Phys.* **37**, 53.

21) Pople, J. A. (1962) *J. Chem. Phys.* **37**, 60.

22) Lazzeretti, P. and Zanasi, R. (1980) *J. Chem. Phys.* **72**, 6768.

23) Lazzeretti, P. and Zanasi, R. (1977) *Inter. J. Quant. Chem.* **12**, 93.

24) McLean, A. D. and Chandler, G. S. (1980) *J. Chem. Phys.* **72**, 5639.

25) Huzinaga, S. (1984) *Gaussian Basis Sets for Molecular Calculations*, Elsevier, New York.

26) Poirier, R., Kari, R. and Csizmadia, I. G. (1985) *Handbook of Gaussian Basis Sets*, Elsevier, New York.

27) Dunning, T. H., Jr. (1970) *J. Chem. Phys.* **53**, 2823.

28) Dunning, T. H., Jr. (1971) *J. Chem. Phys.* **55**, 3958.

29) Roos, B. and Siegbahn, P. (1970) *Theor. Chim. Acta* **17**, 199.

30) Parkanyi, L., Hencsei, P., Bihatsi, L. and Muller, T. (1984) *J. Organomet. Chem.* **269**, 1.

31) Parkanyi, L., Bihatsi, L. and Hencsei, P. (1978) *Cryst. Struct. Comm.* **7**, 435.

32) Lazzeretti, P., Cadioli, B. and Pincelli, U. (1976) *Internat. J. Quant. Chem.* **10**, 771.

33) Tossell, J. A. and Lazzeretti, P. (1985) *J. Chem. Phys.* **84**, 369.

34) Chesnut, D. B. (1989) *Ann. Rep. NMR Spectros.* **21**, 51.

35) Kutzelnigg, W., Fleischer, U. and Schindler, M. (1990) *NMR Basic Principles and Progress* **21**, 165.

36) Fleischer, U., Schindler, M. and Kutzelnigg, W. (1987) *J. Chem. Phys.* **86**, 6337.

37) Hanson, A. E. and Bouman, T. D. (1985) *J. Chem. Phys.* **82**, 5035.

38) Arrighini, G. P., Maestro, M. and Moccia, R. (1970) *J. Chem. Phys.* **52**, 6411.

39) Lazzeretti, P. and Tossell, J. A. (1987) *J. Phys. Chem.* **91**, 800.

40) Arrighini, G. P., Maestro, M. and Moccia, R. (1968) *J. Chem. Phys.* **49**, 882.

41) Lazzeretti, P. and Zanasi, R. (1967) *Phys. Rev.* **1**, 242.

42) Chesnut, D. B. and Foley, C. K. (1986) *J. Chem. Phys.* **84**, 852.

43) Schindler, M. (1987) *J. Am. Chem. Soc.* **109**, 5950.

44) Pestunovich, V. A., Tandura, S. N., Voronkov, M. G., Engelhardt, G., Lippmaa, E., Pehk, T., Sidorkin, V. F., Zelchan, G. I. and Baryshok, V. P. (1978) *Dokl. Akad. nauk SSSR* **240**, 914.

45) Williams, E. A. (1983) *Ann. Rep. NMR Spectros.* **15**, 235.

46) Tossell, J. A. and Lazzeretti, P. (1985) *J. Chem. Phys.* **84**, 369.

AB INITIO IGLO STUDIES OF THE CONFORMATIONAL DEPENDENCE OF THE γ-EFFECT IN THE ^{13}C NMR SPECTRA OF CYCLIC HYDROCARBONS

Michael Barfield
Department of Chemistry
University of Arizona
Tucson, Arizona 85721

ABSTRACT.   Ab initio IGLO (individual gauge for localized molecular orbital) methods of SCF-MO theory are used to study and analyze the mathematical form of the angular dependence of the γ-effects which are often observed in the ^{13}C NMR spectra of cyclic hydrocarbons. Isotropic ^{13}C shielding calculations were performed as a function of dihedral angle for propane, n-butane, i-butane, 2-methylbutane, and 2,3-dimethylbutane. Since γ-effects are usually determined as the chemical shift differences for a molecule in which a hydrogen is replaced by a substituent (e. g., methyl), three models are obtained by taking these molecules in pairs. The angular dependencies arising from each model are compared with the experimental data which have been reported for a series of rigid bicyclic molecules. Only the most general trends are followed by the simplest γ-effect   model   obtained   from   the   n-butane/propane   data.   Better correspondence with the experimental data is found for models based on the 2-methylbutane and 2,3-dimethylbutane ^{13}C shielding data.   The asymmetry in the γ-effect as a function of dihedral angle is in agreement with the experimental data and shows that the γ-effect depends both on dihedral angle and stereochemistry.

1.   INTRODUCTION

Substituent effects [1] in the ^{13}C NMR spectra of aliphatic and alicyclic compounds are of major importance in applying NMR techniques to structural and conformational studies.   Of particular interest are the angularly dependent γ-substituent effects on ^{13}C chemical shifts.   A commonly observed situation is that in which the C4 methyl is _gauche_ to C1 as in 1. The shift of the C1 resonance to lower frequency (higher field on some

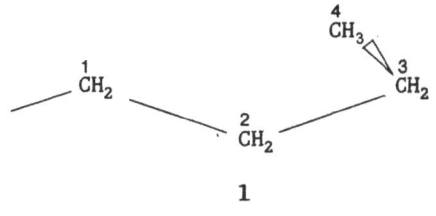

1

J. A. Tossell (ed.), Nuclear Magnetic Shieldings and Molecular Structure, 523–537.

CW NMR spectrometers) has been extensively used in stereochemical studies and ^{13}C chemical shift assignments [2]. Considerable effort has been expended in trying to sort out the electronic factors which are involved [3 - 6]. Several groups attempted to establish an empirical mathematical form for the angular dependence of the γ-effect, but found little indication of systematic relationships [7-9].

The ab initio IGLO method for chemical shielding was used in a previous study by Barfield and Yamamura [10] to describe the ^{13}C chemical shift conformational dependencies at the α-, β-, and γ-positions of aliphatic and alicyclic hydrocarbons. The results for the angularly dependent γ-effect were in accord with the general experimental observations and earlier semiempirical results [5], but did not include a detailed comparison with the experimental data as a function of dihedral angle. Moreover, it is now clear that the standard geometrical model, which was adopted in the earlier work [10], could be a source of error. Of less concern for a study of this type was the use of double-ζ and non-standard basis sets as the *angularly* dependent parts of the substituent effects were shown to be relatively insensitive to the size of the basis sets. This study presents a more detailed examination of the conformational dependence of the γ-effect for hydrocarbons and compares the calculated results with experimental data over a wide range of dihedral angles.

In an attempt to sort out the electronic factors which are responsible for the conformational γ-effects, localized paramagnetic shielding contributions associated with the bonds (PBC) of n-butane, were examined as a function the dihedral angle about the C2-C3 bond [10]. The ^{13}C shielding arises almost entirely from diamagnetic and paramagnetic contributions associated with the four bonds on carbon. However, *the torsional dependence of the C1 chemical shift in* $\underline{n}$-butane is almost entirely determined by the paramagnetic bond contributions (PBC) for the C1-C2 bond. Shielding calculations based on a triple-ζ basis set showed [10] that the PBC for the C1-C2 bond in 1 has a dependence on dihedral angle φ about the C2-C3 bond, which is given approximately by the expression $\sigma_p(\text{PBC}) = 1.3 \cos \varphi - 16.5$ ppm. Although the constants in this expression are somewhat basis set dependent, it is now clear that $\cos\varphi$ is the most important trigonometic form controlling the conformational dependence of the γ-effect.

2. SHIELDING CALCULATIONS

All shielding calculations [11,12] in this study were based on the IGLO (individual gauge for localized orbitals) formulation of Kutzelnigg and Schindler [13]. These methods have been applied with good success to a large number of molecules and include studies of the conformational dependencies of ^{13}C chemical shifts [10, 14]. In the IGLO method localized MO's, which are associated with inner shells, bonding orbitals and lone pairs, have unique origins for the calculations of diamagnetic and paramagnetic terms. Distributed origins methods provide a satisfactory description of ^{13}C chemical shielding using modest basis sets [11-14].

Basis sets for shielding calculations are at least of double-ζ (DZ) quality, e.g., Basis Set I is a (7,3/3) set in the contraction (4111;21/21). The resulting isotropic ^{13}C shielding results for the relevant carbons in several molecules are entered in Table 1. Some of the shielding calculations were also performed with a (9,5/5) Huzinaga set [15] contracted to a (51111;311/2111) set with $\underline{d}$-type polarization functions on C, and a double-ζ set on H. This is Basis Set II' in the notation of the Bochum group [13]. The Basis Set II' results for relevant carbons of the molecules of this study are given in Table 2. The basis set dependence of the ^{13}C isotropic shielding in hydrocarbons has been described elsewhere and will only be mentioned here in the context of conformational γ-effects. The IGLO results were obtained with a modified FORTRAN computer program which was developed by Kutzelnigg and Schindler [13]. All calculations were performed on Digital Equipment Corporation VAX workstations, CONVEX 220, and CONVEX 240 computers.

Previously, it was noted [10] that the ^{13}C isotropic shielding data $\sigma(\varphi)$ could be represented to good precision by means of Fourier series expansions of the forms

$$\sigma(\varphi) = \sum_{n=1}^{6} \underline{A}_n \cos \underline{n}\varphi + \underline{B}, \tag{1}$$

where the coefficients $\underline{A}_n$, etc., are determined by linear regression analyses of the calculated data. In those cases where good precision is obtained this procedure will be used to reduce the number of data tables.

All molecular structures in the model aliphatic compounds of this study are fully optimized ones (subject only to one dihedral angle constraint) at the HF/3-21G* level using the Gaussian 86 code [16] operating on DEC workstations or the Gaussian 90 codes [17] running on a CONVEX 220 computer. This is probably a reasonable compromise in the basis set choice where many computations are required. Results [18] from these laboratories have shown that the shielding data for hydrocarbons are much less sensitive to optimization level in comparison with molecules having heteroatoms, e.g., F-, OH, and -CN substituents.

3. RESULTS AND DISCUSSION

3.1. CONFORMATIONAL DEPENDENCIES OF THE γ-EFFECT BASED ON ALIPHATIC COMPOUNDS AS MODELS FOR RIGID SYSTEMS

Of interest here is the possibility of using calculated ^{13}C shielding data for aliphatic model compounds to interpret the angular dependence of ^{13}C substituent effects in rigid molecules. Carbon-13 shielding calculations were performed as a function of dihedral angle for the series of molecules propane 2, n-butane 3, i-butane 4, 2-methylbutane 5, and 2,3-dimethylbutane 6. In the following sections these molecules will be taken in pairs to describe the conformational dependence of the γ-effect.

3.1.1. γ-Effects Inferred from n-butane and Propane Shielding Data. The molecular structures of propane 2 and n-butane 3 were optimized at 30°

$$\begin{array}{ccc} 1 & 2 & 3 \\ CH_3 - CH_2 - CH_3 \end{array} \qquad\qquad \begin{array}{cccc} 1 & 2 & 3 & 4 \\ CH_3 - CH_2 - CH_2 - CH_3 \end{array}$$

$$\textbf{2} \qquad\qquad\qquad\qquad \textbf{3}$$

intervals of the dihedral angles C1-C2-C3-H3 and C1-C2-C3-C4 at the HF/3-21G* level. The isotropic ^{13}C IGLO shielding results (Basis Set II') were obtained for all carbons in the two molecules as a function of the dihedral angle φ measured about the C2-C3 bond. Analysis of the ^{13}C isotropic shielding data of the C1 carbon of propane via eq 1 shows that the conformational behavior is reproduced by a simple trigonometric form having only a six-fold term,

$$\sigma_{C1}(\varphi) = 0.1 \cos 6\varphi + 180.2 \quad ppm, \tag{2}$$

which has a standard deviation of 0.05 ppm in the estimates of σ_{C1}. In contrast to previous IGLO results [10] which used a standard geometrical model and a smaller basis set, a cos3φ term does not occur. Because the propane C1 shielding in eq 2 is almost independent of dihedral angle, the conformational dependence of the γ-effect in this model is almost entirely determined by the n-butane C1 (C4) shielding. Shielding data for the C1 carbon of n-butane are reproduced by the expression

$$\sigma_{C1}(\varphi) = 2.79 \cos\varphi + 0.8 \cos 2\varphi - 0.3 \cos 3\varphi + 0.2 \cos 4\varphi$$

$$+ 0.2 \cos 5\varphi + 182.8 \quad ppm, \tag{3}$$

with a standard deviation of 0.1 ppm. The "γ-effect" is defined here to be the *chemical shift* difference $\Delta\delta_{C1}$ between the C1 carbon of n-butane 3 and the C1 carbon of propane 2, which is the same as the difference between the calculated *shielding data* for propane and n-butane in eqs 2 and 3, respectively,

$$\Delta\delta_{C1}(\varphi) = -2.8 \cos\varphi - 0.8 \cos 2\varphi + 0.3 \cos 3\varphi - 0.2 \cos 4\varphi$$

$$- 0.2 \cos 5\varphi + 0.1 \cos 6\varphi - 2.6 \quad ppm. \tag{4}$$

This result is very similar to the that reported previously [10]. The cosφ term is still the most important angularly dependent term. The use of optimized molecular geometries (and, possibly, larger basis sets) replaces cos3φ by cos2φ as the next most important term. The $\Delta\delta_{C1}(\varphi)$ data from eq 4 are plotted (solid line) in Figure 1 as a function of the dihedral angle φ (0° to 180°). From eq 4 it can be seen that the conformational dependence of the γ-effect is dominated by the negative cosφ term which gives a minimum at 0° and maximum at 180°. The relative minimum at 180° in Figure 1, associated the negative cos2φ term in eq 4, is a feature which does not appear in the models to be described in the next sections.

The number of experimental ^{13}C NMR reports of γ-effects in hydrocarbons is enormous! In this first attempt at a detailed comparison between calculated results and experimental data, it seemed appropriate to select a manageable set of data which provide the simplest examples consistent with the model compounds. It is also important to consider relatively rigid molecules such as **7 - 16**, which include methyl-substituted bicyclo[2.2.1]heptanes, bicyclo[2.2.2]-octanes, and *trans*-decalins. These are all examples of methyl-disubstituted molecules containing the CH_3-CH-CH-CH_3 moiety (from which γ-effects are deduced by taking the differences in chemical shift with those molecules having one less methyl group). All experimental data were taken from the compilation of Whitesell and Minton [19]. Dihedral angles for the CH_3-CH-CH-CH_3 moieties of compounds **7 - 16** were obtained by molecular mechanics optimization using the MMX force field which is used in the PCMODEL program [20]. A recent study of a number of bicyclic hydrocarbons [21] showed reasonably good agreement between MMX geometries and ab initio results (at the HF/3-21G* or 6-31G** levels).

For comparison with this model, experimental ^{13}C γ-effect data for compounds **7 - 16** are also plotted in Figure 1 as a function of the optimized dihedral angles φ. The latter were rather arbitrarily taken to be accurate to ± 5°. For gauche arrangements of methyl groups, which have dihedral angles near 60° in Figure 1, the relatively large range of experimental values (-6.5 to -1.8 ppm) precludes the possibility of

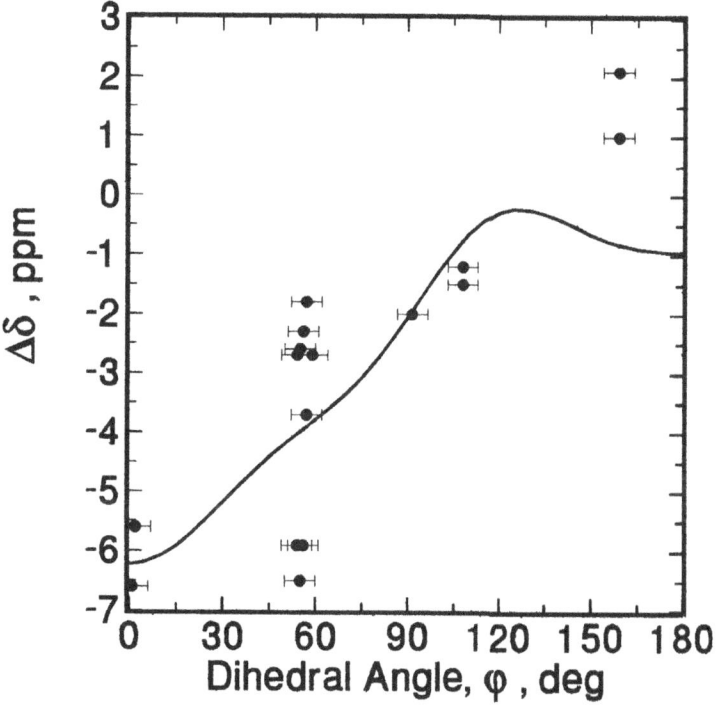

Figure 1. A plot of the calculated (Basis Set II') angular dependence of the γ-effect $\Delta\delta$ versus the dihedral angle φ. The solid line is based on differences between the ^{13}C isotropic shielding data for the Cl carbons of n-butane **3** and propane **2**. Included for comparison are the experimental results (filled circles) for the series of substituted bicyclic compounds **7 - 16**.

528

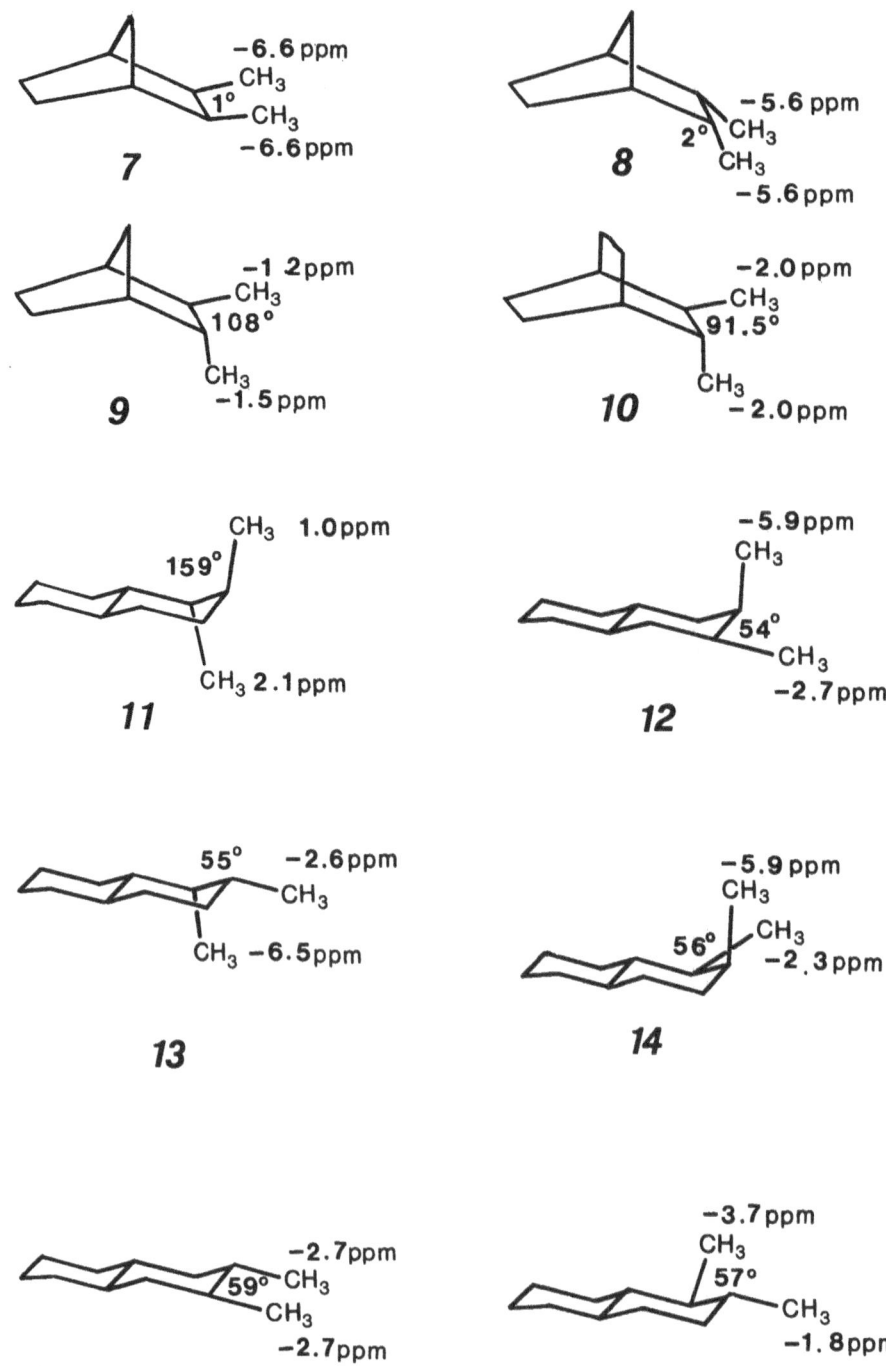

agreement with this model. Furthermore, for the nearly trans arrangement of methyl groups with a dihedral angle near 170°, the calculated and experimental $\Delta\delta$ values do not have the same sign. The experimental data in Figure 1 have a pattern which is very similar to the empirical plots reported by Lambert and Vagenas [9], leading them to conclude that a linear dependence of the γ-effect on dihedral angle was as good as any other. Since the distributed origins methods for ^{13}C shielding generally give better agreement, a more sophisticated model is required.

3.1.2. γ-Effects Inferred from 2-Methylbutane and i-Butane Shielding Data.

A series of IGLO calculations were performed for i-butane **4** and 2-methylbutane **5a** as a function of the dihedral angle φ in the range 0° to 330°.

$$\underset{1}{CH_3}\!\!-\!\!\underset{2}{CH}\!\!-\!\!\underset{3}{CH_3} \qquad \underset{1}{CH_3}\!\!-\!\!\underset{2}{CH_2}\!\!-\!\!\underset{3}{CH}\!\!-\!\!\underset{4}{CH_3}$$
$$\underset{4}{CH_3} \qquad\qquad\qquad \underset{5}{CH_3}$$
$$\mathbf{4} \qquad\qquad\qquad\qquad \mathbf{5a}$$

The calculated data made use of molecular structures which had been optimized at 30° intervals of the H-C1-C2-C3 and C1-C2-C3-C4 dihedral angles. The Basis Set I and II' shielding data (C3 and C4 carbons of **4** and the C4 and C5 carbons of **5**) are given in Tables 1 and 2, respectively, along with the dihedral angles φ(C1-C2-C3-C4) and φ(C1-C2-C3-C5) for 2-methylbutane. Although the ^{13}C shielding data for Basis Set II' are substantially better than for Basis Set I, the calculated γ-effects (differences between shielding values) are only somewhat improved. Because of this improvement (the constant term is more negative by about 1 ppm), Basis Set II' results are used where feasible in preference to the smaller basis set.

Consider three representative 2-methylbutane conformations **5b** - **5d** which

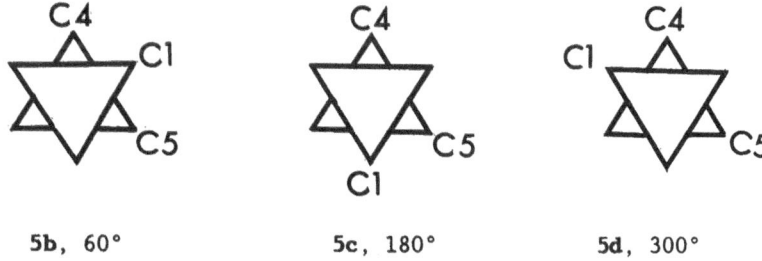

5b, 60°　　　　　　**5c, 180°**　　　　　　**5d, 300°**

are obtained by rotating the C1 methyl clockwise about the C2 - C3 bond. The C1 and C4 methyls are gauche in both **5b** and **5d**, but the ^{13}C chemical shifts are different (for the C5 carbon data in Table 2, for example, there is a 4.1 ppm difference between the calculated shieldings) in these two conformations. For most cyclic molecules of interest an asymmetry in the γ-effect curve is to be expected.

530

Table 1. Calculated IGLO ^{13}C Isotropic Shielding Data (Basis Set I) and Dihedral Angles for Selected Carbons of 2,3-Dimethylbutane **6**, 2-Methylbutane **5** and *i*-Butane **4**.[a]

Dihedral Angles[b]			2,3-Me$_2$butane		2-Methylbutane		*i*-Butane	
φ(C1-C4)	φ(C6-C4)	φ(C6-C5)	C4	C5	C4	C5	C3	C4
0.0	124.9	0.0	202.3	202.3	203.4	197.0	197.7	197.7
30.0	157.6	35.9	200.7	203.6	203.0	198.8	198.1	197.6
60.0	185.3	61.1	200.1	206.5	202.3	202.1	198.8	198.7
90.0	212.5	86.7	199.6	205.4	200.1	203.5	198.4	199.0
120.0	244.0	119.9	197.5	203.4	197.1	203.6	197.7	197.7
150.0	274.2	153.1	198.5	201.3	197.1	202.1	198.1	197.6
180.0	301.6	180.0	200.5	200.5	198.0	201.9	198.8	198.7
210.0	331.1	206.9	202.0	199.7	198.4	200.9	198.4	199.0
240.0	4.2	240.2	203.4	197.4	198.2	198.4	197.7	197.7
270.0	35.8	273.3	204.6	198.9	199.9	197.8	198.1	197.6
300.0	64.2	298.9	206.7	200.4	202.2	198.2	198.8	198.7
330.0	91.1	324.1	204.8	201.7	203.1	198.3	198.4	199.0

[a] Angles are given in degrees and shielding data are given in ppm.
[b] The dihedral angles are those from the HF/3-21G* energy optimized structures with constrained φ(C1-C4) angles.

For this model it is not necessary to treat separately the IGLO γ-effect results for C4 and C5 carbons. Analyzing the C5 carbon shielding data in Table 2 with a 123.8° phase difference in eq 1, gives data which are nearly identical to those for the C4 carbon over the whole dihedral angle range. To obtain a reasonable fit of the calculated γ-effects results to eq 1, it was necessary to analyze separately the regions 0° - 180° and 180° -360°,

$$\Delta\delta(\varphi) = -3.7 \cos\varphi - 0.5 \cos2\varphi + 0.4 \cos3\varphi - 0.4 \cos4\varphi - 2.2 \quad \text{ppm,}$$

$$0° < \varphi < 180° \quad (5a)$$

$$\Delta\delta(\varphi) = -3.0 \cos\varphi - 0.4 \cos2\varphi - 0.2 \cos3\varphi - 2.6 \quad \text{ppm,}$$

$$180° < \varphi < 360° \quad (5b)$$

which have standard deviations 0.1 and 0.3 ppm, respectively. These data are plotted (solid line) in Figure 2 as a function of the dihedral angle φ in the range 0° - 360°. The experimental ^{13}C γ-effect data for the bicyclic compounds **7** -**16** are also plotted in Figure 2 as a function of the dihedral angle φ. The experimental data are divided into two groups corresponding to *threo* and *erythro* carbons. For reasons to be discussed in the next section, these arrangements are identified with γ-effects at the C4 and C5 carbons, respectively. The agreement between calculated and experimental data is better than for the n-butane results in the previous section. Not only are positive values now predicted for the trans arrangement, difficulties with the wide range of gauche values are substantially less severe because the 60° and 300° conformations correspond to distinct physical situations. Furthermore, the asymmetry in the calculated curve in Figure 2 is qualitatively consistent with the experimental trends.

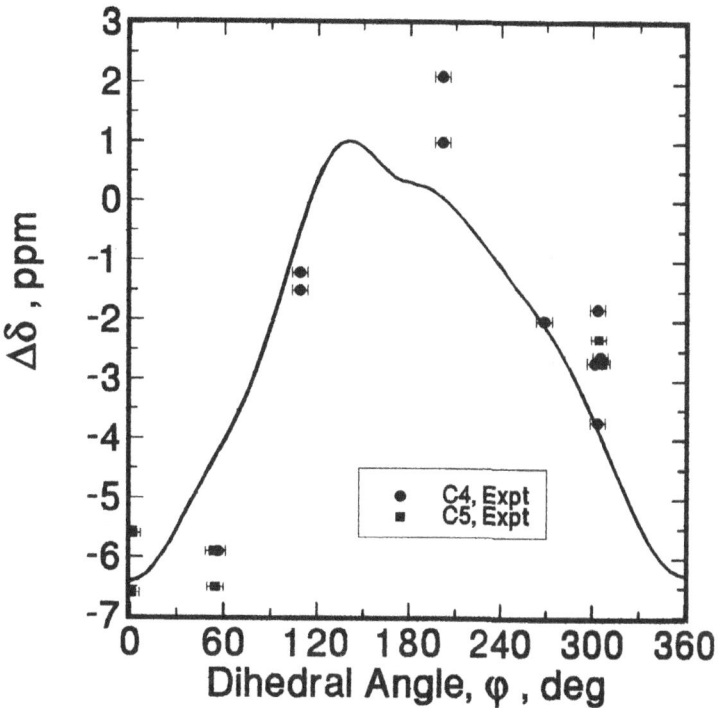

Figure 2. A plot of the calculated (Basis Set II') angular dependence of the γ-effect Δδ versus the dihedral angle φ. The solid line is based on differences between the ¹³C isotropic shielding data for the C4 (C5) carbon of 2-methylbutane 5 and the C3 (C4) carbon of i-butane 4. Included for comparison are the experimental results for a series of substituted bicyclic compounds 7 - 16. The threo (C4) and erythro (C5) arrangements are depicted by filled circles and filled squares, respectively.

3.1.3. γ-Effects Inferred from 2,3-Dimethylbutane and 2-Methylbutane Shielding Data. A series of IGLO calculations (Basis Sets I) were performed for 2,3-dimethylbutane 6a. The calculated data were based on molecular structures which had been optimized at 30° intervals of the C1-C2-C3-C4 dihedral angle between 0° and 330°. Basis Set I shielding

TABLE 2. IGLO ^{13}C Isotropic Shielding Data (Basis Set II') and Dihedral Angles for Selected Carbons of 2-Methylbutane **5** and *i*-Butane **4**.[a]

Dihedral Angles[b]		2-Methylbutane		*i*-Butane	
φ(C1-C4)	φ(C1-C5)	C4	C5	C3	C4
0.0	123.8	178.3	171.4	171.9	171.9
30.0	92.1	177.8	173.4	172.4	171.8
60.0	64.6	177.1	176.9	173.0	173.0
90.0	35.9	174.7	178.3	172.6	173.3
120.0	3.9	171.4	178.6	171.9	171.9
150.0	330.9	171.6	176.8	172.4	171.8
180.0	301.6	172.6	176.6	173.0	173.0
210.0	273.6	172.9	175.5	172.6	173.3
240.0	243.2	172.9	173.0	171.9	171.9
270.0	211.2	174.5	172.3	172.4	171.8
300.0	182.4	176.8	172.8	173.0	173.0
330.0	154.7	177.7	172.8	172.6	173.3

[a] Angles are given in degrees and shielding data are given in ppm.
[b] The dihedral angles are those from HF/3-21G* energy optimized structures with constrained φ(C1-C4) angles.

data for the C4 and C5 carbons of **6a** are given in Table 1 along with the

$$\overset{1}{C}H_3 \text{—} \overset{2}{C}H \text{—} \overset{3}{C}H \text{—} \overset{4}{C}H_3$$

with CH$_3$ (6) below C2 and CH$_3$ (5) below C3

6a

dihedral angles φ(C1-C2-C3-C4), φ(C6-C2-C3-C4), and φ(C6-C2-C3-C5) of 2,3-dimethylbutane. A number of conformations such as **6b** - **6e** can be drawn for 2,3-dimethylbutane. Structures **6b** - **6d**, correspond to clockwise rotation of the

6b   **6c**   **6d**   **6e**

C1 methyl about the C2-C3 bond. The 2,3-dimethylbutane molecule has another set of which **6e** is one member. If C5 and C6 denote methyl groups and C1 and C4 are different (for example, the methylene groups of a cyclohexane ring), conformations **6b** - **6d** and **6e** are *erythro* and *threo*, respectively. If C4 and C6 denote methyl groups the designations are reversed. Accordingly, the calculated and experimental results for the *threo* and the *erythro* conformations are associated with the C4 and C5 carbons, respectively.

The calculated γ-effect data were obtained as differences between the shielding data for the C4 and C5 carbons of 2-methylbutane and the corresponding 2,3-dimethylbutane data in Table 1. These data for C4 (filled circles) and C5 (filled squares), which are plotted in Figure 3, exhibit some differences, particularly in the 120° to 240° range. Differences in the γ-effect for *threo* (C4) and *erythro* (C5) arrangements were not expected. Presumably, differences arise because the calculated shielding data are sensitive to the details of molecular geometry. These effects will be particularly pronounced for the γ-effect, which corresponds to small difference between much larger chemical shift values.

To obtain a reasonable fit of these data to eq 1, for the C4 carbon, only, it was necessary to analyze separately the data in the ranges 0° - 180° and 180° - 360°,

$$\Delta\delta_{C4}(\varphi) = -4.2\ \cos\varphi + 0.3\ \cos2\varphi + 0.6\ \cos3\varphi - 1.7\ \text{ppm},$$

$$0° < \varphi < 180° \qquad (6a)$$

$$\Delta\delta_{C4}(\varphi) = -2.9\ \cos\varphi + 0.2\ \cos2\varphi - 0.5\ \cos3\varphi - 1.3\ \text{ppm},$$

$$180° < \varphi < 360° \qquad (6b)$$

$$\Delta\delta_{C5}(\varphi) = -3.4\ \cos\varphi - 0.4\ \cos2\varphi - 1.6\ \ \text{ppm}, \qquad (6c)$$

where the standard deviations are 0.3, 0.5, and 0.3 ppm, respectively. The $\cos\varphi$ terms still dominate the conformational dependencies, but there are some differences especially for the C4 carbon in which the coefficient of the $\cos3\varphi$ term requires a different sign in eq 6a and 6b to reproduce the skewing of the calculated data. The constant terms occurring in eqs 6a-c (Basis Set I) are more positive by about 1 ppm than the Basis Set II' results of the previous sections. The data from eqs 6a-c are plotted in Figure 3 as a function of the dihedral angle; the solid line and the dot-dash line correspond to the γ-effects at the C4 and C5 carbons, respectively.

The calculated γ-effect data from eqs 6a-c are also plotted in Figure 4 for comparison with the experimental data of compounds **7** - **16**. It appears that the agreement is somewhat improved. Again, the asymmetry of the γ-effect curves accounts very well for the 3 - 4 ppm differences for gauche γ-effects depending on whether the dihedral angles are close to 60° or 300°. The disparities (ca. 1 ppm) in the region between 0° and 60° may be due, in part, to the use of the smaller basis set. However, the experimental data do not permit a distinction between the calculated results for *erythro* (C5) and *threo* (C4) arrangements.

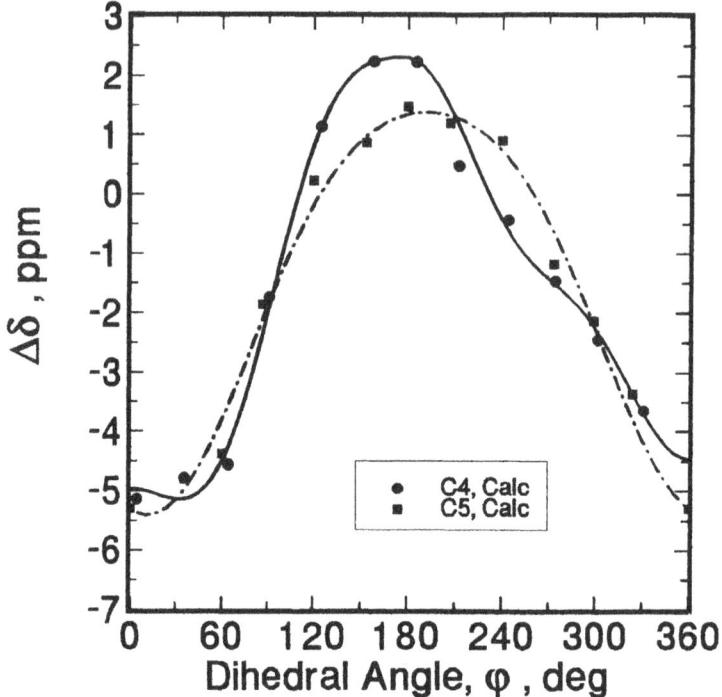

Figure 3. A plot of the calculated (Basis Set I) IGLO γ-effect data $\Delta\delta$ versus the dihedral angle φ for the C4 carbons (filled circles) and the C5 carbons (filled squares) based on differences between the ^{13}C isotropic shielding data for 2,3-dimethylbutane and 2-methylbutane. The solid line and dot-dash lines were obtained by linear regression analyses of the calculated γ-effect data for C4 and C5, respectively.

Consider the changes in the ^{13}C chemical shift of the C4 methyl of a CH_3-CH-CH-CH_3 moiety on removing the C1 methyl group based on the dihedral angle information and the various arrangements of methyl groups (axial-axial, axial-equatorial, equatorial-axial, and equatorial-equatorial) of methyls in the *trans*-decalin molecules.

i.   If both are axial, then $\Delta\delta$ is predicted to be 2.2 ppm (double-ζ)compared with the experimental results of 1.0 and 2.1 ppm based on compound 12.

ii. Removing an axial methyl should shift an equatorial methyl by -2.5 to -2.7 ppm (double-ζ). The experimental results for this situation are -2.7, -2.6, and -2.3 ppm in compounds 12, 13, and 14, respectively.

iii. If the C1 methyl is equatorial and the C4 methyl is axial, then $\Delta\delta_{C4}$ is predicted to be -4.4 ppm (double-ζ). The experimental results are -5.9, -6.5, and -5.9 ppm for the *trans*-decalins **12**, **13**, and **14**.

iv. Removal of a methyl when both are equatorial should give a shift of -2.7 ppm (double-f level). The experimental values are -2.7 ppm in **15**; -3.7 and -1.8 ppm in **16**.

If there is an actual difference in the γ-effect curves for *erythro* and *threo* arrangement, in principle, a single measurement of a γ-effect in the ^{13}C NMR spectra is sufficient to uniquely identify the stereochemical configuration. The situation presented here is not unequivocal because the curves for situations *ii* and *iv* cross at about 300°.

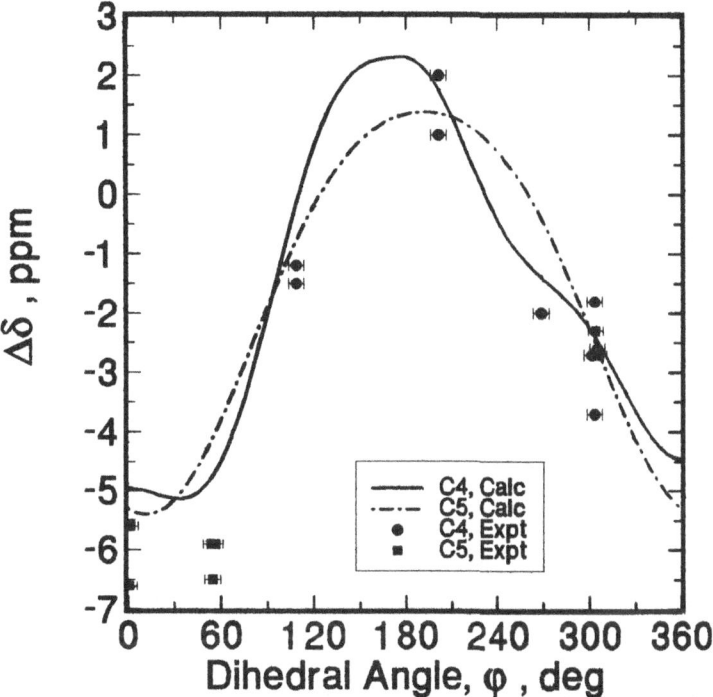

Figure 4. The calculated (Basis Set I) γ-effect data Δδ from eqs 6a-c are plotted as a function of the dihedral angle φ. Included for comparison are the experimental results for a series of substituted bicyclic compounds **7** - **16**. The *threo* (C4) and *erythro* (C5) arrangements are depicted by filled circles and filled squares, respectively.

4. CONCLUSIONS

For the interesting cases of γ-effects in cyclic hydrocarbons, it is shown that the model based on *n*-butane/propane is too simplistic. The best that can be said is that the chemical shift differences between calculated and experimental results exhibit some similar trends. Substantial improvement is noted between the calculated and experimental results if the γ-effect model is based on 2-methylbutane and n-butane. The most important aspect of this model is the distinction between dihedral angles for axial and equatorial methyls as the γ-effects differ by several ppm. The use of the calculated data for 2,3-dimethylbutane as the model compound gives reasonable agreement between the calculated and experimental results over the whole range of dihedral angles. These results indicate that the ^{13}C γ-effect depends both on dihedral angle and stereochemical configuration.

Since the various empirical plots for the γ-effect reported by Lambert and Vagenas have much the same pattern, it seems likely that analysis presented here will prove useful for conformation dependencies including those which have more general applicability in conformational analysis.

5. ACKNOWLEDGMENTS

We wish to extend our appreciation to Professor W. Kutzelnigg and Dr. M. Schindler of the Ruhr-University Bochum for permission to use the IGLO program. Thanks are also extended to Ms. Susan Yamamura for assistance with IGLO code modifications. Appreciation is expressed to the National Science Foundation for Chemical Instrumentation Grants to assist in the purchase of a CONVEX 220 computer.

6. REFERENCES

[1]   Grant, D. M., Paul, E. G. (1964) J. Am. Chem. Soc., 86, 2984-2990.

[2]   Stothers, J. B. (1972) Carbon-13 NMR Spectroscopy, Academic Press, New York.

[3]   Grant, D. M. and Cheney, B. V. (1967) J. Am. Chem. Soc., 89, 5315-5318.

[4]   Woolfenden, W. R. and Grant, D. M. (1966) J. Am. Chem. Soc., 88, 1496-1502.

[5]   Seidman, K. and Maciel, G. E. (1977) J. Am. Chem. Soc., 99, 659-671.

[6]   Li, S. and Chesnut, D. B. (1985) Magn. Reson. Chem., 23, 625-638.

[7]   Lippmaa, E., Pehk, T., and Paasivirta (1973) Org. Magn. Reson., 5, 277-283.

[8]   Chanon, F., Rajzmann, M., Chanon, M., Metzger, J., and Pouzard, G. (1980) Can. J. Chem., 58, 599-603.

[9]   Lambert, J. B. and Vagenas, A. R. (1981) Org. Magn. Reson., 17, 265-270.

[10] Barfield, M. and Yamamura, S. (1990) J. Am. Chem. Soc., 112, 4747-4758.

[11] For reviews of the theory of shielding see, for example, Jameson, C. J. (1991) Nuclear Magnetic Resonance, Specialist Periodical Reports, The Chemical Society London, Burlington House, London, No. 20, and previous chapters in this series.

[12] For a review of ab initio methods in shielding calculations see

Chesnut, D. B. (1989) in Annual Reports on NMR Spectroscopy, Vol. 21, pp. 51-97.

[13] Kutzelnigg, W. (1980) Israel J. Chem., 19, 193-200. Schindler, M. and Kutzelnigg, W. (1982) J. Chem. Phys., 76, 1919-1933. For a review of IGLO results see Kutzelnigg, W. (1989) J. Mol. Struct., 202, 11-61, and Kutzelnigg, W., Fleischer, U., and Schindler, M. (1990) in P. Diehl, E. Fluck, and R. Kosfeld (eds.), NMR Basic Principles and Progress, Springer, Berlin, Vol. 23, p 165.

[14] Jiao, D., Barfield, M., Combariza, J. E, and Hruby, V. J. (1992) J. Am. Chem. Soc. 114, 3639-3643.

[15] Huzinaga, S. (1984) Gaussian Basis Sets for Molecular Calculations, Elsevier, New York.

[16] Gaussian 86, M. J. Frisch, J. S. Binkley, H. B. Schlegel, K. Raghavachari, C. F. Melius, R. L. Martin, J. J. P. Stewart, F. W. Bobrowicz, C. M. Rohlfing, L. R. Kahn, D. J. Defrees, R. Seeger, R. A. Whiteside, D. J. Fox, E. M. Fleuder, and J. A. Pople, Carnegie-Mellon Quantum Chemistry Publishing Unit, Pittsburgh PA, 1984.

[17] Gaussian 90, Revision I, M. J. Frisch, M. Head-Gordon, G. W. Trucks, J. B. Foresman, H. B. Schlegel, K. Raghavachari, M. Robb, J. S. Binkley, C. Gonzalez, D. J. Defrees, D. J. Fox, R. A. Whiteside, R. Seeger, C. F. Melius, J. Baker, R. L. Martin, L. R. Kahn, J. J. P. Stewart, S. Topiol, and J. A. Pople, Gaussian, Inc., Pittsburgh PA, 1990.

[18] Barfield, M. (1992) unpublished results.

[19] Whitesell, J. K. and Minton, M. A. (1987) Stereochemical Analysis of Alicyclic Compounds by C-13 NMR Spectroscopy, Chapman Hall, New York.

[20] Gajewski J. J., Gilbert, K. E., and McKelvie, J. (1990) in D. Liotta, (ed.), Advances in Molecular Modeling, JAI Press, Greenwich, CT.

[21] Barfield, M. and Smith, W. B. (1992) J. Am. Chem. Soc., 114, 1574-1581.

Gas Phase Measurement and *Ab Initio* Calculations of 77Se and 113Cd Chemical Shifts.

Paul D. Ellis*, Jerome D. Odom*, Andrew S. Lipton, Qingping Chen, and James M. Gulick

Department of Chemistry and Biochemistry
University of South Carolina
Columbia, South Carolina 29208

Abstract: The gas phase 77Se and 113Cd chemical shifts have been measured for H_2Se, $HSeCH_3$, $Se(CH_3)_2$, and $Cd(CH_3)_2$ and $Cd(C_2H_5)_2$, respectively. Further, we have computed these chemical shifts using modern *ab initio* theories of shielding. In contrast to other calculations, we demonstrate that these chemical shifts can not be calculated quantitatively. As a result of a Natural Bond Orbital (NBO) analysis it is speculated that the reason for this lack of quantitation is due to the quality of the uncorrelated wave function used in the calculation.

1. Introduction

It is well known that NMR chemical shifts can provide valuable information concerning the structure and bonding in molecules. The clarity of such a relationship is manifested in the ^{1}H and ^{13}C NMR chemical shifts of simple organic molecules.[1] However, the correlation between the chemical shift and structure and bonding becomes less evident for the heavier nuclides in the periodic table.[2] The main reason for this lack of correlation is typically related to the complexity of the molecule or its environment. For example, the molecule may not have the same structure in solution as it has in the solid state or it may interact with different solvents in different ways. Further, the application of molecular theory to such systems is often complicated by the inherently large number of electrons, the quality of the basis sets utilized in such calculations, and the possibility of electron correlation and/or relativistic effects.[3] The former signifies that meaningful correlations between theory and experiment may be difficult to obtain and the latter means that such correlations, if they existed, may be difficult to rationalize. The present work is directed towards a better definition of both aspects of these problems.

We have selected as our experimental systems some examples from the areas of 77Se and 113Cd NMR spectroscopy. These two nuclides were chosen

J. A. Tossell (ed.), Nuclear Magnetic Shieldings and Molecular Structure, 539–555.

because of the success of both as surrogate probes for sulfur (and oxygen)[4] and zinc (and calcium)[5] sites, respectively in biological systems. To date, the surrogate probe strategy has worked well for both systems. The experimental analysis of the ^{113}Cd shielding[6] has progressed further than that for ^{77}Se. This progress has come about as a result of single crystal measurements[5g] where a comparison between structure and shielding can be examined. Further, a larger body of experimental data exists for systems where Cd^{2+} has replaced Zn^{2+} and/or Ca^{2+} in metal dependent or metalloproteins.[6] However, it is important to see if the patterns already observed for these systems can be predicted by modern *ab initio* theories of chemical shielding. The ^{77}Se nuclide represents a challenge from another perspective, namely the sensitivity of the ^{77}Se chemical shift to subtle changes in environment. Classic examples are the ^{2}H isotope effect on the ^{77}Se chemical shift of H_2Se which is 7.02 ppm[7] and the incredible sensitivity of the ^{77}Se chemical shift to diastereomers.[8] Odom et. al.[8] have utilized a selenium-containing chiral auxiliary for detection of enantiomeric excesses in optically active carboxylic acids via ^{77}Se NMR spectroscopy. With this method they have been able to clearly resolve different resonances for diastereomers even when the chiral center is separated from the ^{77}Se nucleus by *eight* bonds![8a] In addition separate selenium resonances are observed when the only difference on the chiral carbon (five bonds from the selenium) is the attachment of a hydrogen or a deuterium atom.[8b] It is not at all clear how a $\Delta\delta$ of 0.119 ppm arises when the chiral carbon is removed from the selenium by eight bonds. An understanding of selenium chemical shifts from a theoretical viewpoint could provide some insight into this remarkable sensitivity. Previously, Odom and coworkers[9] have addressed ^{77}Se substituent-induced chemical shifts in selenols, selenides, and diselenides in a qualitative manner. The data are consistent with the importance of the polarizability of the substituent.

Previous computational efforts on ^{113}Cd chemical shifts have been limited to work by Nakatsuji and coworkers[10] and ourselves.[11] Nakatsuji and coworkers[10] in their work stated that ^{113}Cd chemical shifts could be calculated quantitatively by *ab initio* coupled-Hartree-Fock (CHF) methods. The experimental systems treated were organocadmium compounds, i.e. dimethylcadmium, diethylcadmium, and ethylmethylcadmium. The chemical shift difference for dimethyl- and diethylcadmium, 99.7 ppm for the neat liquids, was difficult for us to explain in our original determination.[5a] We noted in that work the chemical shift difference may result from self association in the neat liquids.

The results of Nakatsuji and coworkers[10] looked excellent; however, on closer examination some puzzling questions arose. The approximate geometry used was based upon Pauling's tetrahedral covalent radii,[12] a C-Cd distance of 2.25Å, and CH bond distances were chosen to be 1.094Å with a CdCH angle of 109.5°. The experimental C-Cd bond length and HCCd angle are known[13] for dimethylcadmium and are 2.112Å and 108.4°, respectively. How sensitive are the results to this choice of geometry? The basis set they used,[14] MIDI-1, needed to be augmented. Two p-polarization functions[15] were added with ζ_p of 0.16 and 0.04094. Further, no d-polarization functions on the Cd were utilized, except the splitting in the occupied 4d functions, i.e. (...|3) to (...|21). A potential problem with the calculation was the nonuniform utilization of the basis set, i.e. MIDI-1 on the cadmium and contiguous atoms, and MINI-1 on the remaining atoms. With this basis set, one of Nakatsuji's[10] conclusions was that d-p metal-ligand interactions are not important for ^{113}Cd chemical shifts. It is interesting to note that the basis set utilized did not allow (except as noted above) such interactions to be fully explored. A natural question at this point is how sensitive are the calculated results to the choice of basis sets?

With these questions in mind we initiated our investigation. To critically examine the success or failure of any *ab initio* method to predict a given molecular property, that property should be experimentally determined in the *same phase* as utilized in the computation. For chemical shift calculations on isolated molecules, this means the experiments should be performed in the gas phase at low pressure. Hence, we will summarize here, our gas phase determination of the ^{113}Cd chemical shifts[11] of $Cd(CH_3)_2$ and $Cd(C_2H_5)_2$. The same strategy will be applied to ^{77}Se chemical shifts. That is, the gas phase chemical shifts of H_2Se, $HSeCH_3$, and $Se(CH_3)_2$ will be presented. The chemical shifts have been computed by several methods with a variety of basis sets. Further, we present the results of calculations for both nuclide chemical shifts that demonstrate even with careful selection of the basis sets and geometry, these shifts cannot be calculated quantitatively. A possible rationale for the quality of the predicted results is presented.

2. Experimental Methods

Synthetic Methods. All solvents were freshly distilled in an inert atmosphere and freeze-thaw degassed prior to use. All samples were handled on a high vacuum system and were purified by trap-to-trap fractionation and/or by the use of a variable temperature cold column.[16] The organocadmium compounds, $Cd(CH_3)_2$ and $Cd(C_2H_5)_2$, were prepared by standard Grignard methods in ether solutions.[17] The purity (>99%) of all samples was confirmed by vapor pressure measurements and multinuclear NMR spectroscopy where applicable.

Hydrogen selenide was obtained commercially (Matheson) and its purity was checked by vapor pressure measurement. The $Se(CH_3)_2$ was prepared as previously described.[18] Methyl selenol was prepared by the reaction of $LiSeCH_3$ with liquid HCl at low temperature. In a typical preparation 15.8 mL of a 1.4M solution of CH_3Li in diethyl ether was added to an evacuated flask at -196 °C containing 1.5g of elemental Se and 30 mL of tetrahydrofuran. The reaction mixture was slowly warmed to room temperature and stirred for two hours. The solution gradually turned milky white indicating the formation of $LiSeCH_3$. All volatile materials were then removed from the flask under dynamic vacuum and excess HCl was condensed onto the solid $LiSeCH_3$ at -196°C. The flask was slowly warmed to -100°C and the reaction mixture was stirred for two hours at that temperature. Removal of volatile materials and vacuum fractionation resulted in pure CH_3SeH.

Sample Preparation. All liquid state NMR samples were prepared by vacuum condensation of the compound and the solvent into a 10 mm NMR tube. Each sample was freeze-thaw degassed before the tube was flame sealed under dynamic vacuum. The concentration of the $Cd(CH_3)_2$ sample was calculated to be 0.088M and that for the $Cd(C_2H_5)_2$ sample was 0.176M. The pressure of all gas phase samples was calculated using the ideal gas equation. For the gas phase ^{113}Cd NMR samples, the appropriate amount of material was condensed under vacuum into an 8mm (7mm I.D. and approximately 5cm in length) insert and then flame sealed under dynamic vacuum. The

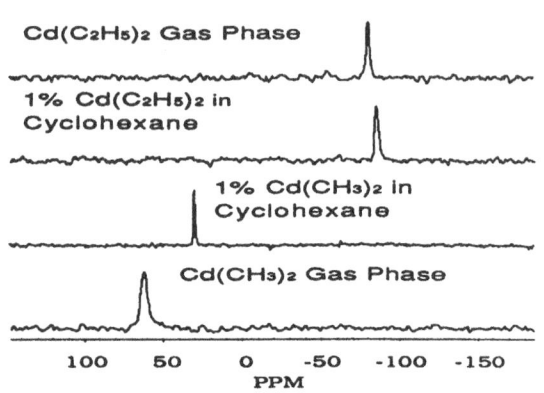

Figure 1 All spectra were Bloch decays run at 88.7 MHz with a spectral width of 30030 Hz zero filled to 8K complex pairs. For the gas samples the $\pi/2$ pulse was 29μs, and for the liquid samples the $\pi/2$ pulse width was 25μs.

8mm sample was placed into a 10mm tube with DMSO-d_6 as a lock solvent. A vortex plug was utilized to keep the 8mm insert from "floating" in the lock solvent. For the selenium gas phase samples the lock solvent was an acidified D_2O solution. Due to the toxicity of the selenium samples, this solution could provide a means for decomposition of the selenium compounds in case of a leak in the inner 8 mm tube.

NMR Spectroscopy. The NMR measurements were made on a Varian XL-400 spectrometer operating at 9.4 Tesla, 88.7 MHz for ^{113}Cd and an XL-300 operating at 7.05 Tesla, 57.3 MHz for ^{77}Se. The probes used (on both instruments) were the standard broad-band, tunable 10mm liquid probes from Varian Associates. The spectra of neat $Cd(CH_3)_2$ and the 1% solutions in cyclohexane were all obtained under unlocked conditions. All of the ^{113}Cd chemical shifts are reported relative to external neat dimethylcadmium. In order to ensure total vaporization of both dimethyl- and diethylcadmium, the gas phase shifts were measured at 97 °C. The chemical shifts of the gas phase sample of dimethylcadmium were independent of temperature, i.e. any variation was within the line width which was typically 3 ppm. Figure 1 summarizes the ^{113}Cd liquid and gas phase NMR spectra. The ^{77}Se NMR spectra were obtained at 25 °C and

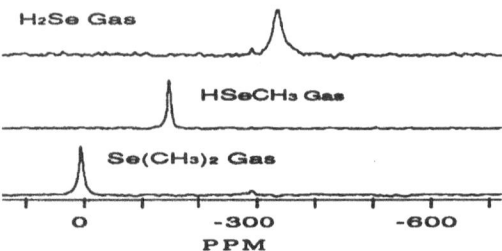

Figure 2 All spectra were recorded at 25 °C at 57.3 MHz. The spectral width was 50 kHz with a $\pi/2$ pulse width of 16μs.

Table I

A Summary of ^{77}Se and ^{113}Cd Chemical Shifts

Sample	Chemical Shift[a]
^{113}Cd Chemical Shifts:	
$Cd(C_2H_5)_2$ ... Neat	-99.7[b]
$Cd(C_2H_5)_2$ ... 1% Solution[c]	-85.23
$Cd(C_2H_5)_2$ ... Gas[d]	-80.52
$Cd(CH_3)_2$ ... Neat	0.0
$Cd(CH_3)_2$ ... 1% Solution[c]	35.32
$Cd(CH_3)_2$ ... Gas[d]	62.09
^{77}Se Chemical Shifts[e]:	
H_2Se ... Gas[f]	-331.70
$HSeCH_3$ ... Gas[f]	-141.62
$Se(CH_3)_2$ ... Gas[f]	13.05

a. The chemical shifts are expressed in ppm with respect to neat dimethylcadmium or a 60% (v/v) solution of dimethylselenide in $CDCl_3$ for ^{113}Cd and ^{77}Se, respectively. A positive chemical shift denotes a resonance to lower shielding.
b. Reference 5a.
c. Approximate concentration in cyclohexane.
d. The sample temperature was 97°C.
e. Luthra, N.P, Dunlap, R.B., Odom, J.D., **J. Magn. Reson.**,52, 318(1983).
f. Sample temperature was 25 °C.

are presented in Figure 2. All of the ^{77}Se chemical shifts are reported to an external sample of 60% (v/v) solution of $Se(CH_3)_2$ in $DCCl_3$. Listed in Table I is a summary of the resulting chemical shifts for cadmium and selenium compounds. In the temperature range of 25 to 75 °C the ^{77}Se chemical shifts were also independent of temperature.

Computational Details. The shielding calculations were performed with an updated VAX version of the RPAC[19] program of Bouman and Hansen[20] capable of utilizing f-orbitals, as well as an AIX version of the TX90 program of Pulay, Hinton and coworkers[21], both at the RHF level. The RPAC program was utilized on a VAX-Station 3540, while the TX90 program was run on an IBM RISC-6000/550 computer. The geometry optimizations were performed with the Gaussian-90 system[22] of programs on both computers. The shielding calculations were performed three ways: the localized orbital/local origin method (LORG) of Hansen and Bouman,[19,20] coupled Hartree-Fock (CHF) method,[23] and the GIAO method as implemented by Pulay and coworkers.[21] The origin dependence of the CHF method, and the sensitivity of the LORG approach to the quality of the basis set is well known[20] and will not be discussed here. We simply present all three methods for comparison purposes. Three varieties of basis sets were employed in the computation, the MIDI-N[14] series (where N ranged from 1 to 5), those of Friedlander, Howell, and Synder[24], and those of Horn and coworkers.[25] These basis sets were used in conjunction with the basis set of Dunning and Hay[26] for carbon and hydrogen. For example, the basis sets MINI-1, MIDI-4, and MIDI-4p2d2f2 for Cd would be designated as (33333|3333|33), (433321|4321|421), and (433321|43311|4311|11), respectively. The Dunning and Hay[26] basis for hydrogen and carbon would be denoted as (31) and (6111|41), respectively. As proved in the previous work,[10] it is necessary to augment the basis functions for chemical shielding calculations. In the present work the exponents for the augmented functions were determined with the Gaussian[22] utility program g90opt. The energy optimized exponents are summarized in Table II. When d- or f-orbital exponents were being optimized, the number of d- or f-Gaussian type orbitals was 5 and 7, respectively. However, in subsequent calculations

Table II
Exponents for Polarization Functions

Atom	ζ_{p1}	ζ_{p2}	ζ_{p1}	ζ_{p2}	ζ_{d1}	ζ_{d2}	ζ_{f1}	ζ_{f2}
Cd[a]	0.9405	0.0755	0.9446	0.0779	0.4113	0.0990	0.9731	0.1193
Cd[b]	0.5138	0.0706	0.4754	0.0733	0.3498	0.1392	0.9594	0.1252
Cd[c]	3.6851	1.3812	3.7141	1.3739	4.2332	1.6709	2.7544	0.7837
Se[d]			1.4026	0.0751	19.618	0.3307		
Se[e]			1.3994		1.0494	0.8614		
Se[f]			2.0066		0.2808			

a. Exponents optimized for the MINI-4 basis set. In the optimization the number of d- and f-functions was limited to 5 and 7, respectively.
b. Exponents optimized for the basis set of Friedlander, Howell, and Synder.[24]
c. Exponents optimized for the AKR/WAG basis set.[27]
d. Optimized for the basis sets of MINI-4.[14]
e. Optimized for AKR/WAG basis set.[27]
f. Optimized for Horn basis set.[25]

using d- or f-orbitals, the number of orbitals was 6 and 10, respectively. For both selenium and cadmium, the exponents were optimized in the triplet state. In the case of cadmium, this was done to simulate the "valence" nature of the added functions.

The first two columns of Table II denote exponents for p-orbital polarization functions when no other polarization functions are to be added to the basis set, i.e. MIDI-4p2. The remaining entries correspond to the case when multiple polarization functions are being added to the basis set, i.e. MIDI-4p2d2 or MIDI-4p2d2f2. The exponents for the p- and d-orbital polarization functions for cadmium did not change upon adding the f-orbital polarization functions.

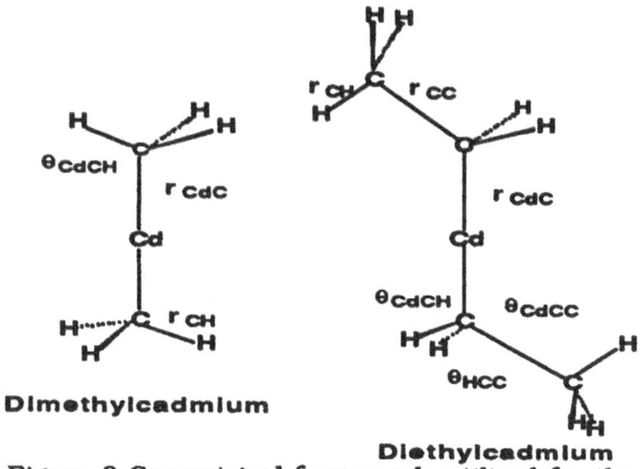

Dimethylcadmium

Diethylcadmium

Figure 3 Geometrical framework utilized for the dimethyl- and diethylcadmium calculations.

Geometries do not play an essential role in the computation of the shielding tensors for dimethyl- and diethylcadmium. However, we have chosen to use theoretical geometries for the simple reason that an experimental geometry for diethylcadmium has not been determined. Hence, a balanced comparison of predicted shielding differences can only be made with optimized theoretical geometries. Figure 3 summarizes the framework geometries utilized for the cadmium compounds. For dimethylcadmium, three parameters were optimized: r_{CdC}, r_{CH}, and Θ_{CdCH}. If

Table III
MP2 Optimized Geometries[a]

$Cd(CH_3)_2$	r_{CdC}	r_{CC}	r_{CH}	Θ_{CdCH}	Θ_{CdCC}	Θ_{HCC}
MIDI-4p2d2f2[b]	2.14966		1.10108	110.10348		
MIDI-4p2d2f2[c]	2.14200		1.10699	110.34688		
fhs-p2d2f2[d]	2.13107		1.10699	110.44172		
$Cd(C_2H_5)_2$						
MIDI4-p2d2f2[b]	2.15816	1.58500	1.10083	107.80597	113.25232	110.74221
MIDI-4p2d2f2[c]	2.14942	1.60144	1.10610	108.42118	112.79217	110.26782
fhs-p2d2f2[d]	2.13808	1.60086	1.10606	108.60664	112.25292	110.40481

Footnotes:
a. All distances are in Angstroms and the angles are in degrees.
b. See footnote b, Table VI for details.
c. See footnote c, Table VI for details.
d. See footnote d, Table VI for details.

one hydrogen from each methyl group, the carbons, and the cadmium lie in a plane, then the angle made by the "out of plane" hydrogens to this plane was constrained to be 60°. The framework of diethylcadmium is more complicated, and as a result only six parameters were optimized: r_{CdC}, r_{CC}, r_{CH}, Θ_{CdCC}, Θ_{CdCH}, and Θ_{HCC}. The basic framework was kept in a "trans-like" geometry and the value of the CCH bond angle for the methyl group was kept at its tetrahedral value of $2 \cos^{-1}(3^{-1/2})$. These parameters are summarized in Table III. Since there are no full experimental geometric parameters available for SeH_2, $HSeCH_3$ and $Se(CH_3)_2$, the geometries were optimized with the MIDI-4p2df2 basis set at the MP2 level.[28] During the geometry optimization $HSeCH_3$ and $Se(CH_3)_2$ were forced to remain in the pseudo-trans conformation. Figure 4 summarizes the framework utilized for the selenium compounds. The optimized geometries are listed in Table IV together with the experimentally obtained partial structural parameters. For ^{77}Se chemical shielding calculations we only report the results of

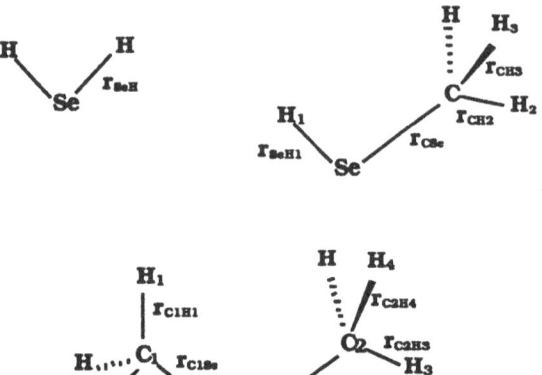

Figure 4 Geometrical framework utilized for the selenium calculations.

GIAO calculations from the TX90 program. The basis sets used were MIDI-4p2d2[14], AKR/WAG-pd2[27] and HORN-pd.[25] The current version of TX90 lacks the capability of handling the orbitals of f-symmetry.

3. Results and Discussion

Cadmium. There are two points that we would like to cover in the present investigation: 1) the question of using chemical shifts from neat liquids as reference points for essentially gas phase calculations and 2) the sensitivity of the calculated shielding tensor to the choice of basis set and geometry. We will address each of these points in turn.

In our original paper on ^{113}Cd NMR spectroscopy,[5a] which dealt in part with the determination of the isotropic chemical shift difference between dimethyl-and diethylcadmium, we noted that the chemical shift of neat dimethylcadmium was due to an associated species. This was clear from the 101.12 ppm chemical shift range observed for 1.0M solutions of dimethylcadmium in a variety of solvents. The most telling data was the shift of 34.67 ppm to lower shielding for the dimethylcadmium in going from the neat liquid to a 1.0M solution in cyclohexane. Hence, it was clear that the observed shift difference of 99.73 ppm for dimethyl- and diethylcadmium was not representative of the isolated molecules. As mentioned above, we have redetermined the shift difference in two ways. First, we have used a solvent which should minimize the apparent self-association in the neat liquids. We chose cyclohexane with concentrations of 0.088M and 0.176M for dimethyl-and diethylcadmium, respectively. These concentrations correspond to approximately 1% (V/V) solutions. The chemical shift difference in this solvent resulted in diethylcadmium being more shielded than dimethylcadmium by 120.6 ppm. Secondly, we have determined the shift

Table IV
Structural Parameters for Hydrogen Selenide, Methyl Selenol and Dimethyl Selenide

parameter[a]	ab initio[b]	exp.	parameter[a]	ab initio[b]	exp.
SeH$_2$			Se(CH$_3$)$_2$		
r(SeH)	1.455	1.460[c]	distance		
α(HSeH)	91.3	90.6[c]	r(C1Se)	1.941	1.945[e]
			r(C2Se)	1.950	
HSeCH$_3$			r(C1H1)	1.094	1.096[e]
distance			r(C1H2)	1.093	1.088[e]
r(CSe)	1.948	1.959[d]	r(C2H3)	1.094	
r(SeH1)	1.457	1.473[d]	r(C2H4)	1.093	
r(CH2)	1.094	1.088[d]	angle		
r(CH3)	1.092		C1SeC2	96.6	96.32[e]
angle			H1C1Se	107.2	105.0[e]
H1SeC	95.2	95.4[d]	H2C1Se	110.2	110.3[e]
H2CSe	106.0		H3C2Se	108.3	
H3CSe	110.6		H4C2Se	110.0	
dih. angle			dih. angel		
H3CSeH2	109.3		H2C1SeH1	109.3	
			H4C2SeH3	109.2	

a. Bond length in Angstrom, bond angle and dihedral angle in degree. For atomic numbering see Figure 4.
b. Optimized with basis set MIDI-4p2df2 at MP2 level.
c. Hehre, W.J., Radom, L., Schleyer, P.R. and Pople, J.A. "Ab Initio Molecular Orbital Theory", pg.139, John Wiley, New York, 1986.
d. Thomas, C.H. **J. Chem. Phys.** 59,70(1973).
e. Pandey, G.K., Dreizler, H. **Z. Naturforsch.** 32a, 482(1977).

difference in the gas phase. The spectra are presented in Figure 1 and the shifts are summarized in Table I. The gas phase chemical shift measured for dimethylcadmium, 62 ppm to lower shielding with respect to neat dimethylcadmium, is larger than the corresponding difference obtained for diethylcadmium, i.e. 19.2 ppm to lower shielding with respect to neat diethylcadmium. These chemical shifts with respect to the corresponding neat liquids are significant. However, one can only speculate at this point as to their origin. We suspect these shifts arise from van der Waals type association, rather than the specific formation of a dimer structure involving bridging methyl groups. However, more experimental work is required to test such speculation.

The chemical shift difference in the gas phase was determined to be 142.6 ppm. To put this chemical shift difference in perspective, the normal range of ^{13}C chemical shifts is typically 200 ppm. Hence, the "simple" replacement of both methyl groups by ethyl groups has led to a gas phase ^{113}Cd chemical shift difference of 142.6 ppm or ~71% of the common ^{13}C chemical shift range! The total range of ^{113}Cd chemical shifts is approximately 900 ppm[5,6] Therefore, the dimethyl-, diethylcadmium gas phase chemical shift represents about 16% of the range of ^{113}Cd chemical shifts. The origin of this chemical shift can only be addressed by calculations of the ^{113}Cd chemical shifts.

Our original data[5a] undoubtedly represented a partial motivation for the pioneering computations of Nakatsuji and coworkers.[10,15] However, as we have pointed out in the introduction, these computations raise some interesting questions. Of particular interest here, is the basis set used in the calculation, i.e. a mixture of MIDI-1 and MINI-1. That is, the MIDI-1 basis was used for cadmium and its contiguous atoms, whereas the remaining atoms used a MINI-1 basis. This type of uneven basis is an example of what has been termed by Chesnut and Moore[29] as a "locally dense" basis set. With such a basis the predicted chemical shift difference was -97 ppm. This result would be considered by anyone to be in excellent agreement with our results for neat liquids, -99.7 ppm, and in modest agreement for the results for the gas phase, i.e. -142.6 ppm. To examine the sensitivity of the previously computed shift

Table V
Basis Set Dependence of the Calculation
of the Chemical Shift Difference of Dimethyl- and Diethylcadmium

	Nakatsuji, et al.[10]	The present work
Cd(CH$_3$)$_2$	σ^{dia} = 4851 ppm	σ^{dia} = 4851 ppm
	σ^{para} = -1090 ppm	σ^{para} = -1092 ppm
	σ^{tot} = 3761 ppm	σ^{tot} = 3759 ppm
Cd(C$_2$H$_5$)$_2$	σ^{dia} = 4896 ppm	σ^{dia} = 4897 ppm
	σ^{para} = -1038 ppm	σ^{para} = -1094 ppm
	σ^{tot} = 3858 ppm	σ^{tot} = 3803 ppm
	δ = -97 ppm	δ = -44 ppm

difference to changes in basis set (for a fixed geometry[10]), we made a subtle change, i.e. namely, the use of MIDI-1 for all of the atoms. The results are outlined in Table V. The results for dimethylcadmium are nearly the same as those reported previously. This near equivalence arises from the subtle differences between MIDI-1 and MINI-1 for hydrogens. However, this is not the case for the diethylcadmium calculation. That is, making a "small" change in the basis set for noncontiguous atoms, i.e. going from MINI-1 to MIDI-1, results in a change in the predicted chemical shift difference of 53 ppm! If we use the energy optimized exponents for a pair of p-functions (see Table II) instead of those employed by Nakatsuji and coworkers,[10,15] the results are equally disappointing. One has to suspect the *quantitative* agreement between theory and experiment reported by Nakatsuji and coworkers[10,15] may be fortuitous.

A reasonable question at this point is whether any basis set can work for these systems? We have addressed this question by performing calculations of the shielding constants with three basis sets, i.e. an augmented MIDI-4 basis on the cadmium and MIDI-4 basis functions on carbon and hydrogen; the same augmented MIDI-4 basis set on the cadmium with the basis set of Dunning and Hay[26] on carbon and hydrogen, and an augmented version of Friedlander, Howell, and Synder's[24] basis function for cadmium in conjunction with the Dunning and Hay[26] basis for carbon and hydrogen. These results are summarized in Table VI.

The first point to make concerning the results summarized in Table VI is a comparison between calculations performed with MIDI-4p2, MIDI-4p2d2, and MIDI-4p2d2f2 basis sets with a geometry optimized at the MP2/MIDI-4p2d2f2 level (the first three entries for $Cd(CH_3)_2$ and $Cd(C_2H_5)_2$ in Table VI). There is a strong dependence of the computed shielding upon the quality of the basis set. Further, there is **no** hint that the dependence is converging. That is, the basis sets are not saturated with respect to their ability to predict the observed shielding. This trend is also reflected in the predicted chemical shifts. The inclusion of f-orbitals is important in describing the chemical shift difference between dimethyl- and diethylcadmium. This is not to say that f-orbitals are essential to describe cadmium chemistry. However, it does say that the calculation must have a better description of the "low lying" virtual orbitals which play such an important role in the coordination chemistry of cadmium.

Clearly the "so-called" paramagnetic term is dominant in describing the shielding of cadmium. However, comparison of the computed shielding by the LORG and CHF methods represents a reminder that only the **sum** of the diamagnetic and the paramagnetic terms are gauge invariant.[1,19,20] That is, comparing the CHF results for the chemical shift difference using the fhs-p2d2f2 basis set would predict that of the -110.84 ppm computed, -45.4 ppm of that shift would arise from the diamagnetic term while -65.4 ppm arises from the paramagnetic term. Comparing the results for the same basis set and geometry, but computed via the LORG approach, one obtains -10.2 ppm and -77.1 ppm, respectively for the diamagnetic and paramagnetic contributions to the computed shift difference.

The overall quality of the calculation is reminiscent of the phrase "...beauty is in the eyes of the beholder...". A worst case comparison would be the -93.52 ppm LORG prediction of the chemical shift difference (i.e. 65.6% of the experimental shift difference). The best case predicted shift difference would correspond to the MIDI-4p2d2f2/D95 CHF results of -115.26 ppm (or 80.8% of the experimental result). However, the results could never be described as quantitative. There are potentially simple reasons for this and they were spelled out in the introduction. That is, there are excellent reasons to believe that relativistic terms, via **L•S** interactions,[3] and/or two-electron contributions (electron-correlation)[30,31] to the shielding should make significant contributions to the observed shielding for heavy atoms such as cadmium. However, before these points can be addressed, the basis set dependence of the shielding must be examined. The **chemical shifts** in Table VI are strongly dependent upon the choice of basis.

Table VI

^{113}Cd Shieldings with MP2 Optimized Geometries[a]

Cd(CH$_3$)$_2$	LORG				CHF			
	σ^{dia}	σ^{para}	σ^{Total}	δ_{LORG}	σ^{dia}	σ^{para}	σ^{Total}	δ_{CHF}
MIDI-4p2[b]	4774.61	-903.72	3870.90	-	4872.16	-944.79	3927.37	-
MIDI-4p2d2[b]	4771.03	-959.61	3811.42	-	4871.97	-1004.45	3867.52	-
MIDI-4p2d2f2[b]	4765.82	-1252.54	3513.28	-	4872.43	-1335.76	3536.68	-
MIDI-4p2d2[c]	4771.89	-931.88	3840.01	-	4872.12	-979.62	3892.50	-
MIDI-4p2d2f2[c]	4765.31	-1266.08	3499.23	-	4872.61	-1346.20	3526.41	-
fhs-p2d2[d]	4762.54	-1143.60	3618.95	-	4875.78	-1226.13	3649.65	-
fhs-p2d2f2[d]	4767.07	-1205.44	3561.63	-	4875.59	-1284.43	3591.15	-
Cd(C$_2$H$_5$)$_2$								
MIDI-4p2[b]	4782.58	-863.33	3919.25	-48.35	4917.22	-923.42	3993.80	-66.43
MIDI-4p2d2[b]	4780.25	-904.52	3875.73	-64.31	4917.04	-967.52	3949.51	-81.99
MIDI-4p2d2f2[b]	4776.43	-1181.53	3594.91	-81.63	4917.46	-1279.09	3638.37	-101.69
MIDI-4p2d2[c]	4780.52	-868.49	3912.02	-72.01	4917.24	-928.57	3988.67	-96.17
MIDI-4p2d2f2[c]	4775.86	-1183.11	3592.75	-93.52	4917.70	-1276.03	3641.67	-115.26
fhs-p2d2[d]	4773.02	-1077.69	3695.33	-76.38	4921.20	-1171.02	3750.19	-100.54
fhs-p2d2f2[d]	4777.32	-1128.34	3648.98	-87.35	4921.01	-1219.02	3701.99	-110.84

a) Optimized at the MIDI- 4p2d2f2 level of basis set. All tensor elements and chemical shifts are expressed in ppm. A negative sign of the **chemical shift** denotes resonances to higher shielding.

b) Geometry optimized with the MIDI-4p2d2f2 basis set on cadmium with a MIDI-4 basis set for carbon and hydrogen. The optimized orbital exponents are given in Table II.

c) Geometry optimized as described in footnote a, except that the basis set of Dunning and Hay[26] was used for carbon and hydrogen.

d) Geometry optimized with the basis set of Friedlander, Howell, and Synder[24] (fhs) for the cadmium augmented with a pair p-, d-, and f-orbitals with a Dunning and Hay[26] basis for the carbon and hydrogens. The values of the orbital exponents used for the fhs basis are given in Table II.

A partial rationalization of this basis set dependence is the potential lack of balance in the basis set. That is, the valence region is probably described reasonably well by the augmented p-, d-, and f-functions. However, it is not clear whether the details (shape and amplitude) of the inner shell orbitals are correct. Clearly, more work is needed, and we are pursuing these points. The basis set is the probable reason as to why the CHF results are closer to experiment than the results predicted by the LORG approach.

In order to better characterize the level of agreement between theory and experiment summarized in Table VI, we have run a set of calculations using GIAO method of Pulay and coworkers[21] with three basis sets, i.e. the augmented AKR/WAG-p2d2 basis set,[27] fhs-p2d2,[24] and the MIDI-4p2d2 basis set.[14] The exponents are described in Table II for cadmium. Further, the D95 basis set[26] utilized on carbon and hydrogen was augmented by a single d- and p-polarization on carbon and hydrogen, respectively. The results of those calculations are summarized in Table VII.

Table VII
GIAO Calculations of ^{113}Cd Chemical Shifts[a]

Cd(CH$_3$)$_2$	σ^d	σ^p	σ^{Total}	δ_{GIAO}
AKR/WAG-p2d2	4814.64	-1211.34	3597.00	-
FHS-p2d2	4805.22	-1356.77	3448.44	-
MIDI-4p2d2	4801.87	-1144.14	3657.73	-
Cd(C$_2$H$_5$)$_2$				
AKR/WAG-p2d2	4807.76	-1114.78	3692.98	-95.97
FHS-p2d2	4805.70	-1260.12	3545.58	-97.14
MIDI-4p2d2	4802.31	-1075.69	3726.62	-68.89

a   The geometries used in all of the calculations are summarized for dimethyl- and diethylcadmium for the optimized MIDI-4p2d2f2 using D95 for carbon and hydrogen, see Table III for the details.

The overall results are essentially the same as depicted in Table VI. That is, the magnitude of the predicted chemical shift difference is underestimated by theory. Even with the GIAO method there is still a strong dependence on the paramagnetic term to the choice of basis. As usual in a GIAO calculation, the diamagnetic term is essentially constant and insensitive to the choice of basis. Within the scope of the present calculation, can an explanation be put forth for this exceptional chemical shift difference observed for dimethyl- and diethylcadmium? To address this point one generally examines differences in structure. However, examination of the results summarized in Table III does not lead one to believe that there are significant structural differences between dimethyl- and diethylcadmium. The exception is molecular symmetry and this point is important but **not causal**. The shift difference between dimethyl- and diethylcadmium is analogous to that of a β-effect in ^{13}C shieldings. A clear physical picture of the origin of the β-effect in ^{13}C shieldings has not been delineated. Given the overall level of agreement between predicted shielding and experiment and the strong basis set dependence manifested in Table VI, it should be concluded that questions relating to the origin of the chemical shift difference between dimethyl- and diethylcadmium is premature at this point. Further, conclusions concerning the importance of ligand orbitals to the observed ^{113}Cd shielding difference are likewise considered premature.

Similar arguments can be made concerning the importance of the geometry utilized in calculations such as is reported here. However, until the level of agreement between theory and experiment improve, such a discussion would seem premature as well.

Selenium. Given its position within the periodic table. relativistic effects[3] should be minimal for selenium. Hence, other than effects due to electron correlation RHF calculations of ^{77}Se chemical shifts should produce satisfactory results between theory and experiment. Table VIII summarizes our calculations using three basis sets. The basis sets utilized are AKR/WAG-pd2,[27] HORN-pd,[25] and MIDI-4p2d2.[14] No f-orbitals were used due to the inability of the present version of the GIAO program to handle these orbitals. For simplicity, this Table only contains the results from GIAO calculations. The results summarized in Table VIII are similar to those predicted for the organocadmium compounds. That is, the chemical shifts are dramatically underestimated. These results are considerably worse than the corresponding calculation for GIAO ^{33}S chemical shifts.[32] The relatively poor agreement between theory and experiment may simply reflect theory's inability to properly describe the lone pairs on selenium.

Natural Bond Orbital Analysis. An important point which can be addressed is whether the computed MO's give rise to the accepted structure for the molecules of interest. That is, are the correct Lewis structures predicted for each molecule? This question is germane given the results of a recent paper of Nemukhin and Weinhold.[33] In this work a natural bond orbital (NBO) analysis was carried out on a series of aluminum oxides. They found for the cases where the electronegativity differences between the atoms was large ($\chi_{Al} = 1.5$ and $\chi_O = 3.5$) that the uncorrelated wave functions did not describe the correct Lewis structures for the aluminum oxides. However, the correct Lewis structures were predicted for the MP2 level or correlated MO's. The electronegativity of cadmium is the *same* as aluminum and $\chi_C = 2.5$. Based on the significant difference in electronegativities for Cd and C, we decided to carry out a series of NBO calculations on dimethyl- and diethylcadmium. At the RHF level the computed wave functions predict the following Lewis structures for dimethyl-and diethylcadmium:

$$H_3C: \quad Cd \rightarrow CH_3 \qquad\qquad H_3CH_2C: \quad Cd \rightarrow CH_2CH_3.$$

That is, both structures are virtually ionic. However, at the MP2 level the computed wave functions predict the following, correct, Lewis structures for dimethyl-and diethylcadmium:

$$H_3C-Cd-CH_3 \qquad\qquad H_3CH_2C-Cd-CH_2CH_3.$$

The bonds in these structures are still polarized towards the carbons, but the amount of covalent character has increased. The results for selenium compounds are not as dramatic. That is, the correct Lewis structures are predicted at the RHF level of theory. However, the population of the lone pair orbitals changes in going from the RHF level to the MP2 level of approximation. The results suggest that performing the shielding calculation at the appropriate correlated level may give rise to unexpected levels of improvement simply arising from a better description of the structure and bonding in these systems.

Table VIII
77Se Chemical Shieldings Calculated With the GIAO Method

	σ^{dia}	σ^{para}	σ^{Total}	δ_{GIAO}	δexp
SeH$_2$					-344.75
midi-4p2d2	2987.06	-836.14	2150.89	-209.07	
AKR/WAG-pd2	2996.79	-847.23	2149.56	-125.03	
Horn-pd	2998.52	-865.12	2133.40	-201.9	
HSeCH$_3$					-154.67
midi-4p2d2	2981.70	-975.50	2006.20	-64.38	
AKR/WAG-pd2	2980.84	-954.79	2041.05	-16.52	
Horn-pd	2998.79	-1026.66	1971.97	-40.47	
Se(CH$_3$)$_2$					0.0
midi-4p2d2	2975.08	-1033.26	1941.82	-	
AKR/WAG-pd2	2994.35	-969.82	2024.53	-	
Horn-pd	2998.99	-1067.49	1931.50	-	

4. Summary and Conclusions

In the present paper we have illustrated several facets of the maturing area of the computation of heavy atom magnetic resonance parameters. We have demonstrated that even with care in geometry and basis set selection, one can not obtain quantitative results for the calculation of [113]Cd or [77]Se chemical shifts. However, the overall trends are apparent from the calculations. In this same vein, care should be exercised in making comments as to what type of orbital interactions will be important in determining [113]Cd chemical shifts. It is suspected that the reasons for the observed lack of quantitative results are two-fold: basis sets and for the case of [113]Cd, not incorporating all of the applicable physics into the calculations. Basis sets are clearly important, but they are not the whole story. Before this area of computational chemistry can advance significantly, the computational theorists must address the unknown importance of spin orbit terms and electron correlation on the chemical shieldings. Finally, we do not suggest that the results reported here be used in some fashion to illustrate which method of calculating the shielding tensor is better, i.e. distributed origin methods, e.g. LORG[19,20] and GIAO[21] or CHF[23] approaches. Such conclusions would be clearly premature.

Acknowledgments

The authors would like to acknowledge the National Institutes of Health via award GM-26295 (PDE) and GM-42907 (JDO) for partial support of this research. Further, we would also like to thank Dr. Dorothy Tudor and Ms. Sibyl Hare of the Computer Services Division at USC for their efforts with regard to the IBM RISC 6000/550 Gaussian calculations. It is a pleasure to acknowledge several stimulating and informative conversations with Tom Bouman as well as his donation of the functional portions of the f-orbital version of the RPAC[19] program to PDE. Sadly, shortly before this conference paper was submitted for publication Tom Bouman died. He will be missed. Finally, PDE would like to acknowledge discussions with Tom Dunning concerning strategies for polarization function development.

References

1. Ditchfield, R.; Ellis, P.D.,in "Topics in Carbon-13 NMR Spectroscopy", Edited by G.C. Levy, Wiley Interscience, Vol 1, pg. 1-51, 1974.

2. Lambert, J.B.; Riddell, F.G., "The Multinuclear Approach to NMR Spectroscopy", NATO ASI Series. Series C. Mathematical and Physical Sciences No. 103.

3. Slater, J.C., "Quantum Theory of Atomic Structure", Vol 1, McGraw-Hill, New York, 1960; Shu, F.H., "The Physics of Astrophysics", Vol 1, University Science Books, California, 1991, pg. 302; Karplus, M. and Porter, R.N., "Atoms & Molecules: An Introduction For Students of Physical Chemistry",W.A. Benjamin, Inc, New York, 1970, pg 210.

4. a) Luthra, N.P., Dunlap, R.B., and Odom, J.D., **Anal. Biochem.**, 117, 94(1981); b) Luthra, N.P., Costello, R.C., Odom, J.D., and Dunlap, R.B., **J. Biol. Chem.**, 257, 1142(1982); c) Boccanfuso, A.M., Griffin, D.W., Dunlap, R.B., and Odom, J.D., **Bioorg. Chem.**, 17, 231(1989), d) House, K.L., Dunlap, R.B., Odom, J.D., Wu, Z.-P, and Hilvert, D., **J. Am. Chem. Soc.**, in press; e) Boles, J.O., Tolleson, W.H., Schmidt, J.C., Dunlap, R.B, and Odom, J.D., **J Biol. Chem.**, in press.

5. a) Cardin, A.D.; Ellis, P.D.; Odom, J.D., and Howard, J.W, **J. Am. Chem. Soc.**, 97, 1672-1679 (1975); b) Ellis, P.D., **Science**, 221, 1141-1146 (1983); c)Ellis, P.D., **J. Biol. Chem.**, 264, 3108-3110 (1989); d) Rivera, E.; Kennedy, M.A.; Adams, R.D.; Ellis, P.D., **J. Am. Chem. Soc.**, 112, 1400-1407 (1990); e) Rivera, E; Kennedy, M.A.; Ellis, P.D., **Adv Magn. Reson.**, 13, 257-274 (1989); f) Marchetti, P.S.; Honkonen, R.S.; Ellis, P.D., **J. Magn. Reson.**, 71, 294-302 (1987); g) Honkonen, R.S. and Ellis, P.D., **J. Am. Chem. Soc.**, 106, 5488-5497 (1984); h) Marchetti, P.S.; Ellis, P.D.; Bryant, R.G., **J. Am. Chem. Soc.**, 107, 8191-8196 (1985).

6. Summers, M.F., **Coord. Chem. Rev.**, 86, 43-134 (1988).

7. Jakobsen, H.J., Zozulen, A.J., Ellis, P.D., and Odom, J.D., **J. Magn. Reson.**, 38, 219 (1980).

8.   a) Silks, L.A. III, Dunlap, R.B., Odom, J.D., **J. Am. Chem. Soc.**, 112, 4979(1990); b) Silks, L.A. III, Peng, J., Odom, J.D., and Dunlap. R.B., **J. Chem. Soc. Perkin, Trans I**, 2495(1991); c) Silks, L.A., III, Peng, J., Odom, J.D., and Dunlap, R.B., **J. Org. Chem.**, 56, 6733(1991).

9.   Luthra, N.P., Boccanfuso, A.M., Dunlap, R.B., and Odom, J.D., **J. Organometal.**, 354, 51 (1988).

10.   Nakatsuji, H; Nakao, T.; Kanda, K., **Chem. Phys.**, 118, 25-32 (1987).

11.   Ellis, P.D., Odom, J.D., Lipton, A.S., and Gulick, J.M., **J. Am. Chem. Soc.**, in press.

12.   Pauling, L., "The Nature of the Chemical Bond", Cornell Univ. Press, 1960.

13.   Rao, K.S.; Stoicheff, B.P.; Turner, R., **Can. J. Phys.**, 38, 1516 (1960).

14.   Tatewaki, H.; Huzinaga, S., **J. Comput. Chem.**, 1, 205 (1980), and a corrected tabulation can be found in Huzinaga, S, "Gaussian Basis Sets for Molecular Calculations", Elsevier, 1984.

15.   Nakatsuji, H.; Kanda, K; Endo, K; Yonezawa, T., **J. Am. Chem. Soc.**, 106, 4653-4660 (1984).

16.   Shiver, D. F., *"The Manipulation of Air-Sensitive Compounds"*, 2nd Ed, McGraw-Hill, New York, N.Y., **1986**.

17.   Gilman, A., and Nelson, J. F., **Rec. Trav.**, 55, 518 (1936);   Sanders, R., and Ashby, E. C., **J. Organomet. Chem.**, 25, 277 (1970).

18.   Bird, M.L.; Challenger, F.; **J. Chem. Soc.**, 1942, 570.

19.   Bouman, T.D. and Hansen, Aa. E.; "RPAC Molecular Properties Package, Version 8.6", Southern Illinois University at Edwardsville, 1990.

20.   Hansen, Aa.E. and Bouman, T.D., **J. Chem. Phys.**, 82, 5035-5047 (1985).

21.   Wolinski, K., Hinton, J.F., Pulay, P.; **J. Am. Chem. Soc.**, 112, 8251-8260 (1990).

22.   Gaussian 90, Revision F, Frish, M.J., Head-Gordon, M., Trucks, G.W., Foresman, J.B., Schlegel, H.B., Raghavachari, K., Robb, M., Brinkley, J.S., Gonzalez, C., Defrees, D.J., Fox, D.J., Whiteside, R.A., Seeger, R., Melius, C.F., Baker, J., Martin, R.L., Kahn, L.R., Stewart, J.J.P., Topiol, S., and Pople, J.A., Gaussian, Inc., Pittsburgh PA, 1990.

23.   An excellent summary of the CHF approach is given by Lipscomb, W.N. in "Advances in Magnetic Resonance", Academic Press, New York, Vol 2, pp 138-176, 1966.

24.   Friedlander, M.E.; Howell, J.M.; Synder,G.; **J. Chem. Phys.**, 77, 1921 (1982).

25.   Schafer, A.; Horn, H.; Ahlrichs, R., **J. Chem. Phys.**, 97, 2571-2577 (1992).

26.   Dunning, T.H. and Hay, P.J., "Modern Theoretical Chemistry", Edited by H.F. Schaefer III, Plenum, 1976, pp. 1-28, Chapter 1.

27.   Rappe, A.K.,; Goddard, W.A., an undocumented feature of the Gaussian-90 system of programs. These basis sets can be found in the directory [g90.basis].

28.   Moller, C., Plesset, M.S.; **Phys. Rev.**, 46, 618 (1934); Binkley, J.S., Pople, J.A.; **Int. J. Quantum Chem.**, 9, 229 (1975).

29.   Chesnut, D.B.; Moore, K.D., **J. Comp. Chem.**, 10, 648-659 (1989).

30.   Oddershede, J. and Geertsen, J., **J. Chem. Phys.**, 92, 6036-6042 (1990).

31.   Bouman, T.D., Hansen, Aa.E., **Chem. Phys. Lett.**, 175, 292-299 (1990).

32.   Wolinski, K.; Hinton, J.F.; Pulay, P.; **J. Am. Chem. Soc.**, 112, 8251 (1990).

33.   Nemukhin, A.V.; Weinhold, F.; **J. Chem. Phys.**, 97, 3420 (1992).

OVERVIEW AND DIRECTIONS FOR THE FUTURE

CYNTHIA J. JAMESON
University of Illinois at Chicago
Department of Chemistry M/C-111
Chicago, Illinois 60680

ABSTRACT. A brief summary of the lectures and discussions is given, with some questions for the future.

1. Overview of the Conference

This is the first conference on nuclear magnetic shielding. It brings together a good mix of theoreticians and experimentalists, all of whom are interested in this magnetic tensor property. This quantity is so sensitive to the electronic environment that sites which can not be distinguished from each other by any other measurement can be easily distinguished by the differences in their NMR shielding. The precision with which such measurements can be carried out allow determinations not only of molecular structure but even detection of subtle changes in geometry such as alteration of bond lengths and bond angles and changes in the number and closeness of neighboring molecules. A few of the questions that have been answered by measurements and calculations of magnetic shielding are: Is the system disordered, averaging dynamically, or does it have small domains of particular well-ordered structures? What is the distribution of molecules in the cavities of a microporous solid? What is the number of bridging oxygen atoms around a phosphorus, aluminum, or silicon atom in a network? Is the molecular structure and geometry proposed on the basis of other evidence correct?

In this conference the nuclear shielding in a broad continuum of systems has been studied, ranging from the one extreme wing of small molecular systems having well-known geometries such as CO, to the other extreme wing comprising of complex systems and networks, some of unknown structure. A variety of theoretical approaches have been used: Post-Hartree-Fock, e.g., SOLO, SOPPA, MC-IGLO, coupled Hartree-Fock or RPA using local origins (IGLO, LORG, GIAO) or using a common origin, Geertsen's gauge invariant method, approximate methods employing a mean excitation energy, and empirical correlations. On the one hand, the largest basis sets and second order electron correlation corrections are being applied to some small molecules such as N_2, CO, HCN. On the other hand, in complex systems, the unit for which the shielding is measured is an extended network of atoms (an aluminosilicate crystallite or a semiconductor alloy) rather than a single molecule having a fixed number of atoms. For these the interpretation is still at the empirical and semi-empirical stages, employing mean excitation energies and qualitative concepts such as ionicity, bond polarity, cation/anion radii. Alternatively, high-quality *ab initio* calculations on model systems to emulate the essential characteristics of the real complex system have been reported here. Examples are the IGLO calculations on the SiO_4^{4-} fragment to simulate the geometry dependence of observed ^{29}Si chemical shifts in polycrystalline silicates (Wolff) and the modeling of the different ^{19}F environments of ^{19}F-labeled phenylalanine in various proteins by using electric field and

557

J. A. Tossell (ed.), Nuclear Magnetic Shieldings and Molecular Structure, 557–560.
© 1993 *Kluwer Academic Publishers.*

electric field gradient derivatives of ^{19}F shielding together with the electric fields generated by point charges representing the electrostatic effects within the protein (Dykstra and Oldfield).

What are some important lessons we can take home from the conference? The damping of errors from incomplete cancellation of long-range contributions due to basis set inadequacies by the use of local origin methods has been well-established. The GIAO method (Pulay, Chesnut), the IGLO method (Kutzelnigg), and the LORG method (Hansen and Bouman) all can provide results of experimental quality or near-experimental quality in the same classes of systems. The first and second row calculations have offered may success stories, but it is quite clear that large basis sets are necessary to obtain near experimental quality results in most cases. When modest basis sets are used in all methods, GIAO appears to do a little better, although an across-the-board comparison with a large number of molecules has yet to be done in the same way that has been done for the LORG vs IGLO comparison. It is now possible to obtain experimental accuracy in many classes of nuclear shieldings, especially for ^{13}C, although a few pathological cases still pose over ten ppm discrepancies (CO, HCN). When a saturated basis set is used the common-origin conventional CHF results (Lazzeretti, Raynes) are indistinguishable from the local-origin results. The gauge origin dependence of shielding calculations is basis-set dependent. (Geertsen, Oddershede). That is, it is not possible to determine that a basis set is saturated simply because the results for a given gauge origin no longer change as the basis set is further improved. A locally-dense basis set may be very useful when only a small number of shieldings are to be calculated in a large molecule (Chesnut). Shielding density maps and current density maps provide stunning visual aids for deriving physical intuition and first-principles understanding of shielding and magnetizability (Lazzeretti, Keith).

Shielding anisotropies provide more complete information than σ_{iso}. There are many examples of systems exhibiting very nearly the same σ_{iso} but very different individual components. (Mason). σ_{iso} being the same offers no clear indication of unchanging electronic environment of the nucleus. Complete experimental shielding tensors [three principal values plus the Euler angles orienting the principal axis system (PAS) of the shielding tensor with respect to the molecular frame] have been obtained unequivocally and accurately in isolated spin pairs (AX systems) by making use of the dipolar tensor (Wasylishen) and in systems of up to 24 ^{13}C nuclei in one molecule, by using ingenious experimental techniques (Grant). *Ab initio* calculations of the shielding tensor are nearly indispensable in obtaining shielding tensor information. Assignment of the PAS seems to be good even when the basis set may not be of sufficient quality to reproduce the individual tensor components, provided the individual σ_{ii} values are sufficiently different from each other.

It is very important to do temperature-dependent experiments because dynamic averaging may give the wrong tensor components (Mason, Kirkpatrick). There is a change in the tensor with physical state/environment. In the condensed phase there may be a) intermolecular effects on the structure of the molecule and the σ tensor reflects this change in geometry (Farrar), b) intermolecular shielding (Jameson), c) hydrogen-bonding (Hansen), or d) electric field effects (Oldfield, Dykstra). The shielding tensor and the molecular structure in solution can both be obtained but (Warning!) in a multi-parameter fit to experiments it is always important to look at the table of correlation coefficients. If the parameters are highly correlated then a unique fit is not obtained. It is important to use more than one set of boundary conditions (new experiments (Farrar).

The dependence of nuclear shieldings on molecular geometry has been a focus point of several papers (Schleyer, Kutzelnigg, Jameson, Raynes, Chesnut, Barfield). For calculating the shielding at the equilibrium geometry, one needs accurate bond lengths and angles because the shielding depends sensitively on these. It is better yet to do vibrational averaging over the nuclear shielding surfaces. This averaging is especially important for comparison with the experimental shielding when the potential energy surface is fairly flat while the shielding surface is not. Rovibrational averaging is also important for the interpretation of the temperature-dependence of the shielding in the isolated molecule and for the interpretation of isotope shifts (such as ^{15}N shielding in ND_3 relative to NH_3).

The intermolecular nuclear shielding surface has a minimum that occurs below the shielding of the infinitely separated molecules (Jameson, Kutzelnigg) and is employed in the interpretation of the density dependence of shielding in the gas phase and of the chemical shifts observed in physisorption. The effects of external electric fields and electric field gradients on shielding can be large and of either sign. (Dykstra, Raynes) These can be used in the interpretation of long-range intermolecular and intramolecular effects on σ (Jameson, Oldfield).

Electron correlation is important for shielding in multiple-bonded systems, especially when lone pairs are also involved (the $n \rightarrow \pi^*$ transitions) and when nearly degenerate or low-lying excited states are involved. For these systems it is necessary to use post-Hartree-Fock methods such as SOPPA (Oddershede), SOLO (Hansen), MC-IGLO (Kutzelnigg). Lone pairs by themselves do not make electron correlation contributions significant; systems with lone pairs can generally be brought into agreement with experiment by saturating the basis set. How large are the actual contributions from electron correlation can not be determined correctly if the basis set is not saturated because part of the difference between the results of a correlated calculation and a calculation at the CHF level may be due to an inadequate basis set. Nevertheless, where correlation effects are known to be significant, it makes sense to do correlated calculations even with a modestly large basis set, specially if basis set errors are damped out by the use of local origins. There appears to be no effects of correlation in ^1H shielding. The sign of the electron correlation effect on shielding is positive for σ_{iso} in most cases, that is, the CHF-level result is usually too deshielded and correlation corrections are positively shielding. Correlation also generally increases $(\sigma_\| - \sigma_\perp)$. There are exceptions, for example, in F_2 correlation makes σ_{iso} more deshielded. For molecules whose electronic ground state cannot be properly described by a single Slater determinant (such as ONNO, O_3, SO_2, NSF, HN=NH) it is important not to use a CHF-level calculation of the shielding for comparison with experiment. Molecules of this type are good candidates for MCSCF calculations.

There are many unanswered questions in shieldings involving 3rd row and below in the Periodic Table, especially for transition metal nuclei, for which there are no absolute shielding scales, no gas phase data; geometries are unknown in solution, and the basis sets are largely untested. While the agreement between calculated shielding differences and observed chemical shifts are good (Nakatsuji), this good agreement may be for-tuitous, since the shielding calculations are shown not to be converged when larger basis sets are used (Ellis). For these it is important to (a) do basis set studies, (b) also calculate the shielding of ligand atoms for which there are absolute shielding scales and com-parison with the experiment can actually be made, and (c) do geometry studies, especially for complex ions where the geometries in solution are not necessarily the same as in the crystalline state, and the theoretical optimized geometry may be sensitive to the counter ions.

2. Directions for the Future

As theorists begin to do calculations for nuclei beyond the 1st and 2nd rows of the Periodic Table, new problems arise:

(1) Basis sets have not been optimized for shielding calculations in 3rd and 4th row atoms; basis set studies are needed. Computation becomes more time and disk-space demanding and it becomes even more important to make sure the CHF - level calculations are converged with respect to basis sets.

(2) Relativistic effects begin to become important. How to do relativistic calculations? Prof. Kutzelnigg led the discussion on this topic and several comments and suggestions were made which can serve as starting points for future work. There is no exact relativistic many-electron Hamiltonian; it is only defined to order c^{-2} (c = velocity of light); singularities appear beyond c^{-2}. This means that we are really speaking of doing calculations only to order c^{-2}. One can do relativistic Hartree-Fock with 4-component spinors but even quantum electrodynamics exact theory cannot go beyond two electrons. So how do we proceed? We can consider deriving a fully relativistic theory with 4-component spinors using a coupled Hartree-Fock scheme, IGLO-version for example. One could do perturbation theory based on the Breit-Pauli Hamiltonian. The most important contribution comes from the spin-orbit coupling terms contributing to the paramagnetic shielding.

(3) Absolute shielding scales are not available. In the first place we no longer have an identity relating the spin rotation tensor and σ^p in relativistic systems, therefore cannot derive an accurate absolute shielding scale this way. I can suggest the following: Find a linear molecule. — A measure of $(\sigma_\| - \sigma_\perp)$ can provide, together with the theoretical $\sigma_\|^d$, the absolute shielding tensor components. Make chemical shift measurements of this molecule (and others) in the gas phase relative to the commonly used reference substance for this nucleus. With this, all the measured chemical shifts can be converted to absolute shielding.

(4) We need more gas phase data for more accurate comparisons of calculations with experiments. We need this even for 1st and 2nd row atoms for example. ^{35}Cl, ^{33}S, ^{29}Si. We also need to carry out vibrational averaging before comparing *ab initio* numbers with experiments since theoretical results are becoming accurate to this level.

5) We need methods of dealing with complex systems which are not yet tractable by *ab initio* methods. We can do high quality *ab initio* calculations on model systems or fragments which emulate the essential characteristics of the real system. This helps to sort out various effects such as charge polarization effects (electric field effects), geometry - dependence (T-O-T angle, etc.), coordination-number-dependence. At the same time we need to think of new clever ways of modelling these complex systems.

(6) We need correlated calculations at higher levels than have already been done. Presently there are several approaches in use: CI, MCSCF (Daborn and Handy, Yeager and Jorgensen), SOPPA (Oddershede), SOLO (Hansen and Bouman), MBPT (Saika), MBPT(2)-GIAO (Gauss), and MC-IGLO (Wülle and Kutzelnigg). These methods have a tendency to overshoot the correlation effects. We can correct for this by using more sophisticated methods, MP4, MCSCF-CC, for example. Since the local origin methods owe their success to removing error terms in the long-range contributions, we can use correlated calculations to improve the long-range contributions. The local correlation effects which are not obtained in a not-saturated basis set might be improved by using locally-dense basis sets in connection with one of the various correlated calculation techniques.

The next conference on this subject will undoubtedly bring new results and understanding along these lines and much more that we do not yet foresee at this time.

The Structure of $B_8H_8^{2-}$ in Solution. Is $B_8H_9^-$ also Involved?

Michael Bühl[+], Alexander M. Mebel[#], Oleg P. Charkin[#], Paul von Ragué Schleyer[+]

[+]Institut für Organische Chemie, Universität Erlangen-Nürnberg, Henkestr. 42, D-8520 Erlangen, Germany.

[#]Institute of New Chemical Problems, Russ. Academy of Sciences, Chernogolovka, Moscow region, 142432, Russia.

Abstract

Ab initio calculations at the MP2/6-31G*//MP2/6-31G* + ZPE(6-31G*) level give relative energies of 0.0, 5.0, and 17.2 kcal/mol for the D_{2d}, C_{2v}, and D_{4d} forms of $B_8H_8^{2-}$, respectively (see figure below). The C_{2v} form is a transient species involved in the fluxional rearrangement of the D_{2d} form. The chemical shift calculations (IGLO) indicate that the fluxional D_{2d} structure corresponds to the species which shows a single δ ^{11}B NMR signal in solution. While no $B_8H_8^{2-}$ form was found to correspond with the 2:4:2 NMR chemical shift pattern, we raise the possibility that the second species observed under certain conditions might be the protonated form, $B_8H_9^-$, rather than a second $B_8H_8^{2-}$ isomer. The observed ^{11}B NMR chemical shifts and the 2:4:2 intensity pattern are consistent with the computed values for B_8H_9 .

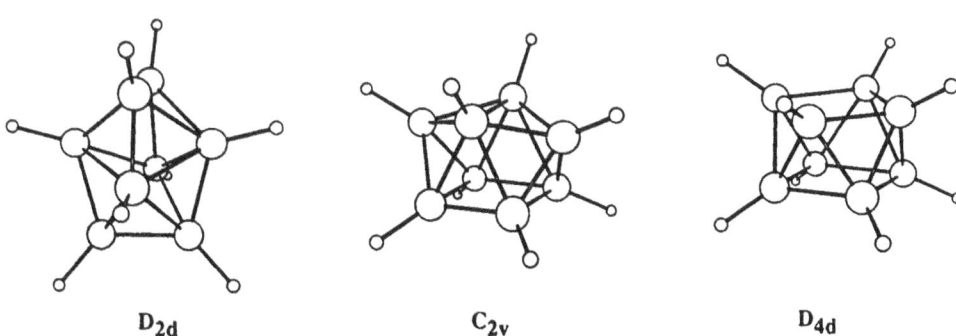

| D_{2d} | C_{2v} | D_{4d} |

J. A. Tossell (ed.), Nuclear Magnetic Shieldings and Molecular Structure, 561–575.

A GENERAL SHAPE FOR INTRA- AND INTERMOLECULAR SHIELDING FUNCTIONS

ANGEL C. de DIOS
University of Illinois at Urbana
Department of Chemistry
Urbana, Illinois 61801

CYNTHIA J. JAMESON
University of Illinois at Chicago
Department of Chemistry M/C-111
Chicago, Illinois 60680

ABSTRACT. The striking similarity between the intermolecular shielding function of ^{39}Ar in the argon dimer and the intramolecular shielding of ^{1}H in H_2^+ prompts us to study the shielding surfaces in other diatomic systems. In this work, the shielding surfaces for ^{39}Ar in Ar-Na$^+$, ^{23}Na in NaH, ^{19}F in HF and F_2, and ^{35}Cl in HCl and ClF are presented. Upon comparison of the shielding functions in these various diatomic systems, a general shape for the shielding function is evident. Although shielding surfaces are not directly observed from experiment, variable temperature and isotope shift measurements indicate that, in general, the shielding is decreasing with internuclear separation. This is consistent with the fact that in these systems the potential function weighs more the region in which the shielding has a negative derivative with respect to the internuclear distance. On the other hand, the experimentally measured second virial coefficients of shielding of rare gas atoms with respect to density are always negative. This is in agreement with having longer distances in the relevant range of separations between the two atoms, the region where the shielding is increasing with internuclear separation. Based on these comparisons, it is clear that the difference between intra- and intermolecular shielding functions lies in what region of the shielding function is sampled as governed by the potential function.

The shielding function of ^{23}Na in NaH provides the unifying picture. At equilibrium separation, the shielding of ^{23}Na in this molecule has a positive derivative with respect to bond length. Upon going to shorter distances, a minimum in the shielding is observed thereby exhibiting the same shape as the other diatomic molecules at short distances. Finally, using the general shape that we propose here, we are able to explain the trends that Chesnut observed in the derivatives of the shielding with respect to bond length in first and second row hydrides.

Experimental Characterization of ^{31}P Chemical Shift Tensors in Phosphine Derivatives

Klaus Eichele and Roderick E. Wasylishen, Department of Chemistry, Dalhousie University, Halifax, Nova Scotia, CANADA B3H 4J3

For an isolated spin, the line shape analysis of the NMR spectrum of a static powder sample gives the three principal components, δ_{ii}, of the chemical shift tensor. However, no information about the orientation of the principal axis system (PAS) with respect to the molecular frame is obtained. Dipolar-chemical shift NMR has proved to be an useful tool to provide two of the three angles required to specify the orientation of the chemical shift tensor. The actual line shape of a spin-½ nucleus A dipolar coupled to a neighbouring nucleus X depends on the principal components of the chemical shift tensor of nucleus A, the dipolar coupling constant R associated with the A,X spin-pair, and the orientation of the dipolar vector, r_{AX}, in the PAS of the shift tensor.

With few exceptions, usually spin-pairs where both nuclei are spin-½ nuclei have been investigated. In the present study, NMR line shapes where X possesses higher spin quantum numbers have been analyzed. Examples are ^{31}P coupled with ^{10,11}B (S = 3 and 3/2) or ^{17}O (S = 5/2) in triphenylphosphine borane and tricyclohexylphosphine oxide, respectively. The orientation of the ^{31}P chemical shift tensors were obtained as well as the sign of the indirect spin-spin coupling constants, J. For both phosphine derivatives, the most shielded principal component δ_{33} deviates from the P-X bond vector, in agreement with the crystal structures for these compounds, and is sensitive for subtle structural differences.

IGLO Calculations of Phosphorus NMR Chemical Shifts

U. Fleischer, W. Kutzelnigg, Lehrstuhl für theoretische Chemie, Ruhr-Universität Bochum, D(W) – 4630 BOCHUM 1, RFA

The IGLO-method /1/ (IGLO stands for Individual Gauge for Localized Orbitals) is a coupled Hartree-Fock (CHF) type method for the calculation of NMR shielding tensors and the magnetic susceptibilities of (closed shell) molecules.
The calculations directly furnish the full information about the principal axes system and the anisotropies of all tensors.

For a wide range of phosphorus compounds the mean deviation of theoretical and experimental data is some 20 to 30ppm (for a shift range of roughly 1000ppm). Deviations for strongly deshielded nuclei are normally larger than those found for shielded ones, since theory at this level has a tendency to overestimate the deshielding contributions.

Unsaturated phosphorus-carbon compounds, i.e. phosphaalkynes, -alkenes, -benzenes, -cyclopenta-dienylanions, are discussed in detail.
The shielding tensors of $2,4,6\text{-t-Bu}_3\text{C}_6\text{H}_2\text{P}=\text{C}(\text{SiMe}_3)_2$ and $2,4,6\text{-t-Bu}_3\text{C}_6\text{H}_2\text{C}\equiv\text{P}$ are known from solid state experiments /2/. Theoretical data for systems with smaller substituents (e.g., C_6H_5, SiH_3) are given.

The orientation of the principal axes of the shielding tensors can be rationalized qualitatively in terms of the matrix elements of the first oder perturbation operators.
Substituent effects in phosphaalkynes are analyzed using the same approach.
The correlation of shieldings and charge densities is discussed.

The shielding tensors of the double bonded systems CH_2PH and CH_2PH_2^+ are compared to those of CH_2PH_3 (and CH_3PH_3^+).

/1/   W. Kutzelnigg   Isr. J. Chem. *1980* **27** *789*
      M. Schindler, W. Kutzelnigg   J. Chem. Phys. *1982* **76** *1919*
      W. Kutzelnigg, U. Fleischer, M. Schindler   NMR – basic principles and progress *1990* **23** *165*
/2/   P.C. Duchamp, M.Pakulski, A.H. Cowley, K.W.Zilm   J. Am. Chem. Soc. *1990* **112** *6803*

^{13}C ORBITAL SHIFT CALCULATION IN GRAPHITE INTERCALATION COMPOUNDS

C. FRETIGNY, M. SAINT JEAN* and M.F. QUINTON

LPQ/CNRS URA 1428, ESPCI, 10 rue Vauquelin 75231 Paris Cedex 05 (France),
*GPS, CNRS-Univ. Paris 6 et 7, Tour 23, 4 Place Jussieu 75251 Paris Cedex 05 (France).

Graphite Intercalation Compounds (GICs) are anisotropic conducting materials obtained by intercalation of atoms or molecules between the graphitic planes (graphenes), the intercalated species being either acceptor ($FeCl_3$, H_2SO_4, AsF_5...) or donor (Li, K, Rb, Cs, Ba...)[1]. Intercalated layers are organized along the crystallographic c-axis : the stage index is defined as the number of graphene between two intercalated layers (pure stages as large as 8 or 9 can be prepared). Due to the semi-metallic character of graphite, the density of states at the Fermi level of GICs ($N(E_F)$) strongly depends on the charge transfer between the intercalated layer and graphene. ^{13}C NMR has been extensively used to understand their electronic properties [2].

Previous analysis of the ^{13}C NMR in GICs lineshift showed that there is a contribution proportional to $N(E_F)$, mainly due to the dipolar interaction with the spin of the conduction electrons [3]. However, two features of the spectra displayed by oriented samples have to be explained. At first, stage 1 donor and acceptor compounds expected to have similar electronic structure and close values of $N(E_F)$ per carbon atom exhibit shift values differing by about 20 ppm when the magnetic field is along the c-axis ($B//c$). Secondly, the orientational dependence of the splitting due to the AB stacking of the graphenes in GICs [4] of stage ≥ 2 can be attributed neither to the contact nor to the dipolar interaction. Consequently, the strongly anisotropic orbital contribution of the conduction electrons must be taken into account.

To investigate this contribution, we applied the formalism recently developed by Kobayashi and Tsukada for the orbital shift in crystals [5], to stage 1 and 2 GICs when $B//c$. We used the model of independent subsystems [6], which is a 2D-LCAO description of the band structure, allowing analytical calculations.

This calculation shows that the sign of the orbital contribution, referred to the shift of pure graphite (macroscopic susceptibility correction done), is different for donors and acceptors, and is found to be consistent with the experimental data. Moreover, this contribution has been found weakly depending on the charge transfer (like the magnetic susceptibility), that is consistent with a total shift varying linearly with $N(E_F)$ in both donor and acceptor compounds. But donor and acceptor GICs cannot be directly compared. For the second stage compounds, the model predicts a splitting of the lines though dipolar contribution does not. The numerical values are sensitively affected by the details of the used atomic orbitals, but a rough estimation of the involved matrix elements leads to lineshifts only one third as great as the experimental ones. Calculations were made at finite temperature and with second neighbor interaction but these parameters do not affect very much the conclusions.

[1] See for example Intercalation in Layered Materials Vol 148 of NATO Advanced study Institute, Series B, edited by M.S. Dresselhaus (Plenum, New York, 1986).

[2] J. CONARD, Ann. Phys. Fr. Colloq. n°2, supp. n°2, vol. 11 (1986) 235.

[3] For a review, see for example C. FRETIGNY, M.F. QUINTON, Intern. Symp. on Magnetic and Electronic Properties of GICs, Grenoble 1991, Chouteaux and Yazami ed.

[4] Y. MANIWA, K. KUME, H. SUEMATSU and S. TANUMA, J. Phys. Soc. Japan 54 (1985) 666.

[5] K. KOBAYASHI and M. TSUKADA, Phys. Rev. B 38 (1988) 8566.

[6] J. BLINOWSKI, NGUYEN HY HAU, C. RIGAUX, J.P. VIREN, R. LE TOULLEC, G. FURDIN, A. HEROLD and J. MELIN, J. Phys (Paris) 41 (1980) 47.

EVALUATION OF CHEMICAL SHIFTS IN SOLID STATE NMR
BY ELECTRONEGATIVITY EQUALIZATION PRINCIPLE

C. Gerardin, M. Henry and F. Taulelle
Chimie de la Matière Condensée, Université P. et M. Curie,
75252 Paris Cedex 05, France

The principle of the chemical shifts evaluation in solid state NMR relies on the atoms charge estimation in a lattice. The way those charges are calculated uses the electronegativity equalization principle[1]:

$$\chi_i = \chi_i^0 + \eta_i q_i + \frac{e^2}{4\pi\varepsilon_0}\sum V_{i,j} q_j = \langle \chi \rangle \quad i \in [1,n], \text{ with the charge conservation, } \sum_{i=1}^{n} q_i = 0 \quad .$$

The electronegativities and hardnesses χ_i^0 and η_i are evaluated from Bratsch[2]. The potential $V_{i,j}$ is due to all the charges of the lattice and is evaluated following Bertaut[3]. Identification of the global hardness to the difference in energy between the HOMO and the LUMO orbitals has been proposed by Pearson[4]. We propose to identify this hardness η to the mean excitation energy ΔE of the second-order perturbation theory of the chemical shift.

This global hardness is given by $\dfrac{1}{\eta} = \dfrac{1}{M} \sum_{i=1}^{n} \dfrac{1}{\eta_i}$ following Yang et al.[5]. Therefore, the system of equations

equalizing the electronegativity can be solved, and provides the charges q_i for each atom, the mean electronegativity $\langle \chi \rangle$, the mean hardness $\langle \eta \rangle = \Delta E$. It is therefore possible to try to explicit the dependancies of the chemical shift as a function of the charges. If we use the Jameson Gutowsky relation of the paramagnetic contribution:

$$\sigma^P = k. \frac{\langle r^{-3} \rangle_p P_u}{\Delta E}$$

then, we can describe some limit cases.

In the case of C3S ($3CaO.SiO_2$), there are 83 inequivalent atoms. The X-ray structure is known and has been used to estimate the charges. The ^{29}Si MAS spectrum shows off the nine different silicons. The chemical shifts have then been plotted as a function of the charge. Actually the δ range is very small, and in such a small domain, this ranking of the chemical shifts shall be legitimate. A linear relation is obtained: $\delta = -96.3q - 13.2$ ppm. The following assignment of the sites can therefore be proposed as Si9, Si1, Si2, Si3, Si7, Si5, Si8 and Si4 in order of decreasing chemical shift with increasing the partial charge.

In the case of a series of oxides the chemical shifts of which have been published[6,7] and the structures known, ΔE has been evaluated. A plot of the chemical shift versus these ΔE values provides the following relationship: $\delta = -2416 + 21970/\Delta E$ (ΔE in eV).

Some other relations have been tested where $\langle r^{-3} \rangle$, P_u and ΔE are expressed as functions of the charge of the atom chosen for chemical shift evaluation[8].

This method provides then the possibility to calculate the charges as a function of the entire lattice and to estimate locally the effect of the charges on each component defining the chemical shift. The only necessary inputs are a correct X-ray structure and an electronegativity scale.

[1]K.V. Genechten, W. Mortier and P. Geerlings *J. Chem. Soc. Chem. Comm.* 1278 (1986)
[2]S.G.. Bratsch *J. Chem. Educ.* 65 34 (1988)
[3]E.J. Bertaut *J. Phys. Radium* 13 499 (1952)
[4]R.J. Pearson *Inorg. Chem.* 27 734 (1988)
[5]W. Yang, C. Lee and S.K. Ghosh *J. Phys. Chem.* 89 5412 (1985)
[6]T.J. Bastow and S.N. Stuart *J. Chem. Phys.* 143 459 (1990)
[7]G.L. Turner, S.E. Chung and E. Oldfield *J. Mag. Res.* 64 316 (1985)
[8]C. Gérardin, M. Henry and F. Taulelle *Proc. Mater. Res. Soc. Symp.* "Better ceramics through chemistry", San Francisco, April 1992

^{129}Xe SHIELDING AS A PROBE OF ZEOLITE STRUCTURE AND DYNAMICS

A. KEITH JAMESON
Loyola University
Department of Chemistry
Chicago, Illinois 60626

CYNTHIA J. JAMESON
University of Illinois at Chicago
Department of Chemistry M/C-111
Chicago, Illinois 60680

ANGEL C. de DIOS
University of Illinois at Urbana
Department of Chemistry
Urbana, Illinois 61801

ABSTRACT. Zeolites are microporous aluminosilicates with cavities which are regular on a molecular scale, with characteristic "diameters" of up to 1.5 nm. They typically contain cations which contribute to interesting and useful chemical properties of great commercial interest. Of the wide variety of physical observables which have been measured, the chemical shift of ^{129}Xe is of particular interest to us. Xenon shielding is a very sensitive probe of the environment. A wide variety of zeolites have previously been studied by ^{129}Xe NMR and variations in chemical shift of 60-250 ppm relative to isolated Xe have been seen, depending on the zeolite, on xenon "loading" (concentration), and on temperature. We presume that it is possible to understand these measurements from a fundamental point of view, in terms of a time average of the ^{129}Xe shielding in a xenon atom as it moves in the potential in which it finds itself inside a zeolite.

Experimentally we have been able to observe adsorbed xenon at equilibrium by using sealed tubes in which the overhead xenon pressure may be as high as 50 atm. This allows us to measure adsorption isotherms at very high loadings entirely by NMR and also observe the ^{129}Xe chemical shift at loadings extending from very low to saturation in variable temperature studies. Xenon in NaA zeolite provides a paradigm for understanding ^{129}Xe chemical shifts in zeolites as a function of loading. In NaA the openings between cages are effectively blocked such that the slow exchange limit exists. This allows the ^{129}Xe chemical shift of individual Xe_n (n = 1-8) and the relative populations of each Xe_n to be measured at various equilibrium loadings. We have been able to utilize the data on NaA to predict reasonably well δ vs. loading at various temperatures for CaA, a closely related zeolite in which the fast exchange limit exists, resulting in a single resonance signal for the adsorbed xenon.

Our goal is to simulate (via Molecular Dynamics (MD) and Grand Canonical Monte Carlo (GCMC) techniques) the nmr observations as well as other physical observations in these systems (isosteric heats, adsorption isotherms). To do this we need several surfaces: the intermolecular potential surface between Xe and the network solid zeolite and between Xe atoms within the zeolite, and the shielding surface for ^{129}Xe as a function of distance from oxygens, Si, Al, cations, and other Xe atoms. For the latter we use *ab initio* ^{39}Ar shielding surfaces scaled to ^{129}Xe. The test of a robust simulation will be to predict observables, including δ(^{129}Xe), in many zeolites with a parameterized potential function which is globally transferable to all zeolites. Our initial GCMC results are very encouraging insofar as the Xe-Xe interactions in NaY zeolites are concerned.

NMR Shielding and Atoms in Molecules

T. A. Keith, Dept. of Chemistry, McMaster Univ., Hamilton, Ontario L8S 4M1, Canada

A method for predicting relatively accurate molecular magnetic response properties from a set of conventional coupled perturbed Hartree Fock (CPHF) calculations for each symmetrically unique atom in a molecule [1] is presented. This method [2] takes advantage of the fact that molecules and their properties are naturally partitioned into atoms and atomic contributions and the fact that a magnetic field induced current deensity distribution within an atom in a molecule is relatively well described by the CPHF method when the gauge origin of the magnetic field's vector potential is placed at its nucleus. An atomic contribution to a molecular nuclear magenetic shielding tensor element or magnetic susceptibility tensor element is determined entirely by the corresponding current density distribution within the atom. The calculation of the atomic contributions to these properties not only affords more accurate predictions than is possible with the conventional single-origin CPHF methhod, but also provides an understanding of these molecular properties at the atomic level. The relative accuracy of the method (IGAIM – Individual Gauges for Atoms in Molecules) is demonstrated by comparing calculated isotropic magnetic susceptibilities and carbon shielding values with experiment and the conventional CPHF method for a wide range of molecules.

References
1. R. F. W. Bader, Atoms in molecules – a quantum theory, Oxford Univ. Press, Oxford, 1990
2. T. A. Keith and R. F. W. Bader, Chem. Phys. Lett., 194, 1, (1992)

Experimental and Calculated Nitrogen Shielding Tensors in the Nitroso Group

Mike D. Lumsden, Gang Wu, Ron D. Curtis and Roderick E. Wasylishen, Department of Chemistry, Dalhousie University, Halifax, N.S., Canada, B3H 4J3

^{15}N solid state NMR spectroscopy has been used to study the nitrogen shielding tensor in the NO fragment of N,N-dimethyl-p-nitroso-^{15}N-aniline (1) and nitrosobenzene-^{15}N$_2$ (2). X-ray diffraction studies have shown that 1 exists as a monomeric nitroso species while 2 shows an azodioxy structure in the solid state. The results show a chemical shift anisotropy (CSA) of 1479 ppm in 1 and 285 ppm in 2. The dipolar - chemical shift NMR spectrum of the dimer has been analyzed to obtain information on the orientation of the principal axis system of the nitrogen chemical shift tensor in the molecular reference frame. The results show the most shielded component to be perpendicular to the molecular plane and the intermediate component is 23° off the N=N bond axis. Using the localized orbital/local origin (LORG) method, the nitrogen chemical shielding tensor has been calculated in two model compounds. The calculations nicely supplement the experimental results, predicting the correct orientation of the shielding tensor in the dimer from the dipolar NMR results and to indicate the orientation in the monomer, where no experimental information is available. Analysis of the molecular orbital contributions to the nitrogen shielding indicate that the source of the large CSA in 1 is predominantly related to low-lying paramagnetic circulations involving the nitrogen lone pair. Unfortunately, large deviations between experiment and theory exist for the principal components, particularly the least shielded component. Finally, a comparison of the nitrogen shift tensor obtained here for 1 with several nitrosyl ruthenium complexes indicates that the least shielded component is sensitive to the X-N=O bond angle, becoming more deshielded as the bond angle decreases.

Calculations of NMR shielding constants using a combination of pseudo-potential and IGLO methods.

V.G.Malkin*#, U.Fleischer and W.Kutzelnigg

*Départment de chimie, Université de Montréal, Québec H3C 3J7
Lehrstuhl für Theor. Chemie, Ruhr-Universität Bochum, Germany

The pseudo-potential approach is one of the most effective methods for reducing the computation time for systems with a significant number of core electrons. However, a brute force combination of the CHF approach with a magnetic field independent nonlocal pseudo-potential in the form

$$V_{pp}(r) = \sum_A V_{pp}^A(r)$$ (1)

$$V_{pp}^A(r) = V_{L+1}^A(r) + \sum_{l=0}^{L} \left[V_l^A(r) - V_{L+1}^A(r) \right] \sum_m |lm\rangle\langle lm|$$ (2)

is inappropriate. In this case the hamiltonian is not invariant with respect to a unitary transformation

$$U = e^{-i\Lambda_K} ; \quad \text{where} \quad \Lambda_K = \frac{1}{2c} \vec{B} \times \left(\vec{R}_K - \vec{R} \right) \cdot \vec{r}$$ (3)

corresponding to a shift of the gauge origin of the vector potential of the external magnetic field. Therefore, we presented here a more general type of pseudo-potential for a system in an external magnetic field in the form

$$V_{pp}(r) = \sum_A e^{i\Lambda_A} V_{pp}^A(r) e^{-i\Lambda_A}$$ (4)

where $V_{pp}^A(r)$ is the same as in Eq. (3). It cares for the invariance of the hamiltonian and has the correct limit in the absence of an external magnetic field.

Results of our calculations of carbon chemical shifts are in good agreement with data of all-electron calculations. Results for silicon chemical shifts are presented and problems concerning the non-transferability of L-shell contribution are discussed.

One of authors (V.G.M) is grateful to the Alexander von Humboldt Foundation for his research fellowship.
Permanent position: Institute of Catalysis, Novosibirsk, Russia

Quantum-chemical calculations of chemical shifts

of adsorbed molecules

V.G.Malkin*+ , U.Fleischer, J.Sauer# and W.Kutzelnigg
Lehrstuhl für Theoret. Chemie, Ruhr-Universität Bochum, Germany
*Département de chimie, Université de Montréal, Quebec, Canada
#Arbeitsgruppe Quantenchemie des Max-Planck-Inst., Berlin, Germany

Resently [1,2], very interesting carbon NMR spectra of products of the reaction of olefins on the H-form of zeolites have been reported. In these spectra, some peaks lie in the region typical for carbocations (in solution). It has been suggested that these peaks are due to adsorbed allyl cation.

For the theoretical interpretation of these spectra, we have studied various clusters that are supposed to simulate the surface structures formed under interaction of an allyl cation and the surface of the H-form of zeolites. Quantum chemical calculations of the geometry and of the carbon chemical shifts of these clusters have been performed.

For the optimization of the geometries the TURBOMOL and MOPAC programs have been used. The chemical shifts have been calculated by means of the IGLO method. The basis set and geometry dependence of the carbon chemical shifts have been studied. We found that the investigated covalent structures cannot be responsible for the down field peaks in carbon NMR spectra. We conclude that only two-center van der Waals complex of allyl cation is likely to be a candidate for the origin of the observed 'cationic' peaks.

One of authors (V.G.M.) is grateful to the Alexander von Humboldt Foundation for his research fellowship.

(1) J.F. Haw, B.R. Richardson, I.S. Oshiro, N.D. Laso, J.A. Speed J. Am. Chem. Soc. 1989, **111**, 2052.
(2) D.-P. Lange, A, Gutsze, J. Alligeir, H.G. Karge Appl. Catal. 1988, **45**, 345.

+ *Permanent address: Institute of Catalysis, Novosibirsk, Russia*

A New Analysis of Proton Chemical Shifts in Proteins

Klara Ösapay and David A. Case

Department of Molecular Biology
Research Institute of Scripps Clinic
La Jolla, California 92037

Abstract. We present an empirical analysis of proton chemical shifts from 17 proteins whose X-ray crystal structures have been determined. The crystal structures are used to estimate the conformation-dependent part of the shift, that is, the difference between the observed shift and that of a "random-coil" linear peptide. The results indicate that a significant improvement over ring-current theories can be made by including the effects of the magnetic anisotropy of the peptide group and estimates of backbone electrostatic contributions. For 5678 protons bonded to carbon, we find a linear correlation coefficient of 0.88 between calculated and observed secondary shifts, with a root-mean-square error of 0.23 ppm; contributions from the peptide group are especially noticeable for protons at the $C\alpha$ position. If we consider only sidechain protons in non-heme proteins, the rms error is 0.18 ppm. New estimates of intensity factors for various ring current contributions are given (including those arising from the heme group) which suggest more nearly equal contributions from various rings than found in earlier studies.

An analysis of the contributions to $H\alpha$ protons in regions of regular seconadary structure confirms expected differences between helices and sheets, and shows that the shifts can be understood primarily in terms of the orientations of neighboring peptide groups and that the ϕ backbone angle is the most important geometrical parameter. Contributions from peptide magnetic anisotropies and electrostatic polarization of the $C\alpha$-H bond have similar geometric dependencies, making it difficult to separate these contributions using the emprical methods employed here.

We have employed these empirical shift calculations as an aid to determining the solution structure of sperm whale myoglobin. Preliminary results indicate that the addition of chemical shift restraints to "standard" NOE-based refinements results in a substantial narrowing of the distribution of acceptable NMR-based structures and that differences between the X-ray and solution structures appear to be small.

ELECTRON SURROUNDINGS OF Cs$^+$ IN THE GRAPHITE INTERCALATION COMPOUND Cs$_{\approx 1}$(THF)$_{\approx 1.5}$C$_{24}$, AS OBSERVED BY ^{133}Cs NMR.

M.-F. QUINTON, F. BEGUIN* and A. P. LEGRAND,

LPQ/CNRS URA 1428, ESPCI, 10 rue Vauquelin, 75231 Paris Cedex 05 (France)
* CRMD - CNRS, 45071 Orléans Cedex 2 (France)

The binary intercalated compounds alkali metal-graphite are known to form ternary intercalated compounds with organic molecules [1] expected to have electronic properties different from the binary ones. In both binary and ternary compounds, the NMR of alkali nuclei is of special interest regarding their electronic properties. Indeed, the investigated nuclear spins mainly interact with the electrons via both quadrupolar and magnetic shielding interactions. ^{133}Cs NMR of the ternary compounds Cesium-Tetrahydrofuran (THF)-Graphite, compared with the host binary ones, leads to interesting information about the modification of the electron surroundings of the Cs ion induced by the THF subsequent intercalation.

The orientational dependence of both quadrupolar splitting and lineshift, observed at room temperature with samples prepared from highly oriented pyrolytic graphite (HOPG) shows that these interactions are axially symmetric about the crystallographic c-axis (perpendicular to the graphitic planes). But when this axial symmetry is originated by the conduction electrons in the host binary compounds Cs$_x$C$_{24}$ (0.8 < x < 1.2), in the ternary Cs$_x$(THF)$_{\approx 1.5}$C$_{24}$ ones, it is due to the electron cloud of THF molecules undergoing a fast anisotropic motion about the c axis [2]. Indeed, we do not observe, in the ternary samples, the strong dependence of the shift on the composition of the intercalated layer, attributed to the special feature of the contact interaction with metallic electrons in the Cs$_x$C$_{24}$ compounds [3]. On the other hand, the THF intercalation reduces the quadrupolar splitting which becomes strongly dependent on the THF concentration, unlike the binary compounds where this splitting was found insensitive to the in-plane Cs concentration.

Between 250 and 300 K, above the 2D phase transition related to the disappearing of the anisotropic motion of THF [2], a small but significant variation of both isotropic and anisotropic shifts is observed. This thermal variation, which can be linearly correlated to the quadrupolar splitting one, confirms that Cs nuclei in the ternary compounds experience the magnetic field due to the moving electron cloud of THF.

[1] See for example R. SETTON, in *Graphite Intercalation Compounds I*, ch. 9, H. ZABEL and S. A. SOLIN ed., Springer Ser. Mat. Sci., Vol. 14 (Springer, Berlin, Heidelberg 1990)
[2] F. BEGUIN, C. LAROCHE, B. GONZALES, M. GOLDMAN, M.-F. QUINTON, *Synth. Met*, 23 (1988) 155.
[3] H. ESTRADE-SZWARCKOPF, M. MALKI, A.-M. FAUGERE, F. FLEURY, P. LAUGINIE, J. CONARD, B. ROUSSEAU, *Proc. of the Sixth Int. Conf. on Intercalation Compounds*, Orléans, France, 1991, to be published.

ON THE CALCULATIONS OF DEUTERIUM LONG RANGE ISOTOPE EFFECTS ON CARBON-13 CHEMICAL SHIFTS

D. VIKIC-TOPIC[1,a], M. HODOSCEK[2,b], A. GRAOVAC[3], E. D. BECKER[1],
G. LODDER[4] and H. ZUILHOF[4]

[1]Laboratory of Chemical Physics, NIDDK, National Institutes of Health, Bethesda, MD 20892
[2]Molecular Graphics and Simulation Laboratory, DCRT, National Institutes of Health, Bethesda, MD 20892
[3]Ruđer Bošković Institute, P.O.Box 1016, HR-41001 Zagreb, Croatia
[4]Gorlaeus Laboratories, Department of Chemistry, P.O.Box 9502, Leiden University, Leiden, The Netherlands

In C-13 NMR spectra of π-systems long range deuterium isotope effects (LRDIE) on chemical shifts, through up to 12 bonds, were observed. Some authors have tried to explain LRDIE by substituent model due to similarities of some LRDIE with substituent effects. This approach assumes that deuterium has its own inductive, hyperconjugative and steric effects. Strictly taken it is contrary to the Born-Oppenheimer approximation (BOA) since it implies different electronic properties of D and H. However in BOA beside the dynamical factor, i.e. vibrational and rotational averaging of nuclear shielding, there is also electronic factor which arises from the small changes in average bond length and average bond angles on isotopic substitution. There are several reports on shortening of the C-D bond compared with the C-H bond ranging from only 0.004Å in the gas phase to even 0.020Å for the molecules dissolved in the liquid crystal phase.

Since calculations of small LRDIE on C-13 shielding in relatively large molecules are still challenging it is interesting to find out the relations of isotopic shifts to other electronic properties of molecule. With respect to it we investigated, by different ab initio calculations, charge/shift relationships in some deuterated aromatic molecules (stilbenes, biphenyls, etc.). We simulated the effect of deuteration on stretching mode by C-H bond reductions for 0.003-0.018Å. We also changed in- and out-of-plane CCH bond angles by 1-2° in attempt to mimic the effect on bending modes. The good correlations between (a) differences of total atomic charges between parent and deuterated molecules (Löwdin population analysis) and measured LRDIE on C-13 chemical shifts (r^2=0.970), and (b) measured C-13 chemical shifts and calculated changes of charges induced by substituents were obtained. However (a) is only valid for bond shortening but not for bond angle changes. The results suggest that D polarizes the π-system as an methyl (electropositive) group but with effects lower by two orders of magnitude. We performed GIAO calculations of shielding with reduced C-H bond and obtained good correlation with LRDIE as well. Preliminar vibrational analysis of few deuterated aromatics performed by PM3 restricted Hartree-Fock calculations showed negligible vibrational coupling of para-D with C-atoms more than five bonds away. The obtained force constants can not be correlated with LRDIE at all. Therefore we claim that LRDIE in π-molecules can be explained, without violating BOA, by shorter C-D than C-H bond, instead by substituent model. Due to the transmissivity of π-electron system the small change of bond length on site of deuteration brings about electronic effects many bonds away.

[a]Permanent address: Ruđer Bošković Institute, Zagreb, Croatia
[b]Permanent address: Institute of Chemistry, Ljubljana, Slovenia

GIAO calculations of the NMR chemical shift for large molecules

K.Wolinski[*], J.F.Hinton, P.Pulay

Department of Chemistry, University of Arkansas

Fayetteville, AR 72701

[*] Faculty of Chemistry, M.C. Sklodowska University

20-031 Lublin, Poland

Recently, we have presented an efficient implementation of the gauge-including atomic orbitals (GIAO) method for NMR chemical shift calculations (1). The following techniques have been used in our program :

1. Elimination of the storage for perturbed two-electron integrals
2. The coupled-perturbed Hartree-Fock (CPHF) equation has been formulated fully in atomic orbital basis set (AO)
3. Three CPHF equations are solved simultaneously by a conjugate gradient type method
4. Neglect of the perturbed two-electron integrals which do not contribute significantly to the NMR shielding tensor

Efficiency

1. small amount of memory required
2. NMR calculations are about 2.5 times longer than energy ones
3. the work which is currently in progress, allows us to belive that performance of integral programs (energy and NMR) can be improved by a factor of 3

Our program can be routinely used for NMR chemical shielding calculations for relatively large molecules i.e. containing 20 or more heavy atoms. Here we present results for ^{13}C in the [5]-, [7]- and [8]-circulene and $C_{30}H_{10}$.

(1) K.Wolinski, J.F.Hinton, P.Pulay, J. Am. Chem. Soc., 112, 8251 (1990)

Is there a viable alternative to the Coulomb gauge for GIAO

calculations of magnetic properties ?

K.Wolinski[*], J.F.Hinton, P.Pulay
Department of Chemistry, University of Arkansas
Fayetteville, AR 72701
[*] Faculty of Chemistry, M.C. Sklodowska University
20-031 Lublin, Poland

In the case of a magnetic field the only physical observable is its strength. The vector potential describing the magnetic field is just a mathematical quantity which can be changed to different forms with the restriction that the resulting field strength satisfies the Maxwell equation (curl H = 0). This freedom of choice of the vector potential is known as a gauge transformation

$$A^{new} = A^{old} + grad\ f$$

where f is an arbitrary function of coordinates.

In the GIAO method the vector potential is taken in the form A = 0.5 (H x r) (sometimes called Coulomb gauge). An interesting question is whether or not the GIAO formalism can be simplified by gauge-transformation performed on this vector potential.
We have considered for example the Landau transformation in which the scalar function f is bi-linear in x,y,z and linear in H. In this case the new vector potential has a simpler form. Consequantly, one-electron operators in perturbational treatment of magnetic properties are transformed into much simpler expressions. For instance, three out of nine components of the matrix operator H_n^{11} (diamagnetic shielding) are zeros, and the H^{20} matrix operator (diamagnetic susceptibility) becomes diagonal. This can be considered as an advantage. However, in order to be invariant, one must transform also the wave function. In the case of the GIAO method such a transformation should be performed on a set of field-dependent basis functions. This introduces orbitals with higher angular momentum than the original ones had.

For the NMR shielding tensor calculations the angular momentum of orbitals is increased by 1 and 2. For the magnetic susceptibility it goes up to 4. In contrast, the traditional Coulomb gauge approach increases angular momentum of basis functions only by 1 for the NMR shielding, and by 1 and 2 for susceptibilities.

Thus, the GIAO method with the Landau gauge requires integrals involving higher orbitals than with the Coulomb gauge. This is obviously more time consuming, especially in the case of two -electron integrals where no advantage can be taken from a simpler form of one-electron operators.

Finally, we conclude that the use of the Landau type gauge in the GIAO method seems to be less efficient than the traditional approach employing the Coulomb gauge.

INDEX

The manufacturer's authorised representative in the EU is Springer
Nature Customer Service Centre GmbH, Europaplatz 3, 69115 Heidelberg,
Germany. If you have any concerns regarding our products, please
contact ProductSafety@springernature.com

Printed and bound by CPI Group (UK) Ltd, Croydon, CR0 4YY

24/04/2026

02096308-0013